Schriften der Mathematisch-naturwissenschaftlichen Klasse
der Heidelberger Akademie der Wissenschaften
Nr. 21 (2010)

ERNST ZERMELO

Collected Works
Gesammelte Werke

Springer

Zermelo in 1907

ERNST ZERMELO

Collected Works
Gesammelte Werke

Responsible for the whole edition
Verantwortlich für die gesamte Edition:

Heinz-Dieter Ebbinghaus,
Craig G. Fraser,
Akihiro Kanamori

ERNST ZERMELO

Collected Works
Gesammelte Werke

VOLUME I
BAND I

Set Theory,
Miscellanea

Mengenlehre,
Varia

VOLUME II
BAND II

Calculus of Variations,
Applied Mathematics,
and Physics

Variationsrechnung,
Angewandte Mathematik
und Physik

ERNST ZERMELO

Collected Works
Gesammelte Werke

VOLUME I
BAND I

Set Theory,
Miscellanea

Mengenlehre,
Varia

Edited by
Herausgegeben von

Heinz-Dieter Ebbinghaus,
Akihiro Kanamori

Springer

Editors
Herausgeber

Heinz-Dieter Ebbinghaus
University of Freiburg, Germany

Craig G. Fraser
University of Toronto, ON, Canada

Akihiro Kanamori
Boston University, MA, USA

ISBN 978-3-642-26209-8 e-ISBN 978-3-540-79384-7
DOI 10.1007/978-3-540-79384-7
Springer Heidelberg Dordrecht London New York

Mathematics Subject Classification (2000): 01A70, 01A75, 03B60, 03E25, 03E30, 03E45, 03E55, 11A41, 91A05

Cover design: WMXDesign GmbH, Heidelberg
Typesetting and production: le-tex publishing services GmbH, Leipzig

Printed on acid-free paper

Springer is part of Springer Science+Business Media (www.springer.com)

Preface to the Zermelo edition

This is a complete edition of the published works of Ernst Zermelo which moreover includes selected correspondence and unpublished manuscripts. Zermelo is generally acknowledged for his pioneering work in axiomatic set theory and for introducing the axiom of choice as a basic principle of mathematics. In contrast, his work in applied mathematics and physics, despite its originality, is hardly recognized or has even been attributed to others. This edition of Zermelo's collected works provides a picture of the entire mathematician. It appears in two volumes. The first volume comprises Zermelo's published papers in set theory and the foundation of mathematics together with isolated papers of an algebraic, analytic, or number-theoretic character. The second volume is dedicated to Zermelo's work in the calculus of variations, mathematical physics, and fluid dynamics. Both volumes are supplemented by selected notes and manuscripts, mainly from Zermelo's *Nachlass*, which throw additional light on his papers, reflect his point of view, or are unpublished continuations of published work. To the best judgment of the editors, the selected notes and manuscripts fully and faithfully represent the essential unpublished writings of Zermelo concerning mathematics. Nevertheless, a possible edition of a third volume comprising further unpublished notes and letters from the *Nachlass* has *expressis verbis* been left open.

Both volumes contain some writings by other authors which include contributions actually written by Zermelo or which react to criticism Zermelo had made. Details are given in the prefaces to the respective volumes.

In order to provide access to a wider audience, the original papers are printed face to face with English translations. As both versions use the same layout, it is easy to go from the translation to the original version and vice versa. The layout itself tries to preserve the appearance of the original papers. For details we refer to the editorial information below.

Each paper or coherent group of papers is preceded by an introductory note which comments on contents, motivation, aims, and influence of the paper(s) concerned. Written by an expert in the field, it came to its final form in discussions with the editors.

Each volume contains a full bibliography of Zermelo together with a schematic *curriculum vitae* which will enable the reader to become acquainted with the personal circumstances from which a paper arose. In addition, Volume I starts with a more detailed biographical sketch of Zermelo's life and work.

Many of these features found their inspiration in the exemplary edition of Kurt Gödel's collected works by Solomon Feferman, John W. Dawson, Jr., and others.

The edition of Zermelo's collected works has a prehistory. Already as early as 1912, at the age of 41 and faced with a serious recurrence of his tuberculosis,

Zermelo conceived plans for an edition of his collected papers, but did not pursue them when his health improved. In 1949, under likewise deplorable personal circumstances, he tried again, this time approaching several publishers, among them Springer-Verlag. But the difficult situation in post-war Germany precluded such an enterprise. Immediately after Zermelo's death, in 1953, the historian of mathematics Helmuth Gericke and the philosopher Gottfried Martin, who had gotten to know Zermelo in the 1930s in Freiburg, started work on a two-volume edition, in 1956 gaining Paul Bernays as a third editor. Support was provided by the Kant-Gesellschaft. However, the plans were not realized; in 1962 work on the edition came to a definite end.

When in early 2004 new plans for an edition of Zermelo's collected works became more concrete, they found the enthusiastic support of Martin Peters of Springer-Verlag. In discussions with him it became clear very quickly that the edition should provide English translations and detailed comments. As Zermelo had been a member of the Heidelberger Akademie der Wissenschaften, the editors turned to the academy for financial support. The application found the warm backing of Hans Günter Dosch, then Sekretar of the class for mathematics and the sciences of the academy. The application was successful. Even more, besides providing generous funding, the academy offered to let the edition appear in its regular series of publications of the class for mathematics and the sciences published by Springer-Verlag.

The editors wish to express deep gratitude to the Heidelberg academy for their ideal, financial support and to Springer-Verlag for their open-minded cooperation. In particular, many thanks go to Hans Günter Dosch and Martin Peters.

Freiburg, Toronto, and Boston
September 2009

Heinz-Dieter Ebbinghaus
Craig G. Fraser
Akihiro Kanamori

Preface to volume I

This first volume of the Zermelo edition focuses on Zermelo's work in set theory and the foundations of mathematics and is supplemented by his papers in pure mathematics. The published papers are accompanied by selected items from Zermelo's *Nachlass* and some letters. Whereas the papers span the time between Zermelo's first encounter with set theory around 1900 in Göttingen and the end of his mathematical research in the mid-1930s in Freiburg, the selected items mainly stem from the early 1930s, from a time when Zermelo's foundational views were in growing opposition to the mainstream developments in mathematical logic.

The primary criterion for the inclusion of an item was the degree of novelty and of insights into Zermelo's foundational views that it provides. Some items may be considered as partly independent drafts of (parts of) a published paper or as offering a variation of (parts of) a published paper (such as *s1931g* with respect to *1932a, 1932b*, and *1935*); some may be viewed as first steps of a continuation of a published paper, for example by elaborating on a feature only touched on there (such as *s1931f* and *s1932d* with respect to *1930a*); some items allow a deeper insight into Zermelo's foundational views (such as *s1929b* and *s1930d*).

The correspondence with Kurt Gödel together with one of Zermelo's letters to Reinhold Baer (*s1931b* to *s1931d*) sheds additional light on Zermelo's foundational views around 1930 which are marked by a strong aversion to finitary approaches to the foundation of mathematics. The last letter *s1941* to Paul Bernays is a revealing testament to Zermelo's mathematical disillusionment and physical deterioration in his final years. The Zermelo part of the correspondence with Reinhold Baer and Arnold Scholz, which could have provided further insights, is lost up to the letter *s1931c* to Baer.

The parts *Landau 1917b* and *D. König 1927b* of *Landau 1917a* and *D. König 1927a*, respectively, are included here because they were written by Zermelo or closely follow a note written by him.

Zermelo was well-educated in and had a continuing enthusiasm for literature and the classics. In the 1920s he translated large parts of Books V to IX of Homer's *Odyssey* into German blank verse. Part of his translation of Book V was published as *1930f*. It is given here together with an appendix which contains his entire translation of Book V.

The introductory notes are a crucial part of this edition. As a matter of fact, all the people we invited to comment on a paper or a group of papers, were instantly ready to join the project and share their experience and knowledge with us and the potential reader. The discussions we had toward securing the most informative and accurate presentations were framed by mutual cooperation and open-mindedness. We particularly appreciate the valuable help

provided by the collaborators in checking bibliographical items and making useful proposals concerning the English translations.

When work on the edition started, the most serious problem we seemed to be faced with was how to get accurate translations. As it turned out, the problem found the best solution imaginable. First, we could use the translations of *1904*, *1908a*, and *1908b* which Stefan Bauer-Mengelberg had provided for Jean van Heijenoort's *From Frege to Gödel.* Furthermore, Warren Goldfarb and R. Gregory Taylor allowed us to use their English translations of *1909a* and *s1931g*, respectively. All the other translations were done by Enzo de Pellegrin. We express our deep gratitude for his extraordinary care and admire his feeling for both languages when handling Zermelo's style with its richness in nuances and its involved sentential structures.

There are many others who have supported us during our work. We express our gratitude to all of them and mention here, in particular, Ruth Allewelt of Springer-Verlag through whom we had smooth cooperation with the publisher; Barbara Hahn who, as the librarian of the Freiburg Mathematical Institute, efficiently provided the literature we needed; and Andrea Köhler of le-tex publishing services who directed the multitude of files and problems with patience and care. We appreciated the supportive interest which Gunther Jost from the Heidelberg Academy showed for the imponderabilities of our work. As ever, Martin Peters from Springer-Verlag was ready to offer valuable help and advice. As a final note, the second-named editor would like to express his special gratitude to the Lichtenberg-Kolleg at Göttingen. Awarded an inaugural fellowship there, he was able to carry out editorial work in a particularly supportive environment at the Gauß Sternwarte, in the city where Zermelo did his best-known work in set theory.

Freiburg and Boston
November 2009

Heinz-Dieter Ebbinghaus
Akihiro Kanamori

Editorial information

Layout. The layout of the texts as well as of the translations mirrors the layout of the originals. Emphasized words, i.e., words in italics or words spaced out or consisting of small capitals, are given in italics. Original pagebreaks are indicated in the texts by "|", and the number of the new original page beginning there is given on the margin.

Editorial annotations. These are set in double square brackets "⟦" and "⟧".

Misprints and errors. Small textual errors in the originals are tacitly corrected; larger ones are corrected with the corrections commented on in editorial annotations.

Wrong words or words missing in the originals have been replaced or added in double square brackets.

Misprints in mathematical expressions in the originals are not corrected in the texts. They are, however, corrected in the translations and noted by an editorial annotation.

Special terminology. Zermelo articulates the following distinction among *Untermenge*, *Teilmenge*, and *Teil*: A set M is a *Untermenge*, *Teilmenge*, *Teil* of the set N if $M \subseteq N$, $M \subsetneq N$, $\emptyset \neq M \subsetneq N$, respectively. However, he evidently deviates from this at times, and in order to truly reflect his usage, *Untermenge*, *Teilmenge*, and *Teil* are always translated as *subset*, *partial set*, and *part*, respectively. This convention has led to changes in the Bauer-Mengelberg translations of the Zermelo papers *1904*, *1908a*, and *1908b*.

Special symbols. Mathematical symbols which are now no longer in use, such as "$\underset{=}{\in}$" for "$\subseteq$", are kept in the texts, but replaced by their modern analogues in the translations.

References. In the texts Zermelo's references to the literature are not altered. Translations as well as introductory notes refer to the main bibliography at the end of the volume and have the form *author(s) year of appearance*, followed by an additional index a, b, $c, \ldots$ if necessary. An example: *Hahn and Zermelo 1904*. If the authors are clear from the context, their names may be omitted; for example, *1904* some lines above is short for *Zermelo 1904*.

References to page numbers are kept in both the texts and the translations; they can be traced via the original pagebreaks and the original page numbers provided in the texts.

Footnotes. Whereas the translations use natural numbers in ascending order as footnote marks, the texts preserve the original marks. It may thus happen that a page of the text may contain identical footnote marks. In such cases the original page numbers on the margin allow for quick correlation of mark and footnote.

Copyright permissions

The editors express their gratitude to the following copyright holders for granting permission for the inclusion of the original texts and their English translations:

Akademie der Wissenschaften zu Göttingen (*1901, Landau 1917b, 1934*);

Springer Verlag (*1904, 1908a, 1908b*, their English translations *1967a,b,c*, and *1909a, 1914, 1932c*);

Universitätsarchiv Göttingen (the quotation from Zermelo's report on Grelling's dissertation in the introductory note to *1909a* and *1909b*);

Cambridge University Press (*1913*);

Universitätsarchiv Freiburg (*s1921, s1929b, s1930d, s1930e, 1930f, Gödel 1931b, s1931e, s1931f, s1931g, s1932d, s1933b, s1937*);

Verlag Walter de Gruyter (*1927*);

Acta Scientiarum Mathematicarum (Szeged) (*D. König 1927b*);

Instytut Matematyczny Polskiej Akademii Nauk (*1929a, 1930a, 1930b, 1935*);

Institute for Advanced Study (Princeton) (*s1931b, s1931d*);

Ulrich Felgner (*s1931c*);

Deutsche Mathematiker-Vereinigung (*1932a*);

Berlin-Brandenburgische Akademie der Wissenschaften (*1932b*);

Archiv der ETH Zürich (*s1941*).

The photographs in this volume are taken from the Zermelo photo collection held in the Abteilung für Mathematische Logik at Freiburg University. The collection will be integrated into the Zermelo *Nachlass* housed in the Universitätsarchiv Freiburg.

Contributors of introductory notes

Oliver Deiser
Fachbereich Mathematik
und Informatik
Freie Universität Berlin
14195 Berlin
Germany
oliver@aleph1.info

Heinz-Dieter Ebbinghaus
Mathematisches Institut
Universität Freiburg
79104 Freiburg
Germany
hde@uni-freiburg.de

Jürgen Elstrodt
Fachbereich Mathematik
und Informatik
Universität Münster
48149 Münster
Germany
elstrod@math.uni-muenster.de

Ulrich Felgner
Mathematisches Institut
Universität Tübingen
72076 Tübingen
Germany
ulrich.felgner@uni-tuebingen.de

Michael Hallett
Department of Philosophy
McGill University
Montreal, Quebec H3A 2T7
Canada
michael.hallett@mcgill.ca

Albert Henrichs
Department of the Classics
Harvard University
Cambridge, MA 02138
USA
henrichs@fas.harvard.edu

Akihiro Kanamori
Department of Mathematics
Boston University
Boston, MA 02215
USA
aki@math.bu.edu

Paul B. Larson
Department of Mathematics
Miami University
Oxford, OH 45056
USA
larsonpb@muohio.edu

Charles Parsons
Department of Philosophy
Harvard University
Cambridge MA 02138
USA
parsons2@fas.harvard.edu

R. Gregory Taylor
Department of Mathematics
and Computer Science
Manhattan College
Riverdale, NY 10471
USA
gregory.taylor@manhattan.edu

Dirk van Dalen
Department of Philosophy
Universiteit Utrecht
3508 TC Utrecht
The Netherlands
dirk.vandalen@phil.uu.nl

Dieter Wolke
Mathematisches Institut
Universität Freiburg
79104 Freiburg
Germany
dieter.wolke@math.uni-freiburg.de

Contents of volume I

Contents of volume II

Ernst Zermelo

Collected Works
Gesammelte Werke

Volume I
Band I

Set Theory,
Miscellanea

Mengenlehre,
Varia

Ernst Zermelo: A glance at his life and work

Heinz-Dieter Ebbinghaus

Ernst Zermelo (1871–1953) is best-known for his great, pioneering work in set theory that transformed the subject. Of focal importance are his introduction of the axiom of choice in 1904 and his axiomatization of set theory in 1908. The axiom of choice led to a methodological enrichment of mathematics, and the axiomatization was the starting point of post-Cantorian set theory. Zermelo also made significant contributions later, particularly in his 1930 introduction of the cumulative hierarchy.

Less known is that Zermelo did significant work in applied mathematics. His dissertation, for example, promoted the Weierstraßian direction in the calculus of variations; in 1912 he wrote the first paper in what is now called the theory of games; and in 1928 he created the pivotal method in the theory of rating systems. His scientific interests included also purely technical questions such as cruise controls for engines. Zermelo also made translations of poetry, which appear here in these collected works.

Two factors are crucial when considering Zermelo's life. First, he had a confrontational personality: he was polemical in his approach to mathematics, especially against those who disagreed with him, and this appeared in his writing and partly stimulated it. Second, he was very ill for a significant part of his middle life, and this slowed down his mathematical activities and ended his academic career when he was in his mid-forties.

Zermelo's confrontational attitude comes together with his striving for truth and objectivity, and the determination with which he stood up for his convictions. Well-educated in and open-minded about philosophy, the classics, and literature, he had the ability of engaging others in a stimulating way.

Zermelo's life and scientific work can naturally be divided into four periods: The time in Berlin, in Göttingen, in Zurich, and, finally, in Freiburg im Breisgau.

The time in Berlin comprises the years from his birth to 1897, leading scientifically to his first disciplines of specialization, the calculus of variations and mathematical physics.

The following years in Göttingen until 1910 are marked by his pioneering work in set theory under the influence and with the encouragement of David Hilbert. Scientifically this was the most fruitful period of Zermelo's life. Personally, however, it was overshadowed by the outbreak of tuberculosis of the lungs which together with a nervous constitution nearly blocked his chances of getting a permanent university position.

In spring 1910 Zermelo obtained a full professorship at the University of Zurich. The years in Switzerland began with the prospect of a fruitful

academic life. But soon they were spoilt by a recurrence of his illness which kept him from any systematic scientific work and enforced his retirement already in 1916.

The time in Freiburg from 1921 until his death in 1953 was sustained by a new academic position—an honorary professorship for mathematics—and was a time of scientific work both in applied mathematics and in the foundations of mathematics. As the latter was directed against the finitary approaches of Kurt Gödel and Thoralf Skolem which were to shape mathematical logic in the 1930s, it was doomed to fail. Moreover, in 1935 Zermelo lost his professorship because of his anti-fascist disposition. The promising new start had come to an end, which left him in a state of resignation.

What follows is an overview of Zermelo's life, one that incorporates his scientific work and development. A detailed analysis of his achievements is reserved for the introductory notes preceding his papers. This overview draws on our full biography (*Ebbinghaus 2007b*) where extensive supporting documentation is provided.

1 Berlin

1.1 Youth

Ernst Friedrich Ferdinand Zermelo was born in Berlin on 27 July 1871 as the second child of the high school teacher ("Gymnasialprofessor") Dr. Theodor Zermelo (1834–1889) and his wife Maria Auguste née Zieger (1847–1878).

Zermelo had five sisters, the one-year older Anna and the younger sisters Elisabeth, Margarete, Lena, and Marie. The letters which he wrote to them throughout his life attest to a close relationship among the siblings, and care and concern on Zermelo's side.

Zermelo's mother Maria Auguste was the only child of the surgeon Dr. Ottomar Hugo Zieger and his wife Auguste née Meißner. Like her parents she suffered from poor health. Exhausted from the strain of multiple pregnancies, she died shortly before Zermelo's seventh birthday. After her death a maid took care of the household.

Zermelo's father Theodor was the son of Ferdinand Zermelo, a bookbinder from Tilsit (today Sovjetsk, Russia) on the river Memel (Neman) and his wife Bertha née Haberland. He studied mainly history and geography in Königsberg (now Kaliningrad, Russia) and in Berlin. Having attained his doctorate at the Friedrich Wilhelm University (now Humboldt University) in Berlin in 1856 with a dissertation in history, he completed his studies with the exams *pro facultate docendi,* i.e., state exams for teaching at secondary schools, which allowed him to teach history, geography, Greek, Latin, and French; he later completed an additional diploma in mathematics. In its evaluation, the examination board found him to have "an adequate knowledge of the elements of geometry and algebra for teaching mathematics in the lower grades." The

board further observed that he had a solid grasp of the basic notions of logic and general grammar and showed an assured competence in their application.

In sum, Ernst Zermelo grew up in a family with an academic background, but not one with a significant orientation towards mathematics and the sciences.

Zermelo had a kind heart, but he was also prone to sharp and polemical reactions and did not refrain from trenchant irony when he was convinced of his opinions. Whereas the first trait might be attributable to the influence of his mother, his determination to speak his mind may be due to the example of his father. In 1875, Theodor Zermelo had published a widely acknowledged paper, *1875*, on the historian and philologist August Ludwig von Schlözer (1735–1809) of Göttingen, the most influential anti-absolutist German publicist of the early age of Enlightenment. It was his aim to raise awareness of those values expressed by Schlözer's plea for free speech, sincerity, tolerance, and humanity on which, in his opinion, the newly founded German Reich should be built. There are indications that the views of his father as reflected in this treatise and the example of von Schlözer had their effect on Zermelo.

Zermelo shared with his father an interest in poetry. As a young teacher, Theodor Zermelo had compiled a large selection of his own translations of poems from England, the United States, France, and Italy. Zermelo was later to translate parts of the Homeric epics, and he enjoyed commenting on daily events in the form of poems. At the age of thirteen, he made "a metric translation from the first book of Virgil's Aeneid" as a birthday present for his father.

In 1880 Zermelo entered the Luisenstädtisches Gymnasium in Berlin, which he finished in March 1889, only one month after the death of his father. His final school report certifies good results in religious education, German language, Latin, Greek, history, geography, mathematics, and physics. The comment in mathematics says that Zermelo followed the instructions with good understanding and that he had reliable knowledge and the ability to use it for skilfully solving problems. In physics it is testified that he was well-acquainted with a number of phenomena and laws. Under the heading "Behaviour and diligence" it is remarked that he followed the lessons with reflection, but that he occasionally showed a certain passivity as a result of physical fatigue, an indication that Zermelo already suffered from poor health during his school days.

1.2 University

In the summer semester 1889 Zermelo matriculated at the Friedrich Wilhelm University in Berlin. During his course of studies he took one semester each at Halle (winter semester 1890/91) and Freiburg (summer semester 1891). After his final examinations he moved to Göttingen (winter semester 1897/98).

When his father died in 1889, he and his five sisters became orphans. As his father's estate was needed to provide for the younger siblings, Zer-

melo applied for grants and succeeded in obtaining scholarships from several foundations which were affiliated with the Friedrich Wilhelm University and established to support gifted students.

In Berlin Zermelo enrolled for philosophy and took courses in mathematics, above all with Johannes Knoblauch, a student of Karl Weierstraß, and with Lazarus Fuchs. Furthermore, he attended courses on experimental physics and heard a course on experimental psychology by Hermann Ebbinghaus.

In Halle he enrolled for mathematics and physics, attending Georg Cantor's courses on elliptical functions and on number theory, Albert Wangerin's courses on differential equations and on spherical astronomy, and a course given by Edmund Husserl on the philosophy of mathematics at the time when Husserl's *Philosophie der Arithmetik* (*1891*) was about to be published. He also took part in a course on logic given by Benno Erdmann, one of the leading philosophical logicians of the time.

In Freiburg he studied mathematical physics with Emil Warburg, analytical geometry and the method of least squares with Jakob Lüroth, experimental psychology with Hugo Münsterberg, and history of philosophy with Alois Riehl. Furthermore he attended a seminar on Heinrich von Kleist.

Having returned to Berlin, he attended several courses by Max Planck on theoretical physics, among them a course on the theory of heat in the winter semester 1893/94. He also attended a course on the principle of the conservation of energy by Wilhelm Wien (summer semester 1893). In mathematics he took part in courses on differential equations by Fuchs, algebraic geometry by Georg Ferdinand Frobenius, and non-Euclidean geometry by Knoblauch. The calculus of variations, one of the central topics of Zermelo's later work, was taught by Hermann Amandus Schwarz in the summer semester 1892. In philosophy he attended "Philosophische Übungen" by the neo-Kantian Friedrich Paulsen and Wilhelm Dilthey's course on the history of philosophy. He also took a course on psychology, again by Hermann Ebbinghaus (winter semester 1893/94). In October 1894 he received his doctorate, and from December 1894 to September 1897 he served as an assistant to Max Planck at the Berlin Institute for Theoretical Physics.

Looking back on Zermelo's university studies, one may emphasize that he acquired a solid and broad knowledge in both mathematics and physics, his main subjects. His advanced studies directed him to his early specialities of research, to mathematical physics and to the calculus of variations, the topic of his dissertation. He attended courses by Georg Cantor, but no course on set theory, at least until his change to Göttingen where Arthur Schoenflies gave such a course in the summer semester 1898. Furthermore, he had no instruction in mathematical logic. Given his teachers in philosophy, Riehl, Paulsen, Dilthey, and Erdmann, it can be assumed that he had a broad, overall knowledge of philosophical theories. In Halle he even became acquainted with Husserl's phenomenological philosophy of mathematics *in*

statu nascendi. His interest in experimental psychology as taught by Hermann Ebbinghaus and Hugo Münsterberg is quite evident.

In February 1897 Zermelo passed his exams *pro facultate docendi.* In philosophy he wrote an essay entitled "What is the significance of the principle of the conservation of energy for the question of the relation between body and mind?" ("Welche Bedeutung hat das Prinzip der Erhaltung der Energie für die Frage nach dem Verhältnis von Leib und Seele?"). According to the reports he was well-informed about the history of philosophy and systematic subjects. His exams in religious education was evaluated as excellent. As to his exams in mathematics the report says that although he was not always aware of the methods in each domain, it was nevertheless evident that he had acquired an excellent mathematical education. The physics part was rated excellent. The exams in geography showed that he was excellent with respect to theoretical explanations, but that he was not that affected by "studying reality." Finally it was certified that he had the knowledge for teaching mineralogy in the high range and chemistry in the middle.

1.3 Scientific work in Berlin

1.3.1 Ph. D. thesis

Zermelo's Ph. D. thesis *Untersuchungen zur Variations-Rechnung* (*Investigations in the calculus of variations*), *1894*, was suggested and guided by Hermann Amandus Schwarz. Schwarz had studied in Berlin with Weierstraß, becoming his most eminent student. In 1892 he had been appointed his teacher's successor. Zermelo became his first Ph. D. student, and work on the thesis may have been inspired by Schwarz's lectures in the summer semester 1892.

Schwarz proposed generalizing methods and results in the calculus of variations, which Weierstraß had obtained for derivations of first order, to higher derivations. The problem concerns variational problems of type

$$J = \int_{t_1}^{t_2} F\big(x, x', \dots x^{(n)}; y, y', \dots y^{(n)}\big)\, dt ,$$

and asks for curves

$$x = \varphi(t) , \quad y = \psi(t)$$

leading to an extremal value of J.

Weierstraß had treated his solution for $n = 1$ in his courses, particularly in a course given in the summer semester 1879 which had been written up under the auspices of the Berlin Mathematical Society. Zermelo studied these lecture notes in the summer of 1892, when he attended Schwarz's course on the calculus of variations. Less than two years later he succeeded in solving the problem Schwarz had set. On 23 March 1894, then 22 years old, Zermelo applied to begin the Ph. D. process. The oral examination took place on 6 October 1894.

In his report of 5 July 1884 Schwarz describes the subject as very difficult; he is convinced that Zermelo provided the very best solution and predicts a lasting influence of the methods Zermelo had developed and of the results he had obtained:

> According to my judgement the author succeeds in generalizing the main investigations of Mr. Weierstraß [...] in the correct manner. In my opinion he thus obtained a valuable completion of our present knowledge in this part of the calculus of variations. Unless I am very much mistaken, all future researchers in this difficult area will have to take up the results of this work and the way they are deduced.

He marked the thesis with the highest degree possible, *diligentia et acuminis specimen egregium*. Co-referee Fuchs shared his evaluation.

Various voices confirm the significance of Zermelo's dissertation for the development of the Weierstraßian direction in the calculus of variations. Examples may be found in Adolf Kneser's monograph on the calculus of variations (*1900*) and in Constantin Carathéodory's similarly influential monograph *1935*. Oskar Bolza (1857–1942) also gives due respect to Zermelo: his epochal monograph *1909* contains numerous quotations of Zermelo's work. Moreover, the quality of the dissertation played a major role when Zermelo was considered for university positions after his *Habilitation*.

1.3.2 Statistical mechanics

The oral examination of the Ph. D. process featured the defence of three theses that could be proposed by the candidate. Zermelo had made the following choice:

I. In the calculus of variations one has to attach importance to an exact definition of maximum or minimum more than has been done up to now.
II. It is not justified to burden physics with the task of reducing all phenomena in nature to the mechanics of atoms.
III. Measurement can be understood as the everywhere applicable means to distinguish and to compare continuously changing qualities.

The first thesis reflects the special care with which Zermelo attends to the definition of maxima and minima in his dissertation in the case of higher derivations. The latter theses, in particular the second one, are clearly aimed against early atomism in physics and the mechanical explanations coming with it. They may have been adopted by Zermelo when attending Planck's lecture on thermodynamics in the winter semester 1893/94 and gained in significance within the next two years when he was an assistant to Planck.

Around 1895 the atomistic point of view, while widely accepted in chemistry, was still under debate in physics, Planck being among the sceptics. For an "atomist", the principles governing heat theory are reducible to the

mechanical behaviour of the particles that constitute the system under consideration. However, as a rule, the number of atoms definitely precludes the possibility of calculating exactly how each of them will behave. To overcome this dilemma, atomists used statistical methods to describe the expected behaviour.

In Boltzmann's presentation the probability $W(s)$ of a system A to be in state s—or, up to a constant, its entropy—is measured by the number (with respect to some suitable measure) of the configurations of A that macroscopically represent s. According to this interpretation the second law now says that physical systems tend toward states of maximal probability or maximal entropy, thereby imposing a direction on the physical processes concerned.

Zermelo published his concerns about statistical mechanics in a note (*1896a*) that took direct aim at Boltzmann, and this provoked a controversy which took place in two rounds in the *Annalen der Physik und Chemie* in 1896/97. Personal invective aside, Zermelo's note led to a re-consideration of foundational questions in physics at a time when probability was emerging and competing with the paradigm of causality. Zermelo observed that the function describing the spatial coordinates of the particles of a bounded system together with their velocities falls under Henri Poincaré's recurrence theorem, which claims that the system in question will again and again approach its initial state provided this state does not belong to a set of exceptional states of measure zero. Thus, as a rule, the entropy of a bounded system cannot steadily increase. His argument became known as *Zermelo's recurrence objection.*[1]

There is evidence that Zermelo was backed by Planck. Boltzmann, aware of the stakes, felt himself compelled to give an immediate answer. In his reply, *1896*, he accepted the application of the recurrence theorem, but defended his theory by arguing that the second law would in itself be subject to probability. The increase of entropy was not absolutely certain, but a decrease or a return to the neighbourhood of the initial state was so improbable and the expected recurrence time so long that one was fully justified in excluding it.

Zermelo answered with further counter-arguments (*1896b*), among them a variant of Josef Loschmidt's reversibility objection, arguing that the notion of probability did not refer to time and, hence, could not be used to impose a direction on physical processes as is the case with the second law. With regard to this point, Boltzmann's defence *1897* brought up the possibility that the universe may be a collection of single worlds, partly in a state of growing entropy and partly in a state of decreasing entropy, thus providing a way to escape the reversibility objection.

Through the next decade, Zermelo maintained interest in statistical mechanics. In his *Habilitation* address, *1900*, given in 1899 in Göttingen, he offered

[1] Three years earlier Poincaré had already formulated the recurrence objection as one of the major difficulties when bringing together experimental results and mechanism (*1893*, 537).

a solution to a problem which he had already mentioned at the end of his rejoinder *1896b*, the problem that on the one hand, Boltzmann applied his statistical view to each state of a system developing in time, whereas, on the other hand, these states were not independent from each other but were intimately linked by the laws of mechanics. Six years later Zermelo published a German translation (*Gibbs 1905*) of Josiah Willard Gibbs' monograph on statistical mechanics (*Gibbs 1902*) which "played an important role in making Gibbs' work known in Germany" (*Uffink 2004*, 1.2). In the preface he wrote in appreciation of Gibbs' book as being the first attempt to develop strictly and on a secure mathematical basis the statistical and the probability theoretical considerations in mechanics. However, despite the growing acceptance of Boltzmann's ideas—even by Planck—Zermelo's reservations concerning the range of statistical mechanics did not get resolved, as described in detail in his review *1906* of Gibb's book. Today Boltzmann's ideas are generally accepted. However, some of Zermelo's questions such as that for a physical explanation why, say, our universe might have been in a state of low entropy when it came to exist, or the question about how Boltzmann's *Einzelwelten* fit the cosmological picture, are still subject to actual research.

2 Göttingen

2.1 Introduction

Already in the summer of 1896, while still an assistant to Planck, Zermelo applied for a position as assistant at the Deutsche Seewarte in Hamburg, the central institution for maritime meteorology of the German Reich. His application was supported by Schwarz and Planck. Apparently, Zermelo was going to give up on an academic career, henceforth dedicating his work to a mathematical treatment of practical metereological problems. For unknown reasons, however, he ultimately decided to pursue the aim of obtaining an academic position. The mathematical treatment of metereological depressions he had started seemed of sufficient worth to be extended to a *Habilitation* thesis, a post-doctoral thesis necessary for obtaining a professorship.

On 19 July 1897 Zermelo turned to Felix Klein in Göttingen for advice, and presumably Klein's answer encouraged him to go to Göttingen. He enrolled there for mathematics on 4 November 1897. Given his special field of competence, Zermelo had made a good choice, as Göttingen was a centre of applied mathematics. It was to become, moreover, the leading centre for research in the foundations of mathematics, a development that began after the turn of the century, the driving force to be David Hilbert. On the initiative of Klein he had been brought from Königsberg University two years before Zermelo arrived.

In 1897 Hilbert was still mainly working in algebraic number theory. He subsequently changed his main field of research, concentrating on the founda-

tions of geometry and the axiomatic method. Zermelo became deeply involved from the first hour on. He grew into the role of Hilbert's most important collaborator in foundational studies in this important early period, slowly moving from applied mathematics and mathematical physics to foundational studies, set theory, and logic. Decades later he would speak of Hilbert as his "first and sole teacher in science" to whom he owes "more than to anybody else with regard to ⟦his⟧ scientific development."

2.2 Applied mathematics in Göttingen

During his first semester in Göttingen Zermelo heard Hilbert's course on irrational numbers and attended the mathematical seminar on differential equations in mechanics directed by Hilbert and Klein. He took exercises in physics with Eduard Riecke and heard thermodynamics with Oskar Emil Meyer. During the next semester he took part in a course on set theory given by Arthur Schoenflies and in Klein's mathematical seminar.

For the time being and in accord to his plans, applied mathematics and physics became Zermelo's main subjects. Hilbert's *Nachlass* contains a manuscript (*Zermelo s1899b*) in which Zermelo studied the movement of an unstretchable material thread in a potential field. In his first Göttingen publication (*1899a*) Zermelo succeeded in generalizing a result of Adolph Mayer (*1899*) on the uniqueness of the solutions of a special differential equation; this equation represents the accelerations of the points of a frictionless system in terms of their coordinates, these restricted by inequalities, and their velocities.

In late 1898, Zermelo completed his *Habilitation* thesis which he had begun already with Planck in Berlin, completing the *Habilitation* process on 4 March 1899 by giving the *Habilitation* address *1900* mentioned above.

The thesis aims at creating a systematic theory of incompressible and frictionless fluids streaming in the sphere. The starting motivation is to get "information about many a process in the development of atmospheric cyclones and ocean currents as far as they concern the whole of the earth and as far as the vertical component of the current can be neglected against its horizontal one." The first two chapters deal with the general theory, while the third chapter deals with the case of finitely many vortices and the last chapter with the special case of three.

Only the first two chapters of the *Habilitation* thesis appeared in print (*1902a*). The *Nachlass* contains the handwritten version of Chapters 3 and 4 (*s1902b, s1902c*). There are no documents that might explain why the last two chapters remained unpublished and whether this was due to a deliberate decision.

The review of the published part in the *Jahrbuch der Fortschritte der Mathematik* (Vol. 33, 0781.01) essentially quotes from the introduction. In later reports on Zermelo's achievements having to do with university position possibilities, his *Habilitation* thesis did not play a major role, whereas his

Ph. D. thesis was highly praised. Perhaps the overall reason for this modest reception may be that for the time being the field of fluid dynamics was shaped by Ludwig Prandtl, professor in Göttingen from 1904 on, in another direction.

In contrast to fluid dynamics, Zermelo remained interested in the calculus of variations. Even when he began to turn his main attention to set theory, he still published several papers in the field. In the first, *1902d*, he provided an intuitive description of some extensions of the problem of shortest lines on a surface, namely for the case of bounded steepness with or without bounded torsion, illustrating them with railroads and roads, respectively, in the mountains.

In the next paper, *1903*, he gave two simplified proofs of a result of Paul du Bois-Reymond (*1879*) which says that, given n and an analytical F, any function y for which $y^{(n)}$ exists and is continuous and which yields an extremum of $\int_a^b F(x, y, \ldots, y^{(n)})dx$ possesses derivatives of arbitrarily high order. The theorem shows that the Lagrangian method for solving the related variational problem which uses the existence of $y^{(2n)}$ and its continuity does not exclude solutions for which $y^{(n)}$ exists and is continuous.

Together with Hans Hahn, one of the initiators of linear functional analysis, Zermelo wrote a continuation of Adolf Kneser's contribution *1904* on the calculus of variations for the *Encyklopädie der Mathematischen Wissenschaften* (*Hahn and Zermelo 1904*), which gives a clear exposition of the Weierstraßian method.

Thanks to his interest in the calculus of variations, Zermelo came into closer contact with Constantin Carathéodory. Carathéodory had also studied in Berlin with Hermann Amandus Schwarz and Max Planck, and was then working in Göttingen with Klein, Hilbert, and Hermann Minkowski, writing his Ph. D. thesis (*1904*) under the final guidance of the latter. His mathematical interests, which extended also into statistical mechanics, led to extensive contacts with Zermelo and developed into a deep, lifelong friendship. In 1906 both were working together on a book on the calculus of variations which, according to Minkowski, "promises to become the best in this field." Probably because of Zermelo's poor state of health then and Carathéodory's move to Bonn in 1908, the book would never be completed.

In his early scientific period Zermelo showed great interest in textbooks and monographs. Already in 1897 he had edited a German translation, *Glazebrook 1897*, of Richard Tetley Glazebrook's elementary textbook on light (*Glazebrook 1894*). In Göttingen he co-published the second edition of the second and third volume of Joseph Alfred Serret's *Lehrbuch der Differential- und Integralrechnung* (*Serret 1899*, *Serret 1904*). A year later he finished his German translation *Gibbs 1905* of Gibb's monograph *1902* on statistical mechanics. This interest corresponds to the care which he will exhibit in preparing his lecture courses. As a matter of fact, however, he was considered an excellent reader only for the more advanced students.

As late as 1909 Zermelo returned once more to physics; together with Ernst Riesenfeld he studied the limit concentration of solvents which are brought together in a horizontal cylinder (*Riesenfeld and Zermelo 1909*).

2.3 Toward set theory

Soon after his arrival in Göttingen, Zermelo turned to foundational questions under the influence of Hilbert. In fact it was his foundational work, in particular his work on set theory, which would lead to his greatest and most influential scientific achievements, but in the end also to a depressing failure. The work is concentrated in two periods which roughly coincide with his decade in Göttingen and the years around 1930 in Freiburg, and corresponds to two specific periods of research on the foundations of mathematics by Hilbert and his collaborators in Göttingen. In the first period Hilbert was articulating his early axiomatic program, and Zermelo's work is clearly along the lines proposed by Hilbert. It is *pro* Hilbert. The second period falls into the years when Hilbert was developing his proof theory which, because of its finitistic character, was flatly rejected by Zermelo. Zermelo's work at this time is *contra* Hilbert.

2.3.1 Hilbert on foundations

In the introductory part of his 1900 Paris address Hilbert considers the general preconditions for the solution of a mathematical problem. For him it is a requirement of mathematical rigour in reasoning that the correctness of a solution can be established "by means of a finite number of inferences based on a finite number of hypotheses."

The second problem of the address concerns the consistency of the axioms for the real numbers, a central topic for subsequent Göttingen research on foundations. Hilbert discusses the basic features of his axiomatic method, stressing for the first time the central role of consistency proofs: The investigation of the foundations of a scientific discipline starts with setting up an axiom system which contains "an exact and complete description of the relations subsisting between the elementary ideas" of that discipline. Given such an axiomatization, "the most important among the numerous questions which can be asked is that of consistency: *To prove [...] that a finite number of logical steps based upon them [the axioms] can never lead to contradictory results.*" Consistency stands out, because it becomes a criterion for mathematical existence: If inconsistent attributes are assigned to a concept, the concept does not exist mathematically, whereas a proof of consistency means a proof of mathematical existence.

It is significant that Hilbert illustrates the prospects for his method with examples from set theory. The proof of the consistency of the arithmetical axioms, he writes, would at the same time prove the mathematical existence

of the complete system of the real numbers ("Inbegriff der reellen Zahlen"), i.e., of the continuum, and he continues that such an existence proof via a consistent axiomatization should also be possible, for example, for Cantor's number classes and cardinals. So there is clearly a desideratum: a consistent axiomatization of set theory.

Hilbert's comments are rooted in his development of an axiomatic system for Euclidean geometry and in problems about inconsistent sets which he had discussed with Cantor already in 1897. Both origins were cast in a new light when the publication of the set-theoretic paradoxes in 1903 suddenly questioned the foundations of mathematics.

There is strong evidence that the problems arising from unlimited comprehension were discussed in the Hilbert group in the late 1890s and that in the course of these discussions Zermelo had come upon the Zermelo-Russell paradox of the set consisting of all sets which are not an element of themselves. However, only Russell's publication of the paradox in 1903 led to the apprehension that the paradoxes presented a severe threat to Hilbert's early consistency programme.

2.3.2 The well-ordering theorem

After the turn of the century, set theory moved into the centre of Zermelo's scientific endeavour. During the winter semester 1900/01 he already gave a course on the subject, centred around the Cantorian notion of cardinal. A first result of his set-theoretic research, *1901*, was presented to the Königliche Gesellschaft der Wissenschaften zu Göttingen at the meeting of 9 March 1901. The main theorem says that if a cardinal $\mathfrak{m}$ remains unchanged under an addition of any cardinal from an infinite series $\mathfrak{p}_1, \mathfrak{p}_2, \mathfrak{p}_3, \ldots$, then it remains unchanged under the addition of the sum of the $\mathfrak{p}_i$. In his argument, Zermelo makes tacit use of the axiom of choice. As a special case of the main result he mentions the equivalence theorem which asserts that two sets are equivalent (i.e., that they can be mapped onto each other by a one-to-one function) if each set is equivalent to a subset of the other.

The year 1904 brought dramatic events. On 9 August, during the Third International Congress of Mathematicians in Heidelberg, Julius König delivered a lecture in which he claimed to have proved that Cantor's continuum conjecture is false, and more, that the continuum is not well-orderable. He thus also claimed to have refuted the well-ordering principle, Cantor's "fundamental law of thought" (*Cantor 1883b*, 550), according to which every set can be well-ordered.

Both Zermelo and Felix Hausdorf very soon and independently suspected an error in König's argument: König's proof rested on an incorrect extrapolation from Felix Bernstein's dissertation *1901*. Only a few weeks later, on 24 September 1904, Zermelo delivered a letter to Hilbert which appeared as *1904* and gives a proof of the well-ordering theorem, the theorem that every set can be well-ordered.

Zermelo based the proof on and for the first time explicitly formulated the axiom of choice. It guarantees for any set S consisting of non-empty sets the existence of a *choice function* on S, a function defined on S and assigning to each set in S one of its elements. As a matter of fact, the axiom had already been used earlier without drawing attention—as we have seen, even by Zermelo himself—and Zermelo was right in speaking of usage "everywhere in mathematical deductions without hesitation." Applications ranged from analysis to set theory itself, some of them avoidable, some of them not. Applications of the latter kind include the fact originating with Cantor (*1878*, 243) that the union of countably many countable sets is countable and the fact originating with Cantor (*1895*, 493) and Émile Borel (*1898*, 12–14) that every infinite set contains a countably infinite subset. Nevertheless, the formulation of the axiom of choice and its use in the well-ordering proof provoked resistance and generated a debate that is unsurpassed in the modern history of classical mathematics, a debate that subsided only after decades.

2.4 Disappointments

On 4 March 1899 Zermelo had become a *Privatdozent*, which meant that he had the right to lecture and was eligible for a professorship, but did not get any regular salary. In spite of Hilbert's unwavering support, eleven years were to pass before Zermelo was offered a first tenure-track position, a full professorship at the University of Zurich. There may be several reasons for this delay: In a letter to Wilhelm Wien of 12 October 1906, Hermann Minkowski remarked that Zermelo's "conspicuous lack of good luck stems from his outer appearance, his nervous haste which comes out in his speaking and conduct." In fact, complaints about Zermelo's nervous state are abundant. Moreover, on 24 January 1910, the board of the Seminar of Mathematics and Physics emphasized that Zermelo had been harmed by criticism of his well-ordering paper *1904* "which ⟦was⟧, according to the experts among us, not justified." After 1905 an additional major reason was Zermelo's tuberculosis, which required years of recuperative stays in southern or alpine regions.

Two years after his *Habilitation*, Zermelo applied for a *Privatdozenten* grant in order to get a livelihood. He was supported by Hilbert who described him as "a gifted scholar with a sharp judgement and a quick intellectual grasp." The application was successful. The grant ensured his economic situation, albeit modestly, until 1907.

In 1903 he was under consideration for a position at the University of Breslau (now Wrocław, Poland) because of his work in applied mathematics and physics. The shortlist consisted of four candidates who were named *aequo loco* in the first place, among them Gerhard Kowalewski and Zermelo. However, two days after this shortlisting, Zermelo was moved to the end of the list. In doing so, the Faculty ignored Hilbert's staunch recommendation. Hilbert had characterized Zermelo as "the real candidate for the Breslau Faculty", praising his "extremely varied" lecturing activity, and characterizing

him as "a modern mathematician who combines versatility with depth in a rare way."

Only after nearly six years as a *Privatdozent* did Zermelo receive the title "Professor". The appointment resulted from an application filed a year earlier by Hilbert. Hilbert had especially emphasized Zermelo's penetrating grasp of the problems of mathematical physics. Very likely, the long time which passed between Hilbert's application and Zermelo's appointment was the result of severe health problems. Zermelo's school certificates had already given evidence of his weak constitution. In the winter of 1904/05 he fell seriously ill, probably of an inflammation of the lungs. In order to recover, he spent the spring and early summer of 1905 in Italy. At the beginning of 1906 he fell ill again, now from pleurisy, and had to take a longer break. Finally, in June 1906, his doctors diagnosed tuberculosis of the lungs and after clinical treatment recommended a longer stay in the mountains. For the time being, Zermelo spent some weeks at the seaside. He returned to Göttingen "in a refreshed state", so Minkowski in a letter to Wien. However, as the recovery did not last, Zermelo had to cancel his course of lectures in the winter semester 1906/07 and took several longer cures in the Swiss Alps lasting until the spring of 1908.

During this time, two further applications for university positions failed: In the autumn of 1906 the application for a professorship at the University of Würzburg, and in spring 1907 the application for a professorship at the Academy of Agriculture at Poppelsdorf near Bonn. The first application had been supported by Minkowski in the warmest way; he described Zermelo as "a mathematician of the highest qualities, of broadest knowledge, of quick and penetrating grasp, of rare critical gift." Commenting on Zermelo's lack of success, Minkowski reported that the physicist Theodor des Coudres, seeing the disparity between Zermelo's deplorable personal situation and his scientific achievements, had called him the prototype of a tragic hero.

2.5 Defence of the well-ordering theorem

In May 1905, during his spring cure in Italy, Zermelo was working on the theory of finite sets. In June he observed that Richard Dedekind's theorem 160 from the *Zahlen* treatise *Dedekind 1888* can be elegantly proved by use of the well-ordering theorem. Dedekind shows that a set is finite in the sense of being equivalent to a proper section of the natural numbers if, and only if, it is Dedekind finite, i.e., not equivalent to a proper subset. For the direction from right to left he makes tacit use of the axiom of choice. Zermelo notes that the argument becomes trivial if one starts with a *well-ordered* Dedekind finite set. He planned to talk about his result at the 77th meeting of the Gesellschaft deutscher Naturforscher und Ärzte at Merano in September 1905 ("about the theory of finite number or something similar"). In the end, he did not do so. However, work continued and led to his papers *1909a* and *1909b* where he gives a set-theoretic definition of finite sets without involving infinite

sets ("Zermelo finiteness"). Viewing natural numbers as finite cardinals, he then shows that the principle of induction becomes a provable fact. He did not extend his investigations by giving a set-theoretic model of arithmetic. This was only done by Kurt Grelling in his dissertation (*1910*) which was officially supervised by Hilbert, but *de facto* by Zermelo.

The work on finite sets, in particular the work on Dedekind's theorem 160, strengthened Zermelo's conviction that the well-ordering principle and the axiom of choice were of fundamental importance: "The theory of finite sets is impossible without the 'principle of choice', and the 'well-ordering theorem' is the true foundation of the whole theory of number", we read in a letter to Hilbert of 29 June 1905. These insights may have triggered off his decision to directly take on the opponents of his well-ordering proof.

In the summer of 1907, while staying in the Swiss Alps, Zermelo worked out a comprehensive answer to his critics in the paper *1908a*. He first wrote of his intention casually to raise some objections; however, besides all their brilliance, some of his remarks demonstrate a high degree of engagement and polemics both in argumentation and tone.

In a clear and definite way Zermelo refutes criticism concerning the technical aspects of his 1904 proof. Concerning the axiom of choice, his arguments reveal quite a few of his personal convictions. For example, in his answer to criticism put forward, for instance, by Émile Borel (*1905a*, 194) which blames him for using the axiom without having proved it, he strongly pleads for evidence as a source of mathematics: The extensive use made hitherto of the axiom "can be explained only by its self-evidence, which, of course, must not be confused with its provability. No matter if this self-evidence is to a certain degree subjective—it is surely a necessary source of mathematical principles."

In order to meet certain misunderstandings of his 1904 proof, Zermelo provides a new proof of the well-ordering theorem. As successive choices had been a focal point of these misunderstandings and in order to support his view that well-orderings have "nothing to do with spatio-temporal arrangement", he bases the new proof on a reformulation of the axiom of choice where choice functions have been replaced by choice sets. He thus arrives at Bertrand Russell's multiplicative axiom (*Russell 1906a*, 48–49).

2.6 Axiomatization of set theory

In order to secure the proof of the well-ordering principle, to further its comprehensibility, and to clearly exhibit the role of the axiom of choice, Zermelo argued about his new 1908 proof that, apart from the axiom of choice, it does not use intricate set-theoretic results but can be based on some "elementary and indispensable set-theoretic principles." By exhibiting these principles more systematically in his axiomatization paper *1908b*, he formulated the first full-fledged axiom system of set theory. Besides this methodological aspect, the "twin papers", the axiomatization paper and the paper *1908a,* are closely connected by mutual references as well as by temporal proximity.

The axiomatization paper is the keystone of Zermelo's set-theoretic work during his first period of foundational research. It is written in the spirit of Hilbert's foundational programme. Hilbert himself belonged to its admirers, characterizing Zermelo in his 1920 course (*1920*, 21–22, 33) as "the person who in recent years has newly founded ⟦set⟧ theory and who has done so, in my view, in the most precise way which is at the same time appropriate to the spirit of the theory [...and] the most brilliant example of a perfected elaboration of the axiomatic method."

When writing down his axiomatization, Zermelo paid due attention to the Zermelo-Russell paradox and the Burali-Forti paradox. They motivate his axiom of separation which restricts unlimited comprehension to comprehension inside a set, allowing only special properties for separation which he calls "definite". The definition of definiteness given is rather vague and was to become a focal point of criticism.

The axioms forming the *Zermelo axiom system* are the axioms of extensionality, elementary sets ($\emptyset$, $\{a\}$, $\{a,b\}$), separation, power set, union, choice, and infinity.

Zermelo's search for axioms started around 1905. A first version of the axiom system was ready in June 1906. The main difference with the final system concerns the axiom of infinity. Whereas the 1906 axiom of infinity ensures the existence of a Dedekind finite set (a set not equivalent to a proper subset), the final version ensures the existence of a nonempty set which is closed under the transition from a set a to its singleton set $\{a\}$. Perhaps this had come to Zermelo from Dedekind's intuitive argument for the existence of an infinite set (*Dedekind 1888*, §5), which conceives an infinite set by considering an object t of one's thoughts, a "thought-thing", and successively conceives new thought-things by the thought of t, the thought of the thought of t, ...

To demonstrate the strength of the system, Zermelo gives a proof of the equivalence theorem, one which turned out to be much like a then still unpublished proof of Dedekind (*Dedekind 1932*, 447–448). The paper culminates in a generalization of König's inequality which is now known as the Zermelo-König inequality. Probably Zermelo had been led to it by the events of the 1904 Heidelberg congress; he had presented it to the Göttingen Mathematical Society already on 24 October 1904. Only in 1921 will Abraham Fraenkel observe a certain weakness of Zermelo's system which will lead to the formulation of the axiom of replacement.

2.7 Last years in Göttingen

In 1907 Zermelo's *Privatdozenten* grant came to an end. As described earlier, various applications for a salaried professorship had failed. Faced with this situation and following Hilbert's initiative, the board of directors of the Seminar of Mathematics and Physics applied for a salaried lectureship for mathematical logic, arguing that foreign countries maintained the leading

role in this scientific discipline and that among the younger generation in Germany only Zermelo could be considered a fully valid representative of this discipline.

The application was successful. Zermelo's appointment marks the birth of mathematical logic as an institutionalized sub-discipline of mathematics in Germany. One might wonder why his teaching position was dedicated to mathematical logic and not, as one could conclude from Zermelo's fields of interest, to set theory. There are clear reasons. After the discovery of the paradoxes, Hilbert's revised axiomatic programme fostered the mutual development of logic and set theory as linked together. Zermelo seemed to be the person suited for enhancing logical competence in Göttingen.

Delayed by his illness, Zermelo's first course on mathematical logic took place in the summer semester 1908. It was attended by 20 students, among them Ludwig Bieberbach, Richard Courant, Kurt Grelling, Ernst Hellinger, Leonard Nelson, and Andreas Speiser. Only a part of the program ("The elementary laws of logic", "The logical form of the mathematical theory (definition, axiom, proof)", "Does arithmetic contain synthetic elements and which are they?") was treated in the course. In particular, Zermelo did not pursue the problem of the paradoxes which Hilbert had so strongly emphasized.

Apparently his involvement with mathematical logic led Zermelo to plans for writing a book on this subject. The project, however, was never realized. Instead and together with Gerhard Hessenberg and Hugo Dingler, he conceived a new plan, aiming at establishing a quarterly journal for the foundations of the whole of mathematics. This project was also dropped when it became clear that Teubner Verlag whom they had approached had both economic concerns and concerns about the character of the journal.

The next opportunities for obtaining a professorship did not meet with success. When Hermann Minkowski died on 12 January 1909, Hilbert did not favour Zermelo, but Edmund Landau as the successor. Hilbert did however use the offer to Landau to have Zermelo's salary increased considerably. In July 1909, the Senate of the University of Würzburg put Zermelo first on the list for a full professorship. Nevertheless, the authorities did not offer the position to him, but to Emil Hilb who had been listed second. Apparently, Zermelo's teaching qualities had been questioned—in spite of Hilbert's judgement that "Zermelo's lecture courses are always very successful."

In January 1910 Felix Bernstein, nine years younger than Zermelo, was offered an extraordinary professorship of insurance mathematics in Göttingen, where he had obtained his Ph. D. degree in 1901 with a thesis in set theory. In order to avoid adverse comparison of Zermelo, the board of directors of the Seminar of Mathematics and Physics applied to make him an extraordinary professor, too, albeit without rising his salary. They argued that "Zermelo is—as far as mathematics is concerned—not only much superior to [...] Bernstein, but [...] one of the most important of all younger German mathematicians." There was no decision on the application how-

ever, because in February 1910 Zermelo was offered a full professorship at the University of Zurich.

3 Zurich

Zermelo's time in Zurich from 1910 to 1916 might be seen as a prosperous period in his life: It was the only time that he held a paid university professorship. On the other hand, it was a dismal time because it was moulded by his tuberculosis and ended with his retirement when he was just 45 years old. The subsequent time in Switzerland until his move to Freiburg im Breisgau in 1921 was one of, in his own words, "a roving life in guest-houses", first in the higher mountains and finally in the mild climate of Locarno on Lago Maggiore. In view of these circumstances, it is not surprising that his scientific work during these years was far from being rich and fruitful.

3.1 The chair in Zurich and its end

In February 1910 Zermelo was offered a full professorship of mathematics at the University of Zurich as the successor of Erhard Schmidt. Schmidt had taken his doctorate under Hilbert's supervision in 1905 in Göttingen and had become a professor in Zurich in 1908. The final choice for the successorship had been between Zermelo and Issai Schur. Obviously impressed by a very positive assessment of Hilbert which had been strongly confirmed by Schmidt, the Faculty had voted 15:1 *pro* Zermelo.

Zermelo was offered the chair initially for a period of six years as was then usual, with the stipulation of representing the field of pure mathematics. The assumption of the position was on 15 April 1910.

In January 1910 Zermelo had undergone a check-up by a Zurich expert for pulmonary diseases who diagnosed tuberculosis in an apparently uncritical state. One year later, however, Zermelo's health had deteriorated to such an extent that he had to take leave for the summer semester 1911 and for the winter semester 1911/12. For the time being he took a cure in a sanatorium for lung patients above Davos. In January 1912 the disease of the right lung had considerably progressed and produced a cavern. The worsening of his health affected his confidence in a final recovery to such a degree that he made definite plans for a publication of his collected works.

The spring of 1913 brought a further depressing event. On 21 February the Senate of the Technical University of Breslau had voted 6 to 2 for the request of the Department for General Sciences to put Zermelo first on the list to be the successor to Carathéodory. The files show that there was resistance to the list because of doubts raised about Zermelo's state of health and about his ability to teach at a technical university. In spite of support from Ernst Steinitz and Gerhard Hessenberg, both professors at the Technical University,

and of Adolf Kneser and Erhard Schmidt, both professors at the University of Breslau, the minister decided for Max Dehn who had been shortlisted in the second place.

In March 1914 Zermelo underwent surgery of the thorax in order to fill the cavern in the right lung with paraffin wax. In spring 1915 there was a second serious recurrence of the disease. It required a longer time off which finally had to be extended until the end of the winter semester 1915/16.

Zermelo being the only full professor of mathematics, the University expressed concern about the situation and confirmed their intention to ensure "an undisruptive, adequate course of classes in the mathematical disciplines through the summer semester 1916." When a medical examination ruled out the possibility of Zermelo lecturing at the beginning of the semester, he was asked to agree to be retired. When he finally did so, albeit "merely because of a probably only temporary inability to work", he was retired as of 15 April.

Zermelo's professorship had come to an end owing to an illness which had interfered with his scientific work for years and very probably had cost him several offers of other university positions. The imposed retirement was to leave its scars. In later years he would feel ill-treated by the "bigwigs", those more successful in their university careers. Moreover, his bitterness resulted in an aversion to Switzerland. There was a widely circulated anecdote, a source for which was a footnote in Abraham Fraenkel's autobiography (*1967*, 149, fn. 55) where he reports on "this ingenious and strange mathematician whose name has held until today an almost magical sound":

> Shortly before the World War he spent a night in the Bavarian Alps. He filled the column "Nationality" in the hotel's registration form with the words: "Not Swiss, thank goodness." Misfortune would have it that shortly after that the head of the Education Department of the Canton Zurich stayed at the same hotel and saw the entry. It was clear that Zermelo could not stay much longer at the University of Zurich.

According to Zermelo himself such an event really happened, probably in the 1920s, in a Black Forest hotel near Freiburg; the entry in the registration form led to inquiries by the Swiss authorities, but had no consequences.

From 1916 to 1919 Zermelo lived in several alpine health resorts, only occasionally returning to Zurich. An exception was his stay in Göttingen during the winter 1916/17. From October 1919 to February 1920 he stayed in Locarno on Lago Maggiore. He spent the spring of 1921 in Southern Tyrol. Evidently his generous pension allowed him to live without financial concerns. This situation was to change considerably when he moved to Germany in 1921; the severe inflation soon to follow would worsen his situation and lead to continual complaints about financial needs.

In October 1916 Zermelo received significant recognition. He was awarded the annual Alfred Ackermann-Teubner prize for the promotion of the mathematical sciences of the University of Leipzig. Later prizewinners include, for example, Emil Artin and Emmy Noether.

3.2 Teaching and scientific activity in Zurich

In the winter semester 1910/11 Zermelo had 39 attendants in his beginners' course on the differential and integral calculus and 20 attendants in his course on differential equations. So he could certainly have gained some gifted students for more intensive mathematical studies, thus establishing his own working group. But the long interruptions of his lecturing activities and his absence from Zurich forced by his illness prevented him from steadily attending to his students. On the other hand, he tried to fulfil his educational duties as well as possible and exercised great care in organizing replacements. Only one dissertation was done in Zurich under his supervision, Waldemar Alexandrow's elementary foundations of measure theory (*1915*).

After having finished his doctorate in 1912 with a number-theoretic thesis supervised by Edmund Landau in Göttingen, Paul Bernays took the opportunity to follow Zermelo to Zurich in order to make his *Habilitation* there. His thesis *1913* on modular elliptic functions was accepted in 1913. He then served as an assistant to Zermelo and as a *Privatdozent* until the spring of 1919. The files document the fruitful collaboration between Zermelo and Bernays. Their personal relationship was marked by mutual esteem. Both continued to exchange letters at least on special occasions, a sign of particular intimacy.

Besides Bernays, Ludwig Bieberbach also made his *Habilitation* under Zermelo's guidance. Bieberbach had accompanied Zermelo from Göttingen to Zurich. Already in 1910 he became a *Privatdozent* there; Zermelo's positive evaluation was shared by Albert Einstein, then an extraordinary professor at the University of Zurich. Four months later, Bieberbach took up a lectureship in Königsberg. Zermelo's relationship to Einstein got closer when Einstein, who had left Zurich in 1911 for the University of Prague, returned to the Eidgenössische Technische Hochschule (ETH) Zurich in 1912. Both met regularly to discuss scientific questions, Einstein appreciating these conversations because of Zermelo's competence in both mathematics and physics.

Contact with Hermann Weyl, whom Zermelo knew from Göttingen and who had obtained a professorship at the ETH Zurich in 1913, was less close. Around that time Weyl had started to favour features of intuitionism. Later Zermelo (*s1930d*) will speak of "a somewhat noisy appearance of the 'intuitionists' who in vociferous polemics announced a 'foundational crisis' of mathematics [...] without being able to replace it by something better." It may have been the deep gulf in foundational views which blocked a fruitful scientific exchange.

Zermelo continued his research as far as his illness would allow. His interests ranged from physics, even engineering techniques, via applied mathematics to algebra and set theory. The dominant feature of his research was diversity, and no topic is pursued in a systematic manner as becomes evident from our discussion below.

In February 1911 Zermelo together with a partner applied for two patents with a Zurich patent attorney. The first patent, "On an oscillation-free reg-

ulator to produce a constant number of revolutions which can be arbitrarily fixed while the machine is running", was for a kind of cruise control for engines. The second patent was for a related means "to produce a constant torque at one of two coupled shafts."

Zermelo was an enthusiastic chess player. His passion motivated two papers. The later one, conceived in 1919, but published only in 1928 (*1928*) provides a method to calculate the outcome of a chess tournament (cf. 4.2). The earlier one (*1913*) represents an address that he gave during the Fifth International Congress of Mathematicians at Cambridge in 1912. The main theorem says that in a chess game, when starting from any position, either White can enforce a win in less than N moves (where N is the number of positions) or Black can enforce a win in less than N moves, or both can enforce a draw. It may be regarded as the first essential result in the theory of games and eventually became known as "Zermelo's theorem". Some fallacies of the proof were corrected by him in *D. König 1927b.*

In late 1913 Zermelo completed an algebraic paper, *1914*, that involves his well-ordering theorem in an essential way. Ten years earlier, Georg Hamel had proved the existence of bases for the vector space of the reals over the field of the rationals (*1905*), defining them along a well-ordering of the reals. Zermelo transfers the method to algebraic independence in order to show the existence of rings whose quotient field is the field of reals and the field of the complex numbers, respectively, and whose algebraic numbers are algebraic integers.

Probably during a stay in a sanatorium in Davos around 1915, Zermelo wrote down a definition of the (von Neumann) ordinals together with their basic properties, attributing some of the results to Bernays. John von Neumann subsequently developed the ordinals in his *1923*. When commenting on Zermelo's anticipation (*1928a*, 321) he remarks that "the fundamental theorem, according to which to each well-ordered set there is a similar ordinal, could not be rigorously proved ⟦by Zermelo⟧ because the replacement axiom was unknown." Zermelo gave a talk on his results in Göttingen on 7 November 1916.

In a letter of 6 May 1921, Abraham Fraenkel asked Zermelo how one can show in the Zermelo axiom system that for an infinite set S the set

$$\{S, \text{ the power set of } S, \text{ the power set of the power set of } S, \ldots\} \qquad (1)$$

is a set. Already on the next day, Zermelo answered that he had missed this point when writing his axiomatization paper. To remedy the shortcoming he proposed a new axiom which is essentially the axiom of replacement with bijective operations instead of arbitrary ones. He instantly drew back because of definiteness issues, but Fraenkel retained the idea, shaping the axiom of replacement with arbitrary operations and proposing it in his *1922b*, which was finished in July 1921.

4 Freiburg

In October 1921, some months after his 50th birthday, Zermelo moved to Freiburg im Breisgau in southwestern Germany, where he subsequently lived for more than three decades until his death in 1953. The mid-1920s saw a resumption of his scientific activity in both applied mathematics and set theory. Furthermore, he was to lecture at the University of Freiburg as an honorary professor. However, there was an evident reversal in the mid-1930s. As his foundational work was directed against the mainstream shaped by inspired members of a new generation such as Kurt Gödel and John von Neumann, Zermelo ended up in a kind of scientific isolation and also in opposition to his mentor David Hilbert. As his political views disagreed with the Nazi ideology, he lost his honorary professorship. In 1935 he was left with deep disappointments and began to live a life of isolation. His last years were brightened by his wife Gertrud whom he had married in 1944.

4.1 A new start

Zermelo's move to Freiburg meant a return to the city where he had studied mathematics and physics during the summer semester of 1891 and experienced the advantages the town might offer to his illness: the mild climate of the upper Rhine valley together with its location at the foothills of the Black Forest. Moreover, the University with its emphasis on arts and humanities accommodated his strong interests in philosophy and the classics.

In the winter semester 1923/24, Zermelo attended Edmund Husserl's course "Erste Philosophie" (cf. *Husserl 1956*) and Jonas Cohn's course "Logik und Erkenntnistheorie". However, after the semester he lost interest in the philosophical trends represented in Freiburg, among them Husserl's phenomenology. This separation deepened when Martin Heidegger became Husserl's successor in 1928. Instead, Oskar Becker, a student of Husserl, who later obtained a professorship for philosophy at the University of Bonn, became Zermelo's main partner in questions of the philosophy of mathematics.

Presumably Zermelo very soon came into contact with the two full professors of mathematics at the Mathematical Institute of the University, Lothar Heffter (1862–1962) and Alfred Loewy (1873–1935). Heffter was a stimulating teacher who had written influential textbooks, e.g. on differential equations and analytic geometry. Loewy likewise worked in the area of differential equations, but also in insurance mathematics, and later mainly in algebra, attracting good students and collaborators to Freiburg, among them Reinhold Baer, Wolfgang Krull, Bernhard H. Neumann, Friedrich Karl Schmidt, Arnold Scholz, and Ernst Witt. Baer (until his move to Halle in 1928) and Scholz (from 1928 to his death in 1941) became Zermelo's closest scientific companions.

Heffter appreciated the contacts with Zermelo so much that he proposed to the Faculty of Sciences and Mathematics that they should offer him an

honorary professorship, an unpaid professorship with the right but not an obligation to lecture. He was deeply convinced of Zermelo's scientific abilities and not concerned that they were sometimes accompanied by nervous and irritable behaviour. In his recollections, (*1952*, 148) he characterizes Zermelo as a researcher of international fame who was personally "extremely strange and fidgety" ("höchst wunderlich und zappelig").

Zermelo accepted the offer with pleasure. Starting with the winter semester 1926, he gave one course in nearly each semester, the subjects ranging from applied mathematics to foundations, sometimes accompanied by exercises or a seminar. Thus he was fully integrated into the teaching programme of the Mathematical Institute to an extent quite unusual for an unpaid honorary professor. After he had lost his position in 1935, he touched on this point with bitterness when he complained about the shameful treatment he had to suffer, adding that this happened "after I had worked at the University for ⟦nine⟧ years without any salary."

The appointment to the honorary professorship coincides with a remarkable increase in Zermelo's scientific activity. It launched his second period of intensive research, about twenty years after his time in Göttingen. One might therefore be tempted to view the professorship as a stimulant. On the other hand, his improved health might also be taken into consideration. The following years until the mid-1930s would see a broad range of activities: Besides his participation in the teaching programme of the Mathematical Institute he first returned to applied mathematics, publishing papers on different topics. As regards the foundations of mathematics, he tried to gain influence on the new developments in mathematical logic and the developments in set theory coming with them. Furthermore, he edited the collected mathematical and philosophical works of Georg Cantor.

Before these projects are presented in greater detail, it should be emphasized that Zermelo's interests went far beyond mathematics and the sciences. Various activities emphasize his broad cultural interests. During his first Freiburg years he started a translation of Homer's *Odyssey*, thereby aiming at "a *liveliness* as immediate as possible" ("möglichst unmittelbare *Lebendigkeit*"). A small part of it was later published as *1930f*.

At the same time Zermelo participated in the educational programme which was organized by the Association of the Friends of the University. In 1932, together with his friend Arnold Scholz, he took part in a cruise to ancient places in Greece, Italy, and Northern Africa. His pleasure at reflecting events and thoughts in little poems and other forms of poetry led him, for example, to an ironic dialogue between Telemachos and Nausicaa, "Tele" and "Nausi", to represent features of Nausicaa as described by Homer and in Goethe's respective fragment.

Despite the diversity of projects which he started and carried out and despite all the personal contacts they involved, Zermelo remained caught in a feeling of loneliness. In many of his letters from this time he mentions his longing for

company ("Gesellschaft") and scientific conversations. Concerning the latter point, his contacts with the Mathematical Institute were not as fruitful and the openness for his logical work not as wide as he had hoped. Loneliness came together with the feeling of financial distress and lack of recognition, rooted in the loss of his Zurich professorship and in cuts in his pension.

4.2 Technical interests and applied mathematics

There are numerous documents evidencing Zermelo's involvement with technical problems during his time in Göttingen and in Zurich. The same is also true for Freiburg. He became interested there in monorail trains and steam and gas turbines. In March 1931 he submitted an—unsuccessful—application to the national German Patent Office, asking for a patent on a construction which used a gyroscope to stabilize bicycles and motorcycles during a stop, at the same time providing energy for a new start. In 1932 he worked out plans for electrodynamic automatic gears, writing down a two-page sketch "On an electrodynamic clutch and braking of motor cars with internal combustion engines" ("Über eine elektrodynamische Kuppelung und Bremsung von Kraftwagen mit Explosionsmotoren"). Apparently he did not prepare an application for the patent office. A success would have been doubtful, as the Daimler company had already equipped a car with hydrodynamic automatic gears as early as 1927.

Shortly after having obtained the honorary professorship, Zermelo began to realize old publication plans. The first paper, *1927*, presented investigations of the measure of point sets that were first described in Waldemar Alexandrow's Ph. D. thesis *1915* written under Zermelo's supervision in Zurich. Then he turned to applied mathematics, thus demonstrating "that [his] old, even though hitherto mostly unhappy love for the 'applications' [had] secretly kept glowing." The resulting papers deal with different topics.

The paper *1928*, already drafted in 1919, treats the problem of how to measure the result of a chess tournament, thereby introducing essential principles of modern tournament rating systems. Zermelo criticizes the practice of defining the relative strength of the participants of a tournament by the number of games they have won: It fails for incomplete tournaments and may lead to inadequate results, for example, in a chess round robin tournament performed by one excellent player and k weak players (here the relative strength of the excellent player differs from that of the next one at most by a factor $\leq 2+\frac{2}{k-1}$). Zermelo uses a maximum likelihood method. The relative strengths of the participants are defined as those winning probabilities which yield the maximal probability for the result of the tournament to occur. His method anticipates what was later called the Bradley-Terry model (*David 1988*, 13).

At the meeting of the Deutsche Mathematiker-Vereinigung in Prague in September 1929 Zermelo gave a talk (*1930c*) that deals with a problem which had come to his mind when the airship "Graf Zeppelin" circumnavigated the

earth in August 1929, namely the question of how an airship moving at a constant speed against the surrounding air has to fly in order to reach a given point Q from a given point P in the shortest time possible. Using methods of the calculus of variations, he was able to solve what is now called the Zermelo navigation problem. The solution illustrates that in changing winds "the helm has to be turned to that side where the ⟦component of the speed of the winds⟧ against the steering direction increases." Later, he generalized his results in *1931a*. The talk and the extended version initiated various alternative methods of addressing the problem, also in different spaces. Concerning the latter point, Zermelo himself made a first step: When Johann Radon was in search of a problem for a Ph. D. student, Zermelo suggested treating navigation on the sphere (cf. *Zita 1931*).

Zermelo's last paper in applied mathematics, *1933a*, is of an elementary character. Having been invited to contribute a paper to a special issue of *Zeitschrift für angewandte Mathematik und Mechanik* in honour of its founder and editor Richard von Mises on the occasion of his 50th birthday, Zermelo accepted and contributed an investigation on splitting lines of ovals with a centre. The subtitle "How does a piece of sugar break?" indicates that his starting point is the question, how a piece of sugar breaks if one grips it at two opposite corners and then tries to divide it. In the introduction he justifies his choice by a double coincidence with the "wise" physicist and philosopher Gustav Theodor Fechner (1801–1887) who had treated the likewise everyday question "Why are sausages not cut straight?" ("Warum wird die Wurst schief durchschnitten?") in a humorous way and published his treatise under the pseudonym "Dr. Mises" as part of his *Kleine Schriften* (*Fechner 1875*).

4.3 Return to the foundations of mathematics

Nine years after his move to Freiburg Zermelo wrote (*s1930d*) that he was brought back to investigations of foundational problems by the "somewhat noisy appearance of the 'intuitionists'." He continued that he did not take sides in the quarrel between intuitionism and formalism, but tried to "help to clarify the questions under consideration [...] as a mathematician by showing objective mathematical connections."

There was a specific reason which brought him back to set-theoretic research: Starting in the early 1910s, his 1908 axiom system had become an object of growing criticism because of the vagueness of the notion of definiteness. This criticism increased in the 1920s, causing him to reconsider this point and publish his results in the definiteness paper *1929a*, the first paper in set theory to appear after a break of more than 15 years.

During the summer semester 1923 and the winter semester 1923/24 Marvin Farber (1901–1980) stayed in Freiburg to study Husserl's phenomenology. Farber, later professor of philosophy at the State University of New York in Buffalo and founder and editor of the journal *Philosophy and Phenomenological Research*, became acquainted with Zermelo probably in Husserl's seminar.

The encounter was the beginning of a longer scientific cooperation and also of a friendship which included Marvin's brother Sidney and Sidney's wife Norma and is reflected in an extended correspondence. In his letters and postcards Zermelo reveals personal biases and his opinion about the people around him. Above all they document the loneliness he felt in Freiburg and draw surprise with many a blunt judgement about his colleagues in the Philosophical Seminar and the Mathematical Institute.

Farber took notes of the discussions he had with Zermelo. They may reflect essential features of Zermelo's views on the philosophy of logic. Central issues re-appear in lectures which Zermelo gave in Warsaw in the spring of 1929 (cf. *s1929b*). The main points are:

- Mathematical reasoning is based on models and, hence, obeys the law of the excludud middle.
- True mathematics is concerned with infinite structures.
- Hilbert's consistency program fails; mathematics can only be justified by its success.

So there is a clear divergence from intuitionism and Hilbert-type formalism.

When Farber returned to Freiburg during the winter semester 1926/27, discussions on logic seemed to become so fruitful that both agreed to write a monograph on logic, Farber being responsible for the first draft. The Farber *Nachlass* contains notes (*Farber 1927*) of the book that were, however, never seen by Zermelo. In 1929, the collaboration came to an end.

Around 1926/27 Zermelo conceived plans to edit a German translation of the works of the Harvard mathematician and philosopher Alfred North Whitehead, an enterprise Whitehead highly appreciated. Finally, however, Zermelo changed his mind. Following a suggestion of members of the Philosophical Seminar, he turned to bringing out an edition of Cantor's collected papers, Springer Verlag being ready to publish them. Work started in early 1927. Zermelo was supported by Reinhold Baer with respect to mathematical questions and by Oskar Becker with respect to philosophical ones. Abraham Fraenkel contributed a biography of Cantor, but because of Zermelo's harsh criticism of an early version, the relations between the two set theorists became estranged. The work, *Cantor 1932,* appeared in the spring of 1932. It was reviewed in about a dozen journals, judgements varying from "sober" to "excellent". Criticism concerning editorial care was made by Ivor Grattan-Guinness (*2000*, 548; *1974*, 134–136).

The second point in the list above is of central importance for Zermelo's view of mathematics. Faced with the insufficiency of his pension, he applied for a fellowship of the Deutsche Forschungsgemeinschaft (then still named "Notgemeinschaft der Deutschen Wissenschaft") and succeeded in obtaining a three-year fellowship for his research project "On the nature and the foundations of pure and applied mathematics and the significance of the infinite in mathematics" ("Wesen und Grundlagen der reinen und der angewandten

Mathematik, die Bedeutung des Unendlichen in der Mathematik"). Scientifically the years of the fellowship form the kernel of Zermelo's Freiburg period of intensive research activities with the focus on foundational questions. His work on set theory and logic was centered around one aim: to provide a foundation for "true" mathematics. According to his firm belief, the emerging formal systems of logic with their inherent weakness in expressibility and range that were to dominate the 1930s went in the wrong direction. Full of energy and without shying away from serious controversies, he set himself against this development.

4.4 Defence of "true" mathematics

After his stay in Warsaw that we have mentioned above, Zermelo spent some weeks at the seaside resort of Sopot near Gdansk where he wrote his definiteness paper *1929a*. Planned already before his visit to Warsaw but finalized during discussions with the Warsaw logicians, in particular with Alfred Tarski, the paper was intended to meet criticism of his 1908 notion of definiteness. Essentially, definiteness now coincides with second-order definability. Second-order quantifiers are viewed as ranging over definite properties—a formal weakness which will later be criticized by Thoralf Skolem.

In his last Warsaw draft (cf. *s1929b*), Zermelo had pleaded for a more convincing conceptual basis for set theory in order to avoid the paradoxes. He had proposed to distinguish between sets and classes—von Neumann's set theory being the example he may have had in mind—and to define sets as domains of structures which are categorically characterizable, i.e., characterizable up to isomorphism, in the universe of classes.

In *s1931f* and *s1932d* he shows that the universe of sets defined in this way satisfies the axioms of set theory. The roughly sketched arguments suffer from the absence of a well-defined language and from his weakness in handling inner models. Both points come together with the refusal to adopt a strict distinction between formulas and the objects they are about, i.e. between syntax and semantics.[2]

Early in 1930 a second proposal for a foundation of set theory was ready for publication: the cumulative hierarchy. In the corresponding paper *1930a* Zermelo bases his investigations on an axiom system which he calls "the extended Zermelo-Fraenkel or ZF system" and which consists of the 1908 axioms without the axiom of infinity (as it would not belong to "general" set theory), together with the axiom of replacement and the axiom of foundation; the axiom of choice is not explicitly included, but taken for granted as a

[2] This distinction or—more adequately—its methodological control, became the watershed that separated the area of the "classical" researchers such as Zermelo and Fraenkel from the domain of the "new" foundations as developed by younger researchers, among them Gödel, Skolem, and von Neumann, and governing the rise of mathematical logic in the 1930s.

"general logical principle". The axioms of separation and replacement are given in a second-order form. By formulating them as first-order schemata according to *Skolem 1923*, one is led to the now usual first-order Zermelo-Fraenkel axiom system ZF (or ZFC, if the axiom of choice is included).

The paper *1930a* contains the first clear statement of the informal interpretation of axiomatic set theory in the cumulative hierarchy. By incorporating his concept of sets as domains of categorically defined structures, Zermelo is led to the assumption of infinitely many strongly inaccessible cardinals, i.e., of an infinite chain of so-called normal domains, of levels of the cumulative hierarchy which are models of set theory, praising this feature as a victory over the Zermelo-Russell paradox: The inconsistent set the paradox provides in one normal domain, is a "perfect" set in the next one. The considerations about large cardinals are only rough sketches; a more detailed version was planned, but was realized merely in a number of approximations (*Zermelo s1931e, s1933b*).

While his *1930a* was in the process of publication, Zermelo was confronted with Skolem's answer *1930* to his definiteness paper where Skolem refers to his first-order approach (in *1923*) to definiteness. According to the Löwenheim-Skolem theorem, there are countable models of set theory, a fact totally alien to Zermelo's idealistic point of view. He now faced a principal shortcoming of finitary axiom systems: Any such system seemed to be unable to describe adequately the Cantorian universe of sets with its endless progression of growing cardinalities. So the decisive role of infinite totalities which he had emphazised so strongly in Warsaw could not be mirrored adequately in a finitary language.

Zermelo's first reaction was to try to refute the existence of countable models of set theory. All his approaches, however, had to fail because they did not leave first-order set theory or suffered from not paying attention to absoluteness. Apart from these *ad hoc* measures he took a more fundamental approach and focused on infinitary languages, maintaining that any adequate language for mathematics had to be infinitary. Of course he knew that his claim asked for a convincing theory of infinitary languages together with some kind of infinitary logic. Indeed, in 1931 he started such an enterprise. The program is succinctly formulated in *s1921*, a one-page note dated 17 July 1921. As there was no hint of infinitary languages in the extensive discussions with Farber and as the Warsaw lectures touch infinitary languages only in one passing remark, it is doubtful whether the note really stems from 1921.

One might wonder why Zermelo did not acknowledge Skolem's approach at least for pragmatic reasons; for Skolem had demonstrated that first-order definiteness was sufficient to carry out all ordinary set-theoretic proofs. Instead, Zermelo developed a strong feeling that Skolem's arguments amounted to a severe attack against mathematics and that he, Zermelo, was in charge of fighting back in order to preserve mathematical science from damage. His former opponents, the intuitionists, could forthwith be included among the adversaries.

In 1931 Zermelo became aware of Gödel's first incompleteness theorem: Under mild conditions which are satisfied in mathematical practice, any consistent finitary axiom system of sufficient number-theoretic strength is incomplete in the sense that it admits propositions which are neither provable nor refutable in it. So Skolem's results about the weakness of finitary axiom systems had gotten a companion, and Zermelo a further opponent.

Zermelo presented his concept of infinitary languages (*1932b*) at the 1931 meeting of the Deutsche Mathematiker-Vereinigung where Gödel also gave a talk on his incompleteness results. Zermelo intended to use this coincidence as a forum for his criticism of both intuitionism and formalism in general and the finitary point of view as shared by Skolem and Gödel in particular. His plans for an extended discussion failed. Instead, there was a "peaceful" meeting (*Taussky-Todd 1987*, 38) with Gödel. After the conference, however, when writing the summary *1932a* of the talk for the news of the Deutsche Mathematiker-Vereinigung, Zermelo obeyed his "particular duty" and extended the presentation of his results by harsh criticism of intuitionism, "Skolemism", and Gödel's finitary approach to incompleteness.

When conceiving his infinitary languages, Zermelo did not clearly distinguish between a proposition and its meaning, between syntax and semantics. Even more, mixing up a proposition and its denotation, he believed he had discovered a flaw in Gödel's argument. Immediately after the Bad Elster meeting he informed Gödel (cf. *s1931c*) about this "essential *gap*" (*s1931b*), locating the alleged mistake in the "finitary prejudice", namely the "erroneous assumption that every mathematically definable notion be expressible by a 'finite' combination of signs." The short correspondence which developed from this letter (*Gödel 1931b*, *Zermelo s1931d*) attests to Gödel's readiness to make clear the technicalities and the meaning of his proof, but also to Zermelo's irreconcilable attitude.

Three years later, Zermelo published the more detailed version *1935* of his infinitary languages and infinitary logic. His hope, however, that in the long run his infinitary approach would gain acceptance was not to be met. It should have been clear very quickly to any well-informed observer that he could not win his fight. First, he did not strive for a presentation of his counter-arguments with the sort of precision exercised by Gödel and Skolem; despite conceptual novelties, his papers on the subject are somewhat vague and, probably for this reason, did not stimulate or influence any further research. Secondly, his epistemological engagement prevented him from getting a technical understanding of the results of Gödel and Skolem in an unprejudiced way and only then pondering their epistemological meaning and utilizing the analyzing power they provide. As it was these results that shaped the discipline of mathematical logic in the 1930s, he placed himself outside the mainstream of mathematical foundations.

In contrast to his increasing isolation in foundational matters, Zermelo received recognition in the German mathematical community. On 18 December

1931, following a suggestion of Richard Courant, he was elected a corresponding member of the mathematical-physical class of the Gesellschaft der Wissenschaften zu Göttingen. On 2 November 1934 he presented his *1934* to the academy where he gave a simple argument for the uniqueness of prime number decomposition which does not use the notions of the greatest common divisor or the least common multiple. On 16 February 1933 he was also elected an extraordinary member of the class for mathematics and the sciences of the Heidelberger Akademie der Wissenschaften; the academy followed a suggestion of the Heidelberg professors Heinrich Liebmann and Artur Rosenthal.

4.5 Loss of the honorary professorship

In March 1935, two years after the Nazis had seized power in Germany, Zermelo lost his honorary professorship due to his resistance against Nazi politics. According to Sanford Segal he thus "provides ⟦a⟧ case of what in Nazi times was reckoned a bold act with damaging results" (*Segal 2003*, 467).

In early 1935 the Mathematical Institute had two full professorships, one held by the applied mathematician Gustav Doetsch, who had followed Lothar Heffter in 1931, the other one by Wilhelm Süss, the successor of Alfred Loewy and later founder of the Mathematical Research Institute Oberwolfach. As a Jew, Loewy had been suspended in April 1933. Arnold Scholz, Zermelo's best friend and scientific partner, had gone to Kiel in 1934, hoping to get a professorship there, but his anti-fascist attitude worked against this.

In January, Doetsch's assistant Schlotter and the student union wrote letters to the Rector of the University and the responsible minister in which they reported maliciously that Zermelo refused to greet with the Hitler salute and that he had "seriously insulted the Führer and the institutions of the Third Reich". They asked that he be removed from the teaching stuff. The rector started an investigation during which Doetsch strongly confirmed the accusations.

When Zermelo was heard on 1 March, he declared that he had tried to follow painstakingly all the legal requirements of the new state; but as he did not receive a salary from the state, he had by no means felt obliged to continuously express convictions of the National Socialist Party which he did not share. Furthermore, he admitted that he had made disparaging remarks about Nazi politics, but had done so only privately. He therefore felt entitled to hold his point of view, but declared himself ready to voluntarily give up an activity which could not give him any pleasure under these circumstances. Already on the next day he wrote a letter to the dean renouncing his honorary professorship. On 26 March he was excluded from the faculty.

The affair thus came to a deplorable end and yet without serious consequences beyond his dismissal. On the one hand, he lost his position under shameful circumstances aggravated by the obstinate and inhuman behaviour of his colleague Doetsch. On the other hand, the Nazi system seemed to have

been satisfied with his dismissal and did not pursue the disparagement of its politics. However, the formal and practically all personal connections between Zermelo and the University had been severed.

4.6 Retreat

The loss of the honorary professorship and the disappointment over the behaviour of some of his former colleagues left Zermelo bitter. Perhaps even more important, the circumstances of the dismissal procedure aggravated his fragile mental constitution. The year 1935 marked the beginning of a decline in his mental energy, a development in reverse of that in 1926 when he was given the honorary professorship. The second active period of his scientific life around 1930 closely coincides with this professorship, and the second half of the thirties comes with a complete withdrawal from the scene of mathematical foundations. Besides letters from Arnold Scholz, there is only one letter in the *Nachlass* written by a well-known researcher: In July 1937 Haskell B. Curry asks for a preprint of *1935*. Apart from the notes on a flawed refutation of the existence of countable models of set theory (*s1937*), there are no notes in the *Nachlass* which stem from the second half of the 1930s and concern foundational questions or non-elementary questions of other areas of mathematics.

In a certain sense, his growing scientific isolation was mirrored by the circumstances of his life. In March 1934 he had left central Freiburg and had moved to Günterstal, a nice suburban village-like residential area in the foothills of the Black Forest. Each day at about noon he walked for nearly a mile to a restaurant to have lunch. As a rule he also went for longer walks in the afternoon. He started to comment on everyday events with little poems, all written with a sense of humour. But he had not lost his sharp irony. In the pamphlet "Cheeky little devil. The golden bigwig-ABC or the art of becoming a perfect bigwig" ("Frech wie Oskar. Das goldene Bonzen-ABC oder die Kunst, ein vollkommener Bonze zu werden") he lists 23 properties which make a perfect bigwig. For instance, a perfect bigwig should be presumptuous like Toeplitz, a charlatan like Weyl, despotic like Klein, obstinate like Hilbert, business-minded like Courant, mad like Brouwer, and careful about his appearance like Doetsch.

Zermelo's retreat from the foundational scene did not mean a retreat from mathematics as a whole or from the diversity of his former activities. The *Nachlass* contains studies on music theory and paper-made slide rules of different kinds to calculate keys and tone intervals, stemming from 1933 and later, and fragments of manuscripts or sketches dealing with various mathematical topics such as puzzle games, permutations of finite and infinite sets, polynomials in division rings, bases of modules, systems of linear equations with infinitely many unknowns, and—probably initiated by the first air attacks of the Second World War—considerations about families of parabolas under the title "Are there places on earth which are protected against bombs?"

("Gibt es Orte der Erdoberfläche, die gegen Fliegerbomben geschützt sind?"). However, none of these elaborations can be considered part of an extended project; they were something like exercises done without contact with actual research.

There are some further plans that did not come to fruition. In 1928 Zermelo had agreed with a publisher to write a book on set theory. Around 1932, perhaps after the edition of Cantor's collected papers had been completed, he elaborated some isolated parts. Apparently, however, work on the book was far from being intensive. In 1942 he gave up, pursuing plans now for a "collection of entertaining exercises for mathematicians and friends of mathematics", "Mathematische Miniaturen", among them several topics which he had treated in papers of his own. After first elaborations of some parts, this project also came to naught.

When Arnold Scholz died on 1 February 1942, Zermelo not only lost a colleague with whom he had intensively discussed mathematics and whose foundational views were close to his own—he also lost his best friend. Each of Scholz's letters conveys warmth and concern for "Dear Zero". Scholz gave advice on everyday problems and was active in the background to help in scientific matters. In the spring of 1941 he had helped to organize a colloquium in Göttingen on the occasion of Zermelo's 70th birthday. Among the speakers were Konrad Knopp and Bartel Leendert van der Waerden. Zermelo gave three talks with topics from his "Miniaturen".

The time that had passed since the loss of the honorary professorship had seen Zermelo working on various enterprises; however, he had been lacking the ability to systematically pursue them. This development may have been natural, given his increasing age and nervous destabilization following the Nazi affair. Concerning the foundations, he felt that his fight against finitism had been lost. Being thus confronted with the success of a direction of mathematical logic which he was not willing to accept, he feared in the end to fall into oblivion. The letter *s1941*, which he sent to Paul Bernays on 1 October 1941 as a response to Bernays' birthday congratulations, candidly brings this out.

4.7 Last years

In the ample country house where Zermelo had rented his apartment, there also lived a younger lady, Gertrud Seekamp. More and more frequently it happened that she spent the evenings together with him. Often these evenings ended with Zermelo reading aloud from classical literature. In the summer of 1944, they decided to get married. The wedding took place on 14 October.

No document in the *Nachlass* shows that Zermelo, now in his seventies, worked on scientific projects. Everyday problems increased during the last years of the war and the time following. It was mainly Gertrud who worked to maintain for them their already modest way of life. Unfortunately, Zermelo's

vision problems—rooted in glaucoma—got more serious. In late 1944 plans for an operation were cancelled because it did not promise success. During the next years his powers of vision deteriorated, and in 1951 he was blind.

With the end of the war in May 1945, southwestern Germany came under the control of a French military government which worked to reinstate the necessary public institutions and infrastructures. On 23 January 1946, Zermelo turned to the Rector's office, requesting the University to reappoint him as a honorary professor. Half a year later, the responsible minister of the subordinate German administration followed through on the request.

In the meanwhile, the financial circumstances of the Zermelo couple had become quite precarious. Since the end of the war the Canton Zurich had stopped the transfer of Zermelo's pension and left the couple without any income. In the next several years Zermelo tried several times to obtain permission to settle in Switzerland in order to get his pension. The Swiss authorities answered negatively; they doubted that the Zurich pension was sufficient for the couple to live in Switzerland. They allowed, however, several stays there in order to procure urgently needed goods. When the transfer of the pension was finally allowed, the circumstances did not change because the modest advance which the German authorities had provided had to be repaid.

The deplorable situation caused by financial need, increasing blindness, and separation from scientific activity Zermelo had to suffer may have intensified his feeling of falling into oblivion. In March 1949 he sent his publication list to the Springer Verlag, suggesting an edition of his collected works under the title *Mathematische und physikalische Abhandlungen 1894–1936*. Springer answered that the current circumstances did not allow for such an edition. Apparently Zermelo made a second unsuccessful attempt and, in January 1950, a third one with the publisher Johann Ambrosius Barth, who had published his translation *Gibbs 1905* of Gibbs' *Elementary Principles in Statistical Mechanics.*

Zermelo's 80th birthday on 27 July 1951 brought congratulations from good friends, among them Heinz Hopf and Reinhold Baer. Baer dedicated his paper *1952* on criteria of finiteness for commutator groups to him, Herbert Bilharz his paper *1951* on the Gaussian method for the approximate calculation of definite integrals. Heinrich Behnke, then editor of the *Mathematische Annalen,* was also among those offering congratulations. He regretted that lack of time had not permitted the preparation of a special issue. Erich Kamke, then president of the Deutsche Mathematiker-Vereinigung, had refused to initiate a special issue of *Mathematische Zeitschrift.* He argued that "for years Zermelo has not distinguished himself mathematically."

Less than two years later and two months before his 82nd birthday, in the morning of 21 May 1953, now totally blind, Zermelo died with no specific illness a cause. On 23 May he was buried in the churchyard of the monastery in the heart of Günterstal. Because of her shortage of money, his widow could only afford a public grave. When it was to expire twenty years later, an initia-

tive supported by the University together with the Faculty of Mathematics took care that he was reburied. He has now found a final resting place in the same churchyard, not far away from the grave of Edmund Husserl. A simple grey stone plate gives his name.

Since the late sixties, many a user of Zermelo's *Nachlass* also paid a visit to his widow Gertrud, enjoying her warm hospitality and open-minded cheerfulness. They listened to what she had to tell about her late husband, and they heard of the mutual care which had shaped their relationship. Gertrud Zermelo died on 15 December 2003, one year after her 100th birthday and more than fifty years after the death of her husband.

Left: Zermelo with his sisters Anna (right) and Elisabeth
Right: Zermelo with his sister Elisabeth around 1907

Zermelo with Marvin Farber around 1925

Zermelo with Warsaw mathematicians, spring of 1929[3]

Zermelo and Bronisław Knaster with Lvov mathematicians, May 1929[4]

---

[3] Front row from left to right: Wacław Sierpiński, Zermelo, Samuel Dickstein, Antoni Przeborski; second row from left to right: Jan Łukasiewicz, Stanisław Leśniewski, Bronisław Knaster, Jerzy Spława-Neyman, Franciszek Leja.

[4] Front row from left to right: Hugo Steinhaus, Zermelo, Stefan Mazurkiewicz; second row from left to right: Kazimierz Kuratowski, Bronisław Knaster, Stefan Banach, Włodzimierz Stożek, Eustachy Żylinski, Stanisław Ruziewicz.

Zermelo in Warsaw, spring of 1929

Zermelo around 1935

Zermelo with his future wife Gertrud in 1935

The Zermelo couple at Lugano in 1948

Gertrud Zermelo in 1999

Zermelo in February 1953

Ernst Zermelo's *curriculum vitae*

Heinz-Dieter Ebbinghaus

1871

27 July: Zermelo is born in Berlin as the second child and the only son of the *Gymnasialprofessor* Theodor Zermelo and his wife Auguste neé Zieger.

1878

3 June: Death of Zermelo's mother.

1880

April: Zermelo enters the Luisenstädtisches Gymnasium in Berlin.

1889

24 January: Death of Zermelo's father.

1 March: Zermelo finishes school. Remarks in his leaving certificate show that he suffers from physical fatigue.

Summer semester – summer semester 1890: Zermelo studies mathematics and physics at the University of Berlin with, among others, Lazarus Fuchs and Johannes Knoblauch.

1890

Winter semester 1890/91: Zermelo studies at the University of Halle-Wittenberg with, among others, Georg Cantor and Edmund Husserl.

1891

Summer semester 1891: Zermelo studies at the University of Freiburg, with his subjects including, as in Berlin and Halle, philosophy and psychology.

Winter semester 1891/92 – summer semester 1897: Zermelo studies again at the University of Berlin with, among others, Ferdinand Frobenius, Max Planck, Hermann Amandus Schwarz, and Wilhelm Wien.

1894

23 March: Zermelo applies to begin the Ph. D. process.

6 October: Zermelo obtains his Ph. D. degree. His dissertation *Untersuchungen zur Variations-Rechnung* was supervised by Hermann Amandus Schwarz.

December – September 1897: Zermelo is an assistant to Max Planck at the Institute for Theoretical Physics of the University of Berlin.

1895

December: Zermelo completes his first paper, *1896a,* which sets out his opposition to Ludwig Boltzmann's statistical theory of heat.

1896

Summer: Zermelo applies for an assistantship at the Deutsche Seewarte in Hamburg, but then decides to pursue an academic career.

15 September: Zermelo completes his second paper, *1896b,* opposing Boltzmann.

1897

German translation *Glazebrook 1897* of *Glazebrook 1894*.

2 February: Zermelo passes the state exam for *Gymnasiallehrer* (high school teachers) that allowed him to teach mathematics and physics as main subjects and chemistry, geography, and mineralogy as additional subjects. According to the reports Zermelo exhibited a broad knowledge in German literature, philosophy, and religion.

19 July: Zermelo asks Felix Klein in Göttingen for support for his *Habilitation.*

Winter semester 1897/98: Zermelo continues his studies at Göttingen with, among others, David Hilbert, Felix Klein, and Arthur Schoenflies.

1899

3 February: David Hilbert presents Zermelo's first paper in applied mathematics, *1899a,* to the Königliche Gesellschaft der Wissenschaften zu Göttingen; it treats differential equations with inequalities.

Zermelo initiates the *Habilitation* process with the *Habilitation* thesis "Hydrodynamische Untersuchungen über die Wirbelbewegungen in einer Kugelfläche" the first part of which is published as *1902a.* The second part (*1902b, 1902c*) remained unpublished; it contains a solution of the 3-vortex problem on the sphere.

4 March: Zermelo gives his *Habilitation* address, *1900,* which proposes an alternative probabilistic approach to Bolzmann's in the latter's work in statistical mechanics. He is granted the *venia legendi* for mathematics at the University of Göttingen.

Around 1900

Beginning of the cooperation with Hilbert on the foundations of mathematics. Zermelo formulates the Zermelo-Russell paradox.

1900

Winter semester 1900/01: Zermelo gives his first course on set theory, the main topic being the Cantorian theory of cardinals.

1901

9 March: David Hilbert presents Zermelo's result on the addition of cardinals, *1901,* to the Königliche Gesellschaft der Wissenschaften zu Göttingen. The proof uses the axiom of choice.

1902

12 May: Zermelo gives a talk on Frege's foundation of arithmetic before the Göttingen Mathematical Society.

Summer semester – winter semester 1906/07: Zermelo receives a *Privatdozenten* grant.

Publication of *1902d,* the first paper on the calculus of variations after the Ph.D. dissertation. It treats shortest lines of bounded steepness with or without bounded torsion.

1903

June: Zermelo is under consideration for an extraordinary professorship of mathematics at the University of Breslau. He is shortlisted in the second position after Gerhard Kowalewski, Franz London, and Josef Wellstein who are shortlisted *aequo loco* in the first position.

1 December: Zermelo completes his second paper on the calculus of variations, *1903.* It gives two simple proofs of a result of Paul du Bois-Reymond on the range of the method of Lagrange.

1904

Beginning of a life-long friendship with Constantin Carathéodory.

Together with Hans Hahn, Zermelo writes a contribution on the calculus of variations, *Hahn and Zermelo 1904,* for the *Encyklopädie der mathematischen Wissenschaften.*

August: Third International Congress of Mathematicians at Heidelberg. Julius König gives a flawed refutation of Cantor's continuum hypothesis. The error is detected by both Zermelo and Felix Hausdorff.

24 September: Zermelo informs Hilbert about his proof of the well-ordering theorem and the essential role of the axiom of choice. The letter is published as *1904.*

15 November: During a meeting of the Göttingen Mathematical Society, Zermelo defends his well-ordering proof against criticism by Julius König, Felix Bernstein, and Arthur Schoenflies.

1905

January: Zermelo falls seriously ill. In order to recover, he spends spring and early summer in Italy.

German translation *Gibbs 1905* of *Gibbs 1902.*

Spring: Zermelo works on the theory of finite sets which finally results in *1909a* and *1909b.*

21 December: Zermelo receives the title "Professor". The application had been filed by Hilbert in December 1904.

1906

Early that year: Zermelo catches pleurisy.

Zermelo works on a book on the calculus of variations together with Carathéodory.

Zermelo publishes a final criticism of Boltzmann's statistical interpretation of the second law of thermodynamics in the review *1906* of *Gibbs 1902.*

Summer semester: Zermelo lectures on "Mengenlehre und Zahlbegriff". He formulates an axiom system of set theory which comes close to the one published by him in 1908.

June: Medical doctors diagnose tuberculosis of the lungs.

Summer: Zermelo spends a longer time at the seaside.

Autumn: Zermelo is under discussion for a full professorship of mathematics at the University of Würzburg. The professorship is given to the extraordinary Würzburg professor Georg Rost. According to Hermann Minkowski Zermelo's difficulties in obtaining a professorship are rooted in his "nervous haste".

Winter 1906/07 – winter 1907/08: Several extended stays in Swiss health resorts for lung diseases.

1907

March: Zermelo applies for a professorship at the Academy of Agriculture in Poppelsdorf without success.

May: During a stay in Montreux Zermelo finishes his paper *1909a.*

14 July and 30 July: During a stay in the Swiss alps Zermelo completes his papers on a new proof of the well-ordering theorem and on the axiomatization of set theory, *1908a* and *1908b,* respectively.

20 August: Following an application by the Göttingen Seminar of Mathematics and Physics, the ministry commissions Zermelo to give lecture courses in mathematical logic and related matters, thus installing the first official lectureship for mathematical logic in Germany.

1908

April: Fourth International Congress of Mathematicians in Rome. Zermelo presents his work on finite sets, *1909b.* He becomes acquainted with Bertrand Russell. Together with Gerhard Hessenberg and Hugo Dingler he conceives plans for establishing a quarterly journal for the foundations of mathematics. The project fails because of diverging views between the group and the Teubner publishing house.

Summer semester: Zermelo gives a course on mathematical logic in fulfilment of his lectureship for mathematical logic and related topics.

1909

July: Zermelo is under consideration for an extraordinary professorship of mathematics at the University of Würzburg. He is shortlisted in the first position. The professorship is given to Emil Hilb shortlisted in the second position.

September: Completion of *Riesenfeld and Zermelo 1909.*

1910

24 January: The board of directors of the Göttingen Seminar of Mathematics and Physics applies to the minister to appoint Zermelo an extraordinary professor.

21 January: Zermelo, being under consideration for a full professorship of mathematics at the University of Zurich, is shortlisted in the first position.

24 February: The *Regierungsrat* of the Canton Zurich approves the choice of Zermelo.

15 April: Zermelo is appointed a full professor at the University of Zurich for an initial period of six years.

1911

28 January: Zermelo applies for leave for the coming summer semester because of a worsening of his tuberculosis.

February and March: Together with a partner, Zermelo applies for several patents concerning, for example, a regulator for controlling the revolutions of a machine.

Zermelo is awarded the interest from the Wolfskehl prize, Hilbert being chairman of the Wolfskehl committee of the Gesellschaft der Wissenschaften zu Göttingen.

Summer semester – winter semester 1911/12: Leave for a cure because of tuberculosis.

1912

January: Serious worsening of tuberculosis diagnosed.

Beginning of the cooperation with Paul Bernays who completes his *Habilitation* with Zermelo in 1913 and stays at the University of Zurich as an assistant to Zermelo and later as a *Privatdozent* until 1919.

August: Fifth International Congress of Mathematicians in Cambridge. Following an invitation by Bertrand Russell, Zermelo gives two talks, one on axiomatic and genetic methods in the foundation of mathematical disciplines and one on the game of chess. The second one results in the paper *1913* which may be considered the first paper in game theory.

Faced with the seriousness of his illness, Zermelo conceives plans for an edition of his collected papers.

1913

Spring: Zermelo is considered for a full professorship in mathematics at the Technical University of Breslau. He is shortlisted in the first position. The professorship is given to Max Dehn, shortlisted in the second position together with Issai Schur.

December: Zermelo completes his paper *1914* on subrings of whole transcendental numbers of the field of the real numbers and the complex numbers, respectively; it makes essential use of the axiom of choice.

1914

Early that year: Zermelo has regular discussions with Albert Einstein.

March: Operation of the thorax by Ferdinand Sauerbruch, the pioneer of thorax surgery.

Around 1915

Zermelo develops a theory of ordinal numbers where the ordinals are defined as by John von Neumann in 1923.

1915

Spring: A new serious outbreak of tuberculosis forces Zermelo to take a one-year leave.

July: Waldemar Alexandrow completes his Ph. D. thesis *Alexandrow 1915.* It is the only thesis guided by Zermelo alone. Kurt Grelling's thesis *Grelling 1910,* which extends Zermelo's theory of finite sets, was officially supervised by David Hilbert, but guided by Zermelo.

Autumn: Several surgical treatments of a tuberculosis of the vocal chords.

1916

21 March: As his illness is expected to extend into the summer semester, Zermelo is urged to agree to retire.

5 April: Zermelo agrees to retire.

15 April: Zermelo retires from his professorship.

31 October: Zermelo is awarded the annual Alfred Ackermann-Teubner prize of the University of Leipzig for the promotion of the mathematical sciences. Later prize winners include, for example, Emil Artin and Emmy Noether.

1 November – February 1917: Zermelo stays in Göttingen.

7 November 1916: Zermelo presents his theory of ordinal numbers to the Göttingen Mathematical Society.

1917

March – October 1919: Zermelo stays in various health resorts in the Swiss alps.

1919

July: First draft of the paper *1928* wherein Zermelo develops a procedure for evaluating the result of a tournament by using a maximum likelihood method.

November – March 1921: Zermelo stays at Locarno, Switzerland.

1921

Spring: Zermelo stays in Southern Tyrol and has correspondence with Abraham Fraenkel.

6 May 1921: Fraenkel informs Zermelo about a gap he has discovered in Zermelo's 1908 axiom system of set theory.

10 May: In his answer to Fraenkel, Zermelo proposes a second-order version of the axiom of replacement in order to close the gap, at the same time criticizing it because of its non-definite character.

17 July (?): Zermelo formulates his "infinity theses" where he describes the aims of his research in infinitary languages and infinitary logic as carried out in the early 1930s.

22 September: Fraenkel announces his axiom of replacement in a talk delivered at the annual meeting of the Deutsche Mathematiker-Vereinigung. Zermelo agrees in principle, but maintains a critical attitude because of a deficiency of definiteness.

1 October: Zermelo settles in Freiburg, Germany.

1923

Winter semester 1923/24: Zermelo attends Edmund Husserl's course "Erste Philosophie".

– 1929: Cooperation with Marvin Farber on the development of a semantically based logic system, in 1927 leading to plans for a monograph on logic.

1924

Summer: Zermelo loses interest in Husserl's phenomenology. Discussions with Marvin Farber on the possibility of obtaining a professorship in the USA.

1926

or earlier: Zermelo starts a translation of Homer's *Odyssey*, one that aims at "liveliness as immediate as possible".

22 April: Zermelo is appointed an *ordentlicher Honorarprofessor* at the Mathematical Institute of the University of Freiburg.

Winter semester – winter semester 1934/35: Zermelo gives regular courses in various fields of mathematics.

– 1932: Zermelo works on the edition of Cantor's collected papers, *Cantor 1932*. He is supported by the mathematicians Reinhold Baer and Arnold Scholz and the philosopher Oskar Becker. The participation of Abraham Fraenkel leads to a mutual estrangement.

1927

12 June: Zermelo completes his paper *1927* on measurability, where he presents results which he had obtained around 1914 and which had first been presented in Alexandrow's thesis *Alexandrow 1915*.

1928

3 August: Zermelo completes his paper *1928* on the evaluation of tournaments.

1929

– 1931: Zermelo receives a grant from the Notgemeinschaft der Deutschen Wissenschaft (Deutsche Forschungsgemeinschaft) for a project on the nature and the foundations of pure and applied mathematics and the significance of the infinite in mathematics.

May and June: Zermelo spends several weeks in Poland, giving talks in Cracow and Lvov and a series of talks in Warsaw. In the latter he presents his

view of the nature of mathematics, arguing strongly against intuitionism and formalism.

11 July: Zermelo completes his paper *1929a* wherein he responds to criticism of his notion of definiteness as put forward by Abraham Fraenkel, Thoralf Skolem, Hermann Weyl, and others.

18 September: At the annual meeting of the Deutsche Mathematiker-Vereinigung in Prague Zermelo gives a talk, *1930c*, on the solution of what is now called the "Zermelo navigation problem". An extension of the result is published as *1931a*.

Arnold Scholz becomes an assistant at the Mathematical Institute of the University of Freiburg, staying there for five years. Until his death on 1 February 1942, he will be Zermelo's closest friend and scientific partner.

1930

13 April: Zermelo completes the paper *1930a* wherein he formulates the second-order Zermelo-Fraenkel axiom system and presents an incisive picture of the cumulative hierarchy.

– 1932: Zermelo's controversy with Skolem and Gödel about finitary mathematics, in particular about Skolem's first-order approach to set theory and Gödel's first incompleteness theorem.

1931

Zermelo develops infinitary languages and an infinitary logic as a response to Skolem and Gödel.

15 September: Zermelo presents his work on infinitary languages and infinitary logic at the annual meeting of the Deutsche Mathematiker-Vereinigung. The talk results in the polemical paper *1932a* and the straightforward *1932b*.

September/October: Correspondence with Gödel about the proof of Gödel's first incompleteness theorem and Zermelo's infinitistic point of view.

18 December: Zermelo is elected a corresponding member of the Gesellschaft der Wissenschaften zu Göttingen on a proposal of Richard Courant.

– 1935: Zermelo continues his research on infinitary languages and infinitary logic which results in the paper *1935*. He works on large cardinals and on a monograph on set theory.

1932

Spring: Zermelo goes on a cruise that visits ancient sites in Greece, Italy, and Northern Africa.

June: Zermelo devises an electrodynamic clutch for motorcars.

July: Zermelo is invited to contribute a paper to a special issue of *Zeitschrift für angewandte Mathematik und Mechanik* in honor of its founder and editor Richard von Mises. He accepts the invitation and contributes the paper *1933a*.

1933

16 February: Zermelo is elected an extraordinary member of the Heidelberger Akademie der Wissenschaften on a proposal of Heinrich Liebmann and Artur Rosenthal.

1934

Zermelo moves to Bernshof, a remote country house in the hilly outskirts of Freiburg where he lives until his death.

2 November: Zermelo presents his paper *1934* on elementary number theory to the Göttingen academy.

1935

2 March: Zermelo resigns his honorary professorhip when denounced for his unwillingness to give the Hitler salute.

– 1940: Smaller scientific projects in various fields of mathematics, further work on a book on set theory and work on a collection of mathematical "miniatures" representing several of his own results.

1937

4 October: Zermelo gives a flawed refutation, *s1937*, of the existence of countable models of set theory.

1941

19 July: Arnold Scholz organizes a colloquium in Göttingen on the occasion of Zermelo's 70th birthday; Zermelo gives three talks that correspond to three items of his collection of mathematical "miniatures". Other speakers include Konrad Knopp and Bartel van der Waerden.

1944

14 October: Marriage to Gertrud Seekamp.

Zermelo suffers from a glaucoma that can no longer be treated adequately and will eventually lead to total blindness.

1946

23 January: Zermelo turns to the Rector's office of the University of Freiburg to request his reappointment as a full honorary professor.

23 July: Zermelo is reappointed an honorary professor. Because of age and increasing blindness, he is unable to lecture.

1947

– 1948: Zermelo stays several times in Switzerland. He can live there on his pension access to which is barred from Germany, leaving the Zermelos in a difficult financial situation.

– 1950: In order to escape financial need, Zermelo tries to move back to Switzerland. His applications fail; the Swiss authorities argue that his Swiss pension does not suffice to provide for his wife as well.

1949

Spring: Zermelo tries to arrange an edition of his collected works. He fails.

1953

21 May: Zermelo, nearly 82 years old, dies at Bernshof in Freiburg.

23 May: Zermelo is buried.

– 1962: Helmuth Gericke and Gottfried Martin work on the edition of Zermelo's collected works. In 1956 Paul Bernays agrees to take part in the edition. The project was not realized.

The Zermelo *Nachlass* is acquired by the University of Freiburg.

2003

15 December: Gertrud Zermelo, 101 years old, dies at Bernshof in Freiburg.

Introductory note to 1901

Oliver Deiser

„Es gibt eine Anzahl mathematischer Sätze, insbesondere auch in der Mengenlehre, die innerhalb eines leicht überschaubaren Begriffssystems formulierbar sind, ohne daß ein Zugang zu einem Beweis auf Anhieb ersichtlich ist. Der eigenartige Reiz der Fragestellung und ihrer Beantwortung sichert solchen Sätzen fortwährendes Interesse. Dazu gehört wohl auch der Cantor-Bernsteinsche Äquivalenzsatz...“ (*Rautenberg 1987*, 71)[1]

1. Introduction

The present paper is Zermelo's first set-theoretic publication. It appeared in the *Nachrichten von der Königlichen Gesellschaft der Wissenschaften zu Göttingen* in 1901. The addendum to the title reads: "Vorgelegt von D. Hilbert in der Sitzung vom 9. März 1901". In the period between 1894 and 1897 Zermelo was scientific assistant to Max Planck in Berlin. In 1897 he moved to Göttingen and wrote his *Habilitationsschrift* in hydrodynamics. Influenced by David Hilbert, he became gradually involved in set-theoretic questions.[2] In the winter semester 1900/01 Zermelo gave a lecture on set theory—one of the first lectures at all on this then rather new subject.[3] [4] We may date the main result of the paper to late winter 1900/01.[5]

The paper discusses topics surrounding the Cantor-Bernstein theorem. Zermelo returned to this theorem again in 1905 and published a new proof

[1] "There is a number of mathematical theorems, in particular also in set theory, which can be formulated in a very clear system of notions, but for which an approach to a proof is not visible at first go. The peculiar appeal of the question and its answer assures those theorems sustained interest. Among them is, to be sure, the equivalence theorem of Cantor-Bernstein." This and all following translations are due to the author, unless otherwise noted.

[2] See *Ebbinghaus 2007b*, 27–50, for Zermelo's turn to set theory and the mathematical environment he found at Göttingen. See also section 6.

[3] The first one might have been given by Arthur Schoenflies in 1898; see section 6.

[4] The manuscript of this lecture (Zermelo *Nachlass*, C 129/150) is not only of historical interest per se, but also relevant to Zermelo's paper. The theorems of the paper appear in the manuscript. But the manuscript—a notebook—contains later additions, corrections and scratch, and it is not always clear what Zermelo presented in his course.

[5] Submissions were published rather quickly at this time, the journal is a local one, and Zermelo handed the paper to Hilbert.

in 1908. We therefore include a broader discussion of this basic theorem and its complicated and intriguing history in sections 2 and 3. The reader who is primarily interested in the contents of the paper can skip to section 4, where we review the usage of cardinal numbers at Zermelo's time. We then look at the results of the paper in detail. The main theorem is a result in cardinal arithmetic, which can be read as a generalization of the Cantor-Bernstein theorem. Interestingly, the proof of the theorem requires the axiom of choice, and thus the principle that would play a pivotal role in Zermelo's work in set theory is already involved—though neither isolated nor emphasized—in his first publication in the subject.

2. The Cantor-Bernstein theorem

Let us start with some basic notions which Cantor had introduced in 1878. For two sets M and N we say that M has *power less than or equal to* N, in symbols $|M| \leq |N|$, if there is an injection from M to N. We say that M and N have *the same power*, in symbols $|M| = |N|$, if there is a bijection between M and N. Then the theorem of Cantor-Bernstein reads:

Cantor-Bernstein theorem
For all sets M and N: If $|M| \leq |N|$ and $|N| \leq |M|$, then $|M| = |N|$.

The theorem is also often called the "equivalence theorem", from the traditional German name "Äquivalenzsatz".[6]

We state several forms of the theorem, which are easily seen to be equivalent. To do this, we introduce some more notions concerning the comparison of powers. For two sets M and N we let $|M| < |N|$, if $|M| \leq |N|$ but $|M| \neq |N|$. We also define $|M| <^* |N|$, if $|M| \leq |N|$ but non$(|N| \leq |M|)$. Finally, we let $|M| \leq^* |N|$, if $|M| <^* |N|$ or $|M| = |N|$.[7]

Equivalent forms of the Cantor-Bernstein theorem
The following are equivalent (by elementary arguments):

(a) For all sets M and N:
If $|M| \leq |N|$ and $|N| \leq |M|$, then $|M| = |N|$.

(b) For all sets $M' \subseteq N \subseteq M$: If $|M'| = |M|$, then $|N| = |M|$.
(inclusion form, "Zwischenmengensatz")

(c) $<$ is transitive.

(d) $\leq$ and $\leq^*$ are identical.

---

[6] Cantor also called two sets of the same power *equivalent*; see *Cantor 1878*, 242.

[7] In the older literature the *-versions are often used instead of our relations $\leq$ and $<$. This matters only as long as the Cantor-Bernstein theorem has not been proved. Cantor used the *-version to define "less than" for cardinals in the first part of his "Beiträge" (see *Cantor 1895*, § 2, 483), and Zermelo is using this definition in his paper. In contrast, Cantor's earlier definition is the version without a star as introduced above, cf. *Cantor 1878*, 242.

The theorem is of central importance in set theory. It is a basic structure theorem about the natural relation $\leq$, of course. But it is also of great value in practice, for it allows proofs of $|M| = |N|$ to be given in two parts: We show $|M| \leq |N|$ and $|N| \leq |M|$.[8] Cantor's striking paper of 1878, where a bijection between $\mathbb{R}$ and $\mathbb{R}^2$ is constructed, is one of many examples showing how the construction in two steps would have been considerably simpler.

Moreover, the theorem is one of the few fundamental facts in the theory of infinite cardinals which can be proved without using the axiom of choice.[9] In fact, two injections $f : M \to N$ and $g : N \to M$ can be definably merged to a bijection $h : M \to N$. The construction of the bijection h is not completely trivial, but lucid. All constructions yield one of two canonical bijections h_1 and h_2 between M and N, and the values $h_1(x)$ and $h_2(x)$ are always either $f(x)$ or $g^{-1}(x)$. We briefly review some proofs, which is also useful for the discussion of the history of the theorem to be given in the next section.[10] [11]

We first sketch the standard proof of the inclusion form, which may be regarded as a slightly simplified version of Bernstein's original argument of 1897. Thus let $M' \subseteq N \subseteq M$, and let $f : M \to M'$ be a bijection. We set $A = M - N$ and define

$$A^* = \bigcup_{n \in \mathbb{N}} f^n[A] .$$

Thus A^* is the orbit of the set A under f. We now let $h_1(x) = f(x)$ for all $x \in A^*$, and $h_1(x) = x$ for all $x \in M - A^*$. It is easy to prove that $h_1 : M \to N$ is indeed a bijection. The mapping h_1 moves A to $f[A]$, $f[A]$ to $f[f[A]]$, etc. All other points are unmoved. The sets $A_n = f^n[A]$, $n \in \mathbb{N}$, are pairwise disjoint, which allows for nice diagrams of the construction. If we define $B = N - M'$ and $B^* = \bigcup_{n \in \mathbb{N}} f^n[B]$, and let $h_2(x) = f(x)$ for $x \notin B^*$ and $h_2(x) = x$ for $x \in B^*$, then $h_2 : M \to N$ is also a bijection.

---

[8] Cf. proofs of "$M = N$" by showing $M \subseteq N$ and $N \subseteq M$, or proofs of "φ iff ψ" by showing that φ implies ψ and that ψ implies φ.

[9] For example, the *comparability theorem*, i.e., that $|M| \leq |N|$ or $|N| \leq |M|$ holds for all sets M and N, is equivalent to the axiom of choice in ZF.

[10] The point of view here is: (1) The theorem is very simple if presented the right way. (2) There are two "equivalence classes" of proofs, one using natural numbers and one not. In each class there are interesting variants. (3) Moreover, the two approaches are not very different, since they construct the same mappings. Seen from today's perspective, they are just one more example for proofs "from below" (in many steps, by recursion) and "from above" (in one step, often by intersection). Cf. the two definitions of the Borel-σ-algebra by transfinite recursion and by intersection; another example is the perfect kernel of a closed set of reals, which can be found in transfinitely many steps using Cantor's derivatives, or in one step using condensation points. Definitions from below have often been considered as finer as they may yield hierarchies.

[11] The following presentation is based on *Deiser 2004*, 71–77.

This proof can easily be modified in order to avoid the use of recursion and natural numbers.[12] The argument, which was independently found by Richard Dedekind (1887), Alwin Korselt (1902), Ernst Zermelo (1905), and Giuseppe Peano (1906), is quite obvious from the modern point of view: The orbit A^* of the above proof can also be defined as the $\subseteq$-least superset of A which is closed under the function f, i.e., we may define

$$A^* = \bigcap\{X \subseteq M \mid A \subseteq X,\ f[X\,] \subseteq X\}\,.$$

In Dedekind's terminology of his famous 1888 treatise "Was sind und was sollen die Zahlen?", A^* is the *chain of* A with respect to the function f. The bijection h_1 is defined as before, and h_2 can be obtained in an analogous way.

An interesting version of Bernstein's argument was given by Julius König in *1906*. Let $f : M \to N$ and $g : N \to M$ be injective. We assume for convenience that M and N are disjoint. Given $x \in M$, we trace its preimages under g and f as long as possible, i.e., we form the series

$$x,\ g^{-1}(x),\ f^{-1}(g^{-1}(x)), \ldots .$$

If this series has a last element which is in M, we call x *special.* For nonspecial x the series either does not terminate at all, or it terminates in N. We can then define a bijection $h_1 : N \to M$ by letting $h_1(x) = f(x)$ if x is special, and $h_1(x) = g^{-1}(x)$ otherwise. The second bijection h_2 is obtained by stipulating $h_2(x) = f(x)$ if x is special or the series of preimages does not terminate, and $h_2(x) = g^{-1}(x)$ otherwise.[13]

Finally, general fixpoint theorems by Knaster-Tarski (1928) and Tarski (1955) can be used to prove the theorem. We give one argument due to Knaster and Tarski.[14] For any set M, a function G: $\mathcal{P}(M) \to \mathcal{P}(M)$ is *monotone*, if for all $X \subseteq Y \subseteq M$ we have that $G(X) \subseteq G(Y)$. A set X is a *fixpoint* of G if $G(X) = X$. Now it is not hard to prove that for every monotone G the sets

$$X_1 = \bigcap\{X \subseteq M \mid G(X) \subseteq X\}$$

and

$$X_2 = \bigcup\{X \subseteq M \mid X \subseteq G(X)\}$$

are the $\subseteq$-least and $\subseteq$-largest fixpoints of G, respectively. In the context of Bernstein's argument above, we define a monotone $G : \mathcal{P}(M) \to \mathcal{P}(M)$ by

---

[12] The use of natural numbers has sometimes been judged as a methodological drawback of Bernstein's proof. See, e.g., *Rautenberg 1987* or Russell's positive reaction to Zermelo's second proof described in *Ebbinghaus 2007b*, 90. See also *Hausdorff 1914*, 50. In particular with respect to Zermelo's 1905 proof, see *Kanamori 2006*, 508.

[13] This proof also appears in *Birkhoff and Mac Lane 1965*, 340, with no reference to König's paper.

[14] See *Knaster 1928*, *Tarski 1955*, footnote 2 on p. 2. See *Tarski 1955*, *Banaschewski and Brümmer 1986*, and *Rautenberg 1987* for more about this approach.

$G(X) = A \cup f[X]$. Then stipulating $h_1(x) = f(x)$ for $x \in X_1$ and $h_1(x) = x$ otherwise, and $h_2(x) = f(x)$ for $x \in X_2$ and $h_2(x) = x$ otherwise, defines our two bijections.

To sum up: The theorem is simple, after all—but only "after all".

3. History of the Cantor-Bernstein theorem

Cantor conjectured the theorem very early. As indicated above, already in 1878 he must have missed the theorem when he constructed a bijection between the line and the plane. The earliest record of the theorem might be a letter to Dedekind dated 5 November 1882, where it appears in its inclusion form. The letter informs us that Cantor discussed the problem with Dedekind when they met in Harzburg in 1882, that Cantor could not see a proof then, but that he now had found a proof using transfinite numbers, and that this would close a "substantial gap in the theory of manifolds".[15] Indeed in 1883, in the fifth part of his series on "Punktmannigfaltigkeiten", Cantor proved the inclusion form for well-ordered sets M of power $\aleph_1$. He stressed that the theorem holds generally "whatever power the set M might have".[16] He announced a proof, which he never gave. It might be that Cantor saw a general proof for well-ordered sets M (for which the theorem is regarded as almost trivial today). He no doubt conjectured that the theorem holds generally, also because he believed that every set can be well-ordered. In his "Beiträge" of 1895, he noted that the theorem is a trivial consequence of the *-version of the trichotomy of cardinals, i.e., it follows from the proposition that $|M| <^* |N|$ or $|M| = |N|$ or $|N| <^* |M|$ holds for all sets M and N.[17] But Cantor never found a simple and general argument. It seems likely that he misjudged the complexity of the problem, and therefore thought that it can only be proved using more advanced tools.

Dedekind found his proof already in July 1887.[18] But, unfortunately for the history of set theory, he hid his light under a bushel, and in particular he did not discuss it with Cantor, who had asked the question five years earlier, and who repeated his interest in the problem in his letters. Dedekind's notes were discovered by J. Cavaillès in his scientific remains, and the proof is published in Dedekind's collected papers.[19] However, something like a sketch of the argument was published by Dedekind as article 63 of his 1888 treatise "Was sind und was sollen die Zahlen?". The theorem of article 63 was not used in the book, and the proof was notably "left to the reader". It yields, read in

---

15 See *Noether and Cavaillès 1937*, 55, and *Cantor 1991*, 85.

16 See *Cantor 1883b*, 581ff.

17 See *Cantor 1897*, small print at p. 484, and also the letter to Philip Jourdain of 4 November 1903, *Cantor 1991*, 434.

18 It is the Dedekind-Korselt-Zermelo-Peano proof given above, which does not use natural numbers. Dedekind constructs the mapping h_1 using the chain A^* of A.

19 See *Dedekind 1932*, 447f.

the context of the proof of the inclusion form given above, the partitions A^*, $M - A^*$ of M and $f[A^*]$, $M - A^*$ of N, from which the bijection h_1 between M and N can be easily read of.[20]

Felix Bernstein presented his proof some time around Easter 1897 at Cantor's seminar in Halle.[21] Borel learned the proof from Cantor at the First International Congress of Mathematicians in Zurich (August 1897).[22] Bernstein's proof was then published in the appendix of Borel's book *Théorie des fonctions* in 1898, and it also appeared in Arthur Schoenflies' "Bericht" of 1900.[23] Because of these two influential publications the theorem and its proof quickly became widely known.

Bernstein met Dedekind in Whitsun 1897, and told him about the theorem he had proved. Dedekind replied, to Bernstein's astonishment, that the theorem is easy to prove using the tools developed in his 1888 treatise. After the meeting Dedekind reconstructed his own proof, which he might have forgotten by that time. We know about this story and the reconstructed proof from a letter by Dedekind to Cantor of 29 August 1899. The proof of the letter was then published by Zermelo in his 1932 edition of Cantor's works.[24] In an annotation Zermelo points out that Dedekind's proof is almost identical to his own proof he published in his 1908 axiomatization paper. He adds: "Why neither Dedekind nor Cantor has then decided to publish this anyway not unimportant proof is not quite understandable today."[25]

[20] We refer the reader to *Ferreirós 1993* and *Ferreirós 1999*, 239ff, for an attempt to explain this irritating state of affairs. Basically Ferreirós argues that Dedekind wanted to test whether Cantor would closely read his works, and the somewhat mysteriously placed article 63 would give him the argument he was looking for. It is well-known that the relationship between the two friends was complicated in some periods, and though this explanation is speculative, it is certainly not absurd. A more conservative explanation is that Dedekind had forgotten in 1887 the problem Cantor had asked in 1882, that he indeed did not really need his theorem of July 1887 in his treatise on numbers, and that he misjudged the general importance of the result. However it may be, the story continues when Bernstein met Dedekind in Whitsun 1897; see below.

[21] See *Borel 1898*, 103f, footnote (3), and *Cantor 1932*, 450.

[22] See again *Borel 1898*, 103f, footnote (3).

[23] See *Borel 1898*, 102f, *Schoenflies 1900*, 16f. See *Grattan-Guinness 2000*, 130, for Schoenflies' "Bericht". Schoenflies mentiones the solution of the problem without proof already in *Schoenflies 1898*, 189.

[24] See *Cantor 1932*, 449, and moreover *Dedekind 1932*, 448.

[25] See *Cantor 1932*, 451, footnotes (2) and (3). See *Zermelo 1908b*, 272f for Zermelo's proof and a footnote about its history. Zermelo states already in 1908 that his proof is based on Dedekind's theory of chains. In 1908 he was not aware that Dedekind had already proved the theorem. Zermelo found his proof in 1905, as we know from a postcard he sent to Hilbert dated 28 June 1905; see *Ebbinghaus 2007b*, 89. Zermelo also sent a letter to Poincaré in January 1906, who published Zermelo's proof (with acknowledgements); see *Poincaré 1906b*, 314. See also *Grattan-Guinness 2000*, 133, footnote, and Zermelo's footnote just quoted.

In 1898, Ernst Schröder published a proof of the theorem, which he found and talked on already in 1894.[26] Because of this, the theorem was—and still is—sometimes called the Schröder-Bernstein theorem.[27] However, Schröder's proof was defective, and so this complicated story is further enriched by a wrong proof. The error was noticed by Schröder himself[28] and independently by Alwin Korselt, a teacher at various German schools, in 1902.[29] Moreover, Korselt found a version of Dedekind's proof. He sent a note to the *Mathematische Annalen* in May 1902, but his proof and his discussion of Schröder's error appeared there only many years later, in 1911. By that time, Zermelo and Peano[30] had independently rediscovered and published the Dedekind-Korselt proof. It adds to the somewhat tragic role Cantor plays in the story that he used to give credit to Schröder for having first proved the theorem in an easy way.[31]

This is a remarkable course of events. Cantor overlooked an easy proof of an important basic set-theoretic theorem which he conjectured in 1882 at the latest. Dedekind proved it in 1887, but for some obscure reasons he did not publish the result, at least not clearly. Schröder proved the theorem in 1894, published it in 1898, became aware of an error, but, for some obscure reasons, did not do anything about it. Korselt found the error in 1902 and moreover he found Dedekind's still unpublished proof by his own. He sent his discoveries to a premier journal, but apparently the paper was rejected. Bernstein presented his proof in 1897 when he was 19 years old, and his proof finds its way to the public through Borel, because Borel talked to Cantor at a congress. Bernstein had a conversation with Dedekind about it, who replied that this was a rather old hat, and in turn sent Cantor a letter with a reconstruction of the proof he had found already ten years ago. Neither Cantor nor Dedekind published or circulated this proof. Peano and Zermelo found it once more, both around 1905, and then in 1911, when the story comes to an end, Korselt was finally allowed to publish his contributions, which should have appeared in 1902. It is a simple theorem—but it is not a simple story.[32]

---

[26] The proof is mentioned in the note *Schröder 1896*, 81, and is given in *Schröder 1898*, 337f. See also *Schoenflies 1900*, 16, and *Ebbinghaus 2007b*, 90.

[27] See, e.g., *Schoenflies 1900*, 16, and *Cantor 1932*, 451.

[28] See *Korselt 1911*, 295, for Korselt's letter to Schröder of 8 May 1902.

[29] Some biographical information about Korselt can be found in *Kreiser 1995*.

[30] See *Peano 1906a*.

[31] See, e.g., *Cantor 1991*, 434.

[32] For some of many discussions of the Cantor-Bernstein theorem and its history see, in chronological order: *Hessenberg 1906*, 13–14, 36–40, *Schoenflies 1913*, 34–39, *Hausdorff 1914*, 47–50, *Fraenkel 1928*, 70–76, *Fraenkel 1961*, 72–79, *Moore 1982*, 42–43, 48–50, *Banaschewski and Brümmer 1986*, *Rautenberg 1987*, *Ferreirós 1999*, 239–241, *Grattan-Guinness 2000*, 132–134, *Kanamori 2006*, 508–512, *Ebbinghaus 2007b*, 73, 89–90.

4. Cardinal numbers in Zermelo's time

The Cantorian definitions comparing the power of two sets just talk about certain relations between sets. Studying the relations $|M| \leq |N|$ and $|M| = |N|$ inevitably leads to some intuitive or formal notion of the *power* or *cardinality* or *cardinal number* of a set M, which is not relational any longer. We will denote it by $|M|$, as it is customary today. In modern set theory, there are basically two definitions of the cardinality of M, which we shall briefly discuss.[33]

In ZFC without the axiom of foundation we can define, for each set M:

$$|M| = \text{“the least ordinal number } \alpha \text{ such that } |M| = |\{\beta \mid \beta < \alpha\}|\text{”.}^{34}$$

What is needed for this definition is that the well-ordering theorem holds *and* that there are definable representatives for all well-orderings having the same length. Thus the axiom of choice as well as the replacement scheme is important, while there is no need for the axiom of foundation.

In ZF, we can still define cardinals by truncating proper classes to sets by what is often called "Dana Scott's method". We here let, for each set M:

$$|M| =$$

$$\{N \mid N \text{ is of minimal rank such that there is a bijection } f : M \to N\} .$$

For this we need to know that the universe can be stratified in a cummulative hierarchy, and thus the axiom of foundation is crucial for this definition.[35]

Axiomatic set theory is today well-known as a framework for all of mathematics: Every mathematical object can be interpreted as a set. It is curious that just a basic notion of naïve set theory—and not of algebra or geometry—is among the most complex from the point of view of axiomatic set theory. It took decades until the naïve Cantorian cardinals were interpreted faithfully as sets. Moreover, Azriel Levy has shown in 1969 that there is no satisfactory definition of the cardinality of M in the theory ZFC without foundation and without choice.[36] One of the two axioms is needed and sufficient. Here a satisfactory definition of cardinal numbers is simply any (definable) class function Card on the universe of sets such that for all sets M and N we have:

$$(\#) \quad \mathrm{Card}(M) = \mathrm{Card}(N) \quad \textit{iff} \quad \text{there is a bijection between } M \text{ and } N.$$

---

[33] See *Deiser 200?* for more on the development of the notion of a cardinal.

[34] Here John von Neumann's precise definition of ordinal numbers of 1923 is used, which was independently found by Zermelo. With this definition of $|M|$, we have that $||M|| = |M|$ holds for all sets M, but this *representation property* is not important here.

[35] This explains from an abstract point of view why Zermelo did not define cardinal numbers in his first introduction of a set-theoretic axiom system in 1908, which is ZFC without foundation and without replacement: The system is not strong enough for either of the two definitions.

[36] See *Levy 1969* and *Kanamori 2006*, 246.

The two precise definitions given above are well beyond the horizon of the year 1901 and much later. But though cardinal numbers were not, by modern standards, really defined, they were used quite successfully in an informal way by Cantor and the following generation. Cantor, in the vein of ancient traditions reaching back to Euclid and Thales, described the cardinality of a set M as a system of units or "ones": There are "as many" units in this system as there are elements in M.[37] In the first part of Cantor's "Beiträge" of 1895 we read:

> *"Mächtigkeit" oder "Kardinalzahl" von M nennen wir den Allgemeinbegriff, welcher mit Hilfe unseres aktiven Denkvermögens dadurch aus der Menge M hervorgeht, daß von der Beschaffenheit ihrer verschiedenen Elemente m und von der Ordnung ihres Gegebenseins abstrahiert wird.* Das Resultat dieses zweifachen Abstraktionsakts, die Kardinalzahl oder Mächtigkeit von M, bezeichnen wir mit $\overline{\overline{M}}$. Da aus jedem einzelnen Elemente M, wenn man von seiner Beschaffenheit absieht, eine "Eins" wird, so ist die Kardinalzahl $\overline{\overline{M}}$ selbst eine aus lauter Einsen zusammengesetzte Menge, die als intellektuelles Abbild oder Projektion der gegebenen Menge M in unserem Geiste Existenz hat.[38]

Cantor's definition of cardinal number was sharply criticized by Frege as being vague, and as demanding an impossible act of abstraction. In particular, Frege objected to Cantor's indistinguishable units.[39]

---

[37] Euclid defined number to be a "set compounded of unities". For the tradition of this definition see, e.g., *Gericke 1970* and *Heath 1931*. This definition was often repeated through the Middle Ages, and Cantor might have found it also in article 87 of Bernard Bolzano's *Wissenschaftslehre*, *Bolzano 1837*.

[38] "*'Power' or 'cardinal number' of M is called that universal notion, which emerges, by means of our active intellectual power, from the set M by abstracting from the character of its different elements and from the ordering of their givenness.* We denote the result of this double act of abstraction, the cardinal number or power of M, by $\overline{\overline{M}}$. Since every single element m becomes, if one abstracts from its character, a "unity", the cardinal number $\overline{\overline{M}}$ itself is a set composed of unities only, which has, as the intellectual effigy or projection of the given set M, its existence in our mind." (See *Cantor 1895*, 481.)

[39] See Frege's 1892 review of Cantor's booklet *Zur Lehre vom Transfiniten* (*1890*), where a version of Cantor's cardinal definition appears, and the polemical draft printed in *Frege 1976*, 76. Frege himself discussed the notions "unit" and "one" in his book *Grundlagen der Arithmetik* of 1884, cf. *Frege 1986*, section III, §§ 29–54. Concerning cardinals, Frege himself used basically the "Russell-style" definition $|M| = \{N \mid |N| = |M|\}$, which has the unpleasant property that $|M|$ is a proper class if and only if M is not empty. Thus Frege throws stones at Cantor living in a glass house.—Cantor defined order types by a similar abstraction process, in which the order of the elements is kept. The treatise *Deiser 2006* can be read as an attempt to express Cantor's definitions of ordinal number and of cardinal number as directly as possible in first-order logic.

Schoenflies in his "Bericht" of 1900 very closely followed Cantor in his definition of cardinal number, but he purges the definition (without any further discussion) and in particular he discards the traditional units:[40]

> Mächtigkeit von M heißt der Allgemeinbegriff, der aus M dadurch hervorgeht, daß von der Beschaffenheit und Ordnung der Elemente abstrahiert wird.[41]

We know from Zermelo's notebook of his 1900/01 lecture on set theory in Göttingen[42] that he followed Cantor without much hesitation as well. The original definition of the notes reads:

> Abstrahiert [man] bei einer vorgelegten Menge von der Beschaffenheit und der Anordnung ihrer Elemente bei festgehaltener Verschiedenheit, so heißt der entstehende Allgemeinbegriff die "Mächtigkeit der Menge".[43]

Zermelo later crossed the definition out and rephrased it as:

> Die Mächtigkeit oder Cardinalzahl einer Menge entsteht durch Abstraktion, indem man absieht von jeder Beschaffenheit und Anordnung der Elemente, bei festgehaltener Verschiedenheit.[44]

In this version the philosophical term "universal notion" has disappeared.

Hausdorff in his famous book of 1914 introduced cardinals by "simply adopting the formal point of view" as certain, but unspecified,—"signs" having the crucial property (#). Now the "act of abstraction" has disappeared as well.[45]

---

[40] Schoenflies' treatise is of particular interest here since it is likely to have influenced Zermelo's paper; Zermelo quotes it in the paper, and it also appears in the references of his 1900/01 lecture notes. See also section 6 of this note.

[41] "The power of M is that universal notion emerging from M by abstracting from the character and ordering of the elements." See *Schoenflies 1900*, 4.

[42] Zermelo *Nachlass*, C 129/150.

[43] "If one abstracts, given a set, from the character and ordering of its elements retaining differentness, then the resulting universal notion is called the 'power of the set'".

[44] "The power or cardinal number of a set emerges from abstraction, by disregarding any character or ordering of the elements, yet retaining differentness."—As we said, the notebook contains later additions, which are hard to date. For example, it mentions Hessenberg's book, which appeared only in 1906.

[45] See *Hausdorff 1914*, 47. Hausdorff had used this definition already in his lectures on set theory; see *Felgner 2002a*, 641. In the rewritten edition of his book of 1927, Hausdorff explicitly leaves the task of finding the "true being" of cardinals to the philosophers—which is a different task than a precise definition of cardinals as certain objects of the theory, which Hausdorff again did not give (see *Hausdorff 1927*, 25). This treatment of cardinal numbers reflects Hausdorff's general attitude not to waste his time on a certain type of foundational difficulties, among them being the paradoxes (see, e. g., the introduction of *Hausdorff 1908*).

So we can observe a continuous reduction of Cantor's original cardinal definition:

Cantor (1895 and before): Abstraction, universal notion, system of units.

Schoenflies (1900), Zermelo (1900/01): Abstraction, universal notion.

Zermelo (1900/01 or somewhat later): Abstraction.

Hausdorff (1914 and earlier): Unspecified assignment of arbitrary symbols such that (#) holds.

Cantor, Schoenflies, Zermelo, Hausdorff, and many others used the new cardinals just the way the more special finite cardinals were used for centuries without a precise definition. After all, the absence of such a definition should not be overestimated, since the crucial property (#) was understood quite well, and emphasized by all authors. And the *laws of cardinal arithmetic* are, to some extent, the only thing that really matters. Once we back up these early texts with a formally satisfying definition of cardinal, they can be seen to have no flaw or inaccuracy.

As the early set theorists did since Cantor's "Beiträge"—and as modern set theorists still do when the axiom of choice is absent—, we shall use the Gothic types ("Fraktur-Buchstaben") $\mathfrak{a}, \mathfrak{b}, \mathfrak{c}, \ldots$ to denote cardinal numbers. Once cardinal numbers have been introduced, the relation $\mathfrak{a} \leq \mathfrak{b}$ as well as the arithmetical operations $\mathfrak{a} + \mathfrak{b}$, $\mathfrak{a} \cdot \mathfrak{b}$, $\mathfrak{a}^{\mathfrak{b}}$ are defined outright:

$$\mathfrak{a} \leq \mathfrak{b} \text{ if } |M| \leq |N|, \quad \mathfrak{a} \leq^* \mathfrak{b} \text{ if } |M| \leq^* |N|, \quad \text{etc.}$$

$$\mathfrak{a} + \mathfrak{b} = |M \times \{0\} \cup N \times \{1\}|, \quad \mathfrak{a} \cdot \mathfrak{b} = |M \times N|,$$

$$\mathfrak{a}^{\mathfrak{b}} = |{}^N M| = |\{f \mid f \text{ is a function from } N \text{ to } M\}|,$$

where M and N are any sets such that $\mathfrak{a} = |M|$ and $\mathfrak{b} = |N|$.[46] It is worth mentioning that these definitions include finite arithmetic. They first appear, with a collection of elementary laws like $\mathfrak{a}(\mathfrak{b} + \mathfrak{c}) = \mathfrak{a}\mathfrak{b} + \mathfrak{a}\mathfrak{c}$, in Cantor's "Beiträge" of 1895.[47]

The definitions of sum and product can be generalized, but in view of Zermelo's paper we shall just need one more definition—depending on the axiom of choice—, namely countable sums of cardinals. We define:

$$\textstyle\sum_{n\in\mathbb{N}} \mathfrak{a}_n = |\bigcup_{n\in\mathbb{N}} M_n \times \{n\}|,$$

[46] Using $M \times \{0\}$ and $N \times \{1\}$ in $\mathfrak{a} \times \mathfrak{b}$ is just a convenient way to ensure disjointness. Concerning axiomatic issues, the definitions make sense with both definitions of $|M|$ given above. For the ZFC definition, M and N can be simply taken to be $\mathfrak{a}$ and $\mathfrak{b}$. For the ZF definition, which lacks the representation property, the product $\mathfrak{a} \cdot \mathfrak{b}$ can be defined to be the single element of $\{|M \times N| \mid M \in \mathfrak{a}, N \in \mathfrak{b}\}$; the same applies to $\mathfrak{a} + \mathfrak{b}$ and $\mathfrak{a}^{\mathfrak{b}}$.

[47] See *Cantor 1895*, 485ff.

where M_n is a set of cardinality $\mathfrak{a}_n$ for all $n \in \mathbb{N}$. The countable axiom of choice AC_ω is needed for this definition.[48] Without this principle, there might be no sequence $\langle M_n \mid n \in \mathbb{N}\rangle$ such that $|M_n| = \mathfrak{a}_n$ for all n, and granted there are such sequences, the value of the infinite sum cannot be shown to be independent of the choice of the sequence.

Cardinal numbers turn out to be very useful in computations, as general laws for cardinals can replace tedious—as well as ingenious—arguments manipulating bijections. The most prominent and quite impressive example, due to Cantor, is the calculation (with $\omega = |\mathbb{N}|$):

$$|\mathbb{R}^2| = (2^\omega)^2 = 2^{2\cdot\omega} = 2^\omega = |\mathbb{R}|\,,$$

proving that there is a bijection between the plane and the line, granted one has already proved that $|\mathbb{R}| = |{}^{\mathbb{N}}\{0,1\}|\ (= |\mathcal{P}(\mathbb{N})|\,)$.[49]

Zermelo uses these little, but quite effective tricks is his paper, and, as we shall see in the next section, he is particularly fond of "doubling" arguments like the following: Assume that $\mathfrak{a} = \mathfrak{b} + \omega$. Then $\mathfrak{a} = \mathfrak{b} + (\omega + \omega) = (\mathfrak{b} + \omega) + \omega = \mathfrak{a} + \omega$. Many such calculations appear in Zermelo's notes of his 1900/01 lecture.

5. The results of the paper

Zermelo proves five theorems. The central result is

Theorem I (Main Theorem)
If $\mathfrak{m} = \mathfrak{m} + \mathfrak{p}_n$ holds for all $n \in \mathbb{N}$, then $\mathfrak{m} = \mathfrak{m} + \sum_{n\in\mathbb{N}} \mathfrak{p}_n$.

We closely follow Zermelo's argumentation, but give a slightly modernized version. Our proof is in ZF + DC, where DC is the principle of Dependent Choices; in fact, the somewhat weaker countable axiom of choice AC_ω is sufficient, as will be discussed below.[50] In ZF + DC, Zermelo's argument is

---

[48] AC_ω is the principle: For every countable set M there is a function g on M such that $g(x) \in x$ for all nonempty $x \in M$. It was studied in *Bernays 1942*, section 10.

[49] See *Cantor 1895*, 488, small print, where Cantor enthusiastically notes that the full content of his 1878 paper can now be algebraically derived "with a few strokes".

[50] The principle DC was introduced by Paul Bernays in 1942; see *Bernays 1942*, 86. It states: If R is a relation such that for every x there is a y with yRx, then there is a sequence $\langle x_n \mid n \in \mathbb{N}\rangle$ such that $x_{n+1}Rx_n$ holds for all n. The basic use is that in recursions on $n \in \mathbb{N}$ we can choose the next object x_{n+1} depending on the choice of x_n. In many cases the use of DC can be weakened to the countable axiom of choice AC_ω, and Zermelo's theorem is one example of this. That this is not always the case, i. e., that DC is stronger than AC_ω was shown by Ronald Jensen; see *Jensen 1966*. The name "dependent choice" is due to Tarski (cf. *Tarski 1948*, 96).

still of interest, while in ZFC Zermelo's main theorem is a trivial consequence of the law $\omega \cdot \mathfrak{m} = \mathfrak{m}$ for all infinite cardinals $\mathfrak{m}$.[51]

We will not use the Cantor-Bernstein theorem in the following proof. In fact, we will prove a generalization of it in ZFC, and then we will prove it in ZF.

We first prove:

(1) If $\mathfrak{m} = \mathfrak{m} + \mathfrak{q}_n$ holds for all $n \in \mathbb{N}$, then $\sum_{n\in\mathbb{N}} \mathfrak{q}_n \leq \mathfrak{m}$.

Let M_0 be a set of cardinality $\mathfrak{m}$. By hypothesis we can, using DC, recursively construct a chain $M_0 \supseteq M_1 \supseteq \ldots \supseteq M_n \supseteq \ldots$ such that:

(i) $|M_n| = \mathfrak{m}$,

(ii) $|M_n - M_{n+1}| = \mathfrak{q}_n$ for all $n \in \mathbb{N}$.

This proves (1), since $\sum_{n\in\mathbb{N}} \mathfrak{q}_n = |\bigcup_{n\in\mathbb{N}}(M_n - M_{n+1})| \leq |M_0| = \mathfrak{m}$.

As a special case of (1) we note:

(2) If $\mathfrak{m} = \mathfrak{m} + \mathfrak{p}$ holds, then $\omega\,\mathfrak{p} = \sum_{n\in\mathbb{N}} \mathfrak{p} \leq \mathfrak{m}$.

But then:

(3) If $\mathfrak{m} = \mathfrak{m} + \mathfrak{p}$ holds, then $\mathfrak{m} = \mathfrak{m} + \omega\,\mathfrak{p}$.

Because by (2) there is a cardinal $\mathfrak{m}'$ such that $\mathfrak{m} = \mathfrak{m}' + \omega\,\mathfrak{p}$. Since $\omega = \omega + \omega$, we have $\mathfrak{m} = \mathfrak{m}' + (\omega + \omega)\,\mathfrak{p} = \mathfrak{m}' + \omega\,\mathfrak{p} + \omega\,\mathfrak{p} = \mathfrak{m} + \omega\,\mathfrak{p}$.

Now assume that $\mathfrak{m} = \mathfrak{m} + \mathfrak{p}_n$ for all n. Then by (3), $\mathfrak{m} = \mathfrak{m} + \omega\,\mathfrak{p}_n$ for all n. Thus by (1) with $\mathfrak{q}_n = \omega\,\mathfrak{p}_n$, there is an $\mathfrak{m}'$ such that $\mathfrak{m} = \mathfrak{m}' + \sum_{n\in\mathbb{N}} \omega\,\mathfrak{p}_n$. But then

$$\mathfrak{m} = \mathfrak{m}' + \textstyle\sum_{n\in\mathbb{N}} (\omega + 1)\,\mathfrak{p}_n = \mathfrak{m}' + \sum_{n\in\mathbb{N}} \omega\,\mathfrak{p}_n + \sum_{n\in\mathbb{N}} \mathfrak{p}_n = \mathfrak{m} + \sum_{n\in\mathbb{N}} \mathfrak{p}_n ,$$

proving the main theorem.

Assertion (3), a stage of the proof, is explicitly stated by Zermelo as:

Theorem II

If $\mathfrak{m} = \mathfrak{m} + \mathfrak{p}$, then $\mathfrak{m} = \mathfrak{m} + \omega\,\mathfrak{p}$.

In our recursive construction of the sets M_n in the proof of (1) we used the principle DC of dependent choices. In fact, the countable axiom of choice AC_ω is sufficient here: Using AC_ω, we choose, for each n, a set $N_n \subseteq M_0$ such that $|N_n| = \mathfrak{m}$, $|M_0 - N_n| = \mathfrak{q}_n$, and we choose, for each n, a bijection

[51] In the main theorem, we may assume that $\mathfrak{m} \geq \omega$, since otherwise all the $\mathfrak{p}_n$ are zero. Since $\mathfrak{m} = \mathfrak{m} + \mathfrak{p}_n$, we have that $\mathfrak{p}_n \leq \mathfrak{m}$ for all n. Thus $\mathfrak{m} \leq \mathfrak{m} + \sum_{n\in\mathbb{N}} \mathfrak{p}_n \leq \sum_{n\in\mathbb{N}} \mathfrak{m} = \omega \cdot \mathfrak{m} = \mathfrak{m}$. The nontrivial result "$\omega \cdot \mathfrak{m} = \mathfrak{m}$ whenever $\mathfrak{m} \geq \omega$" used here was not known at Zermelo's time. Moreover, it is equivalent to the addition law "$\mathfrak{m} + \mathfrak{m} = \mathfrak{m}$ whenever $\mathfrak{m}$ is infinite", which is not derivable from DC; see *Howard and Rubin 1998*. See *Deiser 2005* for more on the history of the $\mathfrak{m} \cdot \mathfrak{m} = \mathfrak{m}$ and $\mathfrak{m} + \mathfrak{m} = \mathfrak{m}$ for infinite $\mathfrak{m}$.

$\varphi_n : M_0 \to N_n$. We now set:

$$M_n = \varphi_0 \varphi_1 \ldots \varphi_{n-1}[M_0] \text{ for each } n \in \mathbb{N} .$$

Then $M_0 \supseteq M_1 \supseteq \ldots M_n \supseteq \ldots$, and $|M_n| = \mathfrak{m}$, $|M_n - M_{n+1}| = |M_0 - \varphi_n[M_0]| = \mathfrak{q}_n$ holds for all $n \in \mathbb{N}$, as is readily verified. The rest is as before. This is in fact the argument used in Zermelo's paper!

On the other hand, the axiom of choice is *not* needed for the proof of theorem II—and it is thus also not needed for theorems III, IV, and V, which are simple corollaries to theorem II. Here we can fix an $N \subseteq M_0$ such that $|N| = \mathfrak{m}$, $|M_0 - N| = \mathfrak{p}$, and a single bijection $\varphi : M_0 \to N$. Then the images $M_n = \varphi^n[M_0]$ form a chain as desired.

As a corollary to theorem II we get the cardinal version of the inclusion form of the Cantor-Bernstein theorem:

Theorem III (Cantor-Bernstein theorem, inclusion form)
If $\mathfrak{m} = \mathfrak{m} + \mathfrak{p} + \mathfrak{q}$, then $\mathfrak{m} = \mathfrak{m} + \mathfrak{p}$.

The proof is given by the following calculation:

$$\mathfrak{m} = \mathfrak{m} + \omega\,(\mathfrak{p} + \mathfrak{q}) = \mathfrak{m} + (\omega + 1)\,\mathfrak{p} + \omega\,\mathfrak{q} = \mathfrak{m} + \omega\,(\mathfrak{p} + \mathfrak{q}) + \mathfrak{p} = \mathfrak{m} + \mathfrak{p} .$$

Thus we proved the Cantor-Bernstein theorem, without using AC. Zermelo explicitly notes that his proof of the Cantor-Bernstein theorem differs from Bernstein's proof only in "Anordnung and Ausdrucksweise", i.e., in "arrangement and wording".[52] Still, Zermelo was the first to cast the argument in an arithmetical form, emphasizing implication (2), from which theorem II as well as the Cantor-Bernstein theorem is easily derived.[53] In this way, the main theorem can indeed be read as a generalization of the Cantor-Bernstein theorem, and Zermelo propagates this view in the paper.

Zermelo also notes another of the equivalences of the Cantor-Bernstein theorem, namely assertion (d) in the list of equivalences given in section 2:

Theorem IV
If $\mathfrak{n} \leq \mathfrak{m}$ then $\mathfrak{n} \leq^* \mathfrak{m}$.

It does not hurt to give the argument in Zermelo's arithmetical language: We let $\mathfrak{m} = \mathfrak{n} + \mathfrak{p}$. It is enough to show that $\mathfrak{m} \leq \mathfrak{n}$ implies that $\mathfrak{m} = \mathfrak{n}$. But if $\mathfrak{n} = \mathfrak{m} + \mathfrak{q}$, then $\mathfrak{m} = \mathfrak{m} + \mathfrak{q} + \mathfrak{p}$, and therefore $\mathfrak{m} = \mathfrak{m} + \mathfrak{q} = \mathfrak{n}$ by theorem III.

52 See page 38 of the paper. Indeed, let $\varphi : M' \cup A \cup B \to M'$ be a bijection, such that M', A, B are pairwise disjoint, $|A| = \mathfrak{p}$, $|B| = \mathfrak{q}$, $|M'| = \mathfrak{m}$. If we analyze Zermelo's argument for this case (with $M_0 = M' \cup A \cup B$, $M_n = \varphi^n[M_0]$), we see that Zermelo uses the sets $A_n = \varphi^n[A]$, $B_n = \varphi^n[B]$ to get the equation $\mathfrak{m} = \mathfrak{m} + \omega\,(\mathfrak{p} + \mathfrak{q})$. We have seen in section 2 that the shifting from A_n to $A_{n+1} = \varphi[A_n]$ is the heart of the matter. In the proof of theorem III, this argument is encapsulated in the cardinal equation $(\omega + 1)\,\mathfrak{p} = \omega\,\mathfrak{p}$.

53 Theorem III can be proved directly from (2): In the situation of theorem III, we can write $\mathfrak{m} = \mathfrak{m}' + \omega\,(\mathfrak{p} + \mathfrak{q})$ by (2). That in fact $\mathfrak{m}' = \mathfrak{m}$ holds here is not needed in the above calculation proving theorem III.

Finally, Zermelo notes:

Theorem V
If $\mathfrak{m} = \mathfrak{m} + \mathfrak{p}$, then $\mathfrak{m}' = \mathfrak{m}' + \mathfrak{p}'$ holds whenever $\mathfrak{m} \leq \mathfrak{m}'$ and $\mathfrak{p}' \leq \mathfrak{p}$.

Indeed, writing $\mathfrak{m}' = \mathfrak{m} + \mathfrak{r}$ and $\mathfrak{p} = \mathfrak{p}' + \mathfrak{q}$, we have

$$\mathfrak{m}' = \mathfrak{m} + \mathfrak{r} = \mathfrak{m} + \mathfrak{p} + \mathfrak{r} = \mathfrak{m}' + \mathfrak{p} = \mathfrak{m}' + \mathfrak{p}' + \mathfrak{q}\ ,$$

and thus the claim follows from theorem III.

At the end of the paper Zermelo looks, given a cardinal $\mathfrak{m}$, at all cardinals $\mathfrak{p}$ having the property $\mathfrak{m} = \mathfrak{m} + \mathfrak{p}$. He summarizes the results of the paper. From a modern point of view we wonder why the addition theorem for infinite cardinals, i. e., that $\mathfrak{m} + \mathfrak{m} = \mathfrak{m}$ holds for all infinite cardinals, is not mentioned as an important open problem in a paper focusing on cardinals which are unchanged by the addition of other cardinals. We note in this context that Zermelo proved in the theory ZF (by theorem II):

$$\text{If } \mathfrak{m} + \mathfrak{m} = \mathfrak{m} \text{ for a cardinal } \mathfrak{m}, \text{ then } \omega\,\mathfrak{m} = \mathfrak{m} + \omega\,\mathfrak{m} = \mathfrak{m}\,.$$[54]

6. Zermelo's set-theoretic background in 1901

Zermelo became accquainted with set theory in Göttingen, where he enrolled in mathematics in November 1897.[55] Hilbert, Schoenflies and Bernstein are the relevant figures in Göttingen with respect to Zermelo's turn to set theory. The root of the unusually high interest at Göttingen in set theory was Felix Klein, who had supported Cantor's publications as editor of the *Mathematische Annalen* already in the 1880s. Klein was also responsible for Hilbert being appointed to a chair at Göttingen in 1895, and moreover he seemed to have encouraged Zermelo to come to Göttingen.[56]

It was Hilbert who excited Zermelo's interest in foundational questions.[57] Hilbert's interest in set theory was stimulated by the correspondence with Cantor, starting in September 1897, about Cantor's distinction between "completed" or "consistent collections" versus "inconsistent collections" ("konsistente Vielheiten" and "inkonsistente Vielheiten").[58] Zermelo in a letter to Heinrich Scholz of 10 April 1936 reports that "around 1900 the set-theoretic antinomies were intensively discussed in the Hilbert circle", and that he had found the antinomy of "the set of all sets that do not contain themselves"

---

[54] This was reproved by Halpern and Howard without awareness of Zermelo's paper; see *Halpern and Howard 1970*, 489.

[55] For Zermelo's time at Göttingen see *Peckhaus 1990a, 1990b*.

[56] See *Ebbinghaus 2007b*, 27.

[57] See *Ebbinghaus 2007b*, 41, and *Moore 2002*, 42. See also *Ebbinghaus 2007b*, 28, for Hilbert's profound influence on Zermelo, in Zermelo's own words.

[58] See *Cantor 1991*, 387ff.

around that time.[59] Apparently Hilbert discussed the contents of Cantor's letters with Zermelo, and arguably the discovery of Russell's paradox around or before 1900 is Zermelo's first important contribution to set theory, though it remained unpublished.

Schoenflies was Professor at Göttingen until 1899, when he accepted a chair in Königsberg. Originally working in geometry, he became fond of Cantor's works, and was asked to contribute a paper about set theory to Felix Klein's new encyclopedia of mathematics.[60] Zermelo seems to have attended a two-hour lecture by Schoenflies with the title "Mengenlehre" in summer 1898.[61] After his "Bericht" appeared in 1900, Schoenflies wrote to Hilbert in a letter of 28 December 1900:

> Wenn Sie Zermelo sehen, so bitten Sie ihn doch, er möchte mir alle Mängel Irrtümer etc. die ihm in meinem Bericht aufstoßen, hierher mitteilen. Er ist ja vielleicht der Einzige, der den Bericht wirklich liest.[62]

Thus Schoenflies appreciated Zermelo as a true and competent reader.

After he proved the equivalence theorem, Felix Bernstein spent a year in Pisa and Rome studying philosophy, archaeology, and art history. He then studied mathematics in Berlin, Halle, Munich, and Göttingen.[63] Bernstein attended courses by Hilbert, Klein and Zermelo in Göttingen in 1899 and

---

[59] See *Ebbinghaus 2007b*, 45f, 277, *Peckhaus 1990b*, 25f, *Moore 2002*, 46, for this and other evidence that Zermelo independently found the "Zermelo-Russell paradox".

[60] See *Schoenflies 1898*.

[61] Zermelo's Göttinger Studienblatt has the following entry for summer 1898: "Über Mengenlehre bei Prof. Schoenflies" (personal communication by Heinz-Dieter Ebbinghaus; see also *Ebbinghaus 2007b*, 10). The Göttinger university calender ("Vorlesungsverzeichnis") has for summer 1898 the entry: "Mengenlehre, Prof. Schönflies, Donnerstag und Freitag 9–10 Uhr" (personal communication by Ulrich Hunger from the University Archive in Göttingen). Nothing else seems to be known about this lecture. But Schoenflies' contribution for the encyclopedia has the title "Mengenlehre", too, and was finished by him before November 1898, as we know from the table of contents of the encyclopedia.

It is tempting to argue that Schoenflies' lecture of 1898 presented the material of this article. Thus Schoenflies seems to have been the first who has given a course on set theory, and Zermelo's 1900/01 lecture would probably have been the second (cf. *Moore 2002*, 44). More importantly, the content of *Schoenflies 1898* might have been Zermelo's first education in set theory apart from his own reading.

See *Fritsch and Fritsch 2001* and *Kaemmel 2006* for more about the life and work of Schoenflies. See also *Grattan-Guinness 2000*, 130ff.

[62] Universitätsarchiv Göttingen, Cod. Ms. David Hilbert 355, Letter 9. "When you see Zermelo, then please ask him to send here ⟦to Königsberg⟧ all flaws, errors, etc., which annoy him in my Bericht. He is probably the only one who is really reading the Bericht." See also *Moore 2002*, 46.

[63] See *Schappacher 2005* for a short biography about Bernstein.

1900.[64] He finished his dissertation in set theory at Göttingen under Hilbert in 1901.[65] One important result of the dissertation appears as "Bernstein's Theorem" in Zermelo's lecture notes of 1900/01.[66]

Apart from personal contacts and local influences, there were the main published sources of set-theoretic knowledge. These were, above all, Cantor's "Punktmannigfaltigkeiten" (*1879–1884*) and "Beiträge" (*1895, 1897*), Dedekind's treatise on numbers (*1888*), Borel's "*Leçons*" (*1898*), and, after the turn of the century, Schoenflies' "Bericht" (*1900*). Zermelo mentions these sources in his lecture notes and in the 1901 paper, and his notation is derived from them.[67]

A full evaluation of Zermelo's lecture notes of 1900/01 would certainly shed more light on Zermelo's understanding of set theory around the turn of the century. This task is encumbered by the untidy and sketchy character of the original notes and the later additions. Moreover, parts of it are written in a forgotten and complicated shorthand.[68]

The reader might grant the author of this note a single paragraph trying a reconstruction. As anyone learning set theory—and in particular as anyone lecturing on the subject—, Zermelo had to think thoroughly about the Cantor-Bernstein question and its solution. The main theorem of the paper appears to be the result of Zermelo's engagement with this topic—not of investigations concerning the arithmetic of infinite cardinals. Like Cantor in the first part of his "Beiträge", Zermelo became fond of the elegance of the arithmetical manipulation of cardinal numbers, which could replace the construction and manipulation of bijections. He successfully applied this calculus to the Cantor-Bernstein problem, and doing so he discovered the main theorem. End of speculation.

---

[64] See *Moore 2002*, 46.

[65] Bernstein's dissertation was published, with some minor updates, in 1905; see *Bernstein 1905*.

[66] See *Bernstein 1905*, 131f. Bernstein proves (tacidly using the axiom of choice): If M and N are sets such that $|M \times N| \leq |M \cup N|$, then $|M| \leq |N|$ or $|N| \leq |M|$. Though the theorem as such was superseded by Zermelo's well-ordering theorem, Bernstein's ingenious argument was not, and it figures prominently in Tarski's proof (*1924b*) showing that the multiplication law "$\mathfrak{m} \cdot \mathfrak{m} = \mathfrak{m}$ for all infinite $\mathfrak{m}$" implies AC in ZF. See also *Deiser 2005*.

[67] Many of Cantor's papers are listed at the beginning of the lectures notes, but only the "Beiträge" are emphasized with a double bar. Dedekind's treatise of 1888 is quoted in the paper, and Zermelo used Dedekind's definition of an infinite set in his lecture and his paper. Borel's book of 1898 and Schoenflies "Bericht" are given as references for the Cantor-Bernstein theorem in the 1901 paper. Borel's book of 1898 and Bernstein's dissertation of 1901 also appear in the references of the lecture notes, but seem to be later additions.

[68] The shorthand is probably a system by Leopold Arends; see *Moore 2002*, 49, footnote 12. See also *Peckhaus 1990b*, 25.

Zermelo's paper is quite original in his use of cardinal numbers, and it is a nice start for someone turning from physics to set theory. But no one could have guessed from this paper that Zermelo's next publication in set theory would be the solution of one of the main open problems of pure set theory: the proof of the well-ordering theorem.

Annotations to the text

[1] Schoenflies used the special symbols $\mathfrak{a}$ for the cardinality of $\mathbb{N}$, $\mathfrak{c}$ for the cardinality of $\mathbb{R}$, and $\mathfrak{f}$ for the cardinality of all real functions. Thus $\mathfrak{c} = 2^{\mathfrak{a}}$ and $\mathfrak{f} = 2^{\mathfrak{c}}$. Here $\mathfrak{a}$ is for "abzählbar", $\mathfrak{c}$ for "Continuum", $\mathfrak{f}$ for "Funktionen". See *Schoenflies 1900*, 5 and 8. This notation can be found quite often in early set theory papers and books. Cantor only used $\aleph_0$ for the power of $\mathbb{N}$ in his "Beiträge", but he used Gothic letters for arbitrary cardinals.

[2] "countable" here means "countably infinite". Since Zermelo uses Dedekind's definition of infinite set, there is no use of the axiom of choice here. It is provable in ZF that the following are equivalent:

(i) M is Dedekind-infinite, i.e., there is a proper subset N of M such that $|M| = |N|$.

(ii) M has a countably infinite subset, i.e., $|\mathbb{N}| \leq |M|$.

For the non-trivial direction from (i) to (ii), let $g : M \to N$ be a bijection, where N is a proper subset of M. Let $x \in M - N$ be arbitrary. Then the orbit of x, i.e., $\{g^n(x) \mid n \in \mathbb{N}\}$, is a countably infinite subset of M. It interesting to note that this argument resembles the argument used to prove the Cantor-Bernstein theorem. Moreover, in cardinal notation we can write the implication just proved as: If $\mathfrak{m} = \mathfrak{m} + 1$, then there is an $\mathfrak{m}'$ such that $\mathfrak{m} = \mathfrak{m}' + \omega \cdot 1$. The doubling argument gives $\mathfrak{m} = \mathfrak{m}' + \omega + \omega = \mathfrak{m} + \omega$, and thus we proved:

$$\text{If } \mathfrak{m} = \mathfrak{m} + 1, \text{ then } \mathfrak{m} = \mathfrak{m} + \omega \cdot 1\ .$$

This should be compared with Zermelo's theorem II.
On the other hand, the equivalence of "M is Dedekind-infinite" and "M is infinite" can only be proved in ZF + AC_ω, where "M is infinite" is defined as usual, i.e., there is no $n \in \mathbb{N}$ such that $|M| = |\{0, 1, \ldots, n-1\}|$.
Dedekind's definition was the preferred definition of "infinite set" until the axiom of choice had been understood. See, e.g., *Schoenflies 1900*, 6. It first appears in *Dedekind 1888*, article 64.

[3] This is a nice cardinal arithmetic motivation of the main theorem, as well as an introduction of the methods that will be used to proof the theorem. As we have seen, the Cantor-Bernstein theorem plays a key role for the paper, which cannot be guessed at this point.

[4] In modern terms, the following proof is in ZF + AC_ω, using the ZF definition of a cardinal number.

[5] Thus the conclusion of the theorem is: $\mathfrak{m} = \mathfrak{m} + \sum_{n \in \mathbb{N}} \mathfrak{p}_n$. In a precise axiomatic environment, AC_ω is needed already at this point.

[6] The notation $A = (B, C)$ means that A is the disjoint union of B and C. Analogously $A = (B_0, B_1, \ldots, B_n, \ldots)$ means that A is the pairwise disjoint union of the sets B_n. This notation was also used by Cantor and Schoenflies; see *Cantor 1895*, 481, and *Schoenflies 1900*, 6. Earlier, Cantor had used the form $\{P_1, P_2, \ldots\}$ to denote pairwise disjoint unions, cf. *Cantor 1880*, 355.

[7] Zermelo uses AC_ω: For each $n \in \mathbb{N}$, he chooses a certain partition (M_n, P_n) of M.

[8] $M \sim N$ is $|M| = |N|$, i.e., there is a bijection between M and N. Again, this notation is used in *Cantor 1895*. In fact, Cantor introduced it in *Cantor 1880*, 356.

[9] Again, Zermelo uses AC_ω: For each $n \in \mathbb{N}$, he chooses a bijection $\varphi_n : M_n \to M$.

Ueber die Addition transfiniter Cardinalzahlen

1901

Vorgelegt von *D. Hilbert* in der Sitzung vom 9. März 1901

Die Summe zweier endlichen Cardinalzahlen, wie überhaupt zweier endlichen positiven Zahlen, ist stets größer als jede von ihnen, also von beiden verschieden. Dieser Satz gilt aber nicht mehr von den transfiniten Cardinalzahlen im Sinne der Cantorschen Mengenlehre [*Cantor*, Beiträge zur Begründung der transfiniten Mengenlehre, Math. Annalen Bd. 46; vgl. auch *Schoenflies*, Bericht über Punktmannigfaltigkeiten Cap. I–IV]. Denn eine unendliche Menge kann einer ihrer Teilmengen aequivalent sein [Cantor ibid. § 6, D, S. 496], und *Dedekind* [„Was sind und was sollen die Zahlen?" § 5] hat eben auf diese Eigenschaft seine Definition der „unendlichen Systeme" gegründet. Somit läßt sich die Cardinalzahl einer unendlichen Menge stets als eine Summe darstellen, deren einer Summandus sie selbst ist. So ist z. B., wenn man unter $\mathfrak{a} = \aleph_0$ die Cardinalzahl der „abzählbaren" Mengen versteht,

$$\mathfrak{a} = \mathfrak{a} + 1 = \mathfrak{a} + \mathfrak{a} = 2\mathfrak{a} ,$$

und da jede unendliche Menge abzählbare Teilmengen besitzt, so ist auch für jede transfinite Cardinalzahl $\mathfrak{m}$

$$\mathfrak{m} = \mathfrak{m}' + \mathfrak{a} = \mathfrak{m}' + \mathfrak{a} + 1 = \mathfrak{m}' + \mathfrak{a} + \mathfrak{a} = \mathfrak{m} + 1 = \mathfrak{m} + \mathfrak{a} .$$

Es kann aber außer $\mathfrak{a}$ noch andere (unendliche) Cardinalzahlen $\mathfrak{p}$ geben, deren Addition die vorgelegte Cardinalzahl $\mathfrak{m}$ nicht ändert:

$$\mathfrak{m} = \mathfrak{m} + \mathfrak{p} ,$$

[10] Here φM is the image of M under φ, i.e., $\varphi M = \varphi[M] = \{f(x) \mid x \in M\}$.
[11] As discussed in section 5, the proofs are in ZF.
[12] The "less than" relation here is $<^*$, and the "less than or equal" relation of theorem IV is $\leq^*$. Zermelo refers to *Cantor 1895*, 483, and *Schoenflies 1900*, 15.
[13] Theorem III is the inclusion form of Cantor-Bernstein, and theorem IV is the nontrivial part of the equivalence of the relations $<^*$ and $<$, i.e., the assertion that $\mathfrak{n} \leq \mathfrak{m}$ implies $\mathfrak{n} \leq^* \mathfrak{m}$ for all cardinals $\mathfrak{n}$ and $\mathfrak{m}$. Zermelo's theorem is basically this implication, since $\mathfrak{n} \leq \mathfrak{m}$ is equivalent to the existence of a cardinal $\mathfrak{p}$ such that $\mathfrak{m} = \mathfrak{n} + \mathfrak{p}$.
[14] Since the Cantor-Bernstein theorem is proved now, $\leq$ and $\leq^*$ agree.

On the addition of transfinite cardinal numbers

1901

Presented by *D. Hilbert* in the session of March 9, 1901

The sum of two finite cardinal numbers, like the sum of any two finite positive numbers, is always greater than either one of them, and hence different from both. But this theorem no longer holds for the transfinite cardinal numbers in the sense of Cantorian set theory [*Cantor 1895;* see also *Schoenflies 1900,* Cap. I–IV]. For an infinite set can be equivalent to one of its partial sets [*Cantor 1895,* §6, D, p. 496], and it is just this property that *Dedekind* [*1888,* §5] chose as a basis for his definition of "infinite systems". Thus the cardinal number of an infinite set can always be represented as a sum one of whose summands is the cardinal number itself. For instance, if we take $\mathfrak{a} = \aleph_0$ to be the cardinal number of the "countable" sets,[1] then

$$\mathfrak{a} = \mathfrak{a} + 1 = \mathfrak{a} + \mathfrak{a} = 2\mathfrak{a} \,,$$

and because every infinite set has countable partial sets,[2] it also holds for every transfinite cardinal number $\mathfrak{m}$ that

$$\mathfrak{m} = \mathfrak{m}' + \mathfrak{a} = \mathfrak{m}' + \mathfrak{a} + 1 = \mathfrak{m}' + \mathfrak{a} + \mathfrak{a} = \mathfrak{m} + 1 = \mathfrak{m} + \mathfrak{a} \,.$$

But besides $\mathfrak{a}$ there can be other (infinite) cardinal numbers $\mathfrak{p}$ whose addition to a given cardinal number $\mathfrak{m}$ does not change it:

$$\mathfrak{m} = \mathfrak{m} + \mathfrak{p} \,,$$

1 ⟦The marks [1], [2],... refer to the annotations at the end of the introductory note.⟧

und die Summe zweier solchen Zahlen $\mathfrak{p}$ und $\mathfrak{q}$ hat wieder dieselbe Eigenschaft. Denn wenn

35 | $$\mathfrak{m} = \mathfrak{m} + \mathfrak{p} = \mathfrak{m} + \mathfrak{q} \,,$$

so ist auch

$$\mathfrak{m} = (\mathfrak{m} + \mathfrak{p}) + \mathfrak{q} = \mathfrak{m} + (\mathfrak{p} + \mathfrak{q}) \,.$$

Wie man unmittelbar sieht, läßt sich der Satz auch auf drei und mehr Summanden $\mathfrak{p}, \mathfrak{q}, \mathfrak{r}$ ausdehnen, er gilt aber, wie wir jetzt zeigen wollen, auch für eine beliebige *abzählbar unendliche* Menge solcher Cardinalzahlen $\mathfrak{p}_1, \mathfrak{p}_2, \mathfrak{p}_3, \ldots$

Satz I. (Hauptsatz) Wenn eine Cardinalzahl $\mathfrak{m}$ *ungeändert bleibt bei der Addition einer beliebigen Cardinalzahl aus der unendlichen Reihe* $\mathfrak{p}_1, \mathfrak{p}_2, \mathfrak{p}_3,$ $\ldots,$ *so bleibt sie auch ungeändert, wenn man alle auf einmal addirt.*

Nach unserer Voraussetzung

$$\mathfrak{m} = \mathfrak{m} + \mathfrak{p}_1 = \mathfrak{m} + \mathfrak{p}_2 = \mathfrak{m} + \mathfrak{p}_3 = \cdots \tag{1}$$

läßt sich nämlich eine Menge M von der Mächtigkeit $\mathfrak{m}$ folgendermaßen in Teilmengen zerlegen

$$M = (P_1, M_1) = (P_2, M_2) = (P_3, M_3) = \cdots \,, \tag{2}$$

wo jede Teilmenge P_λ von der Cardinalzahl $\mathfrak{p}_\lambda$ mit ihrer complementären M_λ kein Element gemeinsam hat and alle M_λ der Menge M selbst aequivalent sind:

$$M \sim M_1 \sim M_2 \sim M_3 \cdots$$

Es existiren also ein-eindeutige Abbildungen $\varphi_1, \varphi_2, \varphi_3, \ldots$ der Menge M auf ihre Teilmengen $M_1, M_2, M_3, \ldots$, so daß

$$M_1 = \varphi_1 M \,, \qquad M_2 = \varphi_2 M \,, \qquad M_3 = \varphi_3 M \,, \qquad \ldots \tag{3}$$

geschrieben werden kann. Für eine beliebige solche Abbildung φ ist aber auch immer

$$\varphi M = (\varphi P_1, \varphi M_1) = (\varphi P_2, \varphi M_2) = (\varphi P_3, \varphi M_3) \,, \ldots \,, \tag{2a}$$

d. h. für alle aequivalenten Mengen gelten die analogen Zerlegungen. Somit ist auch successive

$$\begin{array}{llll}
M & & = (P_1, M_1) & \\
M_1 & = \varphi_1 M & = (\varphi_1 P_2, \varphi_1 M_2) & = (P_2', M_2') \\
M_2' & = \varphi_1 \varphi_2 M & = (\varphi_1 \varphi_2 P_3, \varphi_1 \varphi_2 M_3) & = (P_3', M_3') \\
M_3' & = \varphi_1 \varphi_2 \varphi_3 M & = (\varphi_1 \varphi_2 \varphi_3 P_4, \varphi_1 \varphi_2 \varphi_3 M_4) & = (P_4', M_4') \\
\multicolumn{4}{c}{\dotfill} \\
M_{\lambda-1}' & = \varphi_1 \varphi_2 \ldots \varphi_{\lambda-1} M & = (\varphi_1 \varphi_2 \ldots \varphi_{\lambda-1} P_\lambda, \varphi_1 \varphi_2 \ldots \varphi_{\lambda-1} M_\lambda) & = (P_\lambda', M_\lambda') \,,
\end{array}$$

and the sum of two such numbers $\mathfrak{p}$ and $\mathfrak{q}$ has again the same property. For if

$$\mathfrak{m} = \mathfrak{m} + \mathfrak{p} = \mathfrak{m} + \mathfrak{q} \,,$$

then also

$$\mathfrak{m} = (\mathfrak{m} + \mathfrak{p}) + \mathfrak{q} = \mathfrak{m} + (\mathfrak{p} + \mathfrak{q}) \,.$$ [3]

As is immediately evident, the theorem can be extended to three or more summands $\mathfrak{p}, \mathfrak{q}, \mathfrak{r}$. But it also holds, as we shall now show, for arbitrary *countably infinite* sets of such cardinal numbers $\mathfrak{p}_1, \mathfrak{p}_2, \mathfrak{p}_3, \ldots$

Theorem I. (Main Theorem) If a cardinal number $\mathfrak{m}$ *remains unchanged upon addition to an arbitrary cardinal number from an infinite series* $\mathfrak{p}_1, \mathfrak{p}_2, \mathfrak{p}_3, \ldots$, *then it also remains unchanged when all of them are added to it at once.*[4] [5]

For according to our assumption

$$\mathfrak{m} = \mathfrak{m} + \mathfrak{p}_1 = \mathfrak{m} + \mathfrak{p}_2 = \mathfrak{m} + \mathfrak{p}_3 = \cdots \tag{1}$$

we can decompose a set M of cardinality $\mathfrak{m}$ into partial sets as follows

$$M = (P_1, M_1) = (P_2, M_2) = (P_3, M_3) = \cdots \,, \tag{2}$$ [6] [7]

such that every partial set P_λ of cardinal number $\mathfrak{p}_\lambda$ has no element in common with its complementary M_λ and all M_λ are equivalent to the set M itself:

$$M \sim M_1 \sim M_2 \sim M_3 \cdots$$ [8]

There then exist one-one mappings $\varphi_1, \varphi_2, \varphi_3, \ldots$ from the set M onto its partial sets $M_1, M_2, M_3, \ldots$ such that we can write

$$M_1 = \varphi_1 M, \qquad M_2 = \varphi_2 M, \qquad M_3 = \varphi_3 M, \ldots \tag{3}$$ [9] [10]

But for an arbitrary mapping φ of this kind it also always holds that

$$\varphi M = (\varphi P_1, \varphi M_1) = (\varphi P_2, \varphi M_2) = (\varphi P_3, \varphi M_3) \,, \ldots \,, \tag{2a}$$

i.e., the analogous decompositions hold for all equivalent sets. Hence it also holds, successively, that

$$\begin{array}{llll}
M & & = (P_1, M_1) & \\
M_1 & = \varphi_1 M & = (\varphi_1 P_2, \varphi_1 M_2) & = (P_2', M_2') \\
M_2' & = \varphi_1 \varphi_2 M & = (\varphi_1 \varphi_2 P_3, \varphi_1 \varphi_2 M_3) & = (P_3', M_3') \\
M_3' & = \varphi_1 \varphi_2 \varphi_3 M & = (\varphi_1 \varphi_2 \varphi_3 P_4, \varphi_1 \varphi_2 \varphi_3 M_4) & = (P_4', M_4') \\
 & \ldots\ldots\ldots & \ldots\ldots\ldots & \\
M_{\lambda-1}' & = \varphi_1 \varphi_2 \ldots \varphi_{\lambda-1} M & = (\varphi_1 \varphi_2 \ldots \varphi_{\lambda-1} P_\lambda, \varphi_1 \varphi_2 \ldots \varphi_{\lambda-1} M_\lambda) & = (P_\lambda', M_\lambda') \,,
\end{array}$$

36 | also

$$M = (P_1, P'_2, P'_3 \dots P'_\lambda; M'_\lambda)\,, \tag{4}$$

wenn man die Abbildungen $\varphi_1, \varphi_2, \varphi_3 \dots$ hinter einander ausführt und die Abkürzungen benutzt:

$$\varphi_1\varphi_2 \dots \varphi_{\lambda-1} P_\lambda = P'_\lambda \sim P_\lambda$$
$$\varphi_1\varphi_2 \dots \varphi_{\lambda-1} M_\lambda = \varphi_1\varphi_2 \dots \varphi_\lambda M = M'_\lambda \sim M\,.$$

Hier sind auch von den Teilmengen P'_λ der ursprünglichen Menge M immer je zwei ohne gemeinsame Elemente, und es wird schließlich:

$$M = (P_1, P'_2, P'_3, \dots; M')\,, \tag{5}$$

d. h. M enthält *alle* Teilmengen P'_λ und somit auch ihre Vereinigungsmenge. Für die entsprechenden Cardinalzahlen haben wir also zunächst

$$\mathfrak{m} = \mathfrak{p}_1 + \mathfrak{p}_2 + \mathfrak{p}_3 + \cdots + \mathfrak{m}'\,, \tag{6}$$

und es bleibt nur noch zu zeigen, daß hier die Cardinalzahl $\mathfrak{m}'$ durch $\mathfrak{m}$ selbst ersetzt werden kann.

Zunächst können wir aber die Formel (6) auf den Fall anwenden, wo

$$\mathfrak{p}_1 = \mathfrak{p}_2 = \mathfrak{p}_3 \cdots = \mathfrak{p}\,, \quad \text{also} \;\; \mathfrak{p}_1 + \mathfrak{p}_2 + \mathfrak{p}_3 + \cdots = \mathfrak{a}\mathfrak{p}\,,$$

und dann folgt aus der *einen* Gleichung

$$\mathfrak{m} = \mathfrak{m} + \mathfrak{p}$$

sofort

$$\mathfrak{m} = \mathfrak{a}\mathfrak{p} + \mathfrak{m}' = 2\mathfrak{a}\mathfrak{p} + \mathfrak{m}' = \mathfrak{m} + \mathfrak{a}\mathfrak{p}\,. \tag{7}$$

Wir haben also:

Satz II. Wenn eine Cardinalzahl $\mathfrak{m}$ *bei der Addition einer anderen* $\mathfrak{p}$ *ungeändert bleibt, so bleibt sie es auch bei unendlich oft wiederholter Addition derselben Cardinalzahl, oder bei der Addition ihres* $\mathfrak{a}$*-fachen.*

Somit können wir unsere Voraussetzung (1) folgendermaßen erweitern:

$$\mathfrak{m} = \mathfrak{m} + \mathfrak{a}\mathfrak{p}_1 = \mathfrak{m} + \mathfrak{a}\mathfrak{p}_2 = \mathfrak{m} + \mathfrak{a}\mathfrak{p}_3 \dots \tag{1a}$$

37 und wir haben nach (6) |

$$\begin{aligned}\mathfrak{m} &= \mathfrak{a}\mathfrak{p}_1 + \mathfrak{a}\mathfrak{p}_2 + \cdots + \mathfrak{m}'' \\ &= 2\mathfrak{a}\,(\mathfrak{p}_1 + \mathfrak{p}_2 + \mathfrak{p}_3 + \cdots) + \mathfrak{m}'' = (\mathfrak{a} + 1)\,(\mathfrak{p}_1 + \mathfrak{p}_2 + \mathfrak{p}_3 + \cdots) + \mathfrak{m}'' \\ &= \mathfrak{m} + \mathfrak{a}\,(\mathfrak{p}_1 + \mathfrak{p}_2 + \mathfrak{p}_3 + \cdots) = \mathfrak{m} + \mathfrak{p}_1 + \mathfrak{p}_2 + \mathfrak{p}_3 + \cdots\,,\end{aligned} \tag{8}$$

womit der Satz I sogar in erweiterter Form bewiesen ist.

therefore

$$M = (P_1, P_2', P_3' \dots P_\lambda'; M_\lambda') , \tag{4}$$

assuming that we perform the mappings $\varphi_1, \varphi_2, \varphi_3, \dots$ one after the other and use the abbreviations:

$$\varphi_1\varphi_2 \dots \varphi_{\lambda-1} P_\lambda = P_\lambda' \sim P_\lambda$$
$$\varphi_1\varphi_2 \dots \varphi_{\lambda-1} M_\lambda = \varphi_1\varphi_2 \dots \varphi_\lambda M = M_\lambda' \sim M .$$

Here we also have among the partial sets P_λ' of the original set M always two without common elements, and, finally, it holds that

$$M = (P_1, P_2', P_3', \dots; M') , \tag{5}$$

i.e., M contains *all* partial sets P_λ' and therefore also their union. Hence, for the corresponding cardinal numbers we first have

$$\mathfrak{m} = \mathfrak{p}_1 + \mathfrak{p}_2 + \mathfrak{p}_3 + \dots + \mathfrak{m}' , \tag{6}$$

and it only remains to show that the cardinal number $\mathfrak{m}'$ can be replaced here with $\mathfrak{m}$ itself.

But first we can apply the formula (6) to the case in which

$$\mathfrak{p}_1 = \mathfrak{p}_2 = \mathfrak{p}_3 \dots = \mathfrak{p} , \quad \text{hence } \mathfrak{p}_1 + \mathfrak{p}_2 + \mathfrak{p}_3 + \dots = \mathfrak{a}\mathfrak{p} ,$$

and it then follows immediately from the *single* equation

$$\mathfrak{m} = \mathfrak{m} + \mathfrak{p}$$

that

$$\mathfrak{m} = \mathfrak{a}\mathfrak{p} + \mathfrak{m}' = 2\mathfrak{a}\mathfrak{p} + \mathfrak{m}' = \mathfrak{m} + \mathfrak{a}\mathfrak{p} . \tag{7}$$

We therefore have:[11]

Theorem II. If a cardinal number $\mathfrak{m}$ *remains unchanged upon addition to another cardinal number* $\mathfrak{p}$*, then it remains unchanged when the same cardinal number is added to it infinitely many times, or upon addition to its* $\mathfrak{a}$*-fold.*

We thus can extend our assumption (1) as follows

$$\mathfrak{m} = \mathfrak{m} + \mathfrak{a}\mathfrak{p}_1 = \mathfrak{m} + \mathfrak{a}\mathfrak{p}_2 = \mathfrak{m} + \mathfrak{a}\mathfrak{p}_3 \dots \tag{1a}$$

and, according to (6), we have

$$\begin{aligned} \mathfrak{m} &= \mathfrak{a}\mathfrak{p}_1 + \mathfrak{a}\mathfrak{p}_2 + \dots + \mathfrak{m}'' \\ &= 2\mathfrak{a}\,(\mathfrak{p}_1 + \mathfrak{p}_2 + \mathfrak{p}_3 + \dots) + \mathfrak{m}'' = (\mathfrak{a} + 1)\,(\mathfrak{p}_1 + \mathfrak{p}_2 + \mathfrak{p}_3 + \dots) + \mathfrak{m}'' \\ &= \mathfrak{m} + \mathfrak{a}\,(\mathfrak{p}_1 + \mathfrak{p}_2 + \mathfrak{p}_3 + \dots) = \mathfrak{m} + \mathfrak{p}_1 + \mathfrak{p}_2 + \mathfrak{p}_3 + \dots , \end{aligned} \tag{8}$$

by which we have even proved an extended version of theorem I.

Der Satz läßt sich aber auch umkehren, in folgender Form:

Satz III. Wenn die Summe zweier Cardinalzahlen, zu einer dritten addirt, diese ungeändert läßt, so gilt dies auch von jedem ihrer Summanden.

Ist nämlich

$$\mathfrak{m} = \mathfrak{m} + \mathfrak{p} + \mathfrak{q} \,,$$

so ist nach II auch

$$\begin{aligned} \mathfrak{m} &= \mathfrak{m} + \mathfrak{a}(\mathfrak{p} + \mathfrak{q}) \\ &= \mathfrak{m} + (\mathfrak{a}+1)\mathfrak{p} + \mathfrak{a}\mathfrak{q} = \mathfrak{m} + \mathfrak{a}\mathfrak{p} + (\mathfrak{a}+1)\mathfrak{q} \\ &= \mathfrak{m} + \mathfrak{p} \qquad\qquad\quad\;\; = \mathfrak{m} + \mathfrak{q} \,, \end{aligned}$$

und das gleiche gilt natürlich auch, wenn die Summe mehr als zwei Summanden besitzt.

Dieser Satz III ist aber der eigentliche Kern des 1896 von *Schröder* und *Bernstein* zuerst bewiesenen „*Aequivalenzsatzes*": Wenn von zwei Mengen jede einem Teile der anderen aequivalent ist, so sind sie unter einander aequivalent. (*Borel,* Théorie des fonctions, p. 103 ff., *Schoenflies,* Bericht S. 16).

Ist nämlich

$$a)\quad \mathfrak{m} = \mathfrak{n} + \mathfrak{p}$$

und

$$b)\quad \mathfrak{n} = \mathfrak{m} + \mathfrak{q} \,,$$

so ist auch

$$\mathfrak{m} = \mathfrak{m} + \mathfrak{p} + \mathfrak{q}$$

und daher

$$\mathfrak{m} = \mathfrak{m} + \mathfrak{q} = \mathfrak{n} \,, \qquad \text{q. e. d.}$$

Gilt dagegen nur die erste Gleichung $a)$, aber keine Gleichung der Form $b)$, so daß die erste Menge M keinem Teile der zweiten aequivalent ist, so ist nach der Cantorschen Definition (*Cantor*, Ann. 46 § 2, *Schoenflies* Cap. III) $\mathfrak{m} > \mathfrak{n}$. Aus $\mathfrak{m} = \mathfrak{n} + \mathfrak{p}$ folgt also immer: $\mathfrak{m} \geqq \mathfrak{n}$, d. h.

Satz IV. Die Summe zweier oder mehrerer Cardinalzahlen ist stets größer oder gleich einer jeden von ihnen.

38 | Diese Sätze III und IV, die dem „Aequivalenzsatze" gleichwertig sind, ergaben sich als unmittelbare Folgen des „Kettensatzes" II, den wir hier als Specialfall von I eingeführt hatten. Natürlich läßt sich Satz II aber auch direkt beweisen, man braucht nur in dem oben gegebenen Beweise von I die Teilmengen $P_1, P_2, P_3 \ldots$ und damit auch die Abbildungen $\varphi_1, \varphi_2, \varphi_3, \ldots$ von vorn herein als identisch zu betrachten, und dann unterscheidet sich der Beweis des Aequivalenzsatzes nur in der Anordnung und Ausdrucksweise von

But we can also invert the theorem, as follows:

Theorem III. If adding the sum of two cardinal numbers to a third cardinal number leaves the latter unchanged, then the same holds true for each of its summands.

For if

$$\mathfrak{m} = \mathfrak{m} + \mathfrak{p} + \mathfrak{q}\,,$$

then, according to II, also

$$\begin{aligned} \mathfrak{m} &= \mathfrak{m} + \mathfrak{a}(\mathfrak{p} + \mathfrak{q}) \\ &= \mathfrak{m} + (\mathfrak{a} + 1)\mathfrak{p} + \mathfrak{a}\mathfrak{q} = \mathfrak{m} + \mathfrak{a}\mathfrak{p} + (\mathfrak{a} + 1)\mathfrak{q} \\ &= \mathfrak{m} + \mathfrak{p} \qquad\qquad\qquad\quad = \mathfrak{m} + \mathfrak{q}\,, \end{aligned}$$

and, of course, the same holds if the sum has more than two summands.

But theorem III is the real core of the "*theorem of equivalence*" that was first proved by *Schröder* and *Bernstein* in 1896: If each of two sets is equivalent to some part of the other, then the two sets are equivalent to one another. (*Borel 1898,* p. 103ff, *Schoenflies 1900,* p. 16).

For if

$$a)\quad \mathfrak{m} = \mathfrak{n} + \mathfrak{p}$$

and

$$b)\quad \mathfrak{n} = \mathfrak{m} + \mathfrak{q}\,,$$

then also

$$\mathfrak{m} = \mathfrak{m} + \mathfrak{p} + \mathfrak{q}$$

and therefore

$$\mathfrak{m} = \mathfrak{m} + \mathfrak{q} = \mathfrak{n}\,, \qquad \text{q.e.d.}$$

If, however, only the first equation *a*) holds, but no equation of the form *b*), so that the first set M is not equivalent to any part of the second set, then, according to the Cantorian definition (*Cantor 1895,* §2, *Schoenflies 1900,* Cap. III), $\mathfrak{m} > \mathfrak{n}$.[12] From $\mathfrak{m} = \mathfrak{n} + \mathfrak{p}$ it therefore always follows: $\mathfrak{m} \geq \mathfrak{n}$, i.e.,

Theorem IV. The sum of two or more cardinal numbers is always greater than or equal to each of them.

We obtained theorems III[13] and IV, which amount to the "theorem of equivalence", as immediate consequences of the "chain theorem" II, which we had introduced here as a special case of I. It is of course also possible to prove theorem II directly; in the proof of I given above we only need to consider the partial sets $P_1, P_2, P_3, \ldots$, and hence also the mappings $\varphi_1, \varphi_2, \varphi_3, \ldots$, as identical from the outset. Then, the proof of the theorem

dem oben citirten *Bernstein*'schen. Dagegen ist der Satz I selbst als eine *Erweiterung* des Aequivalenzsatzes aufzufassen, die bisher meines Wissens noch unbewiesen war und sich aus diesem allein nicht ableiten läßt.

Beachten wir nun, daß alle Cardinalzahlen $\mathfrak{p} \geqq \mathfrak{p}'$ und nur diese in der Form $\mathfrak{p} = \mathfrak{p}' + \mathfrak{q}$ und alle Zahlen $\mathfrak{m}' \geqq \mathfrak{m}$ in der Form $\mathfrak{m}' = \mathfrak{m} + \mathfrak{r}$ enthalten sind, so können wir aus der Annahme

$$\mathfrak{m} = \mathfrak{m} + \mathfrak{p} = \mathfrak{m} + \mathfrak{p}' + \mathfrak{q}$$

nach III schließen

$$\mathfrak{m} = \mathfrak{m} + \mathfrak{p}' ,$$

und dann ist auch:

$$\begin{aligned} \mathfrak{m}' = \mathfrak{m} + \mathfrak{r} = \mathfrak{m} + \mathfrak{p} &+ \mathfrak{r} = \mathfrak{m} + \mathfrak{p}' + \mathfrak{r} \\ = \mathfrak{m}' + \mathfrak{p} \, , \quad &= \mathfrak{m}' + \mathfrak{p}' \, , \end{aligned}$$

und wir haben den Satz:

Satz V. Wenn eine Cardinalzahl $\mathfrak{p}$ *eine andere* $\mathfrak{m}$ *bei der Addition ungeändert läßt, so läßt auch jede Zahl* $\mathfrak{p}' \leqq \mathfrak{p}$ *jede andere* $\mathfrak{m}' \geqq \mathfrak{m}$ *bei der Addition ungeändert.*

Die Cardinalzahlen $\mathfrak{p}$, die eine gegebene Cardinalzahl $\mathfrak{m}$ bei der Addition ungeändert lassen, bilden also eine Art „Gruppe" von Zahlen, die sich reproduciren

1) bei ihrer Verkleinerung (Satz V),
2) bei ihrer Multiplication mit $\mathfrak{a} = \aleph_0$ (Satz II),
3) bei ihrer Addition in endlicher oder abzählbar unendlicher Summanden-Anzahl (Satz I).

of equivalence differs from the proof by *Bernstein* cited above only with respect to its arrangement and terminology. By contrast, we have to conceive of theorem I itself as an *extension* of the theorem of equivalence, which, to my knowledge, had not been proved until now and cannot be derived from it alone.

If we now consider that all cardinal numbers $\mathfrak{p} \geq \mathfrak{p}'$ [14] and only those are contained in the form $\mathfrak{p} = \mathfrak{p}' + \mathfrak{q}$ and that all numbers $\mathfrak{m}' \geq \mathfrak{m}$ are contained in the form $\mathfrak{m}' = \mathfrak{m} + \mathfrak{r}$, then from the assumption

$$\mathfrak{m} = \mathfrak{m} + \mathfrak{p} = \mathfrak{m} + \mathfrak{p}' + \mathfrak{q}$$

we can conclude, according to III, that

$$\mathfrak{m} = \mathfrak{m} + \mathfrak{p}',$$

and then it also holds that

$$\begin{aligned} \mathfrak{m}' = \mathfrak{m} + \mathfrak{r} &= \mathfrak{m} + \mathfrak{p} + \mathfrak{r} = \mathfrak{m} + \mathfrak{p}' + \mathfrak{r} \\ &= \mathfrak{m}' + \mathfrak{p}\,, \quad = \mathfrak{m}' + \mathfrak{p}'\,, \end{aligned}$$

and we have the theorem:

Theorem V. If a cardinal number $\mathfrak{p}$ *leaves another number* $\mathfrak{m}$ *unchanged upon addition, then every number* $\mathfrak{p}' \leq \mathfrak{p}$ *leaves every other* $\mathfrak{m}' \geq \mathfrak{m}$ *also unchanged upon addition.*

The cardinal numbers $\mathfrak{p}$ that leave a given cardinal number $\mathfrak{m}$ unchanged upon addition therefore form a kind of "group" of numbers, which again yield such numbers

1) upon their decrease (theorem V),
2) upon their multiplication with $\mathfrak{a} = \aleph_0$ (theorem II),
3) upon their addition in a finite or countably infinite number of summands (theorem I).

Introductory note to 1904 and 1908a

Michael Hallett[†]

Introduction

On 24 September 1904, Zermelo sent a letter from "Münden in Hann." to Hilbert.[1] This letter contained a proof of the famous well-ordering theorem (WOT) from what was later called by Zermelo "the axiom of choice [*Axiom der Auswahl*]" (AC), a principle formulated for the first time in this letter. Shortly after, there appeared in volume 59 of the *Mathematische Annalen* a paper by Zermelo (no date is given for its receipt) of slightly less than three pages which sets out this proof. The title of the paper is "Beweis, daß jede Menge wohlgeordnet werden kann", and it carried as a sub-title in brackets "Aus einem an Herrn Hilbert gerichteten Briefe". The paper (*Zermelo 1904*) was destined to become one of the most celebrated in the history of modern mathematics, and the choice principle it contains one of modern mathematics' most discussed and controversial principles. Unfortunately, Zermelo's actual letter is not preserved among Hilbert's correspondence or Zermelo's papers, and its fate is not known. There do exist in Zermelo's hand some fragments from a draft of this letter, but the bulk of the proof is missing.[2] Thus, all that we really have is the published paper. Zermelo's first proof of WOT provoked some immediate reactions, many of them published in the following volume (Volume 60) of the *Mathematische Annalen*, reactions related both to the use of Zermelo's choice principle itself and also to the general

† I wish to thank Andrew Granville, Alison Laywine and Stephen Menn for their help and advice on various matters, and above all the editors of this volume, Heinz-Dieter Ebbinghaus and Akihiro Kanamori, both for sharing their wide knowledge, and for their indulgence and patience. I also wish to acknowledge the generous support of the Social Sciences and Humanities Research Council of Canada over many years. All passages quoted from French and German in this introductory note have been given in English translation. In most cases, these translations are my own. The exceptions are passages from papers included in the present volume, which have been quoted in the translations used here.

1 "Münden in Hann." is better known today as "Hannoversch Münden" or simply "Hann. Münden". It is a small medieval town, beautifully restored and very picturesque, and lies about 25 kilometers from Göttingen, set at the confluence of the rivers Werra and Fulda, thus giving birth to the Weser which flows into the North Sea.

2 The fragments are preserved in the Zermelo Nachlass in the Universitätsbibliothek in Freiburg, and there are photo-reproductions of one of them in Ebbinghaus's biography of Zermelo, *Ebbinghaus 2007b*, 56–57. The passages which *are* preserved are more or less identical to corresponding sections in the published paper.

set-theoretical context in which that principle was deployed. Zermelo subsequently published a second paper (again with no receipt date, but marked at the end "Chesières, 14 July 1907"[3]). This paper was entitled "Neuer Beweis für die Möglichkeit einer Wohlordnung" (*Zermelo 1908a*) and gives, as its title suggests, a new proof of WOT, again using a version of the choice principle, this time eventually formulated as an "axiom". The new proof (in §1 of the paper) occupied four and a half pages; it was followed in §2 by close to seventeen pages setting out detailed replies to some of the objections to the first proof. Shortly after this, Zermelo published his celebrated axiomatisation of set theory (*Zermelo 1908b*, dated 30 July 1907, also from Chesières); this axiomatisation is mentioned in the second well-ordering paper from just before, and must be considered part of the overall reply which this second paper constitutes, though very little will be said about the axiomatisation in this introductory note. These three papers together mark the beginning of the systematisation of modern, abstract set theory as we know it today.[4]

1. The importance and rôle of well-ordering before Zermelo

In one of the fundamental papers in the genesis of set theory (*Cantor 1883b*), Cantor isolated the notion of *well-ordering* on a collection as one of the central conceptual pillars on which number is built. Cantor observed that the notion of counting number must be based on an underlying ordering of the set of things being counted, an ordering in which there is a first element counted, and, following any collection of counted elements, there must be a *next* element counted, assuming that there are elements still uncounted. This kind of ordering he called a "well-ordering", which we now conceive of as a total-ordering with an extra condition, namely that any non-empty subset has a least element in the ordering. Cantor recognised that each distinct well-ordering of the elements gives rise to a distinct counting number, what he originally called an "*Anzahl* [enumeral]", later an "*Ordnungszahl* [ordinal number]", numbers which are conceptually quite different from *cardinal numbers* or *powers*, meant to express just the *size* of collections.[5] There are two features which make this distinction hard to perceive at first sight. Before

[3] Chèsieres is in Switzerland, a ski resort near Montreux.

[4] The books *Ebbinghaus 2007b*, *Moore 1982*, *Peckhaus 1990a* and the author's own *Hallett 1984* provide indispensable background for understanding Zermelo's set-theoretical work, and in particular his work on well-ordering. The author has also been aided enormously by the papers of Akihiro Kanamori, especially *Kanamori 1997* and *Kanamori 2004*.

[5] Two sets M and N are said to have the same size or power or cardinality if they can be put into one-to-one onto correspondence. In this case we say that $M \simeq N$. If M has the same power as a subset of N, we say that $M \preceq N$; if in addition to this $M \simeq N$ fails, we say that $M \prec N$, and that N has greater power than M.

Cantor and the rise of the modern theory of transfinite numbers, the only counting numbers known were the ordinary finite numbers. And, crucially, for finite collections, it turns out that any two orderings of the same underlying elements, which are certainly well-orderings in Cantor's sense, are *order-isomorphic*, i.e., not essentially distinct.[6] This means that one can in effect identify a number arrived at by counting (an ordinal number) with the cardinal number of the collection counted. Thus, the ordinary natural numbers appear in two guises, and it is possible to determine the size of a finite collection directly by counting it. Cantor observed that this ceases to be the case in rather dramatic fashion once one considers infinite collections; here, the same elements can give rise to a large variety of distinct well-orderings. However, Cantor noticed that if one collects together all the countable ordinal numbers, i.e., numbers representing well-orderings of the set of natural numbers,[7] this collection, which Cantor called the second number-class (the first being the set of natural numbers), must be of greater cardinality than that of the collection of natural numbers itself. Moreover, this size is the cardinal successor to the size of the natural numbers in the very clear sense that any infinite subset of the second number-class is either of the power of the natural numbers or of the power of the whole class itself; thus, there can be no cardinal size which is strictly intermediate. The process generalises: collect together all the ordinal numbers representing well-orderings of the second number-class to form the third number-class, and this must be the immediate successor in size to that of the second number-class, and so on. In this way, Cantor could use the ordinal numbers to generate an infinite sequence of cardinalities or powers. This sequence was later (in *Cantor 1895*) called the aleph-sequence, $\aleph_0$ (the size of the natural numbers), $\aleph_1$ (expressing the size of the second number-class), $\aleph_2$ (expressing the size of the third number-class), and so on. Since the intention was that ordinal numbers could be generated arbitrarily far, then so too, it seems, could the alephs.

The whole development of Cantor's theory of the transfinite was dominated by the continuum hypothesis (CH). In 1878, Cantor had shown that the real numbers cannot be put into one-to-one correspondence with the natural numbers, and thus that there must be at least two infinite powers represented in the ordinary number systems. He conjectured, however, that the real numbers have the property that every infinite set of real numbers is either countable or of the power of the whole set of real numbers, a conjecture which came to be known as CH. This corresponds exactly to the property of

6 Two ordered sets A_1 with ordering R_1 and A_2 with ordering R_2 are said to be order-isomorphic if there is a function f which puts them into one-to-one onto correspondence and in addition is such that, for any two elements a, b in A_1, aR_1b if and only if $f(a)R_2f(b)$.

7 A set is said to be countable if it can be put into one-to-one onto correspondence with the set of natural numbers.

the second number-class discovered in 1883, thus suggesting CH in the form that the continuum has the power of the second number-class.

The development of the number-classes (alephs) provided a clear framework in which to pose CH. In work after 1883, Cantor attempted to show that the alephs actually represent a scale of infinite cardinal *number*. For instance, it is clear that the ordinal numbers are comparable, i.e., for any two ordinal numbers α, β, either $\alpha < \beta$, $\alpha = \beta$ or $\alpha > \beta$. Comparability therefore transfers to the alephs, too. Also necessary for this was the definition of clear and appropriate arithmetical operations of addition, multiplication and exponentiation, generalising the notions for finite collections, and the statement and proof of general laws concerning these. For instance, it seemed clear that the union of a countable collection of countable sets must itself be countable; this generalises to any aleph, and in fact $\aleph_\alpha + \aleph_\alpha = \aleph_\alpha \cdot \aleph_\alpha = \aleph_\alpha$. With the development of transfinite cardinal arithmetic, CH assumes its modern form, namely as the answer to a question about exactly where the power of the continuum fits in the scale of infinite numbers so characterised. Indeed, CH becomes a conjecture about the exponentiation operation in the generalised cardinal arithmetic, for it can be expressed in the form $2^{\aleph_0} = \aleph_1$. This of course generalises, too. In 1883, Cantor had assumed (without remark) that the set of all real functions has the size of the third number-class. Given CH, this then becomes the conjecture that $2^{\aleph_1} = \aleph_2$. More generally, this suggests that taking the power set of an infinite set corresponds to moving up just one level in the aleph scale.

This arithmetical framework for CH carries some clear presuppositions. For one thing, it is clear that any collection in well-ordered form (given that it is represented by an ordinal) must have an aleph representing its size, so clearly the aleph-sequence represents the powers of all the well-ordered sets. But can *any* set be put into well-ordered form? A particular question of this form concerns the continuum itself: if the continuum is equivalent to the second number-class, then clearly it can be well-ordered, which is therefore a necessary condition for showing that the continuum is represented at all in the scale. But *can* it be well-ordered? More generally, to assume that *any* cardinality is represented in the aleph scale is to assume in particular that *any* set can be well-ordered. And to assume that the aleph-sequence is *the* scale of infinite cardinal number is to assume that sets generally can be compared cardinally; i.e., that for any M, N, either $M \preceq N$ or $N \preceq M$, COMP for short. But is this correct?

When introducing the notion of well-ordering in 1883, Cantor expressed his belief (*Cantor 1883b*, 550) that the fact that any manifold can be well-ordered is "a law of thought [*Denkgesetz*]", putting forward what for convenience we can call the well-ordering hypothesis (WOH):

> The concept of *well-ordered set* reveals itself as fundamental for the theory of manifolds. That it is always possible to arrange any *well-defined* set in the form of a *well-ordered* set is, it seems to me, a very

> basic law of thought, rich in consequences, and particularly remarkable in virtue of its general validity. I will return to this in a later memoir.

Cantor says nothing about what it might mean to call the well-ordering hypothesis a "law of thought", but in some ways, when one considers the subsequent developments, this claim is key. It could mean that Cantor considered WOH as something like a logical principle.[8] This, however, is not particularly clear, especially since the set concept itself was new and rather unclearly delimited. Another suggestion is that well-orderability is intrinsic to the way that "well-defined" sets are either presented or conceived, e.g., that it is impossible to think of a collection's being a set without at the same time allowing that its elements can be arranged "discretely" in some way, or even that such arrangement can be automatically deduced from the "definition". Thus, if one views sets as necessary for mathematics, and the concept of set itself necessarily involves the discrete arrangement of their elements, then WOH might appear necessary, too. But all of this is imprecise. One clear implication of Cantor's remark is that WOH is something which does not require proof. Nonetheless, not long after he had stated this, Cantor clearly had doubts both about the well-orderability of the continuum and about cardinal comparability.[9] All of this suggested that WOH and the associated aleph hypothesis (AH), that the alephs represent the scale of infinite cardinality, do require proof, and cannot just be taken as "definitional".

Work subsequent to 1884 suggests that Cantor felt the need to supply arguments for well-ordering. For instance, in his *1895*, to show that every infinite set T has a countable subset (and thus that $\aleph_0$ is the smallest cardinality), Cantor had to *define* a subset of T well-ordered like the natural numbers. The key point is that Cantor felt it necessary to exhibit a well-ordered subset, and did not proceed by first assuming (by appeal to his "*Denkgesetz*") that M can be arranged in well-ordered form. He exhibits such a subset in the following way (loc. cit., 493):

> *Proof.* If one has removed from T a finite number of elements $t_1, t_2, \ldots, t_{\nu-1}$ according to some rule, then the possibility always remains of extracting a further element t_ν. The set $\{t_\nu\}$, in which ν denotes an arbitrary finite, cardinal number, is a subset of T with the cardinal number $\aleph_0$, because $\{t_\nu\} \sim \{\nu\}$.

In his later editorial comments to this proof in his edition of Cantor's collected papers, Zermelo comments as follows (in *Cantor 1932*, 352):

[8] Recall that the phrase "laws of thought [*Gesetze des Denkens*]" is used by Frege in *Frege 1879* to refer to logical laws, laws which "transcend all particulars [*über allen Besonderheiten erhaben*]" (*Vorwort*).

[9] See *Moore 1982*, 44.

> The "proof" of Theorem A, which is purely intuitive and logically unsatisfactory, recalls the well-known primitive attempt to arrive at a *well-ordering* of a given set by successive removal of arbitrary elements. We arrive at a correct proof only when we *start from* an already *well-ordered* set, whose smallest initial segment in fact has the cardinal number $\aleph_0$ sought.

The second context in which an argument was given was an attempt by Cantor (in correspondence first with Hilbert and then Dedekind) to show that every set must have an aleph as a cardinal.[10] What Cantor attempts to show, in effect, is the following. Assume that Ω represents the sequence of all ordinal numbers, and assume (for a *reductio* argument) that V is a "multiplicity" which is not equivalent to any aleph. Then Cantor argues that Ω can be "projected" into V, in turn showing that V must be what he calls an "inconsistent multiplicity", i.e., not a legitimate *set*. It will follow that all *sets* have alephs as cardinals, since they will always be "exhausted" by such a projection by some ordinal or other, in which case they will be cardinally equivalent to an ordinal number-class.[11] Zermelo's dismissal of this attempted proof is no surprise, given the comments just quoted. But he also comments further here exactly on this "projection" (in *Cantor 1932*, 451):

> The weakness of the proof outlined lies precisely here. It is *not* proved that the whole series of numbers Ω can be "projected into" any multiplicity V which does not have an aleph as a cardinal number, but this is rather taken from a somewhat vague "intuition". Apparently *Cantor* imagines the numbers of Ω successively and arbitrarily assigned to elements of V in such a way that every element of V is only used once. *Either* this process must then come to an end, in that all elements of V are used up, in which case V would then be coordinated with an *initial segment* of the number series, and its power consequently an aleph, contrary to assumption; *or* V would remain inexhaustible and would then contain a component equivalent to the whole of Ω, thus an inconsistent component. Here, the intuition of time [*Zeitanschauung*] is being applied to a process which goes beyond all intuition, and a being [*Wesen*] supposed which can make *successsive* arbitrary choices and thereby define a subset V' of V which is not definable by the conditions given.[12]

If it really is "successive" selection which is relied on, then it seems that one must be assuming a subset of instants of time which is well-ordered and

---

[10] The letters can all be found in *Cantor 1991*.

[11] Hardy (in *Hardy 1904*, §2) and Jourdain (in *Jourdain 1905b*, among other places) later endorsed a similar "selection" method to show, respectively, that the continuum has a subset of the power of the second number-class and that every cardinal number is an aleph.

[12] "Component" was Cantor's original term for a subset.

which forms a base ordering from which the "successive" selections are made. In short, what is really presupposed is a well-ordered subset of temporal instants which acts as the basis for a recursive definition. Even in the case of countable subsets, if the "process" is actually to come to a conclusion, the "being" presupposed would presumably have to be able to distinguish a (countably) infinite, discrete sequence of instants within a finite time, and this assumption is, as is well known, a notoriously controversial one. In the general case, the position is actually worse, for here the question of the well-orderability of the given set depends at the very least on the existence of a well-ordered subset of temporal instants of arbitrarily high infinite cardinality. This appears to go against the assumption that time is an ordinary continuum, i.e., of cardinality $2^{\aleph_0}$, unless of course the power set of natural number itself is too "big" to be counted by any ordinal, in which case much of the point of the argument would be lost, for one of its aims is presumably to show that the power of the continuum is somewhere in the aleph-sequence.[13]

Part of what is at issue here, at least implicitly, is what constitutes a proof. It seems obvious that if a set is non-empty, then it must be possible to "choose" an element from it (i.e., there must exist an element in it). Indeed, the obviousness of this is enshrined in the modern logical calculus by the inference principle of existential instantiation (EI): from $\exists x Px$ one can always infer Pc, where c is a new constant. Furthermore, it is clear how this extends to finite sets or the finite extensions of properties by stringing together successive inferential steps. But how can such an inferential procedure be extended to infinite sets, if at all?

Some evidence of the centrality of WOH is provided by Problem 1 on Hilbert's list of mathematical problems in his famous lecture to the International Congress of Mathematicians in Paris in 1900. Hilbert does not mention the transfinite numbers in this context, but states CH rather in its original, 1878 form. He notes Cantor's conviction of the correctness of this conjecture and its "great probability", and then goes on to mention another "remarkable assertion" of Cantor's, namely his belief that the continuum, although not (in its natural order) in well-ordered form, can be rearranged as a well-ordered set. He goes on (*Hilbert 1900b*, 264):

> It seems to me highly desirable *to achieve a direct proof of this remarkable assertion*, perhaps by actually specifying an ordering of the numbers, whereby in each partial system [subset] a first number can be exhibited [*aufgewiesen*].

Russell, writing at roughly the same time, expressed doubts about precisely this (*Russell 1903*, 322–323):

[13] One could in consequence say that Zermelo's criticism reveals something like the same circularity as is revealed in many of the "proofs" of the Euclidean parallel postulate.

> Cantor assumes as an axiom that every class is the field of some well-ordered series, and deduces that all cardinals can be correlated with ordinals .... This assumption seems to me unwarranted, especially in view of the fact that no one has yet succeeded in arranging a class of 2^{α_0} terms in a well-ordered series.

He goes on (loc. cit., 323):

> We do not know that of any two different cardinal numbers one must be the greater, and it may be that 2^{α_0} is neither greater nor less that α_1 and α_2 and their successors, which may be called well-ordered cardinals because they apply to well-ordered series.[14]

And recall that, at the International Congress of Mathematicians in Heidelberg in 1904, König had given an apparently convincing proof that the continuum *cannot* be an aleph. König's argument, as we know, turned out to contain fatal flaws, but in any case, the confusion over it is instructive.[15]

In short, the clear impression in the immediate period leading up to Zermelo's paper was that only WOH would provide a solid foundation on which to build a reasonable notion of infinite cardinal number as a proper framework for tackling CH, and furthermore that it required justification.

2. The 1904 proof

2.1. The proof itself

The steps in Zermelo's proof are very clear. (We use Zermelo's notation.)

(1) Let M be an arbitrarily given set, and let M be its power set. Assume given what Zermelo calls a "covering" of M, i.e., a function γ from non-empty elements of M to M such that $\gamma(X) \in X$, in other words, what would now be called a *choice function*. The argument then shows that such a γ determines a unique well-ordering of M.[16]

(2) Using a fixed such γ, Zermelo then defines the so-called γ-sets M_γ. These satisfy the following conditions:

(a) $M_\gamma \subseteq M$;
(b) M_γ is well-ordered by some ordering $\prec$ specific to M_γ;
(c) If $a \in M_\gamma$, then a must determine an initial segment A of M_γ under $\prec$ in such a way that $a = \gamma(M - A)$, i.e., a is the "distinguished element" (as Zermelo calls it) of the complement of A in M.

---

[14] We, following Cantor, would put 'ℵ' in place of Russell's 'α'.

[15] For an account of König's lecture, its aftermath and Zermelo's rôle and comments, see *Ebbinghaus 2007b*, 50–53.

[16] Zermelo attributes to Erhard Schmidt the idea that each covering gives rise to a well-ordering; see *Zermelo 1904*, 516.

(3) There clearly are γ sets: $\{m_1\}$ is one such, where $m_1 = \gamma(M)$ and where we take the trivial well-ordering. The set $\{m_1, m_2\}$ is also a γ-set, where again $m_1 = \gamma(M)$, $m_2 = \gamma(M - \{m_1\})$, and $\{m_1, m_2\}$ is given the ordering which places m_2 after m_1. Note that $\{m_1, m_2\}$ with the *other* ordering would not be a γ-set. In fact, it is easy to see that if $M' \subseteq M$ is to be a γ-set, then condition (2)(c) means that $\prec$ is uniquely (one is tempted to say, recursively) determined.

(4) Indeed, following this, Zermelo shows that of any two distinct γ-sets, one is identical to an initial segment of the other, and the well-ordering of the latter extends the well-ordering of the former.

(5) Zermelo now considers the set L_γ, which is the union taken over all the γ-sets. It is not difficult to see that L_γ itself must be a γ-set, indeed, the largest such. By definition, $L_\gamma \subseteq M$; but Zermelo shows that equality must hold. If not, then $M - L_\gamma$ would be a non-empty subset of M, in which case we can consider $\gamma(M - L_\gamma) = m'_1$. Now form $L'_\gamma = L_\gamma \cup \{m'_1\}$, and supply it with the well-ordering which is the same as that in L_γ, except that we extend it by fixing that $x \prec m'_1$ for any $x \in L_\gamma$. Clearly now L'_γ is a γ-set, but one which properly extends L_γ, which is a contradiction. Thus $L'_\gamma = M$, and so M can be well-ordered.

As Zermelo points out (p. 516 of his paper), WOT establishes a firm foundation for the theory of infinite cardinality; in particular, it shows, he says, that every set ("for which the totality of its subsets etc. has a sense") can be considered as a well-ordered set "and its power considered as an aleph". Later work of Hartogs (see *Hartogs 1915*) showed that, not only does WOT imply COMP as Zermelo shows, but that COMP itself implies WOT, and thus in turn Zermelo's choice principle. Thus, it is not just COMP which is necessary for a reasonable theory of infinite cardinality, but WOT itself.[17] Despite Zermelo's endorsement here, the correctness of AH is a more complicated matter, for it involves the claim that every set is actually equivalent to an initial segment of the ordinals, and not just well-orderable. In axiomatic frameworks for sets, therefore, the truth of AH depends very much on which ordinals are present as sets in the system.

[17] Hartogs showed that, from COMP, it follows from Zermelo's principles of set existence (minus the choice principle) that for every set M, there is a well-ordered set L which must be cardinally strictly bigger than M; thus, any set M is certainly equivalent to some well-ordered set and thus well-orderable. The core of the result is the lemma (which uses neither COMP nor the choice principle) that for any set M, there is a well-ordered set L such that L cannot be cardinally smaller than M or cardinally equivalent to it (loc. cit., 441–442). This, independently of the question of well-orderability, settles the question of whether or not the continuum, for example, can be bigger than all well-ordered sets, and thus bigger than every aleph.

2.2. The assumptions behind the proof

Zermelo's 1904 proof is remarkable, and remarkably radical, for many reasons. Let us try to set out three of these.

In the first place, very much in the newly established Hilbert tradition, Zermelo begins to set a very clear and circumscribed context for the proof. We start from what is assumed to be a perfectly good (in Cantor's terminology, "well-defined") set, M. Along with this set is the power set M of M, which itself forms the operating domain of the proof, for all of the M_γ are in M. (We also have the union operation, but that is taken across a subset of M.) Moreover, although it seems as if we are considering quite arbitrary well-orderings of subsets of M, in fact the conditions set out in (2) above mean that what we are considering in fact are just certain "chains" in M which are all related in a certain way.

Secondly, and following on from this, Zermelo does not operate with a pre-existing well-ordered set which, in some way, is to be "projected into" M. He builds up the well-ordered set on which he relies. If we go back to Zermelo's argument, it looks as if Zermelo relies from the beginning on the fact that his γ is (what we would call) a choice function. But in fact, as Kanamori has pointed out, that γ is a choice function becomes important only at the end of the proof. Indeed, let γ be *any* function from M to M. (Of course, there are such maps: consider the trivial one which takes every element of M to some fixed element $m \in M$.) Then we can show that there is a unique set $(W, \prec)$ determined by γ such that $W \subseteq M$, $\prec$ is a well-ordering of W, where for each $x \in W$, $x = \gamma(\{y \in W : y \prec x\})$ and where $\gamma(W) \in W$.[18] For instance, if we take the trivial γ just specified, then W is $\{m\}$ with the trivial well-ordering. As a second example, let m_1, m_2 be two distinct elements of M, assuming that M is non-empty and not a singleton. Now consider the map $\gamma : \mathsf{M} \to M$ given by:

$$\gamma(A) = \begin{cases} m_1 : & \text{if } m_1 \notin A, \\ m_2 : & \text{if } m_1 \in A. \end{cases}$$

If we now consider $\{m_1, m_2\}$ with the ordering $\prec$ determined by $m_1 \prec m_2$, then $\gamma(\{y \in W : y \prec m_2\}) = m_2$, and so $\{m_1, m_2\}$ is the well-ordered set determined by this γ.

In proving this, the relevant steps are really just those which Zermelo uses. It is only the *final* stage, i.e., showing that W must be the *whole* set M, which calls on the fact that γ is a choice function, turning this proof into Zermelo's proof of WOT. Moreover, Kanamori's reconstruction makes it obvious that Zermelo *generates* the well-ordered sets he uses, and does not implicitly rely on their being given.[19] The risk of circularity pointed out above (fn. 13) is thus avoided.

---

[18] See *Kanamori 1997*, 292 and *Kanamori 2004*, 493.

[19] For a more generalised setting, based on a general notion of "type-reducing function" (of which one example is a Zermelo choice function), see *Bell 1995*.

As was pointed out, Zermelo uses the term "covering" for what we now call a choice function. The term "covering" seems to have been introduced first by Cantor in his *Cantor 1895*, where the term is used to refer to functions from a set N to a set M (loc. cit., 486):

> By a "*covering of a set N with elements of the set M*", or, more simply expressed, by a "*covering of N with M*", we mean a law [*Gesetz*] by which to every element of N a definite element of M is associated, whereby one and the same element of M can be used repeatedly. The element of M associated with n is, in effect, a unique function of n and can perhaps be denoted by $f(n)$. This function is to be called the "*covering function of n*". The corresponding covering will be called $f(N)$.

Cantor uses such "coverings" to define the exponential function on powers; the power of M raised to the power of N is just the cardinality of the set of all these "coverings of N with M", called by Cantor the "*Belegungsmenge von N mit M*" (p. 487, respectively p. 288), what we would now denote as M^N, but which Cantor denotes by $(N|M)$ or $\{f(N)\}$ (ibid.). Cantor says that these functions are given by "laws", which of course immediately makes us think that the correspondences are all defined ones. Cantor gives a few examples of functions from N to M which do have particularly simple definitions. But one example of a "covering set" which Cantor goes on to explain is that of the covering of a countable set by a two-element set, giving rise to the exponential cardinal $2^{\aleph_0}$, the cardinality of the real numbers in binary expansion. Here there are clearly more functions than can be explicitly defined. Now it could be that either Cantor was not yet aware of the possibility of imprecision here, imprecision which came to the fore in the Richard antinomy, or it could be that "law" is used in a loose sense, to mean something more like what we would call "correspondence". In any case, Zermelo's special use of the term "covering" is a further step away from the notion of "law" looked at in any modern sense: with his covering functions there is no explicit definition of the correspondence, and thus no explicit definition of the function. What Zermelo relies on is just the *existence* of the choice coverings. Note, however, that there is still a psychologistic element in the way in which Zermelo describes such a function: "*Imagine* [denke man sich] *that with every subset M'* [of M] *there is associated an arbitrary element m_1' that occurs in M' itself: let m_1' be called the 'distinguished' element of M'*. This yields a 'covering' γ ... " (*Zermelo 1904*, 514).

Thirdly, what Zermelo gives is a *proof* in the sense which was then being emphasised by Hilbert. Cantor (originally) regarded WOH as a "law of thought"; for Zermelo it is *proved*, and indeed proved in a "finite number of inferences" (unlike, say, the argument from "successive selection"), i.e., as something which conforms very much to the modern form of a proof.

The Hilbertian perspective on these matters is a very important part of the context of Zermelo's work. Ebbinghaus's book on Zermelo makes it

very clear how embedded he was in the Hilbert foundational circle in the early years of the century. (See *Ebbinghaus 2007b*, in particular 36–47. See also *Peckhaus 1990a*.) More particularly, there are two elements of Zermelo's proof which fit very well with Hilbert's foundational approach at that time. The first element concerns what might be called the *programmatic* element of Hilbert's approach to the foundations of mathematics which emerged in the later 1890s, and especially the notion of mathematical existence. And the second concerns more *proof analysis*, a highly important part of Hilbert's work on Euclidean geometry and geometrical systems generally. These matters are intricate, and cannot be discussed adequately here.[20] But it is important for understanding Zermelo's work fully that a rough account be given. Let us begin with the programmatic elements.

First, Hilbert adopted the view that a mature presentation of a mathematical theory must be given according to the axiomatic method. This requires several things:

1. The postulation of the existence of a domain, of a "system of things" (or "systems of things" in the case of many-sorted theories);
2. The presentation of a finite list of axioms;
3. The requirement of finite proofs, which (in principle) begin with the axioms (and only with the axioms) and proceed from these to a conclusion by a "finite number of inferences" (i.e., acceptable inferential steps);
4. The provision of a consistency proof for these axioms, which will show that no contradiction is derivable by such a proof in the system given.

It is important to add some comments on these elements. For one thing, Hilbert was very clear (especially in his unpublished lectures on geometry) that all that is known about the objects in the domain is what is given to us by the axioms and what can be derived from these through "finite proof". In other words, while a domain is postulated, nothing is taken to be known about the things in it independently of the axioms laid down. The basic example was given by geometrical systems of points, lines and planes; although the geometrical domain is made up of these things, nothing can be assumed known about them (in particular no "intuitive" geometrical knowledge from whatever source) other than what is given in the axioms or which can be derived from them by legitimate inference. Secondly, a consistency proof for the axioms is taken to establish the mathematical existence of the domain or (correspondingly) of the system set out by the axioms. Thus, to take the prime example, the "existence" of Euclidean geometry (or since there can be many models, it might be better to say "Euclidean geometries") is shown by the consistency proof given through analytic geometry, and likewise for all the

[20] For a discussion, see both *Hallett 2008* and *Hallett 2010*.

other geometries.[21] Thus, the unit of consistency is the *system* of axioms as a whole, and different systems necessarily give accounts of different primitives. In any case, the expectation is that when a domain is axiomatised, attention will turn (at some point) to a consistency proof, and *this* will deal with the question of mathematical existence. Thus, the task of showing existence is a mathematical one and there is no further ontological or metaphysical mystery to be solved once the axioms are laid down. This view was stated in a radical way by Hilbert in his 1902 lectures on the foundations of geometry: the axioms "create" the domains, and the consistency proofs justify their existence. As he puts it (*Hilbert 1902*, 47; in *Hallett and Majer 2004*, 563):

> We must now show the *freedom from contradiction of these axioms taken together*; ....
> In order to *facilitate* the *understanding* of this, we begin with a remark:
> The things with which mathematics is concerned are defined through axioms, *brought into life.*
> The axioms can be taken quite arbitrarily. However, if these axioms contradict each other, then no logical consequences can be drawn from them; the system defined then does not exist for the mathematician.

This notion of "definition through axioms" can be seen in various writings of Hilbert's from around 1900. For instance, consider this from Hilbert's paper on the axiomatisation of the reals (*Hilbert 1900a*, 184):

> The objections which have been raised against the existence of the totality of all real numbers and infinite sets generally lose all their justification once one has adopted the view stated above [the axiomatic method]. By the set of the real numbers we do not have to imagine something like the totality of all possible laws governing the development of a fundamental series, but rather, as has been set out, a system of things whose mutual relations are given by the *finite and closed* systems of axioms I–IV [for complete ordered fields] given above, and about which statements only have validity in the case where one can derive them via a finite number of inferences from those axioms.[22]

A word now about proof analysis. A great deal of Hilbert's work on geometry concerned the analysis of proofs, of what can, or cannot, be derived from what. Much of Hilbert's novel work on geometry involved the clever use of (arithmetical) models for geometrical systems to demonstrate a succession of

[21] Note that, while the notion of consistency itself is stated in *syntactic* terms, such proofs are always given in the geometrical work as relative consistency proofs effected by *semantic* arguments.

[22] See also Hilbert's famous paper on mathematical problems from 1900, *Hilbert 1900b*, 265–266.

independence results, which, among other things, often show how finely balanced various central assumptions are. But one straightforward kind of proof analysis was designed to reveal what assumptions there are behind accepted "theorems", and this is clearly pertinent in the case of the choice principle and WOT. What Zermelo shows, in effect, is that the choice principle is a necessary and sufficient condition for WOT; and he shows this by furnishing a Hilbertian style proof for the theorem, i.e., a conclusion which follows from (fairly) clear assumptions by means of a finite number of inferential steps. The subsequent evolution of the Hilbert approach to foundations demands that genuine theorems be proved by formal proofs in much the same way that we think of formal proof today. For WOT, it is Zermelo's work which makes this feasible. The operative word here is "feasible"; Zermelo does not give a fully axiomatic proof of WOT, since the background assumptions about the existence of sets are not explicitly set out. Nevertheless, the way is prepared for an axiomatic approach; the purpose of the 1904 proof is to reveal the choice principle as necessary, and to show how, using it, a "finite proof" can be given. Moreover, the central Hilbertian attitude to existence is key: "laws" for functions (and set definitions generally) are shunned in favour of postulates. What Zermelo's proof shows further is that the choice principle is fundamental if there is to be any serious Cantorian theory of infinite cardinal number.

It is worth remarking that the subsequent work showing the independence of AC from the other axioms of set theory vindicates Zermelo's pioneering work; in this respect, it puts Zermelo's revelation of the choice principle in a similar position as that which Hilbert ascribes to the parallel postulate in Euclid's work. He claims that Euclid must have realised that to establish certain "obvious" facts about triangles, rectangles etc. an entirely *new* axiom (Euclid's parallel postulate) was necessary, and moreover that Gauß was the first mathematician "for 2100 years" to see that Euclid had been right. This is related to the "pragmatic attitude" on display in Zermelo's second paper on well-ordering of 1908, and it became, in effect, the reigning attitude towards the choice principle: If certain problems are to be solved, then the choice principle must be adopted.[23] Zermelo later brings out this parallel explicitly (*Zermelo 1908a*, 115):

> Banishing fundamental facts or problems from science merely because they cannot be dealt with by means of certain prescribed principles would be like forbidding the further extension of the theory of parallels in geometry because the axiom upon which this theory rests has been shown to be unprovable.

Zermelo does not in 1904 call the choice principle an axiom, though the assumption is formulated as Axiom IV in *Zermelo 1908a*, 110. In *Zermelo*

[23] For the relevant passages in his lectures where Hilbert discusses Euclid, see *Hallett and Majer 2004*, 261–263 and 343–345.

1908b, which followed shortly after, Zermelo adopts this axiom as Axiom VI, and gives it there the name by which it has become famous, the "Axiom der Auswahl [axiom of choice]". On the other hand, in the 1904 paper, the choice principle is designated a "logical principle". What Zermelo has to say by way of an explanation is very short (*Zermelo 1904*, 516):

> This logical principle cannot, to be sure, be reduced to a still simpler one, but it is applied without hesitation everywhere in mathematical deduction.

It is not clear from this whether he thinks of the choice principle as a "law of thought", as the term "logical principle" might suggest, or whether he thinks it is just intrinsic to mathematical reasoning whenever sets are involved, a position suggested by the phrase "is applied without hesitation everywhere in mathematical deduction [wird aber in der mathematischen Deduktion überall unbedenklich angewendet]". By the time of his second well-ordering paper of 1908, Zermelo seems to have moved away from the idea of AC as a "logical" principle in the sense of a logical law, and appears to put the emphasis more on the view of it as intrinsic to the subject matter. In later work on the set axioms, however, Zermelo again calls AC a "logical principle". (See, e.g., *Zermelo 1930a*, 31.)

In saying that Hilbert's early work on the axiomatic method is an important part of the context of Zermelo's proof, it is not meant to suggest that Zermelo adopted Hilbert's approach to the foundations of mathematics. Indeed, Zermelo developed his own, distinctive approach to foundational matters which was very different from Hilbert's, something which emerges quite clearly from his later work. Nevertheless, as far as the proof of WOT goes, there are striking parallels with Hilbert's axiomatic method, especially in the treatment of mathematical existence. This will emerge more clearly in the 1908 treatment of set existence and well-ordering, to which we now turn.

3. The 1908 proof

3.1. The background to the proof

In 1908, Zermelo published a rather different (but, as we shall see, related) proof of WOT, which turned out to be extremely fruitful. Why did Zermelo bother to give a *new* proof? Of course, giving different proofs can often be heuristically stimulating, and is in any case a common mathematical phenomenon. But in this case there was an extra reason; it was partly in response to certain of the criticisms which had been directed against the 1904 proof, unjustified as Zermelo himself saw these. Although there are several interesting points to be made in comparing the 1904 and 1908 proofs, an important one is this: while the 1904 proof "builds up" (by recursion) to a well-ordering of M beginning with M's distinguished element, the 1908 proof uses a way

of "cutting down" (again by a kind of recursion), this time beginning with M itself. In doing so, Zermelo attempts to make clear that only certain axioms of set existence are required, and that these avoid all the known antinomies. In particular, it is part of his purpose to show that WOT can be proved while bypassing the general abstract theory of well-ordering and its association with the Cantorian ordinals, and therefore also bypassing "the set W" (as it was widely known) of *all* Cantorian ordinals, and consequently the Burali-Forti antinomy.

"Cutting down" as opposed to "building up" was central to this. The step in the 1904 proof showing that $L_\gamma = M$ was one of the steps which was heavily criticised. The argument uses the extendability principle for well-ordered sets, namely "every well-ordered set can be extended to an ordinally greater one" (or every ordinal β has a successor $\beta + 1$), and thereby uses the central principle involved in generating the Burali-Forti antinomy. (Note that in both cases it is used to derive a contradiction.) Thus if L_γ was already the set of all ordinals, or some similar set, the whole proof would be vitiated. This was Bernstein's criticism (*Bernstein 1905b*, 193):

> The possibility that, for a definite set, for example the continuum, the set L_γ of γ-elements could be similar to W [the set of all ordinals] is not disproved. The conclusion that $M = L_\gamma$ is, moreover, only permissible *if* $L_\gamma \neq W$.[24]

The criticism is clear: the proof takes no precautions against the intrusion of the Burali-Forti contradiction. There is a sense in which the Cantor proof (using successive selection) that every set has an aleph as cardinal number, and which is based on the distinction between "consistent" and "inconsistent" sets, *does* take the possible intrusion of the Burali-Forti into account, for the Burali-Forti contradiction is used in the argument by *reductio*, just as it is in modern textbooks. Indeed, Jourdain sees such an argument using W directly as superior to Zermelo's, since it establishes, not just that any consistent set can be well-ordered, but that such sets all have alephs as cardinals. Zermelo, he says, only shows that, since every set can be well-ordered, *if* it has a cardinal, then this cardinal must be an aleph (see *Jourdain 1905b*, 468–469).

Zermelo's replies are dismissive. There is indeed much confusion in this whole nexus of criticism and challenge, so Zermelo's impatience is, to a certain extent, understandable. But nevertheless the underlying point being made is a good one. Whatever Zermelo's *intention*, there is no *explicit* attempt to exclude the possibility that $L_\gamma = W$ and thus the suggestion that antinomy might threaten. Of course, Zermelo, referring to critics who "base their objections upon the 'Burali-Forti antinomy'", declares that this antinomy "*is without significance* for my point of view, since the principles I employed *exclude* the existence of a set W [of all ordinals]" (*Zermelo 1908a*, 128), and

---

[24] See also *Schoenflies 1905*.

hints earlier (pp. 118–119) that the real problem is with the "more elementary" Russell antinomy. It is also true that at the end of the 1904 paper, Zermelo states that the argument holds for those sets M "for which the totality of subsets, and so on, is meaningful [... *für welche die Gesamtheit der Teilmengen usw. einen Sinn hat, ...*]", which, in retrospect is clearly a hint at important restrictions on set formation. Even so, the reply is unfair. It could be that the remark about "the totality of subsets etc." is an indirect reference to difficulties with the comprehension principle, but even so the principle is not repudiated explicitly in the 1904 paper, neither does Zermelo put in its place another principle for the conversion of properties to sets, which is what the *Aussonderungsaxiom* of the 1908 axiomatisation does. Moreover, he does not say that the existence principles on which the proof is based are the *only* set existence principles, and he does not divorce the proof of the theorem from the Cantorian assumptions about well-ordering and ordinals. Indeed, Zermelo assumes that "every set can be well-ordered" is equivalent to the Cantorian "every cardinality is an aleph" (*Zermelo 1904*, 141). And despite his later claim (*Zermelo 1908a*, 119), he does *appear* to use the ordinals and the informal theory of well-ordering in his definition of γ-sets, where a γ-set is "any well-ordered M_γ ...", without any specification of how "well-ordered set" is to be defined. What assurance is there that *this* can all be reduced to Zermelo's principles?

The new proof is designed to make the demarcations much clearer. As Zermelo himself says (*Zermelo 1908a*, 119):

> ... I succeeded in completing my new proof without even the device of rank-ordering, and I hope thereby to have definitively cut off every possibility of introducing W.

In short, Zermelo adopts what we can call a "reductionist" position in the new proof, an attempt to show that WOT can be carried through using just "pure" set-theoretical principles, in effect a principle of subset formation from a given legitimate set, and some specific set existence principles, notably the power set axiom. Zermelo states in his preamble to the 1908 paper that the second proof is designed to "bring out more clearly than the first proof did, the purely formal character of the well-ordering, which has nothing to do with spatio-temporal arrangement" (loc. cit., 107).

As background to understanding this development, we should recall that, while we now take for granted that the system of set theory based on Zermelo's work allows the unproblematic and entirely familiar development of the notions of ordered pair, relation, ordered set, function and the full development of the transfinite numbers via the (von Neumann) ordinals and the alephs, this was by no means the case when Zermelo first developed this theory; many of these notions were not in fact introduced and simplified ("reduced" to the notions of set and set membership) until the 1920s, and even then required an essential extension of Zermelo's framework. Hence, the "reductionist" (or "formal") programme followed by Zermelo and others was

very much concerned with showing how much of the theory (in this case, the theory of well-ordering) can be carried through without recourse to the full panoply of informal notions which became available in axiomatised set theory only much later.

3.2. The basic structure of the new proof

Zermelo's new proof clearly has its origins in Dedekind's notion of "chain [*Kette*]", the central building block of his theory of "simply infinite sequences" (the natural numbers). (See *Dedekind 1888*, especially item 37ff.) Suppose there is a mapping φ on a set K. Consider $\varphi[K]$, which Dedekind calls simply K'; K is called a "chain" (more strictly, a "chain under φ") if $K' \subseteq K$. It follows that the union and intersection of chains are also chains. Dedekind also then discusses the notion of "the chain of A". Suppose that $A \subseteq K$; now consider the intersection of all those chains contained in K which also contain A, denoted by A_0. This is then the basis of the (well-ordered!) set of natural numbers N, this being put equal to $\{1\}_0$, where "1" signifies the base element of N with $1 \notin N'$. Zermelo's definition of a "Θ-chain" can be seen as a "transfinite" variant of this, with provision for behaviour at limits. Suppose M is the set given, $\mathfrak{U}M$ is its power set, and φ is a choice function on $\mathfrak{U}M$, where Zermelo denotes by A' the set $A - \varphi(A)$. $\mathfrak{U}M$ possesses the following three properties: (1) $M \in \mathfrak{U}M$; (2) for each $A \in \mathfrak{U}M$, $A' \in \mathfrak{U}M$; (3) for any $\mathsf{A} \subseteq \mathfrak{U}M$, we have $\bigcap \mathsf{A} \in \mathfrak{U}M$. Zermelo now calls a subset Θ of $\mathfrak{U}M$ which satisfies (1)–(3) a "Θ-chain"; since $\mathfrak{U}M$ is clearly a Θ-chain, such chains exist. The key construction is then the intersection M[25] of all Θ-chains, which, as Zermelo says (p.108), forms "a well-defined subset of $\mathfrak{U}\mathfrak{U}M$". This is again a Θ-chain, and indeed (as is standard with intersection constructions, like A_0) the smallest such. It is easy to see, intuitively, that this intersection is just the set $\{M, M', M'', \ldots\}$ closed under intersections. This chain M is actually well-ordered by reverse inclusion, something we will look at more closely in a moment; Zermelo uses this fact to show that M "corresponds" to a well-ordering of the original set M, but officially the only orderings which are used are inclusion orderings. The key to this correspondence is that the condition stated in the theorem (p. 108) holds, namely that for every subset P of M there is a unique element P_0 in M such that $P \subseteq P_0$ and $\varphi(P_0) \in P$. It shows (in effect) that there is a map $f : \mathfrak{U}M \to \mathsf{M}$ such that $\forall X \in \mathfrak{U}M[X \subseteq f(X) \wedge \varphi(f(X)) \in X]$, where φ is the choice function in question. If we focus now on the subset of $\mathfrak{U}M$ of singletons $\mathsf{M}' = \{\{x\} : x \in M\}$, then $f \restriction \mathsf{M}'$ takes M' one-one onto M, i.e., $\mathsf{M} = \{f(\{x\}) : x \in M\}$. Given this, the well-ordering of M is mirrored on M' by the function f^{-1}; the induced well-ordering on M' can then be mapped onto M itself by the function g which takes $\{x\}$ to x. Thus, it is not so much

[25] Note the change of notation from the 1904 proof, where 'M' stands for the power set of M.

the map f which is important, but rather the existence of the derived map $f' = f \restriction \mathsf{M}'$.

The idea (and indeed the motivation) behind the proof is hinted at towards the end of Zermelo's proof. Suppose a set M is in fact well-ordered by an ordering relation $\prec$. Call the set $\mathfrak{R}(a) = \{x \in M : a \preceq x\}$ the "remainder [*Rest*]" determined by a and the ordering $\prec$. Consider now the set of "remainders" in this ordering, i.e., $\{\mathfrak{R}(x) : x \in M\}$. This set is in fact well-ordered by reverse inclusion, where the successor remainder to $\mathfrak{R}(a)$ is just the remainder determined by a's successor a' under $\prec$, and where intersections are taken at the limit elements (the intersection of a set of remainders is again a remainder). But not only is this set well-ordered by reverse inclusion, the ordering is isomorphic to the ordering $\prec$ on M. That is:

$$a \prec b \quad \text{if and only if} \quad \mathfrak{R}(b) \subset \mathfrak{R}(a).$$

Zermelo's Θ-chains are now meant to define a "remainder set" directly without detour through some $\prec$; the resultant inclusion ordering is then "mirrored" on M. (In fact, $f' \circ g^{-1}$ takes x in M to a "remainder" $\mathfrak{R}(x)$.) Using "initial segments" (and unions instead of intersections) with the inclusion relation, not reverse inclusion, would also work just as well, but taking remainders has the advantage of being able to "build down" from M instead of "up" from $\emptyset$. We have spoken of functions here; in fact, Zermelo avoids such talk. He defines (p. 111) M as being "well-ordered" when each element in M "corresponds" uniquely to such a "remainder". This shows, says Zermelo, that the theory of well-ordering rests "exclusively upon the elementary notions of set theory", and that "the uninformed are only too prone to look for some mystical meaning behind Cantor's relation $a \prec b$" (ibid.).

3.3. Maximal inclusion chains and well-ordering

What are the general conditions for a subset M of $\mathfrak{U}M$ to be a set of remainders of a total- or well-ordering on M? These were first worked on by Hessenberg in *Hessenberg 1906*, 680ff, conditions which were later simplified by others, Hartogs, Fraenkel, and most notably Kuratowski in *Kuratowski 1921* and *Kuratowski 1922*. We will use Kuratowski's analysis to show more clearly *why* the M which Zermelo defines gives rise to a well-ordering of M and why Zermelo proves what he does about it. This will explain in particular the rôle of the choice principle in Zermelo's proof, and the origin of Kuratowski's celebrated set definition of ordered pair.

Hessenberg's conditions are designed to show that M is an inclusion chain of subsets which always contains certain prescribed elements. Kuratowski's analysis shows in effect that these extra conditions are really to ensure that the chain is *maximal*. Kuratowksi in effect shows:

(A) A set M represents the set of remainders of a *total*-ordering on a set M
if and only if
M is a maximal, reverse inclusion chain in $\mathfrak{U}M$

and

(B) A set M represents the set of remainders of a *well*-ordering on a set M
if and only if
M is a maximal, *well-ordered*, reverse inclusion chain in $\mathfrak{U}M$.

(See *Kuratowski 1921*, 161–168.[26]) According to (B), to explain why Zermelo's proof works it is enough to show that Zermelo's M is a maximal, well-ordered, reverse inclusion chain in $\mathfrak{U}M$.

1. M is *well-ordered.* To explain this, we turn to *Kuratowski 1922*. In that paper, Kuratowski was looking for a way of eliminating the necessity for transfinite recursion over the ordinals as part of an attempt to justify the reliance on just Zermelo set theory. Kuratowski openly acknowledged that transfinite numbers had been of enormous historical and heuristic importance. Applications of transfinite numbers, he says, "... have contributed time and time again to progress in various domains of mathematics" (*Kuratowski 1922*, 76). However (loc. cit., 77)

> ... in reasoning with transfinite numbers one implicitly uses an axiom asserting their *existence*; but it is desirable both from the logical and mathematical point of view to pare down the system of axioms employed in demonstrations. Besides, this reduction will free such reasoning from a foreign element, which increases its æsthetic value.

Kuratowski then undertakes to *prove* that the transfinite numbers can be dispensed with for a significant class of applications.[27] Invariably, application of the ordinals in analysis, topology, etc. had focused on definitions by transfinite recursion over the ordinals. Kuratowski succeeds in showing that in a large class of cases of this kind, the ordinals can be avoided by using purely set-theoretic methods which are reproducible in Zermelo's system. As he notes (loc. cit., 77):

> From the viewpoint of Zermelo's axiomatic theory of sets, one can say that the method explained here allows us to deduce theorems

[26] The term Kuratowski in fact uses is not "maximal", but "saturated", i.e., M in $\mathfrak{U}\mathfrak{U}M$ is *saturated* with respect to a property if it possesses that property and is not properly included in any other set in $\mathfrak{U}\mathfrak{U}M$ which has the property. See *Kuratowski 1921*, 163.

[27] Thus, Kuratowski takes the view of transfinite numbers as strictly speaking unnecessary. Before the general acceptance of the von Neumann ordinals in the later 1920s and the consequent "reduction" of the full theory of ordinals to "pure" set theory, there was a vigorous debate about the independent importance of the transfinite numbers. We cannot go into this here.

of a certain well-determined general type *directly* from Zermelo's axioms, that is to say, without the introduction of any independent, supplementary axiom about the existence of transfinite numbers.

Kuratowski's "general method" is justified by his proof of the equivalence of the following two definitional procedures. Both start from a fixed set E, a fixed subset M of E, and a set function G defined on $\mathfrak{U}E$ with values in $\mathfrak{U}E$ and such that $\forall X \in \mathfrak{U}E[G(X) \subseteq X]$.

Procedure 1 is as follows. Define a Θ_E-chain to be any set Z satisfying the conditions:

(i) $\mathsf{Z} \subseteq \mathfrak{U}E$
(ii) $M \in \mathsf{Z}$
(iii) if $X \in \mathsf{Z}$ then $G(X) \in \mathsf{Z}$
(iv) if $\mathsf{Z}' \subseteq \mathsf{Z}$ then $\bigcap \mathsf{Z}' \in \mathsf{Z}$.

Now let $\mathsf{M}(M)$ be the intersection of all Θ_E-chains; the set of Θ_E-chains is non-empty since $\mathfrak{U}E$ itself is one. Now $\mathsf{M}(M)$ is itself a Θ_E-chain, the smallest. This, clearly, is a generalization of Zermelo's 1908 definition of his M.

Procedure 2 defines the following sequence by transfinite recursion over the ordinals:

(i) $M_0 = M$
(ii) $M_{\alpha+1} = G(M_\alpha)$
(iii) $M_\lambda = \bigcap_{\beta<\lambda} M_\beta$ for limit λ.

Put $\mathsf{A}(M) = \{M_\alpha : \alpha \text{ an ordinal}\}$.

Kuratowski now shows that $\mathsf{M}(M)$ and $\mathsf{A}(M)$ represent precisely the same set. The argument is quite simple. Certainly $\mathsf{A}(M)$ must be a Θ_E-chain, so $\mathsf{M}(M) \subseteq \mathsf{A}(M)$. Assume now that $\mathsf{A}(M) - \mathsf{M}(M) \neq \emptyset$, and let M_α be the set with the least ordinal index in $\mathsf{A}(M) - \mathsf{M}(M)$. If α is a limit ordinal, all M_β with $\beta < \alpha$ belong to $\mathsf{M}(M)$; hence by clause (iv) of Procedure 1, $M_\alpha \in \mathsf{M}(M)$. If α is a successor ordinal $\beta + 1$, then $M_\beta \in \mathsf{M}(M)$, and, by clause (iii) of Procedure 1, $M_\alpha \in \mathsf{M}(M)$. Thus, we have a contradiction, and so $\mathsf{A}(M) - \mathsf{M}(M) = \emptyset$.

It follows from this that the Zermelo-style Procedure 1 is a perfectly adequate replacement for the recursion in Procedure 2.[28] More to the point here, the equivalence immediately shows why Zermelo's M is well-ordered. To obtain Zermelo's M, let M and E be given, and put $G(X) = X$ if $X = \emptyset$, and $G(X) = X - \{\varphi(X)\}$ (φ being the given choice function) whenever $X \neq \emptyset$. By Kuratowski's result, M is equal to $\mathsf{A}(M)$ defined by recursion, and $\mathsf{A}(M)$ is well-ordered under the ordering of the indices. So, arguing informally, M

[28] Kuratowski (*Kuratowski 1922*, 81) actually proves an analogue of the transfinite induction theorem for the $\mathsf{M}(M)$, namely: Let ψ be a set property; if $\psi(M), \forall X \subseteq E[\psi(X) \to \psi(G(X))]$, and $\forall \mathsf{Y} \subseteq \mathfrak{U}E[\forall X \in \mathsf{Y}[\psi(X)] \to \psi(\bigcap \mathsf{Y})]]$, then $\forall X \in \mathsf{M}(M)[\psi(X)]$.

must be well-ordered by the same ordering. This also gives us the clue as to the *final* ordering of M as it would look from inside Zermelo's set theory, deprived as it is of the presence of the ordinals. We know $M_\alpha <_{\mathsf{A}(M)} M_\beta$ if and only if $M_\alpha \supset M_\beta$, so M is well-ordered by reverse inclusion.

2. M is *maximal*. The maximality can be characterised as follows: A well-ordered, reverse inclusion chain M in $\mathfrak{U}M$ is maximal if and only if the chain contains M and $\emptyset$, successors in the chain differ by only a single element, and any limit element is just the intersection of its predecessors.

It is clear that if the chain M contains neither M nor $\emptyset$, or if the successor of a member differs from it by *more* than one element, then it would be possible to *extend* M by "squeezing in" another element, thus showing that M cannot be maximal. So the conditions specified are certainly necessary. Now assume that M is a well-ordered, reverse inclusion chain in $\mathfrak{U}M$ containing $\emptyset$ and M, such that successors differ by only a single element, with any limit element equal to the intersection of its predecessors. Assume also that M is not maximal, i.e., that there is a well-ordered, reverse inclusion chain M′ in $\mathfrak{U}M$ which properly includes M. Let B be the least element in M′ − M. The initial segment of M′ determined by B is isomorphic to some initial segment of M (possibly M itself) under identity. Now B is either a limit element, or a successor. If it is a successor, then $B = G(X)$, for some $B \subset X$. But this X must therefore be in M, and so therefore is $G(X) = B$. Suppose B is a limit element, i.e., $B = \bigcap\{X : X \in \mathsf{M} \wedge B \subset X\}$. But all these sets X are in M, thus B is too. Either way, we have a contradiction. Thus, M cannot be properly included in any well-ordered, reverse inclusion chain, and so must be maximal.

Given this equivalence it is now easy to show that Zermelo's M is maximal. Certainly it contains M (we shall deal with $\emptyset$ in a moment), and we saw from the analysis of the ordering on M that any limit element is the intersection of its predecessors. The interesting condition is that governing successors, for it is *precisely to guarantee this* that the choice principle is used. In the Kuratowski Procedures 1 and 2 above, the successor of any element X in $\mathsf{M}(M)$ is $G(X)$. In the particular Zermelo case, the shift from X to its successor $G(X)$ involves in general only the "shedding" of a *single element*, the distinguished element $\varphi(X)$ of X, which means that it is the choice principle which guarantees that nothing can be "squeezed in" between succeeding elements.

The above analysis makes it quite clear that Zermelo's M is a maximal, well-ordered reverse inclusion chain in $\mathfrak{U}M$. Close examination of Zermelo's proof, and particularly its much more perspicuous rendering by Hausdorff (see *Hausdorff 1914*, 136–138), shows that it follows more or less the steps set out above. The advantage of the detour through Kuratowski's work is that it shows *why* the proof should take precisely these steps.

Why can the condition that M contain $\emptyset$ be dropped? The reason is that it actually follows from a result of Kuratowski's on the existence of fixed points for the function $G(X)$ in Procedures 1 and 2. Kuratowski proves $\exists M_0 \in$

$\mathsf{M}(M)[G(M_0) = M_0]$.[29] The proof is easy. Since $\mathsf{M}(M) \subseteq \mathsf{M}(M)$ then by definition $\bigcap \mathsf{M}(M) \in \mathsf{M}(M)$. Let $M_0 = \bigcap \mathsf{M}(M)$; since $M_0 \in \mathsf{M}(M)$, we must have $G(M_0) \in \mathsf{M}(M)$. But $M_0 \subseteq X$ for all $X \in \mathsf{M}(M)$, which means that $M_0 \subseteq G(M_0)$. However, by definition of G, $G(M_0) \subseteq M_0$, so $G(M_0) = M_0$. Hence M_0 is the required fixed point. Now, in the Zermelo case $G(X) = X$ just in case $X = \emptyset$, and $X - \{\varphi(X)\}$ otherwise. Since there must be a fixed point, and it cannot be any non-empty set by definition, it must be the empty-set, thus $\emptyset \in \mathsf{M}$. Consequently, in the Zermelo case, we do not actually have to postulate as an independent condition that $\emptyset \in \mathsf{M}$.

This analysis also shows why the 1904 and the 1908 proofs are very closely related. Suppose we change Zermelo's 1908 definition to make M the intersection of all subsets Z of $\mathfrak{U}M$ satisfying

(i) $\emptyset \in \mathsf{Z}$
(ii) if $X \in \mathsf{Z}$ then $X \cup \{\varphi(M - X)\} \in \mathsf{Z}$
(iii) if $\mathsf{Z}' \subseteq \mathsf{Z}$ then $\bigcup \mathsf{Z}' \in \mathsf{Z}$

Suppose also that we replace (A) and (B) (see the beginning of the current subsection) by their analogues for segments, and analogues for Procedures 1 and 2 (for which, see *Kuratowski 1922*); the conditions on the maximality of a well-ordered inclusion chain are identical, except that now any limit element must be the *union* of its predecessors.[30] Given this, the proof that M well-orders M can be carried through in much the same way.[31] However, it is not difficult to show that, defined in this new way, M is just the collection of all segments which go to make up L_γ in the 1904 proof. In fact, $\bigcup \mathsf{M} = L_\gamma$. In the case of both the 1904 and the 1908 procedures, the choice principle is required, not to guarantee that M is well-ordered, but to guarantee that it is *maximal*.

As was pointed out above, there was no general set-theoretically defined notion of function available to Zermelo, and he avoids functions, even in the new 1908 statement of the choice principle, which asserts the existence, not of a "covering" or a choice *function*, but rather of a choice *set*. The avoidance of functions is not completely satisfactory. At one point in his discussion of Jourdain, Zermelo says (p. 120) that order-types and cardinal numbers are

[29] See *Kuratowski 1922*, 82–83. See also *Kanamori 1997*, 300–302. Kanamori shows how this result fits in with the theorem (above, p. 89) which generalises Zermelo's 1904 proof. In addition, he points out (301) how Kuratowski uses this result to establish a "maximal principle", a version of Zorn's lemma, though *Hausdorff 1909* had already demonstrated a similar proposition. The development of maximal principles was an important step to the establishment of the choice principle in wider mathematical settings. Kanamori also discusses Bourbaki's generalisations of these results. It should be underlined here that all of this work stems from Zermelo's papers of 1904 and 1908.

[30] Kuratowski's fixed-point theorem also has a natural analogue. See *Kanamori 1997*, 300.

[31] *Hausdorff 1935* proves the well-ordering theorem by this method.

just convenient ways of expressing the notions of similarity and equivalence. But be that as it may, these notions themselves do require some notion of function or correspondence. In his paper on the axiomatisation of set theory (§2), Zermelo does begin to develop a theory of equivalence based on a notion of one-to-one mapping of one set onto another. But this is far from the general notion of function, and the lack of this is a barrier, among other things, to any hope of rebuilding classical analysis within the Zermelo framework.

Despite this, Zermelo's treatment of ordering via inclusion chains does in fact point the way to the definition of the Kuratowski ordered pair, and thus to a general treatment of the notion of function. Consider the ordered pair (a, b). This can be considered as the unordered pair $M = \{a, b\}$, together with a relation $a < b$. Treat this relation now via the theory of inclusion chains. The *only* maximal inclusion chains in $\mathfrak{U}M$ are $\{\emptyset, \{a\}, \{a, b\}\}$ and $\{\emptyset, \{b\}, \{a, b\}\}$. Using the definition of the ordering "<" derived from a maximal inclusion chain M which says that $x < y$ iff $\exists X \in \mathsf{M}[x \in X \wedge y \notin X]$, then these chains must correspond to the orderings $a < b$ and $b < a$ on $\{a, b\}$ respectively. If $\emptyset$ is ignored, the resulting chain $\{\{a\}, \{a, b\}\}$ is associated with the relation $a < b$, and thus with the ordered set (pair) (a, b). Kuratowski then, quite naturally, *defines* (a, b) as $\{\{a\}, \{a, b\}\}$. (See *Kuratowski 1921*, 170–171.)

3.4. The status of the choice principle in *Zermelo 1908a*

The 1904 paper states expressly that the axiom is a "logical principle"; the 1908 paper, however, contains no such claim. Indeed, Zermelo's procedure is altogether more cautious. The central theorem (*Zermelo 1908a*, 108) is based on the assumption that, for a given set M, there is a "law" associating with every non-empty subset of M one of its members as a "distinguished element". After he has proved this theorem, Zermelo then says this (loc. cit., 110):

> Now in order to apply our theorem to arbitrary sets, we require only the additional assumption that *a simultaneous choice of distinguished elements is in principle always possible for an arbitrary set of sets*, or, to be more precise, that the same consequences always hold as if such a choice were possible.

The insistence on "simultaneous choices" is no doubt intended to underline that there is no reliance on anything like the procedure of "successive choices" which traces back to Cantor, and thus Zermelo's repudiation of any assimilation of his procedure to Cantor's.[32] In fact, Zermelo goes further, for he then distances himself altogether from the use of the term "choice". He points out that, formulated with this term, the principle "appears to be tainted with subjectivity [*subjektiv gefärbt*]" and exposed to "misinterpretation [*Mißdeutungen*]". He then proceeds to formulate a pure existence axiom which postulates the existence of what is now termed a "choice set" for any set S of

[32] Such an assimilation was explicitly made by Borel. See section 4.1 below.

disjoint non-empty sets, an axiom which does not use either the term "choice" or the term "distinguished element". The choice principle, says Zermelo, can be "reduced" to this axiom, which evidently has a "purely objective character". In his paper on the axiomatisation of set theory of just two weeks later, the axiom is stated in much the same purely existential way, this time without preamble, and with just a footnote referring to the polemical section of the 1908 well-ordering paper for "justification".[33] The axiom is called there the "axiom of choice", but the idiom of choice only figures after the statement of the axiom, where Zermelo says (*Zermelo 1908b*, 266):

> We can also express this axiom by saying that it is always possible to *choose* a single element from each element $M, N, R, \ldots$ of T and to combine all the chosen elements $m, n, r, \ldots$ into a set S_1.

In private, Zermelo was very explicit about the move away from the idiom of choice. In a letter to Fraenkel from 31 March 1921 which refers to the axiom of choice from his 1908 axiomatisation (there Axiom VI on p. 266), he writes:[34]

> I designate [it] the "axiom of choice" only in an improper sense and just for the sake of the usual way of speaking. There is no sense in which my theory deals with a *real* "choice". The statement on p. 266 "We can also express this axiom etc." should be taken as only a remark, not as something which touches the theory. For me Axiom VI is a pure *existence axiom*, nothing more and nothing less, ... The "simultaneous choice" of elements and the gathering of these into a set ... is for me only a way of envisaging matters which renders the (psychological) necessity of my axiom *intuitive*.[35]

Thus, with the procedure adopted in the 1908 paper on well-ordering, Zermelo moves fully to an axiomatic viewpoint: the choice principle is replaced by an axiom, which means, in effect, that only sets are considered for which this axiom holds. This, it seems, is what Zermelo means by saying the same consequences will hold "as if" choice (or the axiom) were possible.

4. Other objections to the 1904 argument

We have already looked briefly at one important line of criticism of Zermelo's 1904 proof concerned with the set-theoretical context and its possible

[33] The immediate context for the introduction of the axiom in the axiomatisation is the discussion of set products. Zermelo states (*Zermelo 1908b*, 266): "In order, now, to obtain the theorem that *the product of several sets can vanish* (that is, be equal to the null set) *only if a factor vanishes* we need a further axiom."

[34] For the original German, see *Ebbinghaus 2007b*, 285.

[35] The central points were reported by Fraenkel in his *1922b*, 232–233.

entanglement with the antinomies. Let us now turn to some other objections to which Zermelo replies in *Zermelo 1908a*. There were basically two forms of objection, those which focus on the choice principle itself, and those which concern certain steps in the argument from the principle to WOT. The first line of objection itself breaks into two sub-classes, one concerned with the unprovability of the choice principle, thus a challenge to its status as a "law of thought" or a "logical principle", and one concerned (broadly speaking) with its non-constructive character. The second line of objection again falls into two sub-classes, concerning (*a*) the set-theoretical context, and (*b*) the use of impredicative procedures. We have briefly discussed the set-theoretical context already; below we will discuss the use of impredicative definitions.

4.1. Objections to the choice principle

The objection that the choice principle is non-constructive was advanced by Borel in *Borel 1905a*. Borel points out that there are two separate problems, Problem ***A*** ("to put M in the form of a well-ordered set"), and Problem ***B*** ("being given *any* subset M' of M to choose from M' in a *determinate*, but otherwise arbitrary, way an element m'", and to assume this choice made for all subsets of M). What Zermelo has shown, he goes on, is that these two problems are equivalent. But, Borel says (loc. cit., 194):

> ...this result, whatever may be its interest, cannot be considered as furnishing a general solution to Problem ***A***.

The reason Borel gives is the following (ibid.):

> In effect, in order for Problem ***B*** to be regarded as solved relative to a given set M, it would be necessary to give at least a theoretical means of determining the distinguished element m' of any subset M'; and this problem appears all the more difficult if one supposes, for determinacy's sake, that M is the continuum.

What Borel says here, in effect, is that Zermelo has not solved ***A***, the well-ordering problem, *as it is stated in Hilbert's Paris lecture*. Recall that Hilbert raises in particular the problem of well-ordering the continuum through what he calls a "direct proof" (*Hilbert 1900b*, 264),

> ...perhaps by actually specifying an ordering of the numbers, whereby in each partial system [subset] a first number can be exhibited [*aufgewiesen*].

In other words, Borel quite rightly points out that Zermelo does not *specify* a well-ordering of M in the sense that he does not give a means of specifying the least element in each non-empty subset in M. The choice principle which Zermelo uses, says Borel, is based on an invalid jump, namely from the assumption that one can make a free choice of distinguished element for

a *given* subset of M, to the assumption that such choices are made for *every* (non-empty) subset. This Borel sees as just as unjustified, particularly for uncountable sets, as the well-ordering argument by successive choices. In particular, then, the postulation of the choice principle by no means solves Problem $\boldsymbol{B}$.

There was an interesting exchange of views on the nature of AC and WOT between four leading French mathematicians (Baire, Borel, Hadamard and Lebesgue), originally meant as comments on Borel's *Mathematische Annalen* paper.[36] Roughly speaking, Baire and Lebesgue supported Borel's standpoint, but Hadamard opposed it, and in doing so, represented something akin to Zermelo's position. According to Hadamard, the central distinction being drawn is between a thing's being *determined* (in Lettre I, he had used the term "defined") and it being *described*. He goes on, writing to Borel (Lettre IV, *Borel 1905b*, 269):

> You [i.e., Baire and Lebesgue, as well as Borel] answer negatively the question put (above, p. 263) by M. Lebesgue: Can one demonstrate the existence of a mathematical object without being able to define it? I respond affirmatively.

He goes on (loc. cit., 270):

> ... the main question, that of knowing if the set given can be well-ordered, evidently does not have the same meaning for me as it does for Baire (or for you and Lebesgue). I would say rather: is the ordering possible? (and not even can *one* order it, for fear of having to imagine who or what this "one" is): Baire would say: Can *we* order it? This is a completely subjective question in my view.

Hadamard then relates his view of matters to established positions in the theory of functions (ibid.), and likens the position of the other three to "the point of view of Kronecker", which, until this point, he had regarded as "peculiar to him [Kronecker]" (loc. cit., 269). Hadamard, in short, defends both the legitimacy (and indeed the mathematical fruitfulness) of the view of "pure existence", and he also defends, moreover, the choice principle, endorsing the legitimacy of the move from the acceptance of a distinguished element for any single arbirary subset to the existence of a choice *function* ranging across all non-empty subsets. (See Lettre I, pp. 262–263.) Hadmard also rejects Borel's assimilation (*Borel 1905a*, 195) of the use of Zermelo's choice principle to the argument by *successive* choices, pointing out that each choice is dependent on the set of choices made before, i.e., on a recursive procedure. It is this, says Hadamard, which makes it doubtful that the "successive choices" procedure

[36] This exchange was published as *Borel 1905b*. Zermelo does not discuss the various views, but he does refer to the publication in his 1908 well-ordering paper, p. 111, fn. **).

can be applied in the transfinite case. Zermelo's procedure, which is based on independent choices, does not suffer from this defect.[37]

Hadamard is perfectly clear that there are in fact two quite separate and distinct well-ordering problems. In Lettre I, Hadamard says (*Borel 1905b*, 262):

> What is certain is that M. Zermelo gives no means of executing *effectively* the operation of which he speaks, and it remains doubtful if anyone in the future could indicate such means. Without doubt, it would have been more interesting to resolve the problem in this form; but the question thus posed (to determine the correspondence sought effectively) is nonetheless completely distinct from that which we examine here (does such a correspondence exist?). The two are separated by precisely that difference of which M. Tannery speaks when he refers to *a correspondence* which can be *defined* and a correspondence which can be *described*. Many important questions in mathematics would change their sense totally, and would also thereby demand different solutions, if one were to substitute the second word for the first.

The fact that there are two problems, and that Zermelo does not solve the problem of the well-ordering of the continuum in the way it was stated by Hilbert, was sharply underlined by a celebrated independence result of Levy's from 1963. Levy (see *Levy 1963*) showed that the statement "There is no definable well-ordering of the continuum" is consistent with the usual axioms of set theory together with AC.

Zermelo's reply to Borel is actually very brief, and, one feels, rather offhand. He sees Borel's point that his way of demonstrating the equivalence of Problems $\boldsymbol{A}$ and $\boldsymbol{B}$ does not amount to a solution of Problem $\boldsymbol{A}$, as asserting just that the choice principle has not been proved. Zermelo remarks that this is precisely why the choice principle is added as an "axiom", i.e., because of its unprovability from "the other principles", thus asserting the similarity of his position to that of Euclid in adopting the parallel postulate: if one wants to do certain things, then AC is indispensable. This point is somewhat different from those which Hadamard makes, for not only does the latter defend the notion of existence which Zermelo in effect adopts, he gives a direct defence of the legitimacy of assuming the existence of choice functions.

Zermelo's reply to Peano is germane to his reply to Borel, and actually leads into the most interesting part of the polemical section of his paper. Peano's objection (see *Peano 1906b*) seems to be that since the choice principle cannot be proved "syllogistically" (i.e., from the principles of his *Formulario*), then it has to be rejected. (Peano does think, however, that finite

[37] This might be seen as an anticipation of Zermelo's later analysis of the problems with the procedure of "successive choice" as applied by Cantor. See the discussion of Cantor's procedure in section 1.

versions of the choice principle are provable, relying essentially on repeated applications of a version for classes of the basic logical principle EI mentioned above in section 1.) Zermelo's reply is the following. Axiom systems like Peano's are constructed so as to be adequate for mathematics; but how does one go about selecting the "basic principles"? One cannot assemble a complete list of adequate principles, says Zermelo, without careful inspection of actual mathematics and thereby a careful assessment of what principles are actually necessary to such a list. He then says that such an inspection would show that the choice principle is surely one such, adding the ironical remark (*Zermelo 1908a*, 111–112):

> They [Borel and Peano] would even have put me in their debt had they now for their part established the unprovability I asserted—namely, that this postulate is logically independent of the others—thereby corroborating my conviction.

Zermelo then puts forward a list of seven problems which "in my opinion, could not be dealt with at all without the principle of choice" (p. 113).[38] In particular he points out that the principle is indispensable for any reasonable theory of infinite cardinality, for only it guarantees the right results for infinite unions and sums, and over and above this it is vital for making sense (via the multiplicative axiom, as Russell calls it) of the very definition of infinite product. That Peano cannot establish the choice principle from his principles, says Zermelo, strongly suggests that his list of principles is not "complete" (*Zermelo 1908a*, 112). Zermelo makes two other points:

1. Zermelo points out in fn. ***) on p. 112 that demonstrating the choice principle for arbitrary finite sets involves induction on the finite numbers, that is, on the ordinal numbers of finite sets. But then one is obliged to show that the notion of finite set is "stable", i.e., the various different definitions are extensionally equivalent. This in turn is only possible using the choice principle, as Zermelo makes clear in the fourth problem he lists.
2. Peano argues that although various results in analysis appear to rely on something like the choice principle, in fact they can be proved without it. Zermelo replies that one can often dispense with the principle in the theory of functions because one is operating in or with closed sets, the point presumably being that one can always rely in these contexts on the existence of limit points.[39] But as Zermelo points out (p. 115), outside of this context, the choice principle is in fact unavoidable.

[38] The enumeration of such problems is clearly associated with Zermelo's remark in the 1904 paper that the choice principle cannot be reduced to anything simpler, and "wird aber in der mathematischen Deduktion überall unbedenklich angewendet" (*Zermelo 1904*, 516).

[39] There is a close analogy here with the proof of the Heine-Borel theorem, first given by use of the ordinals of the second number-class, which were later dispensed with. The details are to be found in *Hallett 1979a* and *Hallett 1979b*.

To summarise, Zermelo proposes that, in seeking "fundamental principles", one does not look for principles which are "evident" or "self-evident", but rather, in assessing the "evidence" for a principle, one has to look carefully at its consequences, at what it enables. As Zermelo stresses, the notion of "evidence" is subjectively coloured, whereas on the other hand the "question [of] whether the principle is *necessary for science* [die Frage nach dem *wissenschaftliche Bedürfnis*]" is an objective one. In other words, the principles one seeks out are fundamental, not in the sense that they are "self-justifying", but rather in the sense that they are *sine qua non* for the area(s) in question, in this case, the theory of sets and all that intimately involves it. "Indeed", says Zermelo (*Zermelo 1908a*, 115),

> ... principles must be judged from the point of view of science, and not science from the point of view of principles fixed once and for all.

4.2. Impredicative definitions

Poincaré (in *Poincaré 1906b*) raises the objection that, in his 1904 proof, Zermelo makes essential use of impredicative definitions. The objection is the familiar one: a set A is defined by referring to all sets with a certain property ψ, and it then transpires that A itself has ψ, which means that it must fall within the range of the quantifier employed in its definition. The central steps in both of Zermelo's proofs involve just such a definition. For example, in the 1904 proof, what Poincaré calls Γ (Zermelo's L_γ) is defined as "the logical sum" (i.e., union) of all the M_γ, and it is then shown that Γ itself is an M_γ (loc. cit., 315). Zermelo himself points out that the same kind of procedure is to be found in the second proof of 1908 in the definition of the maximal Θ-chain. Poincaré, like Russell, regards such definitions of objects A as circular, potentially viciously so, and declares that the key universal quantifier in the definition should only range over those sets whose definitions do not involve A (loc. cit., 315); only then presumably does the definition become "predicative", i.e., in the original usage, succeed in singling out a legitimate class or set.

Zermelo has three replies to this form of criticism. The first is that procedures like this are common both in set theory and in wider mathematics as a whole, and he cites in this instance Cauchy's "well-known proof" of the fundamental theorem of algebra, saying that "up to now it has not occurred to anyone to regard this as something illogical" (p. 117).[40] The second reply is that Poincaré's recipe for making such definitions "predicative" is itself

[40] Zermelo is almost certainly referring to a proof of the fundamental theorem of algebra (in the form every complex polynomial $f(z) = a_n z^n + a_{n-1} z^{n-1} + \ldots + a_1 z + a_0$, with $a_n \neq 0$, has at least one root in the complex numbers $\mathbb{C}$) which goes back to Argand and Cauchy, and which is presented in Chapter 10 of *Cauchy 1821*. The proof shows first that there must be a $z_0 \in \mathbb{C}$ for which $|f(z)|$ achieves a minimum value, and then shows that this minimum value must be 0; z_0 must be a root of $f(z)$. The proof that a minimum exists is a pure existence proof, and it is somewhat hazily dealt with by Cauchy. In modern versions, it is based on

circular, for one would have to have the concept being defined *beforehand* in order to know what is to be excluded in its definition. The third reply is the most important and the most interesting. First, says Zermelo, the question of whether an "arbitrarily given object" falls under a definition must be decidable by means of an objective criterion independently of the concept to be defined, and such criteria *are* given in the cases of his proofs, i.e., the notion of γ-set and Θ-chain respectively. And given this, says Zermelo, there is nothing to prevent the fact that some of the objects falling under the definition stand in a "special relationship" to that concept, and are singled out in some way from the others, for example, as a minimum ("the least number such that ..." or "the smallest set such that ..."). Second, says Zermelo, an object is not created by such a "determination [*Bestimmung*]", and there are in principle lots of different ways of "determining" an object. What Zermelo is doing here is underlining the "existential axiomatic" standpoint of Hilbert discussed above in section 2.2, a position to be adopted officially in the axiomatisation of set theory submitted for publication some two weeks later. In effect, as Zermelo's axiomatisation (*Zermelo 1908b*) postulates explicitly, there is a domain of objects ("things") $\mathfrak{B}$ which include the sets, and nothing is assumed known about these "things" which is not explicitly set out in the axioms. To show that some particular thing "exists" is to show that it is in $\mathfrak{B}$, i.e., to show by means of a finite proof from the axioms that it exists in $\mathfrak{B}$. What "exists", then, is really a matter of what the axioms, taken as a whole, determine. If the separation, power set and choice principles are axioms, then for a given M in the domain, there will be choice functions/sets on the subsets of M, consequently well-orderings, and so forth; if these principles are *not* included as axioms, then such demonstrations of existence will not be forthcoming. From this point of view, defining within the language deployed is much more like what Zermelo calls "determination", since definitions, although in a certain sense arbitrary, have to be supported by existence proofs, and of course in general it will turn out that a given extension can be picked out by several, distinct "determinations". In short, Zermelo's view is that def-

---

the following lemma, unknown to Cauchy: If $g(x, y)$ is a continuous function from $\mathbb{R}^2$ to $\mathbb{R}$, and A is a compact subset of the range of g, then there must be a point $(x_0, y_0) \in A$ for which $g(x_0, y_0)$ is a minimum in A. As far as the polynomial $f(z)$ goes, one can focus on a disc $\{z : |z| \leq r\}$ in $\mathbb{C}$, where r is chosen large enough for the $|z^n|$ term in $|f(z)|$ to dominate the other terms when $|z| > r$, which shows that $|f(z)|$ keeps on growing as the points in $\mathbb{C}$ radiate out beyond the boundary of this disc. The lemma can then be applied to the disc. The fact that the minimum must be 0 follows from Argand's inequality, which says that if $f(z)$ is a non-constant polynomial, then, for every $z \in \mathbb{C}$ for which $f(z) \neq 0$, there is a $z' \in \mathbb{C}$ such that $|f(z')| < |f(z)|$. (For this method of proof, see *Remmert 1991*, especially pp. 111–115.) The definition to which Zermelo alludes is clearly that of the minimum for the function $|f(z)|$, involving, of course, the notion of a greatest lower bound. The first fully constructive proof, showing how to construct a sequence which has z_0 as its limit, was given in *H. Kneser 1940*.

initions pick out (or determine) objects from among the others in the domain being axiomatised; they are not themselves responsible for showing their *existence*, for their "creation". Precisely this view about impredicative definitions was put forward in *Ramsey 1926*, 368–369, and then later in Gödel's 1944 essay on Russell's mathematical logic as part of his analysis of the various things which could be meant by Russell's ambiguously stated vicious circle principle. (See *Gödel 1944*, 136, or *Gödel 1990*, 127–128.) With a little charitable interpretation of the terms "*définie*" (or "*determiné*") and "*décrite*", the same position is also hinted at in Hadamard's letters of 1905.

Poincaré clearly has a very different view of mathematical existence from Zermelo. For one thing, he takes the view that "there is no actual infinite" (*Poincaré 1906b*, 316), and it is the "belief in the actual infinite which has given birth to these non-predicative definitions" (ibid.). Secondly, Poincaré bases his attitude towards these definitions primarily on his analysis of Richard's antinomy, the antinomy about decimals "definable in a finite number of words". Suppose E is the countable list of *all* such definitions; then one can define in a finite number of words a decimal which is not in this list. But, says Poincaré, endorsing what he refers to as Richard's own solution, E is really the set of all such definitions which can be given without invoking the set E itself (*Poincaré 1906b*, 307):

> Without this, the definition of E would contain a vicious circle; one cannot define E by the set E itself.

He goes on:

> But although it is true that we have defined **N** [the new decimal] by a finite number of words, this was only by basing ourselves on the notion of the set E. This is why **N** does not form a part of E.

Poincaré claims that "the same explanation is valid for all the other antinomies" (ibid.).[41] Consequently, he endorses two theses about mathematical existence:

The linguistic specification thesis: A mathematical object (in particular, a set) cannot be said to exist until there is a specification (definition) of it.

[41] Note that Russell analyses the paradoxes according to the following general scheme (*Russell 1906a*, 36): "...there are what we may call *self-reproductive* processes and classes. That is, there are some properties such that, given any class of terms all having such a property, we can always define a new term also having the property in question. Hence we can never collect *all* the terms having the said property into a whole; because whenever we hope we have them all, the collection we have immediately proceeds to generate a new term also having the said property." Such properties are therefore "non-predicative" for Russell. This fits exactly with Poincaré's discussion of the Richard antinomy; Poincaré knew of this paper of Russell's, since he discusses it in his *Poincaré 1906b*.

This is also combined with:

The individual specification thesis: Before a set A can be said to exist, we must be in possession of specifications of all of its potential members; needless to say, to avoid circularity, these specifications cannot make reference, either direct or indirect, to the set A.

It follows that a set A not only cannot be defined by direct or indirect reference to itself, but also that it cannot contain members which are defined by reference to A either.

The position is constructivist in the sense that the existence of sets is clearly tied to their definition. There is no actually infinite domain of sets; rather, we go on adding sets (by coming up with new specifications) *ad indefinitum*, and the specifications are hierarchical, in the sense that a new specification can rely only on specifications given hitherto. The contrast with Zermelo's position could not be starker.

Interestingly, the difference between the two positions was later put nicely by Poincaré himself (*Poincaré 1912*, 4):

> The Pragmatists adopt the point of view of extension, and the Cantorians the point of view of comprehension. For finite sets, the distinction can only be of interest to formal logicians, but it appears to us much more profound where infinite sets are concerned. Adopting the extensional viewpoint, a collection is constituted by the successive addition of new members. By combining old objects, we can construct new objects, and then, with these, newer objects; if the collection is infinite, it is because there is no reason for stopping.
>
> On the other hand, from the point of view of comprehension we start from a collection where there are pre-existent objects, objects which appear to us indistinct at first, but some of which we finally recognise because we attach labels to them and arrange them in drawers. But the objects precede the labelling, and the objects will exist, even though there may not be a curator to classify them.

Poincaré, of course, sees himself as a pragmatist, while Zermelo is clearly a Cantorian.[42]

Concluding remarks

Let us return to the question of the nature of the axiom of choice. As noted above, Zermelo called the choice principle "a logical principle" both in his

[42] This only touches the surface of Poincaré's views, for alongside the position sketched here goes another view, namely that what Poincaré repudiates are sets which are, in modern terminology, not absolute. Particularly important here are the papers which clearly have Zermelo's axiomatisation of set theory in view, i.e., *Poincaré 1909b*, *Poincaré 1910b*, as well as *Poincaré 1912*.

1904 paper and in his later work (*1930a*) on axiom systems for sets. But is it right to deem it logical? It has certainly not been accepted as such, primarily because it depends on the existence of certain sets, and because we know how to construct models in which it is false. Is it close to a logical principle? For finite collections, it seems to be little more than a set-theoretically coloured version of the logical EI principle (see section 1 above). Indeed, Fraenkel in the 1920s[43] thought that, given this, the axiom of choice is *provable*, once the axiom of replacement is available to gather all the individual "choices" into a set. This is not right, although there is a sense in which Fraenkel is "close" to being right: if one adopts Hilbert's ε-operator, with its "transfinite axiom" $A(x) \rightarrow A(\varepsilon A)$ for any predicate A, and then adapts this to the language of set theory, AC does become provable using replacement. However, it is not so surprising that AC follows. The ε-operator acts as a selector across the sets. Given this, one can construct a functional term with a set as domain which chooses elements from each non-empty set in a given set, and replacement then guarantees that these "choices" can be gathered into a set. In this way, the restriction to the "finite number of choices" permitted by a suitable version of EI can be circumvented. Hilbert saw the "transfinite axiom" as a replacement for the standard quantifiers, often calling it the "logical choice axiom". It is, of course, a way of dealing with general *object representatives* falling under predicates without being able to pick out such objects by an explicit description, and this is important to Hilbert's idea of showing that any inconsistency proved by use of the classical quantifiers can be reproduced in a proof which does not use these quantifiers. But the point here is that the "transfinite axiom" does much of the work done by AC.[44]

If not a logical truth, is the choice principle at least true about sets? Is there evidence for its truth? Russell was one philosopher-logician who wrote very thoughtfully about this matter. One problem he identifies is the following (*Russell 1911*, 32):

> As for the supposed evidence for this [multiplicative] axiom, it seems to me that the imagination always presents to us a finite number of classes, even when we wish to speak about an infinite number. But, in the case of finite numbers, we do not require an axiom, since the possibility of selections is easily proved. Therefore, the apparent evidence for the axiom tends to disappear on reflection.

The same problem arises with the axiom of choice and distinguished elements, and with WOH, for instance, when one tries to imagine whether sets can always be "presented" in well-ordered form. The difficulty Russell underlines here is surely connected to the underlying temptation to think that principles obvious when considering finite numbers of things maintain their correctness

---

[43] See *Hallett 1984*, 161–163.

[44] Note that, in certain constructive settings, AC implies the law of excluded middle. See *Bell 2009*, §5.

when one passes to the infinite case. But "evidence" from the finite case does not provide evidence in the infinite case, which, indirectly, was one of Borel's objections to what he perceived as the motivation for Zermelo's AC.

Can there perhaps be indirect evidence in a weaker sense of the obviousness of the choice principle? Innumerable propositions have turned out to be equivalent to AC, as Zermelo showed WOT to be, as Russell and Hartogs showed for the multiplicative axiom and COMP respectively. Some of these equivalent propositions seem "obvious"; does the equivalence with AC make it so? The answer is simply that it does not. Perhaps the correct reaction is again that of Russell expressed almost a hundred years ago (loc. cit., 33):

> It is also necessary to say that Zermelo's theorem, namely that if the axiom is true, then any class can be well-ordered, gives reason to believe that the axiom is false. For, it is scarcely credible that every class is well-orderable. Many able mathematicians have endeavoured to find a well-ordered sequence of real numbers, but no-one has been able to find one. Such arguments do not carry much weight, but one should grant them a certain value.

Russell, as we saw, was sceptical about WOH; hence, the fact that it follows from AC makes Russell sceptical about AC itself. What is true here about

Beweis, daß jede Menge wohlgeordnet werden kann

1904

(Aus einem an Herrn Hilbert gerichteten Briefe.)

... Der betreffende Beweis ist aus Unterhaltungen entstanden, die ich in der vorigen Woche mit Herrn *Erhard Schmidt* geführt habe, und ist folgender.

1) Es sei M eine beliebige Menge von der Mächtigkeit $\mathfrak{m}$, deren Elemente mit m bezeichnet werden mögen, M' von der Mächtigkeit $\mathfrak{m}'$ eine ihrer Teilmengen, welche mindestens ein Element m enthalten muß, aber auch alle Elemente von M umfassen darf, und $M - M'$ die zu M' „komplementäre" Teilmenge. Zwei Teilmengen gelten als verschieden, wenn eine von beiden irgend ein Element enthält, das in der anderen nicht vorkommt. Die Menge aller Teilmengen M' werde mit M bezeichnet.

2) *Jeder Teilmenge M' denke man sich ein beliebiges Element m_1' zugeordnet, das in M' selbst vorkommt und das „ausgezeichnete" Element von M'*

scepticism is also true of "obviousness", which also will not transfer straightforwardly across provable equivalence. Thus, on showing that an "obvious" proposition is equivalent to AC one might well conclude that the "obvious" proposition is not, after all, obvious, just as well as one can conclude that AC inherits the obviousness. Russell's scepticism inclined him to think that AC (or the multiplicative axiom) is doubtful, and therefore probably false (ibid.):

> It is possible that we shall find later a *reductio ad absurdam* which will show that the axiom is false. But, for the moment, it seems to me that it is only doubtful. It is possible that it will turn out to be true, but the evidence is lacking, and it has surprising consequences.

Gödel's result that AC is consistent relative to the other set-theoretic principles rules out the possibility of a *reductio*. Many propositions in modern abstract mathematics, both "obvious" and "surprising", have turned out to involve "choice assumptions". Perhaps in the end, all that can be done is to reiterate what seems to have been the position Zermelo developed in his work of 1904 and then more explicitly in 1908: if certain problems are going to be tackled at all, AC is intrinsic to a modern theory of infinite, abstract sets.

Proof that every set can be well-ordered

1904

(From a letter addressed to Mr. Hilbert.)

... The proof in question grew out of conversations that I had last week with Mr. *Erhard Schmidt,* and it is as follows.

1) Let M be an arbitrary set of cardinality $\mathfrak{m}$, let m denote an arbitrary element of it, let M', of cardinality $\mathfrak{m}'$, be a subset of M that contains at least one element m and may even contain all elements of M, and let $M - M'$ be the subset "complementary" to M'. Two subsets are regarded as distinct if one of them contains some element that does not occur in the other. Let the set of all subsets M' be denoted by M.

2) *Imagine that with every subset M' there is associated an arbitrary element m'_1 that occurs in M' itself; let m'_1 be called the "distinguished" element*

genannt werden möge. So entsteht eine „Belegung“ γ der Menge M mit Elementen der Menge M von besonderer Art. Die Anzahl dieser Belegungen γ ist gleich dem Produkte $\Pi\mathfrak{m}'$ erstreckt über alle Teilmengen M' und ist daher jedenfalls von 0 verschieden. Im folgenden wird nun eine beliebige Belegung γ zu grunde gelegt und aus ihr eine bestimmte Wohlordnung der Elemente von M abgeleitet.

3) *Definition.* Als „γ-Menge“ werde bezeichnet jede wohlgeordnete Menge M_γ aus lauter verschiedenen Elementen von M, welche folgende Beschaffenheit besitzt: ist a ein beliebiges Element von M_γ und A der „zugehörige“ Abschnitt, der aus den vorangehenden Elementen $x \prec a$ von M_γ besteht, so ist a immer das „ausgezeichnete“ Element von $M - A$.

4) *Es gibt γ-Mengen innerhalb M.* So ist z. B. m_1, das ausgezeichnete Element von $M' = M$, selbst eine γ-Menge, ebenso die (geordnete) Menge $M_2 = (m_1, m_2)$, wo m_2 das ausgezeichnete Element von $M - m_1$ ist.

5) *Sind M'_γ und M''_γ irgend zwei verschiedene γ-Mengen* (die aber zu derselben ein für allemal gewählten Belegung γ gehören!), *so ist immer eine von beiden identisch mit einem Abschnitte der anderen.*

515 | Es sei nämlich M'_γ die eine der beiden wohlgeordneten Mengen, welche auf die andere, M''_γ, oder einen ihrer Abschnitte ähnlich abbildbar ist. Dann müssen je zwei bei dieser Abbildung einander entsprechende Elemente miteinander identisch sein. Denn das erste Element *jeder* γ-Menge ist m_1, da der zugehörige Abschnitt A kein Element enthält, also $M - A = M$ ist. Wäre nun m' das *erste* Element von M'_γ, welches von dem entsprechenden Elemente m'' verschieden wäre, so müssten die zugehörigen Abschnitte A' und A'' noch miteinander identisch sein, mithin auch die Komplementärmengen $M - A'$ und $M - A''$ und als deren ausgezeichnete Elemente m' und m'' selbst, gegen die Annahme.

6) *Folgerungen.* Haben zwei γ-Mengen ein Element a gemeinsam, so haben sie auch den Abschnitt A der vorangehenden Elemente gemein. Haben sie *zwei* Elemente a, b gemein, so ist in *beiden* Mengen entweder $a \prec b$ oder $b \prec a$.

7) Bezeichnet man als „γ-Element“ jedes Element von M, das in irgend einer γ-Menge vorkommt, so gilt der Satz: *Die Gesamtheit L_γ aller γ-Elemente läßt sich so ordnen, daß sie selbst eine γ-Menge darstellt, und umfaßt alle Elemente der ursprünglichen Menge M.* Die letztere ist damit selbst wohlgeordnet.

I) Sind a, b zwei beliebige γ-Elemente und M'_γ und M''_γ irgend zwei γ-Mengen, denen sie angehören, so enthält nach 5) die größere der beiden γ-Mengen beide Elemente und bestimmt die Ordnungsbeziehung $a \prec b$ oder $b \prec a$. Diese Ordnungsbeziehung ist nach 6) unabhängig von der Wahl der verwendeten γ-Menge.

II) Sind a, b, c drei beliebige γ-Elemente und $a \prec b$ und $b \prec c$, so ist immer $a \prec c$. Denn jede c enthaltende γ-Menge enthält nach 6) auch b und mithin a, und da sie einfach geordnet ist, so folgt in ihr aus $a \prec b$ und $b \prec c$ in der Tat $a \prec c$. Die Menge L_γ ist also *einfach geordnet.*

of M'. This yields a "covering" γ of the set M by certain elements of the set M. The number of these coverings γ is equal to the product $\Pi \mathfrak{m}'$ taken over all subsets M' and is therefore certainly different from 0. In what follows we take an arbitrary covering γ and derive from it a definite well-ordering of the elements of M.

3) *Definition.* Let us apply the term "γ-set" to any well-ordered set M_γ that consists entirely of elements of M and has the following property: whenever a is an arbitrary element of M_γ and A is the "associated" segment, which consists of the elements x of M such that $x \prec a$, a is the "distinguished" element of $M - A$.

4) *There are γ-sets included in M.* Thus, for example, [[the set containing just]] m_1, the distinguished element of M' when $M' = M$, is itself a γ-set; so is the (ordered) set $M_2 = (m_1, m_2)$, where m_2 is the distinguished element of $M - m_1$.

5) *Whenever M'_γ and M''_γ are any two distinct γ-sets* (associated, however, with the same covering γ chosen once for all!), *one of the two is identical with a segment of the other.*

For, of the two well-ordered sets, let M'_γ be the one for which there exists a similar mapping onto the other, M''_γ, or onto one of its segments. Then any two elements corresponding to each other under this mapping must be identical. For the first element of *every* γ-set is m_1, since the associated segment A contains no element and therefore $M - A = M$. If now m' were the *first* element of M'_γ that differs from the corresponding element m'', the associated segments A' and A'' would still have to be identical, consequently also the complements $M - A'$ and $M - A''$, and thus their distinguished elements m' and m'' themselves, contrary to assumption.

6) *Consequences.* If two γ-sets have an element a in common, they also have the segment A of the preceding elements in common. If they have *two* elements a and b in common, then either in *both* sets $a \prec b$ or [[in *both* sets]] $b \prec a$.

7) If we call any element of M that occurs in some γ-set a "γ-element", the following theorem holds: *The totality L_γ of all γ-elements can be so ordered that it will itself be a γ-set, and it contains all elements of the original set M.* M itself is thereby well-ordered.

I) If a and b are two arbitrary γ-elements and if M'_γ and M''_γ are any two γ-sets to which they respectively belong, then according to 5) the larger of the two γ-sets contains both elements and determines whether the order relation is $a \prec b$ or $b \prec a$. According to 6) this order relation is independent of the γ-sets selected.

II) If a, b and c are three arbitrary γ-elements and if $a \prec b$ and $b \prec c$, then always $a \prec c$. For according to 6) every γ-set containing c also contains b, hence also a, and then, since it is simply ordered, within the set, $a \prec c$ indeed follows from $a \prec b$ and $b \prec c$. The set L_γ is therefore *simply ordered.*

III) Ist L'_γ eine beliebige Teilmenge von L_γ und a eines ihrer Elemente, das der γ-Menge M_γ angehören möge, so enthält M_γ nach 6) alle Elemente $\prec a$, also auch die Teilmenge L''_γ, welche aus L'_γ durch Weglassung aller auf a folgenden Elemente entsteht, und L''_γ besitzt als Teilmenge der wohlgeordneten Menge M_γ ein *erstes* Element, das zugleich erstes Element von L'_γ ist. L_γ ist also auch *wohlgeordnet.*

IV) Ist a ein beliebiges γ-Element und A die Gesamtheit *aller* vorangehenden Elemente $x \prec a$, so ist A nach 6) der zu a gehörige Abschnitt in jeder Menge M_γ, welche a enthält, und a ist mithin nach 3) das ausgezeichnete Element von $M - A$. *Also ist L_γ selbst eine γ-Menge.*

V) Gäbe es ein Element von M, das *keiner* γ-Menge angehörte, also Element von $M - L_\gamma$ wäre, so gäbe es auch ein ausgezeichnetes Element m'_1
516 | von $M - L_\gamma$, und die geordnete Menge (L_γ, m'_1), in der jedes γ-Element dem Element m_1' voranginge, wäre nach 3) selbst eine γ-Menge. Also wäre auch m'_1 ein γ-Element gegen die Annahme, und es ist in Wirklichkeit $L_\gamma = M$, also M selbst eine *wohlgeordnete Menge.*

Somit entspricht jeder Belegung γ eine ganz bestimmte Wohlordnung der Menge M, wenn auch nicht zwei verschiedenen Belegungen immer verschiedene. Jedenfalls muß es *mindestens eine* solche Wohlordnung geben, und jede Menge, für welche die Gesamtheit der Teilmengen usw. einen Sinn hat, darf als eine wohlgeordnete, ihre Mächtigkeit als ein „Alef“ betrachtet werden. So folgt also für jede transfinite Mächtigkeit

$$\mathfrak{m} = 2\mathfrak{m} = \aleph_0\mathfrak{m} = \mathfrak{m}^2 \text{ usw.},$$

und je zwei Mengen sind miteinander „vergleichbar“, d. h. es ist immer die eine ein-eindeutig abbildbar auf die andere oder einen ihrer Teile.

Der vorliegende Beweis beruht auf der Voraussetzung, daß Belegungen γ überhaupt existieren, also auf dem Prinzip, daß es auch für eine unendliche Gesamtheit von Mengen immer Zuordnungen gibt, bei denen jeder Menge eines ihrer Elemente entspricht, oder formal ausgedrückt, daß das Produkt einer unendlichen Gesamtheit von Mengen, deren jede mindestens ein Element enthält, selbst von Null verschieden ist. Dieses logische Prinzip läßt sich zwar nicht auf ein noch einfacheres zurückführen, wird aber in der mathematischen Deduktion überall unbedenklich angewendet. So kann z. B. die Allgemeingültigkeit des Satzes, daß die Anzahl der Teile, in die eine Menge zerfällt, kleiner oder gleich ist der Anzahl aller ihrer Elemente, nicht anders bewiesen werden, als indem man sich jedem der betrachteten Teile eines seiner Elemente zugeordnet denkt.

Die Idee, unter Berufung auf dieses Prinzip eine *beliebige* Belegung γ der Wohlordnung zu grunde zu legen, verdanke ich Herrn *Erhard Schmidt*; meine Durchführung des Beweises beruht dann auf der Verschmelzung der verschiedenen möglichen „γ-Mengen“, d. h. der durch das Ordnungsprinzip sich ergebenden wohlgeordneten Abschnitte.

Münden i. Hann., den 24. September 1904.

III) If L'_γ is an arbitrary subset of L and a is one of its elements, belonging, say, to the γ-set M_γ, then according to 6) M_γ contains all elements preceding a, hence includes the subset L''_γ that is obtained from L'_γ when all elements following a are removed; L''_γ, being a subset of the well-ordered set M_γ, possesses a *first* element, which is also the first element of L'_γ. L_γ is therefore also *well-ordered.*

IV) If a is an arbitrary γ-element and A the totality of *all* preceding elements $x \prec a$, then according to 6), in every set M_γ containing a, A is the segment associated with a; according to 3), consequently, a is the distinguished element of $M - A$. *Therefore L_γ is itself a γ-set.*

V) If there existed an element of M that belonged to *no* γ-set, that consequently was an element of $M - L_\gamma$, there would also exist a distinguished element m'_1 of $M - L_\gamma$, and the ordered set (L_γ, m'_1), in which every γ-element precedes the element m'_1, would itself according to 3) be a γ-set. Then m'_1 too would be a γ-element, contrary to assumption; so really $L_\gamma = M$, and thus M is itself a *well-ordered set.*

Accordingly, to every covering γ there corresponds a definite well-ordering of the set M, even if the well-orderings that correspond to two distinct coverings are not always themselves distinct. There must at any rate exist *at least one* such well-ordering, and every set for which the totality of subsets, and so on, is meaningful may be regarded as well-ordered and its cardinality as an "aleph". It therefore follows that, for every transfinite cardinality,

$$\mathfrak{m} = 2\mathfrak{m} = \aleph_0\mathfrak{m} = \mathfrak{m}^2\,, \quad \text{and so forth;}$$

and any two sets are "comparable"; that is, one of them can always be mapped one-to-one onto the other or one of its parts.

The present proof rests upon the assumption that coverings γ actually do exist, hence upon the principle that even for an infinite totality of sets there are always mappings that associate with every set one of its elements, or, expressed formally, that the product of an infinite totality of sets, each containing at least one element, itself differs from zero. This logical principle cannot, to be sure, be reduced to a still simpler one, but it is applied without hesitation everywhere in mathematical deduction. For example, the validity of the proposition that the number of parts into which a set decomposes is less than or equal to the number of all of its elements cannot be proved except by associating with each of the parts in question one of its elements.

I owe to Mr. *Erhard Schmidt* the idea that, by invoking this principle, we can take an *arbitrary* covering γ as a basis for the well-ordering; the proof, as I carried it through, then rests upon the fusion of the various possible "γ-sets", that is, of the well-ordered segments resulting from the ordering principle.

Münden i. Hann., on the 24th of September 1904.

Neuer Beweis für die Möglichkeit einer Wohlordnung

1908a

Obwohl ich meinen im Jahre 1904 veröffentlichten „Beweis, daß jede Menge wohlgeordnet werden kann“* gegenüber den verschiedenen im § 2 ausführlich zu besprechenden Einwendungen noch heute vollkommen aufrecht erhalte, dürfte doch der hier folgende neue Beweis desselben Theorems nicht ohne Interesse sein, da er einerseits keine speziellen Lehrsätze der Mengentheorie voraussetzt, andererseits aber den rein formalen Charakter der Wohlordnung, die mit räumlich-zeitlicher Anordnung gar nichts zu tun hat, deutlicher als der erste Beweis hervortreten läßt.

§ 1.
Der neue Beweis

Die Voraussetzungen und Schlußformen, deren ich mich bei dem Beweise des nachstehenden Theorems bediene, lassen sich auf die folgenden Postulate zurückführen.

I. Alle diejenigen Elemente einer Menge M, denen eine für jedes einzelne Element wohldefinierte Eigenschaft $\mathfrak{E}$ zukommt, bilden die Elemente einer zweiten Menge $M_{\mathfrak{E}}$, einer „Untermenge“ von M.

Jeder Untermenge M_1 von M entspricht somit eine „komplementäre Untermenge“ $M - M_1$, welche alle in M_1 nicht vorkommenden Elemente von M umfaßt und sich für $M_1 = M$ auf die (leere) „Nullmenge“ reduziert.

II. Alle Untermengen einer Menge M, d. h. alle diejenigen Mengen M_1, deren Elemente gleichzeitig Elemente von M sind, bilden die Elemente einer durch M bestimmten Menge $\mathfrak{U}(M)$.

Aus dem Postulat I ergibt sich leicht der Satz

III. Alle diejenigen Elemente, welche den sämtlichen Mengen $A, B, C, \ldots$, den Elementen einer höheren Menge T, gemeinsam sind, bilden die Elemente
108 einer Menge $Q = \mathfrak{D}(T)$, die als der „Durchschnitt“ | oder als der „gemeinsame Bestandteil“ der Mengen $A, B, C, \ldots$ bezeichnet werden soll.

Theorem. *Ist durch irgend ein Gesetz jeder nicht verschwindenden Untermenge einer Menge M eines ihrer Elemente als „ausgezeichnetes Element“ zugeordnet, so besitzt die Menge $\mathfrak{U}(M)$ aller Untermengen von M eine und nur eine Untermenge M von der Beschaffenheit, daß jeder beliebigen Untermenge P von M immer ein und nur ein Element P_0 von M entspricht, welches P als Untermenge und ein Element von P als ausgezeichnetes Element enthält. Die Menge M wird durch M wohlgeordnet.*

---

* Math. Annalen, Bd. 59, p. 514.

A new proof of the possibility of a well-ordering

1908a

⟦The introductory note just before *1904* also addresses *1908a.*⟧

Although I still fully uphold my "Proof that every set can be well-ordered",[1] in the face of the various objections that will be thoroughly discussed in § 2, the new proof that I give below of the same theorem may yet be of interest, since, on the one hand, it presupposes no specific theorems of set theory and, on the other, it brings out, more clearly than the first proof did, the purely formal character of the well-ordering, which has nothing at all to do with spatio-temporal arrangement.

§ 1
The new proof

The assumptions and forms of inference that I use in the proof of the theorem below can be reduced to the following postulates.

I. All elements of a set M that have a property $\mathfrak{E}$ well-defined for every single element are the elements of another set, $M_{\mathfrak{E}}$, a "subset" of M.

Thus to every subset M_1 of M there corresponds a "complementary subset", $M - M_1$, that contains all elements not occurring in M_1 and, when $M_1 = M$, reduces to the (empty) "null set".

II. All subsets of a set M, that is, all sets M_1 whose elements are also elements of M, are the elements of a set $\mathfrak{U}(M)$ determined by M.

Postulate I easily yields the following proposition:

III. All elements that are common to all of the sets $A, B, C, \ldots$, these being elements of a higher set T, are the elements of a set $Q = \mathfrak{D}(T)$, which will be called the "intersection" or the "common component" of the sets $A, B, C, \ldots$.

Theorem. *If with every nonempty subset of a set M an element of that subset is associated by some law as "distinguished element", then $\mathfrak{U}(M)$, the set of all subsets of M, possesses one and only one subset M such that to every arbitrary subset P of M there always corresponds one and only one element P_0 of M that includes P as a subset and contains an element of P as its distinguished element. The set M is well-ordered by M.*

[1] *Zermelo 1904.*

Beweis. Ist A irgend eine nicht verschwindende Untermenge von M, also Element von $\mathfrak{U}(M)$ und $a = \varphi(A)$ ihr ausgezeichnetes Element, so sei $A' = A - \{a\}$ diejenige Teilmenge von A, welche durch Unterdrückung des ausgezeichneten Elementes entsteht. Nun besitzt die Menge $\mathfrak{U}(M)$ aller Untermengen von M folgende drei Eigenschaften:

1) sie enthält das Element M,

2) sie enthält mit jedem ihrer Elemente A auch das zugehörige A',

3) sie enthält mit jeder ihrer Untermengen $\mathsf{A} = \{A, B, C, \ldots\}$ auch den zugehörigen Durchschnitt $Q = \mathfrak{D}(\mathsf{A})$ als Element.

Wird jetzt eine solche Untermenge Θ von $\mathfrak{U}(M)$, welcher diese drei Eigenschaften ebenfalls zukommen, als eine „Θ-*Kette*" bezeichnet, so ergibt sich unmittelbar, daß der Durchschnitt mehrerer Θ-Ketten immer selbst eine Θ-Kette darstellt, und der Durchschnitt M aller existierenden Θ-Ketten, welche ja gemäß I und II die Elemente einer wohldefinierten Untermenge von $\mathfrak{U}\mathfrak{U}(M)$ bilden, ist somit die kleinste mögliche Θ-Kette; so daß keine echte Teilmenge von M eine Θ-Kette mehr sein kann.

Es sei nun A ein solches Element von M, daß in bezug auf A alle übrigen Elemente X von M in zwei Klassen zerfallen: 1) in Elemente U_A, welche Teilmengen von A sind, und 2) Elemente V_A, welche, wie z. B. M selbst, die Menge A als Teil umfassen. Dann ist, wie wir zeigen wollen, jedes U_A immer von der Beschaffenheit W_A, nämlich eine Untermenge von $A' = A - \{\varphi(A)\}$. In der Tat ist jedes V'_A, da es kein U_A sein kann und doch Element von M sein muß, entweder A selbst oder ein V_A, und jeder Durchschnitt mehrerer V_A wieder ein V_A oder A. Andererseits ist A' sowie jedes W'_A wieder ein W_A, und ebenso jeder Durchschnitt mehrerer W_A sowie der Durchschnitt einiger W_A und einiger V_A oder A wieder ein W_A. Somit bilden die W_A mit den V_A und A zusammen schon eine Θ-Kette, sie erschöpfen also die kleinste Θ-Kette M, und jedes U_A ist wirklich ein W_A, d. h. Untermenge von A'. Hieraus folgt aber unmittelbar, daß auch A' dieselbe Eigenschaft hat wie A, d. h. daß alle anderen Elemente von M entweder Teile von A' sind oder A' als Teil enthalten. Ist endlich Q der Durchschnitt mehrerer $A, B, C, \ldots$ von
109 der soeben für A vorausge- | setzten Beschaffenheit und X irgend ein anderes Element von M, so sind nur zwei Fälle möglich: entweder enthält X eine der Mengen $A, B, C, \ldots$ und damit auch Q als Teil, oder X ist in allen $A, B, C, \ldots$ und damit auch in Q als Untermenge enthalten, d. h. auch Q besitzt die genannte Eigenschaft von A. Da endlich M sämtliche Elemente von M als Untermengen umschließt und daher selbst ein A darstellt, so bilden die wie A beschaffenen Elemente von M wieder eine Θ-Kette, nämlich M selbst, und für zwei beliebige Elemente A und B von M gilt die Alternative, daß entweder B Untermenge von A' oder A Untermenge von B' sein muß.

Jetzt sei P eine beliebige Untermenge von M, und P_0 der Durchschnitt aller solchen Elemente von M, welche P als Untermenge enthalten, und zu denen jedenfalls das Element M gehört. Dann ist auch P_0 Element von M, und das ausgezeichnete Element p_0 von P_0 muß ein Element von P sein, weil sonst auch ${P_0}' = P_0 - \{p_0\}$ alle Elemente von P enthielte und doch nur ein Teil

Proof. If A is any nonempty subset of M and hence an element of $\mathfrak{U}(M)$, and if $a = \varphi(A)$ is its distinguished element, let $A' = A - \{a\}$ be the part of A that results when the distinguished element is removed. Now $\mathfrak{U}(M)$, the set of all subsets of M, possesses the following three properties:

1) It contains the element M;

2) Along with each of its elements A it also contains the corresponding A';

3) Along with each of its subsets $\mathsf{A} = \{A, B, C, \ldots\}$ it also contains the corresponding intersection $Q = \mathfrak{D}(\mathsf{A})$ as an element.

If now a subset Θ of $\mathfrak{U}(M)$ that also has these three properties is called a "Θ-*chain*", it immediately follows that the intersection of several Θ-chains is itself always a Θ-chain, and the intersection M of all existing Θ-chains, which according to I and II are the elements of a well-defined subset of $\mathfrak{U}\mathfrak{U}(M)$, is therefore the smallest possible Θ-chain; therefore no proper partial set of M can be a Θ-chain any longer.

Now let A be an element of M such that all other elements X of M fall into two classes with respect to A: 1) elements U_A that are partial sets of A, and 2) elements V_A that include the set A as a partial set, as for instance M itself does. Then, as we shall now show, every U_A always has the property of being a W_A; that is to say, it is a subset of $A' = A - \{\varphi(A)\}$. In fact, every V_A', since it cannot be a U_A and yet must be an element of M, is either A itself or a V_A, and every intersection of several V_A is again a V_A or A. On the other hand, A', as well as every W_A', is again a W_A; and likewise every intersection of several W_A, as well as the intersection of some W_A and some V_A or A, is again a W_A. Thus the W_A together with the V_A and A already form a Θ-chain; they therefore exhaust the smallest Θ-chain M, and every U_A is actually a W_A, that is, a subset of A'. But from this it immediately follows that A', too, has the same property as A, namely, that all other elements of M are either partial sets of A' or include A' as a partial set. If, finally, Q is the intersection of several $A, B, C, \ldots$ that have the property just assumed of A and if X is any other element of M, then only two cases are possible: either X includes one of the sets $A, B, C, \ldots$, and therewith also Q, as a partial set, or X is included in all of the sets $A, B, C, \ldots$, and therewith also in Q, as a subset, that is, Q too possesses the above-mentioned property of A. Since, finally, M includes all elements of M as subsets and therefore is itself an A, the elements of M that are constituted like A again form a Θ-chain, namely, M itself, and for two arbitrary ⟦distinct⟧ elements A and B of M the alternative holds that either B must be a subset of A' or A a subset of B'.

Now let P be an arbitrary subset of M,[2] and let P_0 be the intersection of all elements of M that include P as a subset and to which at least the element M belongs. Then P_0 also is an element of M, and the distinguished element p_0 of P_0 must be an element of P, since otherwise $P_0' = P_0 - \{p_0\}$ also would contain all elements of P and would still only be a partial set of P_0. Every other element P_1 of M that includes P as a subset must then

[2] ⟦Zermelo erroneously writes "M" instead of "M".⟧

von P_0 wäre. Jedes andere, P als Untermenge enthaltende Element P_1 von M muß dann P_0 als Teil umfassen, d. h. P_0 ist nach dem soeben Bewiesenen eine Untermenge von $P_1{}'$, und das ausgezeichnete Element p_1 von P_1 kann, da es in $P_1{}'$ und somit auch in P_0 nicht vorkommt, kein Element von P sein. Es gibt also in der Tat nur ein einziges Element P_0 von M, welches P als Untermenge und ein Element von P als ausgezeichnetes Element enthält.

Wählt man hier für P eine Menge der Form $\{a\}$, wo a irgend ein Element von M ist, so ergibt sich im besonderen, daß jedem Elemente a von M ein einziges Element A von M entspricht, in welchem a ausgezeichnetes Element ist, und welches mit $\mathfrak{R}(a)$ bezeichnet werden möge. Sind a, b irgend zwei verschiedene Elemente von M, so ist entweder $\mathfrak{R}(a)$ oder $\mathfrak{R}(b)$ das der Menge $P = \{a, b\}$ entsprechende Element P_0 von M, d. h. entweder enthält $\mathfrak{R}(a)$ das Element b oder $\mathfrak{R}(b)$ das Element a, aber niemals tritt beides gleichzeitig ein. Sind endlich a, b, c irgend drei Elemente von M, und ist etwa b Element von $\mathfrak{R}(a)$ und c Element von $\mathfrak{R}(b)$, so kann nur $\mathfrak{R}(a)$ das der Menge $P = \{a, b, c\}$ entsprechende Element P_0 sein, d. h. es ist auch c Element von $\mathfrak{R}(a)$. Schreibt man also $a \prec b$ für den Fall, wo b Element von $\mathfrak{R}(a)$ und $a \neq b$ ist, und sagt dann, das Element a „gehe dem Element b voran", so ergibt sich die Trichotomie:

$$a \prec b\,, \quad a = b\,, \quad b \prec a$$

für irgend zwei Elemente a, b, und aus

$$a \prec b \quad \text{und} \quad b \prec c$$

folgt immer $a \prec c$.

Die Menge M wird also vermittels der Menge M „einfach geordnet" und zwar im Cantorschen Sinne „wohlgeordnet"; denn jeder Untermenge P von
110 M entspricht ein „erstes Element", nämlich das ausgezeichnete | Element p_0 von $P_0 = \mathfrak{R}(p_0)$, welches allen übrigen Elementen p von P „vorangeht", weil alle diese p Elemente von P_0 sind.

Ist umgekehrt die Menge M auf irgend eine Weise wohlgeordnet, so entspricht jedem Elemente a von M eine bestimmte Untermenge $\mathfrak{R}(a)$ von M, welche außer a alle „auf a folgenden" Elemente enthält und als der zu a gehörende „Rest" bezeichnet werden möge. Unterdrückt man in einem solchen Reste $\mathfrak{R}(a)$ das „erste Element" a, so verbleibt der „Rest" des „nächstfolgenden" Elementes a'. Ebenso ist der gemeinsame Bestandteil oder „Durchschnitt" mehrerer Reste immer wieder ein Rest, und schließlich ist die ganze Menge M der Rest $\mathfrak{R}(e)$ ihres ersten Elementes. Somit stellt die Gesamtheit aller Reste in dem oben angegebenen Sinne eine Θ-Kette dar, in welcher das erste Element jedes Restes als „ausgezeichnetes Element" figuriert. Besäße nun $\mathfrak{U}(M)$ außer M eine zweite Untermenge M_1 von der im Theorem geforderten Beschaffenheit, so bestimmte auch M_1 eine Wohlordnung von M mit denselben ausgezeichneten Elementen und müßte daher als Θ-Kette den Durchschnitt M aller Θ-Ketten als Bestandteil enthalten. Bedeutet dann z_0

include P_0 as a partial set; that is, P_0, according to what has just been proved, is a subset of P_1', and the distinguished element p_1 of P_1 cannot be an element of P, since it does not occur in P_1' and hence not in P_0 either. So there really exists only a single element P_0 of M that includes P as a subset and contains an element of P as its distinguished element.

If we here choose for P a set of the form $\{a\}$, where a is any element of M, it follows in particular that to every element a of M there corresponds a single element A of M in which a is the distinguished element; let this element A be denoted by $\mathfrak{R}(a)$. If a and b are any two distinct elements of M, then either $\mathfrak{R}(a)$ or $\mathfrak{R}(b)$ is the element P_0 of M corresponding to the set $P = \{a, b\}$, that is, either $\mathfrak{R}(a)$ contains the element b or $\mathfrak{R}(b)$ the element a, but never both. If, finally, a, b, and c are any three ⟦distinct⟧ elements of M and if, say, b is an element of $\mathfrak{R}(a)$ and c an element of $\mathfrak{R}(b)$, then only $\mathfrak{R}(a)$ can be the element P_0 corresponding to the set $P = \{a, b, c\}$, that is, c also is an element of $\mathfrak{R}(a)$. Therefore, if we write $a \prec b$ when b is an element of $\mathfrak{R}(a)$ and $a \neq b$ (we then say that the element a "precedes the element b") the trichotomy

$$a \prec b\,, \quad a = b\,, \quad \text{or} \quad b \prec a$$

obtains for any two elements a and b, and from

$$a \prec b \quad \text{and} \quad b \prec c$$

it always follows that $a \prec c$.

The set M, therefore, is "simply ordered" by means of the set M, and, moreover, it is "well-ordered" in the sense of Cantor; for to every subset P of M there corresponds a "first element", namely, the distinguished element p_0 of $P_0 = \mathfrak{R}(p_0)$, which "precedes" all other elements p of P, since all these p are elements of P_0.

If, conversely, the set M is well-ordered in any way, then to every element a of M there corresponds a certain subset $\mathfrak{R}(a)$ of M that contains, besides a, all elements "following a"; let us call it the "remainder" associated with a. If from such a remainder $\mathfrak{R}(a)$ we remove the "first element" a, what is left is the "remainder" of the "next" element a'. Likewise, the common component, or "intersection", of several remainders is always again a remainder, and, finally, the entire set M is the remainder $\mathfrak{R}(e)$ of its first element. Thus the totality of all remainders, in the sense specified above, forms a Θ-chain, in which for every remainder the first element is the "distinguished element". If now $\mathfrak{U}(M)$ were to include, besides M, a second subset M_1 constituted as required by the theorem, then M_1 also would determine a well-ordering of M with the same distinguished elements and would therefore, as a Θ-chain, include the intersection M of all Θ-chains as a component. If z_0 then

das ausgezeichnete Element eines Elementes Z von $\mathsf{M}_1 - \mathsf{M}$, so wäre z_0 ausgezeichnetes Element in zwei Elementen von M_1, außer in Z nämlich noch in dem durch M bestimmten $\mathfrak{R}(z_0)$, und dies widerspräche der vorausgesetzten Beschaffenheit von M_1. In Wirklichkeit ist also die Wohlordnung M durch die Wahl der ausgezeichneten Elemente eindeutig bestimmt, und der behauptete Satz ist in allen seinen Teilen bewiesen.

Um nun unser Theorem auf beliebige Mengen anzuwenden, bedürfen wir nur noch der Voraussetzung, daß *die gleichzeitige Auswahl der ausgezeichneten Elemente für eine beliebige Menge von Mengen prinzipiell immer möglich ist*, oder präziser, daß immer dieselben Folgerungen gelten, als ob diese Auswahl möglich wäre. In dieser Formulierung erscheint das zugrunde liegende Prinzip freilich immer noch etwas subjektiv gefärbt und Mißdeutungen ausgesetzt. Da man aber, wie ich an anderer Stelle ausführlicher darlegen werde, vermittels der elementaren und unentbehrlichen mengentheoretischen Prinzipien eine beliebige Menge T' von Mengen $A', B', C', \ldots$ immer durch eine Menge T unter sich elementenfremder Mengen $A, B, C, \ldots$ ersetzen kann, die den Mengen $A', B', C', \ldots$ bezüglich äquivalent sind, so läßt sich das allgemeine „Prinzip der Auswahl" auf das folgende Axiom zurückführen, dessen rein objektiver Charakter unmittelbar einleuchtet.

IV. Axiom. *Eine Menge S, welche in eine Menge getrennter Teile $A, B, C, \ldots$ zerfällt, deren jeder mindestens ein Element enthält, besitzt mindestens eine Untermenge S_1, welche mit jedem der betrachteten Teile $A, B, C, \ldots$ genau ein Element gemein hat.*

111 | Unter Anwendung dieses Axioms ergibt sich somit wie in meiner Note von 1904 der allgemeine Satz, *daß jede Menge einer Wohlordnung fähig ist.*

Die unserem neuen Beweise zugrunde liegende Definition der Wohlordnung, wie sie bereits bei der Formulierung des „Theorems" in Erscheinung trat, hat den Vorzug, ausschließlich auf den Elementarbegriffen der Mengenlehre zu beruhen, während bei der üblichen Darstellung, wie die Erfahrung lehrt, Unkundige nur allzu geneigt sind, hinter der unvermittelt auftretenden Cantorschen Beziehung $a \prec b$ irgend einen mystischen Inhalt zu suchen. Unsere Definition möge hier nochmals ausdrücklich formuliert werden, wie folgt:

Definition. *Eine Menge M heißt „wohlgeordnet", wenn jedem ihrer Elemente a eine Untermenge $\mathfrak{R}(a)$ von M als „Rest" eindeutig entspricht, und wenn jede nicht verschwindende Untermenge P von M ein und nur ein „erstes Element" d. h. ein solches Element p_0 enthält, dessen Rest $\mathfrak{R}(p_0)$ die Menge P als Untermenge umfaßt.*

denotes the distinguished element of an element Z of $\mathsf{M}_1 - \mathsf{M}$, z_0 would be the distinguished element of two elements of M_1—namely, of Z as well as of the $\mathfrak{R}(z_0)$ determined by M—and this would contradict the property assumed of M_1. Therefore, the well-ordering M is indeed uniquely determined by the choice of distinguished elements, and the theorem asserted is proved in its entirety.

Now in order to apply our theorem to arbitrary sets, we require only the additional assumption that *a simultaneous choice of distinguished elements is in principle always possible for an arbitrary set of sets*, or, to be more precise, that the same consequences always hold as if such a choice were possible. In this formulation, to be sure, the principle taken as fundamental still appears to be somewhat tainted with subjectivity and liable to misinterpretation. But since, as I shall show in more detail elsewhere,[3] we can, by means of elementary and indispensable set-theoretic principles, always replace an arbitrary set T' of sets $A', B', C', \ldots$ by a set T of mutually disjoint sets $A, B, C, \ldots$ that are equivalent to the sets $A', B', C', \ldots$, respectively, the "general principle of choice" can be reduced to the following axiom, whose purely objective character is immediately evident.

IV. Axiom. *A set S that can be decomposed into a set of disjoint parts $A, B, C, \ldots$, each containing at least one element, possesses at least one subset S_1 having exactly one element in common with each of the parts $A, B, C, \ldots$ considered.*

Then the application of this axiom, just as in my note of 1904, yields the general theorem that *every set can be well-ordered.*

The definition of well-ordering that has already appeared in the formulation of the "Theorem" and forms the basis of our new proof has the advantage that it rests exclusively upon the elementary notions of set theory, whereas experience shows that, with the usual presentation, the uninformed are only too prone to look for some mystical meaning behind Cantor's relation $a \prec b$, which is suddenly introduced. Let us now once more formulate our definition explicitly, as follows.

Definition. *A set M is said to be "well-ordered" if to any element a of M there corresponds a unique subset $\mathfrak{R}(a)$ of M, the "remainder" of a, and every nonempty subset P of M contains one and only one "first element", that is, an element p_0 such that its remainder $\mathfrak{R}(p_0)$ includes the set P as a subset.*

---

3 ⟦*Zermelo 1908b*, p. 273, no. 28.⟧

§ 2.
Diskussion der Einwände gegen den früheren Beweis

Seit 1904 sind gegen meinen damaligen „Beweis, daß jede Menge wohlgeordnet werden kann“, eine Reihe von Einwendungen gemacht und Kritiken veröffentlicht worden, die bei dieser Gelegenheit einmal im Zusammenhange zur Sprache kommen mögen.

a. Einwände gegen das Auswahlprinzip

An erster Stelle stehen hier diejenigen Einwände, welche sich gegen das oben formulierte „Auswahlpostulat“ richten und somit meine beiden Beweise in gleicher Weise treffen. Ihnen kann ich insofern eine relative Berechtigung einräumen, als ich dieses Postulat, wie ich am Ende meiner Note ausdrücklich hervorhob[*], eben nicht *beweisen* und daher niemand apodiktisch zu seiner Anerkennung zwingen kann. Indem also die Herren *E. Borel*[**] und *G. Peano*[***] in ihren Kritiken den Mangel eines Beweises konstatierten, haben sie sich lediglich auf meinen eigenen Standpunkt gestellt. Sie hätten mich sogar
112 zu Dank verpflichtet, wenn sie | die von mir behauptete Unbeweisbarkeit, d. h. die logische Unabhängigkeit dieses Postulates von den übrigen, nun ihrerseits bewiesen und damit meine Überzeugung bestätigt hätten.

Nun ist *Unbeweisbarkeit* auch in der Mathematik bekanntlich keineswegs gleichbedeutend mit *Ungültigkeit*, da doch eben nicht alles bewiesen werden kann, sondern jeder Beweis wieder unbewiesene Prinzipien voraussetzt. Um also ein solches Grundprinzip zu verwerfen, hätte man seine Ungültigkeit in besonderen Fällen oder widersprechende Konsequenzen feststellen müssen; aber hierzu hat keiner meiner Gegner einen Versuch gemacht.

Auch dem „Formulaire“[*] des Herrn *Peano*, welcher die gesamte Mathematik auf „Syllogismen“ (im aristotelisch-scholastischen Sinne) zurückführen soll[**], liegen eine ganze Anzahl unbeweisbarer Prinzipien zugrunde, und eines darunter, welches dem „Auswahlprinzip“ für eine einzige Menge äquivalent ist und dann syllogistisch auf eine beliebige endliche Anzahl von Mengen ausgedehnt werden kann[***]. Aber das allgemeine Axiom, das ich mir nach dem

---

[*] Math. Annalen, Bd. 59, p. 516. „Dieses logische Prinzip läßt sich zwar nicht auf ein noch einfacheres zurückführen —“.

[**] Math. Annalen, Bd. 60, p. 194; vergl. aber auch *Hadamard*, *Borel*, *Baire*, *Lebesgue* „Cinq lettres sur la théorie des ensembles“, Bulletin de la Société Mathématique de France, t. 33, p. 261.

[***] Rivista di Matematica VIII, Nr. 5, p. 145 ff.

[*] „Formulaire de Mathématiques publié par la Revue de Mathématiques“ Tome II, Turin 1897.

[**] Rivista di Matematica VIII, Nr. 5, p. 147.

[***] ibid. p. 145–147. Übrigens gelingt dieser Beweis nur durch „vollständige Induktion“, ist also nur bindend, wenn man die endlichen Zahlen in Peanoscher Weise durch ihren Ordnungstypus definiert. Legt man dagegen die Dedekind-

§ 2
Discussion of the objections to the earlier proof

Since 1904, the date of my "Proof that every set can be well-ordered", a number of objections have been made to it and various critiques of it have been published. Let me take this opportunity to discuss them together.

a. Objections to the principle of choice

In first place we here consider the objections that are directed against the "postulate of choice" formulated above and therefore strike at both of my proofs in the same way. I concede that they are to some extent justified, since I just cannot *prove* this postulate, as I expressly emphasized at the end of my note,[4] and therefore cannot compel anyone to accept it apodictically. Hence if Mr. *E. Borel*[5] and Mr. *G. Peano*[6] in their critiques note the lack of a proof, they have merely adopted my own point of view. They would even have put me in their debt had they now for their part established the unprovability I asserted—namely, that this postulate is logically independent of the others—thereby corroborating my conviction.

Now even in mathematics *unprovability*, as is well known, is in no way equivalent to *nonvalidity*, since, after all, not everything can be proved, but every proof in turn presupposes unproved principles. Thus, in order to reject such a fundamental principle, one would have had to ascertain that in some particular case it did not hold or to derive contradictory consequences from it; but none of my opponents has made any attempt to do this.

Even Mr. *Peano*'s *Formulaire*,[7] which is an attempt to reduce all of mathematics to "syllogisms" (in the Aristotelian–Scholastic sense),[8] rests upon quite a number of unprovable principles; one of these is equivalent to the "principle of choice" for a single set and can then be extended syllogistically to an arbitrary finite number of sets.[9] But the general axiom that, following

---

[4] *Zermelo 1904*, p. 516. "This logical principle cannot, to be sure, be reduced to a still simpler one—".

[5] *Borel 1905a*, but see also *Borel 1905b*.

[6] *Peano 1906b*, pp. 145–148.

[7] *Peano 1897*.

[8] See *Peano 1906b*, p. 147.

[9] See *Peano 1906b*, pp. 145–147. This proof, incidentally, can be carried out only by mathematical induction; hence it is binding only if we define the finite numbers in Peano's way by means of their order type. If on the other hand we take as a basis Dedekind's definition of a finite set as one that is not equivalent to any of its parts, no proof is possible even for finite sets, since the reduction of the two definitions to each other, as we shall show below (example 4), again requires the principle of choice. In this sense, therefore, *Poincaré*'s remark (*1906b,* p. 313) is justified.

Vorgange anderer Forscher in diesem neuen Falle auf beliebige Mengen anzuwenden erlaubte, findet sich eben nicht unter den Peanoschen Prinzipien, und Herr *Peano* versichert selbst, daß er es auch nicht aus ihnen herleiten könne. Er begnügt sich damit, dies festzustellen, und das Prinzip ist für ihn erledigt. Der Gedanke, daß möglicherweise sein Formulaire gerade in diesem Punkte unvollständig sein könnte, liegt doch nahe, und da es in der Mathematik keine unfehlbaren Autoritäten gibt, so haben wir auch mit dieser Möglichkeit zu rechnen und sie nicht ohne objektive Prüfung von der Hand zu weisen.

Zunächst, wie gelangt denn Herr *Peano* zu seinen eigenen Grundprinzipien und wie rechtfertigt er ihre Aufnahme in den Formulaire, da er sie doch gleichfalls nicht beweisen kann? Offenbar durch Analyse der historisch als gültig anerkannten Schlußweisen und durch den Hinweis auf die anschauliche Evidenz der Prinzipien und auf das wissenschaftliche Bedürfnis — alles
113 Gesichtspunkte, die sich für das bestrittene Prinzip ebenso- | gut geltend machen lassen. Daß dieses Axiom, ohne gerade schulmäßig formuliert zu sein, auf den verschiedensten mathematischen Gebieten, besonders aber in der Mengenlehre von *R. Dedekind*, *G. Cantor*, *F. Bernstein*, *A. Schoenflies*, *J. König* u. a. mit Erfolg sehr häufig angewendet worden ist, ist eine unbestreitbare Tatsache, welche durch die frühere gelegentliche Opposition einiger logischen Puristen nur bestätigt wird. Eine so weitgehende Anwendung eines Prinzips ist nur erklärlich durch seine *Evidenz*, welche mit Beweisbarkeit natürlich nicht verwechselt werden darf. Mag diese Evidenz auch bis zu einem gewissen Grade subjektiv sein, so ist sie doch jedenfalls eine notwendige Quelle mathematischer Prinzipien, wenn auch kein Gegenstand mathematischer Beweise, und die Behauptung *Peanos*,* daß sie mit Mathematik nichts zu tun habe, wird offenbaren Tatsachen nicht gerecht. Was sich aber objektiv entscheiden läßt, die Frage nach dem *wissenschaftlichen Bedürfnis*, möchte ich hier in der Weise der Beurteilung unterbreiten, daß ich eine Reihe von elementaren und grundlegenden Sätzen und Problemen vorlege, welche meines Erachtens ohne das Auswahlprinzip überhaupt nicht erledigt werden könnten.

1) Wenn eine Menge in getrennte Teile $A, B, C, \ldots$ zerfällt, so ist die Menge dieser Teile einer Untermenge von M äquivalent, oder anders ausgedrückt: die Menge der Summanden ist immer von kleinerer oder der gleichen Mächtigkeit wie die Summe.

Zum Beweise muß man sich jedem dieser Teile eines seiner Elemente zugeordnet denken.**

---

sche Definition der endlichen Menge als einer solchen, welche keinem ihrer Teile äquivalent ist, zugrunde, so ist ein Beweis auch für endliche Mengen unmöglich, da die Zurückführung der beiden Definitionen aufeinander, wie wir unten (Beispiel 4) zeigen werden, wieder das Auswahlprinzip erfordert. In diesem Sinne ist also *Poincarés* Bemerkung (Revue de Métaphysique et de Morale 14, p. 313) gerechtfertigt.

* Rivista di Matematica VIII, Nr. 5, p. 147.

** Daß hier ein besonderes Schlußprinzip zugrunde liegt, wurde anläßlich eines Bernsteinschen Beweises wohl zuerst von Herrn *Beppo Levi* ausgesprochen

other researchers, I permitted myself to apply to arbitrary sets in this new case just is not to be found among Peano's principles, and *Peano* himself assures us that he could not derive it from them either. He is content to note this fact, and that finishes the principle for him. The idea that possibly his Formulaire might be incomplete in precisely this point does, after all, suggest itself, and, since there are no infallible authorities in mathematics, we must also take that possibility into account and not reject it without objective examination.

First, how does Mr. *Peano* arrive at his own fundamental principles and how does he justify their inclusion in the Formulaire, since, after all, he cannot prove them either? Evidently by analyzing the modes of inference that in the course of history have come to be recognized as valid and by pointing out that the principles are intuitively evident and necessary for science—considerations that can all be urged equally well in favor of the disputed principle. That this axiom, even though it was never formulated in textbook style, has frequently been used, and successfully at that, in the most diverse fields of mathematics, especially in set theory, by *R. Dedekind, G. Cantor, F. Bernstein, A. Schoenflies, J. König,* and others is an indisputable fact, which is only corroborated by the opposition that, at one time or another, some logical purists directed against it. Such an extensive use of a principle can be explained only by its *self-evidence*, which, of course, must not be confused with its provability. No matter if this self-evidence is to a certain degree subjective—it is surely a necessary source of mathematical principles, even if it is not a tool of mathematical proofs, and *Peano*'s assertion[10] that it has nothing to do with mathematics fails to do justice to manifest facts. But the question that can be objectively decided, whether the principle is *necessary for science*, I should now like to submit to judgment by presenting a number of elementary and fundamental theorems and problems that, in my opinion, could not be dealt with at all without the principle of choice.

1) If a set M can be decomposed into disjoint parts, $A, B, C, \ldots$, the set of these parts is equivalent to a subset of M, or, in other words, the set of summands always has a cardinality lower than, or the same as, that of the sum.

To prove this we must mentally associate with each of these parts one of its elements.[11]

---

[10] *Peano 1906b,* p. 147.

[11] That a particular principle of inference is used here was probably first stated by Mr. *Beppo Levi* (*Levi 1902*, p. 863) in connection with a proof by Bernstein. According to Mr. *F. Bernstein* (*Bernstein 1905b,* p. 193), however, the "hypothesis" that a choice is possible is said to be "dispensable" in all similar cases, for instance also in my proof, if one employs the notion "multivalued equivalence" that he introduces. According to him (*Bernstein 1904*) two sets M and N are said to stand in the relation of multivalued equivalence if an entire set A of one-to-one mappings $\varphi, \chi, \psi, \ldots$, "among which none is distinguished", is given for them instead of just a single one. Hence a pure relational notion such as "dis-

2) Die Summen äquivalenter Mengen sind wieder äquivalent, vorausgesetzt, daß alle Summanden unter sich elementenfremd sind, ein Satz, auf dem der ganze Kalkül mit Mächtigkeiten beruht.

114 | Hier ist es erforderlich, ein System von Abbildungen zu betrachten, welche *gleichzeitig* je zwei äquivalente Summanden aufeinander beziehen; man hat also aus den sämtlichen möglichen Abbildungen, welche zu je zwei äquivalenten Summanden gehören, jedesmal eine einzige auszuwählen.

3) Das Produkt mehrerer Mächtigkeiten kann nur verschwinden, wenn ein Faktor verschwindet, d. h. die Cantorsche „Verbindungsmenge" mehrerer Mengen $A, B, C, \ldots$, deren jede mindestens ein Element enthält, muß gleichfalls mindestens ein Element enthalten. Da aber jedes solche Element eine Menge ist, welche mit jeder der Mengen $A, B, C, \ldots$ gerade ein Element gemein hat, so ist der Satz nur ein anderer Ausdruck des Auswahlpostulates für elementenfremde Mengen (IV. Axiom § 1 fin.).

4) Eine Menge, welche keinem ihrer Teile äquivalent ist, läßt sich immer so ordnen, daß jede Untermenge sowohl ein erstes, als auch ein letztes Element besitzt.

Diesen Satz, auf dem die Theorie der *endlichen* Mengen beruht, beweist man am einfachsten mittelst meines Wohlordnungstheorems. Herr *R. Dedekind* bewies den logisch gleichwertigen Satz, daß eine Menge, welche keinem Abschnitte seiner „Zahlenreihe" äquivalent ist, einen der ganzen Zahlenreihe äquivalenten Bestandteil enthalten muß,* durch simultane Abbildung eines Systems äquivalenter Mengenpaare, also wie hier in 2) gleichfalls mit Hilfe des Auswahlprinzipes.** Weitere Beweise sind mir nicht bekannt.

5) Eine abzählbare Menge von endlichen oder abzählbaren Mengen besitzt immer eine abzählbare Summe.

(Lomb. Ist. Rend. (2) 35, 1902, p. 863). Nach Herrn *F. Bernstein* (Math. Annalen Bd. 60, p. 193) soll allerdings in allen ähnlichen Fällen, z. B. auch in meinem Beweise, die „Hypothese" der möglichen Auswahl „entbehrlich" sein, wenn man den von ihm eingeführten Begriff der „vielwertigen Äquivalenz" benutzt. Zwei Mengen M, N sollen (Gött. Nachr. Math. Phys. 1904, Heft 6) „vielwertig äquivalent" heißen, wenn für sie statt einer einzigen eine ganze Menge A von ein-eindeutigen Abbildungen $\varphi, \chi, \psi, \ldots$ gegeben ist, „unter denen keine ausgezeichnet ist". Hier wird also ein reiner Beziehungsbegriff wie „ausgezeichnet" ohne ergänzende Bestimmung oder Erklärung wie ein absolutes Merkmal verwendet, und die versuchte Unterscheidung von der gewöhnlichen Äquivalenz ist logisch nicht durchführbar. In den betrachteten Beispielen handelt es sich aber auch gar nicht um die „Multiplizität", d. h. um die *Mächtigkeit* der Abbildungsmenge A, sondern lediglich um die Frage, ob *mindestens eine* solche Abbildung φ existiert, eine Frage, die hier durch keine Definition umgangen sondern nur durch ein Axiom entschieden werden kann.

* „Was sind und was sollen die Zahlen?" Nr. 159.

** Diese vielfach übersehene Tatsache wird auch von Herrn *G. Hessenberg* im Vorworte seiner „Grundbegriffe der Mengenlehre" (Göttingen 1906) ausdrücklich anerkannt.

2) The sums of equivalent sets are again equivalent, provided all terms are mutually disjoint, a theorem upon which the entire calculus of cardinalities rests.

Here it is necessary to consider a system of mappings that *simultaneously* correlate any two equivalent summands with each other; thus from all the possible mappings associated with each pair of equivalent summands we must choose a single one, and we must make this choice for every pair.

3) The product of several cardinalities can vanish only if one factor vanishes, that is, Cantor's "connection set" of several sets $A, B, C, \ldots$, each containing at least one element, must likewise contain at least one element. But since every such element is a set having exactly one element in common with each of the sets $A, B, C, \ldots$, the theorem is merely another expression of the postulate of choice for disjoint sets (the axiom in IV, end of § 1 above).

4) A set that is not equivalent to any of its parts can always be ordered in such a way that every subset possesses a first as well as a last element.

This theorem, upon which the theory of *finite* sets rests, is most simply proved by means of my well-ordering theorem. *R. Dedekind*[12] proved a logically equivalent theorem—that a set not equivalent to any segment of his "number sequence" must have a component equivalent to the entire number sequence—by simultaneously mapping a system of equivalent pairs of sets, hence, as in (2) above, also by using the principle of choice.[13] I do not know of any other proof.

5) A denumerable set of finite or denumerable sets always possesses a denumerable sum.

---

tinguished" is used here, without supplementary determination or definition, as an absolute characteristic, and the attempt to differentiate between multivalued and ordinary equivalence is logically not realizable. In the examples considered, however, we are not at all concerned with the "multiplicity", that is, the *cardinality* of the set A of mappings, but merely with the question whether *at least one* such mapping φ exists, a question that cannot be evaded here by any definition and can be settled only by means of an axiom.

[12] *Dedekind 1888,* art. 159.

[13] This fact, which has frequently been overlooked, is expressly acknowledged also by *G. Hessenberg* (*1906*, preface).

Auf diesem Satz beruht die Theorie der abzählbaren Mengen und der „zweiten Zahlenklasse"; er läßt sich aber nur beweisen, indem man die sämtlichen betrachteten endlichen oder abzählbaren Mengen *gleichzeitig* nach dem Normaltypus ordnet.

6) Gibt es eine „Basis aller reellen Zahlen", d. h. ein System von reellen Zahlen, zwischen welchen keine linearen Relationen mit einer endlichen Anzahl ganzzahliger Koeffizienten bestehen, und aus welchen sich alle übrigen linear mit endlichvielen ganzzahligen Koeffizienten zusammensetzen lassen?

7) Gibt es unstetige Lösungen der Funktionalgleichung

$$f(x+y) = f(x) + f(y)\,?$$

115 Diese beiden letzten Fragen sind von Herrn *G. Hamel**** auf | Grund der möglichen Wohlordnung des Kontinuums in bejahendem Sinne entschieden worden.

Die *Cantor*sche Theorie der Mächtigkeiten bedarf also jedenfalls unseres Postulates, ebenso die *Dedekind*sche Theorie der endlichen Mengen, welche die Grundlage der Arithmetik bildet. Die Tatsache, daß man in der Funktionentheorie seine Anwendung gewöhnlich umgehen kann, erklärt sich einfach durch den Umstand, daß man es dort in der Regel mit „abgeschlossenen" Mengen zu tun hat, bei welchen die eindeutige Definition ausgezeichneter Elemente keine Schwierigkeit bietet. Wo dies nicht der Fall ist, also namentlich in der Theorie der durchweg unstetigen Funktionen, wird das Prinzip oft unentbehrlich, wie unser letztes Beispiel zeigt.

Solange nun die hier vorgelegten relativ einfachen Probleme den Hilfsmitteln *Peanos* unzugänglich bleiben, und solange andererseits das Auswahlprinzip nicht positiv widerlegt werden kann, wird man die Vertreter der produktiven Wissenschaft nicht hindern dürfen, sich dieser „Hypothese", wie man es meinetwegen nennen möge, fernerhin zu bedienen und ihre Konsequenzen im weitesten Umfange zu entwickeln, zumal ja doch nur auf diesem Wege etwaige Widersprüche eines Standpunktes aufgedeckt werden könnten. Dabei genügt es, diejenigen Theoreme, welche das Axiom notwendig erfordern, von denen zu trennen, bei welchen es entbehrt werden kann, um auch die gesamte *Peano*sche Mathematik als einen besonderen Zweig, als eine gewissermaßen künstlich verstümmelte Wissenschaft mit zu umfassen. Fundamentale Tatsachen oder Probleme einfach aus der Wissenschaft zu weisen, weil sie sich mit gewissen vorgeschriebenen Prinzipien nicht erledigen lassen, wäre ebenso, als wollte man in der Geometrie den weiteren Ausbau der Parallelentheorie verbieten, weil das betreffende Axiom als unbeweisbar nachgewiesen ist. In der Tat müssen die Prinzipien aus der Wissenschaft, nicht die Wissenschaft aus ein für allemal feststehenden Prinzipien beurteilt werden. Die Geometrie hat existiert vor den *Euklid*ischen „Elementen", ebenso die Arithmetik und Mengenlehre vor dem *Peano*schen „Formulaire", und beide werden noch jeden solchen Versuch einer schulmäßigen Systematisierung unzweifelhaft überleben.

*** Math. Annalen, Bd. 60, p. 459.

Upon this theorem rests the theory of denumerable sets and of the "second number class"; but it can be proved only if we *simultaneously* order all the finite or denumerable sets in question like the natural numbers ⟦nach dem Normaltypus⟧.

6) Does there exist a "base for all real numbers", that is, a system of real numbers such that no linear relation with a finite number of integral coefficients holds between them and that any other real number can be obtained from them by a linear relation with a finite number of integral coefficients?

7) Are there discontinuous solutions of the functional equation

$$f(x+y) = f(x) + f(y)\,?$$

The last two questions were answered affirmatively by *G. Hamel*[14] on the assumption that the continuum can be well-ordered.

Cantor's theory of cardinalities, therefore, certainly requires our postulate, and so does *Dedekind*'s theory of finite sets, which forms the foundation of arithmetic. That in the theory of functions we can usually circumvent its use is to be explained simply by the fact that there as a rule we deal with "closed" sets, for which distinguished elements can be unambiguously defined without any difficulty. Where this is not the case, hence especially in the theory of everywhere discontinuous functions, the principle is often indispensable, as our last example shows.

Now so long as the relatively simple problems mentioned here remain inaccessible to *Peano*'s expedients, and so long as, on the other hand, the principle of choice cannot be definitely refuted, no one has the right to prevent the representatives of productive science from continuing to use this "hypothesis"—as one may call it for all I care—and developing its consequences to the greatest extent, especially since any possible contradiction inherent in a given point of view can be discovered only in that way. We need merely separate the theorems that necessarily require the axiom from those that can be proved without it in order to delimit the whole of *Peano*'s mathematics as a special branch, as an artificially mutilated science, so to speak. Banishing fundamental facts or problems from science merely because they cannot be dealt with by means of certain prescribed principles would be like forbidding the further extension of the theory of parallels in geometry because the axiom upon which this theory rests has been shown to be unprovable. Actually, principles must be judged from the point of view of science, and not science from the point of view of principles fixed once and for all. Geometry existed before *Euclid*'s "Elements", just as arithmetic and set theory did before *Peano*'s "Formulaire", and both of them will no doubt survive all further attempts to systematize them in such a textbook manner.

---

[14] *Hamel 1905.*

Freilich hätte Herr *Peano* noch ein einfaches Mittel, die in Frage stehenden Sätze wie noch viele andere aus seinen eigenen Prinzipien zu beweisen. Er brauchte nur von der neuerdings viel erörterten „*Russell*schen Antinomie" Gebrauch zu machen, da sich aus widersprechenden Prämissen bekanntlich alles beweisen läßt. In der Tat schließen die Prinzipien des Formulaire, welche zwischen „Menge" und „Klasse" keinen Unterschied machen, diesen Widerspruch nicht aus. Dagegen sind, wie ich demnächst an anderer Stelle zeigen
116 werde, die Vertreter der Mengenlehre als einer | rein mathematischen Disziplin, welche nicht auf die Grundbegriffe der traditionellen Logik beschränkt ist, durchaus in der Lage, durch geeignete Spezialisierung ihrer Axiome alle bisher bekannten „Antinomieen" zu vermeiden. Während also der Bereich der Peanoschen Prinzipien, wie wir eben zeigten, zu eng ist, um die Wissenschaft in ihrer vollen Schönheit zu entwickeln, ist er andererseits zu weit, um sie von inneren Widersprüchen frei zu halten; und solange die Antinomieen dieses Systems nicht beseitigt sind, wird man in ihm wohl kaum die endgültige Grundlegung der mathematischen Wissenschaft suchen dürfen.

b. Einwand der nicht-prädikativen Definition

Den hier vertretenen Standpunkt einer in letzter Linie auf Intuition beruhenden produktiven Wissenschaft hat neuerdings auch Herr *H. Poincaré* der Peanoschen „Logistik" gegenüber in einer Reihe von Aufsätzen* geltend gemacht, in denen er auch dem Auswahlprinzip, das er für ein unbeweisbares, aber unentbehrliches Axiom ansieht, durchaus gerecht wird.** Dabei ist er aber, weil seine Gegner sich vorzugsweise der Mengenlehre bedienten, im Angriff soweit gegangen, die ganze *Cantor*sche Theorie, diese ursprüngliche Schöpfung genialer Intuition und spezifisch mathematischen Denkens, mit der von ihm bekämpften Logistik zu identifizieren und ihr ohne Rücksicht auf ihre positiven Leistungen lediglich auf Grund der noch ungeklärten „Antinomieen" jede Existenzberechtigung abzusprechen.*** Kam es ihm nur darauf an, in den Grundlagen der Arithmetik „synthetische Urteile a priori" nachzuweisen, zu denen er zunächst das „Prinzip der vollständigen Induktion" glaubte rechnen zu dürfen, so hätte es den mengentheoretischen Beweisen dieses Prinzips gegenüber genügt, den Grundsätzen, auf denen diese Beweise beruhen, einen synthetischen Charakter zuzuschreiben, und auch die Vertreter der Mengenlehre hätten dies gelten lassen können, da die Unterscheidung von „synthetisch" und „analytisch" dann eine rein philosophische wäre und die Mathematik als solche nicht berührte. Statt dessen hat er es unternommen,

---

* „Les mathématiques et la logique", Revue de Métaphysique et de Morale t. 13; t. 14, p. 17, p. 294, p. 866.

** ibid. 14, p. 311–313: „C'est donc un jugement synthétique a priori sans lequel la théorie cardinale serait impossible, aussi bien pour les nombres finis que pour les nombres infinis."

*** ibid. 14, p. 316: „Il n'y a pas d'infini actuel; les Cantoriens l'ont oublié, et ils sont tombés dans la contradiction."

Of course, there remains to Mr. *Peano* a simple way of proving the theorems in question, as well as many others, from his own principles. He need only use "*Russell*'s antinomy", lately much discussed, since, as is well known, everything can be proved from contradictory premisses. Indeed the principles of the Formulaire, which make no distinction between "set" and "class", do not exclude this contradiction. On the other hand, as I shall soon show elsewhere ⟦*1908a*⟧, those who champion set theory as a purely mathematical discipline that is not confined to the basic notions of traditional logic are certainly in a position to avoid, by suitably restricting their axioms, all "antinomies" discovered until now. Thus while the domain of Peano's principles, as we have just shown, is too narrow to permit the development of our science in its full beauty, it is on the other hand too wide to exclude internal contradictions; and, so long as the antinomies of this system are not eliminated, one is hardly justified in seeking in it the definitive foundation for the science of mathematics.

b. Objection concerning nonpredicative definition

The point of view maintained here, that we are dealing with a productive science resting ultimately upon intuition, was recently urged, in opposition to Peano's "logistic", by Mr. *Poincaré*, too, in a series of essays;[15] in these he also does full justice to the principle of choice, which he considers an unprovable but indispensable axiom.[16] However, he presses his attack so far—since his opponents made use chiefly of set theory—as to identify all of *Cantor*'s theory, this original creation of specifically mathematical thought and the intuition of a genius, with the logistic that he combats and to deny it any right to exist, without regard for its positive achievements, solely on the ground of antinomies that have not yet been resolved.[17] If he was concerned only to show that in the foundations of arithmetic there are "synthetic judgments a priori", among which, he believed, he could reckon the "principle of mathematical induction" above all, it would have sufficed, so far as the set-theoretic proofs of this principle are concerned, to ascribe a synthetic character to the fundamental propositions upon which these proofs rest; even the champions of set theory could have accepted that, since the distinction between "synthetic" and "analytic" would then be a purely philosophical one and not touch mathematics as such. Instead, he undertook to combat mathematical proofs

[15] *Poincaré 1905, 1906a, 1906b, 1906c.*

[16] *Poincaré 1906b*, pp. 311–313: "Hence this is a synthetic judgment a priori; without it the theory of cardinals would be impossible, for finite numbers as well as for infinite ones."

[17] *Poincaré 1906b*, p. 316: "There is no actual infinite; the Cantorians forgot that, and they fell into contradiction."

mathematische Beweise mit den Waffen der formalen Logik zu bekämpfen, und sich damit auf ein Feld begeben, auf dem seine logistischen Gegner ihm überlegen sind.

117 Um die Auffassung *Poincarés* zu verdeutlichen, ist es wohl am | einfachsten, ein Beispiel zu wählen, das dem im § 1 dieses Artikels vorausgeschickten Beweise entnommen ist. Dort habe ich eine besondere Klasse von Mengen definiert, die ich als „Θ-Ketten" bezeichnete, und habe dann nachgewiesen, daß der gemeinsame Bestandteil M aller dieser Θ-Ketten selbst eine Θ-Kette darstellt. Dieses Verfahren ist derjenigen „Ketten"-Theorie nachgebildet, auf welche *R. Dedekind** seine Theorie der endlichen Zahlen gründet, und ist auch sonst in der Mengenlehre gebräuchlich. Nach Herrn *Poincaré*** soll aber eine Definition nur dann „prädikativ" und logisch allein zulässig sein, wenn sie alle solchen Gegenstände *ausschließt*, welche von dem definierten Begriffe „abhängig" sind, d. h. durch ihn irgendwie bestimmt werden können. Demnach hätte in dem hier angeführten Beispiele die Menge M, welche selbst erst durch die Gesamtheit der Θ-Ketten bestimmt ist, von der Definition dieser Ketten ausgeschlossen werden müssen, und meine Definition, welche M selbst als Θ-Kette rechnet, wäre „nicht-prädikativ" und enthielte einen circulus vitiosus. In zwei ganz analogen Fällen, deren letzterer sich auf die „γ-Mengen" meines Beweises von 1904 bezieht, wird dies ausdrücklich als Kritik meines Beweisverfahrens ausgeführt.* * *

Nun ist aber einerseits diese logische Form eines Beweises keineswegs auf die Mengenlehre beschränkt, sondern findet sich genau so in der Analysis überall, wo das Maximum oder Minimum einer vorher definierten „abgeschlossenen" Zahlenmenge Z zu weiteren Folgerungen benutzt wird. Dies geschieht z. B. in dem bekannten Cauchyschen Beweise für den „Fundamentalsatz der Algebra", ohne daß es bisher jemand eingefallen wäre, etwas Unlogisches darin zu erblicken. Andererseits enthält gerade die als „prädikativ" bezeichnete Form der Definition etwas Zirkelhaftes; denn ohne den Begriff schon zu haben, kann man noch gar nicht wissen, welche Gegenstände sich durch ihn einmal bestimmen lassen und deswegen auszuschließen wären. In Wahrheit muß natürlich die Frage, ob ein beliebig vorgelegter Gegenstand unter eine Definition fällt, unabhängig von dem erst zu definierenden Begriffe durch ein *objektives* Kriterium entscheidbar sein. Ist aber ein solches Kriterium einmal gegeben, wie dies in den meinen Beweisen entlehnten Beispielen tatsächlich überall der Fall ist, so hindert nichts, daß einige der Gegenstände, welche unter die Definition fallen, zu demselben Begriffe noch in einer besonderen Beziehung stehen und dadurch — als Minimum oder als gemeinsamer Bestandteil — vor den übrigen ausgezeichnet oder bestimmt werden können.
118 Durch eine | solche „Bestimmung" wird ein Gegenstand ja nicht erst geschaffen, sondern jeder Gegenstand kann auf sehr verschiedene Weisen bestimmt

* „Was sind und was sollen die Zahlen?" § 4.
** Rev. d. Mét. e. d. Mor. 14, p. 307.
*** ibid. p. 314 und 315.

with the weapons of formal logic, thus venturing upon a territory in which his logistic opponents are his betters.

To make *Poincaré*'s conception clearer, the simplest thing would probably be to choose an example from the proof given in § 1 of the present paper. There I defined a special class of sets, which I called "Θ-chains", and then proved that the common component M of all these Θ-chains is itself a Θ-chain. This procedure is modeled upon the theory of "chains" on which *R. Dedekind*[18] bases his theory of finite numbers, and it is also customary elsewhere in set theory. But according to Mr. *Poincaré*[19] a definition is "predicative" and logically admissible only if it *excludes* all objects that are "dependent" upon the notion defined, that is, that can in any way be determined by it. Accordingly, in the example cited here the set M, which itself is determined only by the totality of the Θ-chains, would have had to be excluded from the definition of these chains, and my definition, which counts M itself as a Θ-chain, would be "nonpredicative" and contain a vicious circle. In two passages that are quite similar, the latter referring to the "γ-sets" of my 1904 proof, this is expressly elaborated as a criticism of my proof procedure.[20]

Now, on the one hand, proofs that have this logical form are by no means confined to set theory; exactly the same kind can be found in analysis wherever the maximum or the minimum of a previously defined "closed" set of numbers Z is used for further inferences. This happens, for example, in the well-known Cauchy proof of the "fundamental theorem of algebra", and up to now it has not occurred to anyone to regard this as something illogical. On the other hand, it is precisely the form of definition said to be "predicative" that contains something circular; for unless we already have the notion, we cannot know at all what objects might at some time be determined by it and would therefore have to be excluded. In truth, of course, the question whether an arbitrary given object is subsumed under a definition must be decidable independently of the notion still to be defined, by means of an *objective* criterion. But once such a criterion is given, as is in fact the case everywhere in the examples drawn from my proofs, nothing can prevent some of the objects subsumed under the definition from having in addition a special relation to the same notion and thus being determined by, or distinguished from, the remaining ones—say, as common component or minimum. After all, an object is not created through such a "determination"; rather, every object can be determined in a wide variety of ways, and these different

[18] *Dedekind 1988*, §4.
[19] *Poincaré 1906b*, p. 307.
[20] *Poincaré 1906b*, pp. 314–315.

werden, und diese verschiedenen Bestimmungen liefern nicht identische, sondern nur „äquivalente“ Begriffe, d. h. solche von gleichem „Umfange“. In der Tat scheint, worauf besonders Herr *G. Peano** in diesem Zusammenhange hinweist, die Existenz äquivalenter Begriffe dasjenige zu sein, was Herr *Poincaré* in seiner Kritik übersehen hat. Eine Definition darf sich sehr wohl auf Begriffe stützen, welche dem zu definierenden äquivalent sind; ja in jeder Definition sind Definierendes und Definiertes äquivalente Begriffe, und die strenge Befolgung der *Poincaré*schen Forderung würde jede Definition und damit jede Wissenschaft unmöglich machen.

c. *Einwände, gegründet auf die Menge W*

Die bisher erörterten Kritiken, welche sich gegen die Prinzipien und Beweismethoden der Mengenlehre überhaupt richten, haben naturgemäß bei denjenigen Mathematikern wenig Anklang gefunden, welche wie die Herren *J. König*, *Ph. Jourdain* und *F. Bernstein* auf diesem Gebiete selbst bereits produktiv tätig gewesen sind und sich dabei von der Unentbehrlichkeit der genannten Hilfsmittel überzeugen konnten. Dagegen scheint bei einigen derselben die neuerdings wieder so vielfach erörterte „*Burali-Forti*sche Antinomie“, welche sich auf die „Menge *W* aller Cantorschen Ordnungszahlen“ bezieht, einen allzu weitgehenden Skeptizismus gegenüber der Theorie der Wohlordnung hinterlassen zu haben. Und doch hätte schon die elementare Form,
119 welche Herr *B. Russell*** den | mengentheoretischen Antinomieen gegeben

* Rivista di Matematica VIII, Nr. 5, p. 152. Es handelt sich hier um einen neuen Beweis des Schröder-Bernsteinschen Theorems über die Äquivalenz der Mengen, den ich im Januar 1906 Herrn *Poincaré* brieflich mitgeteilt hatte. Diesen Beweis hatte der letztere im Maihefte der Revue de Métaphysique et de Morale (14, p. 314–315) inhaltlich richtig wiedergegeben und zum Gegenstande einer Kritik gemacht, auf die sich die angeführte Stelle Peanos (zum Teil wörtlich zitierend) bezieht. Herr Peano knüpft aber, ohne diesen Tatbestand zu erwähnen, seine Ausführung an eine vorhergehende Reproduktion seines eigenen, mit dem meinigen wesentlich übereinstimmenden, nur in Begriffsschrift gefaßten Beweises, dessen erste Mitteilung (Rendiconti del Circolo Matematico di Palermo XXI, adunanza 8 Apr. 1906) Herrn Poincaré bei der Abfassung seines Artikels offenbar noch nicht vorgelegen hatte. Warum vermeidet Herr Peano hier, wo er mit mir übereinstimmt, die Nennung meines Namens, um unmittelbar darauf seine Bekämpfung des Auswahlprinzipes so ausdrücklich an meine Adresse zu richten? Es dürfte doch einleuchten, daß nicht die mathematischen Prinzipien, welche Gemeingut sind, sondern lediglich die auf sie gegründeten Beweise Eigentum des einzelnen Mathematikers sein können. Übrigens hatte ich am Schluß meiner Note ausdrücklich bemerkt, daß ich die Heranziehung des Auswahlprinzipes zur Bildung einer „γ-Belegung“ einem Vorschlage des Herrn *Erhard Schmidt* (jetzt in Bonn) verdanke.

** „The Principles of Mathematics“, vol. I (Cambridge 1903), p. 366–368. Indessen hatte ich selbst diese Antinomie unabhängig von Russell gefunden und sie schon vor 1903 u. a. Herrn Prof. Hilbert mitgeteilt.

determinations do not yield identical but merely "equivalent" notions, that is, notions having the same "extension". In fact, the existence of equivalent notions seems to be what Mr. *Poincaré* has overlooked in his critique, as Mr. *Peano*[21] especially emphasizes in this connection. A definition may very well rely upon notions that are equivalent to the one to be defined; indeed, in every definition *definiens* and *definiendum* are equivalent nations, and the strict observance of *Poincaré*'s demand would make every definition, hence all of science, impossible.

c. *Objections based upon the set W*

The criticisms discussed so far, which are directed against the principles and proof methods of set theory in general, naturally met with little approval among the mathematicians who, like Mr. *J. König*, Mr. *Ph. Jourdain*, and Mr. *F. Bernstein*, already have themselves been productively active in this field and thus had the opportunity of convincing themselves that the devices mentioned are indispensable. On the other hand, the "*Burali-Forti* antinomy", recently again so much the subject of discussion, concerning the "set W of all Cantor ordinals" seems to have instilled in some of them an all too pervasive skepticism toward the theory of well-ordering. And yet, even the elementary form that Mr. *B. Russell*[22] gave to the set-theoretic antinomies could have

---

21 *Peano 1906b*, p. 152. At issue here is the new proof of the Schröder–Bernstein theorem on the equivalence of sets that I had communicated to Mr. *Poincaré* by letter in January 1906. He had reproduced this proof, correctly as to content, on pages 314–315 of the May number of the *Revue de métaphysique et de morale* ⟦*Poincaré 1906b*⟧ and had made it the object of a critique, to which the cited passage from Peano refers (in part quoting literally). But Mr. Peano, without mentioning this state of affairs, links his presentation to a previous version of his own proof, which essentially coincides with mine, except that it is formulated in his ideography; Mr. Poincaré apparently had not yet seen this first form (*Peano 1906a*, session of 8 April 1906) of the proof when he wrote his article. Why does Mr. Peano avoid mentioning my name here, where he agrees with me, and then direct his opposition to the principle of choice immediately afterward so expressly at me? It would seem obvious, after all, that not mathematical principles, which are common property, but only the proofs based upon them can be the possession of an individual mathematician. I had expressly stated at the end of my note, incidentally, that I owe to a suggestion by Mr. *Erhard Schmidt* (now in Bonn) the idea of invoking the principle of choice in order to form a "γ-covering".

22 *Russell 1903*, pp. 366–368. I had, however, discovered this antinomy myself, independently of Russell, and had communicated it prior to 1903 to Professor Hilbert among others.

hat, sie überzeugen können, daß die Lösung dieser Schwierigkeiten nicht in der Preisgabe der Wohlordnung, sondern lediglich in einer geeigneten Einschränkung des Mengenbegriffes zu suchen ist. Im Hinblick auf solche Bedenken hatte ich bereits in meinem Beweise von 1904 nicht nur alle irgendwie zweifelhaften Begriffe, sondern sogar die Verwendung der „Ordnungszahlen" überhaupt vermieden und mich sichtlich auf solche Prinzipien und Hilfsmittel beschränkt, welche für sich allein bisher noch zu keinen „Antinomieen" Anlaß gegeben haben. Wenn nun trotzdem einige Kritiker diese ominöse „Menge W" gegen meinen Beweis ins Feld geführt haben, so mußten sie dieselbe erst künstlich hinein interpretieren, und alle aus dem widerspruchsvollen Charakter dieser „Menge" geschöpften Argumente wenden sich gegen ihre Urheber zurück. In meinem neuen Beweise bin ich nun vollends bis zuletzt sogar ohne das Hilfsmittel der Rangordnung ausgekommen und habe damit, wie ich hoffe, jede Möglichkeit zur Einführung von W definitiv abgeschnitten.

Diesem W-Standpunkte scheint Herr *J. König* wenigstens nicht fern zu stehen. Denn obwohl er sich noch in seinem Heidelberger Vortrage* selbst auf das Auswahlprinzip stützte, die wesentlichste Voraussetzung meines Theorems also anerkennt, hat er auch in seinen späteren Publikationen** die Frage
120 nach der möglichen Wohlordnung des Kontinuums | ohne Rücksicht auf meine bereits erschienene Note als ein ungelöstes Problem behandelt, hat es aber bisher unterlassen, seinen Bedenken gegen irgend einen bestimmten Schritt meines Beweises öffentlichen Ausdruck zu geben.

---

* Bericht des III. Internationalen Mathematiker-Kongresses zu Heidelberg, 1904, p. 144: Zum Kontinuum-Problem.

** Math. Annalen Bd. 60, p. 177, Bd. 61, p. 156. Über die „Antinomie Richard", die Herr *König* in dem letzten Artikel hier heranzuziehen versucht, vergl. *G. Peano*, Riv. d. Mat. VIII, Nr. 5, p. 148–157, sowie namentlich *G. Hessenberg*, „Grundbegriffe der Mengenlehre" XXIII („Die Paradoxie der endlichen Bezeichnung"), wo der vorliegende Fehlschluß m. Er. treffend aufgedeckt wird. Der Begriff „endlich definierbar" ist kein absoluter sondern ein Relativbegriff und bezieht sich immer auf die gewählte „Sprache" oder „Bezeichnungsweise". Der Schluß, daß alle endlich definierbaren Gegenstände abzählbar sein müssen, gilt aber nur, wenn für alle ein und dasselbe Zeichensystem verwendet werden soll, und die Frage, ob ein einzelnes Individuum überhaupt einer endlichen Bezeichnung fähig ist oder nicht, ist an und für sich gegenstandslos, da man jedem Dinge nötigenfalls willkürlich eine beliebige Bezeichnung zuordnen kann. Übrigens hat die Wohlordnung des Kontinuums mit dieser Antinomie im Grunde nicht viel mehr zu tun wie jeder andere Satz, den man unter Benutzung eines Widerspruches gleich gut beweisen und widerlegen kann. In der Tat hat auch Herr *F. Bernstein* mit Hilfe der endlichen Definierbarkeit einmal beweisen wollen (Deutsche Math.-Vereinigung 1905, p. 447), daß das Kontinuum der zweiten Zahlenklasse, also einer wohlgeordneten Menge, äquivalent sein müsse; er ist sonach von demselben Begriffe ausgehend zu dem entgegengesetzten Resultate gelangt wie Herr *König*. Die versprochene Ausführung dieses „Beweises" ist allerdings niemals erschienen.

persuaded them that the solution of these difficulties is not to be sought in the surrender of well-ordering but only in a suitable restriction of the notion of set. Already in my 1904 proof, having such reservations in mind, I avoided not only all notions that were in any way dubious but also the use of "ordinals" in general; I clearly restricted myself to principles and devices that have not yet by themselves given rise to any "antinomy". If some critics nevertheless deploy this ominous "set W" against my proof, they must first project it into that proof artificially, and all arguments drawn from the inconsistent character of this "set" turn back upon their authors. Now I succeeded in completing my new proof without even the device of rankordering, and I hope thereby to have definitively cut off every possibility of introducing W.

Mr. *J. König*, it seems, is at least not far from this W point of view. For although he himself still relied upon the principle of choice in his Heidelberg lecture,[23] thus accepting the most essential assumption of my theorem, he treated the question whether the continuum can be well-ordered as an unsolved problem in later publications,[24] without regard for my already published note; but up to now he has abstained from giving public expression to his reservations about any specific step in my proof.

---

[23] *J. König 1905a.*

[24] *J. König 1905b, 1905c.* Concerning the "Richard antinomy", which König tries to adduce in the second of these papers, see *Peano 1906b*, pp. 148–157, as well as, especially, *Hessenberg 1906*, chap. 23, "Die Paradoxie der endlichen Bezeichnung", where the present fallacy is in my opinion appropriately exposed. "Finitely definable" is not an absolute notion but only a relative one, and it is always related to the chosen "language", or "notation". The conclusion that ⟦the set of⟧ all finitely definable objects must be denumerable holds only if one and the same system of signs is to be used for all of them, and the question whether a single individual can or cannot have a finite designation is in and of itself meaningless, since to any object we could, if necessary, arbitrarily assign any designation whatever. That the continuum can be well-ordered, incidentally, does not have much more to do with this antinomy, basically, than does any other proposition; all can equally well be proved and refuted by the use of a contradiction. In fact, it is precisely by means of finite definability that Mr. *F. Bernstein* once (*1905a*) wanted to prove that the continuum is equivalent to the second number class, hence to a well-ordered set; thus, starting from the same notion as Mr. *König*, he arrived at the opposite conclusion. To be sure, the promised realization of this "proof" was never published.

Herr *Ph. Jourdain** behauptet zwar, den Satz von der Wohlordnung schon vor mir, und zwar einfacher und vollständiger, bewiesen zu haben, gibt ihm aber mit Rücksicht auf W eine Auslegung, nach welcher, wie er a. a. O. S. 469 ausdrücklich bemerkt, nicht einmal das Kontinuum ein Aleph zu sein brauchte. Nach ihm sollen nur „konsistente" Mengen, d. h. solche, welche keinen Bestandteil ähnlich W enthalten, Ordnungstypen und Kardinalzahlen besitzen, und gerade in dieser Zulassung „inkonsistenter" Mengen erblickt er die größere „Vollständigkeit" seines Resultates. Da nun aber „Ordnungstypen" und „Kardinalzahlen" in der Cantorschen Theorie nichts anderes sind als bequeme *Ausdrucksmittel*, um die Mengen in bezug auf Ähnlichkeit oder Äquivalenz ihrer Teile miteinander zu vergleichen, so kann ich der Aussage, einer wohlgeordneten Menge komme kein Ordnungstypus oder keine Kardinalzahl zu, keinen verständlichen Sinn abgewinnen, und dieser Versuch, unter Beibehaltung von W die Antinomie zu lösen, scheint mir nur auf ein Spiel mit Worten hinauszukommen. Willkürlich kann man freilich, etwa von der zweiten oder dritten Zahlenklasse an, alle höheren Ordnungstypen ignorieren, nicht mehr als solche anerkennen und erhält dann ein W vom Typus ω bezw. Ω, welches unter den gemachten Annahmen, und soweit man von seinem „Ordnungstypus" absieht, gewiß widerspruchsfrei ist. Nur bleibt es so *völlig unbestimmt* und vor allem ist es eben nicht *dasjenige* W, um das es sich in der Antinomie handelt, nämlich eine Menge von der Beschaffenheit, daß jeder beliebigen wohlgeordneten Menge ein Element von W als Ordnungstypus entspricht. Ein ähnliches Bedenken scheint auch Herrn *Jourdain* nachträglich gekommen zu sein und ihn, wenn auch nicht zum Verzicht auf sein W, so doch zur Einführung einer *zweiten* gleichfalls wohlgeordneten Menge $\mathfrak{W}$
121 veranlaßt zu haben, welche als „absolut un- | endlich" wie das weiter unten zu besprechende *Bernstein*sche W keiner „Fortsetzung" mehr fähig sein soll.

Was nun endlich den *Beweis* betrifft, den Herr *Jourdain* in der genannten Annalen-Note* dem meinigen als den „einfacheren" gegenüberstellt, so ist das von ihm vorgeschlagene Verfahren zur Wohlordnung einer beliebigen Menge M das folgende. Man nehme ein beliebiges Element als erstes, dann noch eines usw., nach einer beliebigen endlichen oder unendlichen Anzahl von Elementen ein beliebiges Element des Restes als nächstfolgendes und fahre so fort,

---

* Math. Annalen Bd. 60, p. 465. In den von ihm zitierten früheren Arbeiten (Phil. Mag. 1904, p. 61, p. 294; 1905, p. 42), auf die er seine Prioritätsansprüche stützt, ist dagegen von einer möglichen Wohlordnung überhaupt nicht die Rede. Vielmehr beschränkt sich in dem ersten dieser Artikel sein „Beweis, daß jede Kardinalzahl ein Aleph ist", lediglich auf einen Versuch, die Möglichkeit von Mächtigkeiten größer als alle Alephs durch den Hinweis auf die „Burali-Fortische Antinomie" auszuschließen. Hier wird also *ohne Beweis* vorausgesetzt, daß eine Menge, deren Kardinalzahl selbst kein Aleph ist, einen der Gesamtheit aller Alephs ähnlichen Bestandteil enthalten müßte; und der bloße Hinweis auf die Methoden und Resultate von Cantor und Hardy, welche sich auf die beiden ersten Mächtigkeiten beziehen, kann diesen Beweis doch unmöglich ersetzen.

* l. c. p. 468.

Mr. *Ph. Jourdain*[25] claims to have proved the well-ordering theorem before I did, and, moreover, in a simpler and more complete way; but, having W in his mind, he interprets the theorem in such a way that, as he expressly states (*1905b*, p. 469), not even the continuum need be an aleph. According to him, only "consistent" sets, that is, those that have no component similar to W, are to possess order types and cardinal numbers, and precisely in thus allowing "inconsistent" sets he sees the greater "completeness" of his result. But now, since in Cantor's theory "order types" and "cardinal numbers" are nothing but convenient *means of expression* for the comparison of sets with respect to the similarity or equivalence of their parts, I cannot extract any intelligible meaning from the proposition that a well-ordered set possesses no order type or cardinal number, and this attempt to resolve the antinomy while retaining W seems to me to amount to a mere word game. One can, to be sure, arbitrarily ignore all higher order types, from, say, the second or third number class on; one can refuse to recognize them as such any longer and thus obtain a W of type ω, or Ω, respectively, which is certainly consistent under the assumptions made, insofar as one disregards its "order type". But then this W remains *completely indeterminate*, and, what is more, it is just not the *same* W as is at issue in the antinomy, namely, a set so constituted that to every arbitrary well-ordered set an element of W corresponds as order type. A similar reservation seems afterward to have occurred to Mr. *Jourdain* himself and to have led him, if not to renounce his W, in any event to introduce a *second*, likewise well-ordered set $\mathfrak{W}$, which being "absolutely infinite" like *Bernstein*'s W, to be discussed below, is claimed to be no longer capable of any "continuation".

Now, finally, concerning the *proof* that Mr. *Jourdain*, in the note cited,[26] opposes to mine as being "simpler", the procedure he proposes for well-ordering an arbitrary set M is the following. Take an arbitrary element as first, then another, and so forth; after an arbitrary finite or infinite number of elements take an arbitrary element of the remainder as the next one; and con-

[25] *Jourdain 1905b*. In the earlier papers cited by him (*Jourdain 1904a, 1904b, 1905a*), upon which he rests his claims to priority, there is, however, no mention at all of the possibility of a well-ordering. Rather, in the first of these articles his "proof that every cardinal number is an aleph" confines itself merely to an attempt to exclude the possibility of cardinalities greater than all alephs by reference to the "Burali-Forti antinomy". Hence it is assumed there *without proof* that a set whose cardinal number is not itself an aleph must have a constituent part similar to the totality of all alephs; a mere reference to the methods and results of Cantor and Hardy, which concern the first two cardinalities, cannot, after all, possibly replace a proof.

[26] *Jourdain 1905b*, p. 468.

bis die ganze Menge erschöpft ist. Diese Idee einer sukzessiven Konstruktion ist nicht neu, sie wurde mir vor längerer Zeit einmal von Herrn *F. Bernstein* mündlich mitgeteilt und geht wahrscheinlich auf Herrn *G. Cantor* zurück, der aber offenbar Bedenken trug, sie als *Beweis* anzuerkennen. Dieselbe Konstruktion empfiehlt auch Herr *E. Borel*,** um sie ohne weitere Begründung sofort zu verwerfen und damit, wie er meint, das Auswahlprinzip ad absurdum zu führen. Das ist ihm aber keineswegs gelungen; denn nicht die unendlich wiederholte *Auswahl* ist es, welche diesem „Beweise" entgegensteht, sondern einfach die Tatsache, daß er nicht zum Ziele führt. Läßt man nämlich das oben erwähnte *Peano*sche Prinzip, welches die Auswahl aus einer einzigen Menge gestattet, einmal gelten, so gibt es keine Grenze mehr für seine wiederholte Anwendung. Aber was *beweist* denn diese ganze Betrachtung? Offenbar nicht mehr, als daß jede wohlgeordnete echte Teilmenge M' von M durch Hinzufügung eines willkürlichen Elementes m' aus dem Reste noch erweitert werden kann; oder vielmehr dies ist die *Voraussetzung*, die — im strikten Gegensatz zur Bernstein-Schoenfliesschen Auffassung — dem ganzen Verfahren zugrunde liegt. Läßt sich nun beweisen, daß unter den wohlgeordneten Bestandteilen von M ein *größter L existiert*, so wie z. B. bei mir die Existenz einer größten γ-Menge L_γ *bewiesen* wird, so muß $L = M$ sein und M ist wohlgeordnet. Genau so will auch Herr Jourdain schließen; nur fehlt ihm dazu als wesentliche Prämisse der Nachweis für die Existenz von L. Diese setzt er vielmehr *unbewiesen* voraus, indem er annimmt, daß sein Verfahren, sofern es die Menge M nicht erschöpft, in einer wohlgeordneten Teilmenge ähnlich $\mathfrak{W}$ seinen Abschluß finden müßte. Die „Einfachheit" dieses Beweises geht also so weit, daß er sich auf einen einzigen Schluß reduziert; allerdings einen Fehlschluß.***

122 | Während Herr *Jourdain*, wie wir sahen, der „inkonsistenten" Menge W immerhin noch zweifelnd gegenübersteht, macht sie Herr *F. Bernstein** bereits zum Gegenstande einer dogmatischen Theorie. Da der widerspruchsvolle Charakter dieser „Menge aller Ordnungszahlen" bekanntlich zutage tritt,

---

** Math. Annalen Bd. 60, p. 194.

*** An derselben Unbestimmtheit wie das *Cantor-Jourdain*sche Verfahren leidet *Hardys* angebliche und u. a. auch von Herrn Schoenflies (l. c. p. 183) ausdrücklich anerkannte „Konstruktion einer Teilmenge des Kontinuums von der zweiten Mächtigkeit" (Quart. Journ. of Math. 1903, p. 87). Er gibt eine Regel, aus einer bereits konstruierten abzählbaren und wohlgeordneten Teilmenge A ein neues Element des Kontinuums abzuleiten, welches von allen vorhergehenden verschieden ist. Da aber diese Regel *nicht eindeutig* ist, sondern von der in weiten Grenzen willkürlichen Darstellung durch „Fundamentalreihen" abhängt, so besitzt sein Verfahren keinen Vorzug vor dem sehr viel einfacheren Cantorschen „Diagonalverfahren", welches nur eine Umordnung nach dem ω-Typus erfordert, und liefert eben so wenig wie dieses eine *bestimmte* Teilmenge von der zweiten Mächtigkeit; sondern bestenfalls einen neuen Beweis der Tatsache, daß die Mächtigkeit des Kontinuums $\geqq \aleph_1$ ist.

* Math. Annalen Bd. 60, p. 187.

tinue in this way until the entire set is exhausted. This idea of a stepwise construction is not new; it was communicated to me orally quite some time ago by Mr. *F. Bernstein*, and it probably goes back to *G. Cantor*, who, however, apparently had reservations about accepting it as a *proof*. Mr. *E. Borel*,[27] too, recommends the same construction, only to reject it immediately without further explanation, thereby, as he believes, reducing the principle of choice to absurdity. But in this he does not succeed at all; for it is not the infinitely repeated *choice* that vitiates this "proof", but simply the fact that the proof does not lead to the goal. For, once we accept the above-mentioned principle of *Peano*, which permits a choice from a single set, there is no longer any limit to its repeated application. But what, then, do all these considerations *prove*? Evidently no more than that every well-ordered proper partial set M' of M can still be extended by the addition of an arbitrary element m' taken from the remainder; or, rather, this is the *assumption* that—in strict contrast to the conception of Bernstein and Schoenflies—forms the basis of the entire procedure. If it can now be proved that among the well-ordered components of M there *exists a largest one*, L, in the same way as, for example, I *prove* the existence of a largest γ-set L_γ, then necessarily $L = M$, and M is well-ordered. Mr. Jourdain, too, wants to make precisely the same inference; only he lacks the essential premiss, the proof of the existence of L. Rather, he presupposes it *without proof*, by assuming that his procedure, insofar as it does not exhaust the set M, would have to terminate in a well-ordered partial set similar to $\mathfrak{W}$. Thus this proof is so "simple" that it reduces to a single inference, albeit a fallacious one.[28]

While for all that Mr. *Jourdain*, as we saw, looks upon the "inconsistent" set W with some doubts, Mr. *F. Bernstein*[29] already makes it the object of a dogmatic theory. Since, as is well known, the contradictory character of this "set of all ordinals" becomes manifest when we add to it a further element, e,

---

[27] *Borel 1905a*.

[28] *Hardy*'s alleged "construction of a partial set of the continuum having the second cardinality" (*1904*), which was expressly accepted also by Mr. Schoenflies (*1905*, p. 183) among others, suffers from the same indefiniteness as the procedure of *Cantor* and *Jourdain*. He gives a rule for deriving, from a previously constructed denumerable and well-ordered partial set A, a new element of the continuum that is distinct from all preceding ones. Since this rule is *not univocal*, however, but depends upon the representation ⟦of an element of the continuum⟧ by means of "fundamental sequences", which is arbitrary within wide limits, there is no reason to prefer his procedure to Cantor's very much simpler "diagonal procedure", which merely requires a reordering according to the type ω; no more than Cantor's does Hardy's procedure yield a *definite* partial set having the second cardinality; rather, at best it yields a new proof of the fact that the cardinality of the continuum is $\geqq \aleph_1$.

[29] *Bernstein 1905b*.

wenn man ihr ein weiteres Element e hinzufügt, welches auf alle Elemente von W folgt, so glaubt Herr *Bernstein* alle Schwierigkeiten beseitigen zu können, indem er eine solche Anhängung eines neuen Elementes als der Definition von W widersprechend für unzulässig erklärt. Die Menge W soll nur die Ordnungstypen aller „fortsetzbaren" wohlgeordneten Mengen oder aller „Abschnitte wohlgeordneter Mengen" umfassen, W selbst aber „nicht fortsetzbar" sein. Von diesem Standpunkte aus kritisiert er dann in meinem Beweise von 1904 den Übergang 7 V von der wohlgeordneten Menge L zu (L, m_1'), weil doch L möglicherweise der Menge W ähnlich sein könnte,[**] und konstruiert mit Hilfe von W eine Menge Z, welche keiner Wohlordnung fähig sein soll.

Freilich ist dies verlorne Liebesmüh. Denn wenn die Menge W einmal existiert, so ist sie, wie Herr *Bernstein* ausdrücklich zugibt, auch wohlgeordnet mit einem bestimmten Ordnungstypus β, und jede andere wohlgeordnete Menge ist entweder W selbst ähnlich oder einem Abschnitte von W. Nach der ein für alle Mal feststehenden *Cantor*schen Definition für das Größer und Kleiner der Ordnungszahlen wäre somit β größer als jede andere Ordnungszahl α, und in der nach der Größe geordneten Menge (W, β) rangierte β *hinter* allen Elementen von W, d. h. W wäre tatsächlich „fortsetzbar", entgegen seiner Definition und trotz aller Verbote. Gegen diese unerwünschte Folgerung weiß Herr *Bernstein* nichts weiter einzuwenden als[***] „daß der
123 Widerspruch nur daraus entsteht, daß β als | auf alle Elemente von W folgend *angenommen* wird. Wenn nur die Vereinigungsmenge $(W; \beta)$ gebildet wird, ohne daß zwischen β und den Elementen von W eine *Ordnungsbeziehung festgesetzt* wird, so führt das zu keinem Widerspruch." Als ob es nur auf das *Wort* „Ordnungsbeziehung" oder die *Schreibweise* $a \prec \beta$ ankäme, und als ob durch Vermeidung eines *Wortes* eine objektive mathematische *Tatsache* beseitigt werden könnte! Die Mathematik wäre keine internationale Wissenschaft, wenn ihre Sätze nicht einen von der Sprache, in der wir sie ausdrücken, unabhängigen objektiven Inhalt besäßen. Bei der Prüfung eines Widerspruches handelt es sich ja gar nicht darum, ob eine bedenkliche Folgerung *wirklich vollzogen* und offiziell *anerkannt*, sondern lediglich, ob sie überhaupt formell *möglich* ist; und diese Möglichkeit allein deswegen zu verneinen, *weil* sie zu einem Widerspruche führt, wäre offenbar eine petitio principii oder ein circulus vitiosus. In der Tat kommt aber das eingeschlagene Verfahren zur Rechtfertigung von W darauf hinaus, den in seiner Definition liegenden Widerspruch nicht zu lösen, sondern zu *ignorieren.* Wenn eine Annahme A nach den allgemeinen Prinzipien zu zwei entgegengesetzten Folgerungen B und B' Anlaß

[**] Herr Bernstein verwirft damit nicht nur einen „Teil", wie er sagt, sondern den gesamten Inhalt des Jourdainschen Beweises, obwohl sein eigenes W dem Jourdainschen $\mathfrak{W}$ genau entspricht. Bei Jourdain kann L, eben weil es fortsetzbar ist, mit $\mathfrak{W}$ nicht ähnlich sein, während Bernstein umgekehrt aus der Ähnlichkeit mit W auf die Nicht-Fortsetzbarkeit schließt. Auch hier werden also aus übereinstimmenden Annahmen entgegengesetzte Folgerungen gezogen.

[***] a. a. O. p. 189. Nur habe ich e durch β ersetzt und einige Worte durch den Druck hervorgehoben.

that follows all elements of W, Mr. *Bernstein* believes that he can overcome all difficulties by declaring that to append such a new element is inadmissible on the grounds that this would contradict the definition of W. The set W is to contain only the order types of all "continuable" well-ordered sets, or of all "segments of well-ordered sets", but W itself is to be "noncontinuable". From this point of view he then criticizes the transition (7 V) from the well-ordered set L to (L, m_1') in my proof of 1904, since, after all, L might possibly be similar to the set W,[30] and by means of W he constructs a set Z that is said not to be capable of any well-ordering.

This, of course, is love's labor lost. For, as Mr. *Bernstein* expressly admits, once the set W exists, it is well-ordered, with a definite order type β, and any other well-ordered set is similar either to W itself or to a segment of W. According to *Cantor*'s definition, which is fixed once and for all, of the relations "greater than" and "less than" for ordinals, β would hence be greater than any other ordinal α, and, in the set (W, β), when it is ordered according to magnitude, β would rank *behind* all elements of W, that is, W would in fact be "continuable", contrary to its definition and in spite of all prohibitions. Mr. *Bernstein* does not know what to say in reply to this undesired consequence other than[31] "that the contradiction comes about only because β is *assumed* to follow upon all elements of W. If only the union $(W; \beta)$ is formed and *no order relation is stipulated* between β and the elements of W, no contradiction is generated". As if it were only a matter of the *term* "order relation" or the *notation* $\alpha \prec \beta$, and as if we could eliminate an objective mathematical *fact* by avoiding a *term*! Mathematics would not be an international science if its propositions did not have an objective content independent of the language in which we express them. In the examination of a contradiction the issue, after all, is not whether a questionable consequence is *actually realized* and officially *accepted*, but merely whether it is formally *possible* at all; and to deny this possibility merely *because* it leads to a contradiction would obviously be begging the question or falling into a vicious circle. In fact, however, the procedure followed in the justification of W amounts to this: the contradiction inherent in its definition is not resolved but *ignored*. If according to universal principles an assumption A yields two contrary consequences B and B', then A must be rejected as untenable. But

---

30 Therewith Bernstein rejects not only a "part", as he says, but the entire content of Jourdain's proof, even though his own W corresponds exactly to Jourdain's $\mathfrak{W}$. For Jourdain, L, precisely because it is continuable, cannot be similar to $\mathfrak{W}$, whereas Bernstein, on the contrary, infers that it is not continuable because it is similar to W. Hence here, too, opposite consequences are drawn from a common assumption.

31 *Bernstein 1905b*, p. 189. I have merely replaced e by β and italicized a few words.

gibt, so ist A als unhaltbar aufzugeben. In dem hier vorliegenden Falle soll es aber erlaubt sein, sich für die eine dieser Folgerungen B zu entscheiden und die andere B', weil sie dann mit B im Widerspruch stände, durch ein Ausnahmegesetz zu verbieten oder durch Namensänderung zu verschleiern. Da dieses Verfahren ersichtlich auf jede beliebige Hypothese A anwendbar wäre, so gäbe es überhaupt keinen Widerspruch; man könnte alles behaupten, aber nichts beweisen, da mit der Möglichkeit eines Widerspruches auch die eines Beweises beseitigt wäre, und eine mathematische Wissenschaft könnte nicht existieren.*

Daß es tatsächlich immer möglich ist, einer *beliebigen* wohlgeordneten Menge M ein weiteres Element u als letztes hinzuzufügen, läßt sich übrigens aus den allgemeinen Prinzipien der Mengenlehre elementar *beweisen*, wenn man nur eine rein formale Definition der Wohlordnung, wie die hier am Schlusse des § 1 gegebene, zugrunde legt. Ist nämlich die Wohlordnung von M gegeben durch das System der „Reste" $\mathfrak{R}(x)$, welche zu den Elementen x
124 von M gehören, so genügt es, jedem dieser Reste (unabhängig von | seiner Anordnung) durch einfache Vereinigung mit der Menge $\{u\}$ das neue Element hinzuzufügen, um dann zusammen mit der Menge $\{u\}$ ein neues Restesystem zu erhalten, welches die verlangte Wohlordnung von $M_1 = M + \{u\}$ leistet. In der Tat ist dann sehr einfach einzusehen, daß jede Untermenge von M_1 wieder ein „erstes Element" in dem a. a. O. definierten Sinne besitzt und daß alle Elemente von M dem Elemente u „vorangehen". Mit diesem Beweise, den ich in meiner Note von 1904 nur wegen seiner trivialen Einfachheit als unnötig weggelassen hatte, ist nach dem Obigen gleichzeitig auch die Nichtexistenz von W gesichert, und alle aus W gezogenen Folgerungen werden hinfällig. Da nun andererseits „Ordnungstypus einer wohlgeordneten Menge" gewiß ein logisch zulässiger *Begriff* ist, so folgt weiter, was allerdings viel einfacher schon aus der „*Russell*schen Antinomie" hervorgeht, daß nicht jeder beliebige Begriffsumfang als Menge behandelt werden darf und daß somit die übliche Mengendefinition zu weit ist. Beschränkt man sich aber in der Mengenlehre auf gewisse feststehende Prinzipien wie die unserem Beweise zugrunde liegenden, einfache Mengen zu bilden und aus gegebenen neue abzuleiten, so lassen sich alle solchen Widersprüche vermeiden.

d. Einwand der speziellen Erzeugungsprinzipien und der Menge Z

Ebenfalls gegen den letzten Schritt 7 V meines Beweises richtet sich vom Standpunkte desselben W-Glaubens aus die Kritik des Herrn *A. Schoen-*

* Über die verschiedenen Versuche zur „Rettung von W" vergl. auch G. Hessenberg „Grundbegriffe der Mengenlehre" XXIV („Ultrafinite Paradoxieen"), wo es am Schlusse von § 98 heißt: „Die Menge W selbst ist übrigens gegen alle Ehrenrettungen im höchsten Grade undankbar. So bemühen sich im 60ten Bande der Mathematischen Annalen gleichzeitig Bernstein und Jourdain um ihre Widerspruchslosigkeit, wobei der erste auf Grund der Eigenschaften von W beweist, daß es Mengen gibt, die nicht wohlgeordnet werden können, während dem zweiten der Beweis des Gegenteils gelingt."

in the present case, we are told, it is permissible to opt for one of these consequences, B, while prohibiting the other, B', by means of some special decree or veiling it by a change of name, lest it yield a contradiction with B. Since this procedure would obviously be applicable to every arbitrary hypothesis A, there would never be any contradiction at all; we could assert everything but prove nothing, since along with the possibility of a contradiction that of a proof would also be eliminated, and no science of mathematics could exist.[32]

That it is in fact always possible to add a further element u as a last element to an *arbitrary* well-ordered set M can, incidentally, be *proved* in an elementary way from the general principles of set theory, provided we adopt a purely formal definition of well-ordering, such as that given here at the end of § 1. For if the well-ordering of M is given by the system of remainders $\mathfrak{R}(x)$ associated with the elements x of M, we need merely add the new element to each of these remainders (independently of its ordering) by means of a simple union with the set $\{u\}$ to obtain then, together with the set $\{u\}$, a new system of remainders that furnishes the desired well-ordering of $M_1 = M + \{u\}$. In fact it is then very easy to see that every subset of M_1 again possesses a "first element" in the sense defined above and that all elements of M "precede" the element u. This proof, which, merely on account of its trivial simplicity, I did not find it necessary to include in my note of 1904, also assures, according to what was said above, the nonexistence of W, and all consequences drawn from W come to nought. Since now, on the other hand, "order type of a well-ordered set" is certainly a logically admissible *notion*, it follows further—as already appears in a much simpler way from "*Russell*'s antinomy", to be sure—that it is not permissible to treat the extension of every arbitrary notion as a set and that therefore the customary definition of set is too wide. But if in set theory we confine ourselves to a number of established principles such as those that constitute the basis of our proof—principles that enable us to form initial sets and to derive new sets from given ones—then all such contradictions can be avoided.

d. Objection concerning particular generating principles and the set Z

The critique of Mr. *A. Schoenflies*,[33] likewise directed against the last step (7 V) of my proof, is inspired by the same belief in W and is therefore

[32] Concerning the various attempts to "save W", see also *Hessenberg 1906*, chap. 24, "Ultrafinite Paradoxieen", where at the end of § 98 we read: "The set W itself is, incidentally, ungrateful in the highest degree for all attempts to redeem its honor. Thus in volume 60 of *Mathematische Annalen* both Bernstein and Jourdain exert themselves on behalf of its consistency, the former proving on the basis of the properties of W that there are sets that cannot be well-ordered, while the latter suceeeds in proving the opposite."

[33] *Schoenflies 1905.*

*flies** und wird somit durch die voraufgehenden Erörterungen mit der *Bernstein*schen gleichzeitig erledigt. Der betreffende Artikel enthält aber noch weitere Irrtümer und Mißverständnisse, die hier nicht unerwähnt bleiben können.

Zunächst unterscheidet Herr *Schoenflies* in der Theorie der wohlgeordneten Mengen „einen allgemeinen und einen speziellen Teil" und behauptet, daß *nur* die Theoreme des „allgemeinen Teiles" auf der bekannten *Cantor*schen *Definition* der Wohlordnung,** die übrigen aber auf „Erzeugungsprinzipien" beruhen. In Wahrheit müssen vielmehr *alle* Theoreme über einen Begriff aus seiner Definition zu beweisen sein, anderenfalls wären sie überhaupt nicht bewiesen. Stehen für einen Begriff zwei Definitionen zur Verfügung, so hat man sich bestimmt für die eine zu entscheiden oder die Äquivalenz der beiden nachzuweisen; jedes Schwanken zwischen zwei Definitionen oder die Ergänzung der einen durch die andere ist logisch völlig unzulässig. — Herr Schoenflies fährt dann fort:

125 | „Will man auf dieser allgemeinen Grundlage den fraglichen Satz beweisen, so hat man zu zeigen, daß eine unendliche Reihe von Mengen M_i, deren Mächtigkeiten abnehmen, nicht existieren kann; jede Reihe der Mächtigkeiten $\mathfrak{m}_1 > \mathfrak{m}_2 > \mathfrak{m}_3 > \cdots$ müßte nach einer endlichen Zahl von Gliedern abbrechen. Dies ist die notwendige *und hinreichende* Bedingung des Satzes." Notwendig aber *nicht* hinreichend, und eben deshalb zu einem Beweise des Theorems nicht zu verwenden. Das angeführte Kriterium bezieht sich lediglich auf die Wohlordnung der nach ihrer Größe geordneten *Mächtigkeiten* und gar nicht auf die Wohlordnung der *Mengen* selbst, um die es sich in meinem Theorem handelt. Sind alle Mächtigkeiten Alephs, so sind sie allerdings auch nach ihrer Größe wohlgeordnet, aber umgekehrt kann man *nicht* schließen, während der Vorschlag des Herrn *Schoenflies* dies augenscheinlich erfordert. Nicht einmal die *Vergleichbarkeit* beliebiger Mengen bezüglich ihrer Mächtigkeit läßt sich nach dieser Methode beweisen, sie liegt ihr vielmehr schon als Voraussetzung zugrunde. „Diesem Weg folgt der Zermelosche Beweis nicht." Allerdings nicht, weil ein Beweis so eben nicht geführt werden kann. „Er operiert mit Hilfsmitteln, die den speziellen Teil der Theorie der wohlgeordneten Mengen, nämlich ihre Erzeugung betreffen." Ganz im Gegenteil beruht mein Beweis ausschließlich auf der klassischen Cantorschen Definition und hat mit „Erzeugungsprinzipien" in dem angenommenen Sinne als selbständigen Beweisquellen nichts zu schaffen.

„Erzeugungsprinzipien kann man zunächst nur *axiomatisch postulieren* und hat dann ihre Berechtigung nachzuweisen. Auch die Einführung der Zahlen der zweiten Zahlenklasse und des Fortgangsprinzips von n auf ω war ursprünglich nur mittels eines solchen Axioms möglich. Der Nachweis seiner Zulässigkeit ist durch die von Herrn *G. Cantor* gegebene ausführliche Theorie dieser Zahlen und ihre ausnahmslose Gesetzmäßigkeit als geführt zu betrach-

* Math. Annalen Bd. 60, p. 181.

** G. Cantor, Math. Annalen Bd. 49, p. 207.

taken care of by the preceding discussion simultaneously with *Bernstein*'s. His paper, however, contains further errors and misunderstandings that cannot be ignored here.

To begin with, Mr. *Schoenflies* distinguishes in the theory of well-ordered sets between "a general and a special part" and claims that *only* the theorems of the "general part" rest upon *Cantor*'s familiar *definition* of well-ordering,[34] while the others rest upon "generating principles". In truth, however, *all* theorems concerning a notion must be provable from its definition; otherwise they would not be proved at all. If there are two definitions at hand for a notion, we must opt definitely for one of them or prove the equivalence of the two; to vacillate in any way between two definitions or to complement one by means of the other is logically completely inadmissible.—Mr. Schoenflies then continues:

"If we want to prove the theorem in question on this general basis, we must show that no infinite sequence of sets M_1 with decreasing cardinalities can exist; every decreasing sequence of cardinalities, $\mathfrak{m}_1 > \mathfrak{m}_2 > \mathfrak{m}_3 > \dots$, would have to terminate after a finite number of terms. This is the necessary *and sufficient* condition for the theorem." Necessary but *not* sufficient, and for precisely this reason not usable for a proof of the theorem. The criterion cited applies merely to the well-ordering of the *cardinalities* ordered according to magnitude, and not at all to the well-ordering of the *sets* themselves, which is the subject of my theorem. If all cardinalities are alephs, then, to be sure, they are also well-ordered according to magnitude, but one *cannot* infer the converse, whereas Mr. *Schoenflies*'s proposal apparently requires that. This method does not even allow one to prove that arbitrary sets *can be compared* with respect to their cardinality; rather, the method already presupposes this as a basis. "Zermelo's proof does not follow this path." Indeed not, since a proof just cannot be carried out in that way. "He operates with devices that concern the special part of the theory of well-ordered sets, namely, their generation." Quite on the contrary, my proof rests exclusively upon Cantor's classical definition and has nothing to do with "generating principles", in the sense assumed, as independent sources of proof.

"Generating principles can at first only be *postulated axiomatically*; then one must demonstrate that they are justified. Even the introduction of the numbers of the second number class and of the principle of passage from n to ω was originally possible only by means of such an axiom. The proof that this axiom is admissible was carried out, as we must recognize, by the detailed theory that Mr. *G. Cantor* gave of these numbers, whose regularity admits

[34] *Cantor 1897*.

ten.“ Vielmehr definiert *G. Cantor** die Zahlen der zweiten Zahlenklasse als die *Ordnungstypen*, welche einer wohlgeordneten abzählbaren Menge zukommen können, und *beweist* alles übrige aus dieser Definition. Vorausgesetzt wird lediglich die Existenz abzählbarer Mengen oder der Typus ω, durch den sie definiert sind; weiter braucht es keines Axioms. Wäre aber die Existenz solcher Ordnungstypen irgendwie zweifelhaft, so könnte auch der schönste Formalismus nicht helfen. In analoger Weise wird jede höhere Zahlenklasse vermittels der vorhergehenden definiert, *ohne* daß es dazu „einer Neuschöpfung, resp. eines neuen Axioms und des Nachweises seiner Berechtigung bedarf“. In meinem Beweise hat nun Herr *Schoenflies* gleichfalls ein neues „Postulat“
126 entdeckt, „das die Erzeugung | wohlgeordneter Mengen betrifft. Es besagt nämlich, daß, wenn L irgend eine wohlgeordnete Menge ist, auch (L, m) eine solche ist“. In Wahrheit gar kein „Postulat“, sondern, wie wir oben sahen, ein *beweisbarer Satz.* „Diese Annahme und insbesondere der Gebrauch, den H. Z. von ihr macht, *schließt* so zu sagen *die sämtlichen möglichen Erzeugungsprinzipien in sich ein.* Sie *enthält aber noch mehr*, und gerade deshalb ist sie unzulässig.“ Auch hier verhält sich alles genau umgekehrt, als Herr *Schoenflies* behauptet. Bei *G. Cantor* dienen die „Erzeugungsprinzipien“ *nicht* zur „Herstellung wohlgeordneter Mengen“, sondern zur systematischen Auffindung sämtlicher Ordnungstypen einer gegebenen Zahlenklasse. Hier ist die Hinzufügung eines einzelnen Elementes an das Ende einer Menge von gegebenem Ordnungstypus ξ, also die Operation $\xi + 1$, das „erste Erzeugungsprinzip“, welches innerhalb einer *jeden* Zahlenklasse neue Ordnungstypen liefert und die *Voraussetzung* aller übrigen Erzeugungsprinzipien bildet. Nur *reicht* es schon in der zweiten Zahlenklasse *nicht aus*, da es, von ω ausgehend, nicht einmal $\omega \cdot 2$ erzeugen könnte, und bedarf daher der Ergänzung durch das „zweite Erzeugungsprinzip“, welches sich auf die „Fundamentalreihen“ vom Typus ω bezieht. In den höheren Zahlenklassen bedürfte man *außer* diesen beiden noch *weiterer* Prinzipien. Niemals aber führt das „erste Erzeugungsprinzip“ $\xi + 1$ aus irgend einer Zahlenklasse hinaus, da die *Mächtigkeit* einer transfiniten Menge, welche durch Definition das unterscheidende Merkmal der verschiedenen Zahlenklassen bildet, durch Hinzufügung eines einzelnen Elementes bekanntlich nicht geändert wird. Dementsprechend wird auch in meinem Beweise die Wohlordnung der Gesamtmenge natürlich *nicht* durch *diese* Operation „hergestellt“, sondern, wie ich am Schlusse ausdrücklich bemerkte, durch die „*Verschmelzung* der verschiedenen möglichen γ-Mengen“. Nur kann eben jede einzelne γ-Menge nach diesem Prinzip „fortgesetzt“ werden, worauf ich dann den Schluß gründe, daß $L_\gamma = M$ sein muß. — Sehr seltsam ist weiter die Behauptung: „*Nirgends* bedarf man sonst einer Annahme, wie sie der Zermelosche Beweis benutzt“ — wo doch die ganze Theorie der Ordnungszahlen und Erzeugungsprinzipien auf dieser Operation $\xi + 1$ basiert.

* Math. Annalen Bd. 49, p. 221.

of no exception." Rather, *G. Cantor*[35] defines the numbers of the second number class as the *order types* that can be associated with well-ordered denumerable sets, and he *proves* everything else from this definition. Only the existence of denumerable sets, or the type ω by which they are defined, is assumed; there is no need of a further axiom. But if the existence of such order types were in any way doubtful, not even the most beautiful formalism would be of help. In an analogous way every higher number class is defined by means of the preceding one, *without* "any need of a new creation, or of a new axiom and the proof of its justification". In my proof, too, Mr. *Schoenflies* discovered a new "postulate", "which concerns the generation of well-ordered sets. For it asserts that, if L is any well-ordered set, so is (L, m)". Actually not a "postulate" at all, but, as we saw above, a *provable proposition.* "This assumption, and in particular the use that Mr. Z⟦ermelo⟧ makes of it, so to speak *includes within itself all possible generating principles.* It *contains even more, however*, and for precisely this reason it is inadmissible." Here, too, everything is just the opposite of what *Schoenflies* asserts. *G. Cantor* does *not* use the "generating principles" for the "production of well-ordered sets" but for the systematic discovery of all order types of a given number class. For him the addition of a single element at the end of a set of given order type ξ, hence the operation $\xi + 1$ is the "first generating principle", which yields new order types within *each* number class and constitutes the *presupposition* of all other generating principles. But already in the second number class it *fails to suffice*, since, beginning with ω, it could not even generate $\omega \cdot 2$ and hence requires supplementation by the "second generating principle", which relates to the "fundamental sequences" of type ω. In the higher number classes one would require still *further* principles *besides* these two. But the "first generating principle" $\xi + 1$ never leads us out of any number class, since the *cardinality* of a transfinite set, which by definition forms the distinguishing characteristic of the various number classes, is, as is well known, not changed by the addition of just a single element. Accordingly, the well-ordering of the entire set is, of course, *not* "produced" by means of *this* operation in my proof either, but, as I expressly noted at the end, by the "*fusion* of the various possible γ-sets". It is only that each single γ-set can be "continued" in accordance with this principle, a fact upon which I then base the inference that necessarily $L_\gamma = M$. Very strange, furthermore, is the assertion that "*nowhere* else is an assumption required such as Zermelo's proof employs"—when in fact the entire theory of ordinals and generating principles is based upon this operation $\xi + 1$.

[35] *Cantor 1897*, p. 221.

Jetzt kommt aber der Hauptpunkt. „Der Begriff der *Gesamtheit* aller überhaupt möglichen Erzeugungsprinzipien wohlgeordneter Mengen ist meines Erachtens ein *wohldefinierter mengentheoretischer Begriff*, ebenso der Begriff der mit ihnen herstellbaren wohlgeordneten Mengen ... in demselben Sinn ... wie die Gesamtheit der ganzen Zahlen.“ Damit wird also das Dogma vom W verkündigt. Auf eine Rechtfertigung dieses widerspruchsvollen Begriffes, die Herr *F. Bernstein* doch wenigstens *versucht* hatte, läßt Herr *Schoenflies* sich gar nicht erst ein, sein „Erachten“ soll uns hier genü-
127 gen. Die nun folgende Unterscheidung zwischen „herstellbaren“ | und „nicht herstellbaren“ wohlgeordneten Mengen ist nicht recht zu verstehen, zumal schon im nächsten Absatze eine „nicht herstellbare“ Menge ohne weiteres auch als „logisch widerspruchsvoll“ bezeichnet wird. „Jedenfalls gelangen wir zu einer ebenfalls *wohldefinierten Gesamtheit* wohlgeordneter Mengen. ... Die dadurch bestimmte wohlgeordnete Menge sei Z. Ihrer Definition nach gibt sie die Grenze an, über die wir bei der wirklichen Herstellung einer wohlgeordneten Menge niemals hinauskommen.“ Offenbar soll dieses Z die Gesamtheit der möglichen *Ordnungstypen* solcher Mengen darstellen, aber diese von *G. Cantor* so sorgfältig getrennten Begriffe einer geordneten „Menge“ und ihres „Ordnungstypus“ werden in dem Schoenfliesschen Artikel überhaupt nicht unterschieden, eine Ungenauigkeit, die, wie wir sogleich sehen werden, auch nicht ohne verhängnisvolle Folgen geblieben ist.

„Nehmen wir zunächst einmal an, daß Z die zweite Zahlenklasse ist, d. h. also, daß alle wohlgeordneten Mengen, die wir ... bilden können, niemals über die zweite Zahlenklasse hinausführen. ... In diesem Falle stellt die Menge (Z, m) einen in sich widerspruchsvollen Begriff dar. ...“ Hiernach scheint also Herr *Schoenflies* den Ordnungstypus von (Z, m) für den *ersten* zu halten, welcher nicht mehr der zweiten Zahlenklasse angehört. Nun hat aber *Cantor* bewiesen, daß *die Gesamtheit der Zahlen der zweiten Zahlenklasse selbst nicht mehr abzählbar* ist, ihr Ordnungstypus also bereits zur nächstfolgenden Zahlenklasse gehört, während man mittelst des „ersten Erzeugungsprinzipes“ $\xi + 1$ immer nur Zahlen derselben Klasse erhält. Sonach wäre also auf Grund der gemachten Annahme nicht erst (Z, m), sondern bereits *Z selbst ein widerspruchsvoller Begriff*. Und in der Tat ist das *Schoenflies*sche Z genau so wie das *Bernstein*sche W, weil es eben seinen eigenen Ordnungstypus nicht als Element enthalten kann, wie wir oben sahen, unter allen Umständen mit inneren Widersprüchen behaftet.

„Auf vorstehender Grundlage“ wird nun eine neue Methode zur Prüfung des fraglichen Satzes vorgeschlagen, welche, wie Herr *Schoenflies* sich nachzuweisen bemüht, gleichfalls zu *keinem Resultate* führt. Gewiß gibt es immer mancherlei Methoden, ein gegebenes Theorem *nicht* zu beweisen, und namentlich verschwommene und widerspruchsvolle Begriffe dürften zur Vermeidung von Beweisen besonders geeignet sein. Nur kann man wirkliche Beweise durch solche Hilfsmittel natürlich auch nicht widerlegen. Das vorgeschlagene Verfahren besteht aber darin, daß der wohlgeordnete Bestandteil L der be-

But now we come to the main point. "The notion of the *totality* of all possible generating principles for well-ordered sets is in my opinion a *well-defined set-theoretic notion*, as is the notion of the well-ordered sets that can be produced with them ... in the same sense ... as the totality of all integers." Thus the dogma of W is proclaimed. Mr. *Schoenflies* does not even venture to justify this inconsistent notion, as Mr. *F. Bernstein* after all had at least *attempted* to do; his "opinion" is to suffice for us here. The distinction that he then makes between well-ordered sets that "can be produced" and those that "cannot be produced" is not really intelligible, especially since already in the next paragraph a set that "cannot be produced" is also characterized without further ado as "logically inconsistent". "In any case we arrive at a *totality* of well-ordered sets that is likewise *well-defined.* ... Let Z be the well-ordered set thus determined. According to its definition, it provides the limit beyond which we never proceed when we actually produce a well-ordered set." Apparently this Z is to be the totality of the possible *order types* of such sets, but the notions of an ordered "set" and its "order type", so carefully differentiated by *G. Cantor*, are not distinguished at all in Schoenflies's paper, an imprecision that, as we shall see immediately, has not remained without fatal consequences either.

"Let us assume for the time being that Z is the second number class, that is, that no well-ordered set that we ... can form ever takes us beyond the second number class. ... In that case the set (Z, m) represents a notion that is inconsistent in itself." Thus according to this passage Mr. *Schoenflies* seems to regard the order type of (Z, m) as the *first* that no longer belongs to the second number class. But now *Cantor* proved that *the totality of the numbers of the second number class is itself no longer denumerable*, hence that its order type already belongs to the next higher number class, whereas when we employ the "first generating principle" $\xi + 1$ we always remain within the same number class. Accordingly, on the basis of the assumption made, not only (Z, m) but already Z would *itself be a contradictory notion.* And in fact *Schoenflies*'s Z, just like *Bernstein*'s W, is afflicted with internal contradictions in any case, precisely because, as we saw above, it cannot contain its own order type as an element.

"On the preceding basis" a new method for the examination of the theorem in question is then proposed, a method that, as Mr. *Schoenflies* strains to show, also leads to *no result.* There are always various methods of *not* proving a given theorem, to be sure, and in particular it would seem that vague and inconsistent notions are especially appropriate when one wants to avoid giving a proof. But then, of course, one cannot refute genuine proofs by means of such devices. The proposed procedure, however, consists in comparing the well-ordered component L of the set M under consideration with the sets W

trachteten Menge M mit den oben charakterisierten Mengen W und Z verglichen wird. Dabei ergeben sich dann verschiedene Fälle, welche alle logisch gleich möglich sein sollen, während nur der eine von ihnen den in meinem
128 Beweise gemachten An- | nahmen entspricht. Das kann natürlich nicht wundernehmen, nachdem Herr *Schoenflies* alle diejenigen Widersprüche, welche zur Ausschließung der übrigen Fälle führen, bereits in seine Voraussetzungen aufgenommen hat.

Wenn also Herr *Schoenflies* am Schlusse seines Artikels bemerkt: „Meines Erachtens sollte man mit Annahmen, die zu widerspruchsvollen Begriffen oder Resultaten führen, auch in der Mengenlehre ebenso verfahren, wie man es sonst zu tun pflegt", so kann man gewiß beistimmen. Dieses Verfahren besteht nämlich darin, daß man solche Annahmen ausschließt und aus widerspruchsvollen Begriffen keine Folgerungen ableitet. In meinem Beweise ist dieses Verfahren auch streng eingehalten, nicht aber in der Kritik des Herrn *Schoenflies.*

e. Zusammenfassung

Die vorstehende Erörterung der gegen meinen Beweis von 1904 gerichteten Opposition läßt sich wohl am einfachsten in die folgenden Sätze zusammenfassen. Während Herr *Poincaré* mit seiner formal-logischen Kritik, welche die Existenz der gesamten Mathematik bedrohen würde, bisher noch auf keiner Seite Zustimmung gefunden hat, lassen sich alle übrigen Gegner in zwei Klassen einteilen. Die einen, welche gegen meine Deduktionen durchaus nichts einzuwenden haben, beanstanden die Anwendung eines unbeweisbaren allgemeinen Prinzipes, ohne zu bedenken, daß solche Axiome jeder mathematischen Theorie zugrunde liegen und daß gerade das von mir herangezogene für den Ausbau der Wissenschaft auch sonst nicht entbehrt werden kann. Die anderen Kritiker dagegen, welche sich durch eingehendere Beschäftigung mit der Mengenlehre von dieser Unentbehrlichkeit überzeugen konnten, gründen ihre Einwände auf die „*Burali-Forti*sche Antinomie", die für meinen Standpunkt tatsächlich *ohne Bedeutung ist*, da die von mir benutzten Prinzipien die Existenz einer Menge W *ausschließen.*

Die verhältnismäßig große Anzahl der gegen meine kleine Note gerichteten Kritiken ist ein Zeugnis dafür, daß dem Satze von der möglichen Wohlordnung beliebiger Mengen offenbar starke Vorurteile im Wege stehen. Die Tatsache aber, daß man in meinem Beweise trotz eingehender Prüfung, für die ich allen Kritikern zu Dank verpflichtet bin, keine *mathematischen Irrtümer* hat nachweisen können, und daß die gegen meine *Prinzipien* erhobenen Einwände einander widersprechen und so sich gewissermaßen gegenseitig aufheben, läßt mich hoffen, daß sich alle diese Widerstände durch genügende Aufklärung mit der Zeit wohl werden überwinden lassen.

Chesières, den 14. Juli 1907.

and Z characterized above. This, then, yields several cases, all of which are said to be logically equally possible, whereas only one of them corresponds to the assumptions made in my proof. This cannot surprise us, of course, after Mr. *Schoenflies* has already incorporated into his assumptions all those contradictions that lead to the exclusion of the other cases.

Thus, when Mr. *Schoenflies* remarks at the end of his paper that "in my opinion we should, in set theory, treat assumptions that lead to inconsistent notions or results just as we are accustomed to treating them elsewhere", we can surely concur. For this means that such assumptions are to be excluded and that no consequences should be derived from inconsistent notions. In my proof this procedure has been strictly observed, too, but not in Mr. *Schoenflies*'s critique.

e. Summary

The preceding discussion of the opposition to my 1904 proof can perhaps be summarized most simply by the following statements. Except for Mr. *Poincaré*, whose critique, based on formal logic—a critique that would threaten the existence of all of mathematics—has hitherto not met with any assent whatsoever, all opponents can be divided into two classes. Those who have no objection at all to my deductions protest the use of an unprovable general principle, without reflecting that such axioms constitute the basis of every mathematical theory and that precisely the one I adduced is indispensable for the extension of the science in other respects, too. The other critics, however, who have been able to convince themselves of this indispensability by a deeper involvement with set theory, base their objections upon the "*Burali-Forti* antinomy", which in fact *is without significance* for my point of view, since the principles I employed *exclude* the existence of a set W.

The relatively large number of criticisms directed against my short note testifies to the fact that, apparently, strong prejudices stand in the way of the theorem that any arbitrary set can be well-ordered. But the fact that in spite of a searching examination, for which I am indebted to all the critics, no *mathematical error* could be demonstrated in my proof and the objections raised against my *principles* are mutually contradictory and thus in a sense cancel each other allows me to hope that in time all of this resistance can be overcome through adequate clarification.

Chesières, on the 14th of July 1907.

Introductory note to 1908b

Ulrich Felgner†

Since antiquity mathematicians have spoken of classes, multitudes, and sets, but set theory as such is a product of the 19th century (Bernard Bolzano, Georg Cantor, Richard Dedekind, et al.). Not until the 20th century did set theory attain an axiomatic foundation (Ernst Zermelo, Thoralf Skolem, Johann von Neumann, Paul Bernays, Wilhelm Ackermann, et al.).

In this essay we discuss Zermelo's paper "Investigations in the foundations of set theory I" (*Investigations* for short). In order fully to appreciate Zermelo's research we must discuss the 19th century emergence of the concept of set as a *terminus technicus* of mathematics as well as the axiomatic method with all its many facets.

§1. The 19th century emergence of the concept of set

In the course of the 19th century mathematicians had to recognize that the continued development of mathematics required deeper foundations. In particular, the desire to accept proofs only on the basis of the logico-conceptual argumentation led increasingly to the inclusion of extensions of concepts (Begriffsumfänge) and then also of sets and classes in mathematical considerations.

In order to construct a theory of sets, it was first necessary clearly to formulate the fundamental concept of "set". This concept is one for which there is almost no intuitive foundation and whose intended content is rather difficult to grasp. In particular, the ontological status of sets had to be clarified. If sets are nothing more than "the mere being together" of various things, then it is questionable whether this coexistence can be treated as an autonomous independent object. In what manner do sets exist?

It was Bernard Bolzano (1781–1848), the Bohemian-Austrian mathematician, philosopher, and theologian, who first systematically studied the concept of set. He began to construct a theory of sets, publishing his results in his *Wissenschaftslehre* (*1837*). Further results appeared posthumously in his *Einleitung in die Größenlehre* (*1975*) and in his booklet *Paradoxien des Unendlichen* (*1851*).

While in previous centuries the word "set" (πλῆθος, multitudo, Vielheit, Menge, aggregate, ...) was used only to name an undetermined accumulation

† The author wishes to thank Hartmut Grandel, Allen Guttmann, and Akihiro Kanamori for their helpful comments on a previous version of the paper. He owes a special debt to Heinz-Dieter Ebbinghaus for making many valuable suggestions for improving the manuscript.

of objects, Bolzano added new meanings to this word. In fact, he added four attributes to the notion of a set. Without these new attributes it would not have been possible to transform the concept of a set from colloquial language into a mathematical *terminus technicus.*

The first attribute is that sets are uniquely determined by their extension, i.e., by the totality of their elements. In contrast to this, structured sets ("Inbegriffe" in Bolzano's terminology) are not determined by their extension alone. Notice also that in general the principle of extensionality is not valid for concepts. Bolzano wrote (*1975*, 152):

> Die Theile, aus denen eine Menge besteht, bestimmen sie, und zwar vollständig und alle auf einerley Art.[1]

About 40 years later Dedekind formulated this principle independently in the first paragraph of his celebrated essay *Was sind und was sollen die Zahlen?*, *1888*. Zermelo introduced the name "Axiom der Bestimmtheit". It is now customary to call this the axiom of extensionality, this being a heritage from Antoine Arnauld, who in 1662 distinguished in his *Logique de Port-Royal* between the extension and the intension of an idea. Cantor in *1878* and Dedekind in *1888* proposed referring to "elements" instead of "parts".

The second attribute is that sets can be constituted of objects from different species. Here the word "species" (εἶδος) should be understood in the Aristotelian sense. Since antiquity it was not usual in mathematics to collect things from different "genera" or "species" into a whole. For example, introducing the concept of number in his *Dictionaire Mathématique* (Amsterdam 1691), Jaques Ozanam (1640–1717) admitted only sets of objects of the same species. He wrote: "Le Nombre est l'assemblage de plusiers choses de même genre" (op. cit., 21). An exception can be found in the work of Gottfried Wilhelm Leibniz (1646–1716). In his *Dissertatio de arte combinatoria* (1667) he postulated the existence of finite multitudes of objects from different species as totalities. He wrote:

> ...liceat quotcunque res simul sumere, et tanquam unum totum supponere.[2]

Bolzano often stressed that sets can consist of objects of different species (cf. *1975*, 101–102).

The third attribute is that sets need not be definable. A set need not be the extension of a concept (Begriffsumfang, étendue de l'idée in the sense of Arnauld's *Logique de Port-Royal*). A set need not be thought of by some person and need not be given by an explicit enumeration of all its elements. In his booklet *Paradoxien des Unendlichen* (*1851*, §14, 17), Bolzano wrote:

[1] The parts out of which a set consists determine the set completely and all in a uniform manner.

[2] ...that it is permitted to comprehend anything whatsoever, and to assume that taken together they form a whole.

> Es gibt also Mengen..., auch ohne daß ein Wesen, welches sie denkt, da ist.[3]

The fourth attribute is that sets are "things" in the sense that they have an immutable form of existence. When previous authors spoke about multitudes or collections, they did not presume that these multitudes or collections were objects which have an independent existence. The word "multitude" (πλῆθος, multitudo, Vielheit, Menge, ...) always meant an undetermined number of several individual objects, but the multitude itself was not considered a substantial object.[4] The following quotation from a letter of Leibniz to Johann Bernoulli from February 21, 1699, may serve as an example for this widespread opinion (cf. *Leibniz 1856*, 575):

> Concedo multitudinem infinitam, sed haec multitudo non facit numerum seu unum totum.[5]

Bolzano defines sets as the mere being together of individual objects, and this coexistence of the objects is itself an object. For him such an object has an extra-mental reality. In his *Einleitung in die Größenlehre* he wrote (*1975*, 100–101):

> ... und ich bemerke, daß die erwähnten Dinge mit Wahrheit nicht zusammengedacht werden könnten, wenn sie nicht auch, ohne daß wir sie noch zusammendenken, schon zusammen wären, schon ... ein Ganzes bildeten, ein *Ding*, welchem gewisse Beschaffenheiten zukommen, die keinem der einzelnen Dinge, seinen Theilen, für sich allein zukommen (müssen). Ich behaupte also, daß Inbegriffe bestehen nicht dadurch, daß wir sie denken, sondern umgekehrt, daß wir nur dann sie mit Wahrheit denken können, wenn sie bestehen auch ohne daß wir sie denken.[6]

Bolzano's position here is comparable to that of the "realists" in the medieval debate on universals (Universalienstreit). However, the term "realism" means one thing with regard to sets (Mengenrealismus) and quite another

---

[3] Hence sets exist ... even if there is no one who thinks of them.

[4] Compare this with the detailed discussion of the word "multitude" (πλῆθος) in Plotinus' treatise *On Numbers* (Περὶ ἀριθμῶν), *Von den Zahlen*; cf. *Plotinus 1964*, section 12, 190–193.

[5] I concede that they (i.e. the terms of an infinite geometric series) do form an infinite multitude, but such a multitude has no number and is not a whole.

[6] ... and I note that the things mentioned above cannot truly be thought of together when they are not yet already together, even before our thinking of them as together, and when they are not yet a whole, a *thing*, whose properties are not (necessarily) properties of its elements. Hence, I claim that sets exist not through our thinking, but conversely, that we are able to think of them only if they also exist without our thinking of them.

with regard to universals (Universalienrealismus), because universals are intensionally determined objects while sets are extensionally determined.

Since then, other mathematicians have also consciously and explicitly asserted the objecthood of sets.[7] They have differed, however, in their conception of what is meant by "object". These differences did not affect the development of set theory in the 19th century, but they became quite important when the paradoxes and antinomies of set theory appeared. We shall come back to this point later.

In the very first paragraph of his essay *1888*, p. 1, Dedekind wrote:

> Ein solches System S (oder ein Inbegriff, eine Mannigfaltigkeit, eine Gesamtheit) ist als Gegenstand unseres Denkens ebenfalls ein Ding.[8]

Similarly, on the very first page of his book *Grundzüge der Mengenlehre* (*1914*), Felix Hausdorff wrote:

> Eine Menge ist eine Zusammenfassung von Dingen zu einem Ganzen, d. h. zu einem neuen Ding.[9]

The word "set" is now much more than a mere manner of speaking. Sets are now understood to have the attribute of objecthood. They may occur as elements in other sets. This was emphasized explicitly by Bolzano (*1975*, 102):

> Die Theile, aus denen ein Inbegriff bestehet, können selbst wieder Inbegriffe seyn.[10]

That new sets can be iteratively generated from existing sets would become a basic feature of set theory after Zermelo.

Bolzano gave a simple real definition ("Real-Definition") for the concept of set. For him sets were the "mere being together" of different objects. Some 60 years later, Cantor proposed a much more manageable definition. He wrote (*1895*, 481):

> Unter einer "Menge" verstehen wir jede Zusammenfassung M von bestimmten wohlunterschiedenen Objekten m unserer Anschauung oder unseres Denkens (welche die "Elemente" von M genannt werden) zu einem Ganzen.[11]

---

[7] This is also emphasized by Michael Hallett in his book *1984*, 34–35. He writes (p. 299): "It was Cantor who first stressed the unity, the objecthood of sets, thereby marking them out from mere aggregations." But Cantor was not the first. Much earlier Bolzano and others pointed to the objecthood and unity of sets.

[8] As an object of our reasoning such a system (or set, multitude, totality) is also a thing.

[9] A set is a comprehension of things into a whole, i.e. into a new thing.

[10] The elements which constitute a set can themselves be sets.

[11] By a "set" we understand any comprehension M of definite, well-differentiated objects m of our intuition or our thought (which are called the "elements" of M) into a whole.

Cantor's definition of the concept of set is widely acknowledged and routinely cited. It is remarkable, however, that Cantor's is a "genetic" definition. It not only indicates what sets are, but also how they can be created. We shall return to this in more detail in §3.

§2. The appearance of the set-theoretic antinomies

Bolzano early on foresaw that working with objects that can theoretically contain anything and everything can cause inconsistencies. In his Mathematical Diary ("Mathematisches Tagebuch", Heft 22, p. 1968), he wrote:[12]

> *Das All der A*, wo *A* eine Vorstellung von beschränktem Umfange bezeichnet, z. B. Mensch — läßt sich recht wohl denken. Sobald man aber für *A* die weiteste aller Vorstellungen, nämlich die eines Etwas überhaupt setzt, so entsteht die Schwierigkeit, daß die Vorstellung *das All der Gegenstände* oder *das All von Allem* oder *das absolute All*, eigentlich auch sich selbst, weil ja dieses All auch wieder *Etwas* ist, umfassen sollte; welches doch ungereimt ist. Ich glaube deshalb, daß man diese Vorstellung in der Tat zu den widersprechenden (imaginären) zählen müsse, gerade wie die Vorstellung von der geschwindesten Bewegung. — Aber auch schon das All der Inbegriffe wäre eine solche sich selbst widersprechende Vorstellung.[13] [14]

It seems that Bolzano had some second thoughts. He added the following note:

> Responsio. Nicht doch! Diese Begriffe sind nicht widersprechend.[15]

Unfortunately, Bolzano did not pursue the antinomy of the universal set (i.e., the set of all sets), which he had thus adumbrated.

George Boole (1815–1864) also encountered the concept of the universal set ("class 1", as he called it) "which comprehends every conceivable class of objects," but he did not recognise its antinomic character (cf. *Boole 1847*, 15). The same is true of Dedekind when he referred to the totality of all things

---

[12] Compare also with *Bolzano 1827/44*, 29.

[13] Quoted in *Sebestik 2000*, 236.

[14] *The universe of all A*, where *A* is an idea of limited extension, e.g. a human being, can be thought of without any difficulties. But as soon as one takes for *A* the largest possible idea, namely the idea of an object in general, then a difficulty arises in that the idea of *the universe of everything*, or *the universe of all*, or *the absolute all*, will contain itself, since this universe is also *something*, which is absurd. Therefore, I think that in fact one should classify this idea along with other self-contradictory (imaginary) ideas, such as the idea of the fastest movement.—But already the universe of all collections would turn out to be such a self-contradictory idea.

[15] Answer. Hold on! These notions are not contradictory.

which might happen to be objects of his thinking ("die Gesamtheit S aller Dinge, welche Gegenstand (seines) Denkens sein können", *Dedekind 1888*, 14, art. 66).

Cantor also encountered the universal set ("Allmenge"). In contrast to Bolzano, Boole, and Dedekind, he acknowledged its antinomic character and tried to expel the concept of a universal set from his set theory. He distinguished in the years 1883 and 1887 three types of sets:

finite sets,
transfinite sets, and
absolutely infinite sets.

Infinite sets are transfinite if they can be embedded within sets which have a greater potency. Their size can be measured by the so-called alephs. In Cantor's opinion, both finite and transfinite sets can be used in mathematical reasoning. He was proud of having secured the domain of transfinite sets for mathematics. His comprehensive two-part treatise, *Cantor 1895* and *1897*, in which he presented his mature work in set theory, carried the highly significant title "Beiträge zur Begründung der transfiniten Mengenlehre" ("Contributions to the foundations of transfinite set theory").

As Cantor explained in a letter of October 5, 1883 to the philosopher Wilhelm Wundt, the absolutely infinite sets are those sets "die nicht mehr vergrößert werden können" ("which cannot be enlarged with regard to their potency"). The size of an absolutely infinite set cannot be measured by an aleph. The adjective "absolute" is meant to indicate that such infinities are so large that they transcend the scale of norms of comparable sizes.[16] Cantor spoke about the "absolute" and the "absolut infinite" for the first time in 1883 in his treatise "Grundlagen einer allgemeinen Mannigfaltigkeitslehre" (cf. *Cantor 1883b*, 556 and 587). He returned to this problem in his "Mitteilungen zur Lehre vom Transfiniten" (*Cantor 1887*, 91).

Cantor thought that the absolute infinite eludes mathematical determination (cf. *Cantor 1887*, 109). In his letter to Grace Chisholm Young of June 6, 1908, he wrote:

> Was über dem Finiten und Transfiniten liegt, ist ... das "Absolute", für den menschlichen Verstand Unfassbare, also der Mathematik gar nicht unterworfene, [das] Unmessbare.[17]

These are highly emotional sentences, but they are not quite accurate. It would have been more appropriate to say that, in the calculus of set theory,

---

[16] The Latin word "absolutus" means "unconditional, freed of all conditions, not depending on anything else, detached". The verb "solvere" means "to solve, to dissolve, to untie, to free".

[17] That which lies beyond the finite and the transfinite, ... is the absolute, for the human intellect ungraspable, and immensurable, hence not a subject of mathematics.

absolute infinities should not be treated in the same way as transfinite infinities. Cantor did not know how to introduce the concepts of finite and transfinite sets so that absolute infinite sets could be excluded from the domain of objects under consideration. Apparently, he would have liked to precondition his set theory with a ban on dealing with absolutely infinite sets, but such prohibitions are prohibited in the composition of a mathematical theory.

Cantor himself did not employ absolutely infinite sets in his mathematical reflections, but other mathematicians came across them and stumbled.

In 1897, Cesare Burali-Forti noticed a problem in Cantor's theory of order types. If Ω is the order type of the set S of all order types of well-ordered sets, then the set $S \cup \{S\}$ has order type $\Omega + 1$. But Ω is the order type of the set of all order types of well-ordered sets, hence nothing can be larger than Ω, a contradiction! Burali-Forti's paper was published in the spring of 1897.[18]

When Cantor became aware of Burali-Forti's paper, he was alarmed.[19] He had realized for quite some time that the domains of all ordinal numbers and of all cardinal numbers (Cantor always spoke of the "set" of all ordinal numbers and the "set" of all cardinal numbers) were absolutely infinite, but now he had to acknowledge that his genetic definition of the concept of set did not differentiate between the creation of a transfinite set and the creation of an absolutely infinite set. During the summer of 1897, he began to modify his definition of the concept of set. We will return to this in §4.

Cantor also noticed contradictions in Dedekind's work on the foundations of arithmetic. Dedekind's "Gedankenwelt" (the totality of the thinkable) is an absolute infinite totality whose size cannot be measured by any cardinal number (cf. *Dedekind 1888*, Theorem 66, and Cantor's letters to Hilbert of November 15, 1899, and January 27, 1900, reprinted in *Cantor 1991*). That this was also apparent to Zermelo can be seen in the account of Dedekind's essay that he wrote for Edmund Landau (see *Landau 1917b*).

It was quite easy for Zermelo to demonstrate that working with Bolzano's "Allmenge", Boole's "class 1", and Dedekind's "Gedankenwelt" leads to inconsistencies. All these classes M have the property that $\mathsf{P}(M) \subseteq M$ where

$$\mathsf{P}(M) = \mathfrak{U}(M) = \{X; X \subseteq M\}$$

---

[18] Burali-Forti (1861–1931) was an assistant to Guiseppe Peano in Turin in the years from 1894 until 1896 and hence familiar with the most recent developments in logic and set theory of his time. Burali-Forti presented his paper on transfinite ordinals in the meeting of the Circolo Matematico di Palermo on March 28, 1897. It appeared in print soon afterwards (cf. *Burali-Forti 1897*). The argument of Burali-Forti is strongly reminiscent of an argument which was given already by the Greek philosopher Plotinus (cf. *Plotinus 1964*, section 17–18, pp. 208–213) and also much later again by Bolzano (see above).

[19] Cantor knew Burali-Forti's paper, since in a letter to Dedekind of August 3, 1899, he talks about the Burali-Forti antinomy and even uses Burali-Forti's notation. However, he nowhere mentions Burali-Forti.

is the power set of M (in Zermelo's *Investigations* always denoted by $\mathfrak{U}(M)$ as the set of all "Untermengen" of M). According to Cantor this is impossible by reasons of cardinality (*Cantor 1890/91*). However, Zermelo realized that neither cardinality arguments nor the introduction of power sets (i.e. higher order objects) are necessary in this context. In fact, he showed that each set M has a subset A which cannot be an element of M. Such a subset is, for example, $A = \{x \in M; x \notin x\}$. If $A \in M$ were valid, it then would follow that $A \in A \Leftrightarrow A \notin A$, which is a contradiction. This argument appears in theorem 10 of the *Investigations.* Zermelo explained this elegant argument in April 16, 1902 to the philosopher Edmund Husserl, one of his Göttingen colleagues (cf. *Husserl 1979*, 399; see also *Rang and Thomas 1981*). Zermelo returned to this argument in *s1932d*, which suggests that Zermelo was quite pleased with it.[20]

In his letter of November 7, 1903 to Frege, Hilbert testified that Zermelo worked out his argument sometime around 1900 and made it public in Göttingen at that time. Hilbert's testinomy is supported by a contemporary document, namely Zermelo's (unpublished) notes for his Göttingen lectures on set theory during the winter semester 1900/1901. In these notes there is an entry which shows that he gave a proof of Cantor's well-known theorem, stated here in terms of power sets rather than as Cantor did in terms of functions:

> Theorem (1890): *For every set M it holds that there is no surjection of M onto its power set* $\mathsf{P}(M)$.

But Zermelo gave a slightly different formulation of the theorem (see the last line on the first page of his §6):

> *For every set M there is no injection of* $\mathsf{P}(M)$ *into M.*

Proof. Suppose that an injective mapping ψ from $\mathsf{P}(M)$ into M existed. Consider $B = \{\psi(x);\ \psi(x) \notin x\}$.

If $\psi(B) \in B$ held, then it would follow that there is an $x \in \mathsf{P}(M)$ such that $\psi(B) = \psi(x)$ and $\psi(x) \notin x$. Since ψ is injective, we get $x = B$, and hence $\psi(B) \notin B$, a contradiction.

If $\psi(B) \notin B$ held, then $\psi(B)$ would satisfy the defining condition of B and hence $\psi(B) \in B$, again a contradiction.

Thus: $\psi(B) \in B \Leftrightarrow \psi(B) \notin B$. The assumption that an injective mapping from $\mathsf{P}(M)$ into M exists is therefore disproved. □

Zermelo communicated his modification of Cantor's theorem to Gerhard Hessenberg as well, and he, following Zermelo's notes and notation, included it in his book *Grundbegriffe der Mengenlehre* (*Hessenberg 1906*, 41–42). This modification is also contained in Zermelo's *Investigations* (see his Theorem 32 on p. 276).

[20] This argument is related to the ancient antinomy of the liar. A detailed account of its historical roots and its effects in later times is given in *Felgner 2009*.

Cantor's proof is faintly reminiscent of Euclid's proof that for any finite number of prime numbers there is always a further prime number (see book 9 of Euclid's *Elements*). Analogously, Cantor proved that for any mapping φ of M into $\mathsf{P}(M)$ there are always some additional subsets beyond the range of φ.

It seems that Zermelo's modification of Cantor's theorem resulted from an attempt to give direct proofs for the inconsistencies of Bolzano's "Allmenge", Boole's "class 1", and Dedekind's "Gedankenwelt" as "sets". In fact, when M is a class such that $\mathsf{P}(M) \subseteq M$, then the identity mapping is an injective mapping from $\mathsf{P}(M)$ into M and the proof of the modified version of Cantor's theorem immediately produces the set of all sets which do not contain themselves as elements, $A = \{x;\, x \notin x\}$. The equivalence $A \in A \Leftrightarrow A \notin A$ which follows immediately from the assumption that A is a set, was also observed by the English logician Bertrand Russell (1872–1970) at about the same time and hence will be called the Zermelo-Russellian antinomy.

It is remarkable that Bolzano had already foreseen that in the presence of some elementary assumptions on the concept of set the concept of universal set ("Allmenge") might lead to inconsistencies. In fact, it does lead to inconsistencies. As we have seen above, the equivalence $A \in A \Leftrightarrow A \notin A$ occurs only because we assumed that $A = \{x;\, x \notin x\}$ belongs to the domain of those objects through which the variable x ranges (i.e. to the domain of all sets). If we do not assume that $\exists x : x = A$ holds, then the argument given above cannot be carried through, since the variable x can no longer be assigned to A and the antinomy disappears. This gives us a hint of how to avoid the antinomy.

(a) In Zermelo's axiomatic set theory and also in the extended version proposed by Zermelo, Thoralf Skolem, and Adolf Abraham Fraenkel, called ZF (or more accurately ZSF), the above-mentioned antinomy can be avoided by introducing expressions of the form

$$\{x;\, \Phi(x)\} \quad (= \text{class of all } x, \text{ that satisfy the property } \Phi)$$

as terms (and hence as "things") only when $\exists y \forall x (x \in y \leftrightarrow \Phi(x))$ is provable from the axioms. (Here $x, y, z, \ldots$ are variables running through the domain of all sets.) When

$$\exists y \forall x (x \in y \leftrightarrow \Phi(x))$$

is not provable from the axioms, then, according to a proposal made by Willard van Orman Quine, expressions of the form $\{x;\, \Phi(x)\}$ are still admitted, but they have to be treated in the strict nominalistic sense. This means that for arbitrary terms t the expressions $t \in \{x;\, \Phi(x)\}$ are admitted as formulas, but $\{x;\, \Phi(x)\}$ alone is not a term and is hence not treated as a *thing* (i.e. a set). In this case $\{x;\, \Phi(x)\}$ is called a "virtual class" (cf. *Quine 1963*, 15–21).

(b) In von Neumann-Bernays-Gödel axiomatic set theory, NBG, the antinomy is avoided by introducing two sorts of variables, variables for sets

$x, y, z, \ldots$ and variables for classes $X, Y, Z, \ldots$, and also by introducing two sorts of terms, set terms and class terms, where all comprehension terms $\{x;\, \Phi(x)\}$ are introduced as class terms. It follows from the axioms of the two-sorted predicate logic that for each class term $\{x;\, \Phi(x)\}$ we have $\exists X : X = \{x;\, \Phi(x)\}$. Accordingly, in NBG set theory class terms can be treated as "things". They are not "sets" but things of a different sort (or genus) and are called "classes".

Our discussion has shown that one should not treat all comprehension terms either as "things" or as "things of one and the same species". The fourth attribute, which was added to the notion of set by Bolzano (see above) makes possible the construction of a hierarchy of transfinite sets, but this attribute can also lead to inconsistencies if applied carelessly. Thus, if one does not want to exclude comprehensions, a careful use is possible, because they can be admitted either

(a) in a strictly nominalistic sense as linguistic objects,

or

(b) as "things" of a higher genus,

thus avoiding in both cases (at least directly) the appearance of the Zermelo-Russellian antinomy. But our discussion has also shown that one should not use ontological assumptions in too unrestricted a manner. A certain parsimony (parsimonia ontologiae) is always necessary. We conclude that the concept of "set" as defined by Bolzano, Cantor, and others is not as clear and unambiguous as they had maintained. One obtains clarity only by introducing the notion of set axiomatically.

§3. Remarks on the axiomatic method

Antiquity has bequeathed us two different methods of theory construction. One of them is the kind of axiomatics found in Euclid's *Elements* (Στοιχεῖα); the other—which is another kind of axiomatics—is the theory of science presented by Aristotle in the first book of his *Posterior analytics* (*Analytica posteriora*, Ἀναλυτικὰ ὕστερα).

In spite of many differences, both forms of axiomatics begin with explicit definitions of fundamental concepts and then proceed to postulates (Euclid's αἰτήματα) or hypotheses (Aristotle's ὑπόθεσεις). While in Euclid the status of the definitions is not very clear, it is quite clear and distinct in Aristotle.

Aristotle distinguishes among three different types of definitions (cf. *Analytica posteriora*, book 2, chap. 10, and *Topica*, books 6 and 7). The most important one is the *definitio essentialis* (ὅρος οὐσιώδης, Wesens-Definition, Essential-Definition), the definition of the essential features of an object. Such a definition names all the properties which an object has to satisfy in order to be defined as such. The properties named in such a definition are the essential properties of the objects under consideration. They define what the thing

"is", what medieval philosophers called its *quiditas*. According to Aristotle the essence of a species of objects is defined when the next higher *genus* is specified (the so-called *genus proximum*) and also the property which isolates the species inside the genus (the *differentia specifica*) is stated.

Concepts for which there are no higher (i.e. more general) concepts cannot have a "Wesens-Definition", a *definitio essentialis*. The usual examples for such undefinable concepts are the concepts of "equality" and "being". Mathematicians differ over whether or not some of the fundamental concepts of mathematics are also undefinable, e.g., basic concepts of geometry such as "point", "space", etc. and of set theory such as "set" and "elementhood".

Blaise Pascal (1623–1662) was perhaps the first thinker who was convinced of the undefinability of fundamental mathematical notions and who carefully argued in favor of his conviction. In an essay written in the years 1655–1658 (published posthumously), conventionally entitled *De l'esprit géométrique*, he outlined his proposals for a consistent development of arithmetic and geometry.

Pascal thought that mathematicians should first of all seek those concepts that are immediately understood by everyone without any attempts to precisely define them. These notions he called "mots primitifs". According to Pascal, examples of such "primitive notions" are the concepts of "space", "time", "equality", "number", "existence", etc. He took them to be trans-subjective self-evident truths immediately comprehensible by everyone capable of speaking a language. This kind of intuitive awareness lies beyond the rational mind, and is in Pascal's words a "*sentiment du coeur*", i.e. a kind of intuitive knowledge of principles of the heart. The "heart" seems here to play the role of the Platonic "Nous" (νοῦς) for the Jansenist Pascal.

On the basis of the "mots primitifs", all other mathematical concepts can be introduced by means of nominal definitions. Starting from the concept of number, which according to Pascal remains undefined, one can define, e.g., the concept of even positive integer, rightly calling it a nominal definition.

The difficult problem of introducing the fundamental concepts of arithmetic and geometry by means of *definitiones essentiales* or any other device was finessed by Pascal with the somehow audacious remark that these concepts cannot be defined and are in any case well-known to everyone. In particular, Pascal was convinced that one need not define the concept of number and that an adequate definition is, in any event, impossible.

Pascal for his time clearly saw the problems connected with the introduction of primitive concepts. However, he erred when he thought that the whole content of primitive concepts is given to us by intuition (*la lumière naturelle*). What is this "intuition", and how can we control it? Pascal does not answer these questions.

The essay *De l'esprit géométrique* made an enormous impression on many mathematicians. As late as the 19th and the 20th centuries there were still adherents of Pascal's method. For example, in his *Leçons sur les fonctions de*

variables réelles (*1905c*) Émile Borel had this to say about the fundamental concept of set:

> L'idée d'ensemble est une notion *primitive* dont nous ne donnerons pas de définition. Citons seulement quelques exemples d'ensembles: l'ensemble des points d'une droite, etc.[21]

In his works on set theory, Hausdorff also adopted Pascal's and Borel's point of view. In the first pages of the second edition of his book on set theory (*Mengenlehre*, *1927*), he wrote:

> Eine Menge entsteht durch Zusammenfassung von Einzeldingen zu einem Ganzen. Eine Menge ist eine Vielheit, als Einheit gedacht. Wenn diese oder ähnliche Sätze Definitionen sein wollten, so würde man mit Recht einwenden, daß sie idem per idem oder gar obscurum per obscurius definieren. Wir können sie aber als Demonstrationen gelten lassen, als Verweisungen auf einen *primitiven*, allen Menschen vertrauten Denkakt, der einer Auflösung in noch ursprünglichere Akte vielleicht weder fähig noch bedürftig ist.[22]

It is clearly unsatisfactory, when called upon to state definitions or axioms to respond that they are unnecessary. The problem of how to demarcate a "primitive act of thinking which is familiar to everyone" remains unsolved, yet such demarcation is necessary if one hopes to avoid the antinomies of set theory.

In contrast to Pascal, other mathematicians and philosophers were convinced that all mathematical disciplines should start from explicit definitions of their fundamental concepts. It may perhaps be advisable to proceed not from *definitiones essentiales* but rather from *causal definitions* ("Kausal-Definitionen" in Aristotle's sense), i.e. from definitions that give the cause for the existence of the objects in question. The problem of whether in mathematics one begins—or should begin—from causal definitions was heatedly discussed by Scholastic and Renaissance mathematicians and philosophers. The utility of such definitions was vigorously challenged by the English philosopher Thomas Hobbes (1588–1679) in his work *De corpore* (London 1655). The German philosopher and mathematician Ehrenfried Walter von Tschirnhaus (1651–1708) took up the problem and published a theory of science based on causal definitions. In his book *Medicina mentis* (Amsterdam 1687, Leipzig

---

[21] The concept of a set is a primitive notion which we do not define here. It suffices to give a few examples of sets: the set of all points on a line, etc.

[22] A set is created when by an act of comprehension individual entities are aggregated into a whole. A set is a multitude, considered as a unit. If such a sentence or a similar sentence were offered as a definition, one could rightly object that it defines idem per idem or even obscurum per obscurius. However, we may accept them as demonstrations, as references to a *primitive* act of thinking which is familiar to everyone, but which is neither capable of nor in need of being resolved into more primitive acts.

1695), he called those definitions *genetic definitions* in which the objects are not only defined by their essential properties, but also by the processes of their generation. In order truly to comprehend an object, one must be able to reconstruct the object mentally. Therefore, the definitions of the fundamental notions of a science must include a statement of the methods of their mental construction or reconstruction.

To construct a theory axiomatically means—according to Tschirnhaus—to give its fundamental concepts in the form of genetic definitions. From the contents of these genetic definitions one extracts the axioms of the theory, and from the axioms one deduces the theorems as usual by applying purely logical principles. Hence, the truth of the axioms is *ex terminis* known for certain, for what is expressed in the axioms has to be clear from that which is deposited in the definitions. In particular, the existence of the objects with which the theory deals, is secured. All this differs greatly from Aristotle's concept of a theory built upon axioms. For Aristotle, the truth of an axiom (he spoke of a hypothesis) must be clear, either immediately or by induction from empirical experience.

In the 18th century the axiomatic method as conceived by Tschirnhaus was brought into a more distinct form by the German philosopher Christian Wolff (1679–1754). In all his mathematical textbooks, he employed this form, which resulted in its widespread diffusion.

A prominent example of the Hobbes-Tschirnhaus-Wolffian concept of axiomatics is Cantorian set theory. At the beginning of his final treatise "Beiträge zur Begründung der transfiniten Mengenlehre" (*1895*, *1897*), Cantor stated his famous definition of the concept of set, which we have already quoted in §1. This definition is obviously an attempt at a genetic definition since on the one hand it indicates what a set "is", namely a multitude of various things, and on the other hand it indicates how a set is generated, namely by an act of comprehension. Cantor, however, does not set up axioms or postulates. After defining the concept of set, he immediately begins to elaborate the content of his definition in a long and admirable sequence of theorems.

In §2 we have already indicated that Cantor's set theory was unable to prevent the appearance of antinomies. The process of generating sets is in general not a finite process. Accordingly, it requires a clarifying metatheory of sets. Thus, genetic definitions of the fundamental concepts of set theory are not very persuasive.

In the course of the 19th century, it became more and more apparent that in geometry as well it is impossible to provide genetic definitions for some of the fundamental concepts such as "point" and "space", although it is certainly possible to give genetic definitions for many other concepts (cf. Heron's book on definitions [23]). This brings out that geometry—like set theory—cannot be based on genetic definitions of fundamental concepts.

---

[23] Cf. *Heronis Alexandrini Opera quae supersunt omnia*, vol. 4: *Definitiones*, edited by Johan Ludvig Heiberg, Teubner Verlag Leipzig 1912.

Indeed, in the 19th century it became increasingly evident that the fundamental concepts of geometry (point, space, ...) and of set theory (set, elementhood, ...) are not explicitly definable because they are not contained in higher concepts. If one extends the Aristotelian table of categories to include mathematical concepts, then these concepts must be listed at the outset.

However, during the 19th century it also became increasingly clear that it is still possible to introduce the fundamental concepts of a theory by some kind of definition even if they are not explicitly definable. They can be introduced via *implicit definitions.*

If one wishes to introduce the fundamental concepts of a theory via implicit definitions, this can be done by first setting up a language in which each fundamental concept is represented by a sign, e.g. by a letter. The axioms of the theory are formulated in this language. The signs representing the fundamental concepts are thus "folded" into the axioms. But the signs are not treated as symbols, i.e., as signs with a prescribed meaning.[24] Only that which can be derived from the axioms determines the meaning of that for which the signs stand. The totality of all axioms limits the possibilities of interpreting the signs and hence of assigning meanings to them. In this way the fundamental notions are implicitly "defined" and delimited.

This was the approach of Johann von Neumann (1903–1957), whose axiomatization of set theory relied upon implicitly defined fundamental concepts. In his paper *1925*, 220, he wrote:

> Man konstruiert eine Reihe von Postulaten, in denen das Wort "Menge" zwar vorkommt, aber ohne jede Bedeutung. Unter "Menge" wird hier ... nur ein Ding verstanden, von dem man nicht mehr weiß und nicht mehr wissen will, als aus den Postulaten über es folgt.[25]

In von Neumann's theory of sets, the whole content of the Bolzano-Cantorian concept of set is neither presupposed nor used. His axiom system is a *formal system* which gives rules for using set theory's fundamental concepts. Von Neumann produced his theory of sets from a *formal* rather than from a *contentual standpoint.*

The formal standpoint was originally developed in algebra in the circle of George Peacock (1791–1858), Duncan Farquharson Gregory (1813–1844), George Boole (1815–1864), and their associates (see *Pycior 1981*), becoming the dominant position only through the axiomatizations of the arithmetic of natural numbers (Richard Dedekind, 1888), of projective geometry (Gino Fano, 1892), of the algebraic concept of a field (Heinrich Weber, 1893), and

---

[24] The Greek word "symbol" (symbolon, σύμβολον) is composed from "syn" (σύν = together) and "bállein" (βάλλειν = to throw). A "symbol" is, hence, a sign to which a certain meaning is attached.

[25] A series of postulates is contructed in which the word "set" appears, but without any meaning. By a "set" nothing else than an object is understood, about which we do not know and do not want to know anything more than what follows from the postulates.

of classical geometry (David Hilbert, 1899). In none of these examples was there any attempt to describe the "nature" of natural numbers, of geometric and algebraic objects, but only to define the respective content of the objects dealt with in these theories.

In the next section, we will have a look at Zermelo's axiom system in order to discern the paradigm which he followed. At first, we will briefly discuss the work of those who prepared the path for axiomatizing set theory.

§4. The axiomatization of set theory

When Cantor in the summer of 1897 began to revise the foundations of his set theory, he thought that he had found a way out of the crisis by distinguishing between "*consistent*" and "*inconsistent*" sets. This pair of concepts had been introduced shortly before by Ernst Schröder in his *Vorlesungen über Algebra der Logik* (*1890*, 211–213). However, Cantor could not find a way out of the crisis by this means because the consistency proofs which are required here were far out of reach.

Some time later, Cantor tried another approach, namely by distinguishing between *finished* (*fertigen*) and *unfinished* (*unfertigen*) sets. In a letter to Hilbert of October 10, 1898, Cantor proposed placing the following definition at the beginning of set theory (cf. *Cantor 1991*, 396; see also Cantor's letter of August 3, 1899, to Dedekind in *Cantor 1932*, 443–444):

> *Definition.* Unter einer *fertigen Menge* verstehe man jede Vielheit, bei welcher alle Elemente *ohne Widerspruch* als zusammenseiend und daher als *ein Ding für sich* gedacht werden können.[26]

In practice, as I have already indicated, this definition by itself is not workable. Nevertheless, Cantor thought it possible to derive the following principles from the content of the above definition. He wrote (cf. *Cantor 1991*, 396):

> Aus d⟦ieser⟧ Definition ... ergeben sich mancherlei Sätze, unter Anderm diese:[27]

This shows once again that Cantor was still following the Tschirnhaus-Wolffian paradigm. The propositions he obtained were the following ones:

(A1) Ist M eine fert. Menge, so ist auch jede Theilmenge von M eine fert. Menge.
(If M is a finished set, then also each subset of M is a finished set.)

[26] *Definition.* By a *finished set* ⟦or *ready set*⟧ one should understand any multitude all of whose elements can be thought of *without contradiction* as being together and thereby forming *a thing in itself.*

[27] From this definition we obtain many propositions, amongst them the following ones:

(A2) Substituirt man in einer fert. M. an Stelle der Elemente fertige Mengen, so ist die hieraus resultierende Vielheit eine fertige M.
(If in a finished set all its elements are substituted by other finished sets then the resulting multitude is a finished set.)

(A3) Ist von zwei aequivalenten Vielheiten die eine eine fert. M., so ist es auch die andere.
(If one of two equivalent multitudes is a finished set, then the other one is one as well.)

(A4) Die Vielheit aller Theilmengen einer fertigen Menge M ist eine fertige Menge.
(The multitude of all subsets of a finished set is a finished set.)

Adding a principle of infinity, he wrote:

> Daß die "*abzählbaren*" Vielheiten fertige Mengen sind, scheint mir ein axiomatisch sicherer Satz zu sein.
> (That all countable multitudes are finished sets seems to me to be an axiomatically irrefutable sentence.)

These are early versions of the axioms of separation, union, replacement, power set and infinity as they later appear in the axiomatic system of Zermelo-Skolem-Fraenkel, ZSF. Only the axiom of union (A2) is not quite correctly formulated. In his letter of August 3, 1899, to Dedekind, Cantor gave the following correct formulation (cf. *Cantor 1991*, 407):

> Jede Menge von Mengen ist, wenn man die letzteren in ihre Elemente auflöst, auch eine Menge.[28]

Émile Borel in his *Leçons sur la théorie des fonctions* (*1898*, 104) reported that in August 1897 he had met Cantor at the International Congress of Mathematicians in Zurich, Switzerland, and that they had discussed set-theoretic topics at great length. Borel's introduction to set theory in his *Leçons* (*1898*, 1–20) is perhaps based on recollections of these discussions. In his *Leçons*, Borel discusses some of the above mentioned axioms. In particular, he discusses the replacement axiom (A3), saying that a multitude which results from a *"given set"* (*"d'une ensemble donné"*) by replacing each element by another element also has to be considered a set. Borel, however, explicitly refused to define the concept of set.

Cantor's definition of a "finished set" is unsuitable as a definition and falls short of its goal. Its inadequacy is apparent. The multitude of all sets can in fact "be thought of without contradiction as being together and hence forming a thing in itself". In the von Neumann-Bernays-Gödel set theory NBG this multitude exists as an object. But this object must not be treated as if it were a true set. However, it is remarkable that it was Cantor who in 1898 first presented a list of set-theoretic axioms.

[28] Each set of sets, when resolved into its elements, is again a set.

Gregory Moore (*1976*) referred to a 1905 paper by the English mathematician A. E. Harward which appeared some years before Zermelo published his axiom system. This paper included some of the Zermeloan axioms, and so Moore asked the "tantalizing question" whether some of Zermelo's ideas had been in the air. Presumably the question has a positive answer, as can be seen by Cantor's letters to Hilbert of October 10, 1898, and to Dedekind of August 3, 1899, and also by Borel's *Leçons*. Zermelo's notebooks also show that he had worked on a formulation of axioms for set theory between 1904 and 1906. Harward in 1905 merely formulated some of Cantor's axioms and also a certain multiplicative axiom. It is not plausible that Harward knew Cantor's letters. Zermelo did not know them either, for otherwise he would not have omitted the axiom of replacement in his system of axioms.

The first persuasive axiomatization of set theory was produced by Zermelo. He submitted his paper in July 1907; it appeared in print in 1908. He had begun preparatory work between 1904 and 1906, and in his lectures on "Set Theory and the Concept of Number" ("Mengenlehre und Zahlbegriff") in the summer semester 1906, he had presented a preliminary version.

Zermelo proposed seven axioms. These are the Cantorian axioms, without the axiom of replacement, but with an axiom of extensionality, an axiom of elementary sets, and an axiom of choice, none of which had been formulated by Cantor. In the following, we shall discuss these axioms in some detail.

An **axiom of extensionality (Axiom I)** was originally formulated by Bolzano in his *Größenlehre*, but this book was published posthumously in 1975 and could not have influenced Zermelo. Independently of Bolzano, Dedekind formulated this axiom in his essay *1888*, §1, art. 2. Presumably, Zermelo took this axiom from Dedekind, calling it "Axiom der Bestimmtheit" ("axiom of decisiveness").

The **existence of an empty set** (also called "null set" ("Nullmenge") by Zermelo and others) is postulated in **Axiom II**. It may be a bit surprising that Zermelo spoke of an *improper set* (*uneigentliche Menge*). But this cautious mode of expression has its cause in Cantor's definition of the concept of set. In fact, it is not clear from Cantor's definition whether the empty class really is a set. Sets, according to Cantor, are generated by an act of aggregating various entities into a whole. An empty class, however, is not generated by such an act. How can an empty class be an object, when it is not created by an act? Bertrand Russell in his *Principles of Mathematics* (*1903*, 74) posed the question: can a non-empty set remain a set when all its elements are removed?[29] He concluded (op. cit., 75) that "*there is no actual null-class*", but, since it is useful to have an empty set at one's disposal, it is nonetheless

[29] Compare this with the cat in Lewis Carroll's *Alice's adventures in wonderland*, London 1865/1866, which "vanished quite slowly, beginning with the end of the tail, and ending with the grin, which remained some time after the rest of it had gone. 'Well! I've often seen a cat without a grin,' thought Alice; 'but a grin without a cat! It's the most curious thing I ever saw in all my life!'."

introduced. From this point of view, it may be understandable that Zermelo spoke of an "improper" set. To emphasize that this was of some importance for Zermelo, as he explained in a 1921 letter to Fraenkel (1891–1965):[30]

> Nebenbei wird mir selbst die Berechtigung dieser "Nullmenge" immer zweifelhafter. Könnte sie nicht entbehrt werden bei geeigneter Einschränkung des "Aussonderungsaxioms"? Tatsächlich dient sie doch nur zur formalen Vereinfachung.[31]

Similarly, Fraenkel in his book *Einleitung in die Mengenlehre* (*1923a*, 15), introduced the concept of an "empty set" for purely formal reasons. The same is true of Hessenberg (*1906*), Hausdorff (*1914*), and many others. In time, however, the empty set assumed the same authority and legitimacy as all the other sets.

The **axiom of separation (Axiom III)** will be discussed in the next section in connection with the difficult problem of "definite properties". Let us mention here only that Zermelo named it "Axiom der Aussonderung".

The **power set axiom (Axiom IV)** postulates that for each set M there is another set whose elements are the (definable and the undefinable) subsets of M. For that reason the size of the power set of a set is almost ungraspable and its power cannot be unambiguously determined. Nonetheless, the power set axiom renders a typical feature of the Bolzano-Cantorian concept of a set (see §1, the third attribute) and also (in conjunction with the axiom of separation in its full strength) makes possible the existence of different transfinite cardinalities.

The **axiom of union (Axiom V)**, Zermelo's "Axiom der Vereinigung", is unproblematic.

The formulation of the **axiom of choice (Axiom VI)** is Zermelo's major creative accomplishment. With the introduction of this axiom, he made a significant step beyond Cantor and Dedekind. The formulation of this axiom took place in conversations with Erhard Schmidt. In the second part of his *Investigations* Zermelo demonstrated that quite a number of set-theoretic problems can be solved with the aid of this axiom. Accordingly, the axiom of choice is indispensable.

In the formulation of the **axiom of infinity (Axiom VII)** Zermelo profited from Dedekind's having provided a definition of infinite set.[32] Dedekind in his essay *1888*, §5, called a set M infinite if there is a one-to-one mapping from M onto a proper subset of M. In his lecture on "Set Theory and the Concept of Number" ("Mengenlehre und Zahlbegriff", summer-semester

[30] Quoted in *Ebbinghaus 2007b*, 285.

[31] By the way, the justification of the "empty set" is becoming more and more doubtful for me. Would it be possible to dispense with it by a suitable limitation of the axiom of separation? Actually, the empty set serves only for some formal simplification.

[32] An account of the history of this finiteness definition is given in the marginal note 20 in *Hausdorff 2002*, 588.

1906), Zermelo formulated the axiom of infinity as follows: there is a set which is equivalent to (i.e. equipollent with) a proper subset of itself. In his *Investigations*, however, Zermelo gave the following more concrete version of the axiom: there is a set Z ("Z" for "Zahlen") with the following properties, $\emptyset \in Z$ and $\{a\} \in Z$ for each $a \in Z$. Here we have $\emptyset \neq \{\emptyset\}$ since $\emptyset$ is empty and $\{\emptyset\}$ is non-empty. We also have $\{\emptyset\} \neq \{\{\emptyset\}\}$, etc.

In Zermelo's system of axioms, an axiom of replacement and an axiom of foundation were missing. Both axioms were introduced shortly after 1920 and added to Zermelo's system.

The **axiom of replacement (Ersetzungsaxiom)**. In a letter to Zermelo of May 6, 1921, Fraenkel had posed the following question (quoted in *Ebbinghaus 2007b*, 136 and 286):

> Es sei Z_0 eine unendliche Menge ... und $\mathfrak{U}(Z_0) = Z_1, \mathfrak{U}(Z_1) = Z_2$, usw. Wie folgt dann aus Ihrer Theorie (Grundl. d. M. I), daß $\{Z_0, Z_1, Z_2, \ldots\}$ eine Menge ist, daß also die Vereinigungsmenge existiert? Würde Ihre Theorie zu einem solchen Beweis nicht genügen, so wäre offenbar z. B. die Existenz von Mengen von der Kardinalzahl $\aleph_\omega$ nicht beweisbar.[33]

In his answer of May 9, 1921, Zermelo wrote (cf. *Ebbinghaus 2007b*, 286) that he had missed this point when writing his *Investigations*, and he continues:

> In der Tat wird da wohl noch ein Axiom nötig sein, aber welches? Man könnte es so versuchen: Sind die Dinge $A, B, C, \ldots$ durch eineindeutige Beziehung den Dingen $a, b, c, \ldots$ zugeordnet, welche die Elemente einer Menge m bilden, so sind auch die Dinge $A, B, C, \ldots$ Elemente einer Menge M.[34]

Zermelo called this axiom "*Zuordnungs-Axiom*" (axiom of assignment, or axiom of coordination), which did not please him very much because he felt that the concept of assignment was not definite enough ("zu wenig definit"). In 1921 (at a meeting, published as *1921*), Fraenkel renamed this axiom the "axiom of replacement" ("Ersetzungsaxiom"), cf. *Fraenkel 1922b*. He adopted Zermelo's formulation almost verbatim, but also allowing replacements which are not one-to-one. However, the nature of these replacements remained unclear. It is remarkable that Skolem in 1922 (at a meeting, published in 1923) too, independently of Zermelo and Fraenkel, also proposed an axiom of replacement (cf. *Skolem 1923*, 145–146). In Skolem's formulation, however, the

[33] Let Z_0 be an infinite set ... and $\mathfrak{U}(Z_0) = Z_1, \mathfrak{U}(Z_1) = Z_2$, etc. How can you show in your theory ... that $\{Z_0, Z_1, Z_2, \ldots\}$ is a set, and, hence, that, e.g., the union of this set exists? If your theory is insufficient to allow such a proof, then obviously the existence of, say, sets of cardinality $\aleph_\omega$ cannot be proved.

[34] Indeed, a new axiom is necessary here, but which axiom? One could try to formulate it in the following manner: If the objects $A, B, C, \ldots$ are assigned to the objects $a, b, c, \ldots$ by a one-to-one relation, and the latter objects form a set, then $A, B, C, \ldots$ are also elements of a set M.

replacements are given by formulas of the basic set-theoretic language. Skolem also emphasized that such an axiom is necessary if one wished to prove the existence of sets of power $\aleph_\omega$.

An axiom of replacement is also needed, e.g. in the proof of the determinacy of all Borel sets of natural numbers, as was shown by Tony Martin in 1975 (cf. *Martin 1975*). Harvey Friedman proved in 1971 that Borel determinacy is not provable in Zermelo's original system of axioms (cf. *Friedman 1971*). An axiom of replacement is also needed, for instance, in the proof of the existence of the set of all hereditarily-finite sets, as was shown by Adrian Mathias in *2001*. It is the axiom of replacement which in general permits the creation of sets via transfinite recursion, as was shown by von Neumann in *1928d* in his axiomatic version of set theory.

Although the axiom of replacement was formulated already in 1898 by Cantor, it was von Neumann who recognized its fundamental importance. This has been pointed out conclusively by Michael Hallett in his book *1984*, 95 and 280–286.

The **axiom of foundation (Fundierungsaxiom)**. In order to exclude sets A which have the property $A \in A$, or $A \in B \in A$ for a suitable B, etc., in 1923 Skolem proposed an axiom of foundation according to which *all descending* $\in$*-chains are finite*. According to this formulation, an axiom of foundation can be introduced only after the introduction of the concepts of finiteness and of function. Von Neumann proposed the following elegant formulation: *each non-empty set M contains an element which has no element in common with M* (cf. *von Neumann 1925*). In the course of the 20th century, other variants of the axiom of foundation were proposed. A discussion of these variants can be found in *Felgner 2002b*.

§5. The problem of "definite" properties

Zermelo's axiomatization made it possible, after a period of uncertainty caused by the appearance of the antinomies, for set theory to be developed further on a solid basis. For that, Zermelo's axiomatization has always been praised. However, some of the Zermeloan axioms met with skepticism. The axiom of choice in particular led to a number of well-known controversies. That need not detain us now. The axiom of separation, however, was also hotly debated because of the notion of "definite property", which occurs in the formulation of that axiom. A few words are required to explain the debate.

In 1882, Cantor spoke of "*well-defined multitudes*" ("wohldefinierte Mannichfaltigkeiten"), meaning sets for which (*1882*, 114)

> auf Grund ihrer Definition und infolge des logischen Prinzips vom ausgeschlossenen Dritten es als intern bestimmt angesehen werden muß, sowohl ob irgendein derselben Begriffssphäre angehöriges Objekt zu der gedachten Mannigfaltigkeit als Element gehört oder nicht,

> als auch, ob zwei zur Menge gehörige Objekte, trotz formaler Unterschiede in der Art des Gegebenseins, einander gleich sind oder nicht.[35]

At first, Zermelo adopted Cantor's terminology. In his notes for his lecture on set theory in the winter semester 1900/1901, which also contain additions from the years 1904–1906, we find in §2 the formulation of some axioms: e.g., an elementary axiom of summation,[36] a weak axiom of foundation,[37] an axiom of power set, and an axiom of separation. In these notes, Zermelo spoke of *well-defined sets* (wohldefinierte Mengen) and *well-defined properties* (wohldefinierte Eigenschaften). In his paper offering a "New proof for the possibility of a well-ordering" (*1908a*) which he finished on July 14, 1907, he formulated the axiom of separation as follows:

> Alle diejenigen Elemente einer Menge M, denen eine für jedes einzelne Element wohldefinierte Eigenschaft $\mathfrak{E}$ zukommt, bilden die Elemente einer zweiten Menge $M_{\mathfrak{E}}$, einer "Untermenge" von M.[38]

But even in his 1906 Göttingen lecture "Set theory and the concept of number" (*Mengenlehre und Zahlbegriff*), we find an axiom in which the word "definite" appears and which was meant to express the fact that all things of the basic domain (Grundbereich) are pairwise distinguishable (cf. *Ebbinghaus 2007b*, 83, and *Peckhaus 1990a*, 96). Zermelo's formulation reads as follows:

> Sind a, b Dinge, so ist entschieden, "definit", ob $a = b$ oder $a \neq b$.[39]

Finally, in his *Investigations* Zermelo formulated the axiom of separation with the less commonly used word "definite" instead of the less precise phrase "well-defined".

Zermelo borrowed the word "definite" from Edmund Husserl, who introduced it shortly after 1900 and discussed it in some detail in a series of two lectures at the Göttingen Mathematical Society in the winter semester 1901/1902. Husserl also discussed it twelve years later in great detail in his *Ideen zu einer reinen phänomenologischen Philosophie* (cf. *Husserl 1950*, 167), noting in a footnote (loc. cit., 168) that the word had entered mathematical discourse without revealing its source. Husserl probably also had Zermelo in mind, for Zermelo had used the word in his axiom of separation without indicating its source.

---

[35] Because of their definition and because of the logical principle of the excluded third one has to consider as internally determined whether or not any object of the same genus belongs to the set, and also whether or not two objects which belong to the set are equal despite formal differences in the way they are given.

[36] It postulates that for each object m the set $\{m\}$ exists whose only element is m, and that in addition for any well-defined set M the set M augmented by the new element m exists and is well-defined.

[37] It states that a well-defined set never contains itself as an element.

[38] All elements of a set M that have a property $\mathfrak{E}$ well-defined for every single element are the elements of another set, $M_{\mathfrak{E}}$, a "subset" of M.

[39] If a and b are objects, then it is determined, "definite", whether $a = b$ or $a \neq b$.

Using the expression "definite", Husserl intended to indicate that "definite properties" are mathematically exhaustive defined properties (cf. *Husserl 1950*, 167, and also §31 in Husserl's *Formale und Transzendentale Logik*, *Husserl 1974*, 98–102). Therefore, a concept is "definite" in the sense of Husserl when it permits a detailed analysis (not of its content, but) of its extension. Zermelo outlined his understanding of the word "definite" roughly as follows. A property $\mathfrak{E}$ is "definite", provided that for each element of a given set M whether $\mathfrak{E}$ holds or not can be decided without any arbitrariness on the basis of the $\in$-relation which holds among the objects of the domain.

Hermann Weyl was among the first who expressed doubts about the clarity and distinctness of the concept of "*Definitheit*". In *1910*, he proposed removing the word "definite" from the formulation of the axiom of separation and admitting it only for those properties which can be formulated

> aufgrund der beiden Beziehungen $=$ and $\in$
> (on the basis of the relations $=$ and $\in$)

as the only non-logical signs. In his pamphlet *Das Kontinuum* (*1918*, 36) he stated more clearly that he was thinking of first-order logic. A few years later Skolem in *1923* made the same proposal and in a rather similar manner von Neumann eliminated the notion of "definiteness". In his letter to Zermelo of August 15, 1923 he wrote that in his axiom system he does not introduce the notion of "definiteness" explicitly but instead states all the admissible schemes for the formation of functions and sets. Von Neumann's letter is reprinted in *Meschkowski 1967*, 271–273.

By requiring "definiteness" Zermelo wanted to prevent the use of properties in the axiom of separation which had led to well-known semantic antinomies such as the antinomy of Jules Richard, 1905, the antinomy of G. G. Berry, 1906, etc. In later years Zermelo came back to the problem of "definiteness". He discussed it in his 1929 essay "Über den Begriff der Definitheit in der Axiomatik" (*1929a*) without, however, reaching a final solution (cf. *Ebbinghaus 2007b*, 179–183).

§6. The background of Zermelo's system of axioms

Zermelo's requirement that the properties which are allowed to occur in the axiom of separation should be "definite" shows that he did not take a formalist position. His position was always the *contentual standpoint* (der inhaltliche Standpunkt). Whether a property is "definite" or not, cannot be decided formally by an application of a general rule, but only by an examination of the content and the meaning of the property.

The fact that Zermelo took the contentual standpoint is apparent everywhere in his work. One sees this, for example, by the way in which he treated the $\in$-sign. Axioms are formulated in everyday language with the minimal

use of signs, namely a sign for the elementhood relation and a few comprehension terms. In Zermelo's axiom system, the $\in$-sign is not an undefined sign whose meaning is defined implicitly by the totality of all axioms, but rather a symbol, i.e., a sign to which the presupposed meaning of true elementhood is attached (see footnote 24).

That Zermelo employed the sign "$\in$" not as an undefined sign but as a symbol can be seen from a remark made by Zermelo in the sequel to his Theorem 10. There he argues that, since the domain $\mathfrak{B}$ of sets itself is not a set, the Zermelo-Russellian antinomy can simply be set aside (Zermelo's *Investigations*, 265). His argument is valid, however, only when the relation "$\in$" is the actual true elementhood relation.

The problem of the "domains" (Bereiche). At the beginning of his essay, Zermelo indicates that set theory is concerned with a "domain" of objects (*einem "Bereich" $\mathfrak{B}$ von Objekten*). In the formulation of some of the axioms, reference is made to these "domains". This provoked considerable confusion and a number of controversies. Is the concept of a "domain" another primitive notion of set theory? In what respect are "domains" different from "sets"? Would it be permissible simply to omit all references to these "domains" or is it necessary to provide axioms for "domains"?

It is striking that only in §1 of his *Investigations* does Zermelo refer to "domains". The word appears there no fewer than eleven times. In the much larger §2, the word "domain" does not appear at all. One gets the impression that the term would be superfluous. This seems in fact to be the case. In his paper "Über Grenzzahlen und Mengenbereiche" (*1930a*), Zermelo formulates all the axioms without any reference to "domains". It becomes clear that speaking of "domains" is only a way of referring informally to subdomains of the world of finite and transfinite sets (probably including urelements) in which all the Zermeloan (or Zermelo-Skolem-Fraenkelian) axioms are valid. Validity in this context means that the $\in$-sign is to be interpreted by the true elementhood relation and the concept of power set by the true power set.

In his axiomatization, Zermelo departs from the content of the concept of set as elaborated by Bolzano and Cantor. However, Zermelo derives from Bolzano and Cantor only those principles (or axioms), which lead to unproblematic sets. He does not start his set theory with an explicit definition of the concept of set, but the Bolzano-Cantor definition of the concept of set is the unmentioned basis for his axiom system. In his paper on the "New proof for the existence of a well-ordering" (*1908a*), Zermelo indicates that in order to avoid the antinomies he will rely upon "specialized versions of Cantor's definition". Until the "correct" definition of the concept of set can be found, the axioms should mention particular characteristics of such a definition. This indicates clearly that Zermelo in his axiomatization is following the classical paradigm (Aristotle, Euclid, Hobbes, Tschirnhaus, Wolff, et al.), but he himself does not yet feel able to present the "correct" definition of what a set is.

As late as 1929–1932 Zermelo was searching for an adequate definition (cf. *Ebbinghaus 2007b*, 213–217).

Although Zermelo worked closely in Göttingen together with Hilbert, and although he was obviously influenced by Hilbert's early axiomatization program, he never adopted Hilbert's formalist standpoint. (In the literature, it is quite frequently claimed wrongly that he did.) Zermelo's standpoint was the contentual standpoint. In fact, he repeatedly criticized the formalist standpoint.

The modification and extension of Zermelo's system proposed by Skolem and Fraenkel is now widely accepted. The system of axioms consists of all of Zermelo's axioms, except for the axiom of choice, supplemented with the axiom of replacement and the axiom of foundation. This system is usually denoted by ZF (or more accurately by ZSF) and is called the Zermelo-Fraenkel (or Zermelo-Skolem-Fraenkel) axiom system. These axioms are formulated in a first-order language whose only non-logical sign is the $\in$-sign. A sign whose intended interpretation is "is a set" is not necessary because all objects referred to in the axioms should be considered "sets".

The question in which sense the sets of the theory ZSF are "things" (as discussed above in §1) has a very liberal answer. In ZSF sets are treated formally as things, which should be taken to mean that they are all included in the domain of values of the individual variables.

The modification of the Zermeloan axiom system proposed by Skolem yielded a change in the underlying philosophical standpoint: the contentual standpoint was replaced by the formalist standpoint. Zermelo did not agree with that change and he expressed his disagreement in harsh polemics.

§7. Cardinal arithmetic in Zermelo's paper

In the second part of his *Investigations* Zermelo treats the notions of finite and transfinite number. In the years around 1900 many aspects of the number concept had become confused and some of them had been threatened by paradoxes and antinomies. It was Zermelo's aim to construct a sound arithmetic of finite and infinite cardinal numbers. To begin with, let us first recall the issues and problems that had emerged with the number concept.

(1) Firstly, we have to mention the concept of natural number. The exposition of *Dedekind 1888* lost credibility because Dedekind posited the existence of an infinite class of all possible thoughts. (This was discussed above in §2, where it was noted that this is an absolute infinite proper class.) Dedekind acknowledged the difficulties connected with his procedure and for quite some time was unwilling to allow a reprint of his essay *1888*.

Frege's approach to the arithmetic of natural numbers was also unsuccessful. He thought, mistakenly, that he could introduce the natural numbers in a canonical way as logical objects, i.e., as classes of concepts related to the

concept of equinumerability. In 1902, when Russell found an inconsistency in Frege's theory, Frege was greatly distressed. He confessed that the basis on which he intended to build arithmetic was shaken.

On the basis of his axiom system, Zermelo was able to postulate the existence of a set Z_0 representing the natural numbers. Z_0 is the intersection of all inductive subsets of a set whose existence is guaranteed by the axiom of infinity. A set M is called *inductive* whenever it contains the empty set as an element, and with each element a also contains the singleton $\{a\}$. The elements of Z_0 are hence $0 = \emptyset, 1 = \{0\}, 2 = \{1\} = \{\{0\}\}, \ldots, n + 1 = \{n\}, \ldots$.

(2) Secondly, we have to mention the concept of the cardinal number $\mathsf{Card}(M)$ of an arbitrary set M. In his construction of cardinal numbers Cantor referred to an act of a twofold abstraction, but he was unable to say what objects are created by this procedure (cf. *Cantor 1895*, 481).

In 1903, Russell defined the cardinality of a set M as the class of all sets which are equipotent with M. He wrote (*1903*, 115):

> This method [...], to define as the number of a class the class of all classes similar to the given class [...], is an irreproachable definition of the number of a class in purely logical terms.

It is incomprehensible why Russell, who displays in full detail the antinomy of the set of all sets which do not contain themselves as elements (*1903*, Chapter X, 101–107), did not realize that his definition of cardinality ran into the same antinomy. (The number 1 e.g. is the set of all singletons and we have hence $\{\{x\}; x \subseteq 1\} \subseteq 1$, which leads to a contradiction in almost the same way as in the Zermeloan argument, see §2.)

When writing his *Investigations*, Zermelo did not yet know how to introduce set-theoretic representatives for the sizes of infinite sets. Therefore, he only treated the notion of equipollence (also called equivalence). Additionally, a suitable set-theoretic representation of the general notion of a function was not yet known to Zermelo (such a definition was first given by Hausdorff in 1914, cf. *Felgner 2002c*), he had to work with equipollent disjoint sets. His theorems 10, 19, and 28 are designed to make this possible.

A successful definition of the notion of ordinal number and of cardinal number was discovered by Zermelo in the years 1913–1915. This is apparent from a footnote in *Bernays 1941*, 6. (See also the careful documentations in *Hallett 1984*, 271–280, and in *Ebbinghaus 2007b*, 133–134.) Independently of Zermelo, von Neumann also introduced the same objects as ordinal numbers. These definitions can be found in all monographs on set theory. A detailed discussion of all the difficulties which are connected with the definition of cardinal numbers can be found in *Felgner 2002a*.

(3) Thirdly, there is the problem of whether or not all cardinal numbers are alephs, i.e., cardinal numbers of well-orderable sets. Cantor repeatedly claimed that this was the case. He tried to prove this with the argument that

the "set" of all alephs is an absolute infinite totality (cf. Cantor's letter of September 26, 1897, to Hilbert, see *Cantor 1991*, 388–390). Philip Jourdain also published a "proof", which failed. It was Zermelo, who in 1904 published the first correct proof of the well-orderability of each set by assuming a new axiom, the axiom of choice. Zermelo's proof was strongly attacked by many mathematicians. In order to make apparent the correctness of his proof, Zermelo sought to make clear the grounds upon which his proof rests.

(4) Fourthly, there is the problem of trichotomy. Cantor made the claim that for any cardinal numbers $\mathfrak{a}$ and $\mathfrak{b}$ one has either $\mathfrak{a} < \mathfrak{b}$, or $\mathfrak{a} = \mathfrak{b}$, or $\mathfrak{a} > \mathfrak{b}$. Cantor announced a proof, but he did not publish it nor reveal to anyone the details of this alleged proof (cf. *Cantor 1895*, 484, *Cantor 1932*, 351, remark 2). The truth of the law of trichotomy was doubted occasionally by skeptics. Russell, for example, wrote in his *Principles of mathematics* (*Russell 1903*, 323):

> [...] and it may be that $2^{\aleph_0}$ is neither greater nor less than $\aleph_1$ and $\aleph_2$ and their successors.

Felix Bernstein (*1901*) and Godfrey Harold Hardy (*1904*) were able to prove $\aleph_1 \leq 2^{\aleph_0}$, but their proofs relied implicitly upon the axiom of choice, which was not yet isolated as a set-theoretic principle. Zermelo formulated the axiom of choice in 1904 and thereby deduced the well-ordering principle from it. Thus, the law of trichotomy follows from the axiom of choice and hence is provable in Zermelo's axiom system.

It should be noted that, conversely, the law of trichotomy implies the well-ordering theorem as was shown by Friedrich Hartogs (*1915*, 436–443). Thus, all three statements—the axiom of choice, the law of trichotomy, and the well-ordering principle—are equivalent on the basis of the remaining axioms.

(5) Fifthly, there is the annoying "result" announced by Julius König in 1904 at the International Congress of Mathematicians in Heidelberg. He offered a "proof" of the non-wellorderability of the continuum. In his "proof" he used the Bernstein aleph-formula $\aleph_\alpha^{\aleph_\beta} = 2^{\aleph_\beta} \cdot \aleph_\alpha$. His "result" was sensational and seemed entirely to topple Cantor's set theory, but, a few days after König's lecture, Hausdorff and Zermelo independently found the mistake in König's "proof". In fact, Bernstein's aleph-formula is valid only for finite ordinals α (cf. *Ebbinghaus 2007a* and *Purkert 2002*, 9–12). In consequence, König withdrew his "proof". The analysis of König's faulty argument led Hausdorff to his recursion formula (Rekurrenz-Formel) and Zermelo to the Zermelo-König inequality, discussed below.

In §2 of the *Investigations* Zermelo aimed at corrections of these misleading and erroneous "results". He sought also to set to rights the arithmetic of finite and transfinite cardinal numbers. He rounded off his paper with the presentation of a few further classical results, for which he was able to present elegant new proofs.

The Dedekind-Bernstein equivalence theorem

Theorems 25–27 present a proof of the important equivalence theorem, which is usually ascribed to Cantor, Bernstein, and Schröder. Cantor, however, in 1882/1883, formulated the theorem only for well-orderable sets. Bernstein found a proof in the winter of 1896/1897 which he included in his PhD dissertation (Göttingen 1901). He published this proof in 1905 (*1905c*) after Borel had incorporated it in his *Leçons sur la théorie des fonctions* (*1898*, 104–105). The logician Ernst Schröder also attempted, in 1898, to prove the theorem, but Alwin Reinhold Korselt, in 1902, found a mistake in Schröder's proof. A totally different proof was given by Dedekind in 1887, but this proof remained unpublished. The manuscript was found by Jean Cavaillès in Dedekind's posthumous papers. (See *Dedekind 1932*, 447–449.) In addition, Dedekind communicated his proof in a letter of August 29, 1899 to Cantor. The letter was published in Cantor's Collected Works (*Cantor 1932*, 449 and 451).

In the first decade of the twentieth century, Zermelo knew only Bernstein's proof. When he lectured on set theory in Göttingen during the winter semester 1900/1901, he included Bernstein's proof of the equivalence theorem. However, he added a little note saying that Bernstein's proof rests heavily on the construction of a chain of the following form $Q, \varphi(Q), \varphi(\varphi(Q)), \varphi(\varphi(\varphi(Q))), \ldots$. We follow Zermelo's notation in the proof of his theorem 25 and consider sets $M' \subseteq M_1 \subseteq M$, a one-to-one mapping φ from M onto M' and put $Q = M_1 - M'$. It is clear that $A_0 = Q \cup \bigcup_{1 \leq n \in \omega} \varphi^n(Q)$ and $\bigcup_{1 \leq n \in \omega} \varphi^n(Q)$ are equipollent, and this simple fact is the basis of Bernstein's proof.

According to Dedekind, such a set A_0 can also be described "from above" as the intersection of all φ-closed sets which contain the initial object Q (as a subset),

$$A_0 = \bigcap \{X \subseteq M;\, Q \subseteq X \;\&\; \varphi(X) \subseteq X\}\,.$$

This opens the possibility of another proof of the equivalence theorem. Zermelo had probably realized this by the spring of 1905. He outlined a new proof on a postcard to Hilbert written at the end of June 1905 (cf. *Ebbinghaus 2007b*, 89 and 279–280). As Zermelo indicated, this new proof rests on Dedekind's theory of chains (cf. *Dedekind 1888, art. 37, 44, 63*). This new and elegant proof is used to establish theorem 25. Much later, Zermelo discovered that Dedekind himself had almost the same proof in 1887/1899.[40] We note in passing that in Zermelo's first proof of the well-ordering theorem (*Zermelo 1904*) the well-ordering is obtained "from below" as a union, and that in his second proof (*Zermelo 1908a*) the well-ordering is obtained "from above" as an intersection. This second proof rests on Dedekind's theory of chains.

---

[40] See the introductory note to *Zermelo 1901* for more on the history of the equivalence theorem.

The multiplication theorem

It is rather surprising that Zermelo does not prove the laws of idempotency for the addition and the multiplication of cardinal numbers, $\aleph_\alpha + \aleph_\alpha = \aleph_\alpha$ and $\aleph_\alpha \cdot \aleph_\alpha = \aleph_\alpha$ for all alephs. This is surprising because all other laws about addition and multiplication easily follow from these laws, and it was Zermelo who first announced these laws without proof at the end of his 1904 paper on the well-ordering theorem. He communicated his proof to Hessenberg and probably also to some other colleagues. In 1906, Hessenberg found his own proof and published it in his book *Grundbegriffe der Mengenlehre* (*1906*). On page 109 he included the following remark:

> In jüngster Zeit hat mir Herr Zermelo einen Beweis mitgeteilt, der von dem hier gegebenen wesentlich verschieden ist und demnächst an anderer Stelle erscheinen wird.[41]

But Zermelo had not yet published his proof. Moreover, in his papers no sketch of a proof has been found so far. (See *Deiser 2005* for more on the history of the multiplication theorem.)

The Zermelo-König inequality

The highlight of the second part of Zermelo's *Investigations* is the proof of the so-called Zermelo-König inequality (theorem 33). From that inequality, the original lemma of König and also the Cantor inequality $\mathfrak{m} < 2^{\mathfrak{m}}$ follow. A diagonal argument which is used in Cantor's proof of his inequality is also used in the proof of the Zermelo-König inequality. From a historical point of view it is remarkable that the first diagonal argument was given by Paul du Bois-Reymond in 1873 in connection with his "Infinitär-Kalkül" (cf. *du Bois-Reymond 1873*).

In his book on transfinite numbers, Ivan Ivanovich Shegalkin independently discovered a proof of the above-mentioned Zermelo-König inequality (cf. *Shegalkin 1907*). However, Zermelo had already proved the inequality in 1904. In his *Investigations* he proudly added (p. 279):

> Das vorstehende (Ende 1904 der Göttinger Mathematischen Gesellschaft von mir mitgeteilte) Theorem ist der allgemeinste bisher bekannte Satz über das Größer und Kleiner der Mächtigkeiten, aus dem alle übrigen sich ableiten lassen.[42]

That the Zermelo-König inequality also plays a fundamental role in the proofs of the laws of the exponentiation of alephs was shown, much later, by Alfred Tarski in *1925*.

---

[41] Recently, Zermelo informed me of his proof, which is essentially different from the one given here and which will appear somewhere else in the near future.

[42] This theorem (communicated by me to the Göttingen Mathematical Society at the end of 1904) is the most general theorem now known concerning the comparison of cardinalities, one from which all the others can be derived.

§8. Concluding remarks

Zermelo wrote his *Investigations* in Chesières, a village situated between Montreux and Martinach (Martigny), near Aelen (Aigle), in the Alpine range at the edge of the Rhône valley. He spent the summer of 1907 there in order to recuperate from his pulmonary illness. Here he also wrote his paper on the "New proof of the possibility of a well-ordering".

Both papers are connected. In order to secure his much-criticized proofs of the well-ordering theorem, he carefully analyzed the grounds upon which his proofs rest. (This has been discussed in detail by Moore in *1978*). But this was surely not Zermelo's main concern. What he wanted to achieve was to secure cardinal arithmetic and in particular to publish his own contributions in this field.

To the title of Zermelo's paper was added the words "Part I", but no Part II has ever been published. In the preface to Part I Zermelo wrote that

Untersuchungen über die Grundlagen der Mengenlehre I

1908b

Die Mengenlehre ist derjenige Zweig der Mathematik, dem die Aufgabe zufällt, die Grundbegriffe der Zahl, der Anordnung und der Funktion in ihrer ursprünglichen Einfachheit mathematisch zu untersuchen und damit die logischen Grundlagen der gesamten Arithmetik und Analysis zu entwickeln; sie bildet somit einen unentbehrlichen Bestandteil der mathematischen Wissenschaft. Nun scheint aber gegenwärtig gerade diese Disziplin in ihrer ganzen Existenz bedroht durch gewisse Widersprüche oder „Antinomieen“, die sich aus ihren scheinbar denknotwendig gegebenen Prinzipien herleiten lassen und bisher noch keine allseitig befriedigende Lösung gefunden haben. Angesichts namentlich der „*Russell*schen Antinomie“ von der „Menge aller Mengen, welche sich selbst nicht als Element enthalten“* scheint es heute nicht mehr zulässig, einem beliebigen logisch definierbaren Begriffe eine „Menge“ oder „Klasse“ als seinen „Umfang“ zuzuweisen. Die ursprüngliche Cantorsche Definition einer „Menge“ als einer „Zusammenfassung von bestimmten wohlunterschiedenen Objekten unserer Anschauung oder

[1] B. Russell, „The Principles of Mathematics“, vol. I, p. 366–368, 101–107.

> ein zweiter Artikel, der die Lehre von der Wohlordnung und ihre[r] Anwendung auf die endlichen Mengen und die Prinzipien der Arithmetik im Zusammenhang entwickeln soll, in Vorbereitung [sei].[43]

The second paper announced here was not written. However Zermelo gave a talk on the intended topic at the International Congress of Mathematicians 1908 in Rome, and published its main contents in the proceedings of the congress, cf. *Zermelo 1909b*. A slightly extended version of that paper appeared in 1909 in the Swedish journal *Acta mathematica* (cf. *Zermelo 1909a*). These two papers may be considered as a substitute for the abandoned second paper.

[43] A second paper, which will develop the theory of well-ordering together with its application to finite sets and the principles of arithmetic, is in preparation.

Investigations in the foundations of set theory I

1908b

Set theory is that branch of mathematics whose task is to investigate mathematically the fundamental notions "number", "order", and "function", taking them in their pristine, simple form, and to develop thereby the logical foundations of all of arithmetic and analysis; thus it constitutes an indispensable component of the science of mathematics. At present, however, the very existence of this discipline seems to be threatened by certain contradictions, or "antinomies", that can be derived from its principles—principles necessarily governing our thinking, it seems—and to which no entirely satisfactory solution has yet been found. In particular, in view of the "*Russell* antinomy" of the set of all sets that do not contain themselves as elements,[1] it no longer seems admissible today to assign to an arbitrary logically definable notion a "set", or "class", as its "extension". Cantor's original definition of a "set" as "a collection, gathered into a whole, of certain well-distinguished

[1] *Russell 1903*, pp. 366–368 and 101–107.

unseres Denkens zu einem Ganzen"[2] bedarf also jedenfalls einer Einschränkung, ohne daß es doch schon gelungen wäre, sie durch eine andere, ebenso einfache zu ersetzen, welche zu keinen solchen Bedenken mehr Anlaß gäbe. Unter diesen Umständen bleibt gegenwärtig nichts anderes übrig, als den umgekehrten Weg einzuschlagen und, ausgehend von der historisch bestehenden „Mengenlehre", die Prinzipien aufzusuchen, welche zur Begründung dieser mathematischen Disziplin erforderlich sind. Diese Aufgabe muß in der Weise gelöst werden, daß man die Prinzipien einmal eng genug einschränkt, um alle Widersprüche auszuschließen, gleichzeitig aber auch weit genug ausdehnt, um alles Wertvolle dieser Lehre beizubehalten.

In der hier vorliegenden Arbeit gedenke ich nun zu zeigen, wie sich die ge-
262 samte von *G. Cantor* und *R. Dedekind* geschaffene Theorie auf | einige wenige Definitionen und auf sieben anscheinend voneinander unabhängige „Prinzipien" oder „Axiome" zurückführen läßt. Die weitere, mehr philosophische Frage nach dem Ursprung und dem Gültigkeitsbereiche dieser Prinzipien soll hier noch unerörtert bleiben. Selbst die gewiß sehr wesentliche „Widerspruchslosigkeit" meiner Axiome habe ich noch nicht streng beweisen können, sondern mich auf den gelegentlichen Hinweis beschränken müssen, daß die bisher bekannten „Antinomieen" sämtlich verschwinden, wenn man die hier vorgeschlagenen Prinzipien zugrunde legt. Für spätere Untersuchungen, welche sich mit solchen tiefer liegenden Problemen beschäftigen, möchte ich hiermit wenigstens eine nützliche Vorarbeit liefern.

Der nachstehende Artikel enthält die Axiome und ihre nächsten Folgerungen, sowie eine auf diese Prinzipien gegründete Theorie der Äquivalenz, welche die formelle Anwendung der Kardinalzahlen vermeidet. Ein zweiter Artikel, der die Lehre von der Wohlordnung und ihre Anwendung auf die endlichen Mengen und die Prinzipien der Arithmetik im Zusammenhange entwickeln soll, ist in Vorbereitung.

§ 1.
Grundlegende Definitionen und Axiome

1. Die Mengenlehre hat zu tun mit einem „*Bereich*" $\mathfrak{B}$ von Objekten, die wir einfach als „*Dinge*" bezeichnen wollen, unter denen die „Mengen" einen Teil bilden. Sollen zwei Symbole a und b dasselbe Ding bezeichnen, so schreiben wir $a = b$, im entgegengesetzten Falle $a \neq b$. Von einem Dinge a sagen wir, es „existiere", wenn es dem Bereiche $\mathfrak{B}$ angehört; ebenso sagen wir von einer Klasse $\mathfrak{K}$ von Dingen, „es gebe Dinge der Klasse $\mathfrak{K}$", wenn $\mathfrak{B}$ mindestens ein Individuum dieser Klasse enthält.

2. Zwischen den Dingen des Bereiches $\mathfrak{B}$ bestehen gewisse „*Grundbeziehungen*" der Form $a \varepsilon b$. Gilt für zwei Dinge a, b die Beziehung $a \varepsilon b$, so sagen wir, „a sei *Element* der Menge b" oder „b enthalte a als Element" oder „besitze

[2] G. Cantor, Math. Annalen Bd. 46, p. 481.

objects of our perception or our thought"[2] therefore certainly requires some restriction; it has not, however, been successfully replaced by one that is just as simple and does not give rise to such reservations. Under these circumstances there is at this point nothing left for us to do but to proceed in the opposite direction and, starting from "set theory" as it is historically given, to seek out the principles required for establishing the foundations of this mathematical discipline. In solving the problem we must, on the one hand, restrict these principles sufficiently to exclude all contradictions and, on the other, take them sufficiently wide to retain all that is valuable in this theory.

Now in the present paper I intend to show how the entire theory created by *G. Cantor* and *R. Dedekind* can be reduced to a few definitions and seven "principles", or "axioms", which appear to be mutually independent. The further, more philosophical, question about the origin of these principles and the extent to which they are valid will not be discussed here. I have not yet even been able to prove rigorously that my axioms are "consistent", though this is certainly very essential; instead I have had to confine myself to pointing out now and then that the "antinomies" discovered so far vanish one and all if the principles here proposed are adopted as a basis. But I hope to have done at least some useful spadework hereby for subsequent investigations in such deeper problems.

The present paper contains the axioms and their most immediate consequences, as well as a theory of equivalence based upon these principles that avoids the formal use of cardinal numbers. A second paper, which will develop the theory of well-ordering together with its application to finite sets and the principles of arithmetic, is in preparation.

§ 1.
Fundamental definitions and axioms

1. Set theory is concerned with a "*domain*" $\mathfrak{B}$ of individuals, which we shall call simply "*objects*" and among which are the "sets". If two symbols, a and b, denote the same object, we write $a = b$, otherwise $a \neq b$. We say of an object a that it "exists" if it belongs to the domain $\mathfrak{B}$; likewise we say of a class $\mathfrak{K}$ of objects that "there exist objects of the class $\mathfrak{K}$" if $\mathfrak{B}$ contains at least one individual of this class.

2. Certain "*fundamental relations*" of the form $a \varepsilon b$ obtain between the objects of the domain $\mathfrak{B}$. If for two objects a and b the relation $a \varepsilon b$ holds, we say "a is an *element* of the set b", "b contains a as an element", or "b possesses

[2] *Cantor 1895*, p. 481.

das Element a". Ein Ding b, welches ein anderes a als Element enthält, kann immer als eine *Menge* bezeichnet werden, aber auch nur dann — mit einer einzigen Ausnahme (Axiom II).

3. Ist jedes Element x einer Menge M gleichzeitig auch Element der Menge N, so daß aus $x \varepsilon M$ stets $x \varepsilon N$ gefolgert werden kann, so sagen wir, „M sei *Untermenge* von N", und schreiben $M \mathrel{\underline{\in}} N$.[1] Es ist stets $M \mathrel{\underline{\in}} M$, und aus
263 $M \mathrel{\underline{\in}} N$ und $N \mathrel{\underline{\in}} R$ folgt immer $M \mathrel{\underline{\in}} R$. „*Elementenfremd*" | heißen zwei Mengen M, N, wenn sie keine „gemeinsamen" Elemente besitzen, oder wenn kein Element von M gleichzeitig Element von N ist.

4. Eine Frage oder Aussage $\mathfrak{E}$, über deren Gültigkeit oder Ungültigkeit die Grundbeziehungen des Bereiches vermöge der Axiome und der allgemeingültigen logischen Gesetze ohne Willkür entscheiden, heißt „*definit*". Ebenso wird auch eine „Klassenaussage" $\mathfrak{E}(x)$, in welcher der variable Term x alle Individuen einer Klasse $\mathfrak{K}$ durchlaufen kann, als „definit" bezeichnet, wenn sie für *jedes einzelne* Individuum x der Klasse $\mathfrak{K}$ definit ist. So ist die Frage, ob $a \varepsilon b$ oder nicht ist, immer definit, ebenso die Frage, ob $M \mathrel{\underline{\in}} N$ oder nicht.

Über die Grundbeziehungen unseres Bereiches $\mathfrak{B}$ gelten nun die folgenden „*Axiome*" oder „*Postulate*".

Axiom I. Ist jedes Element einer Menge M gleichzeitig Element von N und umgekehrt, ist also gleichzeitig $M \mathrel{\underline{\in}} N$ und $N \mathrel{\underline{\in}} M$, so ist immer $M = N$. Oder kürzer: jede Menge ist durch ihre Elemente bestimmt.

(Axiom der Bestimmtheit.)

Die Menge, welche nur die Elemente $a, b, c, \ldots, r$ enthält, wird zur Abkürzung vielfach mit $\{a, b, c, \ldots, r\}$ bezeichnet werden.

Axiom II. Es gibt eine (uneigentliche) Menge, die „*Nullmenge*" 0, welche gar keine Elemente enthält. Ist a irgend ein Ding des Bereiches, so existiert eine Menge $\{a\}$, welche a und nur a als Element enthält; sind a, b irgend zwei Dinge des Bereiches, so existiert immer eine Menge $\{a, b\}$, welche sowohl a als b, aber kein von beiden verschiedenes Ding x als Element enthält.

(Axiom der Elementarmengen.)

5. Nach I sind die „Elementarmengen" $\{a\}$, $\{a, b\}$ immer eindeutig bestimmt, und es gibt nur eine einzige „Nullmenge". Die Frage, ob $a = b$ oder nicht, ist immer definit (Nr. 4), da sie mit der Frage, ob $a \varepsilon \{b\}$ ist, gleichbedeutend ist.

6. Die Nullmenge ist Untermenge jeder Menge M, $0 \mathrel{\underline{\in}} M$; eine gleichzeitig von 0 und M verschiedene Untermenge von M wird als „*Teil*" von M bezeichnet. Die Mengen 0 und $\{a\}$ besitzen keine Teile.

---

[1] Dieses „Subsumptions"-Zeichen wurde von *E. Schröder* („Vorlesungen über Algebra der Logik" Bd. I) eingeführt, Herr *G. Peano* und ihm folgend *B. Russell*, *Whitehead* u.a. brauchen dafür das Zeichen Ↄ.

the element a". An object b may be called a *set* if and—with a single exception (Axiom II)—only if it contains another object, a, as an element.

3. If every element x of a set M is also an element of the set N, so that from $x \varepsilon M$ it always follows that $x \varepsilon N$, we say that "M is a *subset* of N" and we write $M \subseteq N$.[3] We always have $M \subseteq M$ and from $M \subseteq N$ and $N \subseteq R$ it always follows that $M \subseteq R$. Two sets M and N are said to be "*disjoint*" if they possess no common element, or if no element of M is an element of N.

4. A question or assertion $\mathfrak{E}$ is said to be "*definite*" if the fundamental relations of the domain, by means of the axioms and the universally valid laws of logic, determine without arbitrariness whether it holds or not. Likewise a "propositional function" $\mathfrak{E}(x)$, in which the variable term x ranges over all individuals of a class $\mathfrak{K}$, is said to be "definite" if it is definite for *each single* individual x of the class $\mathfrak{K}$. Thus the question whether $a \varepsilon b$ or not is always definite, as is the question whether $M \subseteq N$ or not.

The fundamental relations of our domain $\mathfrak{B}$, now, are subject to the following "*axioms*", or "*postulates*".

Axiom I. If every element of a set M is also an element of N and vice versa, if, therefore, both $M \subseteq N$ and $N \subseteq M$, then always $M = N$; or, more briefly: Every set is determined by its elements.

(Axiom of extensionality.)

The set that contains only the elements $a, b, c, \ldots, r$ will often be denoted briefly by $\{a, b, c, \ldots, r\}$.

Axiom II. There exists a (fictitious) set, the "*null set*", 0, that contains no element at all. If a is any object of the domain, there exists a set $\{a\}$ containing a and only a as element; if a and b are any two objects of the domain, there always exists a set $\{a, b\}$ containing as elements a and b but no object x distinct from both.

(Axiom of elementary sets.)

5. According to Axiom I, the "elementary sets" $\{a\}$ and $\{a, b\}$ are always uniquely determined and there is only a single "null set". The question whether $a = b$ or not is always definite (No. 4), since it is equivalent to the question whether or not $a \varepsilon \{b\}$.

6. The null set is a subset of every set M: $0 \subseteq M$; a subset of M that differs from both 0 and M is called a "part" of M. The sets 0 and $\{a\}$ do not have parts.

---

[3] ⟦Zermelo uses the sign "$\models\!\!\in$" for inclusion. He comments here:⟧ This sign ⟦$\models\!\!\in$⟧ of inclusion was introduced by *Schröder* (*1890*). *Peano* and, following him, *Russell*, *Whitehead*, and others use the sign $\supset$ instead.

Axiom III. Ist die Klassenaussage $\mathfrak{E}(x)$ definit für alle Elemente einer Menge M, so besitzt M immer eine Untermenge $M_{\mathfrak{E}}$, welche alle diejenigen Elemente x von M, für welche $\mathfrak{E}(x)$ wahr ist, und nur solche als Elemente enthält.

(Axiom der Aussonderung.)

Indem das vorstehende Axiom III in weitem Umfange die Definition neuer Mengen gestattet, bildet es einen gewissen Ersatz für die in der Einleitung angeführte und als unhaltbar aufgegebene allgemeine Mengendefinition, von der es sich durch die folgenden Einschränkungen unterscheidet: Erstens dürfen mit Hilfe dieses Axio-
264 mes | niemals Mengen *independent definiert*, sondern immer nur als Untermengen aus bereits gegebenen *ausgesondert* werden, wodurch widerspruchsvolle Gebilde wie „die Menge aller Mengen“ oder „die Menge aller Ordinalzahlen“ und damit nach dem Ausdrucke des Herrn *G. Hessenberg* („Grundbegriffe der Mengenlehre“ XXIV) die „ultrafiniten Paradoxieen“ ausgeschlossen sind. Zugleich muß zweitens das bestimmende Kriterium $\mathfrak{E}(x)$ im Sinne unserer Erklärung Nr. 4 immer „definit“ d. h. für jedes einzelne Element x von M durch die „Grundbeziehungen des Bereiches“ entschieden sein, und hiermit kommen alle solchen Kriterien wie „durch eine endliche Anzahl von Worten definierbar“ und damit die „Antinomie Richard“ oder die „Paradoxie der endlichen Bezeichnung“ (*Hessenberg* a. a. O. XXIII, vergl. dagegen *J. König*, Math. Ann. Bd. 61, p. 156) für unseren Standpunkt in Wegfall. Hieraus folgt aber auch, daß, streng genommen, vor jeder Anwendung unseres Axioms III immer erst das betreffende Kriterium $\mathfrak{E}(x)$ als „definit“ nachgewiesen werden muß, was denn auch in den folgenden Entwicklungen bei jeder Gelegenheit, wo es nicht ganz selbstverständlich ist, immer geschehen soll.

7. Ist $M_1 \mathrel{\underline{\in}} M$, so besitzt M immer eine weitere Untermenge $M - M_1$, die „*Komplementärmenge* von M_1“, welche alle diejenigen Elemente von M umfaßt, die *nicht* Elemente von M_1 sind. Die Komplementärmenge von $M - M_1$ ist wieder M_1. Die Komplementärmenge von $M_1 = M$ ist die Nullmenge 0, die Komplementärmenge jedes „Teiles“ M_1 von M (Nr. 6) ist wieder ein „Teil“ von M.

8. Sind M, N irgend zwei Mengen, so bilden nach III diejenigen Elemente von M, welche gleichzeitig Elemente von N sind, die Elemente einer Untermenge D von M, welche auch Untermenge von N ist und alle M und N *gemeinsamen* Elemente umfaßt. Diese Menge D wird der „gemeinsame Bestandteil“ oder der „*Durchschnitt*“ der Mengen M und N genannt und mit $[M, N]$ bezeichnet. Ist $M \mathrel{\underline{\in}} N$, so ist $[M, N] = M$; ist $N = 0$ oder sind M und N „elementenfremd“ (Nr. 3), so ist $[M, N] = 0$.

9. Ebenso existiert auch für mehrere Mengen $M, N, R, \ldots$ immer ein „Durchschnitt“ $D = [M, N, R, \ldots]$. Ist nämlich T irgend eine Menge, deren Elemente selbst Mengen sind, so entspricht nach III jedem Dinge a eine gewisse Untermenge $T_a \mathrel{\underline{\in}} T$, welche alle diejenigen Elemente von T umfaßt, die a als Element enthalten. Es ist somit für jedes a definit, ob $T_a = T$ ist, d. h. ob a gemeinsames Element aller Elemente von T ist, und ist A ein beliebiges Element von T, so bilden alle Elemente a von A, für welche $T_a = T$

Axiom III. Whenever the propositional function $\mathfrak{E}(x)$ is definite for all elements of a set M, M possesses a subset $M_{\mathfrak{E}}$ containing as elements precisely those elements x of M for which $\mathfrak{E}(x)$ is true.

(Axiom of separation.)

By giving us a large measure of freedom in defining new sets, Axiom III in a sense furnishes a substitute for the general definition of set that was cited in the introduction and rejected as untenable. It differs from that definition in that it contains the following restrictions. In the first place, sets may never be *independently defined* by means of this axiom but must always be *separated* as subsets from sets already given; thus contradictory notions such as "the set of all sets" or "the set of all ordinal numbers", and with them the "ultrafinite paradoxes", to use Mr. *G. Hessenberg*'s expression (*1906*, chap. 24), are excluded. In the second place, moreover, the defining criterion must always be "definite" in the sense of our definition in No. 4 (that is, for each single element x of M the "fundamental relations of the domain" must determine whether it holds or not), with the result that, from our point of view, all criteria such as "definable by means of a finite number of words", hence the "Richard antinomy" and the "paradox of finite denotation" (*Hessenberg 1906*, chap. 23; on the other hand, see *J. König 1905c*) vanish. But it also follows that we must, prior to each application of our Axiom III, prove the criterion $\mathfrak{E}(x)$ in question to be "definite", if we wish to be rigorous; in the considerations developed below this will indeed be proved whenever it is not altogether evident.

7. If $M_1 \subseteq M$, then M always possesses another subset, $M - M_1$, the "*complement* of M_1", which contains all those elements of M that are *not* elements of M_1. The complement of $M - M_1$ is M_1 again. If $M_1 = M$, its complement is the null set, 0; the complement of any "part" (No. 6) M_1 of M is again a "part" of M.

8. If M and N are any two sets, then according to Axiom III all those elements of M that are also elements of N are the elements of a subset D of M; D is also a subset of N and contains all elements *common* to M and N. This set D is called the "common component", or "*intersection*", of the sets M and N and is denoted by $[M, N]$. If $M = N$, then $[M, N] = M$; if $N = 0$ or if M and N are "disjoint" (No. 3), then $[M, N] = 0$.

9. Likewise, for several sets $M, N, R, \ldots$ there always exists an "intersection" $D = [M, N, R, \ldots]$. For, if T is any set whose elements are themselves sets, then according to Axiom III there corresponds to every object a a certain subset T_a of T that contains all those elements of T that contain a as an element. Thus it is definite for every a whether $T_a = T$, that is, whether a is a common element of all elements of T; if A is an arbitrary element of T, all elements a of A for which $T_a = T$ are the elements of a subset D of A that

ist, die Elemente einer Untermenge D von A, welche alle diese gemeinsamen Elemente umfaßt. Diese Menge D wird „der zu T gehörende Durchschnitt" genannt und mit $\mathfrak{D}T$ bezeichnet. Besitzen die Elemente von T keine gemeinsamen Elemente, so ist $\mathfrak{D}T = 0$, und dies ist z. B. immer der Fall, wenn ein Element von T keine Menge oder die Nullmenge ist.

10. *Theorem.* Jede Menge M besitzt mindestens eine Untermenge M_0, welche nicht Element von M ist.

265 | *Beweis.* Für jedes Element x von M ist es definit, ob $x \varepsilon x$ ist oder nicht; diese Möglichkeit $x \varepsilon x$ ist an und für sich durch unsere Axiome nicht ausgeschlossen. Ist nun M_0 diejenige Untermenge von M, welche gemäß III alle solchen Elemente von M umfaßt, für die *nicht* $x \varepsilon x$ ist, so kann M_0 nicht Element von M sein. Denn entweder ist $M_0 \varepsilon M_0$ oder nicht. Im ersteren Falle enthielte M_0 ein Element $x = M_0$, für welches $x \varepsilon x$ wäre, und dieses widerspräche der Definition von M_0. Es ist also sicher *nicht* $M_0 \varepsilon M_0$, und es müßte somit M_0, wenn es Element von M wäre, auch Element von M_0 sein, was soeben ausgeschlossen wurde.

Aus dem Theorem folgt, daß nicht alle Dinge x des Bereiches $\mathfrak{B}$ Elemente einer und derselben Menge sein können; d. h. *der Bereich* $\mathfrak{B}$ *ist selbst keine Menge*, — womit die „*Russell*sche Antinomie" für unseren Standpunkt beseitigt ist.

Axiom IV. Jeder Menge T entspricht eine zweite Menge $\mathfrak{U}T$ (die „*Potenzmenge*" von T), welche alle Untermengen von T und nur solche als Elemente enthält.

(Axiom der Potenzmenge.)

Axiom V. Jeder Menge T entspricht eine Menge $\mathfrak{S}T$ (die „*Vereinigungsmenge*" von T), welche alle Elemente der Elemente von T und nur solche als Elemente enthält.

(Axiom der Vereinigung.)

11. Ist kein Element von T eine von 0 verschiedene Menge, so ist natürlich $\mathfrak{S}T = 0$. Ist $T = \{M, N, R, \ldots\}$, wo die $M, N, R, \ldots$ sämtlich Mengen sind, so schreibt man auch $\mathfrak{S}T = M + N + R + \cdots$ und nennt $\mathfrak{S}T$ die „*Summe* der Mengen $M, N, R, \ldots$", ob einige dieser Mengen $M, N, R, \ldots$ nun gemeinsame Elemente besitzen oder nicht. Es ist immer $M = M + 0 = M + M = M + M + \cdots$.

12. Für die soeben definierte „Addition" der Mengen gilt das „kommutative" und das „assoziative" Gesetz:

$$M + N = N + M\,, \quad M + (N + R) = (M + N) + R\,.$$

contains all these common elements. This set D is called "the intersection associated with T" and is denoted by $\mathfrak{D}T$. If the elements of T do not possess a common element, $\mathfrak{D}T = 0$, and this is always the case if, for example, an element of T is not a set or if it is the null set.

10. *Theorem.* Every set M possesses at least one subset M_0 that is not an element of M.

Proof. It is definite for every element x of M whether $x \varepsilon x$ or not; the possibility that $x \varepsilon x$ is not in itself excluded by our axioms. If now M_0 is the subset of M that, in accordance with Axiom III, contains all those elements of M for which it is *not* the case that $x \varepsilon x$, then M_0 cannot be an element of M. For either $M_0 \varepsilon M_0$ or not. In the first case, M_0 would contain an element $x = M_0$ for which $x \varepsilon x$, and this would contradict the definition of M_0. Thus M_0 is surely *not* an element of M_0, and in consequence M_0, if it were an element of M, would also have to be an element of M_0, which was just excluded.

It follows from the theorem that not all objects x of the domain $\mathfrak{B}$ can be elements of one and the same set; that is, *the domain* $\mathfrak{B}$ *is not itself a set*, and this disposes of the "*Russell* antinomy" so far as we are concerned.

Axiom IV. To every set T there corresponds another set $\mathfrak{U}T$, the "*power set*" of T, that contains as elements precisely all subsets of T.

(Axiom of the power set.)

Axiom V. To every set T there corresponds a set $\mathfrak{S}T$, the "*union*" of T, that contains as elements precisely all elements of the elements of T.

(Axiom of the union.)

11. If no element of T is a set different from 0, then, of course, $\mathfrak{S}T = 0$. If $T = \{M, N, R, \ldots\}$, where $M, N, R, \ldots$ all are sets, we also write $\mathfrak{S}T = M + N + R + \cdots$ and call $\mathfrak{S}T$ the "*sum*" of the sets $M, N, R, \ldots$, whether some of these sets $M, N, R, \ldots$ contain common elements or not. Always $M = M + 0 = M + M = M + M + \cdots$.

12. For the "addition" of sets that we have just defined, the "commutative" and "associative" laws hold:

$$M + N = N + M\,, \quad M + (N + R) = (M + N) + R\,.$$

Endlich gilt für „Summen“ und „Durchschnitte“ (Nr. 8) auch das „distributive“ Gesetz in doppelter Form:

$$[M + N, R] = [M, R] + [N, R]\,,$$
$$[M, N] + R = [M + R, N + R]\,.$$

Den Beweis führt man mit Hilfe von I, indem man zeigt, daß jedes Element der linksstehenden Menge zugleich Element der rechtsstehenden Menge ist und umgekehrt.[1]

266 | 13. *Einführung des Produktes.* Ist M eine von 0 verschiedene Menge und a irgend eines ihrer Elemente, so ist nach Nr. 5 definit, ob $M = \{a\}$ ist oder nicht. *Es ist also immer definit, ob eine vorgelegte Menge aus einem einzigen Element besteht oder nicht.*

Es sei nun T eine Menge, deren Elemente $M, N, R, \ldots$ lauter (untereinander elementenfremde) Mengen sein mögen, und S_1 irgend eine Untermenge ihrer „Vereinigungsmenge“ $\mathfrak{S}\,T$. Dann ist für jedes Element M von T definit, ob der Durchschnitt $[M, S_1]$ aus einem einzigen Element besteht oder nicht. Somit bilden alle diejenigen Elemente von T, welche mit S_1 genau ein Element gemein haben, die Elemente einer gewissen Untermenge T_1 von T, und es ist wieder definit, ob $T_1 = T$ ist oder nicht. Alle Untermengen $S_1 \Subset \mathfrak{S}\,T$, welche mit jedem Elemente von T genau ein Element gemein haben, bilden also nach III die Elemente einer Menge $P = \mathfrak{P}\,T$, welche nach III und IV Untermenge von $\mathfrak{U}\mathfrak{S}\,T$ ist und als die „zu T gehörende *Verbindungsmenge*“ oder als „das *Produkt* der Mengen $M, N, R, \ldots$“ bezeichnet werden soll. Ist $T = \{M, N\}$, oder $T = \{M, N, R\}$, so schreibt man abgekürzt $\mathfrak{P}\,T = MN$ oder $= MNR$.

Um nun den Satz zu gewinnen, daß *ein Produkt mehrerer Mengen nur dann verschwinden* (d. h. der Nullmenge gleich sein) *kann, wenn ein Faktor verschwindet*, brauchen wir ein weiteres Axiom.

Axiom VI. Ist T eine Menge, deren sämtliche Elemente von 0 verschiedene Mengen und untereinander elementenfremd sind, so enthält ihre Vereinigung $\mathfrak{S}\,T$ mindestens eine Untermenge S_1, welche mit jedem Elemente von T ein und nur ein Element gemein hat.

(Axiom der Auswahl.)

Man kann das Axiom auch so ausdrücken, daß man sagt, es sei immer möglich, aus jedem Elemente $M, N, R, \ldots$ von T ein einzelnes Element $m, n, r, \ldots$ *auszuwählen* und alle diese Elemente zu einer Menge S_1 zu vereinigen.[1]

---

[1] Diese vollständige Theorie dieser „logischen Addition und Multiplikation“ findet sich in *E. Schröders* „Algebra der Logik“, Bd. I.

[1] Über die Berechtigung dieses Axiomes vgl. meine Abhandlung Math. Ann. Bd. 65, p. 107–128, wo im § 2 p. 111 ff. die bezügliche Literatur erörtert wird.

Finally, for "sums" and "intersections" (No. 8) the "distributive law" also holds, in the two forms:

$$[M + N, R] = [M, R] + [N, R]$$

and

$$[M, N] + R = [M + R, N + R] .$$

The proof is carried out by means of Axiom I and consists in a demonstration that every element of the set on the left is also an element of the set on the right, and conversely.[4]

13. *Introduction of the product.* If M is a set different from 0 and a is any one of its elements, then according to No. 5 it is definite whether $M = \{a\}$ or not. *It is therefore always definite whether a given set consists of a single element or not.*

Now let T be a set whose elements, $M, N, R, \ldots$, are various (mutually disjoint) sets, and let S_1 be any subset of its "union" $\mathfrak{S}T$. Then it is definite for every element M of T whether the intersection $[M, S_1]$ consists of a single element or not. Thus all those elements of T that have exactly one element in common with S_1 are the elements of a certain subset T_1 of T, and it is again definite whether $T_1 = T$ or not. All subsets S_1 of $\mathfrak{S}T$ that have exactly one element in common with each element of T then are, according to Axiom III, the elements of a set $P = \mathfrak{P}T$, which, according to Axioms III and IV, is a subset of $\mathfrak{U}\mathfrak{S}T$ and will be called the "*connection set* associated with T" or "the *product* of the sets $M, N, R, \ldots$". If $T = \{M, N\}$, or $T = \{M, N, R\}$, we write $\mathfrak{P}T = MN$, or $\mathfrak{P}T = MNR$, respectively, for short.

In order, now, to obtain the theorem that *the product of several sets can vanish* (that is, be equal to the null set) *only if a factor vanishes* we need a further axiom.

Axiom VI. If T is a set whose elements all are sets that are different from 0 and mutually disjoint, its union $\mathfrak{S}T$ includes at least one subset S_1 having one and only one element in common with each element of T.

(Axiom of choice.)

We can also express this axiom by saying that it is always possible to *choose* a single element from each element $M, N, R, \ldots$ of T and to combine all the chosen elements, $m, n, r, \ldots$, into a set S_1.[5]

[4] The complete theory of this "logical addition and multiplication" can be found in *Schröder 1890.*

[5] For the justification of this axiom see my *1908a*, where in § 2, pp. 111 ff, the relevant literature is discussed.

Die vorstehenden Axiome genügen, wie wir sehen werden, um alle wesentlichen Theoreme der allgemeinen Mengenlehre abzuleiten. Um aber die Existenz „unendlicher" Mengen zu sichern, bedürfen wir noch des folgenden, seinem wesentlichen Inhalte von Herrn *R. Dedekind*[2] herrührenden Axiomes.

Axiom VII. Der Bereich enthält mindestens eine Menge Z, welche die Nullmenge als Element enthält und so beschaffen ist, daß jedem ihrer
267 | Elemente a ein weiteres Element der Form $\{a\}$ entspricht, oder welche mit jedem ihrer Elemente a auch die entsprechende Menge $\{a\}$ als Element enthält.

(Axiom des Unendlichen.)

14_{VII}.[1] Ist Z eine beliebige Menge von der in VII geforderten Beschaffenheit, so ist für jede ihrer Untermengen Z_1 definit, ob sie die gleiche Eigenschaft besitzt. Denn ist a irgend ein Element von Z_1, so ist definit, ob auch $\{a\} \varepsilon Z_1$ ist, und alle so beschaffenen Elemente a von Z_1 bilden die Elemente einer Untermenge $Z_1{}'$, für welche definit ist, ob $Z_1{}' = Z_1$ ist oder nicht. Somit bilden alle Untermengen Z_1 von der betrachteten Eigenschaft die Elemente einer Untermenge $T \Subset \mathfrak{U} Z$, und der ihnen entsprechende Durchschnitt (Nr. 9) $Z_0 = \mathfrak{D} T$ ist eine Menge von der gleichen Beschaffenheit. Denn einmal ist 0 gemeinsames Element aller Elemente Z_1 von T, und andererseits, wenn a gemeinsames Element aller dieser Z_1 ist, so ist auch $\{a\}$ allen gemeinsam und somit gleichfalls Element von Z_0.

Ist nun Z' irgend eine andere Menge von der im Axiom geforderten Beschaffenheit, so entspricht ihr in genau derselben Weise wie Z_0 dem Z eine kleinste Untermenge $Z_0{}'$ von der betrachteten Eigenschaft. Nun muß aber auch der Durchschnitt $[Z_0, Z_0{}']$, welcher eine gemeinsame Untermenge von Z und Z' ist, die gleiche Beschaffenheit wie Z und Z' haben und als Untermenge von Z den Bestandteil Z_0, sowie als Untermenge von Z' den Bestandteil $Z_0{}'$ enthalten. Nach I folgt also, daß $[Z_0, Z_0{}'] = Z_0 = Z_0{}'$ sein muß, und daß somit Z_0 *der gemeinsame Bestandteil aller möglichen wie* Z *beschaffenen Mengen* ist, obwohl diese nicht die Elemente einer Menge zu bilden brauchen. Die Menge Z_0 enthält die Elemente $0, \{0\}, \{\{0\}\}$ usw. und möge als „*Zahlenreihe*" bezeichnet werden, weil ihre Elemente die Stelle der Zahlzeichen vertreten können. Sie bildet das einfachste Beispiel einer „abzählbar unendlichen" Menge (Nr. 36).

---

[2] „Was sind und was sollen die Zahlen?" § 5 Nr. 66. Der von Herrn Dedekind hier versuchte „Beweis" dieses Prinzips kann nicht befriedigen, da er von der „Menge alles Denkbaren" ausgeht, während für unseren Standpunkt nach Nr. 10 der Bereich $\mathfrak{B}$ selbst *keine* Menge bildet.

[1] Die Indizes VI oder VII an der Nummer eines Theorems sollen ausdrücken, daß hier das Axiom VI oder VII explizit oder implizit zur Anwendung kommt.

The preceding axioms suffice, as we shall see, for the derivation of all essential theorems of general set theory. But in order to secure the existence of infinite sets we still require the following axiom, which is essentially due to Mr. *R. Dedekind.*[6]

Axiom VII. There exists in the domain at least one set Z that contains the null set as an element and is so constituted that to each of its elements a there corresponds a further element of the form $\{a\}$, in other words, that with each of its elements a it also contains the corresponding set $\{a\}$ as an element.

(Axiom of infinity.)

14_{VII}.[7] If Z is an arbitrary set constituted as required by Axiom VII, it is definite for each of its subsets Z_1 whether it possesses the same property. For, if a is any element of Z_1, it is definite whether $\{a\}$, too, is an element of Z_1, and all elements a of Z_1 that satisfy this condition are the elements of a subset $Z_1{}'$ for which it is definite whether $Z_1{}' = Z_1$ or not. Thus all subsets Z_1 having the property in question are the elements of a subset T of $\mathfrak{U}Z$, and the intersection (No. 9) $Z_0 = \mathfrak{D}T$ that corresponds to them is a set constituted in the same way. For, on the one hand, 0 is a common element of all elements Z_1 of T, and, on the other, if a is a common element of all of these Z_1, then $\{a\}$ is also common to all of them and is thus likewise an element of Z_0.

Now if Z' is any other set constituted as required by the axiom, there corresponds to it a smallest subset Z_0' having the same property, exactly as Z_0 corresponds to Z. But now the intersection $[Z_0, Z_0{}']$, which is a common subset of Z and Z', must be constituted in the same way as Z and Z'; and just as, being a subset of Z, it must contain the component Z_0, so, as a subset of Z', it must contain the component $Z_0{}'$. According to Axiom I it then necessarily follows that $[Z_0, Z_0{}'] = Z_0 = Z_0{}'$ and that Z_0 thus is the *common component of all possible sets constituted like* Z, even though these need not be elements of a set. The set Z_0 contains the elements 0, $\{0\}$, $\{\{0\}\}$, and so forth, and it may be called the "*number sequence*", because its elements can take the place of the numerals. It is the simplest example of a "denumerably infinite" set (below, No. 36).

---

[6] *Dedekind 1888*, §5, art. 66. The "proof" that Mr. Dedekind there attempts to give of this principle cannot be satisfactory, since it takes its departure from "the set of everything thinkable", whereas from our point of view the domain $\mathfrak{B}$ itself, according to No. 10, does *not* form a set.

[7] The subscript VI, or VII, on the number of a section indicates that explicit or implicit use has been made of Axiom VI, or VII, respectively, in establishing the theorem of that section.

§ 2.
Theorie der Äquivalenz

Die „Äquivalenz“ zweier Mengen[2] läßt sich für unseren Standpunkt zunächst nur für den Fall definieren, wo die Mengen „elementenfremd“ (Nr. 3) sind, und kann erst nachträglich auf den allgemeinen Fall ausgedehnt werden.

15. *Definition A.* Zwei elementenfremde Mengen M und N heißen „*unmittelbar äquivalent*“, $M \sim N$, wenn ihr Produkt MN (Nr. 13) mindestens eine solche Untermenge Φ besitzt, daß jedes Element von $M + N$ in einem
268 | und nur einem Elemente $\{m, n\}$ von Φ als Element erscheint. Eine Menge $\Phi \mathrel{\underline{\in}} MN$ von der betrachteten Beschaffenheit heißt eine „*Abbildung von M auf N*“; zwei Elemente m, n, welche in einem Elemente von Φ vereinigt erscheinen, heißen „aufeinander abgebildet“, sie „entsprechen einander“, das eine ist „das Bild“ des anderen.

16. Ist Φ irgend eine Untermenge von MN, also Element von $\mathfrak{U}(MN)$, und x irgend ein Element von $M + N$, so ist es immer definit (Nr. 4), ob die x enthaltenden Elemente von Φ eine Menge bilden, die aus einem einzigen Element besteht (Nr. 13). Somit ist auch definit, ob *alle* Elemente x von $M + N$ diese Eigenschaft besitzen, d. h. ob Φ eine „Abbildung“ von M auf N darstellt oder nicht. Die sämtlichen Abbildungen Φ bilden also nach III die Elemente einer gewissen Untermenge Ω von $\mathfrak{U}(MN)$, und es ist definit, ob Ω von 0 verschieden ist oder nicht. *Für zwei elementenfremde Mengen M, N ist es also immer definit, ob sie äquivalent sind oder nicht.*

17. Sind zwei äquivalente elementenfremde Mengen M, N durch Φ aufeinander abgebildet, so entspricht auch jeder Untermenge $M_1 \mathrel{\underline{\in}} M$ eine äquivalente Untermenge $N_1 \mathrel{\underline{\in}} N$ vermöge einer Abbildung Φ_1, welche eine Untermenge von Φ ist.

Denn für jedes Element $\{m, n\}$ von Φ ist es definit, ob $m \,\varepsilon\, M_1$ ist oder nicht, und alle in dieser Weise zu M_1 gehörenden Elemente von Φ bilden somit die Elemente einer Untermenge $\Phi_1 \mathrel{\underline{\in}} \Phi$. Bezeichnet man nun mit N_1 den Durchschnitt (Nr. 8) von $\mathfrak{S}\Phi_1$ mit N, so erscheint jedes Element von $M_1 + N_1$ nur in einem einzigen Elemente von Φ_1 als Element, weil es sonst auch in Φ mehrfach vorkommen würde, und es ist nach Nr. 15 in der Tat $M_1 \sim N_1$.

18. Sind zwei elementenfremde Mengen M und N einer und derselben dritten Menge R gleichzeitig elementenfremd und äquivalent, oder ist $M \sim R, R \sim R', R' \sim N$, wobei je zwei aufeinander folgende Mengen elementenfremd sein sollen, so ist auch immer $M \sim N$.

---

[2] G. Cantor, Math. Annalen Bd. 46, p. 483.

§ 2.
Theory of equivalence

From our point of view, the "equivalence" of two sets[8] cannot be defined at first except for the case in which the sets are "disjoint" (No. 3); it is only afterward that the definition can be extended to the general case.

15. *Definition A.* Two disjoint sets M and N are said to be "*immediately equivalent*", $M \sim N$, if their product MN (No. 13) possesses at least one subset Φ such that each element of $M + N$ occurs as an element in one and only one element $\{m, n\}$ of Φ. A subset Φ of MN thus constituted is called a "*mapping of M onto N*"; two elements m and n that occur together in one element of Φ are said to be "mapped onto each other"; they "correspond to each other", or one is the "image" of the other.

16. If Φ is any subset of MN and therefore an element of $\mathfrak{U}(MN)$ and if x is any element of $M + N$, it is always definite (No. 4) whether the elements of Φ that contain x form a set consisting of a single element (No. 13). Thus it is also definite whether *all* elements x of $M + N$ possess this property, that is, whether Φ represents a "mapping" of M onto N or not. According to Axiom III, all of the mappings Φ therefore are the elements of a certain subset Ω of $\mathfrak{U}(MN)$, and it is definite whether Ω differs from 0 or not. *It is therefore always definite for two disjoint sets M and N whether they are equivalent or not.*

17. If two equivalent disjoint sets M and N are mapped onto each other by Φ, there also corresponds to each subset M_1 of M an equivalent subset N_1 of N under a mapping Φ_1, that is a subset of Φ.

For it is definite for every element $\{m, n\}$ of Φ whether $m \varepsilon M_1$ or not, and therefore all elements of Φ thus associated with M_1 are the elements of a subset Φ_1 of Φ. If we now denote by N_1 the intersection (No. 8) of $\mathfrak{S}\Phi_1$ with N, each element of $M_1 + N_1$ occurs as an element in only a single element of Φ_1, since otherwise it would occur more than once in Φ as well; and, according to No. 15, we in fact have $M_1 \sim N_1$.

18. If two disjoint sets M and N are disjoint from and equivalent to one and the same third set, R, or if $M \sim R$, $R \sim R'$, and $R' \sim N$, where each of these pairs of equivalent sets is assumed to be disjoint, then always also $M \sim N$.

[8] *Cantor 1895*, p. 483.

Es seien $\Phi \mathrel{\underline{\in}} MR$, $\mathrm{X} \mathrel{\underline{\in}} RR'$, $\Psi \mathrel{\underline{\in}} R'N$ drei „Abbildungen" (Nr. 15), welche bezw. M auf R, R auf R' und R' auf N abbilden. Ist dann $\{m, n\}$ irgend ein Element von MN, so ist definit, ob es ein Element $r \mathrel{\varepsilon} R$ und ein Element $r' \mathrel{\varepsilon} R'$ gibt, so daß gleichzeitig $\{m, r\} \mathrel{\varepsilon} \Phi$, $\{r, r'\} \mathrel{\varepsilon} \mathrm{X}$ und $\{r', n\} \mathrel{\varepsilon} \Psi$ ist. Alle Elemente $\{m, n\}$ von dieser Beschaffenheit bilden somit die Elemente einer Menge $\Omega \mathrel{\underline{\in}} MN$, welche eine Abbildung von M auf N darstellt. Ist nämlich etwa m irgend ein Element von M, so entspricht ihm immer ein einziges Element $r \mathrel{\varepsilon} R$, ein einziges $r' \mathrel{\varepsilon} R'$ und somit auch ein einziges $n \mathrel{\varepsilon} N$ von der verlangten Beschaffenheit; das Analoge gilt für jedes Element n von N. Jedem Elemente von $M + N$ entspricht also in der Tat ein einziges Element $\{m, n\}$ von Ω, und es ist wirklich $M \sim N$.

269 | 19. *Theorem.* Sind M und N irgend zwei Mengen, so gibt es immer eine Menge M', welche der einen M äquivalent und der anderen N elementenfremd ist.

Beweis. Es sei $S = \mathfrak{S}(M + N)$ gemäß V die Menge, welche die Elemente der Elemente von $M + N$ umfaßt, und r gemäß Nr. 10 ein Ding, welches nicht Element von $M + S$ ist. Dann sind die Mengen M und $R = \{r\}$ elementenfremd, und das Produkt $M' = MR$ besitzt die im Theorem verlangte Eigenschaft. In der Tat ist dann jedes Element von M' nach Nr. 13 eine Menge der Form $m' = \{m, r\}$, wo $m \mathrel{\varepsilon} M$ ist, und niemals Element von $M + N$, weil sonst r Element eines Elementes von $M + N$ und damit gemäß V Element von S wäre gegen die Annahme. Also ist M' beiden Mengen M und N elementenfremd.

Ferner entspricht jedem Element m von M ein und nur ein Element $m' = \{m, r\}$, und umgekehrt enthält jedes m' nur ein einziges Element m von M als Element, da r kein Element von M sein sollte. Jedem Element von $M + M'$ entspricht also ein einziges Element $\{m, m'\}$ von MM', für welches $m' = \{m, r\}$ ist, und wenn man alle so beschaffenen Paare $\{m, m'\}$ zu einer Untermenge $\Phi \mathrel{\underline{\in}} MM'$ rechnet, so ist nach Nr. 15 Φ eine Abbildung von M auf M' und $M \sim M'$.

Aus unserem Satze folgt, daß *die sämtlichen Mengen, welche einer nicht verschwindenden Menge M äquivalent sind, nicht die Elemente einer Menge T bilden können*; denn ist T eine beliebige Menge, so gibt es immer eine Menge $M' \sim M$, welche der Vereinigung $\mathfrak{S}T$ elementenfremd und daher *nicht* Element von T ist.

20. Sind M und N irgend zwei Mengen, so ist es immer definit, ob es eine Menge R gibt, welche beiden Mengen M und N gleichzeitig elementenfremd und äquivalent ist.

Es sei nämlich M' gemäß Nr. 19 eine Menge, welche M äquivalent und $M + N$ elementenfremd ist. Dann ist nach Nr. 16 definit, ob $M' \sim N$ ist oder nicht. Im ersteren Falle ist $R = M'$ eine Menge von der verlangten Beschaffenheit, im entgegengesetzten Falle kann es eine solche Menge R überhaupt nicht geben, da nach Nr. 18 aus $M' \sim M$, $M \sim R$ und $R \sim N$ immer $M' \sim N$ folgen müßte gegen die Annahme.

Let the subset Φ of MR, the subset X of RR', and the subset Ψ of $R'N$ be three "mappings" (No. 15) that map M onto R, R onto R', and R' onto N, respectively. If then $\{m, n\}$ is any element of MN, it is definite whether there exist an element r of R and an element r' of R' such that $\{m, r\} \varepsilon \Phi$, $\{r, r'\} \varepsilon \mathrm{X}$, and $\{r', n\} \varepsilon \Psi$. All elements $\{m, n\}$ for which this is the case therefore are the elements of a subset Ω of MN, which represents a mapping of M onto N. For, if m is any element of M, there always correspond to it a single element r of R, a single element r' of R', and therefore also a single element n of N that satisfy the required condition; an analogous statement holds for each element n of N. Therefore to each element of $M + N$ there actually corresponds a single element $\{m, n\}$ of Ω, and we in fact have $M \sim N$.

19. *Theorem.* If M and N are any two sets, there always exists a set M' that is equivalent to one, M, and disjoint from the other, N.

Proof. Let $S = \mathfrak{S}\{M + N\}$, in accordance with Axiom V, be the set that contains the elements of $M + N$, and let r, in accordance with No. 10, be an object that is not an element of $M + S$. Then the sets M and $R = \{r\}$ are disjoint, and the product $M' = MR$ possesses the property required by the theorem. Indeed, every element of M' is then, according to No. 13, a set m' of the form $\{m, r\}$ (where $m \varepsilon M$) but never an element of $M + N$, since otherwise r would be an element of an element of $M + N$, hence, by Axiom V, an element of S, contrary to the assumption. Thus M' is disjoint from both sets, M and N.

Further, there corresponds to each element m of M one and only one element $m' = \{m, r\}$, and conversely each m' contains as an element only a single element m of M, since r was assumed not to be an element of M. To each element of $M + M'$, therefore, there corresponds a single element $\{m, m'\}$ of MM' for which $m' = \{m, r\}$, and if all pairs $\{m, m'\}$ that are so constituted are assumed to form a subset Φ of MM', then, according to No. 15, Φ is a mapping of M onto M', and $M \sim M'$.

It follows from our theorem that *it is not possible for all sets equivalent to a nonempty set M to be the elements of a set T*; for if T is an arbitrary set, there always exists a set M', equivalent to M, that is disjoint from the union $\mathfrak{S}T$ and therefore *not* an element of T.

20. If M and N are any two sets, it is always definite whether there is a set R that is simultaneously disjoint from and equivalent to both sets, M and N.

For let M', in accordance with No. 19, be a set that is equivalent to M and disjoint from $M + N$. Then, according to No. 16, it is definite whether $M' \sim N$ or not. If $M' \sim N$, then $R = M'$ is a set constituted as required; otherwise, such a set R cannot exist at all, since, according to No. 18, it would always necessarily follow from $M' \sim M$, $M \sim R$, and $R \sim N$ that $M' \sim N$, contrary to the assumption.

Das vorstehende Theorem in Verbindung mit Nr. 18 berechtigt uns jetzt zu der folgenden Erweiterung unserer Definition A:

21. *Definition B.* Zwei beliebige (nicht elementenfremde) Mengen M und N heißen „*mittelbar äquivalent*", $M \sim N$, wenn es eine dritte Menge R gibt, welche ihnen beiden elementenfremd und im Sinne der Definition A beiden „unmittelbar äquivalent" ist.

Eine solche durch R „vermittelte" Äquivalenz zweier Mengen M und N
270 wird gegeben durch *zwei simultane* „Abbildungen" $\Phi \mathrel{\overline{\in}} MR$ und $\mid \Psi \mathrel{\overline{\in}} NR$, und zwei Elemente $m \,\varepsilon\, M$ und $n \,\varepsilon\, N$ heißen „entsprechend" oder „aufeinander abgebildet", wenn sie einem und demselben dritten Elemente $r \,\varepsilon\, R$ entsprechen, so daß gleichzeitig $\{m, r\} \,\varepsilon\, \Phi$ und $\{n, r\} \,\varepsilon\, \Psi$ ist. Auch bei einer solchen vermittelten Abbildung entspricht wie in Nr. 17 jeder Untermenge M_1 von M eine äquivalente Untermenge R_1 von R und somit wieder eine äquivalente Untermenge $N_1 \mathrel{\overline{\in}} N$.

Wegen Nr. 18 kann diese Definition B auch auf elementenfremde Mengen M, N angewendet werden, und nach Nr. 20 ist es immer *definit, ob zwei beliebige Mengen* im Sinne dieser Definition *äquivalent sind oder nicht.*

22. Jede Menge ist sich selbst äquivalent. Sind zwei Mengen M, N einer dritten R äquivalent, so sind sie einander selbst äquivalent.

Ist nämlich gemäß Nr. 19 M' eine Menge, welche M elementenfremd und äquivalent ist, so ist gleichzeitig $M \sim M'$ und $M' \sim M$, also nach Nr. 21 wirklich $M \sim M$.

Ist ferner die Äquivalenz der Mengen M und R vermittelt durch M', sowie die Äquivalenz von R und N vermittelt durch N', wobei M' zu M und R, sowie N' zu N und R elementenfremd sein soll, so wählen wir gemäß Nr. 19 eine sechste Menge R', welche $\sim R$ und der Summe $M + N + R$ elementenfremd ist, und haben dann wegen Nr. 18

$$M \sim M' \sim R \sim R', \quad \text{also} \quad M \sim R'$$

und

$$N \sim N' \sim R \sim R', \quad \text{also} \quad N \sim R',$$

so daß nach Nr. 21 die Äquivalenz von M und N durch R' vermittelt ist.

23. Die Nullmenge ist nur sich selbst äquivalent. Jede Menge der Form $\{a\}$ ist jeder anderen Menge $\{b\}$ derselben Form und keiner sonstigen Menge äquivalent.

Denn da das Produkt $0 \cdot M$ immer $= 0$ ist, so kann keine Menge $M \neq 0$ im Sinne der Nr. 15 der Nullmenge (unmittelbar) und somit auch keine Menge M' im Sinne von Nr. 21 ihr „mittelbar" äquivalent sein.

The preceding theorem, in combination with No. 18, now justifies the following extension of our Definition A:

21. *Definition B.* Two arbitrary (not disjoint) sets M and N are said to be "*mediately equivalent*", $M \sim N$, if there exists a third set, R, that is disjoint from both and "immediately equivalent" to both in the sense of Definition A.

Such an equivalence, "mediated" by R, of two sets M and N is given by means of *two simultaneous* mappings, a subset Φ of MR and a subset Ψ of NR, and two elements, m of M and n of N, are said to "correspond" or "be mapped onto each other" if they correspond to one and the same element r of R, so that both $\{m, r\} \varepsilon \Phi$ and $\{n, r\} \varepsilon \Psi$. In the case of such a mediated mapping, too, there corresponds to each subset M_1 of M, as in No. 17, an equivalent subset R_1 of R, and consequently again an equivalent subset N_1 of N.

On account of No. 18, Definition B may also be applied to disjoint sets M and N, and according to No. 20 *it is always definite whether two arbitrary sets are equivalent or not* in the sense of this definition.

22. Every set is equivalent to itself. If two sets, M and N, are equivalent to a third, R, they are equivalent to each other.

For if, in accordance with No. 19, M' is a set that is disjoint from and equivalent to M, both $M \sim M'$ and $M' \sim M$; therefore, according to No. 21, we in fact have $M \sim M$.

If, furthermore, the equivalence of the sets M and R is mediated by M', and that of R and N by N', where M' is assumed to be disjoint from M and R, and N' to be disjoint from N and R, then we choose, in accordance with No. 19, a sixth set, R', equivalent to R and disjoint from the sum $M + N + R$, and we now have, on account of No. 18,

$$M \sim M' \sim R \sim R', \quad \text{therefore} \quad M \sim R',$$

and

$$N \sim N' \sim R \sim R', \quad \text{therefore} \quad N \sim R',$$

so that according to No. 21 the equivalence of M and N is mediated by R'.

23. The null set is equivalent only to itself. Every set of the form $\{a\}$ is equivalent to all other sets $\{b\}$ of the same form, and to no other set.

For, since the product $0 \cdot M$ is always equal to 0,[9] no set $M \neq 0$ can be (immediately) equivalent to the null set in the sense of No. 15, and therefore no set M' can be "mediately" equivalent to it in the sense of No. 21.

---

9 ⟦Up to this point Zermelo has used only juxtaposition for the product; from here on he occasionally uses a dot.⟧

Ist ferner $\{a\}$ elementenfremd zu M, d. h. a nicht $\varepsilon\, M$, so sind alle Elemente des Produktes $\{a\}\, M$ von der Form $\{a, m\}$, und wenn M außer m noch ein weiteres Element p enthielte, so wären $\{a, m\}$ und $\{a, p\}$ nicht elementenfremd, wie in Nr. 15 für jede „Abbildung“ $\Phi \mathrel{\underline{\in}} \{a\}M$ gefordert. Dagegen ist $\{a\} \cdot \{b\} = \{a, b\}$ stets eine Abbildung von $\{a\}$ auf $\{b\}$.

24. *Theorem.* Ist $M \sim M'$ und $N \sim N'$, während M und N einerseits, M' und N' andererseits einander elementenfremd sind, so ist immer

$$M + N \sim M' + N' .$$

Beweis. Wir betrachten zunächst den Fall, wo $M + N$ und $M' + N'$ elementenfremd sind. Dann ist auf beide Äquivalenzen $M \sim M'$ und $N \sim N'$
271 die Definition A Nr. 15 anwendbar, und es gibt zwei | Abbildungen $\Phi \mathrel{\underline{\in}} MM'$ und $\Psi \mathrel{\underline{\in}} NN'$, deren Summe $\Phi + \Psi$ die verlangte Abbildung von $M + N$ auf $M' + N'$ darstellt. Ist nämlich $p\, \varepsilon\, (M + N)$, so ist entweder $p\, \varepsilon\, M$ oder $p\, \varepsilon\, N$, aber wegen $[M, N] = 0$ nicht beides gleichzeitig, und im einen Falle enthält Φ, im anderen Ψ ein einziges Element der Form $\{p, q\}$. Ebenso entspricht auch jedem Elemente q von $M' + N'$ ein und nur ein Element $\{p, q\}$ in $\Phi + \Psi$.

Sind $M + N$ und $M' + N'$ nicht selbst elementenfremd, so gibt es gemäß Nr. 19 eine Menge $S'' \sim M' + N'$, welche der Summe $M + N + M' + N'$ elementenfremd ist, und bei einer Abbildung X von $M' + N'$ auf S'' mögen wegen Nr. 17 den beiden Teilen M' und N' die äquivalenten und elementenfremden Teile M'' und N'' von S'' entsprechen. Dann ist $M \sim M' \sim M''$ sowie $N \sim N' \sim N''$ und, da jetzt $M + N$ und $M'' + N''$ elementenfremd sind, nach dem soeben Bewiesenen

$$M + N \sim M'' + N'' = S'' \sim M' + N' ,$$

also wieder

$$M + N \sim M' + N' .$$

25. *Theorem.* Ist eine Menge M einem ihrer Teile M' äquivalent, so ist sie auch jedem anderen Teile M_1 äquivalent, welcher M' als Bestandteil enthält.

Beweis. Es sei

$$M \sim M' \mathrel{\underline{\in}} M_1 \mathrel{\underline{\in}} M \quad \text{und} \quad Q = M_1 - M' .$$

Wegen der vorausgesetzten Äquivalenz $M \sim M'$ gibt es gemäß Nr. 21 eine Abbildung $\{\Phi, \Psi\}$ von M auf M', vermittelt etwa durch M''. Ist nun A eine beliebige Untermenge von M, so entspricht ihr bei der betrachteten Abbildung eine bestimmte Untermenge A' von M', und es ist definit, ob $A' \mathrel{\underline{\in}} A$ ist oder nicht. Somit bilden alle solchen Elemente A von $\mathfrak{U}\, M$, für welche gleichzeitig $Q \mathrel{\underline{\in}} A$ und $A' \mathrel{\underline{\in}} A$ ist, nach III die Elemente einer gewissen Menge $T \mathrel{\underline{\in}} \mathfrak{U}\, M$, und es ist namentlich M selbst Element von T. Der gemeinsame Bestandteil $A_0 = \mathfrak{D}\, T$ aller Elemente von T (Nr. 9) besitzt nun die folgenden

If, furthermore, $\{a\}$ is disjoint from M, that is, if a is not an element of M, then all elements of the product $\{a\}M$ are of the form $\{a, m\}$ and, if M were to contain, besides m, another element, p, then $\{a, m\}$ and $\{a, p\}$ would not be disjoint, and this, according to No. 15, would prevent any subset Φ of $\{a\}M$ from being a "mapping". On the other hand, $\{a\} \cdot \{b\} = \{a, b\}$ is always a mapping of $\{a\}$ onto $\{b\}$.

24. *Theorem.* If $M \sim M'$ and $N \sim N'$, while M and N on the one hand, and M' and N' on the other, are mutually disjoint, then always

$$M + N \sim M' + N'.$$

Proof. We first consider the case in which $M+N$ and $M'+N'$ are disjoint. Then Definition A (No. 15) is applicable to both of the equivalences $M \sim M'$ and $N \sim N'$, and there are two mappings, the subset Φ of MM' and the subset Ψ of NN', whose sum $\Phi + \Psi$ represents the required mapping of $M + N$ onto $M' + N'$. For, if $p \,\varepsilon (M + N)$, either $p \,\varepsilon\, M$ or $p \,\varepsilon\, N$, but, since $[M, N] = 0$, not both; and Φ in one case, and Ψ in the other, contains a single element of the form $\{p, q\}$. Likewise there corresponds to each element q of $M' + N'$ one and only one element $\{p, q\}$ of $\Phi + \Psi$.

If $M+N$ and $M'+N'$ are not themselves disjoint, there exists, according to No. 19, a set S'' equivalent to $M' + N'$ and disjoint from the sum $M + N + M' + N'$; and, given a mapping X of $M' + N'$ onto S'', equivalent and disjoint parts M'' and N'' of S'' will, according to No. 17, correspond to the two parts M' and N'. Then $M \sim M' \sim M''$, as well as $N \sim N' \sim N''$, and, since now $M + N$ and $M'' + N''$ are disjoint,

$$M + N \sim M'' + N'' = S'' \sim M' + N'$$

according to what has just been proved; therefore again

$$M + N \sim M' + N'.$$

25. *Theorem.* If a set M is equivalent to one of its parts, M', it is also equivalent to any other part M_1, that includes M' as component.

Proof. Let

$$M \sim M' \subseteq M_1 \subseteq M \quad \text{and} \quad Q = M_1 - M'.$$

Because of the equivalence $M \sim M'$ that has been assumed, there exists, according to No. 21, a mapping $\{\Phi, \Psi\}$ of M onto M', mediated by, say, M''. If now A is an arbitrary subset of M, a certain subset A' of M' will correspond to it under the mapping in question, and it is definite whether $A' \subseteq A$ or not. Thus all elements A of $\mathfrak{U} M$ for which we have both $Q \subseteq A$ and $A' \subseteq A$ are, according to Axiom III, the elements of a certain subset T of $\mathfrak{U} M$, and, in particular, M is itself an element of T. The common component $A_0 = \mathfrak{D} T$

Eigenschaften: 1) $Q \mathrel{\overline{\in}} A_0$, weil Q eine gemeinsame Untermenge aller $A \,\varepsilon\, T$ ist, 2) $A_0{}' \mathrel{\overline{\in}} A_0$, weil jedes Element x von A_0 gemeinsames Element aller $A \,\varepsilon\, T$ und somit auch sein Bild $x' \,\varepsilon\, A' \mathrel{\overline{\in}} A$ gemeinsames Element aller A ist. Wegen 1) und 2) ist also auch $A_0 \,\varepsilon\, T$. Endlich ist 3) $A_0 = Q + A_0{}'$. Da nämlich $A_0{}' \mathrel{\overline{\in}} A_0$ und gleichzeitig $\mathrel{\overline{\in}} M' \mathrel{\overline{\in}} M - Q$ ist, so ist einmal $A_0{}' \mathrel{\overline{\in}} A_0 - Q$. Andererseits ist aber auch jedes Element r von $A_0 - Q$ ein Element von $A_0{}'$ und daher $A_0 - Q \mathrel{\overline{\in}} A_0{}'$. In der Tat, wäre r nicht $\varepsilon\, A_0{}'$, so würde auch $A_1 = A_0 - \{r\}$ noch $A_0{}'$ und a fortiori $A_1{}'$ als Bestandteil enthalten und, da es immer noch Q enthält, selbst Element von T sein, während es doch nur ein *Teil* von $A_0 = \mathfrak{D}\, T$ ist. Es ist also

$$M_1 = Q + M' = (Q + A_0{}') + (M' - A_0{}') = A_0 + (M' - A_0{}'),$$

272 | wo die beiden Summanden rechts keine Elemente gemein haben, weil Q und M' elementenfremd sind. Da nun aber $A_0 \sim A_0{}'$ und $M' - A_0{}'$ sich selbst äquivalent ist, so folgt nach Nr. 24

$$M_1 \sim A_0{}' + (M' - A_0{}') = M' \sim M,$$

d. h. wie behauptet, $M_1 \sim M$.

26. *Folgerung.* Ist eine Menge M einem ihrer Teile M' äquivalent, so ist sie auch jeder Menge M_1 äquivalent, welche aus M durch Fortlassung oder Hinzufügung eines einzelnen Elementes entsteht.

Es sei

$$M \sim M' = M - R \quad \text{und} \quad M_1 = M - \{r\},$$

wo $r \,\varepsilon\, R$ sein möge. Dann ist

$$M' = M - \{r\} - (R - \{r\}) \mathrel{\overline{\in}} M - \{r\} = M_1$$

und nach dem vorigen Satze $M \sim M_1$.

Ist ferner

$$M_2 = M - \{a\}, \quad \text{wo} \quad a \,\varepsilon\, M' = M - R$$

ist, so sei

$$M_0 = M - \{a, r\},$$

und wir haben nach Nr. 23 und 24

$$M_2 = M_0 + \{r\} \sim M_0 + \{a\} = M_1 \sim M,$$

also auch

$$M_2 \sim M.$$

Ist endlich

$$M_3 = M + \{c\},$$

wo c nicht $\varepsilon\, M$ ist, so folgt aus $M \sim M'$ wieder nach Nr. 24

$$M_3 = M + \{c\} \sim M' + \{c\} = M - R + \{c\} = M_3 - R,$$

(No. 9) of all elements of T now possesses the following properties: 1) $Q \subseteq A_0$, since Q is a common subset of all elements A of T; 2) $A_0{}' \subseteq A_0$, because every element x of A_0 is a common element of all elements A of T and its map $x' \,\varepsilon\, A' \subseteq A$ is thus also a common element of all A. On account of 1) and 2), therefore, also $A_0 \,\varepsilon\, T$. Finally we have 3) $A_0 = Q + A_0{}'$. For, since $A_0{}' \subseteq A_0$ and also $A_0{}' \subseteq M' \subseteq M - Q$, on the one hand $A_0{}' \subseteq A_0 - Q$. On the other hand, however, every element r of $A_0 - Q$ is also an element of $A_0{}'$, and therefore $A_0 - Q \subseteq A_0{}'$. Indeed, if r were not an element of A_0', then $A_1 = A_0 - \{r\}$ would still have $A_0{}'$, and a fortiori $A_1{}'$, as a component, and, since it still includes Q, it would itself be an element of T, whereas it is in fact only a *part* of $A_0 = \mathfrak{D}T$. Therefore

$$M_1 = Q + M' = (Q + A_0{}') + (M' - A_0{}') = A_0 + (M' - A_0{}')\,,$$

where the two summands on the right have no element in common, since Q and M' are disjoint. But now, since A_0 is equivalent to $A_0{}'$ and $M' - A_0{}'$ is equivalent to itself, it follows according to No. 24 that

$$M_1 \sim A_0{}' + (M' - A_0{}') = M' \sim M\,;$$

that is, $M_1 \sim M$ as asserted.

26. *Corollary.* If a set M is equivalent to one of its parts, M', it is also equivalent to any set M_1 that is obtained from M when a single element is removed or added.

Let

$$M \sim M' = M - R \quad \text{and} \quad M_1 = M - \{r\}\,,$$

where r is some element of R. Then

$$M' = M - \{r\} - (R - \{r\}) \subseteq M - \{r\} = M_1\,,$$

and, according to the previous theorem, $M \sim M_1$.

If, furthermore,

$$M_2 = M - \{a\}\,, \quad \text{where} \quad a \,\varepsilon\, M' = M - R\,.$$

let

$$M_0 = M - \{a, r\}\,,$$

and we have, according to Nos. 23 and 24,

$$M_2 = M_0 + \{r\} \sim M_0 + \{a\} = M_1 \sim M\,;$$

therefore also

$$M_2 \sim M\,.$$

If, finally,

$$M_3 = M + \{c\}\,,$$

where c is not an element of M, it follows from $M \sim M'$, again according to No. 24, that

$$M_3 = M + \{c\} \sim M' + \{c\} = M - R + \{c\} = M_3 - R\,,$$

und nach dem vorher Bewiesenen weiter

$$M = M_3 - \{c\} \sim M_3 \,,$$

womit der Satz in allen seinen Teilen bewiesen ist.

27. *Äquivalenzsatz.* Ist jede von zwei Mengen M, N einer Untermenge der anderen äquivalent, so sind M und N selbst äquivalent.

Es sei $M \sim M' \mathrel{\underline{\in}} N$ und $N \sim N' \mathrel{\underline{\in}} M$. Dann entspricht wegen Nr. 21 der Untermenge M' von N eine äquivalente Untermenge $M'' \mathrel{\underline{\in}} N' \mathrel{\underline{\in}} M$, und es ist $M \sim M' \sim M''$, also nach dem Theorem Nr. 25 auch $M \sim N' \sim N$, q. e. d.[1]

273 | 28. *Theorem.* Ist T eine beliebige Menge, deren Elemente M, N, R, ..., sämtlich Mengen sind, so kann man sie alle gleichzeitig abbilden auf äquivalente Mengen M', N', R', ..., welche die Elemente einer neuen Menge T' bilden und unter sich sowohl wie einer gegebenen Menge Z elementenfremd sind.

Beweis. Es sei $S = \mathfrak{S}T = M + N + R + \cdots$ nach V die Summe aller Elemente von T, und gemäß Nr. 19 sei T'' eine Menge, welche T äquivalent und der Summe $T + S + \mathfrak{S}(S + Z)$ elementenfremd ist, so daß vermöge einer Abbildung Ω jedem Elemente $M, N, R, \ldots$ von T ein bestimmtes Element $M'', N'', R'', \ldots$ von T'' entspricht. Ein beliebiges Element des Produktes ST'' (Nr. 13) ist dann von der Form $\{s, M''\}$, wo $s \,\varepsilon\, S$ und $M'' \,\varepsilon\, T''$ ist, und für jedes solche Element ist es definit (Nr. 4), ob $s \,\varepsilon\, M$ ist, wo M das dem M'' vermöge Ω entsprechende Element von T, d. h. gemäß Nr. 15 $\{M, M''\} \,\varepsilon\, \Omega$ sein soll. Alle so beschaffenen Elemente des Produktes bilden somit wegen III die Elemente einer Untermenge S' von ST'', und diese Menge S' ist $S + Z$ elementenfremd, weil sonst ein $M'' \,\varepsilon\, T''$ als Element von $\{s, M''\}$ Element eines Elementes von $S + Z$, also wegen V Element von $\mathfrak{S}(S + Z)$ wäre gegen die über T'' gemachte Annahme. Ist ferner M ein beliebiges Element von T, und M'' das entsprechende von T'', so bilden diejenigen Elemente $\{s, M''\}$ von S', welche M'' als Element enthalten, nach III eine gewisse Untermenge $M' \mathrel{\underline{\in}} S'$, und es ist $M' \sim M$ vermöge einer Abbildung $\mathsf{M} \mathrel{\underline{\in}} MM' \mathrel{\underline{\in}} SS'$, in

[1] Der hier in den Nrn. 25 und 27 gegebene Beweis des „Äquivalenzsatzes" (auf Grund meiner brieflichen Mitteilung vom Jan. 1906 zuerst publiziert von Herrn *H. Poincaré* in der Revue de Métaphysique et de Morale t. 14, p. 314) beruht lediglich auf der *Dedekind*schen Kettentheorie (Was sind und was sollen die Zahlen? § 4) und vermeidet im Gegensatz zu den älteren Beweisen von *E. Schröder* und *F. Bernstein*, sowie zu dem letzten Beweise von *J. König* (Comptes Rendus t. 143, 9 VII 1906) jede Bezugnahme auf geordnete Reihen vom Typus ω oder das Prinzip der vollständigen Induktion. Einen ganz ähnlichen Beweis veröffentlichte ungefähr gleichzeitig Herr *G. Peano* („Super Teorema de Cantor-Bernstein", Rendiconti del Circolo Matematico XXI sowie Revista de Mathematica VIII, p. 136), wo in der letztgenannten Note zugleich auch der von Herrn *H. Poincaré* gegen meinen Beweis gerichtete Einwand erörtert wird. Vgl. meine Note Math. Ann. Bd. 65, p. 107–128, § 2 b.

and furthermore, according to what was proved previously,

$$M = M_3 - \{c\} \sim M_3\,;$$

thus the theorem is proved in its entirety.

27. *Equivalence theorem.* If each of the two sets M and N is equivalent to a subset of the other, M and N are themselves equivalent.

Let $M \sim M' \subseteq N$ and $N \sim N' \subseteq M$. Then on account of No. 21 there corresponds to the subset M' of N an equivalent set M'' such that $M'' \subseteq N' \subseteq M$, and we have $M \sim M' \sim M''$; therefore, according to the theorem of No. 25, also $M \sim N' \sim N$, q. e. d.[10]

28. *Theorem.* If all the sets $M, N, R, \ldots$ are elements of an arbitrary set T, they can all be simultaneously mapped onto ⟦respectively⟧ equivalent sets $M', N', R', \ldots$ that are the elements of a new set, T', and are disjoint from one another as well as from a given set Z.

Proof. Let $S = \mathfrak{S}T = M + N + R + \cdots$, in accordance with Axiom V, be the sum of all elements of T, and, in accordance with No. 19, let T'' be a set equivalent to T and disjoint from the sum $T + S + \mathfrak{S}(S + Z)$, so that under a mapping Ω to each element of T there corresponds a certain element of T'': $M'', N'', R'', \ldots$ to $M, N, R, \ldots$, respectively. An arbitrary element of the product ST'' (No. 13) will then have the form $\{s, M''\}$, where $s \,\varepsilon\, S$ and $M'' \,\varepsilon\, T''$, and for every such element it is definite (No. 4) whether $s \,\varepsilon\, M$, where M is assumed to be the element of T corresponding to M'' under Ω, that is, according to No. 15, to be such that $\{M, M''\} \,\varepsilon\, \Omega$. All elements of the product that are so constituted then are, on account of Axiom III, the elements of a subset S' of ST'', and this set S' is disjoint from $S + Z$, since otherwise an element M'' of T'', being an element of $\{s, M''\}$, would be an element of an element of $S + Z$ and thus, on account of Axiom V, also an element of $\mathfrak{S}(S + Z)$, contrary to the assumption made about T''. If, furthermore, M is an arbitrary element of T and M'' the corresponding one of T'', those elements $\{s, M''\}$ of S' that contain M'' as an element form, according to Axiom III, a certain subset M' of S', and $M' \sim M$ by virtue of a mapping M ($\mathsf{M} \subseteq MM' \subseteq SS'$) under which to each element m of M there

[10] The proof of the "equivalence theorem" given here in Nos. 25 and 27 (first published by Mr. *H. Poincaré* (*1906b*, p. 314) on the basis of a letter that I wrote in January 1906) rests solely upon *Dedekind*'s chain theory (*1888*, §4) and, unlike the older proofs by *E. Schröder* and *F. Bernstein* as well as the latest proof by *J. König* (*1907*), avoids any reference to ordered sequences of order ω or to the principle of mathematical induction. At approximately the same time Mr. *G. Peano* (*1906a*) published a proof that was quite similar; the paper containing that proof also contains a discussion of the objection directed by Mr. *H. Poincaré* against my proof. See § 2b in my *1908a*.

welcher jedem Elemente m von M ein Element $m' = \{m, M''\}$ von M' entspricht und umgekehrt. Ebenso gehört auch zu jedem anderen Element $N \varepsilon T$ eine äquivalente Untermenge $N' \in S'$ und eine Abbildung $\mathsf{N} \in NN' \in SS'$, durch welche jedem Elemente n von N ein Element $\{n, N''\}$ von N' entspricht. Die beiden Untermengen M' und N', welche zu zwei verschiedenen Elementen M und N von T gehören, sind aber immer elementenfremd, denn wäre etwa

$$\{m, M''\} = \{n, N''\}$$

ein gemeinsames Element von M' und N', so müßte M'' als Element von $\{n, N''\}$ entweder $= N''$ oder $= n$ sein, und im ersten Falle wäre auch $M = N$, im zweiten aber wären T'' und S nicht elementenfremd, gegen die Annahme. Die Untermengen $M', N', R', \ldots$ von S', welche vermöge der Abbildungen
274 $\mathsf{M}, \mathsf{N}, \mathsf{P}, \ldots$ den Elementen $M, N, R, \ldots$, von | T äquivalent sind, sind also in der Tat sowohl unter sich als auch, weil S' es ist, der Menge Z elementenfremd. Endlich ist von jeder Untermenge ${S_1}' \in S'$, welche ein Element $\{s, M''\}$ enthält, immer definit, ob sie mit der entsprechenden Menge M' identisch ist oder nicht, und alle diese $M', N', R', \ldots$ bilden gemäß III und IV die Elemente einer gewissen Menge $T' \in \mathfrak{U}S'$; der Satz ist also in allen seinen Teilen bewiesen.

29_{VI}. *Allgemeines Auswahlprinzip.* Ist T eine Menge, deren Elemente $M, N, R, \ldots$ sämtlich von Null verschiedene Mengen sind, so gibt es immer Mengen P, welche nach einer bestimmten Vorschrift jedem Element M von T eines seiner Elemente $m \varepsilon M$ eindeutig zuordnen.

Beweis. Man wende auf T' das in der vorhergehenden Nr. 28 angegebene Verfahren an, wobei $Z = 0$ gesetzt werden kann, und hat dann alle Mengen $M, N, R, \ldots$ gleichzeitig abgebildet auf die äquivalenten Mengen $M', N', R', \ldots$, welche unter sich elementenfremd sind und die Elemente einer Menge T' bilden. Ist nun P gemäß VI eine solche Untermenge von $\mathfrak{S}\,T'$, welche mit jedem Element von T' genau ein Element gemein hat, so leistet P die verlangte Zuordnung. Ist nämlich M irgend ein Element von T und ist M' das entsprechende Element von T', so enthält P nur ein einziges Element m' von M', und diesem entspricht wieder ein ganz bestimmtes Element m von M.

30_{VI}. *Theorem.* Sind zwei äquivalente Mengen T und T', deren Elemente M, N, $R, \ldots$ bzw. M', N', $R', \ldots$ unter sich elementenfremde Mengen sind, so aufeinander abgebildet, daß jedem Element M der einen Menge eine äquivalente Menge M' als Element der anderen entspricht, so sind auch die zugehörigen Summen $\mathfrak{S}\,T$ und $\mathfrak{S}\,T'$, sowie die entsprechenden Produkte $\mathfrak{P}\,T$ und $\mathfrak{P}\,T'$ einander äquivalent.

Beweis. Wir beweisen den Satz zunächst unter der Annahme, daß $S = \mathfrak{S}\,T$ und $S' = \mathfrak{S}\,T'$ einander elementenfremd sind, in welchem Falle auch jedes Element von T jedem Element von T' elementenfremd sein muß. Aus $M \sim M'$ folgt dann gemäß Nr. 15, daß $\mathfrak{U}(MM')$ eine von 0 verschiedene Untermenge A_M besitzt, welche die sämtlichen möglichen Abbildungen $\mathsf{M}, \mathsf{M}', \mathsf{M}'', \ldots$

corresponds an element $m' = \{m, M''\}$ of M' and conversely. Likewise, with any other element N of T there are associated an equivalent subset N' of S' and a mapping N ($\mathsf{N} \subseteq NN' \subseteq SS'$) under which to each element n of N there corresponds an element $\{n, N''\}$ of N'. The two subsets M' and N', which correspond to two different elements M and N of T, are, however, always disjoint, for, if, say,

$$\{m, M''\} = \{n, N''\}$$

were a common element of M' and N', then M'', being an element of $\{n, N''\}$, would have to be equal either to N'' or to n, and in the first case M would also be equal to N, and in the second T'' and S would not be disjoint, contrary to the assumption. The subsets $M', N', R', \ldots$ of S', which by virtue of the mappings $\mathsf{M}, \mathsf{N}, \mathsf{P}, \ldots$ are equivalent to the elements $M, N, R, \ldots$ of T, are thus indeed disjoint from one another, and they are also disjoint from the set Z since S' is. Finally, it is always definite for every subset ${S_1}'$ of S' containing an element $\{s, M''\}$ whether that subset is identical with the corresponding set M', and all of those among $M', N', R' \ldots$ for which this is the case are, according to Axioms III and IV, the elements of a certain subset T' of $\mathfrak{U}S'$. The theorem is therefore proved in its entirety.

29_{VI}. *General principle of choice.* If T is a set whose elements $M, N, R, \ldots$ all are sets different from the null set, there always exist sets P that, according to a certain rule, uniquely correlate with each element M of T one element m of that M.

Proof. Apply to T[11] the procedure specified in No. 28 above, letting $Z = 0$. This yields simultaneous mappings of all sets $M, N, R, \ldots$ onto the equivalent sets $M', N', R', \ldots$, which are mutually disjoint and form the elements of a set T'. If now P, in accordance with Axiom VI, is a subset of $\mathfrak{S}T'$ that has exactly one element in common with each element of T', then P provides the desired correlation. For, if M is any element of T and M' the corresponding element of T', P contains only a single element m' of M', and to this there again corresponds a well-determined element m of M.

30_{VI} *Theorem.* Let T and T' be two equivalent sets containing as elements the mutually disjoint sets $M, N, R, \ldots$ and the mutually disjoint sets $M', N', R', \ldots$, respectively. If T and T' are mapped onto each other in such a way that to each element M of one set there corresponds an equivalent set M' as an element of the other, the associated sums $\mathfrak{S}T$ and $\mathfrak{S}T'$, as well as the corresponding products $\mathfrak{P}T$ and $\mathfrak{P}T'$, are also equivalent.

Proof. We first prove the theorem on the assumption that $S = \mathfrak{S}T$ and $S' = \mathfrak{S}T'$ are mutually disjoint, in which case each element of T would have to be disjoint from each element of T', too. It then follows from $M \sim M'$, according to No. 15, that $\mathfrak{U}(MM')$ possesses a subset A_M, different from 0, that contains as elements all possible mappings $\mathsf{M}, \mathsf{M}', \mathsf{M}'', \ldots$ of M onto M'.

[11] [Zermelo erroneously writes "T′" instead of "T".]

von M auf M' als Elemente enthält. Ebenso entspricht jedem anderen Element N von T eine Menge $\mathsf{A}_N \mathrel{\underline{\in}} \mathfrak{U}(NN')$, welche die sämtlichen Abbildungen von N auf N' umfaßt, und auch A_N ist $\neq 0$. Alle diese Abbildungsmengen $\mathsf{A}_M, \mathsf{A}_N, \mathsf{A}_R, \ldots$ sind Untermengen von $\mathfrak{U}(SS')$ und bilden daher wegen III und IV die Elemente einer gewissen Untermenge $\mathsf{T} \mathrel{\underline{\in}} \mathfrak{U}\mathfrak{U}(SS')$. Da nun die Elemente von T sämtlich von 0 verschiedene Mengen und unter sich elementenfremd sind (weil aus der Elementenfremdheit von MM' und NN' auch die ihrer Untermengen folgt), so ist nach Axiom VI auch das Produkt
275 $\mathfrak{P}\mathsf{T} \neq 0$, und ein beliebiges Ele- | ment Θ von $\mathfrak{P}$ ist eine Menge der Form $\Theta = \{\mathsf{M}, \mathsf{N}, \mathsf{P}, \ldots\}$, welche von jeder der Mengen $\mathsf{A}_M, \mathsf{A}_N, \mathsf{A}_R, \ldots$ genau ein Element enthält. Die Existenz einer solchen „kombinierten Abbildung“ hätten wir kürzer auch aus dem Theorem Nr. 29 schließen können. Bilden wir nun gemäß V die Vereinigung

$$\Omega = \mathfrak{S}\Theta = \mathsf{M} + \mathsf{N} + \mathsf{P} + \cdots \mathrel{\underline{\in}} SS',$$

so liefert Ω die verlangte Abbildung von S auf S'. Denn jedes Element s von S muß einem und nur einem Elemente von T, etwa M, als Element angehören und daher in einem einzigen Element der entsprechenden Abbildung M als Element erscheinen, während in allen übrigen Summanden $\mathsf{N}, \mathsf{P}, \ldots$ kein Element von M mehr vorkommt. Das analoge gilt auch für jedes Element s' von S', und nach der Definition Nr. 15 ist somit in der Tat $S \sim S'$.

Durch dasselbe Ω und seine Untermengen wird wegen Nr. 17 auch jede Untermenge p von S auf eine äquivalente Untermenge p' von S' abgebildet, und ist insbesondere $p = \{m, n, \ldots\}$ gemäß Nr. 13 ein Element von $P = \mathfrak{P}T$, so ist die ihm entsprechende Untermenge $p' = \{m', n', \ldots\}$ von S' auch ein Element von $\mathfrak{P}T'$. Ist nämlich M' ein beliebiges Element von T', und M das entsprechende Element von T, so enthält p ein und nur ein Element $m \,\varepsilon\, M$ und p' das entsprechende Element $m' \,\varepsilon\, M'$, aber auch kein weiteres Element von M', da ein solches auch einem zweiten Elemente von M in p entsprechen müßte. Ebenso entspricht jedem Elemente $p' \,\varepsilon\, \mathfrak{P}T'$ ein und nur ein Element $p \,\varepsilon\, \mathfrak{P}T$, und wir erhalten in der Tat eine bestimmte Untermenge $\Pi \mathrel{\underline{\in}} \mathfrak{P}T \cdot \mathfrak{P}T'$ als Abbildung von $\mathfrak{P}T$ auf $\mathfrak{P}T'$, so daß auch diese beiden Produkte einander äquivalent sind.

Sind nun aber S und S' nicht mehr elementenfremd, so können wir gemäß Nr. 19 eine dritte Menge S'' einführen, welche S' äquivalent und $S + S'$ elementenfremd ist. Dann entspricht wegen Nr. 17 einer Untermenge $M' \mathrel{\underline{\in}} S'$ eine äquivalente Untermenge $M'' \mathrel{\underline{\in}} S''$, und da die $M', N', R', \ldots$ untereinander elementenfremd sind, so gilt das gleiche auch von den entsprechenden $M'', N'', R'', \ldots$. Da ferner jedes Element s'' von S'' einem Elemente s' von S' entspricht, welches einer der Mengen $M', N', R', \ldots$ angehört, so ist S'' die *Summe* aller dieser $M'', N'', R'', \ldots$, welche die Elemente einer gewissen Untermenge $T'' \mathrel{\underline{\in}} \mathfrak{U}S''$ bilden. Nun haben wir aber $M \sim M' \sim M''$, $N \sim N' \sim N'', \ldots$; es ist also jedes Element M von T dem entsprechenden Element M'' von T'' äquivalent, und da jetzt $S'' = \mathfrak{S}T''$ beiden Summen

Likewise, to any other element N of T there corresponds a subset A_N of $\mathfrak{U}(NN')$ that contains all mappings of N onto N', and A_N is also different from 0. All these mapping-sets $\mathsf{A}_M, \mathsf{A}_N, \mathsf{A}_R, \ldots$ are subsets of $\mathfrak{U}(SS')$ and therefore are, according to Axioms III and IV, the elements of a certain subset T of $\mathfrak{U}\mathfrak{U}(SS')$. Since, now, all elements of T are sets that are different from 0 and mutually disjoint (because from the fact that MM' and NN' are disjoint it follows that their subsets are disjoint), according to Axiom VI the product $\mathfrak{P}\mathsf{T}$ is also different from 0, and an arbitrary element Θ of $\mathfrak{P}\mathsf{T}$ is a set of the form $\{\mathsf{M}, \mathsf{N}, \mathsf{P}, \ldots\}$ that contains exactly one element of each of the sets $\mathsf{A}_M, \mathsf{A}_N, \mathsf{A}_R, \ldots$. We could also have inferred the existence of such a "combined" mapping from the theorem of No. 29, and more quickly at that. If we now form, in accordance with Axiom V, the union

$$\Omega = \mathfrak{S}\Theta = \mathsf{M} + \mathsf{N} + \mathsf{P} + \cdots \subseteq SS',$$

then Ω provides the required mapping of S onto S'. For each element s of S must belong as an element to one and only one element of T, say M, and must therefore occur as an element in a single element of the corresponding mapping M, while no element of M occurs in any of the remaining summands $\mathsf{N}, \mathsf{P}, \ldots$. An analogous statement holds for every element of s' of S', and according to Definition A (No. 15) we thus have $S \sim S'$.

By means of the same Ω and its subsets, each subset p of S is also mapped, on account of No. 17, onto an equivalent subset p' of S', and if in particular $p = \{m, n, \ldots\}$ is, in accordance with No. 13, an element of $P = \mathfrak{P}T$, the subset $p' = \{m', n', \ldots\}$ of S' corresponding to it is an element of $\mathfrak{P}T'$. For, if M' is an arbitrary element of T' and M the corresponding element of T, p contains one and only one element m of M and p' contains the corresponding element m' of M' but no other element of M', since such an element would also have to correspond to a second element of M in p. Likewise, there corresponds to each element p' of $\mathfrak{P}T'$ one and only one element p of $\mathfrak{P}T$, and we indeed obtain a certain subset $\sqcap$ of $\mathfrak{P}T \cdot \mathfrak{P}T'$ as a mapping of $\mathfrak{P}T$ onto $\mathfrak{P}T'$, so that these two products are mutually equivalent.

But if now S and S' are no longer assumed to be disjoint, we can, according to No. 19, introduce a third set, S'', that is equivalent to S' and disjoint from $S + S'$. Then, on account of No. 17, there corresponds to a subset M' of S' an equivalent subset M'' of S'', and, since $M', N', R', \ldots$ are mutually disjoint, the same holds of the corresponding $M'', N'', R'', \ldots$. Since, furthermore, every element s'' of S'' corresponds to an element s' of S' belonging to one of the sets $M', N', R', \ldots$, S'' is the *sum* of all these $M'', N'', R'', \ldots$, which are the elements of a certain subset T'' of $\mathfrak{U}S''$. But now we have $M \sim M' \sim M''$, $N \sim N' \sim N'', \ldots$; thus every element M of T is equivalent to the corresponding element M'' of T'', and, since now $S'' = \mathfrak{S}T''$ is disjoint from both of the sums $S' = \mathfrak{S}T'$ and $S = \mathfrak{S}T$, it follows according to what

$S' = \mathfrak{S}T'$ und $S = \mathfrak{S}T$ elementenfremd ist, so folgt nach dem oben Bewiesenen:

$$\mathfrak{S}T \sim \mathfrak{S}T'' \sim \mathfrak{S}T' \quad \text{und} \quad \mathfrak{P}T \sim \mathfrak{P}T'' \sim \mathfrak{P}T',$$

womit der Satz in voller Allgemeinheit bewiesen ist.

276 31. *Definition.* Ist eine Menge M einer Untermenge der Menge N | äquivalent, aber nicht umgekehrt N einer Untermenge von M, so sagen wir, M sei „von *kleinerer Mächtigkeit* als N“, und schreiben abgekürzt $M < N$.

Folgerungen. a) Da es nach Nr. 21 für irgend zwei Mengen definit ist, ob sie einander äquivalent sind oder nicht, so ist es auch definit, ob M mindestens einem Element von $\mathfrak{U}N$, sowie ob N irgend einem Element von $\mathfrak{U}M$ äquivalent ist. *Es ist also immer definit, ob $M < N$ ist oder nicht.*

b) Die drei Beziehungen $M < N$, $M \sim N$, $N < M$ schließen einander aus.

c) Ist $M < N$ und $N < R$ oder $N \sim R$, so ist immer auch $M < R$.

d) Ist M einer Untermenge von N äquivalent, so ist entweder $M \sim N$ oder $M < N$. Dies ist eine Folge des „Äquivalenzsatzes“ Nr. 27.

e) Die Nullmenge ist von kleinerer Mächtigkeit als jede andere Menge, ebenso jede aus einem einzigen Elemente bestehende Menge $\{a\}$ von kleinerer Mächtigkeit als jede Menge M, welche echte Teile besitzt. Vergl. Nr. 23.

32. *Satz von Cantor.* Ist M eine beliebige Menge, so ist immer $M < \mathfrak{U}M$. *Jede Menge ist von kleinerer Mächtigkeit als die Menge ihrer Untermengen.*

Beweis. Jedem Element m von M entspricht eine Untermenge $\{m\} \mathrel{\underline{\in}} M$. Da es nun für jede Untermenge $M_1 \mathrel{\underline{\in}} M$ definit ist, ob sie nur ein einziges Element enthält (Nr. 13), so bilden alle Untermengen der Form $\{m\}$ die Elemente einer Menge $U_0 \mathrel{\underline{\in}} \mathfrak{U}M$, und es ist $M \sim U_0$.

Wäre umgekehrt $U = \mathfrak{U}M$ äquivalent einer Untermenge $M_0 \mathrel{\underline{\in}} M$, so entspräche vermöge einer Abbildung Φ von U auf M_0 jeder Untermenge $M_1 \mathrel{\underline{\in}} M$ ein bestimmtes Element m_1 von M_0, so daß $\{M_1, m_1\} \varepsilon \Phi$ wäre, und es wäre immer definit, ob $m_1 \varepsilon M_1$ ist oder nicht. Alle solchen Elemente m_1 von M_0, für welche *nicht* $m_1 \varepsilon M_1$ ist, bildeten also die Elemente einer Untermenge $M' \mathrel{\underline{\in}} M_0 \mathrel{\underline{\in}} M$, welche gleichfalls Element von U wäre. Dieser Menge $M' \mathrel{\underline{\in}} M$ kann aber kein Element m' von M_0 entsprechen. Wäre nämlich $m' \varepsilon M'$, so widerspräche dies der Definition von M'. Wäre aber m' *nicht* $\varepsilon M'$, so müßte nach derselben Definition M' auch dieses Element m' enthalten, widersprechend der Annahme. Es ergibt sich also, daß U keiner Untermenge von M äquivalent sein kann, und in Verbindung mit dem zuerst Bewiesenen, $M < \mathfrak{U}M$.

Der Satz gilt für *alle* Mengen M, z. B. auch für $M = 0$, und es ist in der Tat

$$0 < \{0\} = \mathfrak{U}(0).$$

Ebenso ist auch für jedes a

$$\{a\} < \{0, \{a\}\} = \mathfrak{U}\{a\}.$$

has been proved above that

$$\mathfrak{S}T \sim \mathfrak{S}T'' \sim \mathfrak{S}T' \quad \text{and} \quad \mathfrak{P}T \sim \mathfrak{P}T'' \sim \mathfrak{P}T',$$

thus the theorem is proved in full generality.

31. *Definition.* If a set M is equivalent to a subset of the set N, but N is not equivalent to a subset of M, we say that M is "*of lower cardinality* than N", and we write $M < N$ for short.

Corollaries. a) Since according to No. 21 it is definite for any two sets whether they are equivalent or not, it is also definite whether M is equivalent to at least one element of $\mathfrak{U}N$, as well as whether N is equivalent to some element of $\mathfrak{U}M$. *It is therefore always definite whether* $M < N$ *or not.*

b) The three relations $M < N$, $M \sim N$, and $M > N$ are mutually exclusive.

c) Whenever we have $M < N$ and either $N < R$ or $N \sim R$, we have $M < R$.

d) If M is equivalent to a subset of N, then either $M \sim N$ or $M < N$. This is a consequence of the "equivalence theorem" (No. 27).

e) The null set is of lower cardinality than any other set; likewise, every set $\{a\}$ consisting of a single element is of lower cardinality than any set M that has strict parts (see No. 23).

32. *Cantor's theorem.* If M is an arbitrary set, then always $M < \mathfrak{U}M$. *Every set is of lower cardinality than the set of its subsets.*

Proof. To each element m of M there corresponds a subset $\{m\}$ of M. Now since it is definite for each subset M_1 of M whether it contains only a single element (No. 13), all subsets of the form $\{m\}$ are the elements of a subset U_0 of $\mathfrak{U}M$, and $M \sim U_0$.

If on the other hand $U = \mathfrak{U}M$ were equivalent to a subset M_0 of M, then under a mapping Φ of U onto M_0 there would correspond to each subset M_1 of M a certain element m_1 of M_0, so that $\{M_1, m_1\}$ would be an element of Φ, and it would always be definite whether $m_1 \varepsilon M_1$ or not. All those elements m_1 of M_0 for which m_1 is *not* an element of M_1 would then be the elements of a set M' ($M' \subseteq M_0 \subseteq M$), which likewise would be an element of U. But no element m' of M_0 could correspond to this subset M' of M. For if m' were an element of M', this would contradict the definition of M'. But if m' were *not* an element of M', then, according to the same definition, M' would also have to contain this element m', contrary to the assumption. Thus it follows that U cannot be equivalent to any subset of M, and, in combination with what was proved first, $M < \mathfrak{U}M$.

The theorem holds for *all* sets M, even, for instance, for $M = 0$, and indeed

$$0 < \{0\} = \mathfrak{U}0\,.$$

Likewise, for every a,

$$\{a\} < \{0, \{a\}\} = \mathfrak{U}\{a\}\,.$$

Aus dem Satze folgt endlich, daß es zu jeder beliebigen Menge T von
277 | Mengen $M, N, R, \ldots$ immer noch Mengen von größerer Mächtigkeit gibt; z. B. die Menge

$$P = \mathfrak{U}\mathfrak{S}T > \mathfrak{S}T \gtrsim M, N, R, \ldots$$

besitzt diese Eigenschaft.

33_{VI}. *Theorem.* Sind zwei äquivalente Mengen T und T', deren Elemente $M, N, R, \ldots$ bzw. $M', N', R', \ldots$ unter sich elementenfremde Mengen sind, so aufeinander abgebildet, daß jedes Element M von T von kleinerer Mächtigkeit ist als das entsprechende Element M' von T', so ist auch die Summe $S = \mathfrak{S}T$ aller Elemente von T von kleinerer Mächtigkeit als das Produkt $P' = \mathfrak{P}T'$ aller Elemente von T'.

Beweis. Es genügt, den Satz für den Fall zu beweisen, wo die beiden Summen $S = \mathfrak{S}\,T$ und $S' = \mathfrak{S}\,T'$ elementenfremd sind. Die Ausdehnung auf den allgemeinen Fall vollzieht sich dann analog wie bei dem Theorem Nr. 30 und mit Hilfe desselben durch Einschaltung einer dritten Menge $S'' \sim S$, welche S' elementenfremd ist.

Zunächst ist zu zeigen, daß S einer Untermenge von P' äquivalent ist. Wegen $M < M'$ existiert eine von 0 verschiedene Untermenge $\mathsf{A}_M \mathrel{⋹} \mathfrak{U}(MM')$, deren sämtliche Elemente $\mathsf{M}, \mathsf{M}', \mathsf{M}'', \ldots$ Abbildungen sind, welche M auf Untermengen $M_1{}', M_2{}', \ldots$ von M' abbilden. Solche Abbildungsmengen $\mathsf{A}_M, \mathsf{A}_N, \mathsf{A}_R, \ldots$ existieren für je zwei entsprechende Elemente $\{M, M'\}$, $\{N, N'\}$, $\{R, R'\}, \ldots$ von T und T', und jedes Element $\Theta = \{\mathsf{M}, \mathsf{N}, \mathsf{P}, \ldots\}$ ihres Produktes $\mathfrak{P}\mathsf{T} = \mathsf{A}_M \cdot \mathsf{A}_N \cdot \mathsf{A}_R \cdots$ liefert, analog wie in Nr. 30, eine simultane Abbildung sämtlicher Elemente $M, N, R, \ldots$ von T auf äquivalente Untermengen $M_1{}'$, $N_1{}'$, $R_1{}'$, $\ldots$ der entsprechenden Elemente von T'. Durch $\Omega = \mathfrak{S}\Theta \mathrel{⋹} SS'$ wird also jedes Element s von S auf ein Element s' von S' abgebildet, wenn auch nicht umgekehrt jedes von S' auf eines von S.

Nun sind aber die Komplementärmengen $M' - M_1{}', N' - N_1{}', R' - R_1{}', \ldots$, welche die Elemente einer Menge $T_1{}' \mathrel{⋹} \mathfrak{U}S'$ bilden, sämtlich von 0 verschieden, weil wegen $M < M'$ der Fall $M \sim M_1{}' = M'$ immer ausgeschlossen ist. Somit ist auch das Produkt $\mathfrak{P}T_1{}' \neq 0$, und es existiert mindestens eine Menge $q \,\varepsilon\, \mathfrak{P}T_1{}'$ von der Form $q = \{m_0{}', n_0{}', r_0{}', \ldots\} \mathrel{⋹} S'$, welche mit jeder der Mengen $M' - M_1{}', N' - N_1{}', \ldots$ genau ein Element gemein hat und daher auch Element von P' ist.

Ist nun s irgend ein Element von S, und s' das vermöge Ω ihm entsprechende Element von S', so entspricht ihnen beiden noch ein Element s_0 von $q \mathrel{⋹} S'$ in der Weise, daß s' und $s_0{}'$ immer einem und demselben Elemente von T' angehören und somit für $s \,\varepsilon\, M$ immer $s_0{}' = m_0{}'$ ist usw. Da aber im Falle $s' \,\varepsilon\, M_1{}'$ stets $s_0{}' \,\varepsilon\, (M' - M_1{}')$ ist, so sind s' und $s_0{}'$ immer voneinander verschieden. Bilden wir nun die Menge

$$q_s = q - \{s_0{}'\} + \{s'\},$$

Finally, it follows from the theorem that for every arbitrary set T of sets $M, N, R, \ldots$ there always exist further sets of higher cardinality; for example, the set

$$P = \mathfrak{U}\mathfrak{S}T > \mathfrak{S}T \gtrsim M, N, R, \ldots$$

possesses this property.

33_{VI}. *Theorem.* Let T and T' be two equivalent sets containing as elements the mutually disjoint sets $M, N, R, \ldots$ and the mutually disjoint sets $M', N', R', \ldots$, respectively. If T and T' are mapped onto each other in such a way that each element M of T is of lower cardinality than the corresponding element M' of T', the sum $S = \mathfrak{S}T$ of all elements of T is also of lower cardinalty than the product $P' = \mathfrak{P}T'$ of all elements of T'.

Proof. It suffices to prove the theorem for the case in which the two sums $S = \mathfrak{S}T$ and $S' = \mathfrak{S}T'$ are disjoint. The extension to the general case is then accomplished by a method analogous to that used for the theorem in No. 30, through interposition of a third set, S'', equivalent to S and disjoint from S', and by means of that theorem.

First it must be shown that S is equivalent to a subset of P'. Because $M < M'$, there exists a subset A_M of $\mathfrak{U}(MM')$, different from 0, all of whose elements $\mathsf{M}, \mathsf{M}', \mathsf{M}'', \ldots$ are mappings that map M onto subsets ${M_1}', {M_2}', \ldots$ of M'. Such mapping-sets $\mathsf{A}_M, \mathsf{A}_N, \mathsf{A}_R, \ldots$ exist for any two corresponding elements $\{M, M'\}, \{N, N'\}, \{R, R'\}, \ldots$ of T and T', and each element $\Theta = \{\mathsf{M}, \mathsf{N}, \mathsf{P}, \ldots\}$ of their product $\mathfrak{P}\mathsf{T} = \mathsf{A}_M \cdot \mathsf{A}_N \cdot \mathsf{A}_R \ldots$ furnishes, just as in No. 30, a simultaneous mapping of all elements $M, N, R, \ldots$ of T onto equivalent subsets ${M_1}', {N_1}', {R_1}', \ldots$ of the corresponding elements of T'. By means of $\Omega = \mathfrak{S}\Theta \subseteq SS'$, therefore, every element s of S is mapped onto an element s' of S', even though not every element of S' is mapped onto one of S.

But now the complements $M' - {M_1}'$, $N' - {N_1}'$, $R' - {R_1}', \ldots$, which are the elements of a subset T_1' of $\mathfrak{U}S'$, all are different from 0, since, because $M < M'$, the case $M \sim {M_1}' = M'$ is always excluded. Thus also the product $\mathfrak{P}{T_1}' \neq 0$, and there exists at least one set $q \,\varepsilon\, \mathfrak{P}{T_1}'$ of the form $\{{m_0}', {n_0}', {r_0}', \ldots\} \subseteq S'$ that has exactly one element in common with each of the sets $M' - {M_1}'$, $N' - {N_1}', \ldots$, and is therefore also an element of P'.

If now s is any element of S, and s' is the element of S' corresponding to it under Ω, there corresponds to both of them yet another element s_0 of the subset q of S', such that s' and ${s_0}'$ always belong to one and the same element of T', and consequently for $s \,\varepsilon\, M$ always ${s_0}' = {m_0}'$, and so forth. But, since ${s_0}' \,\varepsilon (M' - {M_1}')$ whenever $s' \,\varepsilon\, {M_1}'$, s' and ${s_0}'$ are always distinct. If we now form the set

$$q_s = q - \{{s_0}'\} + \{s'\},$$

278 | welche aus q entsteht, indem wir das eine Element s_0' durch das andere s' ersetzen, so erhalten wir wieder ein Element von P', nämlich eine Untermenge von S', welche mit jeder der Mengen $M', N', R', \ldots$ genau ein Element gemein hat. Diese Elemente q_s von P', welche die Elemente einer Untermenge $P_0' \mathrel{\underline{\in}} P'$ bilden, sind aber sämtlich voneinander verschieden. Denn sind etwa m_1 und m_2 zwei verschiedene Elemente *derselben* Menge $M \varepsilon T$, so sind auch die entsprechenden Elemente m_1' und m_2' von M_1', welche an die Stelle von s' treten, voneinander verschieden, und somit auch

$$q_{m_1} = q - \{m_0'\} + \{m_1'\} \neq q - \{m_0'\} + \{m_2'\} = q_{m_2},$$

da q außer m_0' kein weiteres Element mit M' gemein hat. Sind aber m und n zwei Elemente von S, welche *verschiedenen* Mengen M und N angehören, so hat $q_m = q - \{m_0'\} + \{m'\}$ mit M' ein Element m' von M_1', dagegen $q_n = q - \{n_0'\} + \{n'\}$ mit M' nur das Element $m_0' \varepsilon (M' - M_1')$ gemeinsam, und beide Mengen sind gleichfalls voneinander verschieden. Somit bilden die Paare $\{s, q_s\}$ die Elemente einer Menge $\Phi \mathrel{\underline{\in}} SP_0'$, welche gemäß Nr. 15 den Charakter einer Abbildung besitzt, und es ist in der Tat $S \sim P_0' \mathrel{\underline{\in}} P'$.

Andererseits kann aber P' keiner Untermenge S_0 von S äquivalent sein. Wäre dies nämlich der Fall, so müßte vermöge einer Abbildung $\Psi \mathrel{\underline{\in}} S_0 P' \mathrel{\underline{\in}} SP'$ jedem Elemente $s \varepsilon S_0$ ein Element $p_s \varepsilon P'$ entsprechen. Betrachten wir insbesondere diejenigen Elemente p_m, welche Elementen m des Durchschnittes $M_0 = [M, S_0]$ entsprechen. Jedes dieser p_m enthält dabei ein Element $m'' \varepsilon M'$, nämlich dasjenige, welches p_m als Element von P' mit M' gemeinsam hat; die zu verschiedenen m gehörenden m'' brauchen aber nicht immer verschieden zu sein. Jedenfalls bilden alle m'', die zu den Elementen m von M_0 gehören, die Elemente einer Untermenge M_2' von M', welche von M' selbst verschieden ist, da sonst M' einer Untermenge von $M_0 \mathrel{\underline{\in}} M$ äquivalent wäre gegen die Voraussetzung $M < M'$.[1] In derselben Weise gehören zu *allen* Elementen $M, N, R, \ldots$ von T gewisse echte Teilmengen $M_2', N_2', R_2', \ldots$ der entsprechenden Elemente $M', N', R', \ldots$ von T'. Die zugehörigen Komplementärmengen $M' - M_2', N' - N_2', R' - R_2', \ldots$ sind also sämtlich von 0 verschieden und bilden die Elemente einer Menge $T_2' \mathrel{\underline{\in}} \mathfrak{U}S'$. Ist nun p_0' irgend ein Element von $\mathfrak{P}\, T_2' \neq 0$, so ist es gleichzeitig auch Element von P', kann aber bei der vorausgesetzten Abbildung Ψ keinem Elemente s von S_0 entsprechen. Wäre nämlich etwa $p_0' = p_m$, entspräche also p_0' einem Elemente von M_0, so müßte es nach der gemachten Annahme mit M' ein Element $m'' \varepsilon M_2'$ gemein haben, während in Wirklichkeit p_0' mit M' kein anderes
279 | Element als eines von $M' - M_2'$ gemein haben kann. Ebensowenig kann p_0' irgend einem Elemente von $N_0, R_0, \ldots$ entsprechen, entspricht also überhaupt keinem Elemente von $S_0 \mathrel{\underline{\in}} S$, und die Annahme $P' \sim S_0$ führt auf einen Widerspruch, womit der Beweis der Behauptung $S < P'$ vollendet ist.

Das vorstehende (Ende 1904 der Göttinger Mathematischen Gesellschaft von mir mitgeteilte) Theorem ist der allgemeinste bisher bekannte Satz über das Größer

---

[1] Auch hier kommt das Auswahlaxiom VI zur Anwendung.

which we obtain from q when we replace one of these elements, s_0', by the other, s', we again obtain an element of P', namely, a subset of S' that has exactly one element in common with each of the sets $M', N', R', \ldots$. But the elements of P' such as q_s which are the elements of a subset P_0' of P', all are distinct. For if, say, m_1 and m_2 are two distinct elements of the *same* element M of T, the corresponding elements m_1' and m_2' of M_1', which take the place of s', are also distinct, and thus

$$q_{m_1} = q - \{m_0'\} + \{m_1'\} \neq q - \{m_0'\} + \{m_2'\} = q_{m_2},$$

since, except for m_0', q has no element in common with M'. But if m and n are two elements of S that belong to *distinct* sets M and N, then $q_m = q - \{m_0'\} + \{m'\}$ has one element m' of M_1' in common with M', while $q_n = q - \{n_0'\} + \{n'\}$ has only the element m_0' of $(M' - M_1')$ in common with M', and the two sets are likewise distinct. Thus the pairs $\{s, q_s\}$ are the elements of a subset Φ of SP_0', which, according to No. 15, possesses the character of a mapping, and we in fact have $S \sim P_0' \subseteq P'$.

On the other hand, P' cannot be equivalent to any subset S_0 of S. For if this were the case, there would have to correspond to every element s of S_0 an element p_s of P' under a mapping Ψ ($\Psi \subseteq S_0P' \subseteq SP'$). Let us consider in particular those elements p_m that correspond to elements m of the intersection $M_0 = [M, S_0]$. Each of these p_m contains, then, an element m'' of M', namely, the one that p_m, as an element of P', has in common with M'; but the m'' belonging to distinct m are not necessarily always distinct. In any case, all m'' belonging to the elements m of M_0 are the elements of a subset M_2' of M' that is distinct from M' itself, since otherwise M' would be equivalent to a subset of M_0, which in turn is a subset of M, contrary to the assumption that $M < M'$.[12] Similarly, there answer to *all* elements $M, N, R, \ldots$ of T certain strict partial sets $M_2', N_2', R_2', \ldots$ of the corresponding elements $M', N', R', \ldots$ of T'. The respective complements $M' - M_2'$, $N' - N_2'$, $R' - R_2', \ldots$, therefore, all are different from 0 and are the elements of a subset T_2' of $\mathfrak{U}S'$. If, now, p_0' is any element of $\mathfrak{P}T_2' \neq 0$, it is also an element of P', but under the mapping Ψ whose existence was assumed it cannot correspond to any element s of S_0. For if, say, p_0' were equal to p_m, if, therefore, p_0' were to correspond to an element of M_0, then according to the assumption made it would necessarily have an element m'' of M_2' in common with M', while actually p_0' cannot have any element in common with M' other than one of $M' - M_2'$. Nor can p_0' correspond to any element of $N_0, R_0, \ldots$; thus it corresponds to no element of the subset S_0 of S at all, and the assumption that $P' \sim S_0$ leads to a contradiction, which completes the proof of the assertion that $S < P'$.

This theorem (communicated by me to the Göttingen Mathematical Society at the end of 1904) is the most general theorem now known concerning the comparison

[12] Here, too, use is made of Axiom VI (axiom of choice).

und Kleiner der Mächtigkeiten, aus dem alle übrigen sich ableiten lassen. Der Beweis beruht auf einer Verallgemeinerung des von Herrn *J. König* für einen speziellen Fall (siehe unten) angewandten Verfahrens.

34_{VI}. *Folgerung* (*Satz von J. König*). Ist eine Menge T, deren Elemente sämtlich Mengen und untereinander elementenfremd sind, in der Weise auf eine Untermenge T' von T abgebildet, daß jedem Elemente M von T ein Element M' von T' von größerer Mächtigkeit ($M < M'$) entspricht, so ist immer $\mathfrak{S}\,T < \mathfrak{P}\,T$, sofern $\mathfrak{P}\,T \neq 0$ ist.[1]

Nach dem Theorem Nr. 33 ist in dem betrachteten Falle immer $\mathfrak{S}T < \mathfrak{P}T'$; es bleibt also nur noch zu zeigen, daß hier $\mathfrak{P}T'$ einer Untermenge von $\mathfrak{P}T$ äquivalent ist. Für $T' = T$ ist dies trivial; im anderen Falle ist aber $\mathfrak{P}(T - T') \neq 0$, weil sonst wegen VI die Nullmenge ein Element von $T - T'$ und gegen die Annahme $\mathfrak{P}T = 0$ wäre. Ist aber q irgend ein Element von $\mathfrak{P}(T - T')$ und $p' \,\varepsilon\, \mathfrak{P}T'$, so ist $p' + q$ Element von $\mathfrak{P}T$, nämlich eine Untermenge von $\mathfrak{S}T' + \mathfrak{S}(T - T') = \mathfrak{S}T$, welche mit jedem Elemente von T' sowohl als von $T - T'$ genau ein Element gemein hat. Somit entspricht bei festgehaltenem q jedem Element p' von $\mathfrak{P}T'$ ein bestimmtes Element $p' + q$ von $\mathfrak{P}T$, und alle diese $p' + q$ bilden die Elemente einer gewissen Untermenge P_q von $\mathfrak{P}T$, welche $\sim \mathfrak{P}T'$ ist.

35. Auch der Cantorsche Satz Nr. 32 läßt sich als besonderer Fall aus dem allgemeinen Theorem Nr. 33 gewinnen.

Es sei M eine beliebige Menge, M' gemäß Nr. 19 eine M äquivalente und elementenfremde Menge und $\Phi \in\!\!\!= MM'$ eine beliebige „Abbildung" von M auf M'. Jedem Element m von M entspricht dann ein bestimmtes Element $\{m, m'\}$ von Φ und es ist immer gemäß Nr. 31e

$$\{m\} < \{m, m'\}\,.$$

Diese Mengen $\{m\}$ bilden offenbar die Elemente einer weiteren Menge $T \sim M$, und es ist nach dem Theorem Nr. 33

$$M = \mathfrak{S}T < \mathfrak{P}\Phi\,.$$

Es bleibt also nur noch zu zeigen, daß $\mathfrak{P}\Phi \sim \mathfrak{U}M$ ist. Nun ist jedes Element von $\mathfrak{P}\Phi$ eine Menge der Form $M_1 + (M' - M_1')$, wo M_1 eine Untermenge
280 von M, und M_1' die entsprechende von M' bedeutet. Somit | entspricht in der Tat jedem Elemente M_1 von $\mathfrak{U}M$ ein und nur ein Element von $\mathfrak{P}\Phi$ und umgekehrt, und es ist, wie behauptet,

$$M < \mathfrak{P}\Phi \sim \mathfrak{U}M\,.$$

[1] *J. König*, Math. Ann. Bd. 60, p. 177 für den besonderen Fall, wo die Elemente von T nach ihrer Mächtigkeit geordnet eine Reihe vom Typus ω bilden.

of cardinalities, one from which all the others can be derived. The proof rests upon a generalization of a procedure applied by Mr. *J. König* in a special case (see below).

34_{VI}. *Corollary* (*J. König's theorem*). If a set T whose elements all are sets that are mutually disjoint is mapped onto a subset T' of T in such a way that to each element M of T there corresponds an element M' of T' of higher cardinality ($M < M'$), then $\mathfrak{S}T < \mathfrak{P}T$ whenever $\mathfrak{P}T \neq 0$.[13]

According to the theorem of No. 33, we always have $\mathfrak{S}T < \mathfrak{P}T'$ in the case under consideration; it therefore remains to be shown only that $\mathfrak{P}T'$ is here equivalent to a subset of $\mathfrak{P}T$. When $T' = T$ this is trivial; but in the other case we have $\mathfrak{P}(T - T') \neq 0$, else, on account of Axiom VI, the null set would be an element of $T - T'$, and, contrary to the assumption, $\mathfrak{P}T$ would equal 0. But if q is any element of $\mathfrak{P}(T - T')$ and p' any element of $\mathfrak{P}T'$, then $p' + q$ is an element of $\mathfrak{P}T$, namely, a subset of $\mathfrak{S}T' + \mathfrak{S}(T - T') = \mathfrak{S}T$ that has exactly one element in common with each element of T' as well as of $T - T'$. Thus for a fixed q there corresponds to each element p' of $\mathfrak{P}T'$ a certain element $p'+q$ of $\mathfrak{P}T$, and all these $p'+q$ are the elements of a certain subset P_q of $\mathfrak{P}T$ that is equivalent to $\mathfrak{P}T'$.

35. Cantor's theorem (No. 32) can also be obtained as a special case of the general theorem of No. 33.

Let M be an arbitrary set; let M' be a set—and according to No. 19 such a set does exist—equivalent to and disjoint from M, and let the subset Φ of MM' be an arbitrary mapping of M onto M'. Then to each element m of M there corresponds a definite element $\{m, m'\}$ of Φ and, according to No. 31e, always

$$\{m\} < \{m, m'\}.$$

These sets $\{m\}$ obviously are the elements of a new set, T, that is equivalent to M, and, according to the theorem of No. 33,

$$M = \mathfrak{S}T < \mathfrak{P}\Phi.$$

It thus remains to be shown only that $\mathfrak{P}\Phi \sim \mathfrak{U}M$. Now every element of $\mathfrak{P}\Phi$ is a set of the form $M_1 + (M' - M_1')$, where M_1 is a subset of M and M_1' the corresponding subset of M'. Then there indeed corresponds to each element M_1 of $\mathfrak{U}M$ one and only one element of $\mathfrak{P}\Phi$ and conversely; and, as asserted,

$$M < \mathfrak{P}\Phi \sim \mathfrak{U}M.$$

---

13 *J. König 1905b* for the special case in which the elements of T, when ordered according to cardinality, form a sequence of type ω.

36_{VII}. *Theorem.* Die „Zahlenreihe“ Z_0 (Nr. 14) ist eine „unendliche“ Menge d. h. eine solche, welche einem ihrer Teile äquivalent ist. Umgekehrt enthält auch jede „unendliche“ Menge M einen Bestandteil M_0, welcher „abzählbar unendlich“, d. h. der Zahlenreihe äquivalent ist.

Beweis. Es sei Z eine beliebige Menge, welche gemäß VII das Element 0 und mit jedem ihrer Elemente a auch das entsprechende Element $\{a\}$ enthält, und diese Menge Z sei durch eine Abbildung $\Omega \Subset ZZ'$ gemäß Nr. 19 abgebildet auf eine ihr äquivalente und elementenfremde Menge Z'. Ist nun $\{z, x'\}$ ein beliebiges Element von ZZ', und $\{x, x'\}$ Element von Ω für dasselbe x', so ist immer definit, ob $z = \{x\}$ ist oder nicht. Alle solchen Elemente $\{\{x\}, x'\}$ von ZZ' bilden also nach III die Elemente einer gewissen Untermenge $\Phi \Subset ZZ'$, und Φ ist eine „Abbildung“ von Z' auf $Z_1 \Subset Z$, wo Z_1 alle Elemente der Form $z = \{x\}$ umfaßt. In der Tat entspricht jedem $x' \varepsilon Z'$ ein bestimmtes $\{x\} \varepsilon Z_1$ und umgekehrt, d. h. jedes Element von $Z_1 + Z'$ erscheint in einem und nur einem Elemente von Φ. Es ist also nach Nr. 21 $Z \sim Z' \sim Z_1$, wo Z_1, weil es das Element 0 nicht enthält, nur ein *Teil* von Z ist; und jede wie Z beschaffene Menge, also auch Z_0 ist „unendlich“.

Um nun auch die zweite Hälfte des Theorems zu beweisen, betrachten wir eine beliebige „unendliche“ Menge M, die wir aber mit Rücksicht auf Nr. 19 unbeschadet der Allgemeinheit als elementenfremd zu Z_0 annehmen können. Es sei also $M \sim M' = M - R$, r ein beliebiges Element von $R \neq 0$ und $\{\Phi, \Psi\}$ gemäß Nr. 21 eine Abbildung, bei welcher jedem Elemente $m \varepsilon M$ ein Element $m' \varepsilon M'$ entspricht und umgekehrt. Ferner sei A eine Untermenge des Produktes MZ_0, welche die folgenden Eigenschaften besitzt: 1) sie enthält das Element $\{r, 0\}$; und 2), ist $\{m, z\}$ irgend ein Element von A, so enthält A auch das weitere Element $\{m', z'\}$, wo m' das dem m entsprechende Element von M', und $z' = \{z\}$ wegen Nr. 14 gleichfalls Element von Z_0 ist. Ist nun $\mathsf{A}_0 = \mathfrak{D}\mathsf{T}$ der gemeinsame Bestandteil aller wie A beschaffenen Untermengen von MZ_0, welche wegen III, IV die Elemente einer gewissen Menge $\mathsf{T} \Subset \mathfrak{U}(MZ_0)$ bilden, so besitzt auch A_0, wie man ohne weiteres erkennt, gleichfalls die Eigenschaften 1) und 2), ist also ebenfalls Element von T. Ferner ist, mit alleiniger Ausnahme von $\{r, 0\}$, jedes Element von A_0 auch von der Form $\{m', z'\}$; denn im entgegengesetzten Falle könnten wir es fortlassen, und der Rest von A_0 besäße immer noch die Eigenschaften 1) und 2), ohne doch, wie alle Elemente von T, den Bestandteil A_0 zu
281 enthalten. | Hieraus folgt zunächst, daß das Element $\{r, 0\}$ allen übrigen Elementen von A_0 elementenfremd ist, da weder $r = m' \varepsilon M'$ noch $0 = \{z\} = z'$ sein und somit kein weiteres Element $\{m', z'\}$ eines der Elemente r oder 0 enthalten kann. Ist ferner ein Element $\{m, z\}$ von A_0 allen übrigen elementenfremd, so gilt das gleiche auch von dem entsprechenden Elemente $\{m', z'\}$, da zu jedem Elemente der Form $\{m', z_1'\}$ oder $\{m_1', z'\}$ ein weiteres Element $\{m, z_1\}$ oder $\{m_1, z\}$ gehören müßte. Alle Elemente von A_0, welche allen übrigen elementenfremd sind, bilden also die Elemente einer Untermenge A_0' von A_0, welche die Eigenschaften 1) und 2) besitzt und daher als Element von T umgekehrt A_0 als Untermenge enthält, d. h. mit A_0 identisch ist. Jedes

36_{VII}. *Theorem.* The "number sequence" Z_0 (No. 14) is an "infinite" set, that is, one that is equivalent to one of its parts. Conversely, also, every "infinite" set M contains a "denumerably infinite" component M_0, that is, one that is equivalent to the number sequence.

Proof. Let Z be an arbitrary set that, in accordance with Axiom VII, contains the element 0 and, for each of its elements a, also the corresponding element $\{a\}$ and let this set Z be mapped by means of the subset Ω of ZZ' onto a set Z' equivalent to and disjoint from it, which, according to No. 19, is possible. Now, whenever $\{z, x'\}$ is an arbitrary element of ZZ' and $\{x, x'\}$ an element of Ω for the same x', it is definite whether $z = \{x\}$ or not. All elements of ZZ' that in fact have the form $\{\{x\}, x'\}$ thus are, according to Axiom III, the elements of a certain subset Φ of ZZ', and Φ is a "mapping" of Z' onto the subset Z_1 of Z that contains all elements z of the form $\{x\}$. In fact, to every element x' of Z' there corresponds a certain element $\{x\}$ of Z_1 and conversely; that is, each element of $Z_1 + Z'$ occurs in one and only one element of Φ. Thus, according to No. 21, $Z \sim Z' \sim Z_1$, where Z_1, since it does not contain the element 0, is only a *part* of Z; and every set constituted like Z, hence also Z_0, is "infinite".

To prove now the second half of the theorem as well, we consider an arbitrary "infinite" set M, which, however, we may in view of No. 19 assume without loss of generality to be disjoint from Z_0. Thus let $M \sim M' = M - R$, let r be an arbitrary element of $R \neq 0$, and let $\{\Phi, \Psi\}$ be a mapping, whose existence is possible according to No. 21, under which there corresponds to each element m of M an element m' of M' and conversely. Furthermore let A be a subset of MZ_0 that possesses the following properties: 1) it contains the element $\{r, 0\}$; and 2) if $\{m, z\}$ is any element of A, then A also contains the further element $\{m', z'\}$, where m' is the element of M' corresponding to m, and $z' = \{z\}$ is likewise an element of Z_0 on account of No. 14. If now $\mathsf{A}_0 = \mathfrak{D}\mathsf{T}$ is the common component of all subsets of MZ_0 constituted like A, which, on account of Axioms III and IV, are the elements of a certain subset T of $\mathfrak{U}(MZ_0)$, then A_0 also possesses properties 1) and 2), as we see immediately, and is thus likewise an element of T. Furthermore, with the sole exception of $\{r, 0\}$, every element of A_0, too, has the form $\{m', z'\}$; for in the contrary case we could remove it, and the remainder of A_0 would still possess properties 1) and 2), without, however, including the component A_0, which all elements of T do include. From this it follows first that the element $\{r, 0\}$ is disjoint from all other elements of A_0, since neither $r = m' \varepsilon M'$ nor $0 = \{z\} = z'$ is possible, and therefore no further element $\{m', z'\}$ can contain one of the elements r or 0. Furthermore, if an element $\{m, z\}$ of A_0 is disjoint from all the remaining ones, the same also holds for the corresponding element $\{m', z'\}$, since to each element of the form $\{m', z_1'\}$ or $\{m_1', z'\}$ there would have to correspond a further element, $\{m, z_1\}$ or $\{m_1, z\}$. All those elements of A_0 that are disjoint from all the others, therefore, are the elements of a subset A_0' of A_0 possessing properties 1) and 2); hence, being an element of T, A_0' now includes A_0 as a subset, that is, is identical with A_0. Every

Element von

$$\mathfrak{S}\mathsf{A}_0 = M_0 + Z_{00} \in M + Z_0\,,$$

wo wir mit M_0 und Z_{00} die gemeinsamen Bestandteile von $\mathfrak{S}\mathsf{A}_0$ mit M, bezw. Z_0 bezeichnen, kann also nur in einem einzigen Elemente von A_0 als Element figurieren, und es ist (wegen Nr. 15) $M_0 \sim Z_{00}$. Nun ist aber Z_{00} eine Untermenge von Z_0, welche das Element 0 und mit jedem ihrer Elemente z auch das zugehörige $z' = \{z\}$ enthält; Z_{00} muß also wegen Nr. 14 die ganze Zahlenreihe Z_0 als Bestandteil enthalten, d. h. es ist $Z_{00} = Z_0$ und, wie behauptet, $Z_0 \sim M_0 \in M$.

Chesières, den 30. Juli 1907.

element of

$$\mathfrak{S}\mathsf{A}_0 = M_0 + Z_{00} \subseteq M + Z_0 \,,$$

where M_0 is the common component of $\mathfrak{S}\mathsf{A}_0$ and M, and Z_{00} that of $\mathfrak{S}\mathsf{A}_0$ and Z_0, can therefore figure as an element in only a single element of A_0, and, on account of No. 15, $M_0 \sim Z_{00}$. But now Z_{00} is a subset of Z_0 containing the element 0 and, for every one of its elements z, also the associated $z' = \{z\}$; Z_{00} must therefore on account of No. 14 contain the entire number sequence Z_0 as a component; that is, we have $Z_{00} = Z_0$ and, as asserted, $Z_0 \sim M_0 \subseteq M$.

Chesières, on the 30th of July 1907.

Introductory note to 1909a and 1909b

Charles Parsons

Of these two papers, the second is a briefer but almost certainly later version of the first.[1] The point of view underlying both papers is succinctly expressed by the opening sentence of *1909b*:

> Anyone wishing to found arithmetic on the theory of natural numbers as the *finite cardinal numbers*, faces the task, first and foremost, of defining *finite set*; for cardinal number is by nature a property of sets, and every statement about finite cardinal numbers can always be expressed as one about finite sets.

1909a is also structured by the issue of Henri Poincaré's famous claim of the primitive character of mathematical induction and the dispute that he carried on with Bertrand Russell and his French ally Louis Couturat, both of whom defended logicist views.

Thus *1909a* begins with the question "Is the principle of complete induction provable or not?" But here too Zermelo regards finite sets as fundamental, and he gives attention to the problem of defining "finite set". Obviously, for Zermelo the proper framework is that of set theory as developed by Cantor and Dedekind. In fact the influence of *Dedekind 1888* is quite visible. Much of the work reported in these papers was done in a stay in Italy in 1905, when Zermelo was much occupied with the topic and with *Dedekind 1888* in particular.[2] These papers offer a set-theoretic foundation of arithmetic that updates Dedekind's in several respects, in particular in not requiring the existence of an infinite set. Like Dedekind, Zermelo refrains from singling out a sequence of sets to be the "natural numbers"; see below.

1909a was probably written before *1908b*, which gives his axiomatization of set theory, but Zermelo had already worked out his axiomatization and

---

[1] *1909a* bears the date May 1907, almost a year before the congress at which *1909b* was presented. The former's publication was probably delayed by Zermelo's orginally sending it to Poincaré with the idea that it would be submitted to the *Revue de métaphysique et de morale*, a leading French philosophical journal. Instead Poincaré sent it to *Acta mathematica.* (See *Ebbinghaus 2007b*, 63–64, and *Poincaré 1909a*, 198.) In a letter to Cantor of 24 July 1908 Zermelo mentions both papers and says of *1909a* that he already has the proofs; of *1909b* he says only that he hopes it will appear soon (*Meschkowski 1967*, 268).

1909b slightly refines the definition of shifting, so that the uniqueness of the initial and final element can be proved. However, one change in *1909a* must have been made after May 1907: noting *1908a* as published, although that paper bears a later date, 14 July 1907.

I am indebted to Heinz-Dieter Ebbinghaus and Akihiro Kanamori for comments and relevant information.

[2] See *Ebbinghaus 2007b*, 61–63.

takes account of axiomatic issues. He notes that he will avoid the assumption of an infinite set but will use the axiom of choice, though only to prove that every Dedekind finite set is finite by his own definition.[3] Thus the argument remains within what was then called general set theory.

The best-known result of both papers is an induction principle for finite sets, which Zermelo evidently regards as the most fundamental form of mathematical induction, although it is purely set-theoretic and does not explicitly mention natural numbers, either as primitive or as defined.

However, Zermelo begins with the question of defining "finite set". He notes that several definitions have been proposed (*1909a*, 185) and proposes to prove them equivalent. Although he does not give references (except to Dedekind), several definitions of finite set had appeared in the literature in the preceding twenty years. An obvious definition would be to say that a set M is finite if it has a natural number as its number of elements, i.e. if there is a one-one correspondence between M and $\{x : x < n\}$ for some natural number n. I will call such a set N-finite ("finite in number"). Zermelo follows the recent literature in offering a purely set-theoretic definition. The pioneering such definition was that of *Dedekind 1888*, art. 64: A set is finite if there is no one-one mapping onto a proper subset. Such sets are to this day called Dedekind finite. Dedekind proves that a set is Dedekind finite if and only it is N-finite (art. 160), but the left-right direction would require the axiom of choice.

In the preface to the second, 1893 edition of his *1888*, Dedekind gives an alternate definition whose equivalence to N-finiteness does not require the axiom of choice.[4] This was apparently the first such set-theoretic definition to be published. Zermelo does not mention it in these papers. Others were proposed in papers published shortly before Zermelo's. *Weber 1906* defines a set as finite if it has an ordering, and every ordering of it has a first and last element. In *Stäckel 1907*, a set is defined as finite if it is doubly well-ordered, that is it has an ordering such that every non-empty subset has a first and last element.

Zermelo begins by presenting his own definition (*1909b*, 8). A shifting (*Verschiebung*) of a set M is a one-one mapping f of a non-empty proper subset P of M properly into M, such that for any partition of M into two non-empty subsets M_1 and M_2, there is an element p of M such that either $p \in M_1$ and $f(p) \in M_2$ or vice versa. If $e \notin M_2$, then it is called an initial

[3] He had already presented the axiomatization in lectures in the summer semester of 1906. About the assumption of an infinite set, Zermelo speaks of "Dedekind's axiom", suggesting that he did not take seriously Dedekind's attempted proof of the existence of an infinite set (*1888*, art. 66). Indeed he objects to it in *1908b*, 266 n.

[4] This was pointed out to me by Saul Kripke, correcting the statement of *Parsons 1987*, 205, that the first such definition was that of *Weber 1906*. However, the proof was already given in §7 of *Tarski 1924a*. Dedekind was of course not conscious at the time of the axiom of choice.

element; if $u \notin M_1$, it is called a final element. After Dedekind, a set mapped into itself in this way is called a *finite chain*. We can best express what Zermelo means by using the term for the structure $\langle M, f \rangle$.

In the formulation of *1909b* Zermelo is able to prove (theorem II) that in a finite chain there is exactly one initial and one final element. (The status of an element as initial or final depends on f, but not on the partition.) It is also shown that the partition condition is equivalent to the statement that any subset P of M containing the initial element e and closed under f is all of M.[5] But that is essentially to say that induction holds inside M. The latter statement is presented somewhat more fully as theorem I of *1909a*. Thus a finite chain is essentially a Dedekindian simply infinite system, except that one element has no successor. I will call a set bearing such a structure Z-finite.

Zermelo leaves the empty set out of account, with the consequence that the partition condition implies that only sets with two or more elements count as finite. This was noted in *Grelling 1910*, 16 n. If one admits the empty set, the natural emendation is to say that a set is finite if it is either empty or has one element or has a shifting.[6]

Zermelo next proves that Z-finiteness is equivalent to having a double well-ordering (theorem II of *1909a* or *1909b*). He does not mention Weber's definition, possibly because it is easily seen to be equivalent to double well-ordering.[7] After proving his induction principle, he proves that a set is Z-finite if and only if it is Dedekind finite (theorem IV). It is of course the right-left direction that requires the axiom of choice.[8] Zermelo is able to give a short proof by applying his well-ordering theorem. Given a set M, let R be a well-ordering of M. If M is not Z-finite, it is not doubly well-ordered, so that M has a subset M_1 that has no last element under R. But then the mapping of an element of M_1 onto the next element under R maps M_1 into itself, but the first element of M_1 cannot be in its range. It follows that M_1, and therefore M, can be mapped one-one properly into itself and so is not Dedekind finite.

---

5 In his proof Zermelo remarks that e is the unique element of M not in the range of f (p. 9); it also follows that u is the unique element of M not in its domain.

6 Grelling actually proposes that a set is also finite if it has no proper subsets. But this works only if one continues to exclude the empty set, as Grelling does. That one-element sets are finite is assumed in Zermelo's proofs. Since a one-element set has no non-empty proper subsets, the condition for f to be a shifting is vacuous. But one would still have to stipulate separately that the single element is both initial and final.

7 Stäckel's paper bears the date June 1907, which makes it not too likely that Zermelo had seen it before writing *1909a*. He could hardly have failed to have seen *Weber 1906*, which appeared in April of that year. It is quite possible that Zermelo thought of the double well-ordering criterion independently of Stäckel.

8 Although Zermelo uses his induction principle for the left-right direction, he remarks that the proof is essentially present in Cantor and Dedekind. Using ordinary induction, the immediate conclusion is that an N-finite set is Dedekind finite.

In *1909b*, 10–11, he suggests that this result could not be proved without the axiom of choice, which was confirmed by later independence results.[9]

Zermelo's induction principle is theorem III of each paper. ("Finite" and "Fin(a)" should be understood as Z-finite.) It states that if a statement $A(x)$ holds for any unit set, and if, for any object x and set M such that $x \in M$, if it holds for $M - \{x\}$ then it holds for M, then it holds for all finite sets. We can write it as a schema as follows:

$$\text{(I1)} \qquad \forall x A(\{x\}) \wedge \forall a \forall x \big[\mathrm{Fin}(a) \wedge x \in a \wedge A(a - \{x\}) \to A(a)\big] \to \forall a \big[\mathrm{Fin}(a) \to A(a)\big].$$

The formulation used later, admitting the empty set, is the following:

$$\text{(I2)} \qquad A(\emptyset) \wedge \forall a \forall x \big[\mathrm{Fin}(a) \wedge A(a) \to A(a \cup \{x\})\big] \to \forall a \big[\mathrm{Fin}(a) \to A(a)\big].^{10}$$

Zermelo's proof is brief and uses the double well-ordering property of a finite set M: If $\leq$ is such an ordering, and for $x \in M$ not the last element u, $f(x)$ is the next element, then f is a shifting. By internal induction, for every $x \in M$, A holds in the segment $\{y : y \leq x\}$, and so it holds in the segment $\{y : y \leq u\}$, for u the last element, but that segment is all of M.

1909a contains one further result (theorem V): Every set T of finite sets has an element of smallest cardinality, and if there is a finite set Z such that every element of T is equivalent to a subset of Z, then T contains an element of greatest cardinality. He adds that if one uses the notion of finite cardinal number, then every set of finite cardinals is well-ordered by size, and every initial segment is doubly well-ordered.

In *1909b*, Zermelo closes by remarking that it seems that his results show that arithmetic does not require the assumption of "actually infinite" sets, but this is no longer true for higher analysis and the theory of functions.[11] Someone serious about rejecting the actual infinite would, to be consistent have to "remain with general set theory and elementary number theory and relinquish modern analysis in its entirety" (p. 11). Intuitionism, which would challenge that statement, was just beginning at the time. A lot of the work in the foundations of mathematics since then has addressed in various ways the question how much "modern analysis" one would have to give up in order to avoid the "actual infinite".

In §8 of *1909a*, Zermelo returns to the issues raised by Poincaré. Although he is clearly challenging Poincaré by offering a purely set-theoretic derivation of mathematical induction, he does not commit himself to the conclusion that the principle is analytic, against Poincaré's claim that it is synthetic a priori.

---

[9] See for example *Jech 1973*, 81.

[10] This formulation occurs in *Tarski 1924a*, 54. Cf. *Levy 1979*, 77.

[11] See also *1909a*, 192–93. In the letter to Cantor cited in note 1, Zermelo says that *1909b* "closes with a sharp point against Poincaré's skepticism", presumably skepticism about set theory.

For Zermelo that depends on the question whether the axioms of set theory are purely logical or are "intuitions of a special sort" (*1909a*, 192). He will not venture to decide that "purely philosophical question".

There was also an exchange with Poincaré concerning impredicative definitions. *1909a* is followed immediately by a paper *Poincaré 1909a* commenting on Zermelo's paper and one by Schoenflies. Poincaré observes that Zermelo's proof of internal induction on a finite set uses an impredicative definition (p. 199). In a supplement to his paper, Zermelo replies that the impredicativity can be avoided by reformulation (*1909a*, 193). I believe the reply is fallacious. Given a shifting f on a set M, Zermelo defines a set M_0 as the intersection of all sets M_1 containing the initial element e and closed under f. He states

> This common part is defined *after* the set E [of all the sets M_1], and there is no *petitio principii* because M_0 is not in general a part of E [*partie de E*].

I am not sure exactly what Zermelo means by "part of E".[12] I think he would have to mean "element of E", but then his claim is surely false. Later writers have maintained that there is something inherently impredicative about mathematical induction.[13]

A feature of these papers that is surprising to the contemporary reader is the absence of any use of a set-theoretic representation of the natural numbers. It is well-known that in *1908b* Zermelo, in formulating the axiom of infinity, mentions the sequence $\emptyset$, $\{\emptyset\}$, $\{\{\emptyset\}\}$, ... (p. 267). He says there that it "may be called the *number sequence*, because its elements can take the place of the numerals". Thus this sequence has come to be referred to as the "Zermelo numbers". But no mention of them is made in *1909a* or *1909b*. Evidently Zermelo thought that an approach that did not presuppose singling out such a sequence is more fundamental, as the remark quoted at the beginning of this note indicates. He may also have thought that such a procedure would yield a stronger case for the claim that induction can be founded on set theory and thus does not have the primitive character that Poincaré claimed.

The matter is pursued further in the dissertation *Grelling 1910*, which shows a strong influence of Zermelo although it was officially done under Hilbert. In fact Zermelo wrote an official evaluation of the dissertation, in which he says he set the problem to Grelling:

> To the complete grounding of genuine arithmetic belongs the transition from finite *sets* to the finite *numbers*; it must be possible to assign to each finite set a quite definite individual as "cardinal

[12] In *1909b*, note 1, he states that he uses the German word *Teil* to mean proper subset. But that is irrelevant.

[13] See *Parsons 2008*, § 50, which cites earlier writing by Michael Dummett, Edward Nelson, and the author.

> number" [*Anzahl*], so that to equivalent sets the same number always corresponds. I had set Grelling the task of working out this thought, with the requirement (which makes it more difficult) of basing all proofs on my first six axioms, again avoiding any infinite set.[14]

Grelling's first two chapters prove some theorems about general set theory and about finite sets respectively. He follows Zermelo in defining finiteness as Z-finiteness and makes frequent use of Zermelo's induction principle. The natural numbers are introduced at the beginning of the third chapter. They are not defined as the Zermelo numbers but rather as initial segments of the sequence of Zermelo numbers. If we set $Z_0 = \emptyset$ and $Z_{n+1} = \{Z_n\}$, then the natural number n is identified with the set $\{Z_0, \ldots Z_{n-1}\}$. Grelling's definition amounts to saying that a natural number is a set Z that is either $\{\emptyset\}$ or such that $\lambda x.\{x\}$ restricted to $\{x \in Z : \{x\} \in Z\}$ is a shifting on Z (p. 20). Although Grelling's procedure requires admitting the empty set, he regards the number sequence as beginning with 1 and continues to follow Zermelo in excluding the empty set from the finite sets. Otherwise he would have had to emend Zermelo's induction principle. In Grelling's construction, as in the now standard von Neumann construction, the natural number n is an n-element set, so that it can be proved that every finite set is equivalent to one and only one number (p. 22).[15]

Two later developments are worth mentioning. Work on definitions of "finite set" did not end with Zermelo's work: A new criterion was proposed in *Principia mathematica*, *120.24, of which the authors remark that it could have been used as a definition.[16] The study of the notion of finite set was continued in the Warsaw school and brought to some sort of completion by the young Tarski in his *1924a*, which presented another definition (close to that of *Principia*) and presented and analyzed the earlier work.

[14] "Gutachten über die Arbeit des Herrn K. Grelling, *Die Axiome der Arithmetik mit besonderer Berücksichtigung der Beziehungen zur Mengenlehre*," UA Göttingen, Promotionsakte Kurt Grelling, Az. Phil. Fak., 1908–1914, G. Vol. II, my translation. (Thanks to Volker Peckhaus for providing a copy of this document. The information about its provenance comes from *Peckhaus 1990a*, 145 n. 420.) *Zermelo 1908b* gives seven axioms, of which the axiom of infinity is the seventh. Zermelo goes on to say that Grelling carried out the task without demanding further help from him; it is still not improbable that he engaged in some informal supervision. See also *Peckhaus 1990*, 145–46.

On the life and work of Kurt Grelling (1886-1942), see ibid., 142–49 and *Peckhaus 1993* and *1994*, the latter of which contains a bibliography.

[15] For further discussion of Grelling's construction, see *Parsons 1987*, 206–07.

[16] *Whitehead and Russell 1912*, 207.

Zermelo's induction principle naturally suggests the idea of a stand-alone theory of hereditarily finite sets, with adding a single element to a set as a primitive operation (which I will write as $\oplus$) and Zermelo's induction schema as an axiom. This idea was not pursued until very late; the earliest such work known to me is the abstract *Givant and Tarski 1977*. Their work brought out the fact that the axiom of foundation, although true in the intended model, is not provable in the theory based on (I 2). However, $\in$-induction is derivable

Sur les ensembles finis et le principe de l'induction complète

1909a

1. Introduction

Le principe de l'induction complète est-il démontrable ou non ? Voilà une question qui dans ces dernières années a préoccupé beaucoup d'esprits. Dans plusieurs articles de la *Revue de Métaphysique et de Morale*[1] M. *Poincaré* a défendu la thèse que ce principe est un *jugement synthétique a priori* ; d'autres auteurs comme MM. *Couturat*, *Russell* et *Withehead* ont soutenu le contraire et présenté des démonstrations du principe en question.

Le principe de l'induction permet de démontrer des théorèmes sur les nombres finis en raisonnant de n à $n + 1$. La question dépend par suite de la façon dont on définit le nombre fini. Or pour moi tout théorème que l'on énonce pour des nombres finis n'est rien d'autre qu'un théorème sur les *ensembles finis* ; il faut donc avant tout définir ce qu'on entend par là.

On a proposé plusieurs définitions des ensembles finis. On peut par exemple avec *Dedekind*[2] prendre pour base la transformation d'un ensemble en luimême ; on peut aussi en se servant des idées de *Cantor*[3] partir de la notion des ensembles bien-ordonnés. Il faudrait montrer que toutes ces définitions peuvent être ramenées l'une à l'autre ; c'est ce que je me suis proposé

[1] 13e Année N:o 6, 14e Année N:o 1 ; N:o 3.
[2] Was sind und was sollen die Zahlen ? Braunschweig 1888.
[3] Mathematische Annalen vol. 49 p. 207.

if one strengthens (I 2) to

(I 3) $A(\emptyset) \wedge (\forall x)(\forall y)\{A(x) \wedge A(y) \rightarrow A(x \oplus y)\} \rightarrow (\forall x)A(x)$.[17]

Theories of this kind have been explored by other writers; see *Previale 1994* and *Kirby 2009*, and for exposition *Parsons 2008*, §§ 33, 38.

[17] This formulation assumes, as do Givant and Tarski, that there are no urelements. Some details were given in *Tarski and Givant 1987*, 223–226. For a version allowing urelements, see *Parsons 2008*, 228.

On finite sets and the principle of mathematical induction

1909a

1. Introduction

Is the principle of mathematical induction provable or not? This question has occupied many minds during the past few years. In several articles in the *Revue de Métaphysique et de Morale*,[1] *Mr. Poincaré* has defended the thesis that this principle is a *synthetic a priori judgment*. Other authors, such as *Mr. Couturat*, *Mr. Russell*, and *Mr. Whitehead*, have supported the contrary position and have presented proofs of the principle in question.

The principle of induction permits the proof of theorems about the finite numbers by reasoning from n to $n+1$. In the end, the question depends on the way in which one defines finite number. Now for me, all theorems that one states about finite numbers are in the end nothing but theorems about *finite sets*. Therefore, it is necessary at the outset to define what is meant by that.

Several definitions of finite set have been proposed. One could with *Dedekind*[2] take as a basis the transformation of a set into itself. Alternatively, using the ideas of *Cantor*[3] one could start with the notion of well-ordered set. It would then be necessary to show that all of these definitions can be derived from one another. That is what I intend to do in this article. Towards this

[1] *Poincaré 1905, 1906a, 1906b.*
[2] *Dedekind 1888.*
[3] *Cantor 1897.*

de faire dans cet article. A cet effet je me suis appuyé sur les notions fon-
186 damentales | de *Dedekind* et de *Cantor* tout en les énonçant encore une fois pour la commodité du lecteur. Je serais heureux si ces considérations pouvaient contribuer à bien mettre en évidence l'utilité qu'ont, pour l'étude des fondements des «vraies mathématiques», les notions et les méthodes de la théorie des ensembles.

J'ai pu éviter dans les démonstrations l'emploi de l'axiome de *Dedekind*[1] qui présuppose l'existence d'ensembles infinis ; mais j'ai cru au contraire pouvoir recourir au principe du «choix arbitraire»[2] pour la démonstration du théorème IV.

2. Définitions fondamentales

Nous appellerons «*chaîne simple*»[3] un ensemble M qui jouit de la propriété suivante : Il existe une correspondance univoque et réciproque entre, d'une part, les éléments de M, sauf peut-être l'un d'entre eux que nous nommerons le dernier et, d'autre part, les éléments d'une partie de M, soit M', qui ne contient pas l'un des éléments de M (le premier) ; cette correspondance ne permettant pas la division de M en parties séparées. Deux parties de M sont dites «séparées» par rapport à une certaine correspondance lorsqu'aucun élément de l'une n'a son image dans l'autre et réciproquement.

Un ensemble est appelé «*fini*», si tous ses éléments font partie d'une chaîne simple contenant un dernier élément. Si au contraire une chaîne simple n'a pas de dernier élément, l'ensemble qui contient tous ses éléments est nommé «*dénombrable*».

La définition proposée pour les ensembles finis exprime d'une façon précise ce que M. *Poincaré* entend, en définissant les nombres finis «par récurrence» ou «par des additions successives». En effet, les éléments d'un ensemble sont définis «*successivement*» lorsque chaque élément (sauf le premier) est déterminé par le précédent. Il faut donc qu'il y ait une correspondance telle qu'à chaque élément — à l'exception du premier ou du dernier — corresponde un autre élément de l'ensemble. Mais cela ne suffit pas ; la correspondance supposée ne servirait à rien, s'il n'y avait pas *enchaînement* entre les diverses parties de la série, ou en termes plus précis, s'il existait ce que nous avons
187 appelé des «parties séparées». |

[1] *Dedekind*, l. c. 66.

[2] *Zermelo*, Math. Ann. vol. 59 p. 514.

[3] La notion de «chaîne» est due à *Dedekind* (l. c. 37), mais sa définition des ensembles finis diffère beaucoup de la mienne. Voir cet artiele N:o 6.

end, I rely on the basic notions of *Dedekind* and *Cantor*; but at the same time I will state them again for the convenience of the reader. I would be happy if these considerations were to help bring to light the utility which the methods of set theory have for the study of the foundations of "true mathematics".

In the proofs, I have been able to avoid the use of *Dedekind*'s axiom[4] that asserts the existence of infinite sets. In contrast, I believed that I could have recourse to the principle of "arbitrary choice"[5] for the proof of Theorem IV.

2. Basic definitions

A set M which has the following property is called a "*simple chain*":[6] There exists a single-valued and one-one correspondence between, on the one hand, the elements of M, except perhaps one among them which we call the last, and, on the other hand, the elements of a part of M (call it M') which does not contain one of the elements of M (the first); and this correspondence does not permit the division of M into separate parts. Two parts of M are said to be "separate" in relation to a certain correspondence when no element of one has its image in the other, and conversely.

A set is called "*finite*" if all of its elements are part of a simple chain containing a last element. If, on the contrary, a simple chain has no last element, the set which contains all of its elements is called "*denumerable*".

The proposed definition of finite set expresses in a precise way what *Mr. Poincaré* means by defining the finite numbers "by recursion" or "by successive additions". In fact, the elements of a set are defined "*successively*" when each element (but the first) is determined by the preceding one. It is therefore necessary that there be a correspondence such that to each element—with the exception of the first or the last—there corresponds another element of the set. But that is not sufficient; the assumed correspondence would not work if there were no *connection* between the different parts of the series, or, in more precise terms, if there existed what we have called "separate parts".

---

[4] *Dedekind 1888*, art. 64. ⟦Zermelo erroneously writes "66" instead of "64".⟧

[5] *Zermelo 1904*.

[6] The notion of "chain" is due to *Dedekind* (*1888*, art. 37), but his definition of finite set differs considerably from mine. See Section 6 of this article.

3. L'induction complète appliquée aux membres d'une chaîne simple

Théorème I. Si M est une chaîne simple, tout sous-ensemble de M contenant le premier élément e ainsi que les images de tous ses éléments est identique à M lui-même.

Il en résulterait que toute propriété du premier élément qui, si elle est vraie pour un élément quelconque est vraie aussi pour son image, s'étend à tous les éléments de l'ensemble.

Démonstration. Soit M_0 la partie commune de tous les sous-ensembles M_1 de M qui contiennent e et les images de chacun de leurs éléments, et soit $R = M - M_0$ l'ensemble complémentaire. Alors tous les éléments de M_0 ont leurs images en M_0 puisque ces dernières sont communes à tous les ensembles M_1. La réciproque est également vraie : à l'exception de e tout élément de M_0 est image d'un autre, car autrement on pourrait supprimer en M_0 un élément différent de e et ne jouissant pas de cette propriété ; l'ensemble restant serait encore un M_1, ce qui est contraire à la définition de M_0. Il en résulte qu'aucun élément de M_0 ne peut être l'image d'un élément de R et réciproquement ; les parties M_0 et R seraient donc séparées à moins que M_0 ne soit identique à M.

C'est la définition de l'ensemble M_0 («la chaîne de l'élément e» d'après *Dedekind* l. c. 44) que M. *Poincaré*[1] a rejetée comme «non-prédicative» dans ma démonstration du théorème de *Bernstein*. Mais MM. *Russell*[2] et *Peano*[3] ont déjà fait à l'argumentation de M. *Poincaré* certaines critiques qui me paraissent justifiées. (Voir le dernier alinéa du N :o 6.)

4. Les ensembles doublement bien-ordonnés

Rappelons qu'un ensemble est dit «ordonné» lorsqu'une prescription permet de distinguer lequel de deux éléments quelconques a et b précède et lequel suit l'autre. Un ensemble est dit «bien-ordonné» lorsqu'en plus chacun de ses sous-ensembles possède un «premier élément» et un seul c'est-à-dire un élément qui précède tous les autres.

188 | *Définition.* Nous dirons qu'un ensemble est «*doublement bien-ordonné*» lorsqu'il est bien-ordonné et lorsque chacun de ses sous-ensembles possède non seulement un premier mais aussi un «dernier» élément c'est-à-dire un élément qui suit tous les autres.

Théorème II. Tout ensemble fini M peut être doublement bien-ordonné et, réciproquement, tout ensemble doublement bien-ordonné est fini.

---

[1] Revue de Métaphysique et de Morale 14e Année p. 315.

[2] Rev. d. Met. e. d. Mor. 14e Année p. 632.

[3] Revista de Matematica VIII N:o 5 p. 152.

3. Mathematical induction applied to the members of a simple chain

Theorem I. If M is a simple chain, then each subset of M containing the first element e as well as the images of all its elements is identical to M itself.

It follows from this that every property of the first element which, if it is true for an arbitrary element is also true for its image, extends to all elements of the set.

Proof. Let M_0 be the common part of all subsets M_1 of M that contain e and contain the images of each of their elements, and let $R = M - M_0$ be the complementary set. Then all the elements of M_0 have their images in M_0, since the latter are common to all the sets M_1. The converse is also true: with the exception of e, every element of M_0 is an image of another, for otherwise one could delete from M_0 an element different from e and not possessing this property; the set would still remain an M_1, which is contrary to the definition of M_0. From this it follows that no element of M_0 can be the image of one in R, and conversely. The parts M_0 and R are thus separate, at least if M_0 is not identical to M.

This is the definition of the set M_0 ("the chain of the element e", as Dedekind put it, op. cit., art. 44) that *Mr. Poincaré* [7] rejected as "non-predicative" in my proof of the theorem of *Bernstein.* But *Mr. Russell*[8] and *Mr. Peano*[9] have already made certain criticisms of *Mr. Poincaré*'s arguments that appear to me to be justified (see the last paragraph of Section 6).

4. Doubly well-ordered sets

Recall that a set is called "ordered" when a rule permits one to distinguish which of any two elements a and b precedes, and which follows, the other. A set is called "well-ordered" when further each of its subsets has one and only one "first element", that is to say, an element which precedes all the others.

Definition. A set is called "*doubly well-ordered*" when it is well-ordered and when each of its subsets possesses not only a first, but also a "last" element, that is, an element which follows all the others.

Theorem II. Every finite set M can be doubly well-ordered, and, conversely, every doubly well-ordered set is finite.

---

[7] *Poincaré 1906b*, p. 315.

[8] *Russell 1906b*, p. 632.

[9] *Peano 1906b*, p. 152.

Démonstration. Supposons que l'ensemble fini M soit une chaîne simple dont le premier élément soit désigné par e et le dernier par u. Nous allons montrer que tout élément a de M définit un ensemble $E(a)$ doublement bien-ordonné commençant par e, finissant par a et tel que chaque élément x' de $E(a)$ qui diffère de e soit l'image de l'élément x immédiatement précédant. En effet le théorème est évident pour $a = e$ et, s'il est vrai pour l'élément quelconque a, il est également vrai, comme nous allons le voir, pour son image a'. A cet effet considérons l'ensemble $E(a)$ qui, par hypothèse, finit par a et ajoutons-y a' comme dernier élément ; nous obtenons de cette façon l'ensemble $E(a')$ exigé, doublement bien-ordonné et finissant par a'. On voit donc, en s'appuyant sur le théorème I, qu'on peut considérer finalement l'ensemble $E(u)$. Cet ensemble $E(u)$ contient tous les éléments de M ; car il contient e, et, s'il contient un x différent du dernier élément u, l'élément immédiatement suivant ne peut pas différer de l'image x' de x. En d'autres termes l'ensemble M est doublement bien-ordonné.

Soit d'autre part un ensemble M doublement bien-ordonné, commençant par e et finissant par u. Alors on peut faire correspondre à chaque élément x, à la seule exception près de u, l'élément immédiatement suivant x' et l'on obtient de cette façon, comme nous allons le voir, une chaîne simple. En effet, si ce procédé conduisait à des parties «séparées», une d'entre elles au moins ne contiendrait pas u et posséderait par hypothèse un dernier élément v, tandis que l'élément immédiatement suivant v' figurerait dans la partie complémentaire.

5. L'induction appliquée aux ensembles finis

Théorème III. Soit une proposition démontrée d'une part pour tout ensemble contenant un seul élément et, d'autre part, pour un ensemble fini quelconque chaque fois qu'elle est vraie pour cet ensemble diminué d'un de ses éléments ; alors la proposition est vraie pour tous les ensembles finis. Voilà ce que l'on appelle le raisonnement de n à $n + 1$.

Démonstration. Étant donné un ensemble M, fini et doublement bien-ordonné, la proposition est tout d'abord vraie, par hypothèse, pour le segment
189 $E(e)$ de M | qui ne contient que le premier élément e. Si d'autre part elle est vraie pour un segment $E(a)$ dont le dernier élément est a, et qui contient tous les éléments précédents, elle sera également vraie pour le segment $E(a')$ qu'on obtient en ajoutant à $E(a)$, comme dernier élément, l'élément a' suivant immédiatement a. En vertu du théorème I la proposition sera exacte pour tous les segments $E(a)$ qui finissent par un élément quelconque de M, et en particulier pour $E(u) = M$ lui-même ; c'est-à-dire pour un ensemble fini quelconque.

Proof. Assume that the finite set M is a simple chain whose first element is designated by e, and whose last by u. We shall show that each element a of M defines a set $E(a)$ that is doubly well-ordered, begins with e, ends with a and is such that each element x' of $E(a)$ which differs from e is the image of the element x immediately preceding it. In fact, the theorem is evident for $a = e$, and, if it is true for an arbitrary element a, it is also true, as we shall see, for its image a'. To obtain this, consider the set $E(a)$, which, by hypothesis, ends with a, and add a' to it as a last element. In this way, we obtain the required set $E(a')$, doubly well-ordered and ending with a'. By relying on Theorem I, one can thus finally consider the set $E(u)$. This set $E(u)$ contains all the elements of M, for it contains e, and if it contains x different from the last element u, then the element immediately following it cannot be different from the image x' of x. In other words, the set M is doubly well-ordered.

For the converse, let M be a doubly well-ordered set beginning with e and ending with u. Then one can make correspond to each element x, with the sole exception of u, the element x' immediately following, and in this way one obtains, as we shall see, a simple chain. In fact, if this procedure led to "separate" parts, then at least one among them would not contain u, and would possess by hypothesis a last element v; whereas the element v' immediately following it would lie in the complementary part.

5. Induction applied to finite sets

Theorem III. Let a proposition be proved on the one hand for all sets containing only one element, and on the other hand for an arbitrary finite set each time it is true for this set reduced by one of its elements. Then the proposition is true for all finite sets. This is what is called reasoning from n to $n+1$.

Proof. Given a finite and doubly well-ordered set M, the property is immediately true, by hypothesis, for the segment $E(e)$ of M that contains only the first element e. For the rest, if it is true for a segment $E(a)$ of which the last element is a and which contains all the preceding elements, then it is equally true for the segment $E(a')$ which is obtained by adding to $E(a)$ as a last element the element a' that immediately follows a. By virtue of Theorem I, the proposition will hold for all segments $E(a)$ that end with any element of M whatsoever, and in particular for $E(u) = M$ itself, that is, for an arbitrary finite set.

6. Caractère fondamental des ensembles finis

Définition. Deux ensembles M, N sont appelés «*équivalents*», si l'on peut établir une correspondance univoque et réciproque entre les éléments de l'un et ceux de l'autre.

Théorème IV. Un ensemble fini n'est équivalent à aucune des ses parties; et réciproquement tout ensemble jouissant de cette propriété est fini.[1]

Démonstration. Pour démontrer la première partie du théorème nous faisons d'abord voir que la proposition est vraie pour un ensemble fini M chaque fois qu'elle l'est pour l'ensemble M_1, que l'on obtient en supprimant dans M un élément a. Supposons en effet qu'on ait établi une correspondance univoque et réciproque entre M et M', partie effective de M ; désignons par a' l'image de a et par M'_1 l'ensemble des images des éléments de M_1.

Au cas où M' ne contient pas a, l'élément a' diffère de a, et M'_1 qui ne contient ni a ni a' est une partie effective de M_1.

Si au contraire a fait partie de M' l'ensemble M_1 contient un élément p différent de a qui ne fait pas partie de M', et il y a encore deux cas à considérer.

1°. a' est identique à a, et M'_1 qui ne contient ni a ni p est une partie proprement dite de M_1.

2°. a' diffère de a et de p, et a est contenu dans M'_1. Remplaçons dans M'_1 a par a' ; nous obtenons de cette façon un ensemble M''_1 qui ne contient ni a ni p et qui est par conséquent une partie effective de M_1. Soit b l'élément de M_1 dont a est l'image; au moyen de notre substitution c'est à présent a' qui en est l'image, et nous avons obtenu une correspondance univoque et réciproque entre M_1 et M''_1.

Donc dans tous les cas, si M est équivalent à une de ses parties, M_1 le sera également. Mais l'impossibilité étant évidente pour un ensemble ne
190 possé- | dant qu'un seul élément, en vertu du théorème III elle s'étend à tous les ensembles finis.[1]

Soit d'autre part M un ensemble quelconque qui ne soit équivalent à aucune de ses parties. Nous pouvons admettre que cet ensemble est bien-ordonné en nous appuyant sur un théorème dont j'ai donné la démonstration.[2] Alors tout sous-ensemble M_1 de M doit contenir non seulement un premier mais aussi un dernier élément. Car autrement on pourrait établir une correspondance entre M_1 et une de ses parties en définissant comme image de chaque élément x de M_1 l'élément suivant x', c'est-à-dire le premier de tous les éléments de M_1 qui suivent x. L'ensemble M est donc doublement bien-ordonné, c'est-à-dire, fini.

---

[1] C'est là la distinction de *Dedekind* (l. c. 64) entre les ensembles finis et infinis.

[1] Cette démonstration est due à *Cantor* (Math. Ann. vol. 46, p. 490, D.).

[2] *Zermelo*, Beweis dass jede Menge wohlgeordnet werden kann. Math. Ann. vol. 59.

6. Basic character of finite sets

Definition. Two sets M and N are called "*equivalent*" if one can establish a single-valued and one-one correspondence between the elements of one and those of the other.

Theorem IV. A finite set is not equivalent to any of its parts. Conversely, all sets having this property are finite.[10]

Proof. To prove the first part of the theorem, we show first that the proposition is true for a finite set M whenever it is true for the set M_1 which is obtained by deleting an element a from M. Now suppose that there is a single-valued one-one correspondence between M and a proper part M' of M. Designate the image of a by a', and the set of images of elements of M_1 by M'_1.

In the case that M' does not contain a, the element a' differs from a, and M'_1, which contains neither a nor a', is a proper part of M_1.

If, on the other hand, a is in M', then the set M_1 contains an element p different from a which is not in M', and there are now two cases to consider:

1. a' is identical to a. So M'_1, which contains neither a nor p, is a proper part of M_1.

2. a' differs from both a and p, and a is contained in M'_1. In M'_1 replace a by a'. In this way, we obtain a set M''_1 which contains neither a nor p, and which is consequently a proper part of M_1. Let b be the element in M_1 whose image is a. By means of our substitution it is now a' which is its image, and we have obtained a single-valued and one-one correspondence between M_1 and M''_1.

Thus, in every case, if M is equivalent to one of its parts, M_1 will also be so. But the impossibility of this condition is evident for a set possessing only one element, so in virtue of Theorem III this impossibility extends to all finite sets.[11]

For the converse, let M be an arbitrary set which is not equivalent to any of its parts. We can take it that this set is well-ordered, by relying on a theorem whose proof I have given.[12] Then every subset M_1 of M must contain not only a first but also a last element. For otherwise one could establish a correspondence between M_1 and one of its parts by defining as the image of each x in M_1 the next element x', that is, the first of all the elements of M_1 that follow x. Hence the set M is doubly well-ordered, that is, finite.

---

[10] This is *Dedekind*'s distinction between finite and infinite sets (*1888*, art. 64).

[11] This proof is due to *Cantor* (*1895,* p. 490, D).

[12] *Zermelo 1904.*

La démonstration en question du théorème «que tout ensemble peut être bien-ordonné» est fondée sur «*l'axiome du choix arbitraire*» que l'on peut facilement ramener au suivant : Quand un ensemble S est divisé en parties $A, B, C, \ldots$, dont aucune n'est nulle, il existe toujours un sous-ensemble S_1 de S au moins qui contient un et un seul élément $a, b, c, \ldots$ de chacune des parties $A, B, C, \ldots$ C'est un axiome assez évident dont on s'est servi jusqu'à ces derniers temps presque sans opposition et qui n'a jamais conduit à un faux résultat. Tout récemment cependant MM. *Borel*[3] et *Peano*[4] l'ont rejeté dans tous les cas où l'ensemble S possède une infinité de parties. Sans doute, le principe en question est *indémontrable*, mais il est *indispensable* à certaines théories mathématiques. Il me semble, par exemple, impossible de démontrer le théorème précédent sans avoir recours à cet axiome explicitement ou implicitement ![5] Et M. *Poincaré* est tout à fait du même avis quand il dit :[6] «L'axiome est «self-évident» pour les classes finies ; mais s'il est indémontrable pour les classes infinies, il l'est sans doute aussi pour les classes finies qu'on n'en a pas encore distinguées à ce stade de la théorie ; c'est donc un jugement synthétique a priori sans lequel la «théorie cardinale» serait impossible, aussi bien pour les nombres finis que pour les nombres infinis.»

Dans le même article, M. *Poincaré* a fait à ma démonstration une autre objection, analogue à celle dont nous avons parlé à l'art. 3, savoir qu'une de mes définitions serait «non-prédicative». J'ai discuté à fond cette critique
191 dans | une note récemment parue.[1] Dans ce travail j'ai de plus réfuté les objections de MM. *Schoenfliess* et *Bernstein* qui n'admettent pas l'addition d'un seul élément à un ensemble bien-ordonné.

7. Le type ordinal des nombres cardinaux finis

Théorème V. Un ensemble T dont tous les éléments sont des ensembles finis contient toujours comme élément au moins un ensemble E, «de plus petite puissance», c'est-à-dire tel que chaque élément X de T possède un sous-ensemble équivalent à E. Et d'autre part : étant donné un ensemble fini Z, un ensemble T dont chaque élément X est un ensemble fini et équivalent à un sous-ensemble de Z contient toujours comme élément au moins un ensemble U «de plus grande puissance», c'est-à-dire tel que chaque élément de T soit équivalent à un sous-ensemble de U.

---

[3] Math. Ann. vol. 60 p. 194.

[4] Revista de Matematica VIII N:o 5 § 1.

[5] *Dedekind* dans sa démonstration du théorème équivalent (l. c. § 14) s'en sert de même en considérant (159) une série de représentations simultanées a_n.

[6] Revue d. Mét. e. d. Mor. 14e Année N:o 3 p. 313.

[1] *Zermelo.* Neuer Beweis für die Möglichkeit einer Wohlordnung. Math. Ann. vol. 65, p. 107–128.

The proof in question of the theorem "that every set can be well-ordered" is based on the "*axiom of arbitrary choice*", which can be easily rendered as follows: If a set S is divided into parts $A, B, C, \ldots$ none of which is empty, then there exists at least one subset S_1 of S which contains one and only one element $a, b, c, \ldots$ from each of the parts $A, B, C, \ldots$. The axiom is evident enough, and it has been used until recently almost without opposition; nor has it led to any false result. Recently, however, *Mr. Borel*[13] and *Mr. Peano*[14] have rejected it in all cases in which the set S has an infinite number of parts. Doubtlessly, the principle in question is *unprovable*, but it is *indispensable* for certain mathematical theories. It seems to me, for example, to be impossible to prove the preceding theorem without having recourse, either explicitly or implicitly, to this axiom![15] And *Mr. Poincaré* is completely in agreement when he says:[16] "The axiom is 'self-evident' for finite classes; but if it is unprovable for infinite classes, then it is doubtlessly also unprovable for finite classes, which are not distinguished from them at this stage of the theory. It is therefore a synthetic a priori judgment, without which the 'theory of cardinals' would be impossible, for finite numbers as well as for infinite numbers."

In the same article, *Mr. Poincaré* made another objection to my proof, analogous to the one of which we have already spoken in Section 3, that is, that one of my definitions was "non-predicative". I have examined this criticism thoroughly in a note that appeared recently.[17] In this work, moreover, I refuted the objections of *Mr. Schoenflies* and *Mr. Bernstein*, who do not allow the addition of a single element to a well-ordered set.

7. The ordinal type of the finite cardinal numbers

Theorem V. A set T all of whose elements are finite sets always contains as an element at least one set E "of smallest cardinality", that is, such that each element X of T possesses a subset equivalent to E. And further: given a finite set Z, a set T each element X of which is a finite set equivalent to a subset of Z contains as an element at least one set U "of greatest cardinality", that is, such that each element of T is equivalent to a subset of U.

---

13 *Borel 1905a.*

14 *Peano 1906b*, § 5.

15 *Dedekind*, in his proof of the equivalent theorem (*1888*, § 14), avails himself of the axiom in considering (159) a series of simultaneous representations α_n.

16 *Poincaré 1906b,* p. 313.

17 *Zermelo 1908a.*

En se servant de la notion des «nombres cardinaux finis» on peut énoncer le théorème comme suit : tout ensemble de nombres cardinaux finis ordonné suivant leur grandeur est bien-ordonné et chaque segment de l'ensemble non identique à l'ensemble total est doublement bien-ordonné. L'ensemble de tous les nombres cardinaux finis est donc «dénombrable».

Démonstration. Considérons un élément A quelconque de T que nous supposerons doublement bien-ordonné et faisons usage du théorème que de deux ensembles bien-ordonnés l'un au moins est «semblable» à un segment de l'autre.[2] J'appelle «segment» d'un ensemble bien-ordonné un sous-ensemble qui, lorsqu'il contient un élément a quelconque, contient en même temps tous les éléments de l'ensemble qui précèdent a. Remarquons que deux ensembles «semblables» sont aussi équivalents. Au cas où il n'y a pas de segment de A équivalent à un autre élément de T l'élément A sera l'ensemble E de plus petite puissance demandé, puisque chaque élément de T possède un segment semblable à A. Dans le cas contraire considérons tous les segments de A équivalents à un ou à plusieurs éléments de T. Les derniers éléments de ces segments forment à leur tour un ensemble fini dont le premier élément correspond à un segment E' équivalent à un élément E de T. Cet ensemble E sera celui de plus petite puissance. En effet soit X un élément quelconque
192 de T. Si X est équivalent à un segment | X' de A, l'ensemble E' qui est un segment de X' sera équivalent à un sous-ensemble de X. Dans le cas contraire A sera équivalent à un sous-ensemble de X, et l'ensemble E' qui n'est qu'un sous-ensemble de A le sera également. Donc dans tous les cas E qui est équivalent à E' est équivalent à un sous-ensemble de X.

Pour démontrer la seconde partie du théorème, supposons que l'ensemble Z soit doublement bien-ordonné, et que chaque élément X de T soit équivalent à un segment X' de Z. Les derniers éléments de tous ces segments X' forment un sous-ensemble Z_0 de Z dont le dernier élément est u'. Soit de plus U' le segment de Z finissant par u' et U un élément de T équivalent à U'. Alors chaque élément X de T est équivalent à un segment X' de U' et par conséquent équivalent aussi à un sous-ensemble de U. Donc U est un ensemble de plus grande puissance parmi tous les éléments de T.

8. Conclusion

Les théorèmes I, III et V expriment le principe de l'induction complète sous les diverses formes qu'on peut lui donner ; le principe est ainsi réduit à la définition des ensembles finis que nous avons donnée ou à une des définitions équivalentes. Mais en résulte-t-il que le principe en question soit un jugement analytique ? Cela dépend de la nature des axiomes sur lesquels repose la théorie des ensembles et que nous avons été contraints d'utiliser dans chacune de nos démonstrations. Si ces axiomes, que je me propose d'énoncer

[2] G. *Cantor*, Math. Ann. 49 p. 215 N.

By using the notion of "finite cardinal number", one could state the theorem as follows: all sets of finite cardinal numbers ordered by their size are well-ordered, and each segment of the set not identical to the whole set is doubly well-ordered. The set of all the finite cardinal numbers is hence "denumerable".

Proof. Consider an arbitrary element A of T, which we assume doubly well-ordered, and make use of the theorem that of two well-ordered sets, one is "similar" to a segment of the other.[18] I call a "segment" of a well-ordered set any subset which, when it contains an arbitrary element a, contains at the same time all the elements of the set which precede a. Note that two "similar" sets are also equivalent. In the case in which there is no segment of A equivalent to another element of T, the element A will be the required set E of smallest cardinality, since each element of T will possess a segment similar to A. In the contrary case, consider all of the segments of A equivalent to one or to several elements of T. The last elements of these segments again form a finite set whose first element corresponds to a segment E' equivalent to an element E of T. This set E will be the one of smallest cardinality. For let X be an arbitrary element of T. If X is equivalent to a segment X' of A, then the set E', which is a segment of X', will be equivalent to a subset of X. If not, A will be equivalent to a subset of X, and the set E', which is nothing but a subset of A, will have the same property. Thus in every case, E, which is equivalent to E', is equivalent to a subset of X.

To prove the second part of the theorem, suppose that the set Z is doubly well-ordered and that each element X of T is equivalent to a segment X' of Z. The last elements of all these segments X' form a subset Z_0 of Z, the last element of which is u'. Additionally let U' be the segment of Z ending with u', and U an element of T equivalent to U'. Then each element X of T is equivalent to a segment X' of U', and consequently also equivalent to a subset of U. Hence U is a set of greatest cardinality among all the elements of T.

8. Conclusion

Theorems I, III, and V express the principle of mathematical induction in the various forms in which it can be put; the principle is thus reduced to the definition of finite set which we have given, or to one of the equivalent definitions. But does it result from this that the principle in question is an analytic judgment? That depends on the nature of the axioms on which the theory of sets rests, and which we have necessarily used in each of our proofs. If these axioms, which I plan to state completely in another article,

[18] *Cantor 1897,* p. 215, N.

complètement dans un autre article, ne sont que des principes purement logiques, le principe de l'induction le sera également; si au contraire ils sont des intuitions d'une sorte spéciale, on peut continuer à regarder le principe d'induction comme un effet de l'intuition ou comme un «jugement synthétique a priori». Quant à moi, je n'oserais pour le moment, décider de cette question purement philosophique.

Pour démontrer les théorèmes annoncés, nous ne nous sommes pas appuyés sur l'hypothèse qu'il existe des ensembles infinis, c'est-à-dire des ensembles équivalents à une de leurs parties, hypothèse fondamentale de *Dedekind.* La théorie des ensembles en est certainement indépendante puisqu'elle énonce et démontre les théorèmes valables pour des ensembles *quelconques.* Si donc l'arithmétique élémentaire, c'est-à-dire la théorie des ensembles finis, que l'on peut fonder sur le principe de l'induction complète n'a pas besoin de «l'infini actuel», l'analyse au contraire et la théorie des fonctions qui ont pour base les notions de nombre irrationel et de limite exigent ab-
193 solument la considération d'ensem- | bles infinis. En effet, quelque définition que l'on donne du nombre irrationnel, il sera toujours défini par une *infinité* de nombres rationnels; car autrement le continu serait dénombrable. De même la notion de «limite» ne peut être obtenue qu'en considérant une infinité de valeurs possibles. C'est précisément cette idée des ensembles infinis due à M. *Cantor* qui est le fondement de cette partie des «vraies mathématiques».

En terminant je tiens à remercier mes amis MM. *Carathéodory* et *Jaccottet* qui ont bien voulu m'aider à la rédaction française de ce mémoire.

Montreux, mai 1907.

Supplément

M. *Poincaré* a fait suivre ma note de quelques «Réflexions» qui montrent que l'illustre géomètre est maintenant tout à fait de mon avis en ce qui concerne les «définitions prédicatives». Le petit détour qu'il emploie (p. 199), en modifiant la démonstration célèbre de *Cauchy-Weierstrass*, pour prouver qu'une équation algébrique a toujours une racine n'est peut-être pas absolument nécessaire. Mais dans la remarque «Plus généralement etc.» sur la limite inférieure d'un ensemble quelconque de nombres réels M. *Poincaré* rend nettement ma propre pensée et justifie ma démonstration du «Théorème I» (p. 187). Il suffit, en effet, de substituer à «l'ensemble E de nombres réels», «l'ensemble E des ensembles M_1» et à «limite inférieure e» «partie commune M_0» ; on obtiendra alors :

«Si nous envisageons un ensemble E d'ensembles M_1 on peut démontrer que cet ensemble possède une partie commune M_0 ; cette partie commune est définie *après* l'ensemble E; et il n'y a pas de pétition de principe puisque M_0 ne fait pas en général partie de E. Dans certains cas particuliers, il peut

are nothing but purely logical principles, then the principle of induction will be one as well. If, on the other hand, they are intuitions of a special sort, one could continue to regard the principle of induction as a result of intuition or as a "synthetic a priori judgment". At the present time I would not dare to pass on this purely philosophical question.

To prove the stated theorems, we did not rely on the hypothesis that there exist infinite sets, that is, sets equivalent to one of their parts, a basic hypothesis of *Dedekind.* The theory of sets is certainly independent of it, since it states and proves theorems valid for *arbitrary* sets. Thus, elementary arithmetic, that is, the theory of finite sets, which can be based on the principle of mathematical induction, does not need "the actual infinite". In contrast, analysis and the theory of functions, which have as basis the notions of irrational number and limit, absolutely require the consideration of infinite sets. In fact, whatever definition of irrational number is given, it will always be in terms of an *infinity* of rational numbers; for otherwise the continuum would be denumerable. Similarly, the notion of "limit" can be obtained only by considering an infinite number of possible values. It is exactly this idea of infinite sets, due to *Mr. Cantor*, which is the foundation of this part of "true mathematics".

In closing, I would like to thank my friends *Mr. Carathéodory* and *Mr. Jaccottet*, who have assisted me in editing the French of this article.

Montreux, May 1907.

Supplement

Mr. Poincaré has followed up my article with several "reflections"[19] which show that the illustrious geometrician is now completely in agreement with me concerning "predicative definitions". The little trick which he uses (p. 199), modifying the famous proof of *Cauchy-Weierstrass* that shows that an algebraic equation always has a root, is perhaps not absolutely necessary. But in the remark "More generally, ..." on the lower limit of an arbitrary set of real numbers, *Mr. Poincaré* plainly brings in my own idea, and justifies my proof of Theorem I (p. 187). For it suffices to substitute "the set E of sets M_1" for "the set E of real numbers", and "common part M_0" for "lower limit e". One then obtains:

"If we envisage a set E of sets M_1, one can demonstrate that this set possesses a common part M_0; this common part is defined *after* the set E; and there is no *petitio principii* since M_0 is not in general part of E. In certain particular cases, it could happen that M_0 is part of E. In these cases, there

19 ⟦*Poincaré 1909a*⟧

arriver que M_0 fasse partie de E. Dans ces cas particuliers, il n'y a pas non plus de pétition de principe puisque M_0 ne fait pas partie de E *en vertu de sa définition*, mais par suite d'une démonstration postérieure à la fois à la définition de E et à celle de M_0.»

Et c'est en invoquant l'autorité de M. *Poincaré* lui-même que l'on peut mettre en évidence la légitimité de ma démonstration.

Ueber die Grundlagen der Arithmetik

1909b

Will man die Arithmetik begründen auf die Lehre von den natürlichen Zahlen als den *endlichen Anzahlen*, so handelt es sich vor allem um die Definition der *endlichen Menge*; denn die Anzahl ist ihrer Natur nach Eigenschaft einer Menge, und jede Aussage über endliche Anzahlen lässt sich immer ausdrücken als eine solche über endliche Mengen. Im folgenden soll nun versucht werden, aus einer möglichst einfachen Definition der endlichen Menge die wichtigste Eigenschaft der natürlichen Zahlen, nämlich das *Prinzip der vollständigen Induktion*, herzuleiten und gleichzeitig zu zeigen, dass die verschiedenen sonst gegebenen Definitionen der hier angenommenen aequivalent sind. Bei diesen Entwickelungen werden wir uns auf die von *G. Cantor* und *R. Dedekind* geschaffenen Grundbegriffe und Methoden der allgemeinen Mengenlehre zu stützen haben. Wir werden aber *nicht*, wie noch *Dedekind* in seiner grundlegenden Schrift: „Was sind und was sollen die Zahlen?" es tut, von der Annahme Gebrauch machen, dass es eine „unendliche Menge" gibt, d. h. eine solche, welche einem ihrer Teile[1] aequivalent ist.

Definition. — Eine Menge M heisst „endlich", wenn ein Teil P von M auf einen anderen Teil P' in der Weise umkehrbar eindeutig abgebildet werden kann, dass bei jeder beliebigen Zerlegung von M in zwei (komplementäre) Teile M_1, M_2 wenigstens einer der beiden Teile ein Element p von P enthält, dessen Bildelement p' dem anderen Teile angehört. Eine in dieser Weise auf sich abgebildete Menge M soll als eine „endliche Kette", jede Abbildung dieser Art als eine „Verschiebung" bezeichnet werden. Ein Element e von M, welches in P' nicht vorkommt, heisst ein „Anfangselement", ein Element u, welches in P nicht erscheint, ein „Endelement" der endlichen Kette.

---

[1] Der Ausdruck „Teil" soll hier überall nur in der Bedeutung einer *echten* Teilmenge, welche mindestens ein Element enthält, verstanden werden. Eine Menge M_1, deren sämtliche Elemente der Menge M angehören, bezeichne ich als eine „Untermenge von M".

is also no *petitio principii*, since M_0 is not part of E *by dint of its definition*, but rather in following from a proof posterior to the definitions both of E and of M_0."

So by invoking the authority of *Mr. Poincaré* himself, I can support the legitimacy of my proof.

On the foundations of arithmetic

1909b

⟦The introductory note just before *1909a* also addresses *1909b*.⟧

Anyone wishing to found arithmetic on the theory of the natural numbers as the *finite cardinal numbers* faces the task, first and foremost, of defining *finite sets*; for cardinal number is by nature a property of sets, and every statement about finite cardinal numbers can always be expressed as one about finite sets. In what follows we shall try to derive the most important property of the natural numbers, namely the *principle of mathematical induction*, from as simple a definition as possible of finite sets, and to show, at the same time, that the various other existing definitions are equivalent to the one assumed here. For these developments we shall have to rely on the basic concepts and methods of general set theory created by *G. Cantor* and *R. Dedekind*. We shall, however, *not* use the assumption that there exists an "infinite set", i.e., one which is equivalent to one of its parts[1] as *Dedekind* still does in his fundamental essay: "Was sind und was sollen die Zahlen?".

Definition.—A set M is called "finite" if a part P of M can be mapped one-to-one onto another part P' so that for every arbitrary decomposition of M into two (complementary) parts M_1, M_2 at least one of the two parts contains an element p of P whose image element p' belongs to the other part. A set M thus mapped onto itself shall be called a "finite chain", and every mapping of this kind a "shifting". An element e of M not occurring in P' is called an "initial element", and an element u not occurring in P a "terminal element" of the finite chain.

[1] The expression "part" ⟦Teil⟧ shall here always be taken to only refer to a *proper* partial set ⟦Teilmenge⟧ containing at least one element. I shall refer to a set M_1 whose elements all belong to the set M as a "subset ⟦Untermenge⟧ of M".

Satz I. — Jede endliche Kette enthält nur ein einziges Anfangselement e und ein einziges Endelement u. Jede Untermenge E von M, welche das Anfangselement e enthält und mit einem beliebigen Elemente p von P zugleich auch das ihm entsprechende Element p' von P' enthält, ist mit M identisch;
9 ebenso auch jede Unter- | menge U von M, welche u und mit einem beliebigen Elemente p' von P' zugleich das entsprechende Element p von P enthält. Innerhalb einer endlichen Kette kann also das Schlussverfahren der „vollständigen Induktion" nach beiden Seiten angewendet werden.

Beweis: Es sei e ein beliebiges Anfangselement und E_0 der gemeinsame Bestandteil aller Untermengen E von M, welche e und mit jedem p zugleich das entsprechende p' enthalten. Dann ist auch E_0 von derselben Beschaffenheit; denn e ist gemeinsames Element aller E, und wenn p allen E angehört, so gilt das gleiche auch von dem entsprechenden p'. Ferner ist jedes Element x von E_0, welches von e verschieden ist, ein Element p' von P', und das entsprechende Element p von P ist gleichfalls Element von E_0. Denn in jedem anderen Falle würde die durch Unterdrückung von x aus E_0 entstehende Teilmenge E_x gleichfalls von der Beschaffenheit E sein, und E_0 wäre nicht der gemeinsame Bestandteil *aller* E. Da somit bei der betrachteten Verschiebung jedes Element von E_0 immer nur einem anderen Elemente von E_0 und niemals einem Elemente der Restmenge $M-E_0$ entsprechen kann, so muss nach der Definition der endlichen Kette diese Restmenge leer, d. h. $E_0 = M$ sein, und jede Menge E, welche E_0 umfasst, ist in der Tat mit M identisch. Da ferner ausser e jedes Element der Menge $E_0 = M$ Element von P' ist, so gibt es ausser e kein Anfangselement. In analoger Weise beweist man die entsprechenden Sätze für die Untermengen von der Beschaffenheit U und für das Endelement u.

Satz II. — Jede endliche Menge M lässt sich als „doppelt-wohlgeordnete" Menge darstellen, d. h. in der Weise ordnen, dass jede Untermenge von M sowohl ein erstes als ein letztes Element enthält; und umgekehrt ist jede doppelt-wohlgeordnete Menge auch „endlich" im Sinne unserer Definition.

Beweis: Es sei M dargestellt als endliche Kette vermöge einer Verschiebung Φ. Dann gibt es Elemente q von M von der Beschaffenheit, dass mindestens eine Untermenge M_q von M, welche e und q enthält, bei einer gewissen Anordnung doppelt wohlgeordnet ist mit e als erstem und q als letztem Element, wobei jedem Element p das entsprechende Element p' unmittelbar folgt. Z. B. die aus dem Anfangselement e allein bestehende Teilmenge (e) besitzt diese Eigenschaft. Ist aber ein q von der betrachteten Eigenschaft verschieden von u und somit Element von P, so besitzt auch das vermöge Φ ihm entsprechende Element q' von P' die verlangte Beschaffenheit; denn aus M_q erhält man eine neue doppelt-wohlgeordnete Menge $M_{q'}$, indem man das Element q' als letztes hinzufügt. Vermöge des vorhergenden Satzes I sind also alle Elemente von M, darunter auch das Endelement u, von der Beschaffenheit q. Die mit u endende doppelt-wohlgeordnete Menge M_u ist aber nach demselben Satze mit M identisch. Denn sie enthält e, und ist x irgend eines ihrer von u verschiedenen Elemente, so gibt es in M_u ein unmittelbar folgen-

Theorem I.—Every finite chain contains a unique initial element e and a unique terminal element u. Every subset E of M containing the initial element e and, along with an arbitrary element p of P, also the element p' of P' corresponding to it, is identical to M; the same holds of every subset U of M containing u and, along with an arbitrary element p' of P', also the corresponding element p of P. It is therefore possible to apply the inference method of "mathematical induction" within a finite chain in both directions.

Proof: Let e be an arbitrary initial element and E_0 the common component of all subsets E of M containing e and, along with every p, also the corresponding p'. Then E_0, too, has the same property; for e is an element common to all E, and if p belongs to all E, then the same holds of the corresponding p'. Furthermore, every element x of E_0 distinct from e is an element p' of P', and the corresponding element p of P is also an element of E_0. For in any other case the partial set E_x which results when x is removed from E_0 would also be constituted like E, and E_0 would not be the common component of *all* E. Since in the shifting under consideration every element of E_0 can therefore only correspond to another element of E_0 but never to an element of the remainder set $M - E_0$, this remainder set, by the definition of finite chains, must be empty, i.e., $E_0 = M$, according to the definition of finite chains, and every set E including E_0 is in fact identical to M. Since, furthermore, every element of the set $E_0 = M$, besides e, is an element of P', there are no initial elements besides e. The corresponding theorems for the subsets constituted like U and for the terminal element u are proved in analogous fashion.

Theorem II.—Every finite set M can be represented as a "doubly well-ordered" set, i.e., ordered so that every subset of M contains both a first and a last element; and, conversely, every doubly well-ordered set is also "finite" in the sense of our definition.

Proof: Let M be represented as a finite chain by dint of a shifting Φ. Then there are elements q of M constituted so that at least one subset M_q of M containing e and q is doubly well-ordered in a certain ordering with e as the first and q as the last element, where every element p is immediately succeeded by the corresponding element p'. E.g., the partial set (e) solely consisting of the initial element e possesses this property. But if a q which possesses the property under consideration is different from u, and hence an element of P, then the element q' of P' corresponding to it by virtue of Φ also possesses the required property; for a new doubly well-ordered set $M_{q'}$ results from M_q when the element q' is added as the last one. By virtue of theorem I above, all elements of M, including the terminal element u, are therefore constituted like q. But the doubly well-ordered set M_u ending with u is, according to the same theorem, identical to M. For it contains e, and if x is any of its elements different from u, then there exists in M_u an immediately succeeding element which, according to the definition of M_q, must be identical

des Element, welches nach der Definition der M_q mit dem x entsprechenden Elemente x' identisch sein muss, sodass auch x' der betrachteten Menge M_u angehört. Somit ist die ganze Menge M doppelt-wohlgeordnet.

Umgekehrt wird jede doppelt-wohlgeordnete Menge verwandelt in eine „endliche Kette", wenn man jedes Element x derselben auf das unmittelbar folgende x' abbildet. Hier umfasst nämlich P alle Elemente von M ausser dem letzten, P' alle | ausser dem ersten, da jedes Element ausser u ein unmittelbar folgendes, jedes Element ausser e ein unmittelbar vorangehendes besitzt. Bei einer beliebigen Zerlegung der Menge M in zwei Teile M_1, M_2 wird nun einer dieser Teile das letzte Element u der Gesamtmenge *nicht* enthalten, und das letzte Element u_1 dieses Teiles M_1 wird bei der Abbildung notwendig einem Elemente u_1' von M_2 entsprechen. Die Abbildung genügt also allen Forderungen unserer Definition.

Satz III (Satz der vollständigen Induktion für endliche Cardinalzahlen). — Gilt eine Eigenschaft E für jede aus einem einzigen Element bestehende Menge, und gilt sie ausserdem jedesmal für eine Menge M, wo sie für eine durch Fortlassung eines einzigen Elementes m_1 aus M hervorgehende Menge M_1 gilt, so gilt sie allgemein für alle endlichen Mengen.

Beweis: Es sei M eine beliebige endliche Menge und dargestellt als eine endliche Kette mit dem Anfangselement e und dem Endelement u. Dann entspricht nach dem vorhergehenden Satze II jedem Element q von M eine doppelt-wohlgeordnete Menge M_q mit e als erstem und q als letztem Elemente von der dort betrachteten Beschaffenheit. Hier gilt für die Menge M_e, die sich auf das eine Element e reduziert, die vorausgesetzte Eigenschaft E, und ferner, wenn sie für M_q gilt, so gilt sie auch für $M_{q'}$, weil M_q durch Fortlassung des einen Elementes q' aus $M_{q'}$ hervorgeht. Somit gilt nach dem Satze I die betrachtete Eigenschaft für alle M_q und daher auch für M selbst, welches ja mit M_u identisch ist.

Satz IV. — Keine endliche Menge ist einem ihrer Teile äquivalent, und umgekehrt, jede Menge, welche keinem ihrer Teile äquivalent ist, lässt sich als endliche Kette darstellen.

Der Beweis für den ersten Teil des Satzes wird geführt durch vollständige Induktion vermöge des vorhergehenden Satzes und findet sich sowohl bei *Cantor* als bei *Dedekind*, sodass eine Wiederholung hier überflüssig wäre. Die zweite Behauptung aber beweist man am einfachsten mit Hilfe meines Theorems von der *möglichen Wohlordnung beliebiger Mengen.* Wir denken uns die Menge M, welche keiner ihrer echten Teilmengen äquivalent ist, in beliebiger Weise wohlgeordnet und beweisen, dass sie dann auch *doppelt*-wohlgeordnet sein muss. Wäre nämlich M_1 eine Untermenge von M ohne letztes Element, so entspräche jedem Element x von M_1 innerhalb M_1 ein und nur ein unmittelbar folgendes Element x'. Dabei könnte aber das erste Element e_1 von M_1 unter den Elementen x' nicht vorkommen. Es wäre also M_1 umkehrbar eindeutig abgebildet auf eine echte Teilmenge M_1', und auch die Gesamtmenge M müsste einem ihrer Teile äquivalent sein im Wider-

to the element x' corresponding to x so that x', too, belongs to the set M_u under consideration. Hence, the entire set M is doubly well-ordered.

Conversely, any doubly well-ordered set is transformed into a "finite chain" by mapping each of its elements x onto the immediately succeeding x'. For in this case P contains all elements of M except for the last one, and P' all elements except for the first one, since every element except for u possesses an immediate successor and every element except for e possesses an immediate predecessor. Now, given an arbitrary decomposition of the set M into two parts M_1, M_2, one of these parts will *not* contain the last element u of the total set, and the last element u_1 of this part M_1 will, by necessity, correspond to an element u_1' of M_2 under the mapping. The mapping therefore meets all requirements of our definition.

Theorem III (Theorem on mathematical induction for finite cardinal numbers).—If a property E holds for every set consisting of a single element, and if, furthermore, it holds for a set M whenever it holds for a set M_1 which results when a single element m_1 is removed from M, then it holds generally for all finite sets.

Proof: Let M be an arbitrary finite set and let it be represented as a finite chain with the initial element e and the terminal element u. Then, according to theorem II above, to every element q of M there corresponds a doubly well-ordered set M_q with e as the first and q as the last element constituted in the way considered there. In this case, the assumed property E holds for the set M_e which reduces to the one element e, and, furthermore, assuming that it holds for M_q, it also holds for $M_{q'}$, since M_q results when the one element q' is removed from $M_{q'}$. According to theorem I, the property under consideration therefore holds for all M_q, and hence also for M itself, which, after all, is identical to M_u.

Theorem IV.—No finite set is equivalent to one of its parts, and, conversely, every set equivalent to none of its parts can be represented as a finite chain.

The proof of the first part of the theorem uses mathematical induction by virtue of the theorem stated above. It can be found both in *Cantor* and in *Dedekind* and does not need repeating here. The second claim, however, is most easily proved by use of my theorem on the *possible well-ordering of arbitrary sets*. Let M be a set that is equivalent to none of its proper partial sets and well-ordered in some arbitrary manner. We prove that it must then also be *doubly* well-ordered. For if M_1 were a subset of M without a last element, then to each element x of M_1 there would correspond within M_1 one and only one element x' immediately succeeding it. But in this case the first element e_1 of M_1 could not occur among the elements x'. Hence M_1 could be mapped one-to-one onto a proper partial set M_1', and the total set M, too, would have to be equivalent to one of its parts, contrary to the assumption. Hence a subset M_1 without a last element cannot exist, but M is indeed

spruche mit der Voraussetzung. Somit kann eine Untermenge M_1 ohne letztes Element nicht existieren, sondern M ist in der Tat doppelt-wohlgeordnet und als endliche Kette darstellbar vermöge unseres Satzes II.

Die beiden von mir gegebenen *Beweise* des „Wohlordnungs-Satzes“[1], auf die sich die vorhergehende Deduktion gründet, benutzen allerdings ganz wesentlich das heute immer noch umstrittene „Auswahlprinzip“. Aber diese Voraussetzung scheint so wie so *unentbehrlich* für die Gültigkeit
11 unseres Satzes IV, auch *Dedekind* bedient sich | ihrer, wenn schon in versteckterer Form, bei seinem sonst ganz anders geführten Beweise des vorliegenden Satzes. Ohne das Auswahlprinzip wäre wahrscheinlich unser Theorem überhaupt *unbeweisbar*, und es müsste dann ausser den endlichen und unendlichen Mengen noch eine *dritte* Gattung von Mengen als möglich zugelassen werden, über die wir gar nichts Positives aussagen könnten.

Bei der Herleitung unserer Sätze haben wir von der Existenz „aktual unendlicher“ Mengen nirgends Gebrauch gemacht, und da sich alle Sätze über endliche Anzahlen mit Hilfe des Prinzips der vollständigen Induktion leicht beweisen lassen, so scheint es, als ob die Arithmetik der genannten Voraussetzung überhaupt nicht bedürfte. Dies ist auch für die „niedere Arithmetik“, welche es nur mit rationalen Zahlen zu tun hat, durchaus zutreffend, es gilt aber nicht mehr für die höhere Analysis und die Funktionentheorie, in welchen Grenzbegriff und Irrationalzahl die führende Rolle spielen. In der Tat ist jede *Irrationalzahl* bestimmt durch einen „Schnitt“, d. h. durch eine *unendliche* Menge rationaler Zahlen, und ebenso kann auch ein *Limes* immer nur durch eine *unendliche* Menge von Argument- und Funktionswerten definiert werden. Wer also wirklich Ernst machen wollte mit der Verwerfung des „Aktual-Unendlichen“ in der Mathematik, der müsste folgerichtig bei der allgemeinen Mengenlehre und niederen Zahlentheorie stehen bleiben und auf die gesammte moderne Analysis Verzicht leisten.

[1] Mathematische Annalen, Bd. 59, p. 514; Bd. 65, p. 107.

doubly well-ordered and can be represented as a finite chain by virtue of our theorem II.

My two *proofs* of the "well-ordering theorem",[2] which are the basis of the preceding deduction, essentially involve, however, the "principle of choice", which is still contested today. But this assumption seems *indispensable* for the validity of our theorem IV in any event. *Dedekind*, too, uses it, albeit in more concealed form, in his proof of the present theorem, which is otherwise carried out in an entirely different way. Without the principle of choice, our theorem may well be *unprovable* altogether, and, in addition to the finite and infinite sets, we would have to allow for a *third* possible kind of sets about which nothing positive could be stated.

In deriving our theorems we have nowhere made use of the existence of "actual infinite" sets, and, since all theorems on finite cardinal numbers can easily be proved by means of the principle of mathematical induction, it seems as if arithmetic did not stand in need of this assumption. This is certainly also true of "lower arithmetic", which is only concerned with rational numbers. But it is no longer true of higher analysis and of complex function theory, in which the concepts of limit and irrational number play significant roles. In fact, every *irrational number* is determined by means of a "cut", i.e., by means of an *infinite* set of rational numbers. Likewise, a *limit*, too, can only be defined by means of *infinite* sets of argument and function values. Anyone who is really serious about casting the "actual infinite" out of mathematics would consequently have to stop at general set theory and lower number theory and relinquish modern analysis in its entirety.

[2] *Zermelo 1904, 1908a.*

Introductory note to 1913 and D. König 1927b

Paul B. Larson[†]

Zermelo's *1913, Über eine Anwendung der Mengenlehre auf die Theorie des Schachpiels,* is an account of an address given at the Fifth International Congress of Mathematicians in Cambridge in 1912. It is often cited as the first mathematical analysis of strategies in games. While the paper claims to be an application of set theory, and while it would have appeared that way to Zermelo's contemporaries, the set-theoretic notions in the paper have since become part of standard mathematical practice, and to modern eyes the arguments in the paper are more combinatorial than set-theoretic.[1] The notion of "Zermelo's Theorem" (usually described as a variant of "in chess, either White or Black has a winning strategy, or both can force a draw") derives from this paper. Although statements of this sort follow from the claims made in the paper, Zermelo's arguments for these claims are incomplete. As we shall see below, there are other gaps in the paper, one of which was fixed by Dénes König in his *1927a*. König's paper also contains two paragraphs, *1927b*, on arguments of Zermelo fixing this gap, using ideas similar to König's.

At the beginning of his *1913*, Zermelo notes that although he will discuss chess, his arguments apply to a wider class of games. Initially he describes this class as those two-player games "of reason" in which chance has no role. In the second paragraph of the paper, he makes the assumption that the game has only finitely many possible positions (or, rather, invokes the fact that this is true of chess, where a position of the game consists of the positions of all the pieces plus the identity of the player to move next and information such as which players have castled[2]), and in the third paragraph he says that the rules of the game allow infinite runs, which should be considered ties. In the first paragraph he mentions that there are many positions in the game of chess for which it is known that one player or the other can force a win in a certain number of moves, and proposes investigating whether such an analysis is possible in principle for arbitrary positions.

---

† The author is supported in part by NSF grant DMS-0401603. Dr. Nicole Thesz of Miami University and Dr. Burkhard Militzer of the University of California at Berkeley helped with some original sources in German. Our historical remarks rely heavily on *Ebbinghaus 2007b*, *Kanamori 2004* and *Schwalbe and Walker 2001.*

[1] As an example of how set-theoretic language was perceived at the time (and even much later), note that von Neumann and Morgenstern spend Sections 8–10 of their *1943* on the importance of set-theoretic notions for studying games.

[2] If one includes the list of previous moves then the set of positions becomes infinite.

Zermelo's analysis begins by letting P denote the set of all possible positions of the game, and letting $P^{\mathfrak{a}}$ denote the set of countable sequences of positions, finite or infinite. Fixing a position q, he lets Q be the set of all sequences in $P^{\mathfrak{a}}$ starting with q such that for each successive pair of positions in the sequence the latter member is obtained by a legal move from the former, and such that the sequence either continues infinitely or ends with a stalemate or a win for one player or the other. Zermelo notes that given a position q and a natural number r (Zermelo is not explicit about whether the case $r = 0$ is to be included), White can force a win from q in at most r moves if and only if there is a nonempty set $U_r(q)$ of members of Q such that each $\mathbf{q}$ in $U_r(q)$ is a continuation of q in which White wins in at most r moves (starting from q), and such that for each $\mathbf{q} \in U_r(q)$ and each position in $\mathbf{q}$ where it is Black's turn to move, and for each possible move for Black at that point, there is a $\mathbf{q}' \in U_r(q)$ which agrees with $\mathbf{q}$ up to this point, and has the position resulting from Black making this move as its next member. In more modern terminology (appearing no later than *Kuhn 1953*), Zermelo has introduced here a *game-tree* with root q of *height* at most $r+1$, in which all terminal nodes are wins for White, and in which all nodes for which it is Black's turn to move have successors corresponding to every move available to Black at that position. The existence of such a tree is indeed equivalent to the existence of a *quasi-strategy*[3] for White guaranteeing a win in r moves or fewer: White simply plays to maintain the condition that the continuation of the game starting from q is an initial segment of a member of $U_r(q)$.

Still fixing q and r, Zermelo notes that the union of all such sets $U_r(q)$ would also satisfy the conditions on $U_r(q)$. He calls this union $\overline{U}_r(q)$, and notes that as r increases the sets $\overline{U}_r(q)$ also increase under $\subseteq$ (though of course they may eventually all be the same, and may all be the empty set). For each q such that $\overline{U}_r(q)$ is nonempty for some natural number r, Zermelo lets ρ_q be the least such r, and he lets $U^*(q)$ denote $\overline{U}_{\rho_q}(q)$. He also lets τ denote the maximum of the set of defined values ρ_q.

Zermelo lets t be the integer such that $t+1$ is the size of P, and presents an argument to the effect that $\tau \leq t$. The idea behind this argument is that if some position is repeated during a play by a winning quasi-strategy for White, then one could adjust the quasi-strategy to play from the first occurrence of this position in the way that one played from the second, thus winning in fewer moves. This argument was later shown by König in *1927a* to be incomplete, as it does not account for all possible sequences of moves for Black; that is, playing with the same strategy does not guarantee the same resulting sequence of moves.

---

[3] A *strategy* for White specifies a move for White in each position obtainable by the strategy; a quasi-strategy merely specifies an acceptable set of moves (see *Kechris 1995*). The distinction is important when the Axiom of Choice fails, but is less important here, since P is finite. Nonetheless, we will use the term "quasi-strategy" for the sets of sequences described by Zermelo in this paper.

For each q, Zermelo lets $U(q)$ denote $\overline{U}_\tau(q)$, and claims that $U(q)$ being nonempty is equivalent to the assertion that q is a winning position for White. In fact, $U(q)$ being nonempty is equivalent to the existence of some natural number r such that White has a quasi-strategy guaranteeing a win in r moves or fewer. Zermelo does not address in this paper what it means for a player to have a quasi-strategy guaranteeing a win without specifying an upper bound on the number of moves needed to win. König's subsequent work *1927a* would show that in games in which each player has only finitely many moves available in each position, having a winning quasi-strategy in this more general sense implies having one with a fixed upper bound for the number of moves needed.

Zermelo then defines sets $V_s(q)$, analogous to $U_s(q)$ except that the corresponding quasi-strategies merely guarantee that White does not lose in fewer than s moves, though they allow that White loses on the s-th move. So each $V_s(q)$ is a set of members of Q such that each $\mathbf{q}$ in $V_s(q)$ is a continuation of q in which White does not lose in fewer than s moves starting from q, and such that for each $\mathbf{q} \in V_s(q)$ and each position in $\mathbf{q}$ where it is Black's turn to move, and for each possible move for Black at that point, there is a $\mathbf{q}' \in V_s(q)$ which agrees with $\mathbf{q}$ up to this point, and has the position resulting from Black making this move as its next member. Again, the union $\overline{V}_s(q)$ of all such $V_s(q)$ satisfies these conditions. Now, however, the sets $\overline{V}_s(q)$ are shrinking as s increases.

Zermelo now remarks that, given q, if $\overline{V}_s(q)$ is empty for any positive integer s, then, letting σ be the maximal s for which $\overline{V}(s)$ is nonempty, $\sigma \leq \tau$ (he also lets $V^*(q)$ denote $\overline{V}_\sigma(q)$ in this case). The argument for this is not given (Zermelo also reiterates here that $\tau \leq t$, which, as we noted above, is not satisfactorily demonstrated in this paper, but that issue does not affect this one.) The first missing claim is that if $\overline{V}_s(q)$ is empty, then Black has a quasi-strategy which guarantees a win in $s - 1$ moves or fewer, starting from q. Modulo precise notions of *game* and *strategy*, this fact is sometimes called *determinacy for fixed finite length games of perfect information*; indeed, this assertion is often called Zermelo's Theorem, referring to the arguments in this paper (a generalization is called "the theorem of Zermelo-von Neumann" in *Kuhn 1953*). Granting this point, one needs to see that if Black has a quasi-strategy guaranteeing a win in $s - 1$ moves or fewer, then he or she has a quasi-strategy guaranteeing a win in τ moves or fewer. Given the definition of τ this is clear for suitably symmetric games,[4] but it need not hold in general. Finally, it is clear that if Black has a quasi-strategy that guarantees a win in τ moves or fewer starting from q, then $\overline{V}_{\tau+1}(q)$ is empty.

[4] That is, games where for every position where it is White's term to move there is a position where is it Black's turn such that the game trees below the two conditions are isomorphic. Strictly speaking this is not true of chess, since it can only be White's turn when the pieces are in their initial position.

Zermelo lets $V(q)$ denote $\overline{V}_{\tau+1}(q)$, and claims that $V(q)$ being nonempty is equivalent to White being able to force a draw from the position q. This claim is missing the same steps as the corresponding claim for $U(q)$ above. Given that the game is suitably symmetric, the statement that $V(q)$ is nonempty is equivalent to the statement that White can delay a loss by any specified amount he or she chooses, which again by the subsequent work of König and the finiteness of chess means being able to delay a loss indefinitely.

The penultimate paragraph of the paper provides a partial summary, and asserts in a roundabout manner that in chess, either one player or the other has a winning strategy, or both players can force a draw. Zermelo notes that in each position q, $U(q) \subseteq V(q)$, and if $U(q)$ is nonempty, then White can force a win from q. If $U(q)$ is empty but $V(q)$ is not, then White can force a draw (as we mentioned in the previous paragraph, this is true but not supported by the arguments in the paper). If both sets are empty, then White can delay a loss until the σ-th move, for the value of σ corresponding to q. Furthermore, the two sets $U^*(q)$ (in the case where White can force a win) and $V^*(q)$ (otherwise) make up the set of "correct" moves for White from the position q. Zermelo notes than an analogous situation holds for Black, so that there exists a subset $W(q)$ of Q consisting of all continuations of the game (starting from q) in which both players can be said to have played correctly.

The final paragraph of the paper notes that the paper gives no means of determining in general which player has a winning strategy from which positions in chess, and that, given such a method, chess would in some sense cease to be a game.

As mentioned above, König's *1927a* points out that Zermelo's argument for the statement $\tau \leq t$ is incomplete. König's proof of this statement uses the following statement, which had appeared in his *1926*:[5] if E_i ($i \in \mathbb{N}$) are nonempty finite sets and R is a binary relation such that for each $i \in \mathbb{N}$ and each $x \in E_{i+1}$ there is a $y \in E_i$ such that $(y, x) \in R$, then there exists a sequence $\langle x_i : i \in \mathbb{N}\rangle$ such that each $x_i \in E_i$ and $(x_i, x_{i+1}) \in R$ for all $i \in \mathbb{N}$ (where $\mathbb{N}$ denotes the set of natural numbers). This principle is now known as *König's lemma*, often rephrased as "every infinite finitely-branching tree has an infinite branch." König uses this principle to prove that if G is a game in which each player has only finitely many available moves at each point, and one player has a winning strategy in this game, then this player has a strategy guaranteeing a win within a fixed number of moves. König credits this application of his lemma to a suggestion of von Neumann.

Before publishing his paper, König wrote to Zermelo, pointing out the gap in Zermelo's argument for $\tau \leq t$, and providing a correct proof. Zermelo then replied with a correct proof of his own. König's *1927b* consists of two paragraphs in the final section pertaining to Zermelo's corrected proof.

---

[5] As translated in *Schwalbe and Walker 2001* from *D. König 1927a*; see *Franchella 1997* for much more on the history of this statement.

The first paragraph was apparently written by König, summarizing Zermelo's argument. It contains a proof that if White can force a win from a given position within some fixed number of moves, then White has a winning strategy that guarantees a win in fewer than t moves, where t is the number of positions in the game where it is White's turn to move (note that the definition of t has changed; this t is smaller than the t from *Zermelo 1913*, as we are counting only the number of moves that White makes). To show this, Zermelo lets m_r, for each positive integer r, be the number of such positions from which White can force a win in at most r moves, but cannot force a win in fewer moves (though Zermelo does not give a name for the set of such positions, let us call it M_r). Since the corresponding sets of positions are disjoint, and since there are only finitely many possible positions in the game, m_r is nonzero for only finitely many values of r. Furthermore, if p is a position from which White can win in at most r moves (for some $r > 1$) by first playing w_1, then there must be a response by Black such that the resulting position is in M_{r-1}, since from every such position White can win in at most $r-1$ many moves, but if he could win in fewer moves from every such position, then White could win in fewer than r moves from the position p. Zermelo concludes that the set of values r such that m_r is positive is an initial sequence of the set of positive integers, so if λ is the largest integer r such that m_r is positive, then $m_r \geq 1$ for all positive integers $r \leq \lambda$. Then $m = \sum_{r=1}^{\lambda} m_r$ is smaller than the number of positions in which it is White's turn to play (since, for instance, Black can force a win from some such positions), so λ must be smaller than this version of t. This establishes that if p is a position from which White can force a win in at most r moves (for White), for some positive integer r, then r is less than the total number of positions in which it is White's turn to move.

In the second paragraph, König quotes Zermelo directly. Zermelo gives a proof that if White has a strategy guaranteeing a win, then he has one guaranteeing a win in a fixed number of moves. This is shown by König using his lemma, and Zermelo's argument uses the same idea (and implicitly includes a proof of the lemma). In brief, suppose that p is a position from which White cannot force a win in a fixed number of moves. Then no matter how White plays, there must be a move for Black such that White cannot force a win in a fixed number of moves from the resulting position (if each resulting position p' were in some $M_{r'}$, then p would be in M_{r+1} for r the supremum of these values r'—this uses the fact that each player has just

finitely many possible available moves at each point). This observation gives a strategy for Black to postpone a loss forever, by always moving to ensure that the resulting position is not in any set M_r, contradicting the assumption that White has a winning strategy.

Kalmár (in *1928/1929*) extended König's analysis to games where there may be infinitely many possible moves at some points. In this paper he proved what is now known as Zermelo's Theorem for these games, the statement that in each position of such a game, either one player or the other has a strategy guaranteeing a win, or both players can force a draw. His proof uses a ranking of nodes in the game tree by transfinite ordinals, which was to become an important method in descriptive set theory (see *Kechris 1995*). Using this method, he was able to show that if a player has a winning strategy in such a game, then he has one in which no position is repeated, thus realizing Zermelo's idea from his *1913*.

Aside from the work of König and Kalmár, Zermelo's *1913* would seem to have been forgotten for several decades after it was written. In the interval between Zermelo's paper and König's, Émile Borel published several notes on game theory (for example, *1921, 1924, 1927*), none of which mentions Zermelo. Von Neumann's work in game theory began during this period, and though he was informed of Zermelo's work by König,[6] he does not cite it in his *1928c*. Zermelo's *1913* is not mentioned in von Neumann and Morgenstern's book *1943*, which is often cited as the birthplace of game theory. Many authors credit the birth of game theory to some combination of Borel, von Neumann and Morgenstern (for instance, *Luce and Raiffa 1957, Zagare 1984, Straffin 1993, Vorob'ev 1994, Jones 2000, Ritzberger 2002, Mendelson 2004*, and among these only *Luce and Raiffa 1957, Zagare 1984*, and *Jones 2000* credit Borel). Aside from the work of König and Kalmár, the earliest citation of Zermelo's *1913* that we have been able to find is Kuhn's *1953*. Kuhn credits Zermelo with proving that "a zero-sum two-person game with perfect information always has a saddle-point in pure strategies." As we have seen, the argument that Zermelo gives for this fact in his *1913* is incomplete. Although Zermelo's focus was on other issues, it seems fair to say that this fact is the most significant contribution of his paper.

[6] According to König in a 1927 letter to Zermelo; see *Ebbinghaus 2007b*.

Über eine Anwendung der Mengenlehre auf die Theorie des Schachspiels

1913

Die folgenden Betrachtungen sind unabhängig von den besonderen Regeln des Schachspiels und gelten prinzipiell ebensogut für alle ähnlichen Verstandesspiele, in denen zwei Gegner unter Ausschluss des Zufalls gegeneinander spielen; es soll aber der Bestimmtheit wegen hier jeweilig auf das Schach als das bekannteste aller derartigen Spiele exemplifiziert werden. Auch handelt es sich nicht um irgend eine Methode des praktischen Spiels, sondern lediglich um die Beantwortung der Frage: kann der Wert einer beliebigen während des Spiels möglichen Position für eine der spielenden Parteien sowie der bestmögliche Zug mathematisch-objektiv bestimmt oder wenigstens definiert werden, ohne dass auf solche mehr subjektiv-psychologischen wie die des „vollkommenen Spielers" und dergleichen Bezug genommen zu werden brauchte? Dass dies wenigstens in einzelnen besonderen Fällen möglich ist, beweisen die sogenannten „Schachprobleme", d. h. Beispiele von Positionen, in denen der Anziehende *nachweislich* in einer vorgeschriebenen Anzahl von Zügen das Matt erzwingen kann. Ob aber eine solche Beurteilung der Position auch in anderen Fällen, wo die genaue Durchführung der Analyse in der unübersehbaren Komplikation der möglichen Fortsetzungen ein praktisch unüberwindliches Hindernis findet, wenigstens theoretisch denkbar ist und überhaupt einen Sinn hat, scheint mir doch der Untersuchung wert zu sein, und erst diese Feststellung dürfte für die praktische Theorie der „Endspiele" und der „Eröffnungen", wie wir sie in den Lehrbüchern des Schachspiels finden, die sichere Grundlage bilden. Die im folgenden zur Lösung des Problems verwendete Methode ist der „Mengenlehre" und dem „logischen Kalkül" entnommen und erweist die Fruchtbarkeit dieser mathematischen Disziplinen in einem Falle, wo es sich fast ausschliesslich um *endliche* Gesamtheiten handelt.

Da die Anzahl der Felder, sowie die der ziehenden Steine endlich ist, so ist es auch die Menge P der möglichen Positionen $p_0, p_1, p_2 \dots p_t$, wobei immer Positionen als verschieden aufzufassen sind, je nachdem Weiss oder Schwarz am Zuge ist, eine der Parteien schon rochiert hat, ein gegebener Bauer bereits verwandelt ist u. s. w. Es sei nun q eine dieser Positionen, dann sind von q aus „Endspiele" möglich $\mathfrak{q} = (q, q_1, q_2 \dots)$, nämlich Folgen von Positionen, die mit q beginnen und im Einklang mit den Spielregeln auf einander folgen, sodass jede Position q_λ aus der vorhergehenden $q_{\lambda-1}$ abwechselnd durch einen zulässigen Zug von Weiss oder Schwarz hervorgeht. Solch ein mögliches End-
502 spiel $\mathfrak{q}$ kann entweder in einer „Matt" oder | „Patt" Stellung sein natürliches Ende finden, oder aber auch — theoretisch wenigstens — unbegrenzt verlaufen, in welchem Falle die Partie zweifellos als unentschieden oder „remis" zu gelten hätte. Die Gesamtheit Q aller dieser zu q gehörenden „Endspiele" $\mathfrak{q}$

On an application of set theory to the theory of the game of chess

1913

The following considerations are independent of the particular rules of the game of chess. They are in principle also valid for all similar games of reason in which two opponents play against each other and from which chance is excluded. But, for concreteness's sake, chess, the most well-known among games of this kind, will be used as an example here. Also, we shall not be concerned with any method for actually playing the game, but only with an answer to the following question: Is it possible to determine or, at least define, in a mathematically objective manner the value of an arbitrary position possible in a game of chess for one of the players as well as the best move possible without invoking more subjectively psychological notions such as that of the "perfect player" and similar ones? That this is at least possible in individual special cases is shown by the so-called "chess problems", i.e., examples of positions for which it is *possible to demonstrate* that the player whose turn it is to move can force checkmate in a specified number of moves. It certainly seems worthwhile to me to investigate whether it is at least theoretically possible, and whether it makes sense at all, to evaluate a position also in cases where it is practically impossible to carry out a precise analysis on account of the unsurveyable complication posed by the possible continuations of the game. Only such a determination would provide a firm foundation for the applied theory of "endgames" and "openings" as it can be found in chess textbooks. In what follows, we shall use a method taken from "set theory" and from the "calculus of logic" in order to solve the problem, thereby showing that these mathematical disciplines can be fruitfully applied in a case in which almost exclusively *finite* totalities are concerned.

Since the number of squares and that of the moving pieces is finite, so is the set P of the possible positions p_0, p_1, p_2, $\ldots$ p_t, where positions are always to be considered distinct depending on whether it is White's or Black's turn to move, whether one of the players has already castled, whether a given pawn has already been promoted, etc. Now let q be one of these positions. Then, proceeding from q, "endgames" are possible $\mathfrak{q} = (q, q_1, q_2, \ldots)$, that is, sequences of positions beginning with q and succeeding one another according to the rules of the game so that each position q_λ results from the preceding one $q_{\lambda-1}$ by alternate legal moves of White and Black. A possible endgame $\mathfrak{q}$ of this kind can either come to a natural end in a "checkmate" or "stalemate" position, or—theoretically, at least—go on indefinitely, in which case the game would undoubtedly have to be considered a draw or "remis". The totality Q of all such "endgames" $\mathfrak{q}$ belonging to q is always

ist stets eine wohldefinierte, endliche oder unendliche Untermenge der Menge $P^{\mathfrak{a}}$, welche alle möglichen abzählbaren Folgen gebildet aus Elementen p von P umfasst.

Unter diesen Endspielen $\mathfrak{q}$ können einige in r oder weniger „Zügen" (d. h. einfachen Positionswechseln $p_{\lambda-1} \longrightarrow p_\lambda$, nicht etwa Doppelzügen) zum Gewinn von Weiss führen, doch wird dies in der Regel auch noch vom Spiel des Gegners abhängen. Wie muss aber eine Position q beschaffen sein, damit Weiss, wie Schwarz auch spielt, in höchstens r Zügen den Gewinn *erzwingen* kann? Ich behaupte, die notwendige und hinreichende Bedingung hierfür ist die Existenz einer nicht verschwindenden Untermenge $U_r(q)$ der Menge Q von folgender Beschaffenheit:

1. Alle Elemente $\mathfrak{q}$ von $U_r(q)$ enden in höchstens r Zügen mit dem Gewinn von Weiss, sodass keine dieser Folgen mehr als $r + 1$ Glieder enthält und daher $U_r(q)$ jedenfalls endlich ist.

2. Ist $\mathfrak{q} = (q, q_1, q_2 \dots)$ ein beliebiges Element von $U_r(q)$, q_λ ein beliebiges Glied dieser Reihe, welches einem ausgeführten Zuge von Schwarz entspricht, also entweder immer ein solches gerader oder eines ungerader Ordnung, je nachdem bei q Weiss oder Schwarz am Zuge ist, sowie endlich q'_λ eine mögliche Variante, sodass Schwarz von $q_{\lambda-1}$ aus ebensogut nach q'_λ wie nach q_λ hätte ziehen können, so enthält $U_r(q)$ noch mindestens ein Element der Form $\mathfrak{q}'_\lambda = (q, q_1, \dots q_{\lambda-1}, q'_\lambda, \dots)$, welches mit $\mathfrak{q}$ die ersten λ Glieder gemein hat. In der Tat kann in diesem und nur in diesem Falle Weiss mit einem beliebigen Elemente $\mathfrak{q}$ von $U_r(q)$ beginnen und jedesmal, wo Schwarz q'_λ statt q_λ spielt, mit einem entsprechenden $\mathfrak{q}'_\lambda$ weiterspielen, also unter allen Umständen in höchstens r Zügen gewinnen.

Solcher Untermengen $U_r(q)$ kann es freilich mehrere geben, aber die Summe je zweier ist stets von derselben Beschaffenheit, und ebenso auch die Vereinigung $\overline{U}_r(q)$ aller solchen $U_r(q)$, welche durch q und r eindeutig bestimmt ist und jedenfalls von 0 verschieden sein, d. h. mindestens ein Element enthalten muss, sofern überhaupt solche $U_r(q)$ existieren. Somit ist $\overline{U}_r(q) \neq 0$ die notwendige und hinreichende Bedingung dafür, dass Weiss den Gewinn in höchstens r Zügen erzwingen kann. Ist $r < r'$ so ist stets $\overline{U}_r(q)$ Untermenge von $\overline{U}_{r'}(q)$, weil dann jede Menge $U_r(q)$ sicher auch die an $U_{r'}(q)$ gestellten Anforderungen erfüllt, also in $\overline{U}_{r'}(q)$ enthalten sein muss, und dem kleinsten $r = \rho$, für welches noch $\overline{U}_r(q) \neq 0$ ist, entspricht der gemeinsame Bestandteil $U^*(q) = \overline{U}_\rho(q)$ aller solchen $\overline{U}_r(q)$; dieser umfasst alle solche Fortsetzungen, mit denen Weiss in der kürzesten Zeit gewinnen muss. Nun besitzen aber diese Minimalwerte $\rho = \rho_q$ ihrerseits ein von q unabhängiges Maximum $\tau \leqq t$, wo $t + 1$ die Anzahl der möglichen Positionen ist, sodass $U(q) = \overline{U}_\tau(q) \neq 0$ die notwendige und hinreichende Bedingung dafür darstellt, dass in der Position q irgend ein $\overline{U}_r(q)$ nicht verschwindet und Weiss überhaupt „auf Gewinn steht." Ist nämlich in einer Position q der Gewinn überhaupt zu erzwingen, so ist er es auch, wie wir zeigen wollen, in
503 höchsten t | Zügen. In der Tat müsste jedes Endspiel $\mathfrak{q} = (q, q_1, q_2 \dots q_n)$

a well-defined finite or infinite subset of the set $P^{\mathfrak{a}}$ comprising all possible countable sequences formed from elements p of P.

Some among these endgames $\mathfrak{q}$ can lead to a win for White in r or fewer "moves" (i.e., simple changes in position $p_{\lambda-1} \to p_\lambda$, never in double moves), whereas this will usually also depend on how the opponent plays. But how must a position q be constituted so that White can *force* a win in at most r moves regardless of how Black plays? I claim that the necessary and sufficient condition for this is the existence of a nonempty subset $U_r(q)$ of the set Q with the following properties:

1. All elements $\mathfrak{q}$ of $U_r(q)$ end with a win for White in at most r moves so that none of these sequences contains more than $r+1$ terms, and hence $U_r(q)$ is certainly finite.

2. Let $\mathfrak{q} = (q, q_1, q_2, \ldots)$ be an arbitrary element of $U_r(q)$, and let q_λ be an arbitrary term of this sequence which corresponds to a move made by Black, and hence is always one of either even or odd order depending on whether it is White's or Black's turn to move at q. Finally, let q'_λ be a possible variant so that Black might as well have moved from $q_{\lambda-1}$ to q'_λ instead of to q_λ. Then $U_r(q)$ still contains at least one element of the form $\mathfrak{q}'_\lambda = (q, q_1, \ldots q_{\lambda-1}, q'_\lambda, \ldots)$ which has the first λ terms in common with $\mathfrak{q}$. In fact, it is in this and only this case that White can begin with an arbitrary element $\mathfrak{q}$ of $U_r(q)$ and, whenever Black plays q'_λ instead of q_λ, continue playing with a corresponding $\mathfrak{q}'_\lambda$, and hence win in at most r moves under all circumstances.

To be sure, there can be several such subsets $U_r(q)$. But the sum of any two of them is always constituted in the same way, and the same holds of the union $\overline{U}_r(q)$ of all such $U_r(q)$, which is uniquely determined by q and r and which, in any case, must be different from 0, i.e., contain at least one element, provided that such $U_r(q)$ exist at all. Thus $\overline{U}_r(q) \neq 0$ is the necessary and sufficient condition for White being able to force a win in at most r moves. If $r < r'$, then $\overline{U}_r(q)$ is always a subset of $\overline{U}_{r'}(q)$ since, in this case, every set $U_r(q)$ certainly meets the demands placed on $U_{r'}(q)$ as well, and hence must be contained in $\overline{U}_{r'}(q)$. And to the least $r = \rho$ for which still $\overline{U}_r(q) \neq 0$ there corresponds the common component $U^*(q) = \overline{U}_\rho(q)$ of all such $\overline{U}_r(q)$; it comprises all such continuations by means of which White has to win in the shortest time possible. But now these minimal values $\rho = \rho_q$ themselves possess a maximum $\tau \leq t$ independent of q, where $t+1$ is the number of the possible positions, so that $U(q) = \overline{U}_\tau(q) \neq 0$ is the necessary and sufficient condition for some $\overline{U}_r(q)$ not vanishing in the position q and for White being in a "winning position" at all. For if it is at all possible in a position q to force a win, then it is also possible to do so, as we will show, in at most t moves. In fact, every endgame $\mathfrak{q} = (q, q_1, q_2, \ldots q_n)$ with $n > t$ would

mit $n > t$ mindestens eine Position $q_\alpha = q_\beta$ doppelt enthalten, und Weiss hätte beim ersten Erscheinen derselben ebenso weiter spielen können wie beim zweiten Male und jedenfalls schon früher als beim n ten Zuge gewinnen, also $\rho \leqq t$.

Ist andererseits $U(q) = 0$, so kann Weiss, wenn der Gegner richtig spielt, höchstens remis machen, er kann aber auch „auf Verlust stehen" und wird dann versuchen, dass „Matt" möglichst hinauszuschieben. Soll er sich noch bis zum s^{ten} Zuge halten können, so muss eine Untermenge $V_s(q)$ von Q existieren von folgender Beschaffenheit:

1. In keinem der in $V_s(q)$ enthaltenen Endspiele verliert Weiss vor dem s^{ten} Zuge.

2. Ist $\mathfrak{q}$ ein beliebiges Element von $V_s(q)$ und in $\mathfrak{q}$ durch einen erlaubten Zug von Schwarz q_λ ersetzbar durch q'_λ, so enthält $V_s(q)$ noch mindestens ein Element der Form

$$\mathfrak{q}'_\lambda = (q, q_1, q_2, \dots q_{\lambda-1}, q'_\lambda \dots),$$

welches mit q bis zum λ^{ten} Gliede übereinstimmt und dann mit q'_λ weitergeht.

Auch diese Mengen $V_s(q)$ sind sämtlich Untermengen ihrer Vereinigung $\overline{V}_s(q)$, welche durch q und s eindeutig bestimmt ist und die gleiche Eigenschaft besitzt wie V_s selbst, und für $s > s'$ wird jetzt $\overline{V}_s(q)$ Untermenge von $\overline{V}_{s'}(q)$. Die Zahlen s, für welche $\overline{V}_s(q)$ von 0 verschieden ausfällt, sind entweder unbegrenzt oder $\leqq \sigma \leqq \tau \leqq t$, da der Gegner, wenn überhaupt, den Gewinn in höchstens τ Zügen müsste erzwingen können. Somit kann Weiss dann und nur dann mindestens remis machen, wenn $V(q) = \overline{V}_{\tau+1}(q) \neq 0$ ist, und im anderen Falle kann er vermöge $V^*(q) = \overline{V}_\sigma(q)$ den Verlust noch mindestens $\sigma \leqq \tau$ Züge hinausschieben. Da jedes $U_r(q)$ gewiss auch den an $V_s(q)$ gestellten Anforderungen genügt, so ist jedes $\overline{U}_r(q)$ Untermenge jeder Menge $\overline{V}_s(q)$, und $U(q)$ Untermenge von $V(q)$. Das Ergebnis unserer Betrachtung ist also das folgende:

Jeder während des Spiels möglichen Position q entsprechen zwei wohldefinierte Untermengen $U(q)$ und $V(q)$ aus der Gesamtheit Q der mit q beginnenden Endspiele, deren zweite die erste umschliesst. Ist $U(q)$ von 0 verschieden, so kann Weiss, wie Schwarz auch spielt, den Gewinn erzwingen und zwar in höchstens ρ Zügen vermöge einer gewissen Untermenge $U^*(q)$ von $U(q)$, aber nicht mit Sicherheit in weniger Zügen. Ist $U(q) = 0$ aber $V(q) \neq 0$, so kann Weiss wenigstens remis machen vermöge der in $V(q)$ enthaltenen Endspiele. Verschwindet aber auch $V(q)$, so kann Weiss, wenn der Gegner richtig spielt, den Verlust höchstens bis zum σ^{ten} Zuge hinausschieben vermöge einer wohldefinierten Menge $V^*(q)$ von Fortsetzungen. Auf alle Fälle sind nur die in U^*, bezw. V^* enthaltenen Partieen im Interesse von Weiss als „korrekt" zu betrachten, mit jeder anderen Fortsetzung würde er, wenn in Gewinnstellung, bei richtigem Gegenspiel den gesicherten Gewinn verscherzen oder verzögern, sonst aber den Verlust der Partie ermöglichen oder beschleunigen. Ganz analoge Betrachtungen gelten natürlich auch für Schwarz, und als „korrekt" zu

have to contain at least one position $q_\alpha = q_\beta$ twice, and White could have continued to play at the first occurrence ⟦of the position⟧ just as well as at the second occurrence and certainly could have already won earlier than at the nth move, and hence $\rho \leq t$.

If, on the other hand, $U(q) = 0$, then White can at most achieve a draw, assuming the opponent plays correctly. But he also can be "in a losing position", in which case he will try to defer a "checkmate" for as long as possible. If he is to survive until the sth move, then there must exist a subset $V_s(q)$ of Q constituted as follows:

1. In none of the endgames contained in $V_s(q)$ does White lose before the sth move.

2. If $\mathfrak{q}$ is an arbitrary element of $V_s(q)$ and if it is possible to replace q_λ by q'_λ in $\mathfrak{q}$ by a legal move by Black, then $V_s(q)$ still contains at least one element of the form

$$\mathfrak{q}'_\lambda = (q, q_1, q_2, \ldots q_{\lambda-1}, q'_\lambda \ldots),$$

which is identical to $\mathfrak{q}$ up to the λth term and then continues with q'_λ.

These sets $V_s(q)$, too, are all subsets of their union $\overline{V}_s(q)$, which is uniquely determined by q and s and possesses the same property as V_s itself. For $s > s'$, $\overline{V}_s(q)$ now becomes a subset of $\overline{V}_{s'}(q)$. The numbers s for which $\overline{V}_s(q)$ differs from 0 are either unlimited or $\leq \sigma \leq \tau \leq t$, since the opponent would have to be able to force a win in at most τ moves, if at all. Thus White can at least make a draw if and only if $V(q) = \overline{V}_{\tau+1}(q) \neq 0$. In the other case, he can postpone defeat for at least $\sigma \leq \tau$ moves by virtue of $V^*(q) = \overline{V}_\sigma(q)$. Since every $U_r(q)$ certainly meets the demands placed on $V_s(q)$, every $\overline{U}_r(q)$ is a subset of every set $\overline{V}_s(q)$, and $U(q)$ a subset of $V(q)$. Our consideration thus has the following result:

To each position q possible during the game there correspond two well-defined subsets $U(q)$ and $V(q)$ from the totality Q of the endgames beginning with q, with the latter including the former. If $U(q)$ differs from 0, then White can force the win regardless of how Black plays, namely in at most ρ moves by virtue of a certain subset $U^*(q)$ of $U(q)$, but not with certainty in fewer moves. If $U(q) = 0$ but $V(q) \neq 0$, then White can at least make a draw by virtue of the endgames contained in $V(q)$. But if $V(q)$ vanishes as well, then, assuming that the opponent plays correctly, White can postpone defeat to no later than the σth move by virtue of a well-defined set $V^*(q)$ of continuations. In any event, only those games contained in U^* and V^* respectively are to be considered "correct" from White's point of view. With any other continuation, White, when in a winning position, would either throw away or delay a certain win, assuming that the opponent plays correctly, or else make possible the loss of the game or accelerate it. Very similar considerations are of course valid for Black, and we would have to consider those games as having been

Ende geführte Partieen hätten diejenigen zu gelten, welche *gleichzeitig* den beiderseitigen Bedingungen entsprechen, sie bilden also in jedem Falle wieder eine wohldefinierte Untermenge $W(q)$ von Q.

504 | Die Zahlen t und τ sind von der Position unabhängig und lediglich durch die Spielregeln bestimmt. Jeder möglichen Position entspricht eine τ nicht überschreitende Zahl $\rho = \rho_q$ oder $\sigma = \sigma_q$, je nachdem Weiss oder Schwarz in ρ bezw. σ Zügen, aber nicht in weniger, den Gewinn erzwingen kann. Die spezielle Theorie des Spiels hätte diese Zahlen, soweit dies möglich ist, zu bestimmen oder wenigstens in Grenzen einzuschliessen, was bisher allerdings nur in besonderen Fällen, wie bei den „Problemen“ oder den eigentlichen „Endspielen“ gelungen ist. Die Frage, ob die Anfangsposition p_0 bereits für eine der spielenden Parteien eine „Gewinnstellung“ ist, steht noch offen. Mit ihrer exacten Beantwortung würde freilich das Schach den Charakter eines Spieles überhaupt verlieren.

ended "correctly" which satisfy the conditions on both sides *simultaneously*. Hence, they certainly form a well-defined subset $W(q)$ of Q again.

The numbers t and τ are independent of the position and solely determined by the rules of the game. To every possible position there corresponds a number $\rho = \rho_q$ or $\sigma = \sigma_q$ not exceeding τ, depending on whether White or Black can force the win in ρ or σ, but not fewer, moves. The special theory of the game would have to determine these numbers to the extent to which this is possible, or at least determine the limits within which they must lie. So far this has been done only in special cases such as the "problems" and the "endgames" proper. The question as to whether the initial position p_0 already is a "winning position" for one of the players is still open. Its precise answer would of course deprive chess of its character as a game.

Introductory note to 1914

Ulrich Felgner

In his work on transcendental numbers Edmond Maillet[1] was led to the question whether it is possible to introduce the notion of an "integral element" in the field $\mathbb{C}$ of all complex numbers in analogy to the notions of a rational integer (i.e. an integral element in the field $\mathbb{Q}$ of all rational numbers) and of an algebraic integer (i.e. an integral element in the field A of all algebraic numbers). More precisely, if $\mathbb{Z} = I_{\mathbb{Q}}$ is the usual ring of (rational) integers and I_A the usual ring of algebraic integers, Maillet asked whether there exists a ring $I_{\mathbb{C}}$ of complex numbers such that

(1.) each complex number can be written in the form $\frac{a}{b}$ (with $a, b \in I_{\mathbb{C}}$),

(2.) $I_{\mathbb{C}} \cap \mathbb{Q} = I_{\mathbb{Q}}$ and $I_{\mathbb{C}} \cap A = I_A$.

If such a ring $I_{\mathbb{C}}$ were to exist, then its elements could rightly be called "integral elements" of $\mathbb{C}$. Maillet called them *integral transcendental numbers.* This terminology is somewhat unfortunate, since $I_{\mathbb{C}}$ also contains algebraic numbers. In fact $I_{\mathbb{C}}$ extends the domain of algebraic integers in such a way that among the algebraic numbers only the algebraic integers are integral elements of $\mathbb{C}$.

In algebraic number theory the ring I_A plays a dominant role since the arithmetic of I_A is not essentially different from the arithmetic of the ring $\mathbb{Z} = I_{\mathbb{Q}}$. We have $I_A \cap \mathbb{Q} = \mathbb{Z}$, and although the prime numbers of $\mathbb{Z}$ are multiplicatively decomposable in I_A, they are not units in I_A. Many problems from additive number theory hence have multiplicative correlates which are usually a bit easier to solve.

In his book *1906*, Chapter XI, Maillet conjectured that the zeros of those power series $\Phi(z) = 1 + c_1 z + c_2 z^2 + \ldots$, with rational coefficients $c_n \in \mathbb{Q}$, for which the set

$$\{n;\ \text{all zeros of } 1 + c_1 z + c_2 z^2 + \ldots + c_n z^n \text{ are algebraic integers}\}$$

is infinite, could serve as the "integral transcendental numbers". However Henry Blumberg[2] was able to show (cf. *Blumberg 1913*) that all real numbers

[1] Edmond Théodore Maillet was born in Meaux, France, on 15 December 1865 and died in Geneva, Switzerland on 11 September 1938. From 1911 to 1928 he held a professorship for Analysis and Mechanics at the École National des Ponts et Chaussées in Paris. In 1918 he was president of the Société Mathématique de France. In 1896 he was awarded the Grand Prix des Sciences Mathématique and in 1912 the Prix Poncelet of the Académie des Sciences de Paris.

[2] Henry Blumberg was born in Zhagori (Žagarė), Lithuania, on 13 May 1886 and died on 28 June 1950. From 1925 until 1950 he was a professor of mathematics at Ohio State University, Columbus, Ohio.

would then be "integral trancendental", contradicting the second of Maillet's conditions. Blumberg concluded his paper with the remark that the problem whether domains of "integral transcendental numbers" satisfying both of Maillet's conditions exist remains unsolved.

This problem was solved affirmatively by Zermelo. Using the assumption that the set of all complex numbers $\mathbb{C}$ has a well-ordering, he was able to construct a subring $I_{\mathbb{C}}$ of $\mathbb{C}$ which satisfies Maillet's conditions. Instead of $I_{\mathbb{C}}$ Zermelo wrote $\mathfrak{G}$. In the sequel we shall follow his notation. Zermelo obtains $\mathfrak{G}$ as the integral closure of a transcendency basis H in $\mathbb{C}$.

The construction of the basis H is analogous to that of the so-called Hamel basis (cf. *Hamel 1905*). Let Ω be the order-type of a well-ordering of $\mathbb{C}$ such that Ω is a cardinal number, and let t_0 be the first element of $\mathbb{C}$ which is not in the subset A of all algebraic numbers. Now proceed by transfinite recursion. Let α be an ordinal number, $\alpha \in \Omega$. Assuming that the elements t_β (for $\beta \in \alpha$) are already defined, let t_α be the first element of $\mathbb{C}$ (with respect to the well-ordering) which is not in the algebraic closure of $A \cup \{t_\beta;\ \beta \in \alpha\}$. Then $\mathsf{H} = \{t_\beta;\ \beta \in \Omega\}$ is a "basis" of $\mathbb{C}$ in the sense of Zermelo.

The construction yields that $\mathbb{C}$ is the algebraic closure of the pure transcendental extension $A(\{t_\beta;\ \beta \in \Omega\})$ of the field A. This implies that each complex number z satisfies a polynomial equation with coefficients from $A(\{t_\beta;\ \beta \in \Omega\})$. Among all these equations for z there is, up to its sign, a unique one of minimal complexity which is called the "principal equation" (Hauptgleichung). Such a principal equation has the form

$$C_0 x^n + C_1 x^{n-1} + \cdots + C_n = 0\,,$$

where the coefficients C_j are polynomials of the t_β (for $\beta \in \Omega$) with integral coefficients. A principal equation is characterized by the stipulation that n is minimal, that for this n, C_0 has minimal dimension m with respect to the basis $\{t_\beta;\ \beta \in \Omega\}$, and that for these n and m, C_0 has the least sum of the absolute values of the coefficients. Here the "dimension" of C_j is the largest number which occurs as the sum of the exponents $k_1 + k_2 + \cdots + k_d$ of a non-zero summand $a \cdot t_{\beta_1}^{k_1} \cdot t_{\beta_2}^{k_2} \cdots t_{\beta_d}^{k_d}$.

Zermelo proves that such an equation is primitive over the ring $\mathfrak{J}_{\mathsf{H}}$ of all linear combinations $q_0 t_{\beta_0} + q_1 t_{\beta_1} + \cdots + q_n t_{\beta_n}$ of elements $t_\beta \in \mathsf{H}$ with rational coefficients q, and also irreducible over the quotient field $\mathfrak{R}_{\mathsf{H}}$ of $\mathfrak{J}_{\mathsf{H}}$. (Zermelo writes $\mathfrak{J}_\eta$ and $\mathfrak{R}_\eta$ instead of $\mathfrak{J}_{\mathsf{H}}$ and $\mathfrak{R}_{\mathsf{H}}$.)

Finally, a complex number z is called H-*integral* (η-*ganz*) when in its principal equation we have $C_0 = \pm 1$. Zermelo shows that the set $\mathfrak{G} = \mathfrak{G}_{\mathsf{H}}$ of all H-integral elements forms a ring which satisfies both of Maillet's conditions. $\mathfrak{G}_{\mathsf{H}}$ is hence a domain of "integral elements" in the field of all complex numbers. But it is clear that the property of being "integral" is not a property which belongs to the essence of the number, but depends on the chosen well-ordering. Zermelo notes that one could choose the well-ordering in such a way that we have either $t_0 = \pi$ or $t_0 = e$.

The terminology in Zermelo's paper is in most cases the classical terminology. The notion of an "algebraic integer" (ganz-algebraische Zahl) was introduced by Richard Dedekind in his supplement XI to Dirichlet's *Vorlesungen über Zahlentheorie* (see *Dirichlet 1871*, Suppl. XI, §159 and §173). Also the term "Körper" (field) is Dedekind's invention. Dedekind mentions that in 1857 he used the term "rationales Gebiet" instead, which Leopold Kronecker in 1882 slightly changed into the term "Rationalitätsbereich" (domain of rationality)—see *Dirichlet 1871*, Suppl. XI, the first footnote in §160. Rings of integral elements had been coined "Integritätsbereiche" (domains of integrity) by Kronecker in his treatise *1882*, §5. The notion of a "primitive polynomial" is due to Carl Friedrich Gauss (*Disquisitiones arithmeticae, 1801*, §42). The term "primitive" (ursprünglich) however was introduced by Heinrich Weber in his *Lehrbuch der Algebra* (*1895*, §2). Zermelo does not use Hilbert's notion of "ring" (from his "Zahlbericht", *1897*). It is perhaps a bit surprising that Zermelo does not refer to Ernst Steinitz's important treatise *1910* on the algebraic theory of fields.

Emmy Noether (1882–1935) continued Zermelo's investigations in her paper *1916*, however under a different point of view. She did not aim at concrete constructions of examples for rings of "integral transcendental numbers", but at a classification of all such rings and the determination of their algebraic structure. The more abstract point of view in her work, which is often praised, is quite apparent here. She does not restrict her investigations to the field $\mathbb{C}$ of complex numbers but considers the case of algebraically closed fields right from the beginning. Instead of working with a basis of algebraically independent transcendental numbers, she prefers to calculate simply with indeterminates. Her main result is the description of the structure of those subrings of algebraically closed fields which could play the role of rings of integral elements.

When K is an arbitrary algebraically closed field of characteristic 0 and if A is the algebraic closure of its prime subfield, $A \subseteq K$, then each subring R of K which satisfies both of Maillet's conditions contains a transcendency basis H of K. Let $\mathfrak{M}(\mathsf{H})$ be the family of all subrings R of K which contain H and satisfy both of Maillet's conditions. Emmy Noether gives a description of the algebraic structure of the elements of $\mathfrak{M}(\mathsf{H})$ and also finds rings which are not of the Zermeloan type, e.g. the integral closure of the set of all "integral rational functionals (in the sense of Kronecker and Weber)" of the basis

elements. Emmy Noether finally proves that the intersection of all integrally closed rings in $\mathfrak{M}(\mathsf{H})$ is a ring of the Zermeloan type.

Emmy Noether must have found her results of such significance that—in order to fulfill a requirement for *Habilitation*—she gave a lecture on Zermelo's and her own results. Her lecture "Über ganze transzendente Zahlen" (On integral transcendental numbers) took place on November 9, 1915, in the Mathematical Institute in Göttingen. In a handwritten note attached to her *Habilitationsgesuch* (request for *Habilitation*) summarizing her research up to that time, she wrote that in her paper "The most general domains of integral transcendental numbers", *1916*, she applies, in addition to algebraic and number-theoretic principles, also those of abstract set theory (cf. *Dick 1970*, 35–36).

In fact, the constructions given by Zermelo and Noether depend on the axiom of choice. Together with the papers *Hamel 1905* and *Steinitz 1910* these were the earliest applications of the axiom of choice in algebra. That Zermelo's theorem on the existence of rings $\mathfrak{G}$ satisfying both of Maillet's conditions is provable only in the presence of the axiom of choice can be shown quite easily using Paul Cohen's forcing method.

Johann von Neumann (1903–1957) gave an explicit construction of a set of algebraically independent numbers of the power of the continuum without using the axiom of choice (cf. *1928b*). However, this set is not a basis of the field $\mathbb{C}$ of all complex numbers.

With the use of methods which are available now, it is quite easy to construct domains of integral transcendental numbers which satisfy both conditions required by Maillet. In fact, if $\mathcal{F}$ is a free ultrafilter on the set $\omega = \mathbb{N}$ of all positive integers, then (by the well-known Łoś's theorem) the ultrapower $A^\omega/\mathcal{F}$ is an algebraically closed field in which the field A of all algebraic numbers is canonically embedded as an elementary substructure. The ultrapower ${I_A}^\omega/\mathcal{F}$ of the ring I_A is integrally closed (ganz-algebraisch abgeschlossen) in $A^\omega/\mathcal{F}$. Obviously we have ${I_A}^\omega/\mathcal{F} \cap A = I_A$. Also $A^\omega/\mathcal{F}$ has the power of the continuum and hence, by the isomorphism theorem of Steinitz (*1910*, §21, Satz 9), is isomorphic with the field $\mathbb{C}$ of all complex numbers. (The axiom of choice is used here.) Let φ be such an isomorphism from $A^\omega/\mathcal{F}$ onto $\mathbb{C}$. Then φ fixes A (en bloc), because A is the algebraic closure of the prime field. It follows that $\varphi({I_A}^\omega/\mathcal{F}) = \{\varphi(x);\, x \in {I_A}^\omega/\mathcal{F}\}$ is a subring of $\mathbb{C}$ which satisfies the conditions required by Maillet.

Über ganze transzendente Zahlen

1914

Das Problem der „ganzen transzendenten Zahlen“, wie es neuerdings von Maillet* u. a. behandelt wurde, kommt darauf hinaus, einen Bereich $\mathfrak{G}$ von reellen oder komplexen Zahlen zu definieren, welcher die folgenden Eigenschaften besitzt:

I. Summe, Differenz und Produkt zweier Zahlen von $\mathfrak{G}$ ist wieder eine Zahl von $\mathfrak{G}$.
II. Jede reelle oder komplexe Zahl ist Quotient zweier Zahlen von $\mathfrak{G}$.
III. Jede ganze rationale (bzw. algebraische) Zahl ist Element von $\mathfrak{G}$.
IV. Keine nicht-ganze rationale (bzw. algebraische) Zahl gehört zu $\mathfrak{G}$.

Da es bisher noch nicht gelungen ist, einen Bereich von dieser Beschaffenheit zu konstruieren, so liegt die Vermutung nahe, als ob es deswegen nicht ginge, weil die vier aufgestellten Postulate miteinander im Widerspruche ständen. Im folgenden soll nun gezeigt werden, daß es sich so nicht verhält, indem die Existenz solcher Bereiche $\mathfrak{G}$ auf die Wohlordnung des Kontinuums zurückgeführt wird.

Zu diesem Zwecke wird nach einem zuerst von G. Hamel** und H. Lebesgue*** benutzten Verfahren auf Grund einer beliebigen Wohlordnung Ω eine „algebraische Basis“ der reellen und komplexen Zahlen definiert, d. h. ein System H von Zahlen η, zwischen denen keine algebraischen Beziehungen bestehen, welche aber alle übrigen Zahlen algebraisch auszudrücken gestatten. Es wird sodann gezeigt, daß jede Zahl einer *eindeutig* bestimmten „Hauptgleichung“ genügt, nämlich einer in H irreduziblen und primitiven algebraischen Gleichung, deren Koeffizienten „ganzzahlige“ Polynome der η sind, d. h. solche
435 mit ganzen rationalen Zahlen- | koeffizienten. Werden nun zu $\mathfrak{G}_\eta$ alle diejenigen Zahlen gerechnet, für welche der Anfangskoeffizient der Hauptgleichung ± 1 ist, so läßt sich zeigen, daß für diesen Bereich $\mathfrak{G}_\eta$ alle vier Bedingungen erfüllt sind. Jeder Wohlordnung Ω des Kontinuums entspricht also eine Basis H und damit ein System $\mathfrak{G}_\eta$ von „η-ganzen“ Zahlen, zu denen (nach III) jedenfalls auch alle ganzen algebraischen Zahlen gehören, zugleich aber auch alle Basiszahlen η selbst sowie alle ganzzahligen Polynome der η, während alle rational-gebrochenen Polynome der η mit ganzzahligen Koeffizienten als „nicht ganz“ zu gelten hätten.

---

* Hierüber vgl. H. Blumberg, Arch. Math. Phys. (3) 20, p. 53–57.

** G. Hamel, Eine Basis aller Zahlen und die unstetigen Lösungen der Funktionalgleichung $f(x+y) = f(x) + f(y)$. Math. Ann. 60, S. 459, wo es sich aber nur um *lineare* Beziehungen handelt.

*** H. Lebesgue, Sur les transformations ponctuelles transformant les plans en plans. Atti Ac. Torino 1906–1907.

On integral transcendental numbers

1914

The problem of the "integral transcendental numbers", as recently discussed by *Maillet*[1] among others, amounts to the problem of defining a domain $\mathfrak{G}$ of real or complex numbers which possesses the following properties:

I. The sum, difference and product of two numbers of $\mathfrak{G}$ are again numbers of $\mathfrak{G}$.
II. Every real or complex number is a quotient of two numbers of $\mathfrak{G}$.
III. Every integral rational (or algebraic) number is an element of $\mathfrak{G}$.
IV. No non-integral rational (or algebraic) number belongs to $\mathfrak{G}$.

Since attempts at constructing a domain thus constituted have failed so far, the hypothesis suggests itself that this failure is due to the fact that the four specified postulates contradict one another. That this is not the case shall be shown in the following by reducing the existence of such domains $\mathfrak{G}$ to the well-ordering of the continuum.

To this end, we define an "algebraic basis" of the real and complex numbers using an arbitrary well-ordering Ω in accordance with a method first employed by *G. Hamel*[2] and *H. Lebesgue.*[3] That is, we define a system H of numbers η among which no algebraic relations obtain but which allow for the algebraic expression of all other numbers. We will then show that every number satisfies a *uniquely* determined "principal equation", namely an algebraic equation irreducible over and primitive in H whose coefficients are "integral" polynomials of the η's, i.e., ones with integral rational number coefficients. If we now assume that all those numbers belong to $\mathfrak{G}_\eta$ for which the leading coefficient of the principal equation is ± 1, then we can show that all four conditions are satisfied for this domain $\mathfrak{G}_\eta$. To every well-ordering Ω of the continuum therefore corresponds a basis H, and hence a system $\mathfrak{G}_\eta$ of "η-integral" numbers to which (according to III) belong at least all integral algebraic numbers, but also all basis numbers η themselves and all integral polynomials of the η's, while all rational polynomials of the η's with integral coefficients would have to be regarded as "non-integral".

[1] On this matter, cf. *Blumberg 1913.*
[2] *Hamel 1905,* which, however, is only concerned with *linear* relations.
[3] *Lebesgue 1907.*

§ 1.
Die Basis

Die Gesamtheit $\mathfrak{C}$ der reellen und komplexen Zahlen sei in beliebiger Weise „wohlgeordnet“, d. h. nach G. Cantor so geordnet, daß die Menge $\mathfrak{K}$ wie jede ihrer Teilmengen ein erstes Element enthält. Eine solche Wohlordnung Ω kann stets und zwar auf eindeutige Weise definiert werden, wenn jeder nicht verschwindenden Untermenge von $\mathfrak{C}$ eines ihrer Elemente zugeordnet ist, sie existiert also auf Grund des „Auswahlaxioms“*, und die Gesamtheit aller möglichen Ω ist eine Menge von der Mächtigkeit $2^{\mathfrak{c}}$, sofern $\mathfrak{c}$ die des Kontinuums bezeichnet.

Für das Folgende genügt es, eine *beliebige* Wohlordnung Ω von $\mathfrak{C}$ zugrunde zu legen. Dann entspricht jeder (reellen oder komplexen) Zahl α als „Abschnitt“ eine bestimmte Untermenge $\mathfrak{C}(\alpha)$ von $\mathfrak{C}$, nämlich die Gesamtheit der dem Element α in Ω vorangehenden Elemente, sowie der durch diese Zahlen bestimmte Körper $\mathfrak{K}(\alpha)$.

Nun kann es vorkommen, daß eine Zahl α von den *vorangehenden* „algebraisch abhängig“ ist, nämlich einer algebraischen Gleichung genügt von der Form

$$f(x, \beta) \equiv f(x, \beta_1, \beta_2, \cdots, \beta_\nu) = 0\,, \tag{1}$$

in welcher alle Koeffizienten dem Körper $\mathfrak{K}_\alpha$ angehören, d. h. ganzzahlige rationale Funktionen von endlich vielen dem α vorangehenden Zahlen $\beta_1, \beta_2, \cdots, \beta_\nu$ sind und der Koeffizient der höchsten Potenz von x nicht verschwindet. Insbesondere erfüllen alle algebraischen Zahlen diese Forderung, weil der natürliche Rationalitätsbereich $\mathfrak{R}$ in jedem Körper $\mathfrak{K}(\alpha)$ enthalten ist.

Genügt dagegen eine Zahl η *keiner* Gleichung von der Form (1), ist η
436 „von den vorangehenden unabhängig“, so heiße η eine „Basiszahl“, und | die Gesamtheit *aller* Basiszahlen H wird als die zu Ω gehörende „Basis“ bezeichnet. So ist mindestens die *erste* in Ω vorkommende *transzendente* Zahl eine Basiszahl. Die aus endlich vielen Basiszahlen η gebildeten Polynome mit ganzen rationalen Koeffizienten bilden dann einen Integritätsbereich $\mathfrak{J}_\eta$ und ihre Quotienten den „Basiskörper“ $\mathfrak{R}_\eta$. Alle diejenigen Basiszahlen, welche einem gegebenen Elemente α in Ω vorangehen, bilden den „Basisabschnitt“ $\mathsf{H}(\alpha)$.

Ist dagegen α keine Basiszahl, sondern von den vorangehenden abhängig, so genügt α, wie wir zeigen wollen, mindestens einer Gleichung der Form

$$g(x, \eta) \equiv g(x, \eta_1, \eta_2, \cdots, \eta_t) = 0\,, \tag{2}$$

wo $\eta_1, \eta_2, \cdots, \eta_t$ zu $\mathsf{H}(\alpha)$ gehören, d. h. α ist *algebraisch abhängig von endlich vielen (vorangehenden) Basiszahlen.*

---

* E. Zermelo, Beweis, daß jede Menge wohlgeordnet werden kann. Math. Ann. 59, S. 514. Derselbe, Neuer Beweis für die Möglichkeit der Wohlordnung. Math. Ann. 65, S. 107.

§ 1.
The basis

Let the totality $\mathfrak{C}$ of the real and complex numbers be arbitrarily "well-ordered", that is, following Cantor, ordered so that the set $\mathfrak{C}$ [4] and each of its partial sets contain a first element. Such a well ordering Ω can always be uniquely defined if to every nonempty subset of $\mathfrak{C}$ one of its elements is assigned. It therefore exists by dint of the "axiom of choice",[5] and the totality of all possible Ω is a set of cardinality $2^{\mathfrak{c}}$, where $\mathfrak{c}$ signifies the cardinality of the continuum.

For what follows it shall suffice to assume an *arbitrary* well-ordering Ω of $\mathfrak{C}$. Then to every (real or complex) number α there corresponds as a "segment" a particular subset $\mathfrak{C}(\alpha)$ of $\mathfrak{C}$, namely the totality of the elements preceding the element α in Ω, and also the field $\mathfrak{K}(\alpha)$ determined by means of these numbers.

It is possible now that a number α is "algebraically dependent" on the *preceding* ones, that is, that it satisfies an algebraic equation of the form

$$f(x, \beta) \equiv f(x, \beta_1, \beta_2, \cdots, \beta_\nu) = 0, \tag{1}$$

in which all coefficients belong to the field $\mathfrak{K}(\alpha)$, i.e., in which they all are integral rational functions of finitely many numbers $\beta_1, \beta_2, \cdots, \beta_\nu$ preceding α, and in which the coefficient of the highest power of x does not vanish. This requirement is met, in particular, by all algebraic numbers since the natural rational domain $\mathfrak{R}$ is contained in every field $\mathfrak{K}(\alpha)$.

On the other hand, if a number η satisfies *no* equation of the form (1), if η is "independent of the preceding ones", then η shall be called a "basis number", and the totality of *all* basis numbers H the "basis" belonging to Ω. Thus at least the *first transcendental* number occurring in Ω is a basis number. The polynomials with integral rational coefficients formed from finitely many basis numbers η then form an integral domain $\mathfrak{J}_\eta$, and their quotients the "basis field" $\mathfrak{R}_\eta$. All those basis numbers preceding a given element α in Ω form the "basis segment" $\mathsf{H}(\alpha)$.

If, on the other hand, α is not a basis number but dependent on its predecessors, then it satisfies, as we will show, at least one equation of the form

$$g(x, \eta) \equiv g(x, \eta_1, \eta_2, \cdots, \eta_t) = 0, \tag{2}$$

where $\eta_1, \eta_2, \cdots, \eta_t$ belong to $\mathsf{H}(\alpha)$, i.e., α is *algebraically dependent on finitely many (preceding) basis numbers.*

---

4 ⟦Zermelo erroneously writes "$\mathfrak{K}$" instead of "$\mathfrak{C}$".⟧

5 *Zermelo 1904, 1908a.*

Nach dem bekannten, für wohlgeordnete Mengen gültigen Induktionsverfahren genügt es, den Satz zu beweisen für eine Zahl α unter der Voraussetzung, daß seine Gültigkeit für alle etwa vorangehenden Zahlen β bereits gesichert sei; denn unter den Zahlen α, für welche er ungültig wäre, müßte es in Ω eine erste α_0 geben, und für die vorangehenden wäre er gültig, also auch für α_0 gegen die Annahme.

Es genüge also $x = \alpha$ einer algebraischen Gleichung

$$f(x, \beta) \equiv f(x, \beta_1, \beta_2, \cdots, \beta_\nu) = 0\,, \tag{1}$$

und jede der Zahlen β_λ, welche nicht selbst Basiszahl ist, genüge einer Gleichung

$$g_\lambda(x, \eta) \equiv g_\lambda(x, \eta_1, \eta_2, \cdots, \eta_t) = 0\,, \qquad (2)_\lambda$$

wo wieder alle Basiszahlen $\eta_1, \eta_2, \cdots, \eta_t$, weil sie den β_λ vorangehen, dem Abschnitte $\mathsf{H}(\alpha)$ angehören. Ist ein β_λ selbst eine Basiszahl, so genügt sie ebenfalls einer solchen Gleichung $(2)_\lambda$, nämlich $x - \beta_\lambda = 0$, wo β_λ gleichfalls zu $\mathsf{H}(\alpha)$ gehört.

Um nun die ν Größen $\beta_1, \beta_2, \cdots, \beta_\nu$ aus den $\nu + 1$ Gleichungen (1), $(2)_\lambda$ zu eliminieren, kann man etwa folgendermaßen verfahren.

Die Gleichung (1) können wir auf die Form bringen

$$f(x, \beta) \equiv A_0 x^n + A_1 x^{n-1} + \cdots + A_n = 0\,, \tag{1a}$$

wo jeder Koeffizient $A_\varrho = A_\varrho(\beta)$ ein ganzzahliges Polynom in $\beta_1, \beta_2, \cdots, \beta_\nu$ ist und $A_0(\beta) \neq 0$. Ebenso können wir uns in $(2)_\lambda$ alle Nenner fortgeschafft denken und diese Gleichungen sämtlich als irreduzibel voraussetzen in dem durch $\eta_1, \eta_2, \cdots, \eta_t$ bestimmten Körper $\mathfrak{k}_t$. Dann genügt jeder Gleichung $(2)_\lambda$ ein System konjugierter Größen $\beta_\lambda, \beta_\lambda{}', \beta_\lambda{}'', \cdots$, deren elementare symmetrische Funktionen (als Koeffizienten von $g_\lambda(x)$ dividiert durch den Anfangskoeffizienten $b_{\lambda 0} \neq 0$) rational sind in $\mathfrak{k}_t$. Ersetzt man | nun in A_ϱ jede der Größen $\beta_1, \beta_2, \cdots, \beta_\nu$ durch ihre sämtlichen konjugierten, in allen Kombinationen, so entstehen die Werte $A_\varrho, A_\varrho{}', A_\varrho{}'', \cdots$ und genügen sämtlich einer Gleichung

437

$$G_\varrho(x) \equiv (x - A_\varrho)(x - A_\varrho{}')(x - A_\varrho{}'') \cdots = 0\,,$$

deren Koeffizienten als ganze, ganzzahlige symmetrische Funktionen der Wurzelkombinationen β wieder rational sein müssen in $\mathfrak{k}_t$. Der in $\mathfrak{k}_t$ irreduzible Faktor von $G_\varrho(x)$, welchem $x = A_\varrho$ genügt, sei $\Phi_\varrho(x)$, sodaß wir haben

$$\Phi_\varrho(A_\varrho) = 0\,.$$

Hier ist in $\Phi_0(x)$, da $A_0 \neq 0$, auch jede andere Wurzel von 0 verschieden, weil sonst das konstante Glied verschwände und die Gleichung reduzibel wäre gegen die Annahme. Ersetzt man also in (1a) jeden Koeffizienten A_ϱ durch seine konjugierten und multipliziert die entstehenden Polynome durch alle

According to the well-known method of induction valid for well-ordered sets it suffices to prove the theorem for a number α under the assumption that its validity has already been demonstrated for all preceding numbers β; for among the numbers α for which it would be invalid there would have to exist in Ω a first α_0, and it would be valid for the preceding ones, and hence also for α_0, contrary to the assumption.

Let $x = \alpha$ therefore satisfy an algebraic equation

$$f(x, \beta) \equiv f(x, \beta_1, \beta_2, \cdots, \beta_\nu) = 0, \tag{1}$$

and let every number β_λ which itself is not a basis number satisfy an equation

$$g_\lambda(x, \eta) \equiv g_\lambda(x, \eta_1, \eta_2, \cdots, \eta_t) = 0, \tag*{(2)$_\lambda$}$$

where all basis numbers $\eta_1, \eta_2, \cdots, \eta_t$ again belong to the segment $\mathsf{H}(\alpha)$ since they precede the β_λ's. If a β_λ is itself a basis number, then it, too, satisfies such an equation $(2)_\lambda$, namely $x - \beta_\lambda = 0$, where β_λ also belongs to $\mathsf{H}(\alpha)$.

Now in order to eliminate the ν quantities $\beta_1, \beta_2, \cdots, \beta_\nu$ from the $\nu + 1$ equations $(1), (2)_\lambda$, we can proceed, e.g., as follows.

We can bring the equation (1) into the form

$$f(x, \beta) \equiv A_0 x^n + A_1 x^{n-1} + \cdots + A_n = 0, \tag{1a}$$

where each coefficient $A_\varrho = A_\varrho(\beta)$ is an integral polynomial in $\beta_1, \beta_2, \cdots, \beta_\nu$ and $A_0(\beta) \neq 0$. We can also consider all denominators eliminated in $(2)_\lambda$ and we can assume that all these equations are irreducible over the field $\mathfrak{k}_t$ determined by means of $\eta_1, \eta_2, \cdots, \eta_t$. Then each equation $(2)_\lambda$ is satisfied by a system of conjugated quantities $\beta_\lambda, \beta_\lambda', \beta_\lambda'', \cdots$ whose elementary symmetric functions (as coefficients of $g_\lambda(x)$ divided by the leading coefficient $b_{\lambda 0} \neq 0$) are rational in $\mathfrak{k}_t$. Replacing now in A_ϱ each of the quantities $\beta_1, \beta_2, \cdots, \beta_\nu$ by all its conjugates, in all combinations, generates the values $A_\varrho, A_\varrho', A_\varrho'', \cdots$ which all satisfy an equation

$$G_\varrho(x) \equiv (x - A_\varrho)(x - A_\varrho')(x - A_\varrho'') \cdots = 0,$$

whose coefficients, being integral symmetric functions with integral coefficients of the root combinations β, must again be rational in $\mathfrak{k}_t$. Let the factor of $G_\varrho(x)$ which is irreducible over $\mathfrak{k}_t$ and satisfied by $x = A_\varrho$ be $\Phi_\varrho(x)$ so that we have

$$\Phi_\varrho(A_\varrho) = 0.$$

In this case, any other root is also different from 0 in $\Phi_0(x)$ because $A_0 \neq 0$, since otherwise the constant term would vanish and the equation would be reducible, contrary to the assumption. If we therefore replace in (1a) each coefficient A_ϱ by its conjugates and if we multiply the resulting polynomials

Kombinationen, so entsteht eine Gleichung

$$\Phi(x) \equiv (A_0 x^n + A_1 x^{n-1} + \cdots)(A_0' x^n + A_1' x^{n-1} + \cdots) \cdots = 0\,,$$

welche für $x = \alpha$ erfüllt ist, deren Koeffizienten in bezug auf x als symmetrische Funktionen der $A_\varrho, A_\varrho', \ldots$ wieder dem Körper $\mathfrak{k}_t$ angehören, und deren Anfangskoeffizient als eine Potenz der Norm $A_0 A_0' A_0'' \cdots$ sicher von 0 verschieden ist. Es genügt also α in der Tat einer Gleichung von der Form (2), und die behauptete Abhängigkeit ist bewiesen.

Dagegen kann zwischen endlich vielen *Basiszahlen allein niemals* eine algebraische Gleichung der Form

$$F(\eta_1, \eta_2, \cdots, \eta_t) = 0$$

mit ganzzahligen Koeffizienten bestehen, ohne daß alle Koeffizienten verschwinden. Denn sonst wäre die in der Wohlordnung *letzte* unter den η, z. B. η_t, welche nicht identisch herausfällt, algebraisch abhängig von den vorangehenden, entgegen unserer Definition. Jede solche algebraische Relation zwischen Basiszahlen mit ganzzahligen Koeffizienten muß daher *identisch* bestehen, also auch für *willkürliche* Werte der Veränderlichen $\eta_1, \cdots, \eta_t$.

§ 2.
Die Hauptgleichung

Jede reelle oder komplexe Zahl α, mag sie zur Basis H gehören oder nicht, genügt nach dem oben Bewiesenen mindestens einer algebraischen Gleichung der Form

$$g(x, \eta) \equiv g(x, \eta_1, \eta_2, \cdots, \eta_t) = 0\,, \tag{2}$$

in welcher g ein Polynom mit ganzzahligen Koeffizienten ist, $\eta_1, \eta_2, \cdots, \eta_t$ Basiszahlen sind und der Exponent der höchsten vorkommenden Potenz von x mindestens $= 1$ ist.

438 | Unter allen Gleichungen dieser Form (2), denen α genügt, gibt es sicher solche von *niedrigster Gradzahl* $n \geqq 1$. Unter allen solchen Gleichungen n^{ten} Grades gibt es wieder solche, in denen die *Dimension* des Koeffizienten von x^n in bezug auf die η, d. h. die höchste Exponentensumme eines nicht verschwindenden Gliedes $\eta_1^{\alpha_1} \eta_2^{\alpha_2} \cdots \eta_t^{\alpha_t}$ möglichst klein ist, nämlich $m \geqq 0$. Unter allen diesen Gleichungen vom Grade n und von der Dimension m des Anfangskoeffizienten muß es endlich eine solche geben, in welcher die Summe der absolut genommenen (ganzzahligen) Zahlenkoeffizienten dieses Anfangskoeffizienten möglichst klein und zwar $s \geqq 1$ ist.

Die so charakterisierte Gleichung n^{ten} Grades

$$\varphi(x, \eta) \equiv C_0 x^n + C_1 x^{n-1} + \cdots + C_n = 0\,, \tag{3}$$

through all combinations, then we obtain an equation

$$\Phi(x) \equiv (A_0x^n + A_1x^{n-1} + \cdots)(A_0'x^n + A_1'x^{n-1} + \cdots)\cdots = 0$$

which is satisfied for $x = \alpha$, whose coefficients with respect to x, being symmetric functions of $A_\varrho, A_\varrho', \cdots$, again belong to the field $\mathfrak{k}_t$, and whose leading coefficient, being a power of the norm $A_0A_0'A_0'' \cdots$, certainly differs from 0. So α really satisfies an equation of the form (2), and the asserted dependence is proved.

By contrast, among finitely many *basis numbers alone* there can *never* obtain an algebraic equation of the form

$$F(\eta_1, \eta_2, \cdots, \eta_t) = 0$$

with integral coefficients without the coefficients all vanishing. For otherwise the *last* among the η's in the well-ordering, such as η_t, which does not drop out identically, would be algebraically dependent on its predecessors, contrary to our definition. Each such algebraic relation among basis numbers with integral coefficients must therefore obtain *identically*, and hence also for *arbitrary* values of the variables $\eta_1, \cdots, \eta_t$.

§ 2.
The principal equation

Every real or complex number α, whether or not it belongs to the basis H, satisfies, according to what has been shown above, at least one algebraic equation of the form

$$g(x, \eta) \equiv g(x, \eta_1, \eta_2, \cdots, \eta_t) = 0\,, \tag{2}$$

in which g is a polynomial with integral coefficients, in which $\eta_1, \eta_2, \cdots, \eta_t$ are basis numbers, and the exponent of the highest extant power of x is at least $= 1$.

Among all equations of this form (2) satisfied by α there is certainly one of the *lowest degree number* $n \geqq 1$. Moreover, among all such equations of the nth degree there are those in which the *dimension* of the coefficient of x^n with respect to the η's, i.e., the highest exponent sum of a non-vanishing term $\eta_1{}^{\alpha_1}\,\eta_2{}^{\alpha_2} \cdots \eta_t{}^{\alpha_t}$, is as small as possible, namely $m \geqq 0$. Finally, among all these equations of degree n with the dimension m there must be one in which the sum of the (integral) number coefficients, taken absolutely, of this leading coefficient is as small as possible, namely $s \geqq 1$.

The nth degree equation thus characterized

$$\varphi(x, \eta) \equiv C_0x^n + C_1x^{n-1} + \cdots + C_n = 0 \tag{3}$$

in welcher alle C_ϱ ganzzahlige Polynome in den Basiszahlen η sind und C_0 die kleinste Dimension m in den η und die minimale Koeffizientensumme s besitzt, ist dann, wie wir beweisen wollen, *irreduzibel* im Körper $\mathfrak{R}_\eta$, *primitiv* im entsprechenden Integritätsbereiche $\mathfrak{J}_\eta$ und endlich durch die Zahl α, die ihr genügt, bis auf das Vorzeichen aller Glieder *eindeutig* bestimmt. Sie werde die zu α gehörige „Hauptgleichung" genannt.

Wäre (3) in $\mathfrak{R}_\eta$ *reduzibel*, z. B.

$$\varphi(x,\eta) = \psi(x,\eta)\chi(x,\eta)\,,$$

so würde α einer Gleichung derselben Form (2), aber von niedrigerem Grade $n' < n$ genügen gegen die Annahme.

Wären alle Koeffizienten $C_\varrho(\eta)$ *teilbar* durch ein ganzzahliges Polynom $T(\eta)$ von der Dimension $\tau \geqq 1$, also

$$C_\varrho(\eta) = T(\eta)C_\varrho{}'(\eta)\,,$$

so müßten diese $n+1$ Beziehungen, da die Basiszahlen, wie oben gezeigt, algebraisch unabhängig sind, in den sämtlichen vorkommenden η *identisch* gelten, und α genügte der Gleichung

$$C_0{}'x^n + C_1{}'x^{n-1} + \cdots + C_n{}' = 0\,, \tag{3'}$$

in welcher die Dimension m' von $C_0{}'$, wieder gegen die Annahme, den Wert hätte $m' = m - \tau < m$.

Ebensowenig können alle *Zahlenkoeffizienten* der C_ϱ durch eine und dieselbe ganze *Zahl* $r > 1$ teilbar sein, weil durch Division mit r auch die Koeffizientensumme s von C_0 um den Faktor r *verkleinert* würde im Widerspruche mit der Definition von s. Somit ist (3) in der Tat irreduzibel und primitiv im Bereiche der η.

Es sei jetzt (2) eine *beliebige* Gleichung der betrachteten Form, welcher α genügt, und (3) die soeben gewonnene irreduzible, so würde bei der Divi-
439 sion von $g(x,\eta)$ durch $\varphi(x,\eta)$ ein Rest $\omega(x,\eta)$ sich ergeben, der | in bezug
auf x von niedrigerem als dem n^{ten} Grade wäre und für $x = \alpha$ ebenfalls den Wert 0 annähme, also wieder zu einer Gleichung der Form (2) von niedrigerem Grade führte, entgegen unserer Definition. Der Rest muß also in x identisch verschwinden, aber auch wegen der vorausgesetzten Unabhängigkeit der Basiszahlen identisch in den η, und wir haben

$$k(\eta)g(x,\eta) = \varphi(x,\eta)\psi(x,\eta)\,, \tag{4}$$

wo $\psi(x,\eta)$ wieder ein Polynom derselben Form ist und $k(\eta)$ ein nicht verschwindendes ganzzahliges Polynom in den η. Es ist also $g(x,\eta)$ in bezug auf x immer teilbar durch $\varphi(x,\eta)$, und $g(x,\eta)$ kann nur dann irreduzibel in $\mathfrak{R}_\eta$ oder vom Grade n sein, wenn auch $\psi(x,\eta) = l(\eta)$ von x unabhängig ist.

in which all C_ϱ are integral polynomials in the basis numbers η and in which C_0 possesses the least dimension m in the η's and the minimal coefficient sum s, is then, as we shall prove, *irreducible* over the field $\mathfrak{R}_\eta$ and *primitive* in the corresponding integral domain $\mathfrak{J}_\eta$. Finally, except for the sign of all terms, it is *uniquely* determined by means of the number α satisfying it. It shall be called the "principal equation" belonging to α.

If (3) were *reducible* over $\mathfrak{R}_\eta$, e.g.,

$$\varphi(x,\eta) = \psi(x,\eta)\chi(x,\eta)\,,$$

then α would satisfy an equation which is of the same form (2) but of a lower degree $n' < n$, contrary to the assumption.

If all coefficients $C_\varrho(\eta)$ were *divisible* by an integral polynomial $T(\eta)$ of dimension $\tau \geqq 1$, thus

$$C_\varrho(\eta) = T(\eta)C_\varrho{}'(\eta)\,,$$

then these $n+1$ relations would have to hold *identically* in all the occurring η's since the basis numbers are, as has been shown above, algebraically independent, and α would satisfy the equation

$$C_0{}'x^n + C_1{}'x^{n-1} + \cdots + C_n{}' = 0 \tag{3'}$$

in which the dimension m' of $C_0{}'$ would have the value $m' = m - \tau < m$, again contrary to the assumption.

Likewise, it is not possible that all *number coefficients* of the C_ϱ are divisible by one and the same integral *number* $r > 1$ since division by r would also *diminish* the coefficient sum s of C_0 by the factor r, contrary to the definition of s. Thus (3) is really irreducible over and primitive in the domain of the η's.

Now let (2) be an *arbitrary* equation of the form under consideration satisfied by α and let (3) be the irreducible equation just obtained. Then division of $g(x,\eta)$ by $\varphi(x,\eta)$ would yield a remainder $\omega(x,\eta)$ which, with respect to x, would be of degree lower than n and which would also assume the value 0 for $x = \alpha$. We would thus again arrive at an equation of the form (2) of a lower degree, contrary to our definition. The remainder must therefore vanish identically in x, but also identically in the η's on account of the independence of the basis numbers that has been assumed, and we have

$$k(\eta)g(x,\eta) = \varphi(x,\eta)\psi(x,\eta)\,, \tag{4}$$

where $\psi(x,\eta)$ is again a polynomial of the same form and $k(\eta)$ is a non-vanishing integral polynomial in the η's. Therefore, $g(x,\eta)$ is always divisible by $\varphi(x,\eta)$ with respect to x, and $g(x,\eta)$ can be irreducible over $\mathfrak{R}_\eta$ or of degree n only if $\psi(x,\eta) = l(\eta)$ is also independent of x.

In diesem Falle erhalten wir aus (4) durch Koeffizientenvergleichung das System der Gleichungen

$$k(\eta)B_\varrho(\eta) = l(\eta)C_\varrho(\eta) \qquad (\varrho = 0, 1, 2, \cdots, n)\,, \tag{5$_\varrho$}$$

welche ebenfalls in den η wieder *identisch* bestehen müssen. Nun gelten aber auch für Polynome mehrerer Veränderlicher die bekannten Zerlegungsgesetze.* Setzen wir also, unbeschadet der Allgemeinheit, die Polynome $k(\eta)$ und $l(\eta)$ als teilerfremd voraus (wir könnten sonst durch jeden gemeinsamen Teiler dividieren), so ergibt sich, daß alle B_ϱ durch $l(\eta)$ und alle C_ϱ durch $k(\eta)$ teilbar sein müssen,

$$B_\varrho(\eta) = l(\eta)B_\varrho{}^*(\eta)\,, \quad C_\varrho(\eta) = k(\eta)C_\varrho{}^*(\eta) \quad \text{für} \quad \varrho = 0, 1, \cdots, n\,.$$

Da aber nach dem oben Bewiesenen $\varphi(x, \eta)$ in $\mathfrak{J}_\eta$ *primitiv* sein muß, so ist $k(\eta)$ jedenfalls $= \pm 1$, und $g(x, \eta)$ kann nur dann ebenfalls primitiv sein, wenn auch $l(\eta) = \pm 1$ ist, d. h. es ist in diesem Falle

$$g(x, \eta) = \pm\varphi(x, \eta)\,,$$

und die „Hauptgleichung“ ist in der Tat durch die Eigenschaften der Irreduzibilität und Primitivität bis auf das Vorzeichen *eindeutig* bestimmt.

§ 3.
Der Integritätsbereich

Eine Zahl α wollen wir dann und nur dann als „η-ganz“ dem Bereiche $\mathfrak{G}_\eta$ zurechnen, wenn in der zugehörigen (irreduziblen und primitiven) „Hauptgleichung“ der Koeffizient der höchsten Potenz von x den Wert ± 1 hat, wenn also diese Gleichung die Form annimmt

$$\varphi(x, \eta) \equiv x^n + C_1 x^{n-1} + \cdots + C_n = 0\,, \tag{3*}$$

wo die Koeffizienten $C_1, C_2, \cdots, C_n$ sämtlich Polynome der Basiszahlen mit ganzzahligen Koeffizienten sein sollen.

440 | Wir beweisen zunächst, daß diese Bedingung immer dann erfüllt ist, wenn α *irgendeiner* Gleichung der Form genügt

$$g(x, \eta) \equiv x^m + B_1 x^{m-1} + \cdots + B_m = 0\,, \tag{2*}$$

in welcher der Anfangskoeffizient $= 1$ ist und alle übrigen ganzzahlige Polynome in den η sind.

* Vgl. H. Weber, Lehrbuch der Algebra, kleine Ausgabe, § 20.

In this case, we obtain from (4) by comparison of coefficients the system of equations

$$k(\eta)B_\varrho(\eta) = l(\eta)C_\varrho(\eta) \qquad (\varrho = 0, 1, 2, \cdots, n)\,, \tag{5$_\varrho$}$$

which also must hold *identically* in the η's. But also polynomials in several variables are subject to the well-known laws of decomposition.[6] Without loss of generality, let us therefore assume that the polynomials $k(\eta)$ and $l(\eta)$ have no common divisor (otherwise, we could divide by any common divisor). It then follows that every B_ϱ must be divisible by $l(\eta)$ and every C_ϱ by $k(\eta)$,

$$B_\varrho(\eta) = l(\eta)B_\varrho{}^*(\eta)\,, \quad C_\varrho(\eta) = k(\eta)C_\varrho{}^*(\eta) \quad \text{for} \quad \varrho = 0, 1, \cdots, n\,.$$

But since, according to what has been proved above, $\varphi(x, \eta)$ must be *primitive* in $\mathfrak{J}_\eta$, $k(\eta)$ is certainly $= \pm 1$, and $g(x, \eta)$ can be primitive only if $l(\eta)$ is also $= \pm 1$, i.e., in this case, we have

$$g(x, \eta) = \pm\varphi(x, \eta)\,,$$

and, except for the sign, the "principal equation" is really *uniquely* determined by means of the properties of irreducibility and primitivity.

§ 3.
The integral domain

We will say that a number α belongs to the domain $\mathfrak{G}_\eta$ as "η-integral" if and only if the coefficient of the highest power of x in the corresponding (irreducible and primitive) "principal equation" has the value ± 1, and hence if this equation takes the form

$$\varphi(x, \eta) \equiv x^n + C_1 x^{n-1} + \cdots + C_n = 0\,, \tag{3*}$$

where the coefficients $C_1, C_2, \cdots, C_n$ are all supposed to be polynomials of the basis numbers with integral coefficients.

We first prove that this condition is always satisfied if α satisfies an *arbitrary* equation of the form

$$g(x, \eta) \equiv x^m + B_1 x^{m-1} + \cdots + B_m = 0 \tag{2*}$$

in which the leading coefficient is $= 1$, and in which all other integral polynomials are in the η's.

[6] *Weber 1912,* § 20.

In der Tat haben wir in diesem Falle, wie im vorigen § 2 bewiesen, die in x und η identische Beziehung

$$k(\eta)g(x,\eta) = \varphi(x,\eta)\psi(x,\eta)\,, \tag{4}$$

wo $\varphi(x,\eta) = 0$ die zu α gehörende „Hauptgleichung" bedeutet. Es muß also $k(\eta)$ ein „*Teiler*" des rechtsstehenden Produktes sein, und da $\varphi(x,\eta)$ „primitiv" ist, zugleich auch ein Teiler von $\psi(x,\eta)$, auf Grund des verallgemeinerten „Gaußschen Satzes" über primitive Funktionen.*

Somit haben wir nach Division mit $k(\eta)$

$$g(x,\eta) = \varphi(x,\eta)\overline{\psi}(x,\eta)\,, \tag{4)*$$

wo auch $\overline{\psi}(x,\eta)$ ein ganzzahliges Polynom in x und η mit dem Anfangskoeffizienten $\overline{B}_0(\eta)$ ist, und durch Vergleichung der Anfangskoeffizienten

$$1 = C_0(\eta)\overline{B}_0(\eta)$$

identisch in den η, also beide Faktoren rechts unabhängig von η und $= \pm 1$, d. h. α genügt in der Tat der gestellten Bedingung (3)*.

Auf Grund dieses Hilfssatzes lassen sich jetzt für den Bereich $\mathfrak{G}_\eta$ alle vier im Eingang aufgestellten Forderungen als erfüllt nachweisen.

I. Es seien α und β irgend zwei η-ganze Zahlen und $\varphi(x,\eta) = 0$, $\chi(x,\eta) = 0$ die entsprechenden „Hauptgleichungen". Ferner sei $\gamma = q(\alpha,\beta)$ ein beliebiges ganzzahliges Polynom von α, β, z. B. $\gamma = \alpha \pm \beta$ oder $\gamma = \alpha\beta$ und $\gamma', \gamma'', \cdots$ die Ausdrücke, welche entstehen, indem man α sowohl wie β durch ihre sämtlichen konjugierten Größen $\alpha, \alpha', \alpha'', \cdots, \beta, \beta', \beta'', \cdots$ ersetzt (nämlich durch die sämtlichen Wurzeln der irreduziblen Gleichungen $\varphi = 0$ und $\chi = 0$). Dann genügen alle diese Werte γ der Gleichung

$$(x-\gamma)(x-\gamma')(x-\gamma'')\cdots = 0 \tag{5}$$

mit dem Anfangskoeffizienten 1, während alle übrigen Koeffizienten als ganze, ganzzahlige symmetrische Funktionen der Wurzeln von $\varphi(x,\eta) = 0$ und $\chi(x,\eta) = 0$ wieder ganze, ganzzahlige rationale Funktionen ihrer Koeffizienten und daher ganzzahlige Polynome in den η sein müssen. Also genügt auch γ einer Gleichung (2)* und gehört nach unserem Hilfssatze zu $\mathfrak{G}_\eta$.

II. Ist α eine *beliebige* reelle oder komplexe Zahl und ihre „Hauptgleichung" von der Form

$$\varphi(x,\eta) \equiv C_0x^n + C_1x^{n-1} + \cdots + C_n = 0\,, \tag{3}$$

441 | so genügt $\gamma = C_0\alpha$ der Gleichung

$$\gamma^n + C_1\gamma^{n-1} + \cdots + C_nC_0{}^{n-1} = 0\,, \tag{6}$$

* H. Weber, a. a. O., § 20, 5.

In fact, in this we have, as proved in § 2 above, the relation identical in x and η

$$k(\eta)g(x,\eta) = \varphi(x,\eta)\psi(x,\eta)\,, \tag{4}$$

where $\varphi(x,\eta) = 0$ signifies the "principal equation" belonging to α. So $k(\eta)$ must be a *"divisor"* of the product on the right-hand side, and, since $\varphi(x,\eta)$ is "primitive", it must also be a divisor of $\psi(x,\eta)$ on account of the generalized "Gauss's theorem" on primitive functions.[7]

We thus obtain upon division by $k(\eta)$

$$g(x,\eta) = \varphi(x,\eta)\overline{\psi}(x,\eta)\,, \tag{4)*$$

where $\overline{\psi}(x,\eta)$, too, is an integral polynomial in x and η with the leading coefficient $\overline{B}_0(\eta)$, and, by comparing the leading coefficients,

$$1 = C_0(\eta)\overline{B}_0(\eta)$$

identical in the η's, and hence both factors on the right-hand side ⟦are⟧ independent of η and $= \pm 1$, i.e., α really satisfies the stated condition (3)*.

Using this lemma we can now show that all four requirements specified in the beginning are satisfied for the domain $\mathfrak{G}_\eta$.

I. Let α and β be any two η-integral numbers and let $\varphi(x,\eta) = 0$, $\chi(x,\eta) = 0$ be the corresponding "principal equations". Furthermore, let $\gamma = q(\alpha,\beta)$ be an arbitrary integral polynomial of α,β, e.g., $\gamma = \alpha \pm \beta$ or $\gamma = \alpha\beta$, and let $\gamma',\gamma'',\cdots$ be the expressions obtained by replacing α and β by all their conjugated quantities $\alpha,\alpha',\alpha'',\cdots,\beta,\beta',\beta'',\cdots$ (namely with all the roots of the irreducible equations $\varphi = 0$ and $\chi = 0$). All these values γ then satisfy the equation

$$(x-\gamma)(x-\gamma')(x-\gamma'')\cdots = 0 \tag{5}$$

with the leading coefficient 1, whereas all other coefficients, being integral symmetric function with integral coefficients of the roots of $\varphi(x,\eta) = 0$ and $\chi(x,\eta) = 0$, must again be integral rational functions with integral coefficients of their coefficients, and hence integral polynomials in the η's. Therefore γ, too, satisfies an equation (2)* and, according to our lemma, belongs to $\mathfrak{G}_\eta$.

II. If α is an *arbitrary* real or complex number and if its "principal equation" is of the form

$$\varphi(x,\eta) \equiv C_0x^n + C_1x^{n-1} + \cdots + C_n = 0\,, \tag{3}$$

then $\gamma = C_0\,\alpha$ satisfies the equation

$$\gamma^n + C_1\gamma^{n-1} + \cdots + C_nC_0{}^{n-1} = 0\,, \tag{6}$$

[7] *Weber 1912,* § 20, 5.

welche die Form (2)* hat, während C_0 der Gleichung $x - C_0 = 0$ von der gleichen Beschaffenheit genügt. Somit ist $\alpha = \frac{\gamma}{C_0}$ der Quotient zweier η-ganzen Zahlen.

III. IV. Ist α eine *algebraische Zahl* und die entsprechende in $\mathfrak{R}$ irreduzible und primitive Gleichung

$$f(x) \equiv a_0 x^m + a_1 x^{m-1} + \cdots + a_m = 0 \tag{7}$$

mit ganzen rationalen Koeffizienten, so ist $f(x)$ im Falle $a_0 = 1$ von der Form (2)* und α in der Tat „η-ganz“. Um aber auch das Umgekehrte (IV) zu beweisen, zeigen wir, daß $f(x)$ noch im Bereiche der η irreduzibel und primitiv, also (7) mit der „Hauptgleichung“ identisch ist.

Nach dem oben Bewiesenen ist nämlich auch hier wie in (4)

$$k(\eta) f(x) = \varphi(x, \eta) \psi(x, \eta),$$

wo $\varphi(x, \eta) = 0$ die Hauptgleichung von α bedeutet. Es ist also wieder $k(\eta)$ Teiler des rechtsstehenden Produktes und, da $\varphi(x, \eta)$ primitiv, auch Teiler von $\psi(x, \eta) = k(\eta)\overline{\psi}(x, \eta)$, d. h. wir haben wie oben

$$f(x) = \varphi(x, \eta)\overline{\psi}(x, \eta).$$

Hier müssen die Koeffizienten von x rechts und links übereinstimmen und zwar, wegen der Unabhängigkeit der Basiszahlen, identisch in den η, also auch dann, wenn man die η sämtlich durch 0 ersetzt. Somit wird

$$f(x) = \varphi(x, \eta)\overline{\psi}(x, \eta) = \varphi(x, 0)\overline{\psi}(x, 0).$$

Es wäre also $f(x)$ in $\mathfrak{R}$ zerlegbar gegen die Annahme, außer wenn einer der beiden Faktoren rechts konstant und der andere vom m^{ten} Grade ist. Das letztere kann bei $\overline{\psi}(x, 0)$ nicht der Fall sein, weil sonst auch $\overline{\psi}(x, \eta)$ vom m^{ten} Grade in x wäre und dann $\varphi(x, \eta)$ konstant gegen die Definition der Hauptgleichung. Also müssen $\varphi(x, 0)$ und $\varphi(x, \eta)$ vom m^{ten} Grade sein und $\overline{\psi}(x, \eta) = l(\eta)$ konstant. Alle (ganzzahligen) Koeffizienten von $f(x)$ sind durch $l(\eta)$ teilbar, also $l(\eta)$ von den η unabhängig und selbst eine ganze Zahl und zwar, weil $f(x)$ im natürlichen Integritätsbereiche primitiv ist, $= \pm 1$, d. h. $f(x) = \pm\varphi(x, \eta)$. Somit kann der Anfangskoeffizient der Hauptgleichung nur dann ± 1 sein, wenn α eine *ganze* algebraische Zahl ist.

Unser aus der Basis H abgeleiteter Integritätsbereich $\mathfrak{G}_\eta$ besitzt demnach alle im Eingange des Artikels gestellten Eigenschaften. Er enthält

1. alle *Basiszahlen* η,
2. alle *ganzen rationalen* Funktionen der η mit ganzzahligen Koeffizienten, darunter alle *ganzen rationalen* Zahlen,
3. alle *ganzen algebraischen* Funktionen der η mit ganzzahligen Koeffizienten, darunter alle *ganzen algebraischen* Zahlen.

which has the form (2)*, while C_0 satisfies the equation $x - C_0 = 0$ constituted alike. Thus $\alpha = \frac{\gamma}{C_0}$ is the quotient of two η-integral numbers.

III. IV. If α is an *algebraic number* and the corresponding primitive equation irreducible over $\mathfrak{R}$

$$f(x) \equiv a_0 x^m + a_1 x^{m-1} + \cdots + a_m = 0 \tag{7}$$

with integral rational coefficients, then $f(x)$ is of the form (2)* whenever $a_0 = 1$, and α is really "η-integral". But in order to prove the converse (IV) as well, we show that $f(x)$ is still irreducible over and primitive in the domain of the η's, and hence that (7) is identical with the "principal equation".

For here, as in (4), we have, according to what has been proved above,

$$k(\eta) f(x) = \varphi(x, \eta) \psi(x, \eta) ,$$

where $\varphi(x, \eta) = 0$ signifies the principal equation of α. So $k(\eta)$ is again divisor of the product on the right-hand side and, since $\varphi(x, \eta)$ is primitive, also divisor of $\psi(x, \eta) = k(\eta)\, \overline{\psi}(x, \eta)$, i.e., we have, as above,

$$f(x) = \varphi(x, \eta) \overline{\psi}(x, \eta) .$$

In this case the coefficients of x on the right- and left-end sides must correspond to one another, namely, on account of the independence of the basis numbers, identically in the η's, and hence also in the case of all the η's being replaced by 0. Thus we obtain

$$f(x) = \varphi(x, \eta) \overline{\psi}(x, \eta) = \varphi(x, 0) \overline{\psi}(x, 0) .$$

It would therefore be possible to decompose $f(x)$ in $\mathfrak{R}$, contrary to the assumption, except when one of the two factors on the right-hand side is constant and the other is of mth degree. The latter cannot be the case as far as $\overline{\psi}(x, 0)$ is concerned, since otherwise $\overline{\psi}(x, \eta)$, too, would be of mth degree in x, and hence $\varphi(x, \eta)$ would be constant, contrary to the definition of the principal equation. Therefore $\varphi(x, 0)$ and $\varphi(x, \eta)$ must be of mth degree and $\overline{\psi}(x, \eta) = l(\eta)$ must be constant. All (integral) coefficients of $f(x)$ are divisible by $l(\eta)$, and hence $l(\eta)$ is independent of the η's and itself an integer, namely, since $f(x)$ is primitive in the natural integral domain, $= \pm 1$, i.e., $f(x) = \pm \varphi(x, \eta)$. Thus the leading coefficient of the principal equation can be ± 1 only if α is an *integral* algebraic number.

Our integral domain $\mathfrak{S}_\eta$ derived from the basis H hence possesses all properties stated in the beginning of this paper. It contains

1. all *basis numbers* η,
2. all *integral rational* functions of the η's with integral coefficients, among them all *integral rational* numbers,
3. all *integral algebraic* functions of the η's with integral coefficients, among them all *integral algebraic* numbers.

442 | Dagegen schließt er aus

1. alle *rationalen gebrochenen* Funktionen der η mit ganzen rationalen Koeffizienten, darunter die *rational-gebrochenen* Zahlen,
2. alle *algebraisch gebrochenen* Funktionen der η mit ganzen rationalen Koeffizienten, darunter alle *algebraisch gebrochenen* Zahlen.

Jede *beliebige* reelle oder komplexe Zahl ist darstellbar als Quotient zweier η-ganzen Zahlen, ja als Quotient einer η-ganzen Zahl und eines ganzzahligen Polynoms der η.

Ob eine vorgelegte transzendente Zahl zum Integritätsbereiche gehört oder nicht, hängt wesentlich von der zugrunde gelegten Basis H und damit auch von der gewählten Wohlordnung Ω ab. Durch geeignete Wahl der letzteren wird man jede *beliebige transzendente* Zahl ω (z. B. die Zahl e oder π) zu einer „ganzen" machen können, indem man sie z. B. in der Wohlordnung an die Spitze stellt; ebenso auch ein beliebiges *System* transzendenter Zahlen, sofern diese nur unter sich algebraisch unabhängig sind. Desgleichen wird man ω willkürlich auch zu einer „Einheit" machen können, d. h. zu einer Zahl, welche gleichzeitig mit ihrer Reziproken „ganz" ist. Man braucht zu diesem Zwecke nur die (ebenfalls transzendente) Zahl $\varrho = \omega + \omega^{-1}$ in die Basis aufzunehmen, da in diesem Falle beide Größen ω und ω^{-1} derselben Gleichung genügen

$$x^2 - \varrho x + 1\,,$$

welche von der Form $(2)^*$ ist.

Unsere Definition der Integrität ist auch nicht bestimmt, die Zahlen ihrer inneren Natur nach zu charakterisieren, sondern soll vorläufig nur die *Widerspruchslosigkeit* der an einen solchen Integritätsbereich zu stellenden Forderungen erhärten.

Zürich, den 28. Dezember 1913.

Nachtrag (April 1914)

Die auf die „Hauptgleichung" bezüglichen Beweise kann man formal etwas abkürzen, wenn man den (hier im Eingang des § 3 eingeführten) „Gaußschen Satz" schon auf die Formel (4) des § 2 anwendet und durch Division mit dem „Teiler" $k(\eta)$ den Satz gewinnt:

Ist $g(x,\eta)$ ein ganzzahliges Polynom, das für $x = \alpha$ verschwindet, und $\varphi(x,\eta) = 0$ die zugehörige Hauptgleichung, so ist stets $g(x,\eta)$ algebraisch teilbar durch $\varphi(x,\eta)$, d. h. identisch in x und η

$$g(x,\eta) = \varphi(x,\eta)\overline{\psi}(x,\eta)\,, \tag{4a}$$

wo auch $\overline{\psi}$ ein ganzzahliges Polynom ist.

On the other hand, it excludes

1. all *rational fractional* functions of the η's with integral rational coefficients, among them the *rational fractions*,
2. all *algebraic fractional* functions of the η's with integral rational coefficients, among them all *algebraic fractions.*

Any *arbitrary* real or complex number can be represented as the quotient of two η-integral numbers, and even as the quotient of an η-integral number and of an integral polynomial of the η's.

Whether or not a given transcendental number belongs to the integral domain essentially depends on the underlying basis H, and hence also on the chosen well-ordering Ω. By suitable choice of the latter, it is possible to turn any *arbitrary transcendental* number ω (e.g., the number e or π) into an "integral" one, e.g., by putting it at the beginning of the well-ordering; the same holds true for an arbitrary *system* of transcendental numbers, assuming that they are algebraically independent of one another. Likewise it is possible to turn ω into a "unit", i.e., into a number which, together with its reciprocal, is "integral". To this end, we only need to introduce the (likewise transcendental) number $\varrho = \omega + \omega^{-1}$ into the basis since in this case the two quantities ω and ω^{-1} satisfy the same equation

$$x^2 - \varrho x + 1 \, ,$$

which is of the form (2)*.

Our definition of integrity is not designed to characterize the numbers according to their inner nature either. For the time being, it is only supposed to substantiate the *consistency* of the requirements we must impose on such an integral domain.

Zurich, on the 28th of December 1913.

Postscript (April 1914)

The proofs concerning the "principal equation" are capable of formal abridgment. By applying "Gauss's theorem" (introduced here in the beginning of § 3) already to formula (4) of § 2 and by dividing by the "divisor" $k(\eta)$, we obtain the theorem:

If $g(x,\eta)$ is an integral polynomial which vanishes for $x = \alpha$, and if $\varphi(x,\eta) = 0$ is the corresponding principal equation, then $g(x,\eta)$ is always algebraically divisible by $\varphi(x,\eta)$, i.e., identical in x and η

$$g(x,\eta) = \varphi(x,\eta)\overline{\psi}(x,\eta) \, , \tag{4a}$$

where $\overline{\psi}$, too, is an integral polynomial.

Introductory note to Landau 1917b

Heinz-Dieter Ebbinghaus

Throughout his life Richard Dedekind was linked to the Göttingen mathematical and scientific institutions. He finished his dissertation at the Göttingen mathematical institute under the guidance of Carl Friedrich Gauß in 1852 and completed his *Habilitation* there in 1854. In 1858 he left Göttingen for Zurich, and in 1862 he accepted a professorship in Brunswick, the town where he was born, living there until his death. Already in 1859 he became a corresponding member and in 1862 an external member of the Königliche Gesellschaft der Wissenschaften zu Göttingen.

Dedekind died on 12 February 1916. On 15 May 1917 Edmund Landau gave a memorial lecture on Dedekind's achievements under the auspices of the academy. The address was published as *Landau 1917a*. Part of it is devoted to a description of the contents and the influence of Dedekind's two treatises "Stetigkeit und Irrationale Zahlen" (*Dedekind 1872*) and "Was sind und was sollen die Zahlen?" (*Dedekind 1888*). The considerations concerning the latter were conceived by Zermelo and taken over literally by Landau as Section 3 of the published address (*Landau 1917b*). The text exhibits the frequent use of quotation marks characteristic of Zermelo's writings.

Landau's decision to offer to Zermelo the writing of the part about Dedekind's *1888* was quite natural: Zermelo was well acquainted with Dedekind's treatise; moreover, during the winter of 1916/1917 he was staying in Göttingen. In turn, Zermelo surely welcomed Landau's invitation. He appreciated Dedekind's contributions to set theory to a very high degree. When formulating his axiom system of set theory in *1908b,* he attributed the creation of set theory to both Georg Cantor and Dedekind (p. 261), repeating his opinion several times (so *1909a,* 186, and *1909b,* 9) with respect to the creation of the basic set-theoretic notions. Of course, Zermelo was fully aware of the uniqueness of Cantor's work,[1] but naming Cantor and Dedekind side-by-side shows that he was conscious of Dedekind's part as well. In the spring of 1905 he had studied Dedekind's *Zahlen* treatise intensively. This study left its traces in the twin papers *Zermelo 1908a, 1908b*. The new well-ordering proof in *1908a* and the proof of the equivalence theorem in *1908b* use Dedekind's chain theory. Moreover, the axiomatization and terminology in *1908b* owe much to Dedekind's "Erklärungen" and notational conventions from the first sections of *1888*.[2] Zermelo's study of the *Zahlen* treatise undoubtedly contributed to his shifting of the

[1] In the preface to *Cantor 1932* he writes that set theory is "due to the creative act of a single individual", namely Cantor.

[2] The similarity extends to peculiarities like the scepticism against the empty set.

emphasis to an abstract view of sets, away from the more specific concerns of Cantor.[3]

In the present item, Zermelo gives a two-page description of Dedekind's treatise. In accordance with the character of Landau's address he abstains from technicalities and avoids emphasizing the two points which he had explicitly criticized on earlier occasions.

(1) Dedekind's set-theoretic foundation of the natural numbers rests on the existence of an infinite set. In order to provide an example of such a set, Dedekind refers to the infinite world of thought, thereby getting ensnarled in the Zermelo-Russell paradox. In his axiomatization paper Zermelo had explicitly criticized Dedekind's argument (*1908b,* 266): "The 'proof' that Dedekind [...] attempts [...] cannot be satisfactory, since it takes its departure from 'the set of everything thinkable', whereas from our point of view the ⟦universe of sets⟧ does *not* form a set". In the memorial address Zermelo merely describes Dedekind's argument as "more philosophical rather than mathematical" and as "quite irrelevant for the further development". When referring in this context to the Zermelo-Russell paradox, he names Russell, but hides his authorship under "among others" in accordance with his attitude not to claim priorities.[4]

(2) Dedekind's proof of the equivalence of the notions of Dedekind finiteness and of finiteness as being equivalent to an initial section of the natural numbers (Theorem 160 of *1888*) makes tacit use of the axiom of choice. In 1905, when working on Dedekind's treatise, Zermelo had strongly emphasized the role of the axiom. In a letter to David Hilbert of 29 June 1905 he had stated very clearly that "the theory of finite sets is impossible without the 'principle of choice', and the 'well-ordering theorem' is the true fundament of the whole theory of number". Moreover, the insight into the necessity of the axiom of choice had strengthened his intention to argue against its opponents—an intention which was realized in his *1908a.* In the memorial address, he does not mention the axiom of choice at all.

Zermelo concludes by quoting from Dedekind's introduction to *1888*; in the programmatic first part, Dedekind expresses his view that the possession of (mathematical) truth or the belief in it is never conferred through inner intuition, but by individual inferences. Twelve years later, in the first sentence of his first Warsaw talk (*1929b,* W1), Zermelo will share this opinion: "Mathematics, if one seeks to exhaust its entire scope, is *not* to be characterized in terms of its subject matter (such as space and time, forms of inner intuition, a theory of counting and measuring), but only in terms of the method peculiar to it, proof. Mathematics is the system ("die Systematik") of the provable."

[3] Cf. *Ebbinghaus 2007b,* Sect. 2.7.

[4] To give another example, Zermelo never publicly claimed priority for the notion of the von Neumann ordinals (*von Neumann 1923*) which he conceived already as early as 1915.

Abschnitt 3

Landau 1917b

Dedekinds zweite Leistung besteht in der, gleichfalls in Buchform erschiene-
nen, dünnen Schrift: „Was sind und was sollen die Zahlen?“, auf die er eben-
56 falls | später nicht zurückgekommen ist, wenn sie auch gleichfalls eine Rei-
he von unveränderten Auflagen erlebte. Diese grundlegende Schrift stellt —
ich verdanke die folgende Analyse einem der besten Kenner dieses Gebietes,
Herrn *Zermelo* — den ersten durchgeführten Versuch dar, den Begriff und die
Grundeigenschaften der natürlichen Zahlen rein mengentheoretisch aus der
bloßen Idee der Abbildung von Systemen zu entwickeln. Ein „System“ (d. h.
eine Gesamtheit, Menge) heißt nach *Dedekind* „unendlich“, wenn es „ähn-
lich“, d. h. elementweise ein-eindeutig auf einen echten (nicht mit dem Ganzen
identischen) Teil von sich abgebildet werden kann; jedes andere System heißt
„endlich“. Die Existenz unendlicher Systeme, auf der seine Theorie der Zah-
lenreihe beruht, will *Dedekind*, anstatt sie einfach axiomatisch zu postulieren,
auf das Beispiel unserer „Gedankenwelt“, d. h. die Gesamtheit alles Denkba-
ren, begründen. Er will die Existenz unendlicher Systeme beweisen, indem er
jedem Element s seines Denkens den Gedanken zuordnet, daß s Gegenstand
seines Denkens ist. Aber es hat sich doch später (durch *Russell* u. a.) gezeigt,
daß diese Gedankenwelt nicht als System im gleichen Sinne gelten kann. Doch
ist diese mehr philosophische als mathematische Begründung seiner Annahme
für die weiteren Entwickelungen durchaus unerheblich. Bei jeder Abbildung
eines Systems S auf einen (echten oder unechten) Teil S' von S gibt es „Ket-
ten“, d. h. Teilsysteme K, welche gleichfalls in sich selbst abgebildet werden,
und jeder beliebige Teil A von S läßt sich zu einer kleinsten Kette A_0 ergän-
zen, welche als Durchschnitt aller A umfassenden Ketten einfach als „die Ket-
te von A“ bezeichnet wird. Ist nun e ein Element, welches bei der Abbildung
des (unendlichen) Systems S auf den Teil S' nicht selbst als Bild erscheint, al-
so in S' nicht vorkommt, so heißt die Kette N dieses Elementes e ein „einfach
unendliches System“ und hat die Eigenschaft, daß jedes Teilsystem N_0 von N,
welches das „Grundelement“ e und mit jedem seiner Elemente n auch dessen
Bild n' enthält, mit N identisch sein muß. Jedes solche einfach unendliche
System N repräsentiert rein formal, d. h. ohne Rücksicht auf die besondere
Beschaffenheit der Elemente betrachtet, die Reihe der natürlichen Ordnungs-
zahlen, in welcher das Grundelement 1 und als Bild jedes Elementes n das
„nächstfolgende“ $n' = n + 1$ erscheint, wobei die definierende Ketteneigen-
schaft als „Gesetz der vollständigen Induktion“ das bekannte Schlußverfahren
von n auf $n + 1$ gestattet. Von zwei verschiedenen Elementen a, b eines Sy-
57 stems N wird gezeigt, daß entweder a in der Kette | von b' oder b in der Kette
von a' enthalten ist, und je nachdem dies oder jenes gilt, schreiben wir im Falle
der Zahlenreihe einfach $a < b$ oder $a > b$. In dieser Bezeichnung lassen sich die
bekannten Eigenschaften der Größenbeziehung als gültig nachweisen, insbe-

Section 3

Landau 1917b

Dedekind's second achievement consists of the essay "Was sind und was sollen die Zahlen?",[1] which, too, was published in book form and received several unchanged editions even though it, too, was never revisited by him in later years. The following analysis is due to one of the foremost experts in this field, *Mr. Zermelo*: This fundamental paper represents the first actual attempt to develop the concept and basic properties of the natural numbers in purely set-theoretic terms starting only with the idea of the mapping of systems. *Dedekind* calls a "system" (that is, a totality, a set) "infinite" if it can be mapped "similarly", that is, elementwise one-to-one onto a proper part (not identical to the whole) of itself; any other system is called "finite". Instead of simply postulating the existence of infinite systems, upon which his theory of the number series is based, as an axiom, *Dedekind* wants to justify it by using the example of our "world of thought", that is, the totality of all that is conceivable. He wants to prove the existence of infinite systems by assigning to each element s of his thought the thought that s is the object of his thought. It has later been shown, however, (by *Russell*, among others) that this world of thought cannot be considered a system in the same sense. But this more philosophical rather than mathematical justification of his assumption is quite irrelevant for the further developments. For every mapping of a system S onto a (proper or improper) part S' of S there are "chains", that is, partial systems K which are being mapped into themselves as well, and any arbitrary part A of S can be complemented to the smallest chain A_0, which, being the intersection of all chains comprising A, is simply called "the chain of A". Now if e is an element which itself does not occur as image under the mapping of the (infinite) system S onto the part S', and hence does not occur in S', then the chain N of this element e is called a "simply infinite system" and possesses the property that every partial system N_0 of N containing the "basic element" e as well as the image n' of any of its elements n must be identical to N. Considered from a purely formal point of view, that is, without taking into account the particular constitution of the elements, every simply infinite system N of this kind represents the series of the natural ordinal numbers whose basic element is 1, and in which for each of its element n there occurs as the image the "succeeding one" $n' = n+1$, where the defining property of chains licences the well-known inference method from n to $n+1$ as "principle of mathematical induction". For any two distinct elements a, b of a system N it is shown that either a is contained in the chain of b' or b in the chain of a'. In the case of the number series, we simply write $a < b$ or $b > a$, depending on which one of the two holds. Using this terminology,

[1] ⟦*Dedekind 1888.*⟧

sondere das Gesetz der „Transitivität“ (aus $a < b$ und $b < c$ folgt $a < c$), sowie auch die Tatsache, daß jedes beliebige Teilsystem N_1 von Zahlen eine kleinste enthält. Der Beweis eines allgemeinen Existenzsatzes über die „Definition durch Induktion“ gestattet es nun, auch die arithmetischen Grundoperationen durch besonders charakterisierte neue Abbildungen der Zahlenreihe in sich einzuführen. So wird die „Addition“ $m = a + n$ definiert durch die Postulate

$$a + 1 = a' \quad \text{und} \quad a + n' = (a + n)' \quad \text{für jedes } n\,,$$

die „Multiplikation“ $m = a \cdot n$ durch die folgenden

$$a \cdot 1 = A \quad \text{und} \quad a \cdot n' = a \cdot n + a\,,$$

sowie schließlich die „Potenzierung“ $m = a^n$ durch

$$a^1 = a \quad \text{und} \quad a^{n'} = a^n \cdot a\,.$$

Mit Hilfe der vollständigen Induktion können dann aus diesen definierenden Eigenschaften die weiteren arithmetischen Gesetze der natürlichen Zahlen abgeleitet werden.

Der Übergang von den bisher betrachteten „Ordnungszahlen“ zu den „Cardinalzahlen“ erfolgt zum Schlusse durch den Beweis des Satzes, daß jedes „endliche“ (d. h. nicht unendliche) System Σ auf einen Abschnitt Z_n der Zahlenreihe, welcher alle Zahlen $\leq n$ umfaßt, ähnlich abgebildet werden kann, wobei dann das entsprechende n einfach als die dem System Σ zukommende „Anzahl“ bezeichnet wird. Dagegen enthält ein unendliches System Σ von jedem beliebigen Z_n einen ähnlichen Bestandteil Σ_n. Zum Schluß werden noch einige Sätze über endliche Systeme bewiesen, darunter der Satz, daß ein aus endlich vielen endlichen Systemen zusammengesetzes System wieder endlich ist.

In der Einleitung sagt *Dedekind*: „Diese Schrift kann Jeder verstehen, welcher Das besitzt, was man den gesunden Menschenverstand nennt; philosophische oder mathematische Schulkenntnisse sind dazu nicht im geringsten erforderlich. Aber ich weiß sehr wohl, daß gar Mancher in den schattenhaften Gestalten, die ich ihm vorführe, seine Zahlen, die ihn als treue und vertraute Freunde durch das ganze Leben begleitet haben, kaum wiedererkennen mag; er wird durch die lange, der Beschaffenheit unseres Treppen-Verstandes entsprechende Reihe von einfachen Schlüssen, durch die nüchterne Zergliede-
58 rung der Gedankenreihen, auf denen die Ge-| setze der Zahlen beruhen, abgeschreckt und ungeduldig darüber werden, Beweise für Wahrheiten verfolgen zu sollen, die ihm nach seiner vermeintlichen inneren Anschauung von vornherein einleuchtend und gewiß erscheinen. Ich erblicke dagegen gerade in der Möglichkeit, solche Wahrheiten auf andere, einfachere zurückzuführen, mag die Reihe der Schlüsse noch so lang und scheinbar künstlich sein, einen überzeugenden Beweis dafür, daß ihr Besitz oder der Glaube an sie niemals unmittelbar durch innere Anschauung gegeben, sondern immer nur durch eine mehr oder weniger vollständige Wiederholung der einzelnen Schlüsse erworben ist“.

it is possible to demonstrate the validity of the well-known properties of the relation of magnitude, and in particular that of the law of "transitivity" (from $a < b$ and $b < c$ it follows that $a < c$), as well as the fact that any arbitrary partial system N_1 of numbers contains a least number. The proof of a general existential statement concerning the "definition by induction" now also allows the introduction of the basic operations of arithmetic by means of new, especially characterized mappings of the number series into itself. Thus the ⟦operation of⟧ "addition" $m = a + n$ is defined by means of the postulates

$$a + 1 = a' \text{ and } a + n' = (a + n)' \text{ for every } n\,,$$

the ⟦operation of⟧ "multiplication" $m = a \cdot n$ by means of the following ⟦postulates⟧

$$a \cdot 1 = a \text{ and } a \cdot n' = a \cdot n + a\,,$$

and, finally, the ⟦operation of⟧ "power" $m = a^n$ by means of

$$a^1 = a \text{ and } a^{n'} = a^n \cdot a\,.$$

Using mathematical induction we then can derive from these defining properties the other laws of arithmetic that govern the natural numbers.

The transition from the "ordinal numbers" considered so far to the "cardinal numbers" is made at the end by proving the theorem that every "finite" (i.e., not infinite) system Σ can be mapped similarly onto a segment Z_n of the number series comprising all numbers $\leq n$. In this case, the corresponding n is then simply called the "number" belonging to the system Σ. By contrast, an infinite system Σ contains, for any arbitrary Z_n, a similar component Σ_n. Finally, a few theorems about finite systems are proved, among which is the theorem that a system composed of finitely many finite systems is itself finite.

Dedekind says in the introduction: "Anyone in possession of what is called good common sense can understand this paper; philosophical or mathematical textbook knowledge is not at all required for it. But I know all too well that many a reader will barely recognize in the shadowy forms exhibited here his numbers, those faithful and familiar friends who have accompanied him throughout his life; he will recoil at the long series of simple inferences, which corresponds to the constitution of our staircase-like mind, and at the sober dissection of series of thoughts upon which the laws of numbers are based, and he will grow impatient at having to track proofs of truths which he considers evident and certain from the outset by dint of his supposed inner intuition. I, on the other hand, consider the very possibility of reducing such truths to other, simpler ones, length and apparent artificiality of the series of inferences notwithstanding, as convincing proof that the possession of such truths or the belief in them is never conferred immediately through inner intuition, but always acquired by a more or less complete repetition of the individual inferences."

Introductory note to s1921

R. Gregory Taylor*

The question of greatest urgency confronting nineteenth- and early twentieth-century mathematicians was arguably that of the status of the infinite within mathematics. Zermelo's *s1921*, comprising five multi-part philosophical "theses", should be understood in that spirit. Despite its brevity, *s1921* is somewhat repetitive. It seems that Zermelo had no intention of publishing it even as part of some longer piece. Instead, *s1921* likely functioned as a personal manifesto; clearly, Zermelo sees himself as breaking new ground here. If conceived in July 1921, in fact, *s1921* would contain the earliest intimation of the theory of systems of infinitely long propositions described in *1932a*, *1932b*, and *1935*; but serious questions regarding the dating of *s1921* have been raised (see *Ebbinghaus 2007b*, 205). In any case, talk of theses implies supplementary elaboration, and *1932a*, *1932b*, and *1935* will provide it.

Thesis I concerns the general nature of mathematical propositions and appears to say that, whereas proposition $\ulcorner \forall nm \in \mathbb{N}\,(n+m=m+n) \urcorner$ is genuinely mathematical, $\ulcorner 7+5=12 \urcorner$ is not. One might try to block this reading by analyzing the latter as

$$\forall nm \in \mathbb{N}\,[\text{Seven}(n) \wedge \text{Five}(m) \rightarrow \text{Twelve}(n+m)]\,.$$

This has the virtue of exposing a hidden reference to an infinite domain, but it obscures any sense in which plausibly elementary propositions are being collected together. Instead, Zermelo should probably be taken at his word: Kant holds that singular propositions such as $\ulcorner 7+5=12 \urcorner$ cannot serve as mathematical axioms, and Zermelo takes the further step, it seems, of denying them full mathematical status. On this reading, Thesis I entails that what Zermelo calls "elementary propositions" are, quite generally, not truly mathematical. However unusual, this view has no important consequences for the remainder of Zermelo's foundational program.

In *s1929b* we read that "true mathematics is infinitistic and rests on the assumption of infinite domains". Such infinite domains are elsewhere referred to as *substrates*, described in *Farber 1927* as "contexts giving rise to judgment". (Regarding the relevance of *Farber 1927*, see the closing paragraphs of the present note.) This suggests a weaker reading of Thesis I whereby elementary $\ulcorner 7+5=12 \urcorner$ is, in fact, mathematical by dint of presupposing the entire sequence of natural numbers. (In *Farber 1927* such identities are said to be general propositions by virtue of expressing "general connections".) As for propositions that are not elementary, Zermelo's use of the term *Zusammenfassung* brings to mind Cantor's 1895 definition of set and suggests that nonelementary propositions are like sets.

---

* It is a pleasure to thank H.-D. Ebbinghaus and A. Kanamori for comments on earlier drafts and J. Stanton for editorial assistance.

Should Thesis I be interpreted to mean only that all *nontrivial* mathematical propositions are infinitary in character? In *s1929b* finitary mathematics, of which ⌈7 + 5 = 12⌉ would be part, is described as "not mathematics in the true sense of the word". However, finitary mathematics, as described there, extends beyond the trivial to encompass any proposition that is "verifiable by examination of some finite model", including, it seems, any number-theoretic statement whose verification does not require mathematical induction. For example, the seemingly nontrivial proposition that a certain natural number with over a million factors and over sixteen million digits is a Carmichael number (Fermat pseudoprime) would not count as truly mathematical. So our "nontrivial" reading is problematic. Quite generally, Zermelo's criterion has bizarre consequences. Since 561 is a Carmichael number, the proposition that Carmichael numbers exist would not count as true mathematics either, despite its expansion as an infinite disjunction. Nor would its negation, provided only that "verifiable" is taken as elliptical for "verifiable or refutable". On the other hand, that negation, had it turned out to be true, would then doubtless have been mathematical in the fullest sense. Finally, Thesis I might seem to demote entire subdisciplines, such as finite game theory or the investigation of finite geometries. (It is possible to avoid this conclusion, however (see *Taylor 2002*, §8).)

Thesis II is an affirmation of the idealistic tendency within the philosophy of mathematics that is known as "Platonism". Cantor had famously invoked Plato in connection with his 1883 definition of set and in a manner that surely influenced Zermelo's formulation of Thesis II. (See the first endnote of *Cantor 1883b*.) Zermelo's use of the term "grasp" here prefigures the philosophical passage with which *1932a* opens and suggests that the issue is infinite sets or manifolds or, more immediately, some attribute they all share. Zermelo stops short of telling us what that attribute is. Presumably he has in mind something such as "continuing without end" or a technical equivalent such as being Dedekind infinite.

The influence of Hilbert's first foundational program, from the turn of the century, with its emphasis on axiomatization and consistency proofs, is evident in Thesis III. Its first half indicates that Zermelo takes infinite conjunctions as paradigmatic. Nonetheless, his claim here is, strictly speaking, false: an infinite disjunction is a logical consequence of any one of its disjuncts; and, within the system of *Principia*, what are tantamount to infinite disjunctions can be derived from elementary propositions in very few steps by applying primitive proposition ∗9.1. Thesis III's first half is Zermelo's idiosyncratic recasting of a precept whose traditional formulation is provided by Russell (*Russell 1918/19*, 199):

> You can never arrive at a general proposition by inference from particular propositions alone. You will always have to have at least one general proposition in your premises.

Russell's use of "general" is restricted, reasonably, to propositions that are universally quantified, and the intended meaning of "infinitary" in Thesis III is perhaps similarly narrow in scope but less reasonably: Zermelo's own axiom of infinity amounts to an infinite disjunction—one disjunct for each member of some strongly inaccessible initial segment of a cumulative hierarchy of sets and urelements, say. A charitable reading of Thesis III would be: since no infinite conjunction is derivable from any finite proposition, the axioms of any mathematical theory that purports to speak about every element of an infinite domain must include at least one infinite conjunction. (So the first half of Thesis III would say essentially what Russell says in the quoted passage.) As for the second half of Thesis III, a system of elementary propositions "corresponding" to a given axiomatic theory and establishing its consistency would be a set of elementary propositions whose collective truth implies the truth of the axioms. (This would be in essence what model theorists now call the "positive diagram" of a relational structure for a given formal language.)

Thesis IV is the only one among the five that concerns logic in a stricter sense. Aristotelian logic of the syllogism, said to be finitary, is contrasted with an infinitary Platonistic logic based on intuition. Presumably the latter does not yet exist, in Zermelo's estimation, at the point of writing. The reference to the axiom of choice suggests that intuition of a realm of sets and choice functions is of a piece with intuition of infinitely long propositions and systems of them. So-called intuitionists should embrace the assumption of just such "infinitary" intuition, says Zermelo in effect. Of course, that they do not do so is a consequence of their demand for effective constructions justifying the sort of existence claim involved in any application of the axiom of choice—a point Zermelo might seem to miss. Our own view is that the real issue dividing Zermelo from his contemporaries is his ultra-liberal conceptions of constructibility, provability, and definability. Regarding definability, consider Zermelo's remark, in *1908a*, concerning impredicative definitions: they are ultimately acceptable because supplantable, since "every object can be determined in a wide variety of ways".

Theses II and IV are closely related. In the latter, Zermelo writes of intuition. Infinite sets and, more to the point, infinite propositions would be the objects at which said intuition is directed. At a minimum, one assumes that infinite propositions are present to the mind in his view, from which it seems to follow that such propositions exist. Why is it then necessary to "posit" the infinite as an "idea in Plato's sense", as asserted in Thesis II? The answer may be that, although present to the mind, infinite propositions are nonetheless not surveyable in their entirety; they can never be fully grasped by us. Consequently, we can have no experience of their infinite character, which must then be posited idealistically.

The set concept reappears in Thesis V, whose first half reiterates an idea introduced in Thesis I: truly mathematical propositions are collections of infinitely many elementary propositions. Thesis V, unlike Thesis I, allows for distinct ways of collecting propositions. Specifically, we now have infinite

disjunctions. Since every proposition is said to be a collection of elementary propositions, as opposed to intuitively *more* elementary propositions, the operant notion of collection is extremely general. (Compare the manner in which an arbitrary set is a collection of all the urelements including $\emptyset$ in its transitive closure.) Similarly, in the second half of Thesis V, Zermelo asserts that proofs are mere "regroupings" of elementary propositions. Presumably it is elementary propositions figuring in axioms or other hypotheses that are reorganized. The suggestion is that mathematics is of tautological character, foreshadowing *1935*; specifically, theorems are truth-functional consequences of axioms. A tempting reading of Theses I and V is that (nonelementary) propositions are themselves sets. The propositions-as-sets interpretation might also illumine Zermelo's talk of sets and choice functions in "logic" Thesis IV, where intuition of infinitely long propositions seems more pertinent.

Finally, the distinction between finitary and infinitary propositions, in a natural language context, has been the subject of debate. Thus, to adapt an example from philosopher Carl Hempel, *prima facie* finitary "561 is a Carmichael number" is logically equivalent to "any number that is not a Carmichael number is distinct from 561", which amounts to an infinite conjunction. This has seemed to blur the distinction in the opinion of some. By assuming the clarity of the distinction in mathematical contexts, Zermelo in *s1921* presupposes the set-theoretic systematization of the notion of proposition carried forth in *1932a*, *1932b*, and *1935*.

Zermelo's term *Elementarsatz* mirrors the terminology of *Principia*, where a judgment is called *elementary* "when it merely asserts such things as 'a has the relation R to b'" (*Whitehead and Russell 1925*, 44). Such a proposition is not "atomic" in the sense of having no constituents.

A final word is in order concerning the status of *Farber 1927*, cited above, with respect to Zermelo's views concerning logic and mathematics. Zermelo's protégé during the twenties was the young American philosopher Marvin Farber, who, as the recipient of a Harvard Traveling Fellowship, spent several semesters between 1923 and 1927 at the university in Freiburg and other German universities, where he enrolled in courses taught by Heidegger, Husserl, and Heinrich Rickert. (Farber would go on to found *Philosophy and Phenomenological Research* [1940–], the successor to Husserl's journal.) It is not known when and under what circumstances Zermelo came to know Farber, but they were meeting regularly by April 1924.

Marvin Farber left some two hundred handwritten pages of notes, in English and German, of conversations with, and occasional lectures by, Zermelo concerning the philosophy of logic and mathematics (see *Farber 1926/27a*). In addition, *Farber 1926/27b* comprises multiple outlines of an envisioned book concerning logic to be co-authored in English by Zermelo and Farber. Finally, *Farber 1927* is Farber's redaction of notes for the first three chapters of that joint work.

We attribute *Farber 1927* to Farber alone. Doubtless he alone wrote it. Moreover, their correspondence, which ended in 1929, makes clear that Farber never forwarded *Farber 1927* to Zermelo despite his emphatic appeals. On the

Thesen über das Unendliche in der Mathematik

s1921

17. Juli 1921

I) Jeder echte mathematische Satz hat "infinitären" Charakter, d. h. er bezieht sich auf einen *unendlichen* Bereich und ist als eine Zusammenfassung von unendlich vielen "Elementarsätzen" aufzufassen.

II) Das Unendliche ist uns in der Wirklichkeit weder physisch noch psychisch gegeben, es muß als "Idee" im Platonischen Sinne erfaßt und "gesetzt" werden.

III) Da aus finitären Sätzen niemals infinitäre abgeleitet werden können, so müssen auch die "Axiome" jeder mathematischen Theorie infinitär sein, und die "Widerspruchslosigkeit" einer solchen Theorie kann nicht anders "bewiesen" werden als durch Aufweisung eines entsprechenden widerspruchsfreien Systems von unendlich vielen Elementarsätzen.

IV) Die herkömmliche "Aristotelische" Logik ist ihrer Natur nach finitär und daher ungeeignet zur Begründung der mathematischen Wissenschaft. Es ergibt sich daraus die Notwendigkeit einer erweiterten "infinitären" oder "Platonischen" Logik, die auf einer Art infinitärer "Anschauung" beruht — wie z. B. in der Frage des "Auswahlaxioms" — aber paradoxerweise gerade von den "Intuitionisten" aus Gewohnheitsgründen abgelehnt wird.

V) Jeder mathematische Satz ist aufzufassen als eine Zusammenfassung von (unendlich vielen) Elementarsätzen, den "Grundrelationen", durch Konjunktion, Disjunktion und Negation, und jede Ableitung eines Satzes aus anderen Sätzen, insbesondere jeder "Beweis" ist nichts anderes als eine "Umgruppierung" der zu grunde liegenden Elementarsätze.

other hand, attribution to Farber alone is a conservative choice: *Farber 1927* follows the relevant portions of *Farber 1926/27a* very closely. Consequently, it seems reasonable to take *Farber 1927* to reflect the views of Zermelo.

Theses concerning the infinite in mathematics

s1921

17 July 1921

I) Every genuinely mathematical proposition is "infinitary" in character, that is, it is concerned with an *infinite* domain and is to be considered a collection of infinitely many "elementary propositions".

II) The infinite is not given to us physically or psychologically in reality, it must be grasped as "idea" in Plato's sense and "posited".

III) Since infinitary propositions can never be derived from finitary ones, the "axioms" of any mathematical theory, too, must be infinitary, and the "consistency" of such a theory can be "proved" by no other means than the presentation of a corresponding consistent system of infinitely many elementary propositions.

IV) Traditional "Aristotelian" logic is, according to its nature, finitary, and hence not suited for the foundation of mathematical science. Whence the necessity of an extended "infinitary" or "Platonic" logic that rests on some kind of infinitary "intuition"—as, e.g., in connection with the question of the "axiom of choice"—but which, paradoxically, is rejected by the "intuitionists" by force of habit.

V) Every mathematical proposition must be considered a collection of (infinitely many) elementary propositions, the "fundamental relations", by means of conjunction, disjunction and negation, and every deduction of a proposition from other propositions, in particular every "proof", is nothing but a "regrouping" of the underlying elementary propositions.

Introductory note to 1927

Jürgen Elstrodt

1. Historical context

Zermelo's investigations leading to *1927* were first published by Waldemar Alexandrow in his Zurich doctoral thesis *1915*. In his report[1] of 9 July 1915 on this thesis, Zermelo criticizes "the truly discouraging effect of the French original memoirs" on the promulgation of the theory of measure and expresses the opinion that "a simple and clearly arranged development of the theory from the fundamentals satisfying at the same time all requirements of scientific rigour [would be] a rewarding project". He points out that in pursuit of this aim he had "put some new theorems [...] including the ideas of proof [...] at the author's disposal for a first publication", and he states that "accordingly, the problem was more of a reporting and compiling nature than of a creative one". Nevertheless, he acknowledges that Alexandrow's work may "be viewed as a valuable scientific achievement". In the introduction to his thesis, Alexandrow likewise points out that "the line of thought of the entire development and the essential results [...] have been communicated to [...] [[the author]] by Herrn Prof. Zermelo for a systematic elaboration". — It is only in 1927 after a long break that Zermelo himself submits his paper *1927* for publication.

2. Contents

In describing the contents of *1927* we use modern standard terminology for easier reading. Zermelo frequently uses his own terminology and notations which occasionally differ from the standard usage of his time. We will explain these in the subsequent text. Apart from the last two pages of *1927* virtually all sets under consideration are assumed to be subsets of a fixed closed bounded interval $\Lambda \subseteq \mathbb{R}^\nu$.

The starting point of Zermelo's investigation are Guiseppe Peano's notions of "distributive" and "collective" (in Peano's terminology "antidistributive") properties.[2] A property $\mathfrak{E}$ of unions of intervals is called *distributive* whenever $\mathfrak{E}$ holds for a union $I \cup J$ of two intervals if and only if $\mathfrak{E}$ is true for at least one of the sets I or J. On the other hand, $\mathfrak{E}$ is called *collective* whenever $\mathfrak{E}$ holds for a union $I \cup J$ if and only if $\mathfrak{E}$ is true for both I and J. Zermelo first recalls the following theorem of Peano which is traced back to Georg Cantor by Peano himself:[3] Let the distributive property $\mathfrak{E}$ be defined for

[1] Universität Zürich, Promotionen, Fakultät Phil. II, 1915, U 110e 19.

[2] See *Genocchi 1899*, Anhang V, §§ 9–11 and 14, or *Peano 1990*.

[3] See *Genocchi 1899*, 379, or *Peano 1990*.

unions of intervals contained in a closed bounded interval Λ, and assume that $\mathfrak{E}$ holds for Λ. Then there exists a point $P \in \Lambda$ such that $\mathfrak{E}$ is true in any interval neighbourhood of P. If a collective property holds in an interval neighbourhood of any point $P \in \Lambda$, then this property also holds for all of Λ. As an application, Zermelo points out that the Bolzano-Weierstraß theorem and the theorem on the existence of a condensation point of any uncountable closed bounded set follow at once from the "distributive" part of this theorem. Similarly, the "collective" part yields the Heine-Borel covering theorem in the following form: Any covering of a closed bounded interval by open intervals has a finite subcovering. The argument extends to any open covering of an arbitrary closed bounded subset of $\mathbb{R}^\nu$ thus giving the non-trivial implication of the familiar Heine-Borel theorem in a particularly lucid way.

In virtually every account of Lebesgue measure theory from Émile Borel's beginnings to contemporary textbooks, the Heine-Borel theorem plays a truly crucial role in the proof of the countable additivity of Lebesgue measure on the set of intervals in $\mathbb{R}^\nu$. (After that, the measure is usually extended to a (countably additive) measure on the σ-algebra of all Lebesgue measurable sets.) Zermelo (*1927*) suggests a somewhat different approach in which the Heine-Borel theorem likewise plays a crucial role: First, he defines the measure $\mathfrak{m}(A)$ for intervals and countable unions A of intervals in the standard way. But he does not show forthwith that this definition really makes sense, that is, that $\mathfrak{m}(A)$ has the same value for all sequences of disjoint intervals with union A. (Note that Zermelo uses the word "extension" ("Ausdehnung") instead of "measure" or "volume".) Then he applies his version of the Heine-Borel theorem to prove a "subsumption theorem" which simply states the monotonicity of the measure on countable unions of intervals and thus implies that the definition of $\mathfrak{m}(A)$ is independent of the choice of a sequence of disjoint intervals whose union is A.

The *outer measure* $\mathfrak{m}(E)$ of any subset $E \subseteq \Lambda$ is defined in § 2 as the infimum of the measures of all countable unions of intervals covering E. (Observe that Zermelo uses the word "measure" ("Maß") instead of "outer measure".) The "general subsumption theorem" states that the outer measure is monotone on the set of all subsets of Λ. Zermelo then proves the so-called "basic formula" ("Grundformel")

$$\mathfrak{m}(E_1 \cup E_2) + \mathfrak{m}(E_1 \cap E_2) \leq \mathfrak{m}(E_1) + \mathfrak{m}(E_2) \qquad (F)$$

($E_1, E_2 \subseteq \Lambda$), which plays a decisive role in the subsequent developments. Formula (F) is not really new; see e.g. *Hausdorff 1914*, 410, (1). But Zermelo suggests a particularly lucid way to discuss the notion of measurability by means of (F): He first defines the "discrepancy" ("Diskrepanz") of a subset $E \subseteq \Lambda$ by

$$\mathfrak{d}(E) = \mathfrak{m}(E) + \mathfrak{m}(\Lambda \setminus E) - \mathfrak{m}(\Lambda) .$$

In Lebesgue's terminology, $\mathfrak{d}(E)$ is equal to the outer measure of E minus the inner measure of E. Hence it is very suggestive to call a set $E \subseteq \Lambda$ *measurable* if and only if $\mathfrak{d}(E) = 0$. (An altogether similar method was used by Guiseppe Vitali in *1904*.) Observe now that (F) evidently also holds for $\mathfrak{d}$ instead of $\mathfrak{m}$. This opens the way for a perspicuous proof of what Zermelo calls the "principal theorem of Lebesgue's theory": The system of measurable subsets of Λ is a σ-algebra. (In particular, all open and all closed subsets of Λ are measurable.) Zermelo's approach also readily yields that the restriction of $\mathfrak{m}$ to the σ-algebra of measurable subsets of Λ is countably additive.

An important ingredient in the context of *1927* is the theorem that any set $E \subseteq \Lambda$ has an "isometric hull" ("maßgleiche Hülle"), that is, a measurable set $A \supseteq E$ such that $\mathfrak{m}(A) = \mathfrak{m}(E)$. (This result was already known from *de la Vallée Poussin 1909*.) By means of this result Zermelo concludes that if $C \subseteq \Lambda$ is a measurable set, then we have for any set $E \subseteq \Lambda$

$$\mathfrak{m}(E) = \mathfrak{m}(E \cap C) + \mathfrak{m}(E \cap (\Lambda \setminus C)) \qquad (C)$$

(see eq. (10)). This is Constantin Carathéodory's famous condition for measurability (*Carathéodory 1914*) which nowadays is frequently used as a definition since it allows measurability to be discussed irrespective of any assumptions on finiteness of the outer measure. Moreover, Carathéodory's approach immediately extends to outer measures on arbitrary abstract spaces. Surprisingly, Zermelo fails to prove that condition (C) is in fact a characterization of measurability although this is observed in *Alexandrow 1915,* 64. Strangely, Zermelo also makes no reference to *Carathéodory 1914, 1918, 1927* although Carathéodory dedicated his important monograph of 1918 to his "friends Erhard Schmidt and Ernst Zermelo". — Zermelo makes further use of the isometric hull to prove that equality holds in (F) whenever at least one of the sets E_1, E_2 is measurable. However, he fails to remark that equality does not hold unrestrictedly in (F) although this was already noted in *Alexandrow 1915,* 68 (simply choose a non-measurable set E_1 and let $E_2 := \Lambda \setminus E_1$). The two final results of § 2 say: Any nonmeasurable set admits of a decomposition into a measurable set and a so-called "minimal set" ("Minimalmenge") whose discrepancy equals its outer measure, that is, whose inner measure vanishes. An isometric hull of any set $E \subseteq \Lambda$ may be obtained by adjoining an appropriate set of accumulation points of E to E.

In his final § 3 Zermelo coins the notions of "measure-containing" ("maßhaltig") sets having strictly positive outer measure, and "discrepant" ("diskrepant") sets which by definition are nonmeasurable. Localizing these notions he further defines "measure-containing points" and "discrepancy points" of a set. Peano's theorem then implies: Whenever a set E is measure-containing or

discrepant in a closed bounded interval λ, there exists a measure-containing point or a discrepancy point in λ, respectively. The main results are: The sets of measure-containing points of E and discrepancy points, respectively, are perfect point sets. The outer measure of E is equal to the outer measure of the intersection of E with the set of measure-containing points, and an analogous result holds for the notion of discrepancy. The final pages give some indications on how to extend the main results to unbounded sets.

3. Further developments

As mentioned above, Zermelo first communicated his results to his doctoral student Waldemar Alexandrow who elaborated on them in his thesis *1915*. There can be no doubt that under the hard conditions of World War I this thesis had limited circulation. A look into the literature relevant to the subject reveals that only very few works evaluated *Alexandrow 1915* or *Zermelo 1927*. There is an expert half-page review of Alexandrow's thesis by Arthur Rosenthal in the *Jahrbuch über die Fortschritte der Mathematik* 46 (1916–1918), 289 (published belatedly in 1923/24). In his review of *Zermelo 1927* in the same *Jahrbuch* 53 (1927), 175 (published in 1931), Rosenthal merely refers to this evaluation. Rosenthal also gives a detailed report on *Alexandrow 1915* in *Zoretti and Rosenthal 1924*; see pp. 882, 885, 974ff, 988ff. The work gives a comprehensive survey of the various approaches to the theory of measure as developed by Georg Cantor, Camille Jordan, Émile Borel, Henri Lebesgue, Guiseppe Vitali, William Henry Young, Charles-Jean de la Vallée Poussin, Constantin Carathéodory et al. In *Carathéodory 1918* Alexandrow's thesis is quoted only by title (see p. 690), whereas in *Carathéodory 1927,* 702 the author remarks that this thesis is an "original account of Lebesgue's theory influenced by E. Zermelo". The textbook *Schlesinger and Plessner 1926* contains two brief quotations of *Alexandrow 1915* on pp. 37, 56. The thesis' influence appears to have been diminished by the almost simultaneous appearance of a number of weighty publications. We mention a few typical examples: *Radon 1913* strives for more generality and develops the theory of Lebesgue-Stieltjes measures and integrals on $\mathbb{R}^n$. *Carathéodory 1918* is a comprehensive treatise on the theory of Lebesgue measure and integration which in its treatment of measurability is so general as to pave the way for axiomatic measure theory on abstract sets. The theory of Lebesgue is also contained in *Hobson 1907* and *de la Vallée Poussin 1909, 1916*. Finally, *Fréchet 1915, 1923, 1924* extensively expound the theory of measure and integration on abstract spaces. When Zermelo finally published his *1927* it may have been too late to have a lasting effect and his contribution apparently soon fell into oblivion.

Über das Maß und die Diskrepanz von Punktmengen

1927

Die vorliegende Arbeit enthält Untersuchungen über das *Lebesgue*sche Maß, die bereits eine Reihe von Jahren zurückliegen, aber abgesehen von einer unter meiner Leitung hergestellten Dissertation von *Waldemar Alexandrow*[1] und einem kurzen Bericht in der Enzyklopädie d. math. Wissensch. (Bd. II, 3, 7, *Zoretti-Rosenthal*, II C 9 a „Die Punktmengen") bisher noch nicht veröffentlicht worden sind. Ihre Bedeutung liegt teils im formalen Aufbau der Theorie, welcher an Stelle des „äußeren" und des „inneren" Maßes nur das erstere sowie der neu eingeführte Begriff der „Diskrepanz" zu Grunde gelegt werden, sowie in weitergehenden Sätzen über die zu einer beliebigen Punktmenge gehörende „Maßmenge" und „Diskrepanzmenge".

§ 1. Die Grundlagen der Theorie des Maßes

Ein „Intervall" λ im ν-dimensionalen Raume ist eine Punktmenge, deren Koordinaten $x^{(i)}$ beschränkt sind durch Bedingungen der Form

$$l^{(i)} \leqq x^{(i)} \leqq l'^{(i)} \qquad (l^{(i)} \leqq l'^{(i)})$$

und heißt „abgeschlossen", wenn die gesamte Begrenzung $(l^{(i)}, l'^{(i)})$ mit zugerechnet wird, „eigentlich", wenn jedes $l^{(i)} < l'^{(i)}$ ist, d. h. wenn es „innere" Punkte $l^{(i)} < x^{(i)} < l'^{(i)}$ enthält. Jedes Intervall λ, das einen Punkt $\mathfrak{p}$ „eigentlich" enthält, heißt eine „Umgebung" dieses Punktes. Ein „abgeschlossenes" Intervall enthält stets mindestens einen Punkt, während ein „offenes" Intervall bei $l^{(i)} = l'^{(i)}$ auch „leer" sein kann. Eine Folge sukzessive ineinander liegender abgeschlossener Intervalle besitzt als gemeinsamen Bestandteil (Durchschnitt) wieder ein abgeschlossenes Intervall, enthält also mindestens einen gemeinsamen Punkt. Eine (endliche oder unendliche) „Intervallsumme" $\alpha + \beta + \gamma + \cdots$ ist eine Punktmenge σ, welche alle Punkte von α, alle von β usw. und nur solche Punkte enthält. Jede Intervallsumme läßt sich auch als „exklusive", d. h. als Summe paarweise „punktfremder" Intervalle darstellen. Auch der „Durchschnitt" $\alpha\beta$ von zwei, sowie der von endlich vielen Intervallen $\alpha_1 \alpha_2 \ldots \alpha_t$ ist wieder eine endliche Intervallsumme.

Eine Punktmenge E heißt „beschränkt", wenn sie ganz in einem „Gesamt-Intervall" Λ enthalten ist, $E \underline{\in} \Lambda$, und hat in bezug auf dieses Gesamtintervall
155 | eine „Komplementärmenge" $\overline{E} = \Lambda - E$. Das Komplement $\overline{\lambda}$ einer beschränkten endlichen Intervallsumme ist selbst eine endliche Intervallsumme.

---

[1] „Elementare Grundlagen für die Theorie des Maßes", Zürich 1915.

On the measure and the discrepancy of point sets

1927

The present paper contains investigations of Lebesgue measure which already date back a number of years but have not been published yet, except for a dissertation by *Waldemar Alexandrow*[1] supervised by myself and a brief report in the Enzyklopädie d⟦er⟧ math⟦ematischen⟧ Wissensch⟦aften⟧ (*Zoretti and Rosenthal 1924*). Its significance lies in part in the formal construction of the theory, which, instead of being based on "outer" and "inner" measure, is only based on the former as well as on the newly introduced concept of "discrepancy". Its significance also lies in further theorems on the "measure set" and the "discrepancy set" belonging to an arbitrary point set.

§ 1. The foundations of the theory of measure

An "interval" λ in a ν-dimensional space is a point set whose coordinates $x^{(i)}$ are bounded by means of conditions of the form

$$l^{(i)} \leqq x^{(i)} \leqq l'^{(i)} \qquad (l^{(i)} \leqq l'^{(i)}),$$

and it is called "closed" if the entire delimitation $(l^{(i)}, l'^{(i)})$ is being included, "proper" if every $l^{(i)} < l'^{(i)}$, i.e., if it contains "inner" points $l^{(i)} < x^{(i)} < l'^{(i)}$. Every interval λ "properly" containing a point $\mathfrak{p}$ is called a "neighborhood" of this point. A "closed" interval always contains at least one point, whereas an "open" interval can also be "empty" for $l^{(i)} = l'^{(i)}$. A sequence of successively nested closed intervals possesses a closed interval as its common component (intersection), and hence it contains at least one common point. A (finite or infinite) "interval sum" $\alpha + \beta + \gamma + \cdots$ is a point set σ containing all points of α, all points of β etc., and only those points. Every interval sum can also be represented as an "exclusive" sum, i.e., as a sum of mutually "pointwise disjoint" intervals. The "intersection" $\alpha\beta$ of two as well as that of finitely many intervals $\alpha_1\alpha_2 \ldots \alpha_t$ are again finite interval sums.

A point set E is called "bounded" if it is entirely contained in a "total interval" Λ, $E \subseteq \Lambda$, and it has a "complementary set" $\overline{E} = \Lambda - E$ with respect to this total interval. The complement $\overline{\lambda}$ of a bounded finite interval sum is itself a finite interval sum.

[1] *Alexandrow 1915.*

Eine für Intervalle und Intervallsummen definierte Eigenschaft $\mathfrak{E}$ heißt (nach *Peano*) „distributiv", falls sie für die Summe $\alpha + \beta$ zweier Intervalle dann und nur dann besteht, wenn sie für mindestens eines derselben gilt:

$$\mathfrak{E}_{\alpha+\beta} \equiv \mathfrak{E}_\alpha \vee \mathfrak{E}_\beta \qquad (\text{„oder"}) .$$

Eine Eigenschaft heißt dagegen „kollektiv" (bei *Peano* „antidistributiv") falls sie für $\alpha + \beta$ dann und nur dann gilt, wenn sie für beide Teilintervalle gilt:

$$\mathfrak{F}_{\alpha+\beta} \equiv \mathfrak{F}_\alpha \wedge \mathfrak{F}_\beta \qquad (\text{„und"}) .$$

Jede „kollektive" Eigenschaft ist die Negation einer „distributiven" und umgekehrt:

$$\mathfrak{F}_\alpha \equiv \overline{\mathfrak{E}}_\alpha , \quad \mathfrak{E}_\alpha \equiv \overline{\mathfrak{F}}_\alpha .$$

Für distributive und kollektive Eigenschaften besteht der

Satz von Peano. *Gilt eine* distributive *Eigenschaft* $\mathfrak{E}$ *in einem (endlichen) abgeschlossenen Intervall* Λ, *so enthält dieses mindestens einen Punkt* $\mathfrak{q}$ *von der Beschaffenheit, daß* $\mathfrak{E}$ *in* jeder *Umgebung* ω *von* $\mathfrak{q}$ *gilt. Gilt eine* kollektive *Eigenschaft* $\mathfrak{F}$ *in einer Umgebung* jedes *Punktes* $\mathfrak{p}$ *des abgeschlossenen Intervalles* Λ, *so gilt sie auch für das Gesamtintervall.*

Eine Anwendung des Satzes von der „distributiven" Eigenschaft ist die Existenz der „Häufungspunkte" für jede beschränkte unendliche Punktmenge, sowie die Existenz der „Kondensations"- oder „Verdichtungspunkte" für eine *nicht abzählbare* beschränkte Punktmenge. Die wichtigste Anwendung der „kollektiven" Eigenschaft bildet der *Heine-Borel*sche

„Überdeckungssatz". *Ist ein abgeschlossenes Intervall* Λ *„eigentlich überdeckbar" durch eine unendliche Menge* Δ *von Intervallen, so daß* jeder *Punkt* $\mathfrak{p}$ *von* Λ *im Innern mindestens eines dieser Intervalle liegt, so ist das gesamte Intervall* Λ *auch „endlich überdeckbar", d. h. eigentlich überdeckbar durch eine* endliche *Teilmenge* Δ' *von* Δ.

Denn die Eigenschaft eines Teilintervalls α von Λ, durch Δ „endlich überdeckbar" zu sein, ist „kollektiv" in angegebenem Sinne, und sie gilt nach Voraussetzung in der „Umgebung" jedes Punktes $\mathfrak{p}$ von Λ, nämlich in jedem (offenen) Intervall δ der Menge Δ, welches den Punkt im Innern enthält. Der Satz läßt sich dann übertragen auf die Überdeckung einer beliebigen beschränkten und abgeschlossenen Punktmenge A.

Ein „*Gebiet*" G ist eine Punktmenge, welche nur „innere Punkte" enthält, d. h. mit jedem ihrer Punkte $\mathfrak{p}$ zugleich auch eine ganze „Umgebung" ω von $\mathfrak{p}$. Die Komplementärmenge $\overline{G}$ eines beschränkten Gebietes in bezug auf das abgeschlossene „Gesamtintervall" Λ, welches G umschließt, ist eine *abgeschlossene* Punktmenge und umgekehrt.

A property $\mathfrak{E}$ defined for intervals and interval sums is called "distributive" (following *Peano*) in case it obtains for the sum $\alpha + \beta$ of two intervals if and only if it holds for at least one of these:

$$\mathfrak{E}_{\alpha+\beta} \equiv \mathfrak{E}_\alpha \vee \mathfrak{E}_\beta \qquad (\text{"or"}).$$

On the other hand, a property $[\![\mathfrak{F}]\!]$ is called "collective" ("antidistributive" in *Peano*) in case it holds for $\alpha + \beta$ if and only if it holds for both partial intervals:

$$\mathfrak{F}_{\alpha+\beta} \equiv \mathfrak{F}_\alpha \wedge \mathfrak{F}_\beta \qquad (\text{"and"}).$$

Every "collective" property is the negation of a "distributive" one, and vice versa:

$$\mathfrak{F}_\alpha \equiv \overline{\mathfrak{E}}_\alpha\,, \quad \mathfrak{E}_\alpha \equiv \overline{\mathfrak{F}}_\alpha\,.$$

Distributive and collective properties are subject to

Peano's theorem. *If a* distributive *property $\mathfrak{E}$ holds in a (finite) closed interval Λ, then the latter contains at least one point $\mathfrak{q}$ constituted so that $\mathfrak{E}$ holds in* every *neighborhood ω of $\mathfrak{q}$. If a* collective *property $\mathfrak{F}$ holds in a neighborhood of* every *point $\mathfrak{p}$ of the closed interval Λ, then it also holds for the total interval.*

The existence of "accumulation points" for every bounded infinite point set is an application of the theorem on the "distributive" property, as is the existence of "condensation" or "compression points" for *uncountable* bounded point sets. The most important application of the "collective" property is the *Heine-Borel*

"Covering theorem". *If a closed interval Λ can be "properly covered" by means of an infinite set Δ of intervals so that* every *point $\mathfrak{p}$ of Λ lies in the interior of at least one of these intervals, then the entire interval Λ can also be "finitely covered", i.e., it can be properly covered by means of a* finite *partial set Δ' of Δ.*

For the property of being "finitely coverable" by means of Δ which a partial interval α of Λ possesses is "collective" in the specified sense, and it holds by dint of the assumption in the "neighborhood" of every point $\mathfrak{p}$ of Λ, namely in every (open) interval δ of the set Δ which contains the point in its interior. The theorem can then be extended to the covering of an arbitrary bounded and closed point set A.

A "*domain*" G is a point set only containing "inner points", i.e., one which contains, along with each of its points $\mathfrak{p}$, also an entire "neighborhood" ω of $\mathfrak{p}$. The complementary set $\overline{G}$ of a bounded domain with respect to the closed "total interval" Λ including G is a *closed* point set, and vice versa.

Satz. *Jedes „beschränkte" Gebiet ist eine (endliche oder unendliche) Intervallmenge.*

Denn es sei $\varepsilon_1, \varepsilon_2, \ldots$ eine gegen 0 abnehmende Folge positiver Zahlen und G_n die Teilmenge von G, deren Abstand von der *abgeschlossenen* Punktmenge $\overline{G} \geqq \varepsilon_n$ ist. Dann ist auch G_n abgeschlossen, wird also von der Menge Δ
156 der Umgebungen, | welche G eigentlich überdecken, nach dem vorigen Satze auch „endlich überdeckt", also etwa durch die endliche Intervallmenge A_n. Die Summe *aller* dieser Intervallsummen A_n überdeckt dann das *ganze* Gebiet G und ist gleichzeitig in ihm enthalten. Also ist $G = \mathsf{A}_1 + \mathsf{A}_2 + \mathsf{A}_3 + \cdots$ selbst eine (abzählbare) Intervallsumme.

Die „*Ausdehnung*" eines Intervalles λ, das durch die Ungleichheiten

$$l^{(i)} \leqq x^{(i)} \leqq l'^{(i)}$$

gegeben ist, wird definiert als Produkt sämtlicher Differenzen $l'^{(i)} - l^{(i)}$ und ist positiv und von 0 verschieden für jedes eigentliche Intervall. Die Ausdehnung einer endlichen Intervallsumme $\mathsf{A} = \alpha_1 + \alpha_2 + \cdots + \alpha_t$ ist gleich der Summe der Ausdehnungen ihrer „exklusiven" Bestandteile:

$$\mathsf{A} = \alpha_1 + \overline{\alpha}_1\alpha_2 + \overline{\alpha}_1\overline{\alpha}_2\alpha_3 + \cdots + \overline{\alpha}_1 \ldots \overline{\alpha}_{n-1}\alpha_n\,,$$
$$\mathfrak{m}\mathsf{A} = \mathfrak{m}\alpha_1 + \mathfrak{m}(\overline{\alpha}_1\alpha_2) + \cdots + \mathfrak{m}(\overline{\alpha}_1 \ldots \overline{\alpha}_{n-1}\alpha_n)\,,$$

und ist unabhängig von der Wahl der benutzten Zerlegungen. Die Ausdehnung einer unendlichen *abzählbaren* Intervallsumme

$$\mathsf{A} = \alpha_1 + \alpha_2 + \alpha_3 + \cdots$$

ist gleich dem Grenzwerte $\lim\limits_{n=\infty} \mathfrak{m}\mathsf{A}_n$, dem die Ausdehnungen $\mathfrak{m}\mathsf{A}_n$ der endlichen Teilsummen

$$\mathsf{A}_n = \alpha_1 + \alpha_2 + \cdots + \alpha_n$$

mit wachsendem Index zustreben. Sie ist endlich und $\leqq \mathfrak{m}\Lambda$, für jede „beschränkte" Intervallsumme $\mathsf{A}_n \Subset \Lambda$ und unabhängig von der Reihenfolge der Summanden. Für alle (endlichen und abzählbaren) Intervallsummen gilt nun der

Subsumptionssatz. *Die Ausdehnung jeder Intervallsumme* A *ist nicht größer als die jeder anderen sie ganz überdeckenden Intervallsumme* B *und ist daher unabhängig auch von der Wahl der sie erzeugenden Intervallmenge.*

Der Satz ist trivial für endliche Intervallmengen. Für jede abzählbare Intervallsumme A und für jedes $\varepsilon > 0$ gibt es aber eine *abgeschlossene endliche* Intervallsumme $\mathsf{A}' \Subset \mathsf{A}$, für welche

$$0 \leqq \mathfrak{m}\mathsf{A} - \mathfrak{m}\mathsf{A}' < \varepsilon\,,$$

Theorem. *Every "bounded" domain is a (finite or infinite) interval set.*

For let $\varepsilon_1, \varepsilon_2, \ldots$ be a sequence of positive numbers decreasing towards 0 and let G_n be the partial set of G whose distance from the *closed* point set $\overline{G}$ is $\geqq \varepsilon_n$. Then G_n, too, is closed, and hence, according to the theorem above, it, too, is "finitely covered" by the set Δ of the neighborhoods properly covering G, thus, e.g., by the finite interval set A_n. The sum of *all* these interval sums A_n then covers the *entire* domain G and is also contained in it. Therefore, $G = \mathsf{A}_1 + \mathsf{A}_2 + \mathsf{A}_3 + \cdots$ is itself a (countable) interval sum.

The "*extension*" of an interval λ given by the inequalities

$$l^{(i)} \leqq x^{(i)} \leqq l'^{(i)}$$

is defined as the product of all the differences $l'^{(i)} - l^{(i)}$, and it is positive and different from 0 for every proper interval. The extension of a finite interval sum $\mathsf{A} = \alpha_1 + \alpha_2 + \cdots + \alpha_t$ is equal to the sum of the extensions of its "exclusive" components:

$$\begin{aligned} \mathsf{A} &= \alpha_1 + \overline{\alpha}_1\alpha_2 + \overline{\alpha}_1\overline{\alpha}_2\alpha_3 + \cdots + \overline{\alpha}_1 \ldots \overline{\alpha}_{n-1}\alpha_n\,, \\ \mathfrak{m}\mathsf{A} &= \mathfrak{m}\alpha_1 + \mathfrak{m}(\overline{\alpha}_1\alpha_2) + \cdots + \mathfrak{m}(\overline{\alpha}_1 \ldots \overline{\alpha}_{n-1}\alpha_n)\,, \end{aligned}$$

and it is independent of the choice of the decompositions employed. The extension of an infinite *countable* interval sum

$$\mathsf{A} = \alpha_1 + \alpha_2 + \alpha_3 + \cdots$$

is equal to the limit $\lim\limits_{n=\infty} \mathfrak{m}\mathsf{A}_n$, to which the extensions $\mathfrak{m}\mathsf{A}_n$ of the finite partial sums

$$\mathsf{A}_n = \alpha_1 + \alpha_2 + \cdots + \alpha_n$$

converge as the index increases. It is finite and $\leqq \mathfrak{m}\,\Lambda$, for every "bounded" interval sum $\mathsf{A}_n \subseteq \Lambda$, and independent of the order of the summands. All (finite and countable) interval sums are now subject to the

Subsumption theorem. *The extension of any interval sum* A *is not greater than that of any other interval sum* B *covering it completely, and hence it is also independent of the choice of the interval set generating it.*

The theorem is trivial for finite interval sets. But for every countable interval sum A and for every $\varepsilon > 0$ there exists a *closed finite* interval sum $\mathsf{A}' \subseteq \mathsf{A}$ for which

$$0 \leqq \mathfrak{m}\mathsf{A} - \mathfrak{m}\mathsf{A}' < \varepsilon\,,$$

und die man erhält, indem man zunächst A durch eine erzeugende endliche Teilsumme A_t approximiert und dann jedes ihrer erzeugenden eigentlichen Intervalle durch ein kleineres abgeschlossenes ersetzt. Diese abgeschlossene Punktmenge A' wird nun zugleich mit A überdeckt durch die abzählbare Intervallsumme B und *eigentlich* überdeckt durch die Summe B' der entsprechend *vergrößerten* Intervalle β', wobei die Vergrößerungen doch so klein gehalten werden können, daß $\mathfrak{m}\mathsf{B}' < \mathfrak{m}\mathsf{B} + \varepsilon$. Nach dem „Überdeckungssatze" (S. 155) ist dann A' auch *endlich* überdeckbar durch B', wird somit bereits überdeckt durch eine *endliche* Teilsumme B'_n von B', und es gilt für die endlichen Intervallsummen A' und B'_n, $\mathfrak{m}\mathsf{A}' \leqq \mathfrak{m}\mathsf{B}'_n$, also

$$\mathfrak{m}\mathsf{A} - \varepsilon < \mathfrak{m}\mathsf{A}' \leqq \mathfrak{m}\mathsf{B}'_n \leqq \mathfrak{m}\mathsf{B}' < \mathfrak{m}\mathsf{B} + \varepsilon\,,$$

d. h. $\mathfrak{m}\mathsf{A} < \mathfrak{m}\mathsf{B} + 2\varepsilon$ für beliebiges $\varepsilon > 0$, d. h. $\mathfrak{m}\mathsf{A} \leqq \mathfrak{m}\mathsf{B}$. Stellen nun A und B durch verschiedene erzeugende Intervallmengen Δ und Δ' dieselbe
157 Punktmenge | dar, so ist gleichzeitig $\mathsf{A} \Subset \mathsf{B}$ und $\mathsf{B} \Subset \mathsf{A}$, also $\mathfrak{m}\mathsf{A} = \mathfrak{m}\mathsf{B}$. Außerdem gelten für beliebige endliche und abzählbare Intervallsummen die Grundformeln

$$\mathfrak{m}(\mathsf{A}+\mathsf{B}) + \mathfrak{m}\mathsf{A}\mathsf{B} = \mathfrak{m}\mathsf{A} + \mathfrak{m}\mathsf{B}\,, \tag{1}$$

wo $\mathsf{A}+\mathsf{B}$ die Vereinigung der Intervallsummen A und B und AB ihren Durchschnitt (der sich ebenfalls als Intervallsumme darstellen läßt) bedeuten. Für exklusive Intervallsummen ($\mathsf{AB} = 0$) wird also

$$\mathfrak{m}(\mathsf{A}+\mathsf{B}) = \mathfrak{m}\mathsf{A} + \mathfrak{m}\mathsf{B}\,. \tag{1a}$$

Allgemein aber gilt für jede endliche oder abzählbare Summe von Intervallsummen

$$\mathfrak{m}(\mathsf{A}_1 + \mathsf{A}_2 + \mathsf{A}_3 + \cdots) \leqq \mathfrak{m}\mathsf{A}_1 + \mathfrak{m}\mathsf{A}_2 + \mathfrak{m}\mathsf{A}_3 + \cdots \tag{2}$$

und für exklusive Intervallsummen $\mathsf{A}_1, \mathsf{A}_2, \ldots$

$$\mathfrak{m}(\mathsf{A}_1 + \mathsf{A}_2 + \cdots) = \mathfrak{m}\mathsf{A}_1 + \mathfrak{m}\mathsf{A}_2 + \cdots\,. \tag{2a}$$

§ 2. Maß und Diskrepanz beschränkter Punktmengen

Definition 1. *Das „Maß" $\mathfrak{m}E$ einer auf das Gesamtintervall Λ beschränkten Punktmenge E ist die untere Grenze für die Ausdehnung $\mathfrak{m}\mathsf{A}$ jeder innerhalb Λ die Menge E umschließenden (abzählbaren) Intervallsumme A.*

Nach dieser Definition und dem „Subsumptionssatze" (§ 1) ist also stets

$$0 \leqq \mathfrak{m}E = \underset{E \Subset A}{\mathfrak{U}}\ \mathfrak{m}\mathsf{A} \leqq \mathfrak{m}\mathsf{A} \leqq \mathfrak{m}\Lambda = I\,.$$

and which is obtained by first approximating A by means of a generating finite partial sum A_t, and by then replacing each of its generating proper intervals by a smaller closed one. This closed point set A' is now covered, along with A, by the countable interval sum B and it is *properly* covered by the sum B' of the correspondingly *enlarged* intervals β', where these increases can still be kept small enough so that $\mathfrak{m}\mathsf{B}' < \mathfrak{m}\mathsf{B} + \varepsilon$. Then, according to the "covering theorem" (p. 155), A' is also *finitely* coverable by B', and hence it is already covered by a *finite* partial sum B'_n of B'. For the finite interval sums A' and B'_n we have $\mathfrak{m}\mathsf{A}' \leqq \mathfrak{m}\mathsf{B}'_n$, and hence

$$\mathfrak{m}\mathsf{A} - \varepsilon < \mathfrak{m}\mathsf{A}' \leqq \mathfrak{m}\mathsf{B}'_n \leqq \mathfrak{m}\mathsf{B}' < \mathfrak{m}\mathsf{B} + \varepsilon\,,$$

i.e., $\mathfrak{m}\mathsf{A} < \mathfrak{m}\mathsf{B} + 2\varepsilon$ for arbitrary $\varepsilon > 0$, i.e., $\mathfrak{m}\mathsf{A} \leqq \mathfrak{m}\mathsf{B}$. If now A and B represent the same point set by means of different generating interval sets Δ and Δ', then both $\mathsf{A} \subseteq \mathsf{B}$ and $\mathsf{B} \subseteq \mathsf{A}$, therefore $\mathfrak{m}\mathsf{A} = \mathfrak{m}\mathsf{B}$. Furthermore arbitrary finite and countable interval sums are subject to the basic formulas

$$\mathfrak{m}(\mathsf{A} + \mathsf{B}) + \mathfrak{m}\mathsf{A}\mathsf{B} = \mathfrak{m}\mathsf{A} + \mathfrak{m}\mathsf{B}\,, \tag{1}$$

where $\mathsf{A} + \mathsf{B}$ signifies the union of the interval sums A and B, and $\mathsf{A}\,\mathsf{B}$ their intersection (which can also be represented as an interval sum). For exclusive interval sums ($\mathsf{A}\,\mathsf{B} = 0$) we therefore have

$$\mathfrak{m}(\mathsf{A} + \mathsf{B}) = \mathfrak{m}\mathsf{A} + \mathfrak{m}\mathsf{B}\,. \tag{1a}$$

Generally, however, we have for every finite or countable sum of interval sums

$$\mathfrak{m}(\mathsf{A}_1 + \mathsf{A}_2 + \mathsf{A}_3 + \cdots) \leqq \mathfrak{m}\mathsf{A}_1 + \mathfrak{m}\mathsf{A}_2 + \mathfrak{m}\mathsf{A}_3 + \cdots \tag{2}$$

and for exclusive interval sums $\mathsf{A}_1, \mathsf{A}_2, \ldots$

$$\mathfrak{m}(\mathsf{A}_1 + \mathsf{A}_2 + \cdots) = \mathfrak{m}\mathsf{A}_1 + \mathfrak{m}\mathsf{A}_2 + \cdots\,. \tag{2a}$$

§ 2. Measure and discrepancy of bounded point sets

Definition 1. *The "measure" $\mathfrak{m}E$ of a point set E bounded with respect to the total interval Λ is the infimum for the extension $\mathfrak{m}\mathsf{A}$ of every (countable) interval sum A including the set E within Λ.*

According to this definition and the "subsumption theorem" (§ 1) we therefore always have

$$0 \leqq \mathfrak{m}E = \underset{E \subseteq A}{\mathfrak{U}}\ \mathfrak{m}\mathsf{A} \leqq \mathfrak{m}\mathsf{A} \leqq \mathfrak{m}\Lambda = I\,.$$

Für jede Intervallsumme A fällt also das „Maß" zusammen mit der „Ausdehnung". Ist $E_1 \Subset E$ eine beliebige Teilmenge von E und A eine beliebige E überdeckende Intervallsumme, so ist auch $E_1 \Subset \mathsf{A}$ also

$$\mathfrak{m}E_1 \leqq \mathfrak{m}\mathsf{A} \qquad \text{und somit auch} \qquad \mathfrak{m}E_1 \leqq \mathfrak{U}\mathfrak{m}\mathsf{A} = \mathfrak{m}E\,.$$

Es gilt also auch für beliebige beschränkte Punktmengen E der

Allgemeine Subsumptionssatz. *Das Maß einer Punktmenge E ist nicht kleiner als das Maß jeder ihrer Teilmengen.*

Da allgemein für den Durchschnitt und die Summe von Punktmengen gilt

$$E_1E_2 \Subset E_1 \Subset E_1 + E_2 \Subset I\,,$$

so ist auch stets

$$0 \leqq \mathfrak{m}E_1E_2 \leqq \mathfrak{m}E_1 \leqq \mathfrak{m}(E_1 + E_2) \leqq I\,.$$

Sind ferner A_1 und A_2 beliebige E_1 bzw. E_2 überdeckende Intervallsummen, so ist wegen

$$E_1 + E_2 \Subset \mathsf{A}_1 + \mathsf{A}_2\,, \qquad E_1E_2 \Subset \mathsf{A}_1\mathsf{A}_2$$

wieder

$$\mathfrak{m}(E_1 + E_2) + \mathfrak{m}E_1E_2 \leqq \mathfrak{m}(\mathsf{A}_1 + \mathsf{A}_2) + \mathfrak{m}\mathsf{A}_1\mathsf{A}_2 = \mathfrak{m}\mathsf{A}_1 + \mathfrak{m}\mathsf{A}_2\,,$$

und da die Summanden rechts sich den Grenzwerten $\mathfrak{m}E_1$ und $\mathfrak{m}E_2$ beliebig nähern können, so gilt die Formel („Grundformel")

$$\mathfrak{m}(E_1 + E_2) + \mathfrak{m}E_1E_2 \leqq \mathfrak{m}E_1 + \mathfrak{m}E_2\,. \tag{3}$$

Definition 2. *Die „Diskrepanz"*[1] *$\mathfrak{d}E$ einer beschränkten Punktmenge E ist die untere Grenze für die Ausdehnung des Durchschnittes AB einer beliebigen die*
158 *| Menge E und einer ihre Komplementärmenge $\overline{E}$ überdeckenden Intervallsumme*

$$0 \leqq \mathfrak{d}E = \underset{E \Subset \mathsf{A},\, \overline{E} \Subset \mathsf{B}}{\mathfrak{U}} \mathsf{m}(\mathsf{AB}) \leqq I = \mathfrak{m}\mathsf{\Lambda}\,.$$

Da mit E auch A und B auf das Gesamtintervall $\mathsf{\Lambda}$ beschränkt werden sollen und zusammen $E + \overline{E} = \mathsf{\Lambda}$ überdecken, so muß $\mathfrak{m}(\mathsf{A} + \mathsf{B}) = \mathfrak{m}\mathsf{\Lambda} = I$ sein und es ist nach (1)

$$\mathfrak{m}(\mathsf{A} + \mathsf{B}) + \mathfrak{m}\mathsf{AB} = \mathfrak{m}\mathsf{A} + \mathfrak{m}\mathsf{B}\,,$$

[1] Bei *Lebesgue* die Differenz des „äußeren" und „inneren Maßes".

Hence, for every interval sum A, the "measure" coincides with the "extension". If $E_1 \subseteq E$ is an arbitrary partial set of E and A an arbitrary interval sum covering E, then we also have $E_1 \subseteq \mathsf{A}$, therefore

$$\mathfrak{m}E_1 \leqq \mathfrak{m}\mathsf{A}\,, \quad \text{and hence also} \quad \mathfrak{m}E_1 \leqq \mathfrak{U}\,\mathfrak{m}\mathsf{A} = \mathfrak{m}E\,.$$

Arbitrary bounded point sets E, too, are therefore subject to the

General subsumption theorem. *The measure of a point set E is not smaller than the measure of each of its partial sets.*

Since for the intersection and the sum of point sets we generally have

$$E_1\,E_2 \subseteq E_1 \subseteq E_1 + E_2 \subseteq \varLambda\,,$$ [2]

we also always have

$$0 \leqq \mathfrak{m}E_1E_2 \leqq \mathfrak{m}E_1 \leqq \mathfrak{m}(E_1 + E_2) \leqq I\,.$$

If, furthermore, A_1 and A_2 are arbitrary interval sums covering E_1 and E_2, we have, on account of

$$E_1 + E_2 \subseteq \mathsf{A}_1 + \mathsf{A}_2\,, \quad E_1E_2 \subseteq \mathsf{A}_1\mathsf{A}_2\,,$$

again

$$\mathfrak{m}(E_1 + E_2) + \mathfrak{m}E_1E_2 \leqq \mathfrak{m}(\mathsf{A}_1 + \mathsf{A}_2) + \mathfrak{m}\mathsf{A}_1\mathsf{A}_2 = \mathfrak{m}\mathsf{A}_1 + \mathfrak{m}\mathsf{A}_2\,,$$

and since the summands on the right side may arbitrarily approach the limits $\mathfrak{m}E_1$ and $\mathfrak{m}E_2$, we have the formula ("basic formula")

$$\mathfrak{m}(E_1 + E_2) + \mathfrak{m}E_1E_2 \leqq \mathfrak{m}E_1 + \mathfrak{m}E_2\,. \tag{3}$$

Definition 2. *The "discrepancy"*[3] $\mathfrak{d}E$ *of a bounded point set E is the infimum for the extension of the intersection* AB *of an arbitrary interval sum covering the set E and of one covering its complementary set $\overline{E}$*

$$0 \leqq \mathfrak{d}E = \mathop{\mathfrak{U}}_{E\subseteq\mathsf{A},\,\overline{E}\subseteq\mathsf{B}} \mathfrak{m}(\mathsf{AB}) \leqq I = \mathfrak{m}\mathsf{\Lambda}\,.$$ [4]

Since, along with E, A and B are also supposed to be bounded with respect to the total interval $\mathsf{\Lambda}$, and since they together cover $E + \overline{E} = \mathsf{\Lambda}$, it must be the case that $\mathfrak{m}(\mathsf{A} + \mathsf{B}) = \mathfrak{m}\mathsf{\Lambda} = I$, and, according to (1), we have

$$\mathfrak{m}(\mathsf{A} + \mathsf{B}) + \mathfrak{m}\mathsf{AB} = \mathfrak{m}\mathsf{A} + \mathfrak{m}\mathsf{B}\,,$$

[2] ⟦Zermelo erroneously writes "I" instead of "$\varLambda$".⟧

[3] In *Lebesgue*, the difference of the "outer" and the "inner measure".

[4] ⟦"$\mathfrak{U}$" stands for "infimum".⟧

also

$$\mathfrak{m}\mathsf{AB} = \mathfrak{m}\mathsf{A} + \mathfrak{m}\mathsf{B} - I\,,$$

und durch Approximation der Grenzwerte

$$\underset{E\Subset\mathsf{A},\,\overline{E}\Subset\mathsf{B}}{\mathfrak{U}} \mathfrak{m}\mathsf{AB} = \underset{E\Subset\mathsf{A}}{\mathfrak{U}} \mathfrak{m}\mathsf{A} + \underset{\overline{E}\Subset\mathsf{B}}{\mathfrak{U}} \mathfrak{m}\mathsf{B} - I\,,$$

d. h.

$$\mathfrak{d}E = \mathfrak{m}E + \mathfrak{m}\overline{E} - I = \mathfrak{d}\overline{E}\,. \tag{4}$$

Eine beschränkte Menge E heißt „*meßbar*“ (*Lebesgue*), wenn ihre Diskrepanz verschwindet $\mathfrak{d}E = 0$, (wenn das „äußere“ gleich dem „inneren Maß“ ist). Sie heißt „*Minimalmenge*“, wenn ihre Diskrepanz gleich ihrem Maß ist: $\mathfrak{d}E = \mathfrak{m}E$, $\mathfrak{m}\overline{E} = I$ (wenn ihr „inneres Maß“ null ist). *Jeder Teil einer Minimalmenge ist wieder eine Minimalmenge.* Denn aus $E_1 \Subset E$ folgt $\overline{E} \Subset \overline{E}_1$, also $\mathfrak{m}\overline{E}_1 \geqq \mathfrak{m}\overline{E} = I$ und $\mathfrak{d}\overline{E}_1 = \mathfrak{d}E_1 = \mathfrak{m}\overline{E}_1 + \mathfrak{m}E_1 - I = \mathfrak{m}E_1$. Wendet man die „Hauptformel“ (3) auf $\overline{E}_1$ und $\overline{E}_2$ an und beachtet, daß $\overline{E_1E_2} = \overline{E}_1 + \overline{E}_2$ und $\overline{E_1 + E_2} = \overline{E}_1\overline{E}_2$ ist, so ergibt sich aus

$$\begin{array}{ll} \mathfrak{m}(E_1 + E_2) + \mathfrak{m}E_1E_2 \leqq \mathfrak{m}E_1 + \mathfrak{m}E_2 & \text{und} \\ \mathfrak{m}(\overline{E}_1 + \overline{E}_2) + \mathfrak{m}(\overline{E}_1\overline{E}_2) \leqq \mathfrak{m}\overline{E}_1 + \mathfrak{m}\overline{E}_2 & \text{durch Addition} \\ \hline \mathfrak{d}(E_1 + E_2) + \mathfrak{d}(E_1E_2) \leqq \mathfrak{d}E_1 + \mathfrak{d}E_2\,, & \end{array} \tag{5}$$

wo immer die *Un*gleichheit gelten muß, wenn sie auch nur in einer der vorstehenden Formeln gelten sollte.

Da $\mathfrak{m}E$ und $\mathfrak{d}E$ nie negativ sein können, so gilt a fortiori

$$\mathfrak{m}(E_1 + E_2) \leqq \mathfrak{m}E_1 + \mathfrak{m}E_2\,, \tag{3a}$$

$$\mathfrak{d}(E_1 + E_2) \leqq \mathfrak{d}E_1 + \mathfrak{d}E_2\,, \tag{5a}$$

und durch vollständige Induktion folgt für jede endliche Summe

$$\mathfrak{m}(E_1 + E_2 + \cdots + E_n) \leqq \mathfrak{m}E_1 + \mathfrak{m}E_2 + \cdots + \mathfrak{m}E_n\,, \tag{6}$$

$$\mathfrak{d}(E_1 + E_2 + \cdots + E_n) \leqq \mathfrak{d}E_1 + \mathfrak{d}E_2 + \cdots + \mathfrak{d}E_n\,. \tag{7}$$

Ferner ist $\mathfrak{d}(E_1E_2) = \mathfrak{d}(\overline{E_1E_2}) = \mathfrak{d}(\overline{E}_1 + \overline{E}_2) \leqq \mathfrak{d}\overline{E}_1 + \mathfrak{d}\overline{E}_2 = \mathfrak{d}E_1 + \mathfrak{d}E_2$ und allgemein

$$\mathfrak{d}(E_1E_2 \cdots E_n) \leqq \mathfrak{d}E_1 + \mathfrak{d}E_2 + \cdots + \mathfrak{d}E_n\,. \tag{8}$$

Die Ausdehnung der Formeln (6), (7) und (8) auf (abzählbar) unendliche Summen bedarf noch eines Beweises, da der Grenzübergang nicht ohne weiteres möglich ist.

Es seien die Punktmengen $E_1, E_2, E_3, \ldots$ sämtlich im Gesamtintervall Λ enthalten und der Reihe nach überdeckt von den (gleichfalls in Λ enthaltenen) Intervallsummen $\mathsf{A}_1, \mathsf{A}_2, \mathsf{A}_3, \ldots$. Dann wird auch

and hence

$$\mathfrak{m}\mathsf{AB} = \mathfrak{m}\mathsf{A} + \mathfrak{m}\mathsf{B} - I\,,$$

and by approximation of the limits

$$\underset{E\subseteq\mathsf{A}\,,\,\overline{E}\subseteq\mathsf{B}}{\mathfrak{U}} \mathfrak{m}\mathsf{AB} = \underset{E\subseteq\mathsf{A}}{\mathfrak{U}} \mathfrak{m}\mathsf{A} + \underset{\overline{E}\subseteq\mathsf{B}}{\mathfrak{U}} \mathfrak{m}\mathsf{B} - I\,,$$

i.e.,

$$\mathfrak{d}E = \mathfrak{m}E + \mathfrak{m}\overline{E} - I = \mathfrak{d}\overline{E}\,. \tag{4}$$

A bounded set E is called "*measurable*" (*Lebesgue*) if its discrepancy vanishes $\mathfrak{d}E = 0$, (if its "outer" is equal to its "inner measure"). It is called a "*minimal set*" if its discrepancy is equal to its measure: $\mathfrak{d}E = \mathfrak{m}E$, $\mathfrak{m}\overline{E} = I$ (if its "inner measure" is zero). *Every part of a minimal set is again a minimal set.* For it follows from $E_1 \subseteq E$ that $\overline{E} \subseteq \overline{E}_1$, and hence $\mathfrak{m}\overline{E}_1 \geqq \mathfrak{m}\overline{E} = I$ and $\mathfrak{d}\overline{E}_1 = \mathfrak{d}E_1 = \mathfrak{m}\overline{E}_1 + \mathfrak{m}E_1 - I = \mathfrak{m}E_1$. If we apply the "basic formula" (3) to $\overline{E}_1$ and $\overline{E}_2$, then, considering that $\overline{E_1E_2} = \overline{E}_1 + \overline{E}_2$ and $\overline{E_1 + E_2} = \overline{E}_1\overline{E}_2$, we obtain from

$$\begin{array}{ll} \mathfrak{m}(E_1 + E_2) + \mathfrak{m}E_1E_2 \leqq \mathfrak{m}E_1 + \mathfrak{m}E_2 & \text{and} \\ \mathfrak{m}(\overline{E}_1 + \overline{E}_2) + \mathfrak{m}(\overline{E}_1\overline{E}_2) \leqq \mathfrak{m}\overline{E}_1 + \mathfrak{m}\overline{E}_2 & \text{by addition} \\ \hline \mathfrak{d}(E_1 + E_2) + \mathfrak{d}(E_1E_2) \leqq \mathfrak{d}E_1 + \mathfrak{d}E_2\,, & \end{array} \tag{5}$$

where the *in*equality must always hold if it holds even in only one of the preceding formulas.

Since $\mathfrak{m}E$ and $\mathfrak{d}E$ can never be negative, we a fortiori have

$$\mathfrak{m}(E_1 + E_2) \leqq \mathfrak{m}E_1 + \mathfrak{m}E_2\,, \tag{3a}$$

$$\mathfrak{d}(E_1 + E_2) \leqq \mathfrak{d}E_1 + \mathfrak{d}E_2\,, \tag{5a}$$

and, by mathematical induction, it follows for every finite sum that

$$\mathfrak{m}(E_1 + E_2 + \cdots + E_n) \leqq \mathfrak{m}E_1 + \mathfrak{m}E_2 + \cdots + \mathfrak{m}E_n\,, \tag{6}$$

$$\mathfrak{d}(E_1 + E_2 + \cdots + E_n) \leqq \mathfrak{d}E_1 + \mathfrak{d}E_2 + \cdots + \mathfrak{d}E_n\,. \tag{7}$$

Furthermore, we have $\mathfrak{d}(E_1\,E_2) = \mathfrak{d}(\overline{E_1\,E_2}) = \mathfrak{d}(\overline{E}_1 + \overline{E}_2) \leqq \mathfrak{d}\overline{E}_1 + \mathfrak{d}\overline{E}_2 = \mathfrak{d}E_1 + \mathfrak{d}E_2$ and, generally,

$$\mathfrak{d}(E_1E_2\cdots E_n) \leqq \mathfrak{d}E_1 + \mathfrak{d}E_2 + \cdots + \mathfrak{d}E_n\,. \tag{8}$$

The extension of the formulas (6), (7) and (8) to (countable) infinite sums still requires a proof since it is not possible to pass to the limit without further ado.

Let us assume that the point sets $E_1, E_2, E_3, \ldots$ are all contained in the total interval Λ and, one by one, covered by the interval sums $\mathsf{A}_1, \mathsf{A}_2, \mathsf{A}_3, \ldots$ (also contained in Λ). Then also

$$E = E_1 + E_2 + E_3 + \cdots$$

159 | $$E = E_1 + E_2 + E_3 + \cdots$$

überdeckt von

$$\mathsf{A}^* = \mathsf{A}_1 + \mathsf{A}_2 + \mathsf{A}_3 + \cdots .$$

Also ist nach dem Subsumptionssatze

$$\begin{aligned} \mathfrak{m}E \leqq \mathfrak{m}\mathsf{A}^* &\leqq \mathfrak{m}\mathsf{A}_1 + \mathfrak{m}\mathsf{A}_2 + \mathfrak{m}\mathsf{A}_3 + \cdots \\ &\leqq (\mathfrak{m}E_1 + \varepsilon_1) + (\mathfrak{m}E_2 + \varepsilon_2) + \cdots , \end{aligned}$$

wobei jede der positiven Größen ε_n unabhängig von den übrigen durch Wahl von A_n beliebig klein gemacht werden kann, also auch ihre Summe, und es ist in der Tat

$$\mathfrak{m}E = \mathfrak{m}(E_1 + E_2 + \cdots) \leqq \mathfrak{m}E_1 + \mathfrak{m}E_2 + \cdots , \tag{6a}$$

was natürlich nur Bedeutung hat, falls die rechts stehende Summe konvergiert.

Ferner sei B_n eine $\overline{E}_n$ überdeckende Intervallsumme innerhalb Λ. Dann wird $\overline{E} = \overline{E}_1\overline{E}_2 \cdots$, der Durchschnitt aller $\overline{E}_n$, gleichfalls überdeckt von

$$\mathsf{B}^* = \mathsf{B}_1\mathsf{B}_2 \cdots , \text{ dem Durchschnitt aller } \mathsf{B}_n ,$$

und es ist jedenfalls $\mathsf{B}^* \in \mathsf{B}_n$. Es wird also

$$\begin{aligned} \mathfrak{d}E = \mathfrak{U}\mathfrak{m}\mathsf{A}\mathsf{B} &\leqq \mathfrak{m}\mathsf{A}^*\mathsf{B}^* = \mathfrak{m}(\mathsf{A}_1\mathsf{B}^* + \mathsf{A}_2\mathsf{B}^* + \cdots) \\ &\leqq \mathfrak{m}\mathsf{A}_1\mathsf{B}^* + \mathfrak{m}\mathsf{A}_2\mathsf{B}^* + \cdots \\ &\leqq \mathfrak{m}\mathsf{A}_1\mathsf{B}_1 + \mathfrak{m}\mathsf{A}_2\mathsf{B}_2 + \cdots . \end{aligned}$$

$$\mathfrak{d}(E_1 + E_2 + \cdots) = \mathfrak{d}E \leqq \mathfrak{d}E_1 + \mathfrak{d}E_2 + \cdots , \tag{7a}$$

da jedes $\mathfrak{m}\mathsf{A}_n\mathsf{B}_n$ unabhängig von allen übrigen das entsprechende $\mathfrak{d}E_n$ approximiert.

Ersetzt man in dieser Formel jedes E_n durch $\overline{E}_n$, also E durch $\overline{E}_1 + \overline{E}_2 + \overline{E}_3 + \cdots$ und $\overline{E}$ durch $E_1E_2E_3 \cdots$, den Durchschnitt aller E_n, und beachtet, daß stets $\mathfrak{d}E = \mathfrak{d}\overline{E}$ ist, so ergibt sich

$$\mathfrak{d}(E_1E_2 \cdots) \leqq \mathfrak{d}E_1 + \mathfrak{d}E_2 + \cdots . \tag{8a}$$

Sind hier alle E_n „meßbar", d. h. $\mathfrak{d}E_n = 0$, so werden auch die linken Seiten in (7a) und (8a) verschwinden, und wir haben den *Hauptsatz* der *Lebesgue*schen Theorie:

Die Summe (Vereinigung) sowie der Durchschnitt von endlich vielen oder abzählbar unendlich vielen meßbaren Mengen ist wieder meßbar.

Sind demnach in der Grundformel (3) E_1 und E_2 beide meßbar, so sind es auch $E_1 + E_2$ und E_1E_2, und es gilt in (5) die *Gleichheit*; sie muß also auch in (3) gelten, d. h.

is covered by

$$\mathsf{A}^* = \mathsf{A}_1 + \mathsf{A}_2 + \mathsf{A}_3 + \cdots .$$

According to the subsumption theorem, we therefore have

$$\begin{aligned} \mathfrak{m}E \leqq \mathfrak{m}\mathsf{A}^* &\leqq \mathfrak{m}\mathsf{A}_1 + \mathfrak{m}\mathsf{A}_2 + \mathfrak{m}\mathsf{A}_3 + \cdots \\ &\leqq (\mathfrak{m}E_1 + \varepsilon_1) + (\mathfrak{m}E_2 + \varepsilon_2) + \cdots , \end{aligned}$$

where, by choice of A_n, each of the positive quantities ε_n can be made arbitrarily small independently of the others, hence also its sum, and we in fact have

$$\mathfrak{m}E = \mathfrak{m}(E_1 + E_2 + \cdots) \leqq \mathfrak{m}E_1 + \mathfrak{m}E_2 + \cdots , \qquad (6a)$$

which, of course, is meaningful only if the sum on the right side converges.

Furthermore, let B_n be an interval sum within Λ covering $\overline{E}_n$. Then $\overline{E} = \overline{E}_1\overline{E}_2\cdots$, the intersection of all $\overline{E}_n$, is also covered by

$$\mathsf{B}^* = \mathsf{B}_1\mathsf{B}_2\cdots , \text{ the intersection of all } \mathsf{B}_n ,$$

and it is certainly the case that $\mathsf{B}^* \subseteq \mathsf{B}_n$. We therefore have

$$\begin{aligned} \mathfrak{d}E = \mathfrak{U}\mathfrak{m}\mathsf{AB} \leqq \mathfrak{m}\mathsf{A}^*\mathsf{B}^* &= \mathfrak{m}(\mathsf{A}_1\mathsf{B}^* + \mathsf{A}_2\mathsf{B}^* + \cdots) \\ &\leqq \mathfrak{m}\mathsf{A}_1\mathsf{B}^* + \mathfrak{m}\mathsf{A}_2\mathsf{B}^* + \cdots \\ &\leqq \mathfrak{m}\mathsf{A}_1\mathsf{B}_1 + \mathfrak{m}\mathsf{A}_2\mathsf{B}_2 + \cdots . \end{aligned}$$

$$\mathfrak{d}(E_1 + E_2 + \cdots) = \mathfrak{d}E \leqq \mathfrak{d}E_1 + \mathfrak{d}E_2 + \cdots , \qquad (7a)$$

since each $\mathfrak{m}\mathsf{A}_n\mathsf{B}_n$ approximates the corresponding $\mathfrak{d}E_n$ independently of all others.

If we replace every E_n in this formula by $\overline{E}_n$, hence E by $\overline{E}_1 + \overline{E}_2 + \overline{E}_3 + \cdots$ and $\overline{E}$ by $E_1E_2E_3\cdots$, the intersection of all E_n, then, considering that always $\mathfrak{d}E = \mathfrak{d}\overline{E}$, we obtain

$$\mathfrak{d}(E_1E_2\cdots) \leqq \mathfrak{d}E_1 + \mathfrak{d}E_2 + \cdots . \qquad (8a)$$

If all E_n are "measurable" here, i.e., $\mathfrak{d}E_n = 0$, then the left sides in (7a) and (8 a) also vanish, and we have the *principal theorem* of *Lebesgue*'s theory:

Both the sum (union) and the intersection of finitely many or countably infinitely many measurable sets are again measurable.

If therefore E_1 and E_2 in the basic formula (3) are both measurable, then so are $E_1 + E_2$ and E_1E_2, and the *equality* holds in (5); and hence it must also hold in (3), i.e.,

$$\mathfrak{m}(E_1 + E_2) + \mathfrak{m}E_1E_2 = \mathfrak{m}E_1 + \mathfrak{m}E_2 \quad (\mathfrak{d}E_1 = \mathfrak{d}E_2 = 0) \qquad (3^*)$$

$$\mathfrak{m}(E_1 + E_2) + \mathfrak{m}E_1E_2 = \mathfrak{m}E_1 + \mathfrak{m}E_2 \quad (\mathfrak{d}E_1 = \mathfrak{d}E_2 = 0) \tag{3*}$$

entsprechend der Formel (1) für Intervallsummen. Sind in diesem Falle E_1 und E_2 zugleich „elementefremd" oder auch nur „maßfremd", d. h. $\mathfrak{m}E_1E_2 = 0$, so folgt

$$\mathfrak{m}(E_1 + E_2) = \mathfrak{m}E_1 + \mathfrak{m}E_2\,, \tag{3a*}$$

und diese Formel läßt sich durch vollständige Induktion auf jede *endliche* Anzahl von Summanden ausdehnen:

$$\mathfrak{m}(E_1 + E_2 + \cdots + E_n) = \mathfrak{m}E_1 + \mathfrak{m}E_2 + \cdots + \mathfrak{m}E_n\,, \tag{6a*}$$

sofern alle $E_1, E_2, \ldots, E_n$ meßbar und unter sich „maßfremd" sind. Hierbei wird nur die Tatsache benutzt, daß wenn zwei Mengen E_1 und E_2 zu einer dritten E „maßfremd" sind:

160 |

$$\mathfrak{m}E_1E = \mathfrak{m}E_2E = 0\,,$$

das gleiche auch von ihrer Summe gelten muß. Denn es ist

$$\mathfrak{m}(E_1 + E_2)E = \mathfrak{m}(E_1E + E_2E) \leqq \mathfrak{m}E_1E + \mathfrak{m}E_2E = 0\,.$$

Das Entsprechende gilt auch für beliebig viele, ja abzählbar unendlich viele Summanden vermöge der Formel (6a).

Nun seien die unendlich vielen auf Λ (das Hauptintervall) beschränkten Mengen $E_1, E_2, \ldots$ sämtlich „meßbar" und unter sich „elementefremd". Dann sind auch die Teilsummen

$$S_n = E_1 + E_2 + \cdots + E_n\,,$$

wie die Reste

$$R_n = E\overline{S}_n = E_{n+1} + E_{n+2} + \cdots$$

sämtlich meßbar und daher

$$\mathfrak{m}E = \mathfrak{m}(S_n + R_n) = \mathfrak{m}S_n + \mathfrak{m}R_n \geqq \mathfrak{m}S_n \quad \text{und}$$
$$\mathfrak{m}S_n = \mathfrak{m}E_1 + \mathfrak{m}E_2 + \cdots + \mathfrak{m}E_n \leqq \mathfrak{m}E\,.$$

Die Reihe positiver Glieder

$$s = \mathfrak{m}E_1 + \mathfrak{m}E_2 + \cdots$$

ist also konvergent und ihr Restglied

$$r_n = \mathfrak{m}E_{n+1} + \mathfrak{m}E_{n+2} + \cdots \geqq \mathfrak{m}R_n$$

nähert sich mit wachsenden n der Null. Es wird somit

$$\mathfrak{m}E = \lim_{n=\infty} \mathfrak{m}S_n = \mathfrak{m}E_1 + \mathfrak{m}E_2 + \cdots = \mathfrak{m}(E_1 + E_2 + \cdots)\,.$$

according to the formula (1) for interval sums. If in this case E_1 and E_2 are at the same time "elementwise disjoint" or even only "measurewise disjoint", i.e., $\mathfrak{m}E_1E_2 = 0$, then it follows that

$$\mathfrak{m}(E_1 + E_2) = \mathfrak{m}E_1 + \mathfrak{m}E_2\,, \tag{3a*}$$

and, by mathematical induction, this formula can be extended to any *finite* number of summands:

$$\mathfrak{m}(E_1 + E_2 + \cdots + E_n) = \mathfrak{m}E_1 + \mathfrak{m}E_2 + \cdots + \mathfrak{m}E_n\,, \tag{6a*}$$

provided that all $E_1, E_2, \ldots, E_n$ are measurable and mutually "measurewise disjoint". Here we only use the fact that if two sets E_1 and E_2 are "measurewise disjoint" from a third E:

$$\mathfrak{m}E_1\,E = \mathfrak{m}E_2\,E = 0\,,$$

then the same must also hold of their sum. For we have

$$\mathfrak{m}(E_1 + E_2)E = \mathfrak{m}(E_1E + E_2E) \leqq \mathfrak{m}E_1E + \mathfrak{m}E_2E = 0\,.$$

The corresponding fact also obtains for arbitrarily many, and even countably infinitely many summands by virtue of the formula (6a).

Let us now assume that the infinitely many sets $E_1, E_2, \ldots$ bounded with respect to Λ (the principal interval) are all "measurable" and mutually "elementwise disjoint" ⟦and let $E = E_1 + E_2 + \cdots$⟧.[5] Then all the partial sums

$$S_n = E_1 + E_2 + \cdots + E_n\,,$$

and all the remainders

$$R_n = E\overline{S}_n = E_{n+1} + E_{n+2} + \cdots$$

are also measurable, and hence we have

$$\begin{aligned} \mathfrak{m}E &= \mathfrak{m}(S_n + R_n) = \mathfrak{m}S_n + \mathfrak{m}R_n \geqq \mathfrak{m}S_n \quad \text{and} \\ \mathfrak{m}S_n &= \mathfrak{m}E_1 + \mathfrak{m}E_2 + \cdots + \mathfrak{m}E_n \leqq \mathfrak{m}E\,. \end{aligned}$$

The series of positive terms

$$s = \mathfrak{m}E_1 + \mathfrak{m}E_2 + \cdots$$

is therefore convergent, and its remainder term

$$r_n = \mathfrak{m}E_{n+1} + \mathfrak{m}E_{n+2} + \cdots \geqq \mathfrak{m}R_n$$

converges to zero as n increases. We thus have

$$\mathfrak{m}E = \lim_{n=\infty} \mathfrak{m}S_n = \mathfrak{m}E_1 + \mathfrak{m}E_2 + \cdots = \mathfrak{m}(E_1 + E_2 + \cdots)\,.$$

[5] ⟦Zermelo fails to provide a definition for E.⟧

Sind dagegen die Summanden $E_1, E_2, \ldots$ nicht mehr untereinander punktfremd, so sind es doch die ihnen entsprechenden Teilmengen

$$E_1^* = E_1\,, \quad E^* = \overline{E}_1 E_2\,, \quad E^* = \overline{E}_1 \overline{E}_2 E_3\,, \ldots, \quad E_n^* = \overline{E}_1 \overline{E}_2 \cdots \overline{E}_{n-1} E_n\,,$$

und es wird wieder

$$S_n^* = E_1^* + E_2^* + \cdots E_n^* = S_n, \quad E = E_1^* + E_2^* + \cdots\,,$$

also nach dem soeben Bewiesenen

$$\begin{aligned} \mathfrak{m}E = \mathfrak{m}E_1^* + \mathfrak{m}E_2^* + \cdots &= \lim_{n=\infty} \mathfrak{m}S_n^* = \lim_{n=\infty} \mathfrak{m}S_n \\ &= \lim_{n=\infty} \mathfrak{m}(E_1 + E_2 + \cdots + E_n)\,. \end{aligned} \tag{6a**}$$

Nun sei

$$P = E_1 E_2 \cdots$$

der Durchschnitt von abzählbar unendlich vielen einzeln *meßbaren* Punktmengen $E_1, E_2, \ldots$ innerhalb des Gesamtintervalls Λ und

$$P_n = E_1 E_2 \cdots E_n$$

der Durchschnitt der ersten n von ihnen. Dann wird

$$\overline{P}_n = \overline{E}_1 + \overline{E}_2 + \cdots + \overline{E}_n\,,$$

und daher nach dem vorigen Satze

$$\mathfrak{m}\overline{P} = \mathfrak{m}(\overline{E}_1 + \overline{E}_2 + \cdots) = \lim_{n=\infty} \mathfrak{m}\overline{P}_n\,,$$

also, da mit $\overline{P}$ auch P meßbar ist,

161 |

$$\mathfrak{m}P = I - \mathfrak{m}\overline{P} = \lim_{n=\infty}(I - \mathfrak{m}\overline{P}_n) = \lim_{n=\infty} \mathfrak{m}P_n\,,$$

d. h.

$$\mathfrak{m}(E_1 E_2 \cdots) = \lim_{n=\infty} \mathfrak{m}(E_1 E_2 \cdots E_n)\,. \tag{9}$$

Da die Summe wie der Durchschnitt von abzählbar vielen meßbaren Punktmengen wieder meßbar ist, sowie auch jedes Intervall (wegen $\mathfrak{m}\lambda + \mathfrak{m}\overline{\lambda} = I$), so ist auch jede (abzählbare) *Intervallsumme* meßbar, also auch jedes *Gebiet* (S. 155) und als Komplement eines Gebietes auch jede *abgeschlossene* (beschränkte) Punktmenge.

On the other hand, if the summands $E_1, E_2, \ldots$ are no longer mutually pointwise disjoint, then the partial sets corresponding to them are still mutually pointwise disjoint

$$E_1^* = E_1, \quad E_2^* = \overline{E}_1 E_2, \quad E_3^* = \overline{E}_1 \overline{E}_2 E_3, \ldots, \quad E_n^* = \overline{E}_1 \overline{E}_2 \cdots \overline{E}_{n-1} E_n,$$ [6]

and we again have

$$S_n^* = E_1^* + E_2^* + \cdots E_n^* = S_n, \quad E = E_1^* + E_2^* + \cdots,$$

therefore, according to what has just been proved,

$$\begin{aligned} \mathfrak{m}E &= \mathfrak{m}E_1^* + \mathfrak{m}E_2^* + \cdots = \lim_{n=\infty} \mathfrak{m}S_n^* = \lim_{n=\infty} \mathfrak{m}S_n \\ &= \lim_{n=\infty} \mathfrak{m}(E_1 + E_2 + \cdots + E_n). \end{aligned} \tag{6a**}$$

Let now

$$P = E_1 E_2 \cdots$$

be the intersection of countably infinitely many individually *measurable* point sets $E_1, E_2, \ldots$ within the total interval Λ and let

$$P_n = E_1 E_2 \cdots E_n$$

be the intersection of the first n of them. We then have

$$\overline{P}_n = \overline{E}_1 + \overline{E}_2 + \cdots + \overline{E}_n,$$

and hence, according to the preceding theorem, we have

$$\mathfrak{m}\overline{P} = \mathfrak{m}(\overline{E}_1 + \overline{E}_2 + \cdots) = \lim_{n=\infty} \mathfrak{m}\overline{P}_n.$$

Since, along with $\overline{P}$, P, too, is measurable, we thus have

$$\mathfrak{m}P = I - \mathfrak{m}\overline{P} = \lim_{n=\infty} (I - \mathfrak{m}\overline{P}_n) = \lim_{n=\infty} \mathfrak{m}P_n,$$

i.e.,

$$\mathfrak{m}(E_1 E_2 \cdots) = \lim_{n=\infty} \mathfrak{m}(E_1 E_2 \cdots E_n). \tag{9}$$

Since the sum, like the intersection, of countably many measurable point sets is again measurable, and since the same holds of every interval (because $\mathfrak{m}\lambda + \mathfrak{m}\overline{\lambda} = I$), every (countable) *interval sum*, too, is measurable, and hence also every *domain* (p. 155), and, being the complement of a domain, also every *closed* (bounded) point set.

[6] ⟦Zermelo erroneously writes "E^*" for "E_2^*" and for "E_3^*".⟧

Ist E eine beliebige (meßbare oder nicht meßbare) Punktmenge im Gesamtintervall und $\varepsilon_1, \varepsilon_2, \ldots$ eine nach 0 abnehmende Folge positiver Zahlen, so gibt es (nach der Definition des Maßes) zu jedem n eine E überdeckende Intervallsumme A_n derart, daß

$$\mathfrak{m}E \leqq \mathfrak{m}\mathsf{A}_n < \mathfrak{m}E + \varepsilon_n \,,$$

und der *Durchschnitt* aller dieser Intervallsummen A_n ,

$$A = \mathsf{A}_1\mathsf{A}_2 \cdots \Subset \mathsf{A}_n \,,$$

ist nach dem soeben Bewiesenen wieder eine *meßbare* E überdeckende Punktmenge, für welche nach dem Subsumptionssatze gilt:

$$\mathfrak{m}E \leq \mathfrak{m}A \leq \mathfrak{m}\mathsf{A}_n < \mathfrak{m}E + \varepsilon_n \,.$$

Also im Grenzfalle

$$\mathfrak{m}E = \mathfrak{m}A \,, \quad \text{wo} \quad E \Subset A \quad \text{und} \quad \mathfrak{d}A = 0 \,, \quad \text{d. h.}$$

Satz. *Jede beschränkte Punktmenge E besitzt eine „maßgleiche Hülle", d. h. eine sie überdeckende meßbare Punktmenge von gleichem Maße. (De la Vallée-Poussin.)*

Ist A eine „maßgleiche Hülle" von E und C eine beliebige „meßbare" Menge, so ist wegen $EC \Subset AC$, $E\overline{C} \Subset A\overline{C}$ nach dem Subsumptionssatze und nach (3a*)

$$\mathfrak{m}E = \mathfrak{m}(EC + E\overline{C}) \leq \mathfrak{m}EC + \mathfrak{m}E\overline{C} \leqq \mathfrak{m}AC + \mathfrak{m}A\overline{C} = \mathfrak{m}A = \mathfrak{m}E \,,$$

also

$$\mathfrak{m}E = \mathfrak{m}EC + \mathfrak{m}E\overline{C} \quad (\mathfrak{d}C = 0) \,, \tag{10}$$

und, weil jetzt für beide Summanden die Gleichheit gelten muß,

$$\mathfrak{m}EC = \mathfrak{m}AC \,, \quad \mathfrak{m}E\overline{C} = \mathfrak{m}A\overline{C} \,.$$

Ferner ist nach (10)

$$\begin{aligned} \mathfrak{m}(E + C) &= \mathfrak{m}(E + C)C + \mathfrak{m}(E + C)\overline{C} = \mathfrak{m}C + \mathfrak{m}E\overline{C} \\ &= \mathfrak{m}C + \mathfrak{m}A\overline{C} = \mathfrak{m}(C + A\overline{C}) = \mathfrak{m}(A + C) \,. \end{aligned}$$

Ist also A eine maßgleiche Hülle von E, so sind auch AC und $A+C$ maßgleiche Hüllen von EC bezw. $E + C$. Sind A und B maßgleiche Hüllen von E bzw. $\overline{E}$, so wird

$$\mathfrak{m}E + \mathfrak{m}\overline{E} = \mathfrak{m}A + \mathfrak{m}B = \mathfrak{m}(A + B) + \mathfrak{m}AB \,,$$

If E is an arbitrary (measurable or not measurable) point set in the total interval and if $\varepsilon_1, \varepsilon_2, \ldots$ is a sequence of positive numbers decreasing towards 0, then for each n there exists (according to the definition of measure) an interval sum A_n covering E such that

$$\mathfrak{m}E \leqq \mathfrak{m}\mathsf{A}_n < \mathfrak{m}E + \varepsilon_n ,$$

and the *intersection* of all these interval sums A_n,

$$A = \mathsf{A}_1\mathsf{A}_2 \cdots \subseteq \mathsf{A}_n ,$$

is, according to what has just been proved, again a *measurable* point set covering E for which, according to the subsumption theorem,

$$\mathfrak{m}E \leq \mathfrak{m}A \leq \mathfrak{m}\mathsf{A}_n < \mathfrak{m}E + \varepsilon_n .$$

In the limit case we therefore have

$$\mathfrak{m}E = \mathfrak{m}A , \quad \text{where} \quad E \subseteq A \quad \text{and} \quad \mathfrak{d}A = 0 , \quad \text{i.e.,}$$

Theorem. *Every bounded point set E possesses an "isometric hull", i.e., a measurable point set which covers it and is of equal measure. (De la Vallée-Poussin.)*

If A is an "isometric hull" of E and if C is an arbitrary "measurable" set, then, on account of $EC \subseteq AC$, we have $E\overline{C} \subseteq A\overline{C}$, according to the subsumption theorem, and, according to (3a*),

$$\mathfrak{m}E = \mathfrak{m}(EC + E\overline{C}) \leq \mathfrak{m}EC + \mathfrak{m}E\overline{C} \leqq \mathfrak{m}AC + \mathfrak{m}A\overline{C} = \mathfrak{m}A = \mathfrak{m}E ,$$

and hence

$$\mathfrak{m}E = \mathfrak{m}EC + \mathfrak{m}E\overline{C} \quad (\mathfrak{d}C = 0) , \tag{10}$$

and, since the equality must now hold for both summands, we have

$$\mathfrak{m}EC = \mathfrak{m}AC , \quad \mathfrak{m}E\overline{C} = \mathfrak{m}A\overline{C} .$$

Furthermore, according to (10), we have

$$\begin{aligned}\mathfrak{m}(E+C) &= \mathfrak{m}(E+C)C + \mathfrak{m}(E+C)\overline{C} = \mathfrak{m}C + \mathfrak{m}E\overline{C} \\ &= \mathfrak{m}C + \mathfrak{m}A\overline{C} = \mathfrak{m}(C + A\overline{C}) = \mathfrak{m}(A+C) .\end{aligned}$$

Thus if A is an isometric hull of E, then AC and $A+C$ are also isometric hulls of EC and $E+C$ respectively. If A and B are isometric hulls of E and $\overline{E}$ respectively, then we have

$$\mathfrak{m}E + \mathfrak{m}\overline{E} = \mathfrak{m}A + \mathfrak{m}B = \mathfrak{m}(A+B) + \mathfrak{m}AB ,$$

also, da $A + B = \Lambda$ und somit $\mathfrak{m}(A + B) = I$ ist,

$$\mathfrak{d}E = \mathfrak{m}E + \mathfrak{m}\overline{E} - I = \mathfrak{m}AB\,; \tag{11}$$

sowie, wenn C eine beliebige meßbare Menge ist, da dann $\mathfrak{m}EC = \mathfrak{m}AC$, $\mathfrak{m}\overline{EC} = \mathfrak{m}(\overline{E} + \overline{C}) = \mathfrak{m}(B + \overline{C})$, also $B + \overline{C}$ maßgleiche Hülle von $\overline{EC}$ ist,

162 |

$$\mathfrak{d}EC = \mathfrak{m}AC(B + \overline{C}) = \mathfrak{m}ABC\,. \tag{11a}$$

Sind jetzt A_1 und B_1 zwei beliebige E bzw. $\overline{E}$ überdeckende meßbare Mengen innerhalb des Gesamtintervalles, so wird

$$\mathfrak{m}E = \mathfrak{m}A \leqq \mathfrak{m}A_1\,, \quad \mathfrak{m}\overline{E} = \mathfrak{m}B \leqq \mathfrak{m}B_1\,, \quad \text{also}$$

$$\mathfrak{d}E = \mathfrak{m}E + \mathfrak{m}\overline{E} - I \leqq \mathfrak{m}A_1 + \mathfrak{m}B_1 - \mathfrak{m}(A_1 + B_1) = \mathfrak{m}A_1B_1\,,$$

d. h.

$$\mathfrak{d}E = \mathfrak{m}AB = \mathfrak{U}\mathfrak{m}A_1B_1$$

die untere Grenze für das Maß des Durchschnittes von je zwei solchen meßbaren Mengen A_1, B_1.

$$\mathfrak{d}EC = \mathfrak{m}ABC = \mathfrak{U}\mathfrak{m}A_1B_1C\,. \tag{11a*}$$

(Verallgemeinerungen der Definitionsformel (2) für die Diskrepanz.) Ferner wird, wenn wieder A, B maßgleiche Hüllen von E und $\overline{E}$ bezeichnen,

$$\begin{aligned}\mathfrak{m}EC + \mathfrak{m}\overline{E}C &= \mathfrak{m}AC + \mathfrak{m}BC = \mathfrak{m}(AC + BC) + \mathfrak{m}AC \cdot BC \\ &= \mathfrak{m}(A + B)C + \mathfrak{m}ABC = \mathfrak{m}C + \mathfrak{d}EC\,, \quad \text{d. h.}\end{aligned}$$

$$\mathfrak{d}EC = \mathfrak{m}EC + \mathfrak{m}\overline{E}C - \mathfrak{m}C \tag{12}$$

für jede meßbare Punktmenge C als Verallgemeinerung der Formel (4), sowie

$$\begin{aligned}\mathfrak{d}(E + C) = \mathfrak{d}(\overline{E + C}) = \mathfrak{d}(\overline{E}\,\overline{C}) &= \mathfrak{m}\overline{E}\,\overline{C} + \mathfrak{m}E\overline{C} - \mathfrak{m}\overline{C} \\ &= \mathfrak{m}E\overline{C} + \mathfrak{m}\overline{E}\,\overline{C} - \mathfrak{m}\overline{C}\end{aligned} \tag{12a}$$

für jede *meßbare* Menge C. Durch Addition folgt nun

$$\begin{aligned}\mathfrak{d}(E + C) + \mathfrak{d}EC &= (\mathfrak{m}EC + \mathfrak{m}E\overline{C}) + (\mathfrak{m}\overline{E}C + \mathfrak{m}\overline{E}\,\overline{C}) - (\mathfrak{m}C + \mathfrak{m}\overline{C}) \\ &= \mathfrak{m}E + \mathfrak{m}\overline{E} - I = \mathfrak{d}E.\end{aligned}$$

Diese Formel

$$\mathfrak{d}(E + C) + \mathfrak{d}EC = \mathfrak{d}E\,, \tag{5*}$$

and hence, since we have $A+B=\Lambda$ and thus $\mathfrak{m}(A+B)=I$, we have

$$\mathfrak{d}E=\mathfrak{m}E+\mathfrak{m}\overline{E}-I=\mathfrak{m}AB\,. \tag{11}$$

We also have

$$\mathfrak{d}EC=\mathfrak{m}AC(B+\overline{C})=\mathfrak{m}ABC\,, \tag{11a}$$

if C is an arbitrary measurable set. For in this case $\mathfrak{m}EC=\mathfrak{m}AC$ and $\mathfrak{m}\overline{E\,C}=\mathfrak{m}(\overline{E}+\overline{C})=\mathfrak{m}(B+\overline{C})$, and hence $B+\overline{C}$ is an isometric hull of $\overline{E\,C}$.

If now A_1 and B_1 are two arbitrary measurable sets covering E and $\overline{E}$ respectively within the total interval, then we have

$$\mathfrak{m}E=\mathfrak{m}A\leqq\mathfrak{m}A_1\,,\quad \mathfrak{m}\overline{E}=\mathfrak{m}B\leqq\mathfrak{m}B_1\,,\quad \text{and hence}$$

$$\mathfrak{d}E=\mathfrak{m}E+\mathfrak{m}\overline{E}-I\leqq\mathfrak{m}A_1+\mathfrak{m}B_1-\mathfrak{m}(A_1+B_1)=\mathfrak{m}A_1B_1\,,$$

i.e.,

$$\mathfrak{d}E=\mathfrak{m}AB=\mathfrak{U}\mathfrak{m}A_1B_1$$

is the infimum for the measure of the intersection of any two such measurable sets A_1,B_1.

$$\mathfrak{d}EC=\mathfrak{m}ABC=\mathfrak{U}\mathfrak{m}A_1B_1C\,. \tag{11a*}$$

(Generalizations of the defining formula (2) for the discrepancy.) Furthermore if A,B again signify isometric hulls of E and $\overline{E}$,

$$\begin{aligned}\mathfrak{m}EC+\mathfrak{m}\overline{E}C&=\mathfrak{m}AC+\mathfrak{m}BC=\mathfrak{m}(AC+BC)+\mathfrak{m}AC\cdot BC\\&=\mathfrak{m}(A+B)C+\mathfrak{m}ABC=\mathfrak{m}C+\mathfrak{d}EC\,,\quad\text{i.e.,}\\ \mathfrak{d}EC&=\mathfrak{m}EC+\mathfrak{m}\overline{E}C-\mathfrak{m}C\end{aligned} \tag{12}$$

for every measurable point set C as a generalization of the formula (4) as well as

$$\begin{aligned}\mathfrak{d}(E+C)=\mathfrak{d}(\overline{E+C})=\mathfrak{d}(\overline{E}\,\overline{C})&=\mathfrak{m}\overline{E}\,\overline{C}+\mathfrak{m}E\overline{C}-\mathfrak{m}\overline{C}\\&=\mathfrak{m}E\overline{C}+\mathfrak{m}\overline{E}\,\overline{C}-\mathfrak{m}\overline{C}\end{aligned} \tag{12a}$$

for every *measurable* set C. By addition it now follows that

$$\begin{aligned}\mathfrak{d}(E+C)+\mathfrak{d}EC&=(\mathfrak{m}EC+\mathfrak{m}E\overline{C})+(\mathfrak{m}\overline{E}C+\mathfrak{m}\overline{E}\,\overline{C})-(\mathfrak{m}C+\mathfrak{m}\overline{C})\\&=\mathfrak{m}E+\mathfrak{m}\overline{E}-I=\mathfrak{d}E\,.\end{aligned}$$

We can also obtain this formula

$$\mathfrak{d}(E+C)+\mathfrak{d}EC=\mathfrak{d}E \tag{5*}$$

erhalten wir auch als Ungleichheit mit $\leq$ durch Addition der allgemein gültigen Formeln

$$\mathfrak{m}(E+C)+\mathfrak{m}EC \leqq \mathfrak{m}E+\mathfrak{m}C$$

$$\mathfrak{m}(\overline{E}+\overline{C})+\mathfrak{m}\overline{E}\,\overline{C} \leqq \mathfrak{m}\overline{E}+\mathfrak{m}\overline{C}$$

mit Rücksicht auf die Beziehungen $\overline{E+C}=\overline{E}\,\overline{C}$, $\overline{EC}=\overline{E}+\overline{C}$ und $\mathfrak{m}C+\mathfrak{m}\overline{C}=I$. Da aber tatsächlich in (5*) die Gleichheit gilt, so gilt sie auch in beiden durch Addition vereinigten Formeln, d. h.

$$\mathfrak{m}(E+C)+\mathfrak{m}EC=\mathfrak{m}E+\mathfrak{m}C\,, \tag{3**}$$

als Verallgemeinerung von (3*), wenn auch nur *eine* der beiden Mengen E_1, E_2 meßbar ist. Dieselbe Formel ergibt sich auch unmittelbar durch Benutzung der maßgleichen Hülle A:

$$\mathfrak{m}(E+C)+\mathfrak{m}EC=\mathfrak{m}(A+C)+\mathfrak{m}AC=\mathfrak{m}A+\mathfrak{m}C=\mathfrak{m}E+\mathfrak{m}C\,.$$

Sind die Punktmengen $C_1, C_2, \ldots$ sämtlich meßbar und unter sich maßfremd $\mathfrak{m}C_rC_s=0$ $(r\neq s)$, so ist auch ihre Summe meßbar und $\mathfrak{m}EC_rC_s \leqq \mathfrak{m}C_rC_s=0$ für jede Menge E. Somit wird nach S. 161 und (6a**)

163 |
$$\begin{aligned}\mathfrak{m}E(C_1+C_2+\cdots)&=\mathfrak{m}A(C_1+C_2+\cdots)=\mathfrak{m}AC_1+\mathfrak{m}AC_2+\cdots \\ &=\mathfrak{m}EC_1+\mathfrak{m}EC_2+\cdots\,,\end{aligned} \tag{13}$$

sowie

$$\mathfrak{d}E(C_1+C_2+\cdots)=\mathfrak{m}AB(C_1+C_2+\cdots)=\mathfrak{d}EC_1+\mathfrak{d}EC_2+\cdots\,. \tag{14}$$

Ist wieder B maßgleiche Hülle von $\overline{E}$ (und zugleich A eine von E), so wird $E=EB+E\overline{B}=E_1+E_0$, sowie

$$\mathfrak{m}E_1=\mathfrak{m}EB=\mathfrak{m}AB=\mathfrak{d}E=\mathfrak{d}EB=\mathfrak{d}E_1 \qquad \text{und}$$

$$\mathfrak{d}E_0=\mathfrak{d}E\overline{B}=\mathfrak{m}AB\overline{B}=0\,, \qquad \text{d. h.:}$$

Jede diskrepante (beschränkte) Punktmenge E läßt sich zerlegen in eine „meßbare“ Menge E_0 und eine „Minimalmenge“ E_1.

Ist E eine beliebige Menge, A eine maßgleiche Hülle und E' ihre Ableitung (die Menge ihrer Häufungspunkte), so ist die Summe $E+E'$ „abgeschlossen“, mithin meßbar, und es wird

$$\mathfrak{m}E=\mathfrak{m}E(E+E')=\mathfrak{m}A(E+E')=\mathfrak{m}(E+AE')\,, \tag{15}$$

also $A_0=A(E+E')=E+AE'$ wegen $E \in A$, und $E \in E+E'$ ist eine maßgleiche Hülle von E, welche außer Punkten von E nur noch solche von E' enthält.

Jede beschränkte Punktmenge kann lediglich durch Punkte ihrer Ableitung zu einer maßgleichen Hülle ergänzt werden.

in the form of an inequality with $\leqq$ by adding the generally valid formulas

$$\mathfrak{m}(E+C)+\mathfrak{m}EC \leqq \mathfrak{m}E+\mathfrak{m}C$$

$$\mathfrak{m}(\overline{E}+\overline{C})+\mathfrak{m}\overline{E}\,\overline{C} \leqq \mathfrak{m}\overline{E}+\mathfrak{m}\overline{C}$$

considering the relations $\overline{E+C}=\overline{E}\,\overline{C}$, $\overline{EC}=\overline{E}+\overline{C}$ and $\mathfrak{m}C+\mathfrak{m}\overline{C}=I$. But since the equality in fact holds in (5*), it therefore also holds in both formulas united by means of addition, i.e.,

$$\mathfrak{m}(E+C)+\mathfrak{m}EC=\mathfrak{m}E+\mathfrak{m}C\,, \tag{3**}$$

as a generalization of (3*), if even only *one* of the two sets E_1, E_2 is measurable. We immediately obtain the same formula also by using the isometric hull A:

$$\mathfrak{m}(E+C)+\mathfrak{m}EC=\mathfrak{m}(A+C)+\mathfrak{m}AC=\mathfrak{m}A+\mathfrak{m}C=\mathfrak{m}E+\mathfrak{m}C\,.$$

If the point sets $C_1, C_2, \ldots$ are all measurable and mutually measurewise disjoint $\mathfrak{m}C_rC_s=0$ $(r\neq s)$, then their sum, too, is measurable and $\mathfrak{m}EC_rC_s \leqq \mathfrak{m}C_rC_s=0$ for every set E. We thus have according to p. 161 and (6a**)

$$\begin{aligned}\mathfrak{m}E(C_1+C_2+\cdots)&=\mathfrak{m}A(C_1+C_2+\cdots)=\mathfrak{m}AC_1+\mathfrak{m}AC_2+\cdots\\&=\mathfrak{m}EC_1+\mathfrak{m}EC_2+\cdots\,,\end{aligned} \tag{13}$$

as well as

$$\mathfrak{d}E(C_1+C_2+\cdots)=\mathfrak{m}AB(C_1+C_2+\cdots)=\mathfrak{d}EC_1+\mathfrak{d}EC_2+\cdots\,. \tag{14}$$

If B is again an isometric hull of $\overline{E}$ (and, at the same time, A one of E), then we have $E=E\,B+E\,\overline{B}=E_1+E_0$, as well as

$$\mathfrak{m}E_1=\mathfrak{m}EB=\mathfrak{m}AB=\mathfrak{d}E=\mathfrak{d}EB=\mathfrak{d}E_1 \qquad \text{and}$$

$$\mathfrak{d}E_0=\mathfrak{d}E\overline{B}=\mathfrak{m}AB\overline{B}=0\,, \qquad \text{i.e.:}$$

Every discrepant (bounded) point set E can be decomposed into a "measurable" set E_0 and a "minimal set" E_1.

If E is an arbitrary set, A an isometric hull and E' its derivation (the set of its accumulation points), then the sum $E+E'$ is "closed", and therefore measurable, and we have

$$\mathfrak{m}E=\mathfrak{m}E(E+E')=\mathfrak{m}A(E+E')=\mathfrak{m}(E+AE')\,, \tag{15}$$

and hence $A_0=A\,(E+E')=E+A\,E'$, since $E\subseteq A$, and $E\subseteq E+E'$ is an isometric hull of E containing, aside from points of E, only points of E'.

Every bounded point set can be complemented to an isometric hull using only points of its derivation.

§ 3. Maßmenge und Diskrepanzmenge. Unbeschränkte Punktmengen

Eine Punktmenge E heiße „maßhaltig“, wenn sie ein von Null verschiedenes Maß besitzt (keine „Nullmenge“ ist), „diskrepant“, wenn sie eine von Null verschiedene Diskrepanz besitzt, also „nicht meßbar“ ist. Sie heißt „in einem Intervall λ maßhaltig bzw. diskrepant“, wenn die im Intervall enthaltene Teilmenge $E\lambda$ maßhaltig bzw. diskrepant ist.

Die Eigenschaft, in einem Intervall maßhaltig bzw. diskrepant zu sein, ist nun aber eine „distributive“ (§ 1). Denn es ist bei jeder Zerlegung $\lambda = \lambda_1 + \lambda_2$ in zwei (getrennte) Teilintervalle, da Intervalle meßbar sind, nach (13) und (14)

$$\mathfrak{m}E\lambda = \mathfrak{m}E(\lambda_1 + \lambda_2) = \mathfrak{m}E\lambda_1 + \mathfrak{m}E\lambda_2 \,,$$

$$\mathfrak{d}E\lambda = \mathfrak{d}E(\lambda_1 + \lambda_2) = \mathfrak{d}E\lambda_1 + \mathfrak{d}E\lambda_2 \,,$$

und die linke Seite wird, da beide Summanden wesentlich positiv sind, dann und nur dann verschwinden, wenn beide Summanden es tun. Nach dem *Peano*schen Satze (§ 1) folgt also:

Satz 1. *In jedem abgeschlossenen Intervalle λ, in welchem eine Punktmenge „maßhaltig“ bzw. „diskrepant“ ist, besitzt sie mindestens einen „Maßpunkt“ $\mathfrak{p}$, in dessen* beliebiger *Umgebung sie noch maßhaltig ist. In jedem abgeschlossenen Intervalle, wo sie diskrepant ist, besitzt sie mindestens einen „Diskrepanzpunkt“ $\mathfrak{q}$, in dessen beliebiger Umgebung sie noch diskrepant ist.*

Die Menge der „Maßpunkte“ $\mathfrak{p}$, die zu einer Menge E gehören, heiße ihre „Maßmenge“ $\mathfrak{M}E$, die Menge ihrer Diskrepanzpunkte $\mathfrak{q}$ heiße ihre „Diskrepanzmenge“ $\mathfrak{D}E$.

164 | Da wegen $\mathfrak{d}E\lambda \leqq \mathfrak{m}E\lambda$ eine Punktmenge in jedem Intervall, wo sie diskrepant ist, auch maßhaltig sein muß, so ist jeder Diskrepanzpunkt auch ein Maßpunkt und $\mathfrak{D}E \Subset \mathfrak{M}E$; *die Diskrepanzmenge ist in der Maßmenge enthalten.* Ist ferner $E_1 \Subset E$ eine Untermenge von E, so gilt für jedes Intervall

$$0 < \mathfrak{m}E_1\lambda \leqq \mathfrak{m}E\lambda \,,$$

und jeder Maßpunkt von E_1 ist zugleich Maßpunkt von E, also

$$\mathfrak{M}E_1 \Subset ME \,.$$

Sind A und B „maßgleiche Hüllen“ von E und $\overline{E}$ innerhalb des Gesamtintervalls Λ, so ist auch in jedem kleineren Intervall $\mathfrak{m}E\lambda = \mathfrak{m}A\lambda$, $\mathfrak{m}\overline{E}\lambda = \mathfrak{m}B\lambda$,

§ 3. The measure set and the discrepancy set. Unbounded point sets

A point set E is called "measure-containing" if it possesses a measure different from zero (if it is not a "null set"). It is called "discrepant" if it possesses a discrepancy different from zero, that is, if it is "not measurable". It is called "measure-containing in an interval λ" and "discrepant in it" if the partial set $E\lambda$ contained in the interval is measure-containing and discrepant, respectively.

But now the property of being measure-containing or discrepant in an interval is "distributive" (§ 1). For, since intervals are measurable, we have, according to (13) and (14), for every decomposition $\lambda = \lambda_1 + \lambda_2$ into two (separate) partial intervals

$$\mathfrak{m}E\lambda = \mathfrak{m}E(\lambda_1 + \lambda_2) = \mathfrak{m}E\lambda_1 + \mathfrak{m}E\lambda_2\,,$$

$$\mathfrak{d}E\lambda = \mathfrak{d}E(\lambda_1 + \lambda_2) = \mathfrak{d}E\lambda_1 + \mathfrak{d}E\lambda_2\,,$$

and, since both summands are essentially positive, the left side vanishes if and only if both summands do. According to *Peano*'s theorem (§ 1), it therefore follows that:

Theorem 1. *In every closed interval λ in which a point set is "measure-containing" or "discrepant",*[7] *it possesses at least one "measure point" $\mathfrak{p}$ in whose* arbitrary *neighborhood it is still measure-containing. In every closed interval where it is discrepant, it possesses at least one "discrepancy point" $\mathfrak{q}$ in whose arbitrary neighborhood it is still discrepant.*

The set of "measure points" $\mathfrak{p}$ belonging to a set E shall be called its "measure set" $\mathfrak{M}E$, and the set of its discrepancy points $\mathfrak{q}$ shall be called its "discrepancy set" $\mathfrak{D}E$.

Since, on account of $\mathfrak{d}E\lambda \leqq \mathfrak{m}E\lambda$, a point set must be measure-containing in every interval where it is discrepant, every discrepancy point is also a measure point, and we have $\mathfrak{D}E \subseteq \mathfrak{M}E$; *the discrepancy set is included in the measure set.* If furthermore $E_1 \subseteq E$ is a subset of E, then for every interval we have

$$0 < \mathfrak{m}E_1\lambda \leqq \mathfrak{m}E\lambda\,,$$

and every measure point of E_1 is also a measure point of E, that is,

$$\mathfrak{M}E_1 \subseteq \mathfrak{M}E\,.^8$$

If A and B are "isometric hulls" of E and $\overline{E}$ within the total interval Λ, then in every smaller interval we also have $\mathfrak{m}E\lambda = \mathfrak{m}A\lambda$, $\mathfrak{m}\overline{E}\lambda = \mathfrak{m}B\lambda$, and,

---

[7] ⟦The words "or "discrepant"," should be deleted.⟧

[8] ⟦Zermelo erroneously writes "ME" instead of "$\mathfrak{M}E$".⟧

sowie nach (11a)

$$\mathfrak{d}E\lambda = \mathfrak{m}AB\lambda\,,$$

d. h. die Menge AB ist maßhaltig in jedem Intervall λ, wo E diskrepant ist, und umgekehrt, also

$$\mathfrak{D}E = \mathfrak{M}(AB)\,.$$

Ist C irgendeine meßbare Menge und λ ein Intervall, z. B. die Umgebung eines Punktes $\mathfrak{p}$, so ist nach (10)

$$\mathfrak{m}E\lambda = \mathfrak{m}EC\lambda + \mathfrak{m}E\overline{C}\lambda$$

und

$$\mathfrak{d}E\lambda = \mathfrak{d}EC\lambda + \mathfrak{d}E\overline{C}\lambda\,;$$

jeder Maßpunkt von E ist also auch Maßpunkt von EC oder $E\overline{C}$ und jeder Diskrepanzpunkt von E auch Diskrepanzpunkt von EC oder $E\overline{C}$, d. h.

$$(16)\quad \mathfrak{M}E = \mathfrak{M}(EC) + \mathfrak{M}(E\overline{C})\,, \qquad (17)\quad \mathfrak{D}E = \mathfrak{D}(EC) + \mathfrak{D}(E\overline{C})\,.$$

Ist $\mathfrak{p}$ irgend ein Maßpunkt der Menge E, also E maßhaltig in *jeder* Umgebung λ, $\mathfrak{m}E\lambda > 0$, so ist $\mathfrak{p}$ natürlich auch Häufungspunkt von E, ja „Kondensationspunkt“ von E, weil sonst $E\lambda$ „abzählbar“ und vom Maße 0 wäre. Zugleich muß aber $\mathfrak{p}$ auch Häufungspunkt von $P = \mathfrak{M}E$ sein. Denn sonst gäbe es eine abgeschlossene Umgebung λ von $\mathfrak{p}$, welche außer $\mathfrak{p}$ keinen Maßpunkt mehr enthielte, und man könnte $\mathfrak{p}$ in ein Intervallgebiet ω von der Ausdehnung $\mathfrak{m}\omega < \varepsilon$ einschließen, so daß der abgeschlossene Restbereich $\lambda\overline{\omega}$ keinen Maßpunkt von E mehr enthielte, also $\mathfrak{m}E\lambda\overline{\omega} = 0$ wäre. Somit würde $\mathfrak{m}E\lambda = \mathfrak{m}E\lambda\omega + \mathfrak{m}E\lambda\overline{\omega} = \mathfrak{m}E\lambda\omega \leqq \mathfrak{m}\omega < \varepsilon$ beliebig klein, also Null und $\mathfrak{p}$ wäre kein Maßpunkt von E. Die Maßmenge P ist also „in sich dicht“ und, da jede Umgebung λ eines Häufungspunktes $\mathfrak{p}'$ von P wieder Maßpunkte von E enthält, also selbst maßhaltig ist, zugleich auch „abgeschlossen“ und damit „perfekt“, gehört also zum „abgeschlossenen Bestandteil“ der Ableitung E' von E. Das gleiche gilt auch von der „Diskrepanzmenge“ $Q = \mathfrak{D}E$, die ja, wie oben gezeigt, als „Maßmenge“ des Durchschnitts AB dargestellt werden kann. Wir haben somit:

Satz 2. *Die Maßmenge $\mathfrak{M}E$ wie die Diskrepanzmenge $\mathfrak{D}E$ jeder Punktmenge E sind perfekte Punktmengen, gehören also zum perfekten Bestandteile ihrer Ableitung E'.*

Weiter erhalten wir als Hauptsatz dieses Abschnittes:

Satz 3. *Das Maß einer beschränkten Punktmenge ist gleich dem Maß ihres Durchschnittes mit ihrer Maßmenge; die Diskrepanz einer Menge gleich der Diskrepanz ihres Durchschnittes mit ihrer Diskrepanzmenge.*

according to (11 a),

$$\mathfrak{d}E\lambda = \mathfrak{m}AB\lambda\,,$$

i.e., the set AB is measure-containing in every interval λ where E is discrepant, and vice versa, that is,

$$\mathfrak{D}E = \mathfrak{M}(AB)\,.$$

If C is any measurable set and λ an interval, e.g., the neighborhood of a point $\mathfrak{p}$, then, according to (10), we have

$$\mathfrak{m}E\lambda = \mathfrak{m}EC\lambda + \mathfrak{m}E\overline{C}\lambda$$

and

$$\mathfrak{d}E\lambda = \mathfrak{d}EC\lambda + \mathfrak{d}E\overline{C}\lambda\,;$$

every measure point of E is therefore also a measure point of EC or $E\overline{C}$, and every discrepancy point of E is also a discrepancy point of EC or $E\overline{C}$, i.e.,

$$(16)\quad \mathfrak{M}E = \mathfrak{M}(EC) + \mathfrak{M}(E\overline{C})\,,\qquad (17)\quad \mathfrak{D}E = \mathfrak{D}(EC) + \mathfrak{D}(E\overline{C})\,.$$

If $\mathfrak{p}$ is any measure point of the set E, and hence, if E is measure-containing in *every* neighborhood λ, $\mathfrak{m}E\lambda > 0$, then $\mathfrak{p}$ is of course also an accumulation point of E, and even a "condensation point" of E, since otherwise $E\lambda$ would be "countable" ⟦for some λ⟧ and of measure 0. But, at the same time, $\mathfrak{p}$ must also be an accumulation point of $P = \mathfrak{M}E$. For otherwise there would exist a closed neighborhood λ of $\mathfrak{p}$ containing, besides $\mathfrak{p}$, no further measure points, and it would be possible to include $\mathfrak{p}$ in an interval domain ω with the extension $\mathfrak{m}\omega < \varepsilon$ so that the closed remainder domain $\lambda\overline{\omega}$ would no longer contain a measure point of E, and hence we would have $\mathfrak{m}E\lambda\overline{\omega} = 0$. Thus $\mathfrak{m}E\,\lambda = \mathfrak{m}E\lambda\omega + \mathfrak{m}E\lambda\overline{\omega} = \mathfrak{m}E\lambda\omega \leqq \mathfrak{m}\omega < \varepsilon$ would become arbitrarily small, and hence zero, and $\mathfrak{p}$ would not be a measure point of E. The measure set P is therefore "dense within itself" and, since every neighborhood λ of an accumulation point $\mathfrak{p}'$ of P again contains measure points of E, and hence is measure-containing itself, also "closed" and thus "perfect". It therefore belongs to the "closed component" of the derivation E' of E. The same also holds of the "discrepancy set" $Q = \mathfrak{D}E$ which, as has been shown above, can be represented as the "measure set" of the intersection AB. We thus have:

Theorem 2. *Both the measure set $\mathfrak{M}E$ and the discrepancy set $\mathfrak{D}E$ of any point set E are perfect point sets, and hence belong to the perfect component of its derivation E'.*

Furthermore we obtain as the principal theorem of this section:

Theorem 3. *The measure of a bounded point set is equal to the measure of its intersection with its measure set; the discrepancy of a set is equal to the discrepancy of its intersection with its discrepancy set.*

165 | Es sei nämlich E wieder eine beliebige auf das Intervall Λ beschränkte Punktmenge und $P = \mathfrak{M}E$ ihre Maßmenge. Ferner sei $\varepsilon_1, \varepsilon_2, \ldots$ eine nach Null konvergierende Folge positiver Zahlen und E_n die Teilmenge des Durchschnittes $E\overline{P}$, deren Punkte von der *abgeschlossenen* Punktmenge P einen Abstand $\geqq \varepsilon_n$ besitzen. Dann gehört jeder Punkt von $E\overline{P}$ einer dieser Mengen E_n an (welche sukzessive ineinander liegen), und es ist

$$E\overline{P} = E_1 + E_2 + \cdots .$$

Jede dieser Teilmengen E_n ist aber eine „Nullmenge". Denn wäre E_n „maßhaltig" in Λ, so müßte nach Satz 1 E_n in Λ mindestens einen Maßpunkt $\mathfrak{p}_n$ besitzen, der zugleich ein *Häufungspunkt* von E_n und wegen $E_n \mathrel{\underline{\in}} E$ auch Maßpunkt von E, also ein Punkt von P wäre und dann einen Abstand $\geqq \varepsilon_n$ von E_n haben müßte. Somit wird in der Tat $\mathfrak{m}E_n = 0$ und nach (6a)

$$\mathfrak{m}E\overline{P} \leqq \mathfrak{m}E_1 + \mathfrak{m}E_2 + \cdots = 0\,,$$

also, da P als abgeschlossene Menge auch meßbar ist, nach (10)

$$\mathfrak{m}E = \mathfrak{m}EP + \mathfrak{m}E\overline{P} = \mathfrak{m}EP = \mathfrak{m}(E\mathfrak{M}E)\,. \tag{18}$$

Ist weiter E_1 eine beliebige Punktmenge zwischen EP und E, d. h. $EP \mathrel{\underline{\in}} E_1 \mathrel{\underline{\in}} E$, so ist nach dem Subsumptionssatze

$$\mathfrak{m}E = \mathfrak{m}EP \leqq \mathfrak{m}E_1 \leqq \mathfrak{m}E\,, \quad \text{also auch} \quad \mathfrak{m}E_1 = \mathfrak{m}E = \mathfrak{m}EP\,.$$

Das Maß einer Menge bleibt also ungeändert bei Weglassung von beliebig vielen Punkten, welche keine Maßpunkte sind. Da ferner die Diskrepanzmenge $Q = \mathfrak{D}E$ als abgeschlossene Menge gleichfalls meßbar ist, so folgt nach (11) und (10)

$$\mathfrak{d}E = \mathfrak{m}AB = \mathfrak{m}ABQ + \mathfrak{m}AB\overline{Q} = \mathfrak{d}EQ + \mathfrak{d}E\overline{Q}\,,$$

wo A und B maßgleiche Hüllen von E und $\overline{E}$ bezeichnen. Weiter ist

$$Q = \mathfrak{D}E = \mathfrak{M}(AB)\,,$$

und $AB\overline{Q}$ ist elementefremd zu $Q = \mathfrak{M}(AB)$, enthält keine Maßpunkte von AB und somit auch keine von $AB\overline{Q} \mathrel{\underline{\in}} AB$, ist mit ihrer eigenen Maßmenge punktfremd und hat daher das Maß Null (nach dem soeben Bewiesenen); d. h. es ist, wie behauptet

$$\mathfrak{d}E = \mathfrak{m}ABQ = \mathfrak{d}EQ = \mathfrak{d}(E\mathfrak{D}E)\,. \tag{19}$$

Eine Ausdehnung des Satzes auf eine zwischen E und EQ liegende Menge E_1 ist hier natürlich nicht möglich, da für die Diskrepanz ein Analogon des Subsumptionssatzes nicht gilt.

For let E again be an arbitrary point set bounded with respect to the interval Λ and let $P = \mathfrak{M}E$ be its measure set. Furthermore let $\varepsilon_1, \varepsilon_2, \ldots$ be a sequence of positive numbers converging to zero and let E_n be that partial set of the intersection $E\overline{P}$ whose points lie at a distance $\geqq \varepsilon_n$ from the *closed* point set P. Then every point of $E\overline{P}$ belongs to one of these sets E_n (which are successively nested), and we have

$$E\overline{P} = E_1 + E_2 + \cdots .$$

But each of these partial sets E_n is a "null set". For if E_n were "measure-containing" in Λ, then, according to theorem 1, E_n would have to contain at least one measure point $\mathfrak{p}_n$ in Λ, which would also be an *accumulation point* of E_n and, since $E_n \subseteq E$, also a measure point of E. Hence it would be a point of P and it would then lie at a distance $\geqq \varepsilon_n$ from E_n. Thus we in fact have $\mathfrak{m}\, E_n = 0$ and according to (6 a)

$$\mathfrak{m}E\overline{P} \leqq \mathfrak{m}E_1 + \mathfrak{m}E_2 + \cdots = 0 ,$$

and hence, since P, being a closed set, is also measurable, according to (10)

$$\mathfrak{m}E = \mathfrak{m}EP + \mathfrak{m}E\overline{P} = \mathfrak{m}EP = \mathfrak{m}(E\mathfrak{M}E) . \tag{18}$$

If, furthermore, E_1 is an arbitrary point set between EP and E, i.e., $EP \subseteq E_1 \subseteq E$, then, according to the subsumption theorem, we have

$$\mathfrak{m}E = \mathfrak{m}EP \leqq \mathfrak{m}E_1 \leqq \mathfrak{m}E , \quad \text{hence also} \quad \mathfrak{m}E_1 = \mathfrak{m}E = \mathfrak{m}EP .$$

The measure of a set therefore remains unchanged when arbitrarily many points that are not measure points are removed. Since, furthermore, the discrepancy set $Q = \mathfrak{D}E$, being a closed set, is also measurable, it follows, according to (11) and (10), that

$$\mathfrak{d}E = \mathfrak{m}AB = \mathfrak{m}ABQ + \mathfrak{m}AB\overline{Q} = \mathfrak{d}EQ + \mathfrak{d}E\overline{Q} ,$$

where A and B are isometric hulls of E and $\overline{E}$. Furthermore

$$Q = \mathfrak{D}E = \mathfrak{M}(AB) ,$$

and $AB\overline{Q}$ is elementwise disjoint from $Q = \mathfrak{M}(AB)$. It contains no measure points of AB, and hence also none of $AB\overline{Q} \subseteq AB$. It is pointwise disjoint from its own measure set, and hence possesses the measure zero (according to what has just been proved); i.e., we have, as stated,

$$\mathfrak{d}E = \mathfrak{m}ABQ = \mathfrak{d}EQ = \mathfrak{d}(E\mathfrak{D}E) . \tag{19}$$

In this case it is of course not possible to extend the theorem to a set E_1 lying between E and EQ, since an analogue of the subsumption theorem is not valid for the discrepancy.

Da die Eigenschaft eines „Maßpunktes“ bzw. eines „Diskrepanzpunktes“ $\mathfrak{p}$ sich nur auf die in der nächsten Umgebung von $\mathfrak{p}$ befindlichen Punkte von E bezieht und von den ferner gelegenen Punkten völlig unabhängig ist (es kommt immer nur der Durchschnitt $E\omega$ mit einer „Umgebung“ ω von $\mathfrak{p}$ in Betracht), so kann die „Maßmenge“ bzw. „Diskrepanzmenge“ ebenso gut wie für beschränkte auch für unbeschränkte Punktmengen E definiert werden, und die Sätze 1–3 behalten auch für solche Mengen ihre Gültigkeit. So ergeben sich als naturgemäße Erweiterungen der ursprünglichen Festsetzungen die Definitionen:

166 | *Eine (beschränkte oder unbeschränkte) Punktmenge E heißt eine „Nullmenge“, wenn sie keine Maßpunkte besitzt, ihr Durchschnitt $E\lambda$ mit jedem endlichen Intervall also das Maß Null hat; eine Menge heißt „meßbar“, wenn sie keine Diskrepanzpunkte besitzt, wenn sie also in keinem endlichen Intervall diskrepant ist.*

Die Sätze über „Nullmengen“ bzw. „meßbare Mengen“ behalten dann sämtlich ihre Gültigkeit, soweit ihre in endlichen Intervallen enthaltenen Teilmengen in Betracht kommen. Nur brauchen nicht mehr ein „Gesamtmaß“ $\mathfrak{m}E$ oder eine „Gesamtdiskrepanz“ $\mathfrak{d}E$ als endliche Zahlen zu existieren. Doch können solche in besonderen Fällen als Grenzwerte definiert werden. Es sei nämlich $\lambda_1, \lambda_2, \ldots$ eine unbegrenzte Folge von (exklusiven oder nicht exklusiven) Intervallen von der Beschaffenheit, daß jeder Punkt von E in mindestens einem dieser Intervalle gelegen ist, so daß die Summe

$$\Sigma = \lambda_1 + \lambda_2 + \cdots ,$$

die sich auch ins Unendliche erstrecken kann, die gesamte Menge E überdeckt. Bezeichnen wir nun mit Σ_n die n-te Teilsumme

$$\Sigma_n = \lambda_1 + \lambda_2 + \cdots + \lambda_n ,$$

so können wir setzen

$$\mathfrak{m}E = \lim_{n=\infty} \mathfrak{m}(E\Sigma_n) , \qquad \mathfrak{d}E = \lim_{n=\infty} \mathfrak{d}(E\Sigma_n) ,$$

insofern diese Grenzwerte existieren.

Da nun für alle Intervallsummen nach (6a) und (7a) jedenfalls

$$\mathfrak{m}E\Sigma_n \leqq \mathfrak{m}E\lambda_1 + \mathfrak{m}E\lambda_2 + \cdots + \mathfrak{m}E\lambda_n$$
$$\mathfrak{d}E\Sigma_n \leqq \mathfrak{d}E\lambda_1 + \mathfrak{d}E\lambda_2 + \cdots + \mathfrak{d}E\lambda_n ,$$

so wird allgemein

$$\mathfrak{m}E = \lim_{n=\infty} \mathfrak{m}(E\Sigma_n) \leqq \mathfrak{m}E\lambda_1 + \mathfrak{m}E\lambda_2 + \cdots \tag{20}$$

$$\mathfrak{d}E = \lim_{n=\infty} \mathfrak{d}(E\Sigma_n) \leqq \mathfrak{d}E\lambda_1 + \mathfrak{d}E\lambda_2 + \cdots . \tag{21}$$

Since the property of a "measure point" and that of a "discrepancy point" $\mathfrak{p}$ refers only to the points of E in the immediate neighborhood of $\mathfrak{p}$ and is entirely independent of the points at greater distance (only the intersection $E\omega$ with a "neighborhood" ω of $\mathfrak{p}$ is being considered), it is possible to define the "measure set" and "discrepancy set", respectively, for unbounded point sets E just as well as for bounded ones, and the theorems 1–3 remain valid for such sets as well. We thus obtain the following definitions as natural expansions of the original stipulations:

A (bounded or unbounded) point set E is called a "null set" if it possesses no measure points, that is, if its intersection $E\lambda$ with every finite interval has the measure zero; a set is called "measurable" if it possesses no discrepancy points, that is, if it is discrepant in no finite interval.

The theorems on "null sets" and on "measurable sets" then all still hold as far as their partial sets contained in finite intervals are concerned. But a "total measure" $\mathfrak{m}E$ or a "total discrepancy" $\mathfrak{d}E$ need no longer exist as finite numbers. It is, however, possible to define them as limits in special cases. For let $\lambda_1, \lambda_2, \ldots$ be an infinite sequence of (exclusive or not exclusive) intervals so constituted that every point of E lies in at least one of these intervals so that the sum

$$\Sigma = \lambda_1 + \lambda_2 + \cdots ,$$

which can also extend to infinity, covers the entire set E. Calling the nth partial sum Σ_n

$$\Sigma_n = \lambda_1 + \lambda_2 + \cdots + \lambda_n ,$$

we can put

$$\mathfrak{m}E = \lim_{n=\infty} \mathfrak{m}(E\Sigma_n) , \qquad \mathfrak{d}E = \lim_{n=\infty} \mathfrak{d}(E\Sigma_n) ,$$

provided that these limits exist.

Since now for all interval sums we certainly have, according to (6 a) and (7 a),

$$\begin{aligned} \mathfrak{m}E\Sigma_n &\leqq \mathfrak{m}E\lambda_1 + \mathfrak{m}E\lambda_2 + \cdots + \mathfrak{m}E\lambda_n \\ \mathfrak{d}E\Sigma_n &\leqq \mathfrak{d}E\lambda_1 + \mathfrak{d}E\lambda_2 + \cdots + \mathfrak{d}E\lambda_n , \end{aligned}$$

we generally have

$$\mathfrak{m}E = \lim_{n=\infty} \mathfrak{m}(E\Sigma_n) \leqq \mathfrak{m}E\lambda_1 + \mathfrak{m}E\lambda_2 + \cdots \tag{20}$$

$$\mathfrak{d}E = \lim_{n=\infty} \mathfrak{d}(E\Sigma_n) \leqq \mathfrak{d}E\lambda_1 + \mathfrak{d}E\lambda_2 + \cdots . \tag{21}$$

Sind aber die Intervalle $\lambda_1, \lambda_2, \ldots$ untereinander exklusiv, so wird nach (13) und (14), da Intervalle immer meßbar sind,

$$\mathfrak{m}E\Sigma_n = \mathfrak{m}E\lambda_1 + \cdots + \mathfrak{m}E\lambda_n$$
$$\mathfrak{d}E\Sigma_n = \mathfrak{d}E\lambda_1 + \cdots + \mathfrak{d}E\lambda_n\,,$$

und somit

$$\mathfrak{m}E = \mathfrak{m}E\lambda_1 + \mathfrak{m}E\lambda_2 + \cdots \tag{20a}$$
$$\mathfrak{d}E = \mathfrak{d}E\lambda_1 + \mathfrak{d}E\lambda_2 + \cdots\,. \tag{21a}$$

Insbesondere ergibt sich hieraus für „Nullmengen“ E stets ein „Gesamtmaß“ 0 und für „meßbare“ Mengen eine „Gesamtdiskrepanz“ 0.

Haben die (unbeschränkten) Mengen E_1 und E_2 endliche Maßzahlen bzw. Diskrepanzen, so ergibt sich aus den Hauptformeln (3) und (5) für jede Intervallsumme Σ_n

$$\mathfrak{m}(E_1 + E_2)\Sigma_n + \mathfrak{m}E_1E_2\Sigma_n \leqq \mathfrak{m}E_1\Sigma_n + \mathfrak{m}E_2\Sigma_n$$
$$\mathfrak{d}(E_1 + E_2)\Sigma_n + \mathfrak{d}E_1E_2\Sigma_n \leqq \mathfrak{d}E_1\Sigma_n + \mathfrak{d}E_2\Sigma_n$$

und, wenn die Grenzwerte $\mathfrak{m}E_1$, $\mathfrak{m}E_2$ bzw. $\mathfrak{d}E_1$, $\mathfrak{d}E_2$ existieren, wieder

167 | $$\mathfrak{m}(E_1 + E_2) + \mathfrak{m}E_1E_2 \leqq \mathfrak{m}E_1 + \mathfrak{m}E_2$$

$$\mathfrak{d}(E_1 + E_2) + \mathfrak{d}E_1E_2 \leqq \mathfrak{d}E_1 + \mathfrak{d}E_2\,,$$

und entsprechend für endliche und abzählbar unendliche Summen

$$\mathfrak{m}(E_1 + E_2 + \cdots) \leqq \mathfrak{m}E_1 + \mathfrak{m}E_2 + \cdots$$
$$\mathfrak{d}(E_1 + E_2 + \cdots) \leqq \mathfrak{d}E_1 + \mathfrak{d}E_2 + \cdots\,,$$

wobei für „meßbare“ und exklusive Summanden wieder das Gleichheitszeichen gelten muß.

Auch der Begriff der „maßgleichen Hülle“ läßt sich ausdehnen auf unbeschränkte Punktmengen vermöge der folgenden Betrachtung. Ist A maßgleiche Hülle von E im „Gesamtintervall“ Λ, und λ irgendein Teilintervall, so ist nach S. 161 auch $A\lambda$ maßgleiche Hülle von $E\lambda$. Besitzt ferner die unbeschränkte Punktmenge E in den exklusiven Intervallen $\lambda_1, \lambda_2, \ldots$ die maßgleichen Hüllen $A_1, A_2, \ldots$, also $\mathfrak{m}E\lambda_n = \mathfrak{m}A_n = \mathfrak{m}A_n\lambda_n$ und ist λ irgendein weiteres (endliches) Intervall, so ist auch

$$\mathfrak{m}E\lambda_n\lambda = \mathfrak{m}A_n\lambda\,,$$

sowie

$$\mathfrak{m}E(\lambda_1 + \lambda_2 + \cdots)\lambda = \mathfrak{m}(A_1 + A_2 + \cdots)\lambda = \mathfrak{m}A\lambda\,,$$

But if the intervals $\lambda_1, \lambda_2, \ldots$ are mutually exclusive, then, since intervals are always measurable, we have, according to (13) and (14),

$$\begin{aligned}\mathfrak{m}E\Sigma_n &= \mathfrak{m}E\lambda_1 + \cdots + \mathfrak{m}E\lambda_n \\ \mathfrak{d}E\Sigma_n &= \mathfrak{d}E\lambda_1 + \cdots + \mathfrak{d}E\lambda_n\,,\end{aligned}$$

and thus

$$\mathfrak{m}E = \mathfrak{m}E\lambda_1 + \mathfrak{m}E\lambda_2 + \cdots \tag{20a}$$

$$\mathfrak{d}E = \mathfrak{d}E\lambda_1 + \mathfrak{d}E\lambda_2 + \cdots\,. \tag{21a}$$

This in particular always yields a "total measure" 0 for "null sets" E and a "total discrepancy" 0 for "measurable" sets.

If the (unbounded) sets E_1 and E_2 possess finite measure numbers or discrepancies, then the principal formulas (3) and (5) yield for each interval sum Σ_n

$$\begin{aligned}\mathfrak{m}(E_1+E_2)\Sigma_n + \mathfrak{m}E_1E_2\Sigma_n &\leqq \mathfrak{m}E_1\Sigma_n + \mathfrak{m}E_2\Sigma_n \\ \mathfrak{d}(E_1+E_2)\Sigma_n + \mathfrak{d}E_1E_2\Sigma_n &\leqq \mathfrak{d}E_1\Sigma_n + \mathfrak{d}E_2\Sigma_n\end{aligned}$$

and, if the limits $\mathfrak{m}E_1, \mathfrak{m}E_2$ and $\mathfrak{d}E_1, \mathfrak{d}E_2$ respectively exist, again

$$\begin{aligned}\mathfrak{m}(E_1+E_2) + \mathfrak{m}E_1E_2 &\leqq \mathfrak{m}E_1 + \mathfrak{m}E_2 \\ \mathfrak{d}(E_1+E_2) + \mathfrak{d}E_1E_2 &\leqq \mathfrak{d}E_1 + \mathfrak{d}E_2\,,\end{aligned}$$

and, accordingly, for finite and countably infinite sums

$$\begin{aligned}\mathfrak{m}(E_1+E_2+\cdots) &\leqq \mathfrak{m}E_1 + \mathfrak{m}E_2 + \cdots \\ \mathfrak{d}(E_1+E_2+\cdots) &\leqq \mathfrak{d}E_1 + \mathfrak{d}E_2 + \cdots\,,\end{aligned}$$

where the equality must again hold for "measurable" and exclusive summands.

It is also possible to extend the concept of the "isometric hull" to unbounded point sets by virtue of the following consideration. If A is an isometric hull of E in the "total interval" Λ, and λ some partial interval, then, according to p. 161, $A\lambda$, too, is an isometric hull of $E\lambda$. If, furthermore, the unbounded point set E possesses in the exclusive intervals $\lambda_1, \lambda_2, \ldots$ the isometric hulls $A_1, A_2, \ldots$, and hence $\mathfrak{m}E\lambda_n = \mathfrak{m}A_n = \mathfrak{m}A_n\lambda_n$, and if λ is some further (finite) interval, then we also have

$$\mathfrak{m}E\lambda_n\lambda = \mathfrak{m}A_n\lambda\,,$$

as well as

$$\mathfrak{m}E(\lambda_1+\lambda_2+\cdots)\lambda = \mathfrak{m}(A_1+A_2+\cdots)\lambda = \mathfrak{m}A\lambda\,,$$

wobei die Summe auf die Intervalle $\lambda_1, \lambda_2, \ldots$ beschränkt werden kann, die zur Überdeckung von λ ausreichen, also

$$\mathfrak{m}E\lambda = \mathfrak{m}A\lambda\,,$$

d. h. die Punktmenge $A = A_1 + A_2 + \cdots$ spielt in *jedem* „Gesamtintervall" λ für E die Rolle einer „maßgleichen Hülle", die dann noch gemäß Satz 3 auf ihre Maßmenge $\mathfrak{M}A = \mathfrak{M}E$ „reduziert" werden kann.

Ebenso kann auch für $\overline{E}$, die Komplementärmenge von E in bezug auf den (unendlichen) Gesamtraum, eine entsprechende „maßgleiche Hülle"

$$B = B_1 + B_2 + \cdots$$

konstruiert werden, so daß in *jedem* Intervall λ

$$\mathfrak{m}\overline{E}\lambda = \mathfrak{m}B\lambda$$

wird. Dann wird auch in jedem Intervall λ

$$\mathfrak{d}(E\lambda) = \mathfrak{m}(AB\lambda)\,,$$

und der (wieder nach Satz 3 auf seine Maßmenge $\mathfrak{M}(AB) = \mathfrak{D}E$ reduzierbare) Durchschnitt

$$C = AB \in \mathfrak{D}E$$

(die „Diskrepanzhülle" von E) hat dann die Eigenschaft, daß ihr Maß in jedem Intervalle die Diskrepanz der „Ausgangsmenge" E darstellt, wiewohl das „Gesamtintervall" jetzt unendlich ist und damit die ursprüngliche Definition der Diskrepanz (als Differenz des „äußeren" und „inneren Maßes" bei *Lebesgue*) gemäß der Formel (4) ihre Bedeutung verliert.

Freiburg i. Br., den 12. Juni 1927.

where it is possible to restrict the sum to the intervals λ_1, $\lambda_2, \ldots$ sufficient for the covering of λ, and hence

$$\mathfrak{m}E\lambda = \mathfrak{m}A\lambda\,,$$

i.e., in *every* "total interval" λ the point set $A = A_1 + A_2 + \cdots$ plays vis-à-vis E the part of an "isometric hull", which, according to theorem 3, can then also be "reduced" to its measure set $\mathfrak{M}A = \mathfrak{M}E$.

Likewise, it is possible to construct also for $\overline{E}$, the complementary set of E with respect to the (infinite) total space, a corresponding "isometric hull"

$$B = B_1 + B_2 + \cdots$$

so that, in *every* interval λ, we have

$$\mathfrak{m}\overline{E}\lambda = \mathfrak{m}B\lambda\,.$$

Then, in every interval λ, we also have

$$\mathfrak{d}(E\lambda) = \mathfrak{m}(AB\lambda)\,,$$

and the intersection (reducible, again, according to theorem 3, to its measure set $\mathfrak{M}(AB) = \mathfrak{D}E$)

$$C = AB \subseteq \mathfrak{D}E$$

(the "discrepancy hull" of E) then possesses the property of its measure representing the discrepancy of the "origin set" E in every interval, although the "total interval" is now infinite, and hence the original definition of discrepancy (as the difference of the "outer" and the "inner measure" in *Lebesgue*) in accordance with the formula (4) loses its meaning.

Freiburg i. Br., on the 12th of June 1927.

Zusatz zu § 5

D. König 1927b

Nachdem ich die vorangehenden Untersuchungen Herrn *Zermelo* zukommen liess, hatte er die Freundlichkeit, den folgenden Beweis seines oben erwähnten Satzes über die Schranke t mir mitzuteilen.

Da die Gesamtheit aller Positionen endlich ist, so ist die Gesamtheit derjenigen Positionen, in denen Weiss am Zuge ist und von denen aus Weiss den Sieg in höchstens r Zügen, aber nicht in weniger Zügen, erzwingen kann, ebenfalls endlich. Diese endliche Anzahl sei m_r $(r = 1, 2, 3, \ldots)$. Aus demselben Grund muss die Summe $\sum m_r = m_1 + m_2 + m_3 + \cdots$, als die Anzahl verschiedener Positionen, endlich sein, d. h. mit einem Glied m_λ abbrechen; so dass $m_\lambda \geqq 1$, aber für $r > \lambda$ stets $m_r = 0$ wird (es ist klar, dass nicht sämtliche m_r verschwinden können). Es sei p eine Position, von der aus Weiss, mit dem Zuge w_1 beginnend, den Sieg in höchstens r Zügen, nicht aber in weniger Zügen erzwingen kann. Aus jedem der durch einen auf w_1 folgenden Zug von Schwarz entstehenden endlichvielen Positionen kann dann Weiss den Sieg in höchstens $r-1$ Zügen erzwingen. Unter diesen endlichvielen Positionen gibt es auch sicher eine, von der aus Weiss in weniger als $r-1$ Zügen den Sieg *nicht* erzwingen kann, da sonst Weiss von p aus in weniger als r Zügen den Sieg erzwingen könnte. Mit m_r ist also auch m_{r-1} von Null verschieden. Da also $m_\lambda \geqq 1$ ist, ist auch $m_{\lambda-1} \geqq 1$, also auch $m_{\lambda-2} \geqq 1$, u. s. w. bis $m_1 \geqq 1$. Hieraus folgt, dass die Anzahl sämtlicher Positionen mit Weiss am Zuge, von denen aus Weiss den Gewinn in einer begrenzten Anzahl von Zügen erzwingen kann

$$m = \sum m_r = m_1 + m_2 + \cdots + m_\lambda$$

grösser oder gleich λ ist. Andererseits ist natürlich m kleiner als die Anzahl t sämtlicher Positionen mit Weiss am Zuge. Also ist $\lambda < t$. Lässt sich also der Gewinn von einer Position aus in einer begrenzten Anzahl von Zügen erzwingen, so lässt er sich auch in weniger als t Zügen. Da aber — wie wir oben gezeigt haben — aus jeder Gewinnstellung der Sieg in einer begrenzten Anzahl von Zügen erreicht werden kann, so ist in der Tat t eine universelle obere Schranke für die nötigen Zugzahlen bei irgendeiner Gewinnstellung.

130 Durch diesen Beweis[1] hat *Zermelo* die oben erwähnte Lücke | in seiner Cambridger Darstellung vollkommen und zwar in elegantester Weise ausgefüllt.

Einer oben ausgesprochenen Vermutung entsprechend benutzt dieser Beweis den *Neumann*schen Satz über die Beschränktheit der Zugzahlen, den wir

[1] Laut früheren mündlichen Mitteilungen des Herrn *J. von Neumann* war ihm ein auf derselben Grundidee beruhender Beweis bekannt.

Addition to § 5

D. König 1927b

⟦The introductory note just before *1913* also addresses *D. König 1927b*.⟧

After having received from me the preceding investigations, Mr. *Zermelo* was kind enough to communicate to me the following proof of his theorem about the bound t mentioned above.

Since the totality of all positions is finite, the totality of those positions in which it is White's turn to move and from which White can force victory in at most, but not less than, r moves is also finite. Let this finite number be m_r $(r = 1, 2, 3, \ldots)$. For the same reason, the sum $\sum m_r = m_1 + m_2 + m_3 + \ldots$, being the number of different positions, must be finite, that is, it must terminate with a member m_λ; so that we have $m_\lambda \geq 1$, but for $r > \lambda$ always $m_r = 0$ (it is obvious that not all m_r can disappear). Let p be a position from which White, starting with the move w_1, can force victory in at most, but not less than, r moves. From each of the finitely many positions arising from a move by Black in immediate succession to w_1 White can then force victory in at most $r - 1$ moves. Among these finitely many positions there certainly is also one from which White can *not* force victory in less than $r - 1$ moves, since, otherwise, White could force victory in less than r moves from p. Hence, along with m_r, m_{r-1}, too, is different from zero. Since we therefore have $m_\lambda \geq 1$, we also have $m_{\lambda-1} \geq 1$, hence also $m_{\lambda-2} \geq 1$, etc., up to $m_1 \geq 1$. From this it follows that the number of all positions in which it is White's turn to move and from which White can force victory in a limited number of moves

$$m = \sum m_r = m_1 + m_2 + \cdots + m_\lambda$$

is greater than or equal to λ. On the other hand, m is of course smaller than the number t of all positions in which it is White's turn to move. Hence, $\lambda < t$. If we can therefore force victory from a position in a limited number of moves, then we can also do so in a number of moves smaller than t. But since it is possible to gain victory—as we have shown above—from every winning position in a limited number of moves, t in fact is a universal upper bound of the number of necessary moves for any winning position.

With this proof,[1] *Zermelo* has completely and most elegantly filled the gap in his Cambridge account.

In accordance with a conjecture put forth above, this proof makes use of *Neumann*'s theorem on the boundedness of the number of moves, which

[1] According to earlier, personal communications with Mr. J. von *Neumann*, a proof was known to him, which rests on the very same basic idea.

oben auf das Lemma A) zurückgeführt haben. Auch für diesen Beschränktheitssatz hat mir Herr *Zermelo* einen Beweis mitgeteilt, der vom Lemma A) explizite keinen Gebrauch macht. Dieser ausserordentlich einfach formulierte Beweis soll — ebenfalls mit der Erlaubniss des Herrn *Zermelo* — hier *wörtlich* mitgeteilt werden.[1]

„Es sei p_0 eine Position, in welcher Weiss am Zuge in *keiner begrenzten* Anzahl von Zügen das Mat erzwingen kann, sondern, je nach dem Spiel des Gegners, vielleicht in unbegrenzt wachsender Anzahl. Dann wird auf *jedem* Zug von Weiss der Schwarze eine Position p_1 herbeiführen können, welche von der *gleichen* Beschaffenheit ist. Denn sonst würde, da die Zahl der möglichen Züge endlich ist, Weiss auch von p_0 aus mit begrenzter Zugzahl zum Ziele kommen. Somit wird, wie Weiss auch spielt, bei richtigem Gegenspiel eine unbegrenzte Folge $p_0, p_1, p_2, \ldots$ von Positionen (mit Weiss am Zuge) entstehen, welche *sämtlich* die Beschaffenheit p_0 haben, also niemals zum Mat führen können. Ist also in einer Position p_0 der Sieg *überhaupt* zu erzwingen, so auch in einer begrenzten Zugzahl.“

Zu diesem Beweis will ich nur folgende Bemerkung hinzufügen. Wenn man diesen Beweis ganz ausführlich darstellt und bis auf die oben gegebene Definition der „Gewinnstellung“ zurückführt, ersieht man, dass er mit dem oben (§ 5) gegebenen übereinstimmt, nur dass er auch den oben (§ 1) gegebenen Beweis des Lemmas A) mitenthält. — Wenn auch die Formulierung des Beweises hierdurch länger wird, scheint mir ein Beweis, der den Beweis des Lemmas A) von den übrigen Überlegungen lostrennt, klarer den logischen Inhalt hervortreten zu lassen. Und zwar gilt dies nicht nur für die Anwendung auf die Theorie der Spiele, sondern auf jede Überlegung (wie die in § 2, 3 oder 4), welche im Wesentlichen auf dem Lemma A) basiert. Die Sache steht hier ähnlich, wie bei vielen Beweisen der Analysis und der Geometrie, wo durch die Isolierung des implizite angewandten *Borel*schen Überdeckungssatzes die Überlegungen durchsichtiger werden.

[1] Herr *Zermelo* beabsichtigt seine Untersuchungen über die Schach-Theorie demnächst in zusammenhängender Darstellung erscheinen ⟦zu⟧ lassen.

was reduced to lemma A) above. Mr. *Zermelo* has communicated to me a proof of this theorem on boundedness as well, which makes no explicit use of lemma A). This proof, which is stated with extraordinary simplicity, shall be communicated here *verbatim*—again with the permission of Mr. *Zermelo*.[2]

"Let p_0 be a position in which it is White's turn to move and in which White can force checkmate in *no limited* number of moves but perhaps, depending on the opponent's play, in an indefinitely increasing number. In succession to *every* move by White, Black will then be able to create a position p_1 constituted *identically*. For otherwise White would reach the goal also from p_0 in a limited number of moves since the number of possible moves is finite. Thus, whatever White's play may be, an infinite sequence $p_0, p_1, p_2, \ldots$ of positions (in which it is White's turn) will arise, *all* of which have the constitution p_0, and hence can never lead to checkmate, assuming the opponent plays correctly. So if it is possible to force victory *at all* in some position p_0, then likewise in a limited number of moves."

To this proof I will only add the following comment. When stated in full detail and reduced to the definition of "winning positions" given above, this proof can be seen to agree with the one stated above (§5), except that it also already contains the proof of lemma A) stated above (§1).—It seems to me that, even at the expense of greater length, a proof that separates the proof of lemma A) from the other considerations brings out more clearly the logical content. This holds true not only for the application to the theory of games, but for applications to any consideration (such as those in § 2, 3, and 4) that essentially rests on lemma A). This case is similar to that of many proofs in analysis and geometry in which considerations are made more transparent by isolating *Borel*'s covering theorem, which is applied implicitly.

[2] Mr. *Zermelo* intends to publish a comprehensive account of his investigations in the theory of chess in the near future.

Introductory note to 1929a

Heinz-Dieter Ebbinghaus

Zermelo's *1929a*, the first set-theoretic paper to appear after a break of more than 20 years, is a reaction against criticism of his 1908 notion of definiteness. The paper provoked a swift response by Thoralf Skolem, *1930*, a response which had a strong impact on Zermelo's foundational views in the 1930s. In the following we first describe the range and nature of the criticism. We then comment on the paper, and in the concluding section we discuss the impact of Skolem's response.[1]

Definiteness between 1908 and 1929

When axiomatizing set theory in his *1908b*, Zermelo had in mind two aims: to have the axiom system suitably strong in order to preserve the scope of naïve set theory, and to have it sufficiently narrow in order to avoid inconsistencies. An essential feature of his system is the axiom of separation. It restricts unlimited comprehension to comprehension within a set and is intended to exclude inconsistent totalities such as the set of all sets, the set of all ordinals, or the set $\{x \mid x \notin x\}$ leading to the Zermelo-Russell paradox. Basically it says that for any set S and (certain) properties $E(x)$ there exists the set $\{x \in S \mid E(x)\}$ of the elements of S having property E. In order to exclude Richard's paradox, Zermelo restricts the propositional functions allowed for defining the properties E to those which he calls "definite", providing the following description (*1908b*, 263): A propositional function is definite if *the fundamental relations of the domain, by means of the axioms and the universally valid laws of logic, determine without arbitrariness whether it holds or not.* The criticism of Zermelo's notion of definiteness is rooted in the vagueness of this definition. We describe, in chronological order, how after raising objections solutions were proposed successively by Hermann Weyl, Arthur Schoenflies, Abraham Fraenkel, and Skolem.

In his *Habilitation* address Weyl states (*1910*, 112) that Zermelo's notion "is in need of greater precision." He sketches an alternative definition of definiteness which in essence coincides with first-order definability where arbitrary sets are allowed as parameters, i.e., with the version proposed by David Hilbert in 1920[2] and by Skolem in 1922 (see below). Later, in *Das*

[1] For more details we refer to section 4.9. of *Moore 1982*, to *Ebbinghaus 2003*, and to sections 2.9.4 and 4.6 of *Ebbinghaus 2007b*.

[2] Hilbert made this precise in his lecture course "Probleme der mathematischen Logik", Göttingen, summer semester 1920, in the section "§5 (Zermelos axiomatische Begründung der Mengenlehre und Analysis)". As Weyl made his proposal while also staying in Göttingen, it may be reasonable to assume that already in

Kontinuum, Weyl characterizes Zermelo's definition as an "apparently unsatisfactory explanation" (*1918*, 36), again proposing a first-order solution and thereby touching a point which plays an important role for Zermelo: The first-order version of definiteness asks for the use of "finitely many" in the inductive definition of first-order formulae; it thus makes use of the notion of natural number. Zermelo refused to accept basic properties of the natural numbers when axiomatizing set theory, as it was set theory that should provide these properties.[3] Weyl distances himself sharply from such reservations, describing his former efforts to avoid the use of the notion of finite number as "hunting a scholastic pseudo-problem" and expresses his conviction that "*the imagination of iteration, of the series of natural numbers, should form the ultimate foundation of mathematical thinking*" and should not itself be subject to a set-theoretic foundation (*1918*, 36ff), thus also sharing a basic conviction with intuitionism.

Schoenflies interprets "definite" as "uniquely determined", namely by the axioms (pointing to Hilbert's axioms of geometry and the "between" relation as an example) and by logic (pointing to the use of terms such as "identical" and "all"). He therefore can state (*1911*, 227, fn. 2) that Zermelo's definition of definiteness is dispensable.

Between 1922 and 1927 Fraenkel criticizes Zermelo's notion several times and in several ways. Early on he calls Zermelo's description (*1922b*, 231) "the weakest point or rather the only weak point" of the axiomatization and (*1923b*, 2) "an essential deficiency". In the second edition of his *Einleitung in die Mengenlehre* (*1923a*, 196ff) he speaks of a mathematically not well-defined concept that might become a new source of concern. In his *Zehn Vorlesungen*, in a subsection entitled "The necessity of eradicating the notion of property", he characterizes Zermelo's explanation as a mere comment that would lead to difficulties "lying on the same level as Richard's paradox", as it referred to the axioms and at the same time was an essential part of their formulation (*1927*, 104). In order to "eradicate this sore spot" he then gives the version of definiteness which he had already published before (*1922a*, 253–254, *1925*, 254) and which he also uses in the second edition of his *Mengenlehre*. It is given by a rather involved inductive definition which does, however, not go beyond first-order definability.

In his address *Skolem 1923* to the 1922 Helsinki Congress of Scandinavian Mathematicians, Skolem calls Zermelo's notion of definiteness "a very

---

the 1910s definiteness was generally understood as first-order definability in the Hilbert circle.

[3] Zermelo never accepted a distinction between formulas and the objects they are about such as numbers, points, or sets. (For a possible exception concerning variables, see *s1931g*.) In particular, he did not distinguish between meta-language, where the use of natural numbers should not matter, and object language, an attitude that did not change even when such a distinction became common in the mid-1930s and indispensable for understanding, for example, Gödel's incompleteness results.

imperfect point". In order to eliminate this deficiency, he proposes identifying definiteness with first-order definability. Taking the axiom of replacement into account as well, he thus gives the Zermelo axioms what was to become their final form in modern set theory.

Zermelo's definiteness paper

Zermelo became acquainted with the criticism of Fraenkel, Schoenflies, and Weyl; he probably did not become aware in the 1920s of Skolem's Helsinki address. In early 1929 Zermelo decided to answer his critics. In May, during a stay in Warsaw, he discussed the matter in particular with Bronisław Knaster, Stanisław Leśniewski, and Alfred Tarski.[4] As a result, he felt encouraged to write the paper in late spring.

In the beginning, he states that, apparently, not everyone had understood his concept of definiteness, that some had found it unclear, and some had tried to avoid it. The latter are divided into two groups.[5]

The first group consists of those who simply drop the concept, find it superfluous, or leave it to general logic. He may be referring here, for example, to Schoenflies. He may also be referring to work in set theory that did not take foundational problems into consideration. Such an attitude is reflected, for instance, in Hausdorff's influential *Grundzüge der Mengenlehre* (*1914*); in the preface of the second edition (*1927*), Hausdorff states that, as in the first edition, he did not see the need to discuss foundational issues.[6]

Zermelo defends his procedure by pointing to his aim to base set theory on assumptions as weak as possible in order to avoid the paradoxes. It would not have been appropriate to give a logistic foundation at a time "where no universally acknowledged mathematical logic existed" and "most mathematicians harboured suspicions against any kind of logistic."

The second group avoiding the concept of definiteness consists of those who restrict it to special propositional formulae. The reference here is Abraham Fraenkel.[7] As stated above, Fraenkel's formulae are inductively defined, a way which Zermelo terms "constructive" or "genetic" and which he decries because of its implicit use of the notion of natural number. He then pleads for the "axiomatic" way as the right one for fixing definiteness, giving credit to John von Neumann for having performed this task in his axiomatization

---

[4] See Zermelo's report to the Notgemeinschaft der deutschen Wissenschaft of 9 December 1929, Universitätsarchiv Freiburg, C 129/140.

[5] In his paper *1908a* Zermelo organized the critics of the axiom of choice into different groups and proceeded to argue against each. Now, he is following a similar approach with definiteness.

[6] For Hausdorff's foundational views, cf. *Purkert 2002*, in particular sect. 1.

[7] Very probably, Zermelo knew about Weyl's first-order approach to definiteness and perhaps also about Hilbert's explicit first-order definition of definiteness in his 1920 lecture course.

of set theory (*1925*), but blaming this approach as overly knotty and difficult to understand and promising a simpler solution in what follows.

Zermelo's distinction between "genetic" and "axiomatic" may go back to discussions with David Hilbert. In *1900a*, 180–181, Hilbert refers to two different ways of defining infinite totalities, namely "genetically" (sometimes also termed "constructively"), e.g., by some inductive definition, or "axiomatically", i.e., by a system of postulates which should be satisfied by the intended domain.[8] Examples of the first kind include the natural numbers as arising from the number zero by finitely often applying the successor operation. Examples of the latter kind include the natural numbers as given by the Peano axioms and the real numbers as given by the axioms for completely ordered fields. Hilbert pleads for the latter kind (ibid.): "Despite the high pedagogic and heuristic value of the genetic method, for the final presentation and the complete logical grounding of our knowledge the axiomatic method deserves the first rank." In 1912 Zermelo gave an address about this dualism at the Fifth International Congresss of Mathematicians in Cambridge entitled "On axiomatic and genetic methods in the foundation of mathematical disciplines" ("Ueber axiomatische und genetische Methoden bei der Grundlegung mathematischer Disziplinen"). Nothing is known about its contents. However, it is plausible that he supported Hilbert's opinion.

After having discussed the different approaches to definiteness, Zermelo presents his own way. Using modern terms, we may describe it for the case of sets as follows: A proposition φ which is in some way "composed" from atomic statements of the form $a \in b$, is definite if, given any set M and any assignment of elements of M to constants, it is determined as either true or false. Atomic statements alien to set theory such as "a is painted green" or "a is definable in English by a finite number of words" (recalling Richard's paradox) are explicitly excluded.

Zermelo does not distinguish between constants and elements; the atomic statements, his "fundamental relations", are formed by the elements themselves. In fact, he will never distinguish explicitly between a sentence and its meaning, a laxity which will separate him from the mainstream of mathematical logic in the 1930s when such a distinction became commonplace.

Having provided the frame of atomic statements, he continues that it suggests itself as a determination of the concept of definiteness to just allow as compositions the logical operations of negation, conjunction, disjunction, and quantification, performed an arbitrary finite number of times. Criticizing once more the use of the notion of finiteness in this genetic approach, he arrives at his favoured axiomatic definition (p. 343):

(1) The set of definite propositions contains the fundamental relations and is closed under the operations of negation, conjunction, disjunction, first-order quantification, and second-order quantification.

[8] For the early discussion of "construction" *vs.* "axiomatization" cf. *Ferreirós 1999*, 119ff. See also §3 of the introductory note to *1908b*.

(2) The set of definite propositions does not possess a proper part satisfying (1).

To be more exact, the second-order quantifications range only over properties defined by definite propositional functions, a problematic point which Skolem will address in his reply.

Zermelo claims (p. 340) that this definition makes precise what he had in mind when writing down his 1908 version. Indeed, the nature of the fundamental relations is clarified. However, the role of (the universally valid laws of) logic amounts to nothing more than providing propositional connectives and first-order and second-order quantifications, and the axioms do not have an explicit role at all.

When motivating his definition of definiteness, Zermelo introduces (p. 341) the notion of a categorical axiom system without making use of it. This passage reflects an approach to the notion of set which he had presented in one of his Warsaw talks (*s1929b*) some weeks earlier and which was eventually to become his favourite one, leading to a reformulation of the axiom of separation.[9] At the end, he announces further investigations which are to concern—in modern terminology—the basics of second-order model theory along the lines drawn in notes W1 and W2 of his Warsaw talks.[10] There are, however, no documents giving evidence that he ever began such investigations.

Skolem's answer and the consequences

In a swift response to Zermelo's definiteness paper, Skolem (*1930*) points out that—except for the second-order quantification—both, Zermelo's genetic and axiomatic definition of definiteness coincide with the first-order approach of his Helsinki address (*1923*). He blames Zermelo for the use of set theory in item (2) of his axiomatic version and sharply criticizes the vagueness concerning second-order quantification. Moreover, by pointing to his Helsinki address, he informs Zermelo about a notable result he had obtained there: Skolem's first-order version of definiteness leads to a

[9] Cf. (the introductory note to) *s1931f* and *s1932d*.
[10] Cf. (the introductory note to) *s1929b*.

first-order version of Zermelo's axiom system. By generalizing Löwenheim's theorem about the existence of countable models from single first-order formulae to countable sets of first-order formulae, Skolem had shown that—granted its satisfiability—this system has a countable model. It was this consequence that roused Zermelo's epistemological resistance. In fact, henceforth his scientific endeavours will be largely directed toward ruling out the existence of a countable model of set theory. Several ways can be distinguished in which he tried to do so:

- He tried to reduce Skolem's approach ad absurdum by proving that there cannot be a countable model of set theory.[11]
- He considerably developed his new approach to the notion of set via categorical definitions.[12]
- He avoided the notion of definiteness in his 1930 axiom system of set theory, choosing a second-order version of separation instead.[13]
- Localizing the deficiency of Skolem's axiomatization in its finitary approach, he emphasized that any genuine axiom system could only be an infinitary one, working out his ideas in several papers on infinitary languages and infinitary logic.[14]

Strangely enough, Zermelo did not try to check Skolem's proof, nor did he acknowledge Skolem's result as an interesting mathematical theorem which may shed light on the scope of the mathematical method—a behaviour in line with his fight against Gödel's incompleteness results.[15] As it was the results and the methods of Gödel and Skolem which shaped the discipline of mathematical logic in the 1930s, Zermelo placed himself outside the mainstream of mathematical foundations, and his campaign failed. In this sense the notion of definiteness and in particular his *1929a* carried the seeds of his eventual scientific fate.

---

[11] Cf. *Ebbinghaus 2004* for early attempts and (the introductory note to) *s1937* for the latest one.

[12] Cf. (the introductory note to) *s1931f* and *s1932d*.

[13] Cf. (the introductory note to) *1930a*.

[14] Cf. (the introductory notes to) *s1921*, *1932a*, *1932b*, and *1935*.

[15] Cf. (the introductory note to) *s1931b*, *s1931c*, *Gödel 1931b*, and *1931d*.

Über den Begriff der Definitheit in der Axiomatik

1929a

In meinen „Untersuchungen über die Grundlagen der Mengenlehre" (Math. Ann. 65. 1908) habe ich auf S. 263 den Terminus „definit" eingeführt für solche Aussagen $E(x)$ „über deren Gültigkeit oder Ungültigkeit die Grundbeziehungen des Bereiches vermöge der Axiome und der allgemeingültigen logischen Gesetze ohne Willkür entscheiden". Dieser Begriff scheint nicht allgemein verstanden worden zu sein; manche finden ihn „unklar", und die meisten neueren Autoren auf dem Gebiete der mengentheoretischen Axiomatik versuchen ohne ihn auszukommen. Die verschiedenen Versuche, den Begriff der „Definitheit" zu vermeiden, lassen sich in die folgenden Gruppen einteilen:

1) Man läßt ihn einfach *weg*, hält ihn für *überflüssig* oder überläßt ihn der allgemeinen Logik bzw. der Logistik, da seine Bedeutung eben nicht spezifisch mengentheoretischer Natur ist. Dieser einfache Behelf, einer Schwierigkeit dadurch aus dem Wege zu gehen, daß man sie ignoriert oder einer anderen Wissenschaft zuschiebt, wäre gewiß auch für mich das Bequemste gewesen und hätte mir manche Angriffe und Kritiken erspart. Aber ist es schon an und für sich nicht die Aufgabe des wissenschaftlichen Forschers, sich die Arbeit bequem zu machen, so gilt dies noch am wenigsten für Untersuchungen, die es mit den „Grundlagen" einer Wissenschaft zu tun haben. Es kam mir darauf an, die Hauptsätze der allgemeinen Mengenlehre aus möglichst geringen Voraussetzungen und mit möglichst beschränkten Hilfsmitteln herzuleiten, und ich erkannte, daß der uneingeschränkte Gebrauch der „Satzfunktionen" hier
340 entbehrlich, ja wegen gewisser „Antinomien" vielleicht sogar | gefährlich war. Eine allgemein anerkannte „mathematische Logik", auf die ich mich hätte berufen können, gab es damals noch nicht — so wenig wie heute, wo jeder Grundlagen-Forscher seine eigene Logistik hat. Eine solche logistische Grundlage in extenso selbständig zu entwickeln, wäre damals schwerlich am Platze gewesen, wo es sich doch zunächst um andere Aufgaben handelte, und zu einer Zeit, wo noch unter den Mathematikern weitgehendes Mißtrauen gegen jede Logistik bestand. Immerhin glaubte ich durch meine Erklärung des fraglichen Begriffes und vor allem durch die von ihm gemachten Anwendungen hinreichend deutlich gemacht zu haben, wie er gemeint war.

2) Man sucht den *allgemeinen* Definitheits-Begriff zu vermeiden, indem man ihn *spezialisiert*, indem man in der Mengenlehre nur Satz-Funktionen von besonderer Form zuläßt, wie dies z. B. *A. Fraenkel* in der zweiten Auflage seiner „Mengenlehre" getan hat. Aber bei der Charakterisierung der zugelassenen Funktionen verfährt er *konstruktiv*, was dem Zweck und Wesen der *axiomatischen* Methode im Grunde widerspricht und außerdem vom Begriffe

On the concept of definiteness in axiomatics

1929a

On p. 263 of my "Investigations in the foundations of set theory" (*1908b*) I introduced the term "definite" for an assertion $E(x)$ "if the fundamental relations of the domain, by means of the axioms and the universally valid laws of logic, determine without arbitrariness whether it holds or not." Apparently, not everyone has understood this concept; some find it "unclear", and a great majority of those who have written on the topic of set-theoretic axiomatics in recent times tries to make do without it. The various attempts at avoiding the concept of "definiteness" can be divided into the following groups:

1) It is simply *dropped*, considered *superfluous* or left to general logic or to logistic entirely, on the grounds that it lacks a specifically set-theoretic meaning. To be sure, it would have been the most convenient thing for me, too, to take this simple makeshift approach to avoiding a difficulty by ignoring it or by foisting it upon another science. I would have been spared many an attack and criticism. But if it is true that taking the easy way out is not a scientific researcher's proper task, then the same holds even more so for those investigations that are concerned with the "foundations" of a science. What I wanted to do was to derive the main theorems of general set theory from the smallest possible number of assumptions and by means of the most restricted expedients. I recognized that in order to do so, the unrestricted use of "propositional functions" would be dispensable or, on account of certain "antinomies", perhaps even dangerous. At the time, a universally acknowledged "mathematical logic" on which I could have relied did not exist—nor does it exist today when every foundational researcher has his own logistic. With my primary tasks being different, however, it hardly would have been appropriate for me to develop in extenso such a logistical foundation, particularly at a time when most mathematicians still harbored suspicions about any kind of logistic. But I believed that my explanation of the concept in question, and in particular its applications, had at least made sufficiently clear how it was meant.

2) One seeks to avoid the *general* concept of definiteness by *restricting* it, by permitting only propositional functions of a special form in set theory. *A. Fraenkel*, e.g., proceeded along such lines in the second edition of his "Set Theory".[1] However, in characterizing the permitted functions, he proceeds *constructively*, which actually contradicts purpose and nature of the *axiomatic* method and also depends on the concept of *finite number* whose

[1] ⟦*Fraenkel 1923a.*⟧

der *endlichen Anzahl* abhängt, dessen Klärung doch gerade eine der Hauptaufgaben der Mengenlehre sein sollte. In der Tat scheint er auch mit diesem Verfahren wenig Nachfolge gefunden zu haben.

3) Die angemessenste Methode, dem Begriffe der „Definitheit" beizukommen und ihn zu präzisieren, scheint mir vielmehr seine *Axiomatisierung*, und das ist denn auch das Verfahren gewesen, das ich selbst, wenn auch ohne ausdrückliche Formulierung, im Auge gehabt und in den Beweisführungen der genannten Arbeit zur Anwendung gebracht habe. Diese Auffassung liegt augenscheinlich auch der „*Axiomatik der Mengenlehre*" *J. v. Neumann's* zu Grunde, in welcher von vornherein neben und vor dem Begriff der „Menge" derjenige der „*Funktion*" eingeführt und axiomatisiert wird. Doch scheint mir das ganze System *v. Neumann's*, so interessant und wertvoll es in mancher Beziehung auch sein mag, doch in seinen Grundlagen allzu verwickelt und schwer verständlich, als daß es sich nicht den Versuch lohnte, mit einfacheren mengentheoretischen Grundbegriffen auszukommen. So soll denn im Folgenden versucht werden, den Begriff der „*definiten*" Eigenschaften oder Aussagen im Rahmen einer allgemeinen Axiomatik als ein spezifisch logi-
341 sches | Problem im besonderen Hinblick, aber ohne Beschränkung auf die Grundlagen der Mengenlehre axiomatisch aufzuklären.

Jedes Axiomen-System A bestimmt als die Gesamtheit seiner Folgerungen ein „*logisch abgeschlossenes System*" d. h. ein System S von Sätzen, welches alle aus ihm rein logisch ableitbaren Sätze bereits enthält. Ist ein solches System „*konsistent*" d. h. widerspruchsfrei, so muß es auch „*realisierbar*" d. h. darstellbar sein durch ein „*Modell*", durch eine *vollständige Matrix* der in den Axiomen bzw. im System vorkommenden „*Grundrelationen*". Zwei ein und dasselbe System S realisierende Modelle M_1 und M_2 heißen „*isomorph*", wenn die Bereiche B_1 und B_2 ihrer Elemente in der Weise ein-eindeutig auf einander abgebildet werden können, daß alle Grundrelationen erhalten bleiben; ein System S, für welches je zwei *beliebige* Modelle isomorph sind, heißt „*kategorisch*".

Ist $r(x, y, z, \ldots)$ irgend eine der in S vorkommenden „Grundrelationen" und M ein beliebiges realisierendes Modell, mit dem Elementen-Bereich B, so gilt für jedes einzelne dem Bereiche B entnommene Werte-System x, y, z entweder r oder seine Negation $\overline{r}$ in dem betrachteten Modell: unser Modell ist bezüglich der Grundrelationen „*disjunktiv*". Das gleiche gilt aber auch für alle aus den Grundrelationen „*ableitbaren*" zusammengesetzten Relationen der Form

$$f(x, y, z, \ldots) ,$$

auch für sie ist in jedem Modell durch die Matrix der Grundrelationen eindeutig entschieden, ob sie in ihm gelten oder nicht. Eine solche *in jedem Modell* durch die Grundrelationen entschiedene Eigenschaft oder Relation ist es eben, die in der genannten Arbeit als „*definit*" bezeichnet werden sollte. „Definit" ist also, was *in jedem* einzelnen *Modell entschieden* ist, aber in

clarification, after all, is supposed to be one of set theory's principal tasks. In fact, his procedure does not seem to have been adopted by many.

3) The method best suited to getting a grip on the concept of "definiteness" and to making it precise lies rather, it seems to me, in its *axiomatization.* This then was the procedure which I had in mind, even if I did not expressly formulate it, and of which I made use when conducting the proofs in the paper mentioned above. This conception appears to form the basis also of *J. v. Neumann*'s "*Axiomatics of Set Theory*",[2] in which the concept of "*function*" is introduced and axiomatized along with, and even before, the concept of "set". But, its interest and value in certain respects notwithstanding, *v. Neumann*'s entire system seems to me overly knotty and difficult to understand, so much so that an attempt to make do with simpler set-theoretic basic concepts seems to me worthwhile. Thus, in what follows, I shall try to axiomatically clarify the concept of "*definite*" properties, or assertions, within the framework of a general axiomatics as a special logical problem. I shall consider it in particular, but not exclusively, in relation to the foundations of set theory.

Every axiom system A determines as the totality of its consequences a "*logically closed system*", that is, a system S of propositions already containing all propositions that are derivable from it by purely logical means. If such a system is "*consistent*", that is, free from contradictions, then it must be "*realizable*" as well, that is, representable by means of a "*model*", by means of a *complete matrix* of the "*fundamental relations*" that occur in the axioms or in the system. Two models M_1 and M_2 realizing one and the same system S are called "*isomorphic*" if it is possible to map the domains B_1 and B_2 of their elements one-to-one onto one another so that all fundamental relations are preserved; a system S for which any two *arbitrary* models are isomorphic is called "*categorical*".

If $r\,(x, y, z, \ldots)$ is any of the "fundamental relations" occurring in S and M an arbitrary realizing model with the domain of elements B, then either r or its negation $\overline{r}$ holds in the model under consideration for every single system of values x, y, z from the domain B: our model is "*disjunctive*" with respect to the fundamental relations. But the same holds also for all composite relations "*derivable*" from the fundamental relations of the form

$$f(x, y, z, \ldots)\ ,$$

For those, too, it is uniquely decided in every model by means of the matrix of the fundamental relations whether or not they hold in it. Such a property or relation decided by means of the fundamental relations *in every model* is just what, in the paper mentioned above, was supposed to be called "*definite*". "Definite" is thus what is *decided in every* single *model*, but may be decided differently in different models; "decidedness" refers to the individual *model*,

[2] ⟦*Neumann 1928d.*⟧

verschiedenen Modellen auch in verschiedener Weise entschieden sein kann; die „Entschiedenheit“ bezieht sich auf das einzelne *Modell*, die „Definitheit“ selbst auf die betrachtete *Relation* und auf das ganze *System*.

Natürlich gibt es auch „*nicht-definite*“ Eigenschaften in jedem System, wie etwa „grün angestrichene“ Mengen oder Irrationalzahlen, „die durch keine *endliche Anzahl von Worten* in einer beliebigen europäischen Sprache definirt werden können“, und es erscheint mir keineswegs überflüssig, auf solche „unsinnigen“ oder „unwissenschaftlichen“ Definitionen Bezug zu nehmen: hat
342 man doch mit | solchen Hilfsmitteln u. a. „die *Unmöglichkeit der Wohlordnung des Kontinuums*“ nachweisen wollen. „*Indefinit*“ sind überhaupt alle Sätze, welche *systemfremde* Relationen enthalten oder solche, die durch die Grundrelationen nicht *eindeutig* bestimmt und festgelegt sind; und dies ist keineswegs immer auf den ersten Blick schon ersichtlich.

Aber welche Sätze und Eigenschaften sind nun wirklich „definit“? Wie kann man entscheiden, ob ein vorgelegter Satz es ist? Oder wie kann man sich vergewissern, keine „indefinite“ Eigenschaft bei der Beweisführung verwendet zu haben? Naheliegend wäre die folgende Begriffs-Bestimmung:

Ein Satz heißt „definit“ für ein gegebenes System, wenn er aus den Grundrelationen des Systems aufgebaut ist ausschließlich vermöge der logischen Elementar-Operationen der Negation, Konjunktion und Disjunktion sowie der Quantifikation, alle diese Operationen in beliebiger aber endlicher Wiederholung und Zusammensetzung.

Diese Bestimmung wäre aber wieder *genetischer, konstruktiver* Natur, sie betrifft nicht den Satz *selbst* sondern seine *Entstehung* oder *Erzeugung*, verwendet den Begriff der *unbestimmten endlichen Anzahl*, liefert in verwickelteren Fällen *keine zweifelsfreie Entscheidung* und ist überhaupt, als dem Wesen der axiomatischen Methode *widerstreitend*, in irgend einer *Axiomatik*, und vollends einer Axiomatik der Mengenlehre, die doch dem Anzahlbegriff vorausgehen soll, m. Er. ebenso wenig am Platze wie die oben besprochene *Fraenkel*’sche Formulierung des Aussonderungs-Axioms. Die hier angedeutete *genetische* Bestimmung muß also durch eine *axiomatische* ersetzt werden, die, wenn sie auch nicht in *allen* Fällen eine praktische Entscheidung ermöglicht, doch wenigstens von den genannten Übelständen frei und einer präzisen Fassung fähig ist.

Wir versuchen also folgende *Axiomatik der Definitheit*:

Es sei gegeben ein Bereich B (oder allgemeiner eine Vielheit von Bereichen $B_1, B_2, \ldots$) sowie ein System R von *Grundrelationen* der Form

$$r(x, y, z, \ldots)\,,$$

wo die Variablen $x, y, z, \ldots$ entsprechend den Bereichen B angehören. Dann sagen wir von einem Satze p, er sei „*definit in Bezug auf R*“ und schreiben

$$D(p)\,, \quad \text{oder einfacher} \quad Dp\,,$$

whereas "definiteness" itself refers to the *relation* under consideration and to the entire *system*.

There are of course also "*non-definite*" properties in every system such as sets which "are painted green" or irrational numbers which "cannot be defined by means of a *finite number of words* in any European language". I certainly do not deem it superfluous to consider such "nonsensical" or "unscientific" definitions: for it is by use of such expedients that some have sought to show, e.g., "*the impossibility of the well-ordering of the continuum*". Generally speaking, "*non-definite*" are all those propositions containing relations *alien to the system* and those containing relations not *uniquely* determined and fixed by the fundamental relations; and whether this is the case is by no means always immediately evident.

But which propositions and properties are now in fact "definite"? How can we decide whether a given proposition is "definite"? Alternatively, how can we be sure not to have made use of any "non-definite" properties when conducting a proof? The following would seem to suggest itself as a determination of the concept:

A proposition is called "definite" for a given system if it is constructed from the fundamental relations of the system only by virtue of the logical elementary operations of negation, conjunction and disjunction, as well as quantification, all these operations in arbitrary yet finite repetition and composition.

But this determination of the concept would again be of a *genetic, constructive* nature. It is not concerned with the proposition *itself* but its *formation* or *generation*. It uses the concept of *non-definite finite number*. In more complicated cases, it yields *no decision beyond all doubt*. It *runs counter* to the nature of the axiomatic method, and therefore it is really, in my opinion, just as out of place in any axiomatics, and in particular in an axiomatics of set theory, which, after all, is supposed to precede the notion of number, as is *Fraenkel*'s formulation of the separation axiom discussed above. The *genetic* determination suggested here must therefore be replaced by an *axiomatic* determination, which, while not allowing for a practical decision in *all* cases, at least does not suffer from the ills mentioned here and lends itself to precise rendition.

Let us try the following *axiomatics of definiteness*:

Consider a domain B (or, more generally, a multiplicity of domains $B_1, B_2, \ldots$) and a system R of *fundamental relations* of the form

$$r(x, y, z, \ldots),$$

where the variables $x, y, z, \ldots$ belong to the corresponding domains B. Then we say of a proposition p that it is "*definite with respect to R*" and write

$$D(p), \quad \text{or, more simply,} \quad Dp,$$

343 | wenn folgendes stattfindet:

I) $Dr(x, y, z, \ldots)$ für jede Relation aus R und jede Variablen-Kombination aus B. *Definit sind zunächst alle Grundrelationen.*

II) *Die Definitheit überträgt sich auf zusammengesetzte Aussagen*, nämlich

1) Gilt Dp so gilt auch immer $D\overline{p}$, wenn $\overline{p}$ die Negation von p bedeutet.
2) Gilt Dp und zugleich Dq, so gilt auch $D(pq)$ sowie auf Grund von 1) auch $D(p + q)$, wo pq bzw. $p + q$ die Verbindungen durch „und" und „oder" bedeuten.
3) Gilt $Df(x, y, z, \ldots)$ für *alle* (erlaubten) Werte-Kombinationen, also

$$\bigcap_{x,y,z} Df(x, y, z, \ldots),$$

so gilt auch $D \bigcap_{x,y,z} f(x, y, z, \ldots)$ und somit auch für die Partikular-Aussage $D \bigcup^{x,y,z} f(x, y, z, \ldots)$.

4) Gilt $DF(f)$ für alle definiten Funktoren $f = f(x, y, z, \ldots)$, so gelten auch $D \bigcap_f F(f)$ und $D \bigcup^f F(f)$.

Die Definitheit überträgt sich auf die Quantifikationen.

Mit diesen Axiomen I) und II) werden zwar manche Sätze als „*definit*" charakterisiert, aber keiner als „*indefinit*", und es könnte nach ihnen allein *jeder beliebige* Satz als „*definit*" gelten; es muß also zur *vollständigen* Charakterisierung noch ein *negatives, einschränkendes* Axiom hinzukommen. Dieses neue, abschließende Axiom ist freilich von etwas anderer Form als die vorangehenden, indem es sich nicht sowohl auf die einzelnen „definiten" Sätze p als auf ihre *Gesamtheit* P bezieht. Diese Gesamtheit P hat nämlich die folgenden Eigenschaften:

1) sie genügt den Postulaten I und II, d. h. sie *enthält* sämtliche Grundrelationen und sie *ergänzt sich selbst* durch die unter II angegebenen logischen Grundoperationen der Negation, Konjunktion, Disjunktion und Quantifikation;
2) *kein echter Teil* P_1 von P besitzt die *gleiche* Eigenschaft, den Postulaten I und II im selben Sinne zu genügen.

In der Tat, hat ein System P ⟦von Sätzen die unter⟧[1] 1) und 2) angegebenen Eigenschaften, so müssen alle Sätze p dieses Systems „*definit*" sein. Denn sonst bildeten die „definiten" Sätze dieses Systems ein *Teilsystem* P_1
344 von der durch 2) ausgeschlossenen Beschaf- | fenheit. Sollen nun *alle übrigen* Sätze (außerhalb P) als „*indefinit*" gelten, so erhalten wir noch das folgende

[1] ⟦In the original text the four words are permuted as follows: "die unter von Sätzen".⟧

if the following is the case:

I) $Dr(x, y, z, \dots)$ for every relation from R and every combination of variables from B. *First, all fundamental relations are definite.*

II) *Definiteness is passed on to composite assertions as follows*
 1) If Dp holds, then also always $D\overline{p}$, assuming that $\overline{p}$ signifies the negation of p.
 2) If Dp holds and, at the same time, also Dq, then $D(p\,q)$ holds as well, as does $D(p + q)$ on account of 1), where $p\,q$ and $p + q$ signify the compositions obtained by means of "and" and "or" respectively
 3) If $Df(x, y, z, \dots)$ holds for *all* (permissible) combinations of values, and hence

$$\bigcap_{x,y,z} Df(x, y, z, \dots)\,,$$

 then $D \bigcap_{x,y,z} f(x, y, z, \dots)$ holds as well, and thus also for the assertion of the particular $D \bigcup^{x,y,z} f(x, y, z, \dots)$.
 4) If $DF(f)$ holds for all definite functors $f = f(x, y, z, \dots)$, then $D\bigcap_f F(f)$ and $D\bigcup^f F(f)$ hold as well.
 Definiteness is passed on to the quantifications.

While some propositions are characterized as "*definite*" by means of axioms I) and II), none is characterized as "*non-definite*", and, according to these axioms alone, *any arbitrary* proposition could count as "*definite*"; in order to provide a *complete* characterization we must therefore add a *negative, restrictive* axiom. This new, complementary axiom differs of course with respect to its form from the axioms mentioned above since it does not refer both to the individual "definite" propositions p and to their *totality* P. For this totality P has the following properties:

1) it satisfies the postulates I and II, that is, it *contains* all fundamental relations and it *supplements itself* by means of the logical basic operations of negation, conjunction, disjunction and quantification specified in II;
2) *no proper part* P_1 of P possesses the *same* property of satisfying the postulates I and II in the same sense.

In fact, if a system P of propositions possesses the properties specified in 1) and 2), then all propositions p of this system must be "*definite*". For, otherwise, the "definite" propositions of this system would form a *partial system* P_1 of the constitution excluded by 2). If, now, *all remaining* propositions (outside of P) are supposed to count as "*non-definite*", then we also obtain the following

Axiom III) *Ist P das System* aller *„definiten" Sätze, oder allgemeiner* irgend ein *System von Sätzen p von der Beschaffenheit Dp, so besitzt es* kein echtes *Untersystem* P_1*, welches gemäß I und II einerseits die sämtlichen* Grundrelationen *aus R enthält, andererseits aber alle Negationen, Konjunktionen, Disjunktionen und Quantifikationen der eigenen Sätze, bzw. Satzfunktionen bereits umfaßt.*

Natürlich ist auch dieses Axiom nur ein anderer Ausdruck für die Forderung, daß alle „definiten" Sätze sich aus den „Grundrelationen" der Form I durch eine *endliche Anzahl von Operationen* der Form II) herleiten lassen. Es kommt mir aber hier eben darauf an, nachzuweisen, daß der Begriff der *„Definitheit"* auch rein *axiomatisch* und ohne explizite Benutzung der *„endlichen Anzahl"* präzisierbar ist. Die weitere Untersuchung des hier charakterisierten *Systems P aller* (in Bezug auf ein Relationssystem *R*) *definiten Sätze* und seiner Bedeutung für die *„logisch abgeschlossenen Systeme"* und die sie realisierenden *„Modelle"* werde einer späteren ausführlichen Darstellung vorbehalten.

Bad Zoppot, den 11-ten Juli 1929.

0 Axiom III) *If P is the system of* all *"definite" propositions, or, more generally,* any *system of propositions p of the constitution Dp, then it has* no proper *subsystem P_1 that, on the one hand, contains all fundamental relations from R, in accordance with I and II, while already including, on the other hand, all negations, conjunctions, disjunctions and quantifications of its own propositions and propositional functions.*

This axiom, too, is of course just another expression of the demand that all "definite" propositions be derivable from the "fundamental relations" of the form I by means of a *finite number of operations* of the form II). But my concern here is to show that it is possible to make the concept of "definiteness" precise by purely *axiomatic* means and without explicit use of "*finite number*". I postpone a further, detailed investigation of the system P characterized here of all propositions that are definite (with respect to a relation system R) and of its significance for both the "logically closed systems" and the "models" realizing them to a later occasion.

Bad Zoppot, on the 11th of July 1929.

Introductory note to s1929b and 1930b

Heinz-Dieter Ebbinghaus

Following an invitation of the Faculty of Mathematics and Sciences of the University of Warsaw, Zermelo gave nine one-hour talks at the Warsaw Mathematical Institute between 27 May and 10 June 1929. On 24 May he gave an additional talk on the logical form of mathematical theories (*1930b*) to the Polish Mathematical Society and a further one on 7 June on reflection in analytical curves.[1] The topics he treated in Warsaw range from a justification of classical logic and the infinite to a critical discussion of intuitionism and Hilbert's proof theory, but he also dealt with specific topics such as a new notion of set or the (von Neumann) ordinals. Taken together, these talks provide a complete picture of his foundational views at this time and served as a platform from which he developed new insights when faced with the pioneering results of Thoralf Skolem and Kurt Gödel. Zermelo's views had evidently been sharpened by intensive discussions with the philosopher Marvin Farber in the 1920s.

In the first section we give details about the Warsaw program. The next sections comment on the drafts (*s1929b*) underlying the talks. *1930b* consists of a one-sentence description of the address of 24 May to the Polish Mathematical Society. Very probably, Zermelo therein gave a more detailed exploration of the thoughts laid down in draft W1 described below.

The Programme

In chronological order the Warsaw talks were entitled as follows:

T1 What is mathematics? Mathematics as the logic of the infinite. (Was ist Mathematik? Die Mathematik als die Logik des Unendlichen.)

T2 Axiom systems and logically complete systems as a foundation of general axiomatics. (Axiomensysteme und logisch vollständige Systeme als Grundlage der allgemeinen Axiomatik.)

T3 On disjunctive systems and the principle of the excluded middle. (Über disjunktive Systeme und den Satz vom ausgeschlossenen Dritten.)

T4 On infinite domains and the importance of the infinite for the whole of mathematics. (Über unendliche Bereiche und die Bedeutung des Unendlichen für die gesamte Mathematik.)

[1] Details concerning the stay in Warsaw and the cooperation with Marvin Farber mentioned below can be found in *Ebbinghaus 2007b*, sect. 4.4, and in sect. 4.3 of the introductory biography. There are no documents with information about the contents of the analytical talk.

T5 On the consistency of arithmetic and the possibility of a formal proof. (Über die Widerspruchslosigkeit der Arithmetik und die Möglichkeit eines formalen Beweises.)

T6 On the axiomatics of set theory. (Über die Axiomatik der Mengenlehre.)

T7 On the possibility of an independent definition of the notion of set. (Über die Möglichkeit einer independenten Mengendefinition.)

T8 The theory of "basic sequences" as a substitute for "ordinal numbers". (Theorie der "Grundfolgen" als Ersatz der "Ordnungszahlen".)

T9 On some basic questions of mathematics. (Über einige Grundfragen der Mathematik.)

There are no notes that provide information about the contents of the talks. However, *s1929b* comprises drafts of six talks with partly deviating titles:[2]

W1 What is mathematics? (Was ist Mathematik?)

W2 Disjunctive systems and the law of the excluded middle. (Disjunktive Systeme und der Satz vom ausgeschlossenen Dritten.)

W3 Finite and infinite domains. (Endliche und unendliche Bereiche.)

W4 How can the assumption of the infinite be justified? (Wie rechtfertigt sich die Annahme des Unendlichen?)

W5 Continuation: Can the consistency of arithmetic be "proved"? (Fortsetzung: Kann die Widerspruchslosigkeit der Arithmetik "bewiesen" werden?)

W6 On Sets, classes, and domains. An attempt at defining the concept of sets. (Über Mengen, Klassen und Bereiche. Versuch einer Definition des Mengen-Begriffs.)

Most likely, draft W1 corresponds to T1, draft W2 to T2 and T3, draft W3 to T4, drafts W4 and W5 to T5, and draft W6 to T6 and T7. Talk T8 presumably treated Zermelo's notion of ordinal as conceived around 1915 in Zurich and identical with von Neumann's definition in *1923*.[3] As earlier correspondence between Zermelo and the Warsaw mathematician Bronisław Knaster reveals, talk T9 may have addressed questions such as that of the relationship between mathematics and intuition, about the role of applied mathematics with its application of the mathematically infinite to the empirically finite, and that of the relation between thinking and being, between sentence and fact, topics which had been favoured in particular by Wacław Sierpiński. Basic features may be inferred from *Zermelo s1930d*: The second appendix lists

[2] The list of six topics at the beginning may represent Zermelo's original plan; cf. *Ebbinghaus 2007b*, 168, fn. 124.

[3] Cf. *Ebbinghaus 2007b*, 133–134, for details.

five topics of intended investigations; the last one, "On the relation of mathematics to *intuition*" ("Über das Verhältnis der Mathematik zur *Anschauung*") states that mathematics starts beyond intuition: "Mathematics begins only with the infinitistic-logical processing of material given in intuition. Thus it can *not* itself be based on 'intuition'. In geometry, too, the advantage of 'Euclidean geometry' is *not* based on its 'being given in intuition', but merely on its logico-mathematical simplicity."

Draft W1: Mathematics as applied logic

The draft starts right away with a clear description of the nature of mathematics: Mathematics cannot be characterized by its objects, but solely by its method which consists in a systematization of the provable, and in this sense is "applied logic". In contrast to applied logic, "pure logic" is concerned with a general theory of logical systems of sentences. A proof is a deduction of a new proposition from other ones which follows general logical rules. An ideal logical system is a system which is closed under all conclusions, i.e. which is logically complete. A typical example is given by the set of sentences deducible from a set A of axioms. As the underlying axiom system may be varied, the complete logical system is something like an invariant of all axiom systems equivalent to A. Zermelo tends to believe that the study of logically complete systems may provide advantages comparable to those obtained by the transition from algebraic equations to algebraic fields.

The draft ends by defining consistency of an axiom system or a logically complete system of sentences as satisfiablity by a model, a model which "exists or at least is conceivable", and with a definition of categoricity which will play a role in drafts W2 and W6.

When reading the explication on proofs, one might think that Zermelo has in mind some system of logical rules underlying the deductions which constitute the proofs. However, in none of the talks does he give any indication about any formal system of logic. Even more, in draft W4 he will strongly doubt that the inferences applied in mathematical reasoning can be codified into a complete system.

Paul Bernays[4] points out that the term "system of sentences" ("Satzsystem") for a mathematical theory can also be found in earlier papers of Paul Hertz (*1922, 1923, 1929*).[5] However, neither Zermelo nor Farber refer to Hertz.

Drafts W2 and W3: The nature of mathematics

Already some years before, in discussions with Marvin Farber, Zermelo's criticism of intuitionism had focused on the principle of the excluded middle.

[4] In a letter to Helmuth Gericke of 2 October 1958.

[5] For Hertz's work cf. *Schroeder-Heister 2002*.

From the point of view of intuitionism, an (intuitionistic) proof of a disjunction $(\varphi \vee \psi)$ consists in either a proof of φ or a proof of ψ, and a proof of a negation $\neg\varphi$ consists in reducing φ to an absurdity. Hence, a proof of a disjunction $(\varphi \vee \neg\varphi)$ requires either a proof of φ or a reduction of φ to an absurdity. If, for example, φ is the Goldbach conjecture (*Every even natural number* ≥ 4 *is the sum of two prime numbers*), then $(\varphi \vee \neg\varphi)$ is not yet proved, because neither a proof nor a refutation of φ has thus far been established in ongoing mathematics. Hence, the principle of the excluded middle cannot be accepted as being universally valid.

Draft W2 argues against this feature of intuitionism: Mathematical reasoning inside a logical system always refers to existing or hypothetical models of the system. As the sentences true in a model form a *disjunctive* system, i.e. a logically complete system which for any sentence φ either contains φ or $\neg\varphi$, mathematicians are right if they use the principle of the excluded middle, provided the system is consistent, i.e. satisfiable. Even more, without this feature, mathematics would lose its true character: It would have to be restricted to disjunctive systems, and, hence, already the general theory of groups and the general theory of fields would be lost.

It is clear from these considerations that for Zermelo consistency in the sense of having a model, i.e. satisfiability, is the basic assumption for any mathematical theory. In notes of Marvin Farber, very probably from 1924, one finds that "Zermelo states that mathematicians must have the courage to do this", to presuppose the existence of a model.

Having stated his basic viewpoint, Zermelo continues with the cryptic statement that consistency "has meaning only when a closed circle of logical operations and principles for the possible inferences is assumed", essential among them the principle of the excluded middle. Farber has that models are "governed" by logical principles. One may judge these statements as giving support to the view that a clear understanding of the relationship between semantic aspects as embodied by the notion of model and syntactic aspects as exemplified in the system of logical rules and principles is still missing. Such an understanding would presuppose a sharp distinction between the two aspects. Despite its dominating role in the development of mathematical logic in the 1930s, this distinction will be completely missed by Zermelo.

Draft W2 ends with a brief discussion of categorical axiom systems, i.e. axiom systems having exactly one model up to isomorphism. Zermelo shows that categorical axiom systems yield a disjunctive set of consequences. Interestingly enough, he also considers the converse direction, trying to conclude that disjunctiveness of a system S implies categoricity of S (a statement which is wrong, for example, in the framework of first-order logic). The intuitive argument runs as follows: Given a model M of S, the question whether a model M' of S is isomorphic to M, cannot be generally answered in the negative (as $M' = M$ is a counterexample) and hence, because of disjunctiveness, should be answered in the positive, thus yielding categoricity. He instantly draws back by stating that a justification of such a conclusion could only be

given in the framework of a general proof theory. However, the inclusion of some kind of meta-statement about categoricity into the range of sentences of the system S—even if only for a moment's reflection—shows that a wide and informal notion of language lies at the root of his considerations. Two years later he will offer a candidate for a new proof theory. It will be an infinitary one, dealing with infinitely long proofs about infinitely long sentences.

Draft W3 strengthens the semantic approach of draft W2 by asserting that "true" ("eigentliche") mathematics is necessarily based on the assumption of *infinite* models. This also holds for theories of finite structures such as the theory of finite groups or the theory of finite fields. As they treat structures of arbitrarily high finite cardinality, they "are, in fact, developed only within the framework of a comprehensive arithmetic, which itself is infinite." On the other hand, "finitistic" mathematics, mathematics where the assertions may be verified in finite models, is no longer "mathematics in the true sense of the word". Taking up his characterisation of mathematics as applied logic in draft W1, he hence characterizes true mathematics as the "logic of the infinite".

Drafts W4 and W5: The justification problem

Zermelo does not deny that the assumption of the existence of (infinite) models may have led to inconsistencies. Hence, there is the task of justifying it. He discusses several possibilities to do so, taking the arithmetic of the real numbers, i.e. analysis, as an example. The possibilities are the following:

1. Giving a proof "in the proper sense of the word" of the existence of an infinite model.
2. Giving an "explicitly specified and ready-made" infinite model.
3. Giving a Hilbert-type consistency proof.

The first possibility is dismissed by a surprising argument (draft W4): A proof in the true sense is impossible as the existence of a model is an axiomatic assumption. Zermelo was surely unaware of Kurt Gödel's 1929 first-order completeness proof and, hence, was unaware of the possibility of proving the existence of a model of a first-order theory by proving its consistency relative to a complete system of rules of inference. Moreover, there are doubts whether he would have accepted such a justification. As we shall see below, he would have argued, for example, that such a proof of consistency merely shifted the problem of consistency from the theory concerned to the logical theory of propositions, to the question whether this theory has an infinite model.

The second possibility is dismissed because "the infinite as such defies, after all, all attempts at making it manifest" (draft W4), it "is not given [...] physically or psychologically in reality, it must be grasped as 'idea' in Plato's sense and 'posited'" (*s1921*, thesis II). So only the third possibility remains, a possibility aimed at by Hilbert's proof theory. As a Hilbert-type consistency proof does not immediately yield the existence of a(n infinite) model, the third

possibility offers only a "second-class justification". But Zermelo believes that even this minor aim cannot be reached; he puts forward two arguments.

(3.1) A consistency proof would have to be based on a "thorough and complete formalization of all the logic relevant to mathematics". In draft W4 he argues that such a formalization can never be secured.

(3.2) A consistency proof for arithmetic, so his argument in draft W5 goes, can only be given on the basis of a suitable proof theory which would have to deal with infinitely many sentences. Therefore a respective result would merely reduce the consistency of the infinite set of real numbers to another infinite domain, that of sentences, and nothing would be gained.

In the end, Zermelo comes to the conclusion that there is no other way than the "simplest" way of accepting the idea of "infinite domains" which "inevitably obtrudes itself on us as we engage in logico-mathematical thinking" (draft W5) and "is capable of justification solely by its success, by the fact that it (and it alone!) has made possible the creation and development of all extant arithmetic" (draft W4).

The arguments (3.1) and (3.2) are remarkable. A year earlier Hilbert and Wilhelm Ackermann had presented a complete formalization of propositional logic in their *Grundzüge der theoretischen Logik* (*Hilbert and Ackermann 1928*) and had asked there for a similar result for first-order logic. Argument (3.1) dismisses the relevance of their completeness proof. Already on 24 August 1928 Zermelo had written to Marvin Farber that the *Grundzüge* are "more than meagre". His doubts about formal logical systems were to become even stronger when he learnt about their shortcomings as exemplified by the existence of countable models of first-order set theory and the consequences of Gödel's first incompleteness theorem.

Argument (3.2) works against Hilbert's proof theory. In the letter to Farber just mentioned Zermelo confesses that he "no longer expects something overwhelming from ⟦Hilbert's⟧ 'foundations of arithmetic' which ⟦Hilbert⟧ announced again and again." Thus during the 1920s Zermelo very clearly distances himself from Hilbert's aims. This is a remarkable development as he had been Hilbert's most important collaborator in mathematical foundations during his time at Göttingen. Still in 1932, in a letter to Richard Courant, he speaks of Hilbert as his "first and sole teacher in science". Nevertheless he was very conscious of his estrangement from Hilbert. When he read an article in the Vossische Zeitung of 28 July 1931 which had appeared on the occasion of his 60th birthday and attributed "highest achievements of modern mathematics" to him—characterizing him as a follower of Hilbert's axiomatic method who only tried to go his own way in the foundation of mathematics in recent years—he underlined the words "only" and "tried", adding an exclamation mark to the latter.

There is a basic reason underlying Zermelo's opposition to Hilbert's proof theory. The theory asks for a formalization of logic and presupposes a certain elementary reasoning as manifest in Hilbert's metamathematics which is clearly distinguished from "real" mathematics. For Zermelo, however, the

constructs of logic had the same epistemological status as the other objects of mathematics such as numbers, points, and sets. Therefore it did not make sense for him "to prop up the formalism on the formalism" (draft W5). This point of view was accompanied by his refusal to adopt a strict distinction between formulas and the objects they are about, i.e. between syntax and semantics, and became apparent again in his refusal to accept basic properties of the natural numbers when axiomatizing set theory.[6]

About two years later, when Zermelo was confronted with the results of Gödel and Skolem that exhibited a principal weakness of finitely formalized mathematics, he became convinced that the richness of mathematics and logic could only be grasped by infinitary languages and an appropriate infinitary logic.[7] The Warsaw drafts touch this point only marginally in the beginning of draft W4, where it is stated that arithmetical sentences comprise an infinite multiplicity of elementary propositions (Einzelaussagen).

Draft W6: An alternative concept of set

Given Zermelo's argument for the impossibility of a consistency proof for arithmetic, a consistency proof for set theory should be impossible as well. But set theory had suffered from paradoxes. So there was no justification by success. To overcome this dilemma, Zermelo pleads for grounding set theory on principles that are conceptually more convincing. In draft W6 he starts

[6] Cf. (the introductory note to) *1929a*.

[7] Cf. (the introductory notes to) *s1921*, *1932a*, *1932b*, and *1935*.

Vortrags-Themata für Warschau 1929*

s1929b

1) Was ist Mathematik? Logik und Anschauung in der Mathematik.

2) Axiomensysteme und logisch vollständige Systeme als Grundlage der allgemeinen Axiomatik.

3) Über disjunktive Systeme und den Satz von ausgeschlossenen Dritten.

4) Über unendliche Bereiche und die Widerspruchslosigkeit der Arithmetik.

* ⟦Page 7 of the typescript is missing.⟧

a search for such principles. Locating the roots of the paradoxes in "Cantor's original definition of a set as a 'well-defined comprehension of objects where for each object it is determined whether or not it belongs to it'" (referring to *Cantor 1882*, 114–115) Zermelo proposes a new definition of set free from such deficiencies. He finds it in a set theory distinguishing between sets and classes—von Neumann's set theory being the example he has in mind—together with a definition of sets which allows their separation from classes without employing the usual variety of set existence principles. Faced with the necessity that sets need to have a cardinality, he proposes taking as sets those classes which allow a categorical definition, i. e., which are the domains of the models of a definition which has exactly one model up to isomorphism and, hence, also fixes the cardinality of its models. Examples are the set of natural numbers and the set of real numbers, categorically defined by the Peano axioms and by Hilbert's axioms for completely ordered fields, respectively. Due to the sketchy character of the draft there are no details concerning the language and the additional notions allowed for the definitions. Later notes taking this approach will not give details either—despite the fact that sets as categorically defined domains will henceforth dominate Zermelo's picture of sets and will play a major role in his later investigations about large cardinals.[8] The concluding question of draft W6 concerning the validity of the set-theoretic axioms in the universe of categorically defined sets will be answered to some extent in *s1931f* and *s1932d* and commented upon in the introductory notes to these papers.

---

[8] Cf. (the introductory notes to) *1930a* and *s1931e*.

Lecture topics for Warsaw 1929

s1929b

1) What is mathematics? Logic and intuition in mathematics.

2) Axiom systems and logically complete systems as foundation of general axiomatics.

3) On disjunctive systems and the law of the excluded middle.

4) On infinite domains and the consistency of arithmetic.

5) Über Mengen und Klassen und den Versuch einer independenten Mengen-Definition.

6) Theorie der "Grundfolgen" als Mengen zweiter Stufe anstelle der Cantorschen "Ordnungszahlen".

2 | W1.[1] Was ist Mathematik?

Die Mathematik ist *nicht* nach ihrem Gegenstande (etwa: Raum und Zeit, Formen der inneren Anschauung, Lehre vom Zählen und Messen u. dergleichen) zu charakterisieren sondern, wenn man ihren ganzen Umfang erschöpfen will, allein durch ihr eigentümliches Verfahren, den Beweis. Die Mathematik ist eine Systematik des Beweisbaren und als solche eine angewandte Logik; sie hat zur Aufgabe die systematische Entwickelung der "logischen Systeme", während die "reine Logik" nur die allgemeine Theorie der logischen Systeme untersucht. Was heißt nun "Beweisen"? Ein "Beweis" ist die nach allgemeinen logischen Regeln oder Gesetzen erfolgende Ableitung eines neuen Satzes aus anderen, vorgegebenen Sätzen, durch deren Wahrheit seine eigene gesichert ist. Das Ideal einer mathematischen Disziplin wäre demnach ein System von Sätzen, welches alle aus ihm rein logisch ableitbaren Sätze bereits in sich enthält d. h. ein "logisch vollständiges System". Ein "vollständiges" System ist z. B. die Gesamtheit aller logischen Folgerungen, die aus einem vorgegebenen System von Grund-Annahmen, einem "Axiomen-System" ableitbar sind. Aber nicht jedes "vollständige System" ist notwendig durch eine endliche Anzahl von Axiomen bestimmt. Ein und dasselbe vollständige System kann durch mehrere, ja durch unendlich viele verschiedene Axiomen-Systeme gegeben sein, z. B. die Euklidische Geometrie oder die Arithmetik der reellen bzw. der komplexen Zahlen. Ein vollständiges System ist also gleichsam die "Invariante" aller äquivalenten Axiomen-Systeme, und die Frage nach der "Unabhängigkeit" der Axiome geht es nichts an. Ein vollständiges System verhält sich zu jedem es bestimmenden Axiomensystem wie ein "Körper" zu seiner "Basis", und die Untersuchung solcher "vollständigen Systeme" verspricht vielleicht ähnliche Vorteile der größeren Allgemeinheit und Übersichtlichkeit wie der Übergang von den algebraischen Gleichungen zu den algebraischen Körpern.

Die bisher entwickelten mathematischen Disziplinen beziehen sich immer auf einen Bereich von "Dingen", Objekten, Gegenständen, zwischen denen gewisse "Grund-Relationen" bestehen — oder nicht bestehen, z. B. die Relation $x + 1 = y$ im System der "Peano'schen Axiome". Ein Axiomensystem oder ein vollständiges System wird dann "realisiert" durch ein "Modell" d. h. durch eine volle "Matrix" eines speziellen Bereiches, durch welche das Bestehen oder Nicht-Bestehen der Grundrelationen zwischen je zwei (oder mehr) Dingen des Bereiches eindeutig entschieden ist, und die gleichzeitig alle Sätze

[1] ⟦The numbers Wn for "nth Warsaw draft" do not appear in Zermelo's notes.⟧

5) On sets and classes and the attempt at an independent definition of sets.

6) Theory of "basic sequences" as sets of second order instead of Cantor's "ordinal numbers".

W1. What is mathematics?

Mathematics, if one seeks to exhaust its entire scope, is *not* to be characterized in terms of its subject matter (such as space and time, forms of inner intuition, a theory of counting and measuring), but only in terms of the method peculiar to it, proof. Mathematics is the system of the provable and, as such, it is an applied logic; its task is the systematic development of the "logical systems", whereas "pure logic" investigates only the general theory of logical systems. Now what does "prove" mean? A "proof" is the derivation of a new proposition from other, given propositions that proceeds in accordance with general logical rules and laws, where the truth of the latter secures that of the former. Accordingly, the ideal of a mathematical discipline would be a system of propositions that already contains all propositions derivable from it by purely logical means, that is, a "logically complete system". A "complete" system is, e.g., the totality of all logical consequences derivable from a given system of basic assumptions, an "axiom system". But not every "complete system" is necessarily determined by means of a finite number of axioms. One and the same complete system can be given by means of several, and even infinitely many, different axiom systems, such as Euclidean geometry or the arithmetic of the reals and that of the complex numbers. A complete system is therefore the "invariant", so to speak, of all equivalent axiom systems, and the question of the "independence" of the axioms does not concern it. A complete system is related to every axiom system determining it like a "field" is related to its "basis", and the investigation of such "complete systems" may hold promise of greater generality and clarity similar to that inherent in the transition from algebraic equations to algebraic fields.

The mathematical disciplines developed so far are always concerned with a domain of "things", entities, or objects among which certain "fundamental relations" obtain—or do not obtain, such as the relation $x + 1 = y$ in the system of "Peano's axioms". An axiom system or a complete system is then "realized" by means of a "model", that is, by means of a full "matrix" of a particular domain by which it is uniquely decided whether or not the fundamental relations obtain among any two (or more) things of the domain, and

unseres Systems erfüllt. Nur wenn ein solches "Modell" existiert oder wenigstens denkbar ist, gilt unser System als "konsistent" d. h. realisierbar; ist die Existenz eines Modelles als logisch unmöglich nachweisbar, so ist damit das System selber "inkonsistent" und "widerspruchsvoll". Sind je zwei Modelle,
2a | welche ein und dasselbe System realisieren, unter einander "isomorph" d. h. ihre Bereiche derart ein-eindeutig auf einander abbildbar, daß die Grundrelationen gleichzeitig gelten und nicht gelten, so heißt das System "kategorisch". Konsistenten aber nicht kategorischen Systemen entsprechen mindestens zwei wesentlich verschiedene d. h. nicht isomorphe Modelle, kategorischen dagegen nur ein einziger "Modell-Typus".

3 | W2. Disjunktive Systeme und der Satz vom ausgeschlossenen Dritten

In einem logisch vollständigen System kann es vorkommen, daß ein beliebiger aus den Grundrelationen aufgebauter Satz S entweder selbst im System enthalten ist oder seine Negation $\underline{S}$. In diesem Falle heißt das System "disjunktiv", in jedem anderen Falle "nicht-disjunktiv". In einem disjunktiven System gilt also der "Satz vom ausgeschlossenen Dritten" im strengsten Sinn. Nun ist aber auch jede Relations-Matrix ihrer Definition nach "disjunktiv", und somit ist auch das System der für ein spezielles Modell M gültigen und aus den Grundrelationen aufgebauten Sätze ebenfalls "disjunktiv", auch wenn das ursprüngliche System, dessen Modell M darstellt, das durch M realisiert wird, selbst nicht-disjunktiv sein sollte. Also kann auch der "Satz vom ausgeschlossenen Dritten" auf ein beliebiges disjunktives oder nicht-disjunktives System bei der *Beweisführung* immer angewendet werden, sofern dieses System nur überhaupt als "konsistent" und damit als "realisierbar" vorausgesetzt wird. Nur gilt eben die vollständige Disjunktion im Allgemeinen immer nur innerhalb eines speziellen, wenn auch sonst willkürlichen "Modelles", das nicht gerade explizit gegeben zu sein braucht, dessen Existenz aber schon durch die bloße Annahme der "Konsistenz" gesichert ist. In der Tat ist es auch immer diese (auf hypothetische Modelle) eingeschränkte Anwendungsform des allgemeinen logischen Prinzipes, die in der Beweisführung der klassischen Mathematik aller Disziplinen eine so wesentliche Rolle spielt und m. Er. auch gar nicht entbehrt werden kann. Eine Mathematik ohne den (richtig verstandenen) Satz vom ausgeschlossenen Dritten, wie sie die "Intuitionisten" fordern und selbst glauben bieten zu können, wäre überhaupt keine Mathematik mehr, und höchstens die Theorie der "disjunktiven" Systeme, dieser eng begrenzten Klasse von Systemen, zu denen schon die allgemeine Gruppentheorie wie die allgemeine Körpertheorie nicht mehr gehören, wäre dann noch zugelassen. Die "Realisierbarkeit" durch Modelle ist eben die Grundvoraussetzung aller mathematischen Theorien, und ohne sie verliert auch die Frage nach der "Widerspruchsfreiheit" eines Axiomen-Systemes ⟦ihre⟧ eigentliche Bedeutung. Denn die Axiome selbst tun einander nichts, bevor sie nicht auf ein und dasselbe (gegebene oder hypostasierte) Modell angewendet wer-

which satisfies all propositions of our system as well. Only if such a "model" exists, or at least is conceivable, does our system count as "consistent", that is, realizable; if the existence of a model can be shown to be logically impossible, then the system itself is "inconsistent" and "contradictory". If any two models realizing one and the same system are mutually "isomorphic", that is, if their domains can be mapped onto one another so that the fundamental relations continue to obtain and not to obtain respectively, then the system is called "categorical". To systems which are consistent but not categorical there correspond at least two essentially different, that is, non-isomorphic models. To categorical ⟦systems⟧, by contrast, ⟦there corresponds⟧ only a single "model-type".

W2. Disjunctive systems and the law of the excluded middle

In a logically complete system, it is possible that for any proposition S constructed from the fundamental relations either it itself or its negation $\underline{S}$ is contained in the system. In this case, the system is called "disjunctive". In any other case, it is called "non-disjunctive". In a disjunctive system, the "law of the excluded middle" holds therefore in the strictest sense. But now, according to its definition, every relation matrix, too, is "disjunctive", and hence the system of propositions valid for a particular model M and constructed from the fundamental relations is "disjunctive" as well, even if the original system whose model represents M, which is realized by means of M, is itself non-disjunctive. Therefore, the "law of the excluded middle" can also always be applied to a system, be it disjunctive or non-disjunctive, when carrying out a *proof*, provided that this system is assumed to be "consistent", and hence "realizable", at all. But the complete disjunction holds in general only within a particular, if otherwise arbitrary, "model", which need not be explicitly specified but whose existence is already secured by the mere assumption of "consistency". It is always, in fact, this way of applying the general logical principle restricted to hypothetical models that plays such an essential and, in my opinion, quite indispensable role in the *proofs* of all disciplines of classical mathematics. A mathematics without the law of the excluded middle (understood correctly) as the "intuitionists" demand it, confident in their ability to provide it, would cease to be mathematics. At best, the theory of "disjunctive" systems would then still be permitted, the theory of this narrowly restricted class of systems to which even general group theory and general field theory no longer belong. The "realizability" by means of models is, after all, the presupposition of any mathematical theory, and without it, the question of the "consistency" of an axiom system loses ⟦its⟧ true meaning. For the axioms themselves do no harm to each other, unless they are applied to one and the same (given or assumed) model. "Consistency" has meaning

den. “Widerspruchslosigkeit” hat erst einen Sinn, wenn ein geschlossener Kreis logischer Operationen und Prinzipien für die möglichen Schlußfolgerungen zugrunde gelegt wird. Zu diesen Prinzipien gehört aber ganz wesentlich eben der Satz vom ausgeschlossenen Dritten.

4 | Sind kategorische Systeme zugleich disjunktiv und umgekehrt?

Ist ein System “konsistent” und zugleich “kategorisch”, so ist es realisierbar durch Modelle $M, M', \ldots$, welche alle unter einander isomorph sind. Ist nun T irgend eine im Modell M aus den Grundrelationen aufgebaute Beziehung, so gilt in M entweder T oder $\underline{T}$ und daher wegen der Isomorphie entsprechend entweder T oder $\underline{T}$ in irgend einem anderen Modell M'. Ist hier T ein von der besonderen Natur des Substrates seiner Bedeutung nach unabhängiger Satz, so gilt er, wenn er in M gilt, auch in jedem anderen Modell M', d. h. er gilt in unserem System S überhaupt, er gehört dem System an, und das gleiche gilt entsprechend auch von $\underline{T}$; d. h. aber entweder T oder $\underline{T}$ gehört zu S, und unser System ist in der Tat disjunktiv. Umgekehrt sei nun S ein disjunktives System und M, M' irgend zwei das System realisierende Modelle. Dann muß auch die Frage entschieden sein, ob ein beliebiges S realisierendes Modell M einem bestimmten M' isomorph ist oder nicht. Da diese Frage, wenn M' selbst zu den realisierenden gehört, nicht allgemein verneint werden kann, so muß sie allgemein bejaht werden, d. h. unser System S ist auch kategorisch. Voraussetzung ist für diese Argumentation einmal, daß S überhaupt realisierbar d. h. konsistent ist, sowie, daß die Möglichkeit der isomorphen Abbildung eines willkürlichen Modells M auf ein bestimmtes M' zu den durch die Grundrelationen entschiedenen und somit nach unserer *Annahme* unter die vollständige Disjunktion fallenden Fragen gehört. Die Zulässigkeit dieser letzten Annahme bedürfte freilich noch eingehenderer Erörterung, die m. E. nur im Rahmen einer allgemeinen “Beweisbarkeitstheorie” geführt werden müßte.

5 | W3. Endliche und unendliche Bereiche

Es kann der Fall eintreten, daß ein logisches System realisierbar ist durch ein Modell, dessen Matrix vollständig und in extenso angegeben werden kann, dessen Bereich dann also jedenfalls auch endlich und von begrenzter Ausdehnung sein muß. Hier kann die Konsistenz des Systems und damit auch die Widerspruchslosigkeit des zugrunde liegenden Axiomensystems *nachgewiesen* werden. So wird das den allgemeinen Gruppen-Begriff charakterisierende Postulaten-System schon realisiert durch ein aus einem einzigen Elemente bestehendes Modell, desgleichen das System der für wohlgeordnete Mengen gültigen Sätze durch eine solche von 2 Elementen, endlich auch das ganze System der Aussage- wie der Subsumptions-Logik durch ein aus 1 bzw. 2 Elementen gebildetes Modell. In den angeführten Beispielen ist aber dieser direkte Nachweis der Konsistenz nur dadurch möglich, daß die betrachteten Systeme nicht-kategorisch sind und deswegen sowohl durch endliche wie durch unendliche Modelle realisiert werden können. Für kategorische Systeme dagegen

only when a closed circle of logical operations and principles for the possible inferences is assumed. But key among these principles is just the law of the excluded middle.

Are all categorical systems disjunctive, and vice versa?

If a system is both "consistent" and "categorical", then it is realizable by means of mutually isomorphic models $M, M', \ldots$. If now T is some relation constructed in the model M from the fundamental relations, then either T or $\underline{T}$ obtains in M, and hence either T or $\underline{T}$ obtains in any other model M' on account of the isomorphism. If, in this case, T is a proposition whose meaning is independent of the particular nature of the substratum, then it also holds in every other model M', assuming that it holds in M, that is, it holds in our system S without qualification, it belongs to the system, and the same holds true for $\underline{T}$ as well; this, however, means that either T or $\underline{T}$ belongs to S, and our system is, indeed, disjunctive. Conversely, now, let S be a disjunctive system and let M, M' be any two models realizing it. Then, the question whether or not an arbitrary model M realizing S is isomorphic to a particular M' must be decided as well. Since this question cannot be generally answered in the negative, assuming that M' itself belongs to the realizing models, it must be generally answered in the affirmative, that is, our system S is categorical as well. What we assume here is, first, that S is realizable at all, that is, that it is consistent. Second, we assume that the question whether it is possible to isomorphically map an arbitrary model M onto a particular M' belongs to those questions that are decided by means of the fundamental relations, and hence to those which, by our *assumption*, fall under the rubric of complete disjunction. Whether this last assumption is permissible, however, stands in need of a more detailed investigation, which, in my opinion, can only be carried out within the framework of a general "theory of provability".

W3. Finite and infinite domains

It is possible that a logical system is realizable by means of a model whose matrix is complete and can be specified in extenso, and hence whose domain must also be finite and of limited extension. In this case, we can *show* the consistency of the system, and hence also the consistency of the underlying axiom system. For instance, the postulate system characterizing the general concept of groups is already realized by means of a model consisting of a single element. The same holds true for the realization of the system of propositions valid for well-ordered sets by means of a model consisting of 2 elements, and, finally, also for the entire system of propositional logic and of the logic of subsumption by means of a model formed from 1 and 2 elements respectively. But this direct demonstration of consistency is only possible in the specified examples on account of the fact that the considered systems are non-categorical, and hence realizable by means of both finite and infinite models. This method of manifest representation fails, by contrast, for

wie z. B. das der Peano'schen Postulate oder überhaupt für alle Systeme, die ausschließlich durch unendliche Modelle realisiert werden können, bei denen also die Unendlichkeit des Bereiches wesentlich ist, versagt dieses Verfahren der sinnfälligen Repräsentation. Nun ist aber die ganze herkömmliche Arithmetik auf die Annahme einer unbegrenzten Zahlenreihe, wie sie etwa durch die P⟦eano⟧'schen Postulate definiert werden kann, gegründet, und selbst die Theorie der endlichen Gruppen, der endlichen Mengen, der Kongruenzen und der endlichen Körper von Primzahl-Charakteristik gewinnen ihre eigentlich mathematische und nicht-triviale Bedeutung allein durch den Umstand, daß sie auf endliche Bereiche von beliebiger Elementen-Zahl angewendet werden können, also tatsächlich immer nur im Rahmen einer umfassenden, selbst unendlichen Arithmetik entwickelt werden. Eine rein "finitistische" Mathematik, in der man eigentlich nichts mehr zu beweisen brauchte, weil doch alles schon am endlichen Modell verifiziert werden könnte, wäre keine Mathematik mehr im wahren Sinne. Die wahre Mathematik ist vielmehr ihrem Wesen nach infinitistisch und auf die Annahme unendlicher Bereiche gegründet; sie kann geradezu als die "Logik des Unendlichen" bezeichnet werden.

6 | W4. Wie rechtfertigt sich die Annahme des Unendlichen?

Die Arithmetik — wie im Grunde auch jede andere mathematische Disziplin — besteht im wesentlichen aus Sätzen, die unendliche Vielheiten von Einzelaussagen umfassen, ihre Axiome beziehen sich auf Bereiche von Dingen, die auf echte Teilbereiche ein-eindeutig (oder ein-vieldeutig) abgebildet werden können, also im eigentlichen Sinne "aktual unendlich" sind. Die Annahme solcher unendlichen Bereiche (es brauchen nicht immer gerade "Mengen" im Sinne der Mengenlehre zu sein!) ist also die Grund-Annahme der gesamten Mathematik, die als solche (nach dem Satze vom Grunde!) gewiß auch einer Rechtfertigung bedarf. Ein "Beweis" im eigentlichen Sinne ist hier, wo es sich um eine axiomatische Annahme handelt, natürlich nicht möglich. Ebenso unmöglich ist aber auch die Realisierung durch ein explizit gegebenes und fertig vorgelegtes Modell, weil das Unendliche eben nirgends als solches sinnfällig aufgewiesen werden kann. Rechtfertigen läßt sich eine solche Annahme lediglich durch ihren Erfolg, durch die Tatsache, daß sie (und sie allein!) die Schöpfung und Entwickelung der ganzen bisherigen Arithmetik, die eben wesentlich eine Wissenschaft des Unendlichen ist, ermöglicht hat. Aber besteht diese Wissenschaft zu Recht? Könnte nicht gerade diese scheinbar so fruchtbare Hypothese des Unendlichen geradezu Widersprüche in die Mathematik hineingebracht und damit das eigentliche Wesen dieser auf ihre Folgerichtigkeit so stolzen Wissenschaft von Grund aus zerstört haben? So paradox dies auch erscheinen möchte bei einer in zweitausendjähriger Entwickelung mit den glänzendsten Erfolgen gekrönten Wissenschaft, so wird doch die Möglichkeit als solche, daß unsere Mathematik auf Widersprüche gebaut sei, nicht von vornherein und ohne nähere Prüfung von der Hand zu weisen sein. Kann nun aber die Widerspruchslosigkeit der ("infinitistischen")

categorical systems such as that of Peano's postulates, and, in general, for all systems which can only be realized by means of infinite models, that is, those systems for which the infinity of domain is essential. But, now, traditional arithmetic rests in its entirety on the assumption of an unlimited number series as can be defined, e.g., by means of P⟦eano⟧'s postulates. Even finite group theory, the theories of finite sets, of congruences and that of finite fields of prime characteristic derive their proper mathematical and non-trivial significance solely from the fact that they can be applied to finite domains of any number of elements, that is, that they are, in fact, developed only within the framework of a comprehensive arithmetic, which itself is infinite. A purely "finitistic" mathematics, in which nothing really requires proof since everything is already verifiable by use of the finite model, would no longer be mathematics in the true sense of the word. Rather, true mathematics is infinitistic according to its nature and rests on the assumption of infinite domains; it may even be called the "logic of the infinite".

W4. How can the assumption of the infinite be justified?

Arithmetic—like basically any other mathematical discipline—consists, generally speaking, of propositions comprising infinite multitudes of particular assertions. Its axioms are concerned with domains of things that can be mapped one-to-one (or one-to-many) onto proper partial domains and hence are really "actually infinite". To assume such infinite domains (which do not always have to be "sets" in the set-theoretic sense of the word!) is therefore to make the basic assumption underlying all mathematics, which, as such, certainly requires some justification as well (in accordance with the principle of sufficient reason!). A "proof", in the proper sense of the word, is of course not possible where an axiomatic assumption is concerned, as is the case here. But equally impossible is the realization by means of an explicitly specified and ready-made model since the infinite as such defies, after all, all attempts at making it manifest. Such an assumption is capable of justification solely by its success, by the fact that it (and it alone!) has made possible the creation and development of all extant arithmetic, which is, in essence, simply a science of the infinite. But is the existence of this science justified? Could it not be the case that this seemingly so fruitful hypothesis of the infinite carries contradictions into mathematics, thereby utterly destroying the real essence of this science, which prides itself so much on the correctness of its inferences? As paradoxical as it may seem in the case of a science that has achieved the greatest successes in the two thousand years of its development, we cannot immediately dismiss without closer consideration the possibility that our mathematics rests on contradictions. But, now, can the consistency

Arithmetik selbst logisch-mathematisch bewiesen werden? Ein "Konsistenz-Beweis" durch Realisierung in einem aufweisbaren Modell oder durch Einbettung der "infinitistischen" in eine "finitistische" Arithmetik kann nach dem Obigen hier nicht in Frage kommen. Es bliebe also nur noch die Möglichkeit nachzuweisen, daß aus den arithmetischen Axiomen auf formal-logischem Wege erkennbare Widersprüche niemals abgeleitet werden können. Ein solcher Nachweis müßte sich, sofern er möglich ist, gründen auf eine durchgehende und vollständige Formalisierung der ganzen für die Mathematik in Betracht kommenden Logik. Jede "Unvollständigkeit" der zugrunde gelegten "Beweis-Theorie", jede etwa vergessene Schlußmöglichkeit würde den ganzen Beweis in Frage stellen. Da nun aber eine derartige "Vollständigkeit" augenscheinlich niemals verbürgt werden kann, so entfällt damit m. E. auch jede Möglichkeit, die Widerspruchslosigkeit formal zu beweisen.

8 | **W5. Fortsetzung: Kann die Widerspruchslosigkeit der Arithmetik "bewiesen" werden?**

Ein anderes Argument gegen Beweis-Versuche der Widerspruchsfreiheit wäre das folgende: Jede formale "Beweistheorie", die einem solchen Beweis-Versuche zugrunde liegen muß, hat die Form einer logisch-mathematischen Theorie, bezogen auf einen Bereich von Sätzen (oder Aussagen), die durch (logische) Grund-Relationen verknüpft sind und gewissen "Axiomen", eben den logischen Prinzipien, unterliegen. Diese Axiome fordern u. a. wesentlich die unbegrenzte Verknüpfungsmöglichkeit der betrachteten Sätze (in beliebiger endlicher Anzahl von Kombinationen), setzen also selbst wieder einen unendlichen Bereich voraus und definieren damit ihrerseits ein "infinitistisches System" in genau demselben Sinne wie das der Arithmetik, dessen Zulässigkeit gerade in Frage steht. Jeder solche "Beweis" setzt also im Grunde das zu Beweisende schon voraus. Gegen diese Argumentation ließe sich nun freilich ein Einwand erheben, durch den, scheint es, der angebliche "Zirkel" vermieden würde. Das unserer "Beweistheorie" zugrunde liegende Axiomen-System ließe sich so einrichten, daß aktual unendliche Bereiche zwar zugelassen, aber nicht (ausdrücklich) gefordert würden, wie das ja in nicht-kategorischen Systemen, wie wir oben ausführten, gewiß vorkommen kann. Dann würde auch jede formell für unbeschränkte Schlußketten geführte Deduktion (welche die Widerspruchslosigkeit erweisen soll) für begrenzte erst recht gelten: wenn durch beliebig lange Schlußketten aus der gemachten Annahme kein Widerspruch hergeleitet werden kann, dann doch auch gewiß nicht durch solche von begrenzter Länge! Aber ist denn eine derartige Fassung der logischen Axiome auch wirklich durchführbar? Müßte nicht vielmehr, wenn in dieser Frage der Widerspruchslosigkeit irgend etwas "bewiesen" werden soll, die unbeschränkte Gültigkeit der logischen Prinzipien vorausgesetzt werden? Und eben diese Grundsätze postulieren gerade unweigerlich "unendliche" d. h. auf echte Teile abbildbare Bereiche. Ist dem aber so, dann ist die Existenz des Unendlichen als logisches Postulat, das jeder "Beweistheorie" zugrunde liegen

of ("infinitistic") arithmetic itself be proved by logico-mathematical means? To provide a "consistency proof" by means of realization in a model capable of exhibition or by embedding the "infinitistic" arithmetic in a "finitistic" one is, according to what was said above, out of the question. What therefore remains is the possibility of showing that contradictions detectable by formal-logical means are not derivable from the arithmetical axioms. Such a demonstration, if it were possible, would have to rest on a thorough and complete formalization of all the logic relevant to mathematics. Any "incompleteness" of the underlying "proof theory" such as a neglected possible inference would jeopardize the entire proof. But, now, since such "completeness" can obviously never be guaranteed, it is, in my opinion, not possible to furnish a formal proof of the consistency.

W5. Continuation: Can the consistency of arithmetic be "proved"?

Another argument against attempts at proving the consistency would run as follows: Any formal "proof theory", which would have to underly such an attempt, assumes the form of a logico-mathematical theory that is concerned with a domain of propositions (or assertions) which are connected by means of (logical) fundamental relations and which are subject to certain "axioms", namely the logical principles. These axioms essentially demand, among other things, the unlimited possibility of connecting the propositions under consideration (in any finite number of combinations). In other words, they themselves again presuppose an infinite domain, thereby defining in turn an "infinitistic" system in precisely the same sense as that of arithmetic, whose very legitimacy is being called into question. Hence, any such "proof" really already presupposes what is to be proved. Now, to this line of reasoning we could of course raise an objection by which the alleged "circle" can be avoided, or so it would appear. The axiom system underlying our "proof theory" could be so designed that actually infinite domains would still be permitted but not (explicitly) demanded, such as it may certainly happen with non-categorical systems, as was explained above. Then, any derivation nominally made for unlimited chains of inference (in order to show the consistency) would hold all the more for those which are limited: if we cannot derive a contradiction from a given assumption by using chains of inference of arbitrary length, then we certainly cannot do so by using chains of limited length either! But, then, is such a rendition of the logical axioms really feasible? Would we not rather have to assume the unlimited validity of the logical principles if anything at all is to be "proved" concerning this question of consistency? And it is precisely these basic laws that inevitably postulate "infinite" domains, that is, domains that can be mapped onto proper parts. But if this is the case, then the existence of the infinite as a logical postulate, which must form the basis

müßte, bereits a priori gesichert und bedarf gar keines Beweises. Überhaupt geht es nun einmal nicht an, den Formalismus wieder auf den Formalismus zu stützen; irgend einmal muß doch wirklich gedacht, muß etwas gesetzt, etwas angenommen werden. Und die einfachste Annahme, die gemacht werden kann und die zur Begründung der Arithmetik (wie auch der gesamten klassischen Mathematik) ausreicht, ist eben jene Idee der "unendlichen Bereiche", die sich dem logisch-mathematischen Denken geradezu zwangsmäßig aufdrängt, und auf die auch tatsächlich unsere ganze Wissenschaft, so wie sie sich historisch entwickelt hat, aufgebaut ist.

9 | **W6. Über Mengen, Klassen und Bereiche. Versuch einer Definition des Mengen-Begriffs**

Cantor's ursprüngliche Definition einer Menge als einer "wohldefinierten Zusammenfassung von Objekten, bei der von jedem Ding entschieden ist, ob es dazu gehört oder nicht", ist längst als *unzureichend* anerkannt, meistens unter Hinweis auf die "Antinomien" der Mengenlehre, die bei dieser Definition unvermeidlich wären. Zur Vermeidung dieser Antinomien hat man dann verschiedene Wege eingeschlagen: einmal die Russell'sche "Typen-Theorie", welche, um die Identität von "Mengen" und "Klassen" aufrecht zu erhalten, genötigt ist, die logischen Grund-Operationen in bedenklichem Maße einzuschränken, und andererseits die rein "axiomatische" Begründung der Mengenlehre, welche die allgemeine Logik unbehelligt läßt, dafür aber grundsätzlich zwischen "Mengen" und "Klassen" unterscheidet. Hier erscheinen nun freilich die einzelnen den Mengen-Begriff einschränkenden "Axiome" leicht als willkürlich und ohne inneren Zusammenhang unter einander, und man möchte versuchen, sie alle aus einem einzigen einheitlichen Prinzipe herzuleiten, das dann als unterscheidendes Merkmal zwischen Mengen und Klassen eine eigentliche "Definition" des Mengen-Begriffes darstellen würde. Vielleicht gelingt dies mit Hilfe der folgenden Überlegung. Eine "Menge" im Sinne der Mengenlehre muß jedenfalls auch eine "Mächtigkeit" besitzen, welche durch ihre Definition eindeutig bestimmt ist, d. h. irgend zwei unter die Definition fallende "Klassen" oder "Bereiche" müssen im Cantor'schen Sinne "aquivalent", d. h. ein-eindeutig auf einander abbildbar sein. Dies ist aber gewiß der Fall für Bereiche, die durch "kategorische" Systeme bestimmt werden, die etwa einem kategorischen Axiomen-Systeme genügen, wie z. B. die "abzählbaren" Mengen dem System der Peano'schen Postulate. Danach liegt es nahe, allgemein die "Mengen" als die "Bereiche kategorischer Systeme" zu definieren. Dann müßte man aber auch nachweisen, daß die der Mengenlehre zugrunde liegenden Axiome dieser "Definition" entsprechen, von den so erklärten "Mengen" erfüllt werden. Aber da entsteht schon gleich eine Schwierigkeit bei dem wichtigsten Axiom der ganzen Mengenlehre, dem "Aussonderungsaxiom". Denn wie könnte gezeigt werden, daß jeder Teilbereich eines so bestimmten Bereiches ebenfalls einem "kategorischen" System von Sätzen oder Axiomen zugehört? Dazu müßten doch in der Regel neue "Grundrelationen"

of any "proof theory", is already guaranteed a priori and does not stand in need of a proof at all. Generally speaking, propping up the formalism on the formalism itself again simply won't do; at some point, there has to be real thought, something has to be posited or assumed. And the simplest assumption that we can make and that suffices for the foundation of arithmetic (as well as for that of classical mathematics in its entirety) is precisely this idea of the "infinite domains". This idea almost inevitably obtrudes itself on us as we engage in logico-mathematical thinking, and, in fact, our entire science has been built upon it throughout its historical development.

W6. On sets, classes and domains. An attempt at defining the concept of set

Cantor's original definition of a set as a "well-defined collection of objects where for each thing it is determined whether or not it belongs to it" has long been recognized as *insufficient*, usually with reference to the "antinomies" of set theory which would necessarily arise with this definition. Different paths have been taken in order to avoid these antinomies: on the one hand, we have Russell's "type theory", which is forced to restrict the basic operations of logic to an alarming extent in order to sustain the identity of "sets" and "classes". On the other hand, we have the purely "axiomatic" foundation of set theory, which, while leaving general logic as it is, distinguishes between "sets" and "classes" as a matter of principle. In this case, the various "axioms" restricting the concept of a set easily appear arbitrary and lacking in inner cohesion, and one is tempted to try deriving them all from a single uniform principle which then, being a mark of difference between sets and classes, would serve as a proper "definition" of the concept of sets. The following consideration may help to achieve this. A "set" in the sense of set theory must under all circumstances have a "cardinality", which is uniquely determined by means of its definition, that is, any two "classes" or "domains" falling under the definition must be "equivalent" in Cantor's sense, that is, it must be possible to map one onto the other one-to-one. But this certainly is the case for domains which are determined by means of "categorical" systems, which, e.g., satisfy a categorical axiom system such as the "countable" sets, which satisfy the system of Peano's postulates. Accordingly, the idea of generally defining "sets" as the "domains of categorical systems" seems to suggest itself. But then one would also have to show that the axioms underlying set theory correspond to this "definition", that they are satisfied by the "sets" so explained. Already at this point, however, a difficulty immediately arises for the most important axiom of all set theory, the "separation axiom". For how could we show that every partial domain of a domain so determined also belongs to a "categorical" system of propositions or axioms? To this end, new "fundamental relations" would usually have to be introduced, e.g., in the case of "Peano's axioms" for

eingeführt werden, z. B. im Falle der "Peano'schen Axiome" für jeden Teilbereich eine besondere Abbildung auf sich selbst, da die ursprüngliche des ganzen Bereiches hier augenscheinlich versagt. Um dieser Schwierigkeit zu entgehen, könnte man nun die ursprüngliche Definition so abändern, daß man sagt: eine "Menge" ist ein Bereich, der wenigstens durch Hinzufügung weiterer Elemente zu einem "kategorisch bestimmten" *ergänzt* werden kann. Damit verzichtet man von vornherein auf den Vorzug, die Mengen allein durch "innere" Eigenschaften zu bestimmen.

each partial domain a special mapping onto itself since the original one of the entire domain obviously fails here. Now, in order to avoid this difficulty, we might modify the original definition as follows: a "set" is a domain which can at least be *complemented* so that it becomes a "categorically determined" ⟦domain⟧ by adding further elements to it. We thus sacrifice from the outset the advantage of determining sets solely by means of "inner" properties.

Introductory note to 1930a

Akihiro Kanamori†

Zermelo in his remarkable *1930a* offered his final axiomatization of set theory as well as a striking, synthetic view of a procession of natural models that would have a modern resonance. Appearing only six articles after *Skolem 1930* in *Fundamenta mathematicae*, Zermelo's article seemed strategically placed as a response, an aspect that we will discuss below, but its dramatically new picture of set theory reflects gained experience and suggests the germination of ideas over a prolonged period. The subtitle, "New investigations in the foundations of set theory", evidently recalls his axiomatization article *1908b,* differing only in the "New" from the title of that article. The new article is a *tour de force* which sets out principles that would be adopted in the further development of set theory and draws attention to the cumulative hierarchy picture, dialectically enriched by initial segments serving as natural models.

In Section 1, Zermelo formulates his axiom system, the "constitutive axioms" of "general set theory", and though the presentation is opaque largely because of a second-order lens, the thrust of ZFC is there. Indeed, Zermelo used the term "Zermelo-Fraenkel" to indicate the result of adding the replacement axiom (which subsumes the separation axiom) to his *1908b* axioms.[1] Zermelo actually proceeded with his "ZF'-system", the result of deleting the axiom of infinity as not being part of general set theory; assuming the axiom of choice as an implicit underlying "general logical principle"; and newly adjoining the axiom of foundation—the term having its source in this article.

Concerning the replacement axiom, Zermelo in a letter of 9 May 1921 to Abraham Fraenkel had formulated a version and aired, though with skepticism, the possibility of adopting it as a new axiom.[2] Fraenkel advocated that adoption in a paper *1922b*, completed on 1 July 1921, through which he would become associated with the axiom. It was however the work of *von Neumann 1928d*, done in the early 1920s, that made evident the importance of adopting replacement, for the formalization of transfinite recursion and the existence of sets defined therewith.

Concerning the foundation axiom, in modern notation the axiom, as is well-known, entails that the universe V of sets is stratified into cumulative ranks V_α, where by transfinite recursion $V_0 = \emptyset$, $V_{\alpha+1} = \mathcal{P}(V_\alpha)$ the power set of V_α, and $V_\delta = \bigcup_{\alpha<\delta} V_\alpha$ for limit ordinals δ—and $V = \bigcup_\alpha V_\alpha$, *the cumulative*

† This note draws heavily on *Kanamori 2004,* with the permission of the Association for Symbolic Logic which holds the copyright.

[1] "Zermelo-Fraenkel" was first invoked by *von Neumann 1928d*, 374, for this purpose.

[2] See *Ebbinghaus 2007b,* 135ff.

hierarchy. Zermelo substantially advanced this schematic, generative picture with his inclusion of foundation in an axiomatization.[3]

In modern set theory, the replacement and foundation axioms focus the notion of set, with the first making possible the means of transfinite recursion and induction, and the second making possible the application of those means to get results about *all* sets. In a notable inversion, what has come to be regarded as the underlying *iterative conception* has become a heuristic for motivating the axioms of set theory generally. It is nowadays almost banal that foundation is the one axiom unnecessary for the recasting of mathematics in set-theoretic terms, but the axiom ascribes to membership the salient feature that distinguishes investigations specific to set theory as an autonomous field of mathematics. Indeed, it can be fairly said that modern set theory is at base a study couched in well-foundedness, the Cantorian well-ordering doctrines adapted to the Zermelian generative conception of sets.

This combined operative effect of replacement and foundation is much in evidence in the workings of Zermelo's article. However, Zermelo nowhere points out how replacement is implicated in definitions by transfinite recursion. He just applies such definitions routinely in his analysis of various cumulative hierarchies, even though their various levels, analogues of the V_α, being sets have their provenance in replacement. This aside, Zermelo's explicit appeals to separation and replacement raise the issue of their applicability. What properties, as given by propositional functions or logical formulas, are to be in the purview of these two axioms? The vagueness of "definite" property for separation had prompted several efforts at remedy including Zermelo's own *1929a,* and Zermelo now writes (p. 31) that "framing the axioms in a suitable fashion" replacement implies separation. The discussion below of Zermelo's models clarifies how these two axioms are to be taken in his second-order context. There is also an issue about Zermelo's specific formulation of foundation, and this too will be addressed in due course.

In Section 2, Zermelo provides formulations now basic to modern set theory, but as with the ZFC axioms the presentation is opaque, here because of Zermelo's insistence on having urelements, with one having been fixed as *the* empty set.[4] Zermelo formulates the von Neumann ordinals, but starting generally from any urelement u: $u, \{u\}, \{u, \{u\}\}, \ldots$. (For convenience, these will be referred to as the *u-ordinals*; the usual (von Neumann) ordinals are a special case, and we shall avail ourselves of them and their usual notation in what follows.) Notably, Zermelo, in unpublished work perhaps as early as

[3] Paul Bernays in a letter of 3 May 1931 to Kurt Gödel (*Gödel 2003,* 105ff) also provided an axiomatization of set theory in which foundation is included, an axiomatization which, significantly, was to be formalizable in first-order logic. This letter would be the conduit to Gödel's use of a version of the axiomatization in his monograph *1940* on the constructible universe.

[4] See the earlier "U)" clause, in which "an arbitrarily chosen 'urelement' $\mathfrak{u}_0$ takes the place of the 'null set'."

1913, may have been the first to sketch the rudiments of the von Neumann ordinals.[5] Nonetheless, it is evident from his presentation that for Zermelo Cantor's ordinal numbers retain a separate autonomy; Zermelo's reductionism interestingly did not extend to identifying the numbers with sets, and the various u-ordinals only "represent" the ordinal numbers.

With urelements in play, Zermelo investigates the various *normal domains*, models of ZF′ when the membership relation is restricted to them. In Zermelo's words, a normal domain has a "width" given by its *basis* consisting of urelements, and a "height" given by its *characteristic*, the supremum of ordinal numbers represented in it. To modern eyes versed in pure set theory, i.e. set theory without urelements and with one universe, Zermelo's width and height seem arcane, but as one reads on in the article, it becomes a crucial feature of the *applicability* of set theory. For Zermelo a mathematical context has basic subject matter and then various possible set-theoretic super-structures built on top for the application of set-theoretic ideas and constructions.

Zermelo's crucial observation is that there are simple set-theoretic conditions on the ordinals that secure his ZF′, conditions that newly underscore how the Zermelian sets are to be an algebraic closure of his axioms. In an inspired move, Zermelo takes the characteristic π of a normal domain P to be again an ordinal number, thereby "resolving" the Burali-Forti paradox by having π outside of P but within set theory. Zermelo's simple conditions are:

(I) π is a regular cardinal, i.e. if $\alpha < \pi$ and $F\colon \alpha \to \pi$ is arbitrary, then $\bigcup F``\alpha < \pi$, and

(II) π is a strong limit cardinal, i.e. if $\beta < \pi$, then $2^\beta < \pi$.

Zermelo initially observes that these conditions are *necessary*; he argues in terms of his representing u-ordinals, but we can proceed directly with the usual (von Neumann) ordinals: To establish (I), suppose that $\alpha < \pi$ and $F\colon \alpha \to \pi$. Since $\alpha \in P$, by a crucial use of replacement $F``\alpha$ is a *set* in $P$. $\bigcup F``\alpha$ is thus a set, in fact an ordinal, in P, and hence $\bigcup F``\alpha < \pi$. To establish (II), suppose that $\beta < \pi$. Then $\mathcal{P}(\beta)$ is a set in $P$ by the power set axiom. If to the contrary $2^\beta \geq \pi$, there would be a $G\colon \mathcal{P}(\beta) \to \pi$ such that $G``\mathcal{P}(\beta) = \pi$. But by another crucial application of replacement $G``\mathcal{P}(\beta)$ must be a set in P, which is a contradiction.

The intended applicability of separation and replacement can be gleaned from these arguments. Zermelo (p. 30) had stated separation in terms of propositional functions and provided the following footnote:

> Like the replacement function in E) [the replacement axiom] the propositional function $\mathfrak{f}(x)$ can be completely *arbitrary* here, and all consequences of it restricted to a particular class of functions cease to apply from the present point of view. I shall consider elsewhere more

[5] See *Ebbinghaus 2007b*, 133.

> thoroughly "the question of definiteness" in connection with my last contribution to this journal (*Zermelo 1929a*) and with the critical "remarks" by Th. Skolem (*Skolem 1930*).

In his *1929a,* Zermelo had proposed that the "definite" properties for separation be given in terms of second-order quantification, and Skolem had responded with alacrity to criticize the vagueness of this proposal and moreover to emphasize the "Skolem paradox", whereby the axioms of set theory cast in first-order terms have countable models even though they entail the existence of uncountable sets. Zermelo found this repugnant and would work against it,[6] but, in any case, there is a *mathematical* necessity for how replacement and separation must be taken in the context of *1930a.*

The above arguments for the necessity of conditions (I) and (II) both require that replacement be applied without restriction. In modern terms, *The replacement axiom should be taken as a single, second-order axiom quantifying over all possibilities*, yielding what we now call second-order ZF. It does not even suffice for replacement to be a schema of second-order axioms, and the reference in the above cited footnote to *Zermelo 1929a*, which would still sanction separation and replacement taken as schemata, is equivocating. Zermelo's exposition is generally less meticulous than it was two decades before, and haphazard on the role of replacement. For the necessity of (II), he does not explicitly associate u-ordinals to the subsets of a set, and when finally he appeals to replacement it is for a limit case made redundant by (I).

In Section 3, Zermelo continues with three "development" theorems for his normal domains. The first states that each normal domain P is indeed stratified according to rank because of foundation: With Q the basis of urelements and π the characteristic of P, Zermelo defines the corresponding cumulative ranks by transfinite recursion:

$$P_1 = Q\,;\;\; P_{\alpha+1} = P_\alpha \cup \mathcal{P}(P_\alpha)\,;\;\text{ and } P_\alpha = \bigcup\nolimits_{\beta<\alpha} P_\beta \text{ for limit ordinals } \alpha,$$

and concludes that

$$P = P_\pi = \bigcup\nolimits_{\alpha<\pi} P_\alpha\,.$$

Zermelo emphasizes the partitioning into disjoint layers $Q_\alpha = P_{\alpha+1} - P_\alpha$ and points out that each such layer contains a u-ordinal, so that the hierarchy is strict. It becomes evident that Zermelo is entertaining all possible subsets in his normal domains, so that the $\mathcal{P}(P_\alpha)$ above has an absolute significance independent of the domain P.[7]

The first development theorem raises an issue about the axioms of foundation and infinity. Zermelo (p. 31) actually formulates foundation both as stipulating that there are no infinite descending $\in$-chains, "Or, what amounts to the same thing: Every partial domain T contains at least one element t_0 that has no element t in T." As is now well-known, the latter form implies

[6] See *Ebbinghaus 2007b*, 196ff and the introductory note to *1929a*.

[7] This is clarified by the first isomorphism theorem mentioned below.

the former only in the presence of substantial axioms, including the axiom of choice. Aside from this, the latter, a second-order form of foundation, was needed in the proof of the first development theorem: It is well-known that usual (set) foundation implies such a strong form assuming *transitive containment*, that every set is a subset of a transitive set.[8] Furthermore, the (first-order) axioms of replacement and infinity do imply transitive containment.[9] However, Zermelo is not assuming the axiom of infinity. In fact, it is a latter-day observation that second-order ZF with only (set) foundation but without infinity does not suffice to establish transitive containment, and in fact has models whose membership relation is ill-founded.[10] Hence, foundation as Zermelo formulated it in second-order terms *is* necessary for his cumulative hierarchy analysis in the absence of infinity, i.e. in case of normal domains with characteristic ω.

The second development theorem addresses *unit* domains, those normal domains P with a single urelement, $Q = \{u\}$, and provides information about their ranks. Zermelo had defined a Beth-type cardinal-valued function as follows: $\Psi(0) = 0$; $\Psi(\xi+1) = 2^{\Psi(\xi)}$; and $\Psi(\alpha) = \sup_{\xi<\alpha} \Psi(\xi)$ for limit ordinals α. Zermelo now proves that each P_α has cardinality $\Psi(\alpha)$ for infinite α.

In remarks following the second theorem Zermelo concludes that unit domains satisfy von Neumann's axiom IV2, that a class is a set exactly when there is no surjection from that class onto the entire universe. Zermelo thus establishes the consistency of IV2 relative to his axioms, proceeding in his second-order context with natural models. Zermelo notes: "Set theory, however, would lose most of its applicability if it were restricted to 'unit domains'." Moreover, Zermelo (p. 45) returns with emphasis to this point to criticize the putative restrictiveness of von Neumann's axiom. For Zermelo, one should be able to start with an unrestricted amount of basic subject matter as urelements and then "apply" set theory by imposing cumulative hierarchy superstructures on top. However, the following result shows that Zermelo's reservations have only to do with the nature and size of the totality of urelements. $|X|$ denotes the cardinality of a set X in the presence of the axiom of choice, i.e. the least ordinal bijective with X.

Proposition. For any normal domain P with basis Q and characteristic π, von Neumann's axiom IV2 holds in P *iff* Q is a set satisfying $|Q| \leq \pi$.

[8] Suppose that T is a non-empty class, say with $x \in T$. Let t be a transitive set such that $\{x\} \subseteq t$ and consider the set $t \cap T$, given by separation. By (set) foundation there is a $t_0 \in t \cap T$ such that $t_0 \cap t \cap T = \emptyset$. But then, since t is transitive, $t_0 \cap T = \emptyset$. This argument was probably first given by Gödel, in his letter of 20 July 1939 to Bernays (*Gödel 2003*, 121).

[9] For any set x and $n \in \omega$, recursively define $x_0 = x$, and $x_{n+1} = \bigcup x_n$. Then $\bigcup_n x_n$ is a transitive set containing x. The use of the axioms of replacement and infinity in this argument was noted by Gödel in the letter cited in the previous footnote.

[10] See *Vopěnka and Hajek 1963* and *Hauschild 1966*.

Proof. Suppose first that von Neumann's axiom holds in P. Since π itself is a proper class of P, π is surjective onto P and hence $|Q| \leq \pi$.

For the converse, suppose that Q is a set satisfying $|Q| \leq \pi$. Every set in P, being well-orderable, is bijective with some u-ordinal in P,[11] and so has cardinality less than π. Hence, one can prove by induction that for the development $P = \bigcup_{\alpha<\pi} P_\alpha$, $|P_\alpha| \leq \bigcup_{\beta<\pi} \pi^\beta = \pi$ for every $\alpha \leq \pi$, the equality following from (I) and (II) for π. So presuming that urelements are exempted, von Neumann's axiom for P amounts to: For any $X \subseteq P$, $X \in P$ *iff* $|X| < \pi$. The forward direction here was already noted. For the converse, if $|X| < \pi$, then applying replacement and (I) to the function $F\colon X \to \pi$ given by $F(x) =$ the least ξ such that $x \in P_{\xi+1}$, it follows that there is an $\alpha < \pi$ such that $X \subseteq P_\alpha$, and so $X \in P_{\alpha+1} \subseteq P$.

The third development theorem provides a more refined hierarchy for a normal domain P based on the Ψ function—to wit the $(\alpha+1)$th cumulative level is now to be formed by adjoining only subsets of the αth level of cardinality at most $\Psi(\alpha)$—and with this hierarchy establishes that conditions (I) and (II) are also *sufficient* for P to be a normal domain. Actually, Zermelo seems to be assuming here that $|Q| \leq \pi$ for the basis Q and characteristic π. By the above proposition, such "canonical" developments can only be given for a normal domain exactly when it satisfies von Neumann's axiom IV2. The implicit generality of Zermelo's theorem seems to work against the alleged restrictiveness of IV2.

Those cardinals π satisfying conditions (I) and (II) are called *boundary numbers* [*Grenzzahlen*] by Zermelo,[12] and when $\pi > \omega$ are now called the *(strongly) inaccessible* cardinals. These cardinals are basic in the theory of large cardinals, a mainstream of modern set theory devoted to the investigation of strong hypotheses and consistency strength, and in fact are the modest beginnings of a natural linear hierarchy of stronger and stronger postulations extending ZFC.[13] It is through Zermelo's *1930a* that inaccessible cardinals became structurally relevant for set theory as the delimiters of natural models. Just 13 articles before Zermelo's in *Fundamenta mathematicae*, *Sierpiński and Tarski 1930* had formulated the inaccessible cardinals arithmetically as those uncountable cardinals that are not the product of fewer cardinals each of smaller power and observed that inaccessible cardinals are regular limit cardinals. The first large cardinals ever considered, the regular limit cardinals had appeared in *Hausdorff 1908*, 443, and following *Hausdorff 1914*, 131, Zermelo calls them the "exorbitant numbers". Be that as it may, in the early

---

[11] This fundamental von Neumann result, a consequence of (first-order) replacement, was pointed out by Zermelo (p. 32).

[12] "Boundary numbers" is how "Grenzzahlen" has been translated, but the more literal "limit numbers" has its connotative advantages as well: Zermelo in his *1930a* refers to Kant's antinomies, and Kant in his *Prolegomena* had distinguished between *Grenzen* and *Schranken*, with the first having entities beyond.

[13] See *Kanamori 2003*.

model-theoretic investigations of set theory the inaccessible cardinals provided the natural models as envisioned by Zermelo. Years later *Shepherdson 1952* provided more formal proofs of Zermelo's results in a first-order context with sets and classes but without urelements, taking account of the relativity of concepts and isolating Zermelo's models as the transitive, super-complete models.[14] Recently *Uzquiano 1999* has investigated models of second-order Zermelo set theory (no replacement axiom but taking the separation axiom as a single second-order axiom) and showed that V_δ for limit ordinals $\delta > \omega$ are by no means the only possibilities and that there is already considerable variation at level ω.

In Section 4, Zermelo proceeds with three isomorphism theorems that establish a second-order categoricity of sorts for his axioms in terms of the cardinal numbers of the bases and characteristics. The first isomorphism theorem states that two normal domains with the same characteristic and bases of the same cardinality are isomorphic, the isomorphisms generated by bijections between the bases and extended through the two cumulative hierarchies as structured by foundation. Unbridled second-order replacement is crucial here as well, this time to establish that the cumulative ranks of the two domains are level-by-level extensionally correlative, though Zermelo does not mention replacement at all. This correlation also brings out how the $\mathcal{P}(P_\alpha)$'s for Zermelo have an absolute significance. The second isomorphism theorem states that two normal domains with different characteristics and bases of the same cardinality are such that one is isomorphic to a cumulative rank of the other. The third isomorphism theorem states that two normal domains with the same characteristic are such that one is isomorphic to a subdomain of the other. Hence, as Zermelo emphasizes, a normal domain is characterized up to isomorphism by its *type*, the pair $\langle \mathfrak{q}, \pi \rangle$ where $\mathfrak{q}$ is the cardinal number of the basis, which can be arbitrary, and π is the characteristic, which must be ω or inaccessible; and given two types $\langle \mathfrak{q}, \pi \rangle$ and $\langle \mathfrak{q}', \pi' \rangle$, isomorphic embeddability is a consequence of $\mathfrak{q} \leq \mathfrak{q}'$ and $\pi \leq \pi'$.

In Section 5, Zermelo concludes with a brief discussion of existence, consistency, and categoricity. Notably, he initially assumes "the existence of domains of set theory that satisfy the ZF-axioms for an arbitrary basis", and concludes forthwith the existence of domains that also satisfy foundation. This was the main thrust of *von Neumann 1929,* which devoted several pages to the result in a formalized axiomatic setting. Speculating then on the possibilities for characteristics, Zermelo points out that ω is a characteristic, as starting with any normal domain P with basis Q one can consider the corresponding

[14] A class is *super-complete iff* any sub*set* is an element. *Shepherdson 1952,* 227, wrote: "[Equivalent results] were obtained by Zermelo although in an insufficiently rigorous manner. He appeared to take no account of the relativity of set-theoretical concepts pointed out by Skolem." Skolem relativity has seemingly become entrenched, but Zermelo of course was deliberately working in a second-order context and decidedly opposed "Skolemism"!

subdomain P_ω of the canonical development of P (cf. the third development theorem—P_ω is in effect the set of hereditarily finite sets). This also suggests why Zermelo deliberately eschews the axiom of infinity, which thus establishes the relative consistency of ZF′. Zermelo proceeds by analogy to the least inaccessible cardinal via the ordinal type of the next normal domain—what he calls the "Cantorian" normal domain—and points out how, like ω, such a cardinal cannot be proved to exist in ZF′. This kind of positing by analogy with the axiom of infinity is now typical in the theory of large cardinals and is resonant with Cantor's own seamless account of number across the finite and the transfinite. Since already it is seen that ω may exist in one model but not another, Zermelo wrote (p. 45): "Our axiom system is *non-categorical* after all, which, in this case, is not a disadvantage, but rather an *advantage.* For the enormous significance and unlimited applicability of set theory rests precisely on this fact."

Zermelo argues more broadly how a characteristic specifies a model with "suitable postulates" that determine it categorically. In a sweeping climax he puts forward the general hypothesis that "*every categorically determined domain can also be conceived of* [*aufgefaßt*] *as a 'set' in some way;* that is, that it can occur as an element of a (suitably chosen) normal domain"[15] and postulated "the *existence of an unlimited sequence of boundary numbers* [*Grenzzahlen*] as a new axiom for the 'meta-theory of sets' ..." The postulation would bijectively correlate the ordinal numbers with the inaccessible cardinals and so provide for an endless procession of models.[16] The open-endedness of Zermelo's original *1908b* axiomatization had been structured by replacement and foundation, but after synthesizing the sense of progression inherent in the new cumulative hierarchy picture and the sense of completion in the inaccessible cardinals, Zermelo advanced a new open-endedness with an eternal return of models. This dynamic view of sets and set theory was a marked departure from Cantor's (and later, Gödel's) focus on a fixed universe of sets. Through means dramatically different and complementary to Cantor's absolute infinite, Zermelo dissolved the traditional antinomies of set theory through a dialectical interplay between the global and the local.[17] Furthermore, not only did Zermelo subsume von Neumann's axiom IV2, that principled means of handling classes too large, but by having such classes be elements in a next normal domain and therefore coming under the purview of his generative axioms like power set, Zermelo dissolved further antinomies like the incompatibility of $2^\pi > \pi$ with π being the cardinal number of the

[15] This anticipates the "closed" domains of the last page of *s1930e*.

[16] Tarski in his *1938* also and later posited arbitrarily large inaccessible cardinals via his axiom of inaccessible sets; he was led to this axiom by cardinality and closure considerations, and he formulated it in such a way that it implies the axiom of choice. In contrast to Zermelo's informal, second-order approach Tarski could be seen to be working in first-order ZF.

[17] *Tait 1998* provides a sophisticated account of this aspect of Zermelo's conception of set theory and draws out large cardinal reflection principles.

universal class, the first "paradox" that Russell had come to, while studying Cantor's work.

Zermelo (p. 47) concludes grandly:

> The two diametrically opposed tendencies of the thinking mind, the idea of creative *progress* and that of summary *completion*, which form also the basis for Kant's "antinomies", find their symbolic presentation as well as their symbolic reconciliation in the transfinite number series that rests upon the notion of well-ordering and which, though lacking in the true completion on account of its boundless progressing, possesses relative way stations, namely those "boundary numbers" separating the higher from the lower model types. Thus, instead of leading to constriction and mutilation, the set-theoretical "antinomies" lead, when understood correctly, to yet still unpredictable development of the mathematical science and its enrichment.

Remarkable and distinctive though *Zermelo 1930a* was, it was historically overshadowed by the epochal work of Kurt Gödel. In the summer of 1930 the young Gödel established his now celebrated incompleteness results. As part of a steady intellectual development, he forthwith moved into set theory as a transfinite extension of the theory of types, and in the later 1930s established the relative consistency of the axiom of choice and of the continuum hypothesis with the inner model L of constructible sets. The approaches of Gödel and of Zermelo to set theory merit comparison with respect to the underlying logic, the emergence of the cumulative hierarchy view, the focus on models of set theory, and subsequent influence.

First and foremost, first-order logic is part and parcel of Gödel's work both in completeness and set theory. Zermelo proceeded in second-order terms and ultimately did not take the linguistic turn, in that he did not develop an uninterpreted formalism. Whereas the Skolem paradox much exercised Zermelo, Gödel subsumed paradox into method by invoking Skolem's analysis to establish the continuum hypothesis in L. Gödel showed how first-order definability can be formalized and used in a transfinite recursive construction to establish striking new mathematical results. This significantly contributed to a lasting ascendancy for first-order logic, which beyond its sufficiency as a logical framework for mathematics was newly seen to have considerable operational efficacy.

Zermelo first adopted the axiom of foundation, and indeed it was thematically central to his *1930a* analysis. Gödel had foundation in his *1940* monograph on L, in an axiomatization adapted from one of Bernays, and indeed, it was axiomatically correlative to his view of set theory as a hierarchy of cumulative types. Gödel came close to *Zermelo 1930a* in his informal sketch *1939* about L when he stated his relative consistency results in terms of the *Zermelo 1908b* axioms as rendered in first-order logic and asserted that L_{Ω}, where Ω is "the first inaccessible number", is a model of Zermelo's axioms together with replacement. Also, making his only explicit reference to *Zermelo 1930a*, Gödel in his *1947*, 520, later gave the existence of inaccessible cardinals as the simplest example of an axiom that asserts still further iterations of the 'set of' operation and can supplement the axioms of set theory without arbitrariness.[18]

Beyond the imprint on Gödel himself, which could be regarded as significant, *Zermelo 1930a* seemed to have had little influence on the further development of set theory, presumably because of its second-order lens and its lack of rigorous detail and attention to relativism. On the other hand, Gödel's work with L with its incisive analysis and use of first-order definability was readily recognized as a signal advance. Issues about consistency, truth, and definability were brought to the forefront, and the continuum hypothesis result established the mathematical importance of a hierarchical analysis. As the construction of L was gradually digested, the sense it promoted of a cumulative hierarchy reverberated to become the basic picture of the universe of sets. Nonetheless, with the assimilation of set-theoretic rigor, increasing confidence in consistency, and the emergence of ZFC as the canonical set theory, there has been of late new appreciation of the sweep of *1930a* especially because of renewed interest in second-order logic.[19]

---

[18] Gödel referenced *Zermelo 1930a* after writing: "[This] axiom, roughly speaking, means nothing else but that the totality of sets obtainable by exclusive use of the processes of formation of sets expressed in the other axioms forms again a set (and, therefore, a new basis for a further application of these processes)." This was just what Zermelo had emphasized; for Gödel there would also be the overlay of truth in the "next higher system".

[19] See for example *Tait 1998* and *Shapiro-Uzquiano 2008*.

Über Grenzzahlen und Mengenbereiche

Neue Untersuchungen über die Grundlagen der Mengenlehre

1930a

In der folgenden Arbeit handelt es sich um die Untersuchung der „Bereiche", bestehend aus Mengen und Urelementen, in denen die „allgemeinen" Axiome der Mengenlehre (die „Zermelo-Fraenkelschen Axiome" mit einer Ergänzung) erfüllt sind, und um den Nachweis, daß ein solcher „Normalbereich" bis auf isomorphe Abbildungen bestimmt ist durch zwei Zahlen, durch die Mächtigkeit seiner „Basis" d. h. der Gesamtheit seiner „Urelemente" (die keine eigentlichen Mengen sind) und durch seine „Charakteristik" d. h. den Ordnungstypus aller in ihm enthaltenen „Grundfolgen" oder aller in ihm durch Mengen vertretenen Ordnungszahlen. Es wird gezeigt, daß diese beiden Zahlen unabhängig von einander beliebig gewählt werden können, sofern die „Charakteristik" den Bedingungen einer „Grenzzahl" genügt, nämlich gleichzeitig eine „Kernzahl" oder „reguläre Anfangszahl" und „Eigenwert" oder „kritische Zahl" einer gewissen „Normalfunktion" zu sein. Die schrankenlose Fortsetzbarkeit der transfiniten Zahlenreihe gestattet danach die Darstellung der Mengenlehre in einer ebenso unbegrenzten Folge wohlunterschiedener „Modelle". Und eben die scharfe Unterscheidung zwischen den verschiedenen Modellen des (nicht-kategorischen!) Axiomensystems sichert uns auch eine befriedigende Aufklärung der „ultrafiniten Antinomien", indem immer die „Unmengen" des einen Modells sich als eigentliche „Mengen" darstellen im nächstfolgenden wie in allen höheren Modellen.

Als Hilfsmittel der Untersuchung bieten sich einmal die „Grundfolgen", nämlich die in jedem Normalbereich vorhandenen einfachsten Vertreter der verschiedenen Ordnungszahlen, und zweitens die „Entwickelung" des Normalbereiches, seine Zerlegung in eine wohlgeordnete Folge getrennter „Schichten",
30 wobei die einer Schicht | angehörenden Mengen immer in den vorangehenden „wurzeln", sodaß ihre Elemente in diesen liegen, und selbst wieder den folgenden als Material dienen.

§ 1. Die konstituierenden Axiome

Das unserer Untersuchung zugrunde liegende Axiomensystem der Mengenlehre ist im Wesentlichen das „Zermelo-Fraenkelsche", nämlich das durch das Fraenkelsche „Ersetzungsaxiom" ergänzte System meiner Axiome von 1908* mit der Abänderung daß einmal mein „Unendlichkeits-Axiom" als nicht zur

* Math. Ann. Bd. 65, S. 261–281.

On boundary numbers and domains of sets
New investigations in the foundations of set theory

1930a

The present paper investigates "domains", consisting of sets and urelements, in which the "general" axioms of set theory (the "Zermelo-Fraenkel axioms" with a supplementation) are satisfied. The paper also provides a demonstration that such a "normal domain" is determined up to isomorphic mappings by two numbers: the cardinality of its "basis", that is, the totality of its "urelements" (which are not proper sets), and its "characteristic", that is, the order type of all "basic sequences" contained in it or of all ordinal numbers represented in it by sets. It is shown that we can choose these two numbers arbitrarily and independently of one another, provided that the "characteristic" satisfies the conditions for being a "boundary number"; that is, being a "core number", or "regular initial number", as well as an "eigenvalue", or "critical number", of a certain "normal function" at the same time. The boundless continuability of the transfinite number sequence therefore permits the presentation of set theory in a likewise unlimited sequence of well-differentiated "models". And it is this sharp distinction between the different models of the (non-categorical!) axiom system that allows us to resolve the "ultrafinite antinomies" to our satisfaction by always presenting the "non-sets" of one model as proper "sets" in the next model and in all higher models.

Two notions suggest themselves as expedients for our investigation: first, the "basic sequences", that is, the simplest representatives of the different ordinal numbers, which occur in every normal domain. Second, the "development" of the normal domain, its decomposition into a well-ordered sequence of separated "layers" where the sets belonging to one layer are always "rooted" in the preceding layers such that their elements lie in those layers, and they themselves, in turn, serve as material for subsequent layers.

§ 1. The constitutive axioms

The axiom system of set theory that forms the basis of our investigation is essentially the "Zermelo-Fraenkel" axiom system, namely, the system that results when my axioms of 1908[1] are supplemented with Fraenkel's "replacement axiom" and modified so that, on the one hand, my "axiom of infinity"

[1] *Zermelo 1908b.*

„allgemeinen“ Mengenlehre gehörig weggelassen und andererseits das „Axiom der Fundierung“ hinzugefügt wird, wodurch „zirkelhafte“ und „abgründige“ Mengen ausgeschlossen werden. Demgemäß bezeichnen wir als „ergänztes ZF-System“ oder abgekürzt als „ZF'-System“ die Gesamtheit der folgenden Axiome:

B) *Axiom der Bestimmtheit:* Jede Menge ist durch ihre Elemente bestimmt, sofern sie überhaupt Elemente besitzt.

A) *Axiom der Aussonderung:* Durch jede Satzfunktion $\mathfrak{f}(x)$ wird aus jeder Menge m eine Untermenge $m_\mathfrak{f}$ ausgesondert, welche alle Elemente x umfasst, für die $\mathfrak{f}(x)$ wahr ist. Oder: jedem Teil einer Menge entspricht selbst eine Menge, welche alle Elemente dieses Teiles enthält[1].

P) *Axiom der Paarung:* Sind a, b irgend zwei Elemente, so gibt es eine Menge, welche beide als Elemente enthält.

U) *Axiom der Potenzmenge:* Jeder Menge m entspricht eine Menge $\mathfrak{U}m$, welche alle Untermengen von m als Elemente enthält, einschließlich der Nullmenge und m selbst. An die Stelle der „Nullmenge“ tritt hier ein beliebig ausgewähltes „Urelement“ u_0.

V) *Axiom der Vereinigung*: Jeder Menge m entspricht eine Menge $\mathfrak{S}m$, welche die Elemente ihrer Elemente enthält.

E) *Axiom der Ersetzung:* Ersetzt man die Elemente x einer Menge m ein-
31 deutig durch beliebige Elemente x' des Bereiches, so ent- | hält dieser auch
eine Menge m', welche alle diese x' zu Elementen hat.

F) *Axiom der Fundierung:* Jede (rückschreitende) Kette von Elementen, in welcher jedes Glied Element des vorangehenden ist, bricht mit endlichem Index ab bei einem Urelement. Oder, was gleichbedeutend ist: Jeder Teilbereich T enthält wenigstens ein Element t_0, das kein Element t in T hat.

Dieses letzte Axiom, durch welches alle „zirkelhaften“ namentlich auch alle „sich selbst enthaltenden“, überhaupt alle „wurzellosen“ Mengen ausgeschlossen werden, war bei allen praktischen Anwendungen der Mengenlehre bisher immer erfüllt, bringt also vorläufig keine wesentliche Einschränkung der Theorie.

---

[1] Die Satzfunktion $\mathfrak{f}(x)$ kann hier ganz *beliebig* sein, wie auch die Ersetzungsfunktion in E), und alle aus ihrer Beschränkung auf eine besondere Klasse von Funktionen gezogenen Folgerungen kommen für den hier angenommen Standpunkt in Wegfall. Eine eingehende Erörterung der „Definitheitsfrage“ in Anschluß an meine letzte Note in dieser Zeitschrift (Fund. Math. T. XIV, S. 339–344) und an die kritischen „Bemerkungen“ des Herrn Th. Skolem (ebendort T. XV, S. 337–341) behalte ich mir vor.

is dropped because it does not belong to "general" set theory and, on the other hand, the "axiom of foundation" is added, whereby "circular" and "non-grounded" sets are excluded. Accordingly, we call the totality of the following axioms the "supplemented ZF-system", or "ZF'-system", for short:

B) *Axiom of extensionality*: Every set is determined by its elements, provided that it has any elements at all.

A) *Axiom of separation*: Every propositional function $\mathfrak{f}(x)$ separates from every set m a subset $m_{\mathfrak{f}}$ containing all those elements x for which $\mathfrak{f}(x)$ is true. Or: To each part of a set there in turn corresponds a set containing all elements of this part.[2]

P) *Axiom of pairing*: If a and b are any two elements, then there is a set that contains both of them as its elements.

U) *Axiom of the power set*: To every set m there corresponds a set $\mathfrak{U}m$ that contains as elements all subsets of m, including the null set and m itself. Here, an arbitrarily chosen "urelement" $\mathfrak{u}_0$ takes the place of the "null set".

V) *Axiom of the union*: To every set m there corresponds a set $\mathfrak{S}m$ that contains the elements of its elements.

E) *Axiom of replacement*: If the elements x of a set m are replaced in a unique way by arbitrary elements x' of the domain, then the domain contains also a set m' that has as its elements all these elements x'.

F) *Axiom of foundation*: Every (decreasing) chain of elements, in which each term is an element of the preceding one, terminates with finite index at an urelement. Or, what amounts to the same thing: Every partial domain T contains at least one element t_0 that has no element t in T.

This last axiom, which excludes all "circular" sets, all "self-membered" sets in particular, and all "rootless" sets in general, has always been satisfied in all practical applications of set theory, and, hence, does not result in an essential restriction of the theory for the time being.

[2] Like the replacement function in E), the propositional function $\mathfrak{f}(x)$ can be completely *arbitrary* here, and all consequences of restricting it to a particular class of functions cease to apply from the present point of view. I shall consider elsewhere more thoroughly "the question of definiteness" in connection with my last contribution to this journal (*1929a*) and with the critical "remarks" by Mr. Th. Skolem (*Skolem 1930*).

Auf die „Unabhängigkeit" der Axiome kommt es uns hier nicht an: bei geeigneter Fassung ließe sich etwa A) aus E) ableiten oder P) aus U) und E). Das „Auswahl-Axiom" ist hier nicht ausdrücklich formuliert, da es einen anderen Charakter hat als die übrigen und nicht zur Abgrenzung der Bereiche dienen kann. Es wird aber unserer ganzen Untersuchung als allgemeines logisches Prinzip zugrunde gelegt, und namentlich wird auf Grund dieses Prinzipes im Folgenden jede vorkommende Menge auch als *wohlordnungsfähig* vorausgesetzt werden.

Dieses Axiomensystem BAPUVEF, das wir als das „ZF'-System" bezeichnen wollen, nehmen wir hier zum Ausgangspunkt und bezeichnen als einen „*Normalbereich*" einen Bereich von „Mengen" und „Urelementen", der in Bezug auf die „Grundrelation" $a \epsilon b$ unserem ZF'-System genügt. „Bereiche" dieser Art, ihre „Elemente", ihre „Unterbereiche", ihre „Summen" und „Durchschnitte" werden wir dabei nach den allgemeinen mengentheoretischen Begriffen und Axiomen genau wie *Mengen* behandeln, von denen sie sich auch in keinem sachlich wesentlichen Punkte unterscheiden, wir werden sie aber immer nur als „Bereiche" und nicht als „Mengen" bezeichnen zur Unterscheidung von den „Mengen" als den Elementen des betrachteten Bereiches.

§ 2. Die Grundfolgen eines Normalbereiches und seine Charakteristik

Als „*Grundfolge*" bezeichne ich *eine wohlgeordnete Menge, in welcher jedes Element (mit Ausnahme des ersten, das ein „Urelement" sein muß), identisch ist mit der Menge aller ihm vorangehenden Elemente.*

32 | So entstammen dem Urelement u die Grundfolgen

$$g_0 = u, \quad g_1 = \{u\}, \quad g_2 = \{u, \{u\}\}, \quad g_3 = \{u, \{u\}, \{u, \{u\}\}\},$$

und so fort nach der Regel

$$g_{\alpha+1} = g_\alpha + \{g_\alpha\} \quad \text{und} \quad g_\alpha = \sum_{\beta<\alpha} g_\beta, \text{ wenn } \alpha \text{ eine Limeszahl ist.}$$

Allgemein ist eine Grundfolge *eine durch die ε-Beziehung geordnete Menge*, die dann wegen F) auch *wohlgeordnet* sein muß, und es gelten für sie u. a. die folgenden leicht zu verifizierenden Sätze:

1) Jedes in einer Grundfolge enthaltene Element ist Element aller folgenden und enthält alle vorangehenden als Elemente.
2) Jedes Element sowie jeder Abschnitt einer Grundfolge ist selbst eine Grundfolge.
3) Aus jeder Grundfolge entsteht eine neue, wenn man zu ihren Elementen die Menge selbst als letztes Element hinzufügt: $g' = g + \{g\}$, wobei ihr Ordnungstypus gerade um 1 vermehrt wird.

We are not concerned here with the "independence" of the axioms: framing the axioms in a suitable fashion, we could, for instance, derive A) from E), or P) from U) and E). We have not explicitly formulated the "axiom of choice" here because it differs in character from the other axioms and cannot be used to delimit the domains. However, we use it as a general logical principle upon which our entire investigation is based; in particular, it is on the basis of this principle that we shall assume in the following that every set is *capable of being well-ordered.*

This axiom system BAPUVEF, which we will call the "ZF'-system", serves as our starting point. We call "*normal domain*" a domain of "sets" and "urelements" that satisfies our "ZF'-system" with regard to the "basic relation" $a \epsilon b$. We shall treat "domains" of this kind, their "elements", their "subdomains", their "sums" and "intersections" according to the general set-theoretic concepts and axioms exactly like *sets*, from which they do not substantially differ anyway. But we shall always call them "domains" rather than "sets" in order to distinguish them from the "sets" that are the elements of the domain under consideration.

§ 2. The basic sequences of a normal domain and its characteristic

I call a *"basic sequence" a well-ordered set in which every element (with the exception of the first, which has to be an urelement) is identical to the set of all elements preceding it.*

From the urelement u thus issue the basic sequences

$$g_0 = u, \quad g_1 = \{u\}, \quad g_2 = \{u, \{u\}\}, \quad g_3 = \{u, \{u\}, \{u, \{u\}\}\},$$

and so forth in accordance with the rule

$$g_{\alpha+1} = g_\alpha + \{g_\alpha\}, \quad \text{and} \quad g_\alpha = \sum_{\beta<\alpha} g_\beta \text{ if } \alpha \text{ is a limit number}.$$

Generally speaking, a basic sequence is *a set that is ordered by the ε-relation*, and which then, on account of F), must also be *well-ordered.* The following easily verifiable theorems, among others, hold for basic sequences:

1) Each element in a basic sequence is contained as an element in all those following it and contains as elements all those preceding it.
2) Each element is itself a basic sequence, as is each segment of a basic sequence.
3) A new basic sequence results from any basic sequence when the set itself is added to its elements as the last element: $g' = g + \{g\}$, when its order type is increased by exactly 1.

4) Aus jeder Menge T von Grundfolgen mit identischem Anfangselement u entsteht durch Vereinigung eine neue Grundfolge $\mathfrak{S}T$, welche die Elemente von T sämtlich als Abschnitte und außer sich selbst als Elemente enthält. Auch hier ist der Ordnungstypus der neuen Grundfolge der auf die der gegebenen nächstfolgende.
5) Von zwei verschiedenen Grundfolgen mit identischem Anfangselement ist immer die eine Abschnitt und Element der anderen. Nämlich immer diejenige vom kleineren Ordnungstypus, die wir dann auch einfach als die „kleinere“ bezeichnen wollen.
6) Ist u ein Urelement und r eine nach dem Typus ϱ wohlgeordnete Menge in einem Normalbereich, so enthält dieser Bereich auch eine der Menge r ähnliche Grundfolge g_ϱ mit u als Anfangselement.
Angenommen nämlich, der Satz sei richtig für alle Ordnungszahlen $\varrho < \alpha$, so gilt er auch für $\varrho = \alpha$. Denn entweder ist $\alpha = \beta + 1$ und g_β hat den Typus β, dann hat nach 3) g' den Typus $\beta + 1 = \alpha$. Oder α ist Limeszahl, dann ist die Vereinigung Σg_β aller g_β für $\beta < \alpha$ nach 4) selbst eine Grundfolge und zwar vom Typus α, da jeder ihrer echten Abschnitte selbst ein $g_\beta < g_\alpha$ ist.
7) Die Gesamtheit *aller* in einem Normalbereich P enthaltenen Grundfolgen g_α mit gemeinsamem Anfangselement u bildet einen wohldefinierten Unterbereich G_u von P, und die entsprechenden Ordnungszahlen α einen wohldefinierten Abschnitt Z_π der Zahlenreihe vom Ordnungstypus π, aber der Bereich P enthält keine „Menge“ w, die alle diese Grundfolgen zu Elementen hätte, und | ebenso wenig eine wohlgeordnete Menge vom Ordnungstypus π, sondern π ist lediglich die *obere Grenze* aller in P durch Mengen vertretenen Ordnungszahlen. Anderenfalls ergäbe sich die bekannte „Burali-Fortische Antinomie“. 33

Die so definierte Ordnungszahl π, die hier als „Grenzzahl“ oder „Charakteristik“ des Normalbereiches bezeichnet werden soll, ist aber nicht willkürlich, sondern muß, um „Grenzzahl-Charakter“ zu haben, gewissen Bedingungen genügen. Es sind dies die folgenden:

I) Jede Grenzzahl hat „*Kernzahl*-Charakter“ d.h. sie ist eine „reguläre Anfangszahl“, nämlich *keiner kleineren „konfinal“*[2].

Wäre nämlich π konfinal $\varrho < \pi$, so enthielte der Abschnitt Z_π der Zahlenreihe eine Teilfolge vom Ordnungstypus ϱ bestehend aus Zahlen $\alpha_\nu < \pi$, die keinem echten Abschnitte $Z_\alpha < Z_\pi$ angehörten. Jeder dieser Zahlen a_ν entspräche dann in P eine Grundfolge g_{α_ν} vom gleichen Ordnungstypus, und auch die Vereinigung aller dieser g_{α_ν} wäre nach 4) wieder eine Grundfolge g_α des Normalbereiches, während doch ihr Ordnungstypus $\alpha = \lim \alpha_\nu = \pi$ sein müßte nach der Annahme. Also ist π eine „Kernzahl“ oder eine „reguläre Anfangszahl“ und zwar, wie wir sehen werden, eine solche „zweiter Art“, eine „exorbitante“ Zahl. (Hausdorff a. a. O. S. 131) Wäre nämlich $\pi = \omega_{\nu+1}$, so

[2] Vergl. *F. Hausdorff*, Grundzüge der Mengenlehre, 1. Aufl. Kap. IV § 4.

4) The union of any set T of basic sequences whose initial elements are identical gives rise to a new basic sequence $\mathfrak{S}T$ that contains all elements of T as segments and, with the exception of itself, also as elements. Once again, the order type of the new basic sequence immediately follows that of the given basic sequences.
5) Of any two different basic sequences with the same initial element, one is always contained in the other, both as an element and as a segment. Because it is always the sequence of the smaller order type that is so contained in the other, we shall simply refer to it as the "smaller" one of the two sequences.
6) If a normal domain contains an urelement u and a well-ordered set r of type ϱ, then it also contains a basic sequence g_ϱ that is similar to the set r and whose initial element is u.
For, assuming that it holds for all ordinal numbers $\varrho < \alpha$, the theorem also holds for $\varrho = \alpha$. For either $\alpha = \beta + 1$ and g_β has type β, and, therefore, g' has type $\beta + 1 = \alpha$, according to 3). Or α is a limit number. In this case, the union Σg_β of all g_β for $\beta < \alpha$ is itself, according to 4), a basic sequence. Moreover, since each of its proper segments is itself a $g_\beta < g_\alpha$, it is a basic sequence of type α.
7) The totality of *all* basic sequences g_α with common initial element u contained in a normal domain P forms a well-defined subdomain G_u of P, and the corresponding ordinal numbers α form a well-defined segment Z_π of the number series of order type π. The domain P, however, does not contain a "set" w that would contain as its elements all these basic sequences. Nor does it contain a well-ordered set of order type π. Rather, π is only the *upper limit* of all ordinal numbers represented in P by sets. Otherwise, the well-known "Burali-Forti antinomy" would follow.

The ordinal number π so defined, which will be called here the "boundary number", or "characteristic", of the normal domain, is, however, not arbitrary. In order to possess the "character of a boundary number" it must satisfy the following conditions:

I) Every boundary number possesses the "character of a *core number*", that is, it is a "regular initial number", it is *"cofinal" with no smaller number.*[3]

For if π were cofinal with $\varrho < \pi$, then the segment Z_π of the number series would contain a partial series of order type ϱ consisting of numbers $\alpha_\nu < \pi$ that would belong to no proper segment $Z_\alpha < Z_\pi$. To each of these numbers α_ν there would then correspond in P a basic sequence g_{α_ν} of the same order type. And, by 4), the union of all these g_{α_ν} would be a basic sequence g_α of the normal domain as well, while, by the assumption made, its order type would have to be $\alpha = \lim \alpha_\nu = \pi$. Therefore, π is a "core number", or "regular initial number", particularly, as we shall see, one of the "second kind", that is, "exorbitant" number (*Hausdorff 1914*, 131). For if $\pi = \omega_{\nu+1}$, then still

[3] Cf. *Hausdorff 1914*, chap. IV § 4.

wäre noch $\omega_\nu < \pi$, und der Bereich enthielte eine Grundfolge g_{ω_ν} von diesem Typus sowie die zugehörige Potenzmenge $m = \mathfrak{U}g_{\omega_\nu}$ von der Kardinalzahl $\mathfrak{m} > \overline{\omega}_\nu$, also $\mathfrak{m} \geq \overline{\omega}_{\nu+1} = \overline{\pi}$ im Widerspruch mit der Definition von π.

Wäre nun die Cantorsche Vermutung erwiesen, daß die Potenzmenge $\mathfrak{U}m$ immer gerade die nächst höhere Mächtigkeit habe, so würde aus $\mathfrak{m} < \overline{\pi}$ auch immer folgen $2^{\mathfrak{m}} < \pi$ und jede „exorbitante" Zahl π wäre auch „Grenzzahl" eines Normalbereiches[3]. Da aber tatsächlich diese Frage noch unentschieden ist, so brauchen wir zur Charakterisierung der „Grenzzahlen" noch eine weitere Bedingung, die hier mit Hilfe einer gewissen „Normalfunktion"[4] hergeleitet werden soll.

Ist ξ eine beliebige im Normalbereich vertretene Ordnungszahl, so enthält
34 dieser außer der Grundfolge g_ξ wegen U) auch eine | Grundfolge mit dem Index $\xi^* = \varphi(\xi)$, der Anfangszahl der Zahlenklasse, die zur Kardinalzahl $2^{\overline{\xi^*}}$ gehört. Diese Funktion $\varphi(\xi)$ ist zwar noch keine Normalfunktion, da verschiedenen Argumenten ξ gleiche Funktionswerte entsprechen können. Wohl aber gelangen wir zu einer solchen durch Iteration von φ in folgender Weise, indem wir festsetzen:

$$1)\quad \psi(0) = 0\,, \quad 2)\quad \psi(\xi+1) = \psi(\xi)^* = \varphi\psi(\xi)\,, \quad 3)\quad \psi(\alpha) = \lim_{\xi<\alpha} \psi(\xi)\,,$$

wenn α eine Limeszahl ist. Hierdurch wird die Funktion ψ für beliebige Argumente ξ eindeutig bestimmt, und auch die Bedingungen einer Normalfunktion sind erfüllt. Denn aus $\alpha < \beta$ folgt immer $\alpha + 1 \leq \beta$ und daher durch transfinite Induktion

$$\psi(\alpha) < \psi(\alpha)^* = \psi(\alpha+1) \leq \psi(\beta)\,,$$

sowie aus 3), daß allgemein $\lim \psi(\alpha_\nu) = \psi(\lim \alpha_\nu)$, die Funktion also auch „stetig" ist. Durch unsere Funktion $\psi(\xi)$ wird aber nicht nur die *ganze* Zahlenreihe ähnlich und stetig abgebildet auf einen Teil derselben, sondern auch jeder einem Normalbereich entsprechende *Abschnitt* Z_π auf einen Teil von sich selbst. Ist nämlich $\alpha < \pi$, so ist auch $\psi(\alpha) < \pi$, wie durch Induktion gezeigt werden soll. Angenommen, es sei stets $\psi(\xi) < \pi$ für alle $\xi < \alpha$, so ist auch $\psi(\xi+1) = \psi(\xi)^* < \pi$, weil doch der Normalbereich mit jeder seiner Mengen m auch ihre Potenzmenge $\mathfrak{U}m$ enthalten soll, also $\psi(\alpha) < \pi$, wenn α von erster Art. Ist aber α eine Limeszahl, so entsprechen den Elementen der Grundfolge g_α, die ja selbst Grundfolgen g_ξ von kleinerem Typus sind, eindeutig die Grundfolgen $g_{\psi(\xi)} < g_\pi$; diese letzteren sind also nach E) selbst Elemente einer Menge in P, und ihre Vereinigung $\Sigma g_{\psi(\xi)}$ ist nach 4) selbst eine Grundfolge g_ϱ des Normalbereiches. Hier ist aber $\varrho = \lim_{\xi<\alpha} \psi(\xi) = \psi(\alpha)$

[3] Vergl. *R. Baer*, Zur Axiomatik der Kardinalzahlarithmetik, Math. Zeitschr. Bd. 29, S. 382 f. und die Fußnote 8) auf S. 382.

[4] *Hausdorff* a. a. O. Kap. V, 3, S. 130. Bezüglich der hier verwendeten besonderen Normalfunktion $\psi(\xi)$ vergl. auch *A. Tarski*, Fund. Math. T. VII, S. 1–15.

$\omega_\nu < \pi$, and the domain would contain both a basic sequence g_{ω_ν} of this type as well as the corresponding power set $m = \mathfrak{U} g_{\omega_\nu}$ of cardinal number $\mathfrak{m} > \overline{\omega}_\nu$, and hence $\mathfrak{m} \geq \overline{\omega}_{\nu+1} = \overline{\pi}$, contrary to the definition of π.

Now, if there existed a demonstration for Cantor's conjecture that the power set $\mathfrak{U}m$ is always of exactly the next higher cardinality, then $2^{\mathfrak{m}} < \pi$ would always follow from $\mathfrak{m} < \overline{\pi}$, and every "exorbitant" number π would also be the "boundary number" of a normal domain.[4] But since this question in fact is still undecided, we need a further condition in order to characterize the "boundary numbers", which we shall provide here using a certain "normal function".[5]

If ξ is an arbitrary ordinal number represented in the normal domain, then, on account of U), the normal domain contains in addition to the basic sequence g_ξ also a basic sequence with index $\xi^* = \varphi(\xi)$, the initial number of the number class belonging to the cardinal number $2^{\overline{\xi}}$.[6] This function $\varphi(\xi)$ is not a normal function yet, since the same functional values may correspond to different arguments ξ. But we can arrive at such a normal function by iteration of φ as follows, using the definitions:

$$1)\quad \psi(0) = 0\,, \quad 2)\quad \psi(\xi+1) = \psi(\xi)^* = \varphi\psi(\xi)\,, \quad 3)\quad \psi(\alpha) = \lim_{\xi<\alpha} \psi(\xi)\,,$$

if α is a limit number. Thus, the function ψ is uniquely determined for arbitrary arguments ξ, and the conditions for normal functions are satisfied as well. For from $\alpha < \beta$ it always follows that $\alpha+1 \leq \beta$, and hence by transfinite induction

$$\psi\,(\alpha) < \psi\,(\alpha)^* = \psi\,(\alpha+1) \leq \psi\,(\beta)\,,$$

and from 3) that, in general, $\lim \psi(\alpha_\nu) = \psi(\lim \alpha_\nu)$, and hence that the function is also "continuous". But not only is the *entire* number series mapped similarly and continuously onto a part of itself by virtue of our function $\psi(\xi)$. Each *segment* Z_π corresponding to a normal domain is mapped onto a part of itself as well. For if $\alpha < \pi$, then also $\psi(\alpha) < \pi$, as we shall show by induction. Suppose that always $\psi(\xi) < \pi$ for all $\xi < \alpha$. Then $\psi(\xi+1) = \psi(\xi)^* < \pi$, since the normal domain is supposed to contain the power set $\mathfrak{U}m$ of each of its sets m, hence $\psi(\alpha) < \pi$, if α is of the first kind. But if α is a limit number, then to the elements of the basic sequence g_α, which themselves are basic sequences g_ξ of smaller type, correspond uniquely the basic sequences $g_{\psi(\xi)} < g_\pi$. The latter are themselves, according to E), elements of a set in P, and their union $\Sigma g_{\psi(\xi)}$ is itself, according to 4), a basic sequence g_ϱ of the normal domain. But in this case $\varrho = \lim_{\xi<\alpha} \psi(\xi) = \psi(\alpha)$, and therefore

[4] Cf. *Baer 1929*, 382f. and footnote 8 on p. 382.

[5] *Hausdorff 1914*, chap. V, § 3, p. 114 ⟦Zermelo erroneously writes "p. 130"⟧. With respect to the special normal function $\psi(\xi)$ used here, cf. also *Tarski 1925*.

[6] ⟦Zermelo erroneously writes "$2^{\overline{\xi^*}}$" instead of "$2^{\overline{\xi}}$".⟧

und daher, wie behauptet, $\psi(\alpha) < \pi$. Wäre nun $\pi < \psi(\pi) = \lim_{\alpha<\pi} \psi(\alpha)$, so gäbe es bereits ein $\alpha < \pi$, für welches $\psi(\alpha) > \pi$ wäre, im Widerspruch mit dem Bewiesenen. Somit ergibt sich denn als *zweite Bedingung*:

II) Jede „Grenzzahl" oder „Charakteristik" eines Normalbereiches ist gleichzeitig ein „Eigenwert" oder „kritische Zahl" unserer oben definierten Normalfunktion $\psi(\xi)$.

Diese beiden Bedingungen, denen jede „Grenzzahl" genügen muß, sind im Wesentlichen von einander unabhängig, sofern man | in I) ausschließlich den Kernzahl-Charakter postuliert. Daß es keine Kernzahl *erster Art* sein kann, folgt dann unmittelbar aus der zweiten Bedingung: für zwei auf einander folgende (transfinite) Anfangszahlen ω_ν und $\omega_{\nu+1}$ ist nämlich $\omega_\nu < \omega_\nu + 1 < \omega_{\nu+1}$ und daher

$$\omega_{\nu+1} \leq \omega_\nu^* \leq \psi(\omega_\nu)^* = \psi(\omega_\nu + 1) < \psi(\omega_{\nu+1}),$$

also $\omega_{\nu+1}$ gewiß kein Eigenwert der Normalfunktion. Dagegen wäre nach der Cantorschen Vermutung, für jede „exorbitante" Zahl, jede „Kernzahl zweiter Art" als solche schon die zweite Bedingung erfüllt[5]. Denn in diesem Falle wäre $\psi(\xi) = \omega_\xi$ für alle transfiniten ξ und mit $\xi < \pi$ auch immer $\omega_\xi < \pi$, die Normalfunktion ω_ξ hätte Eigenwerte $< \pi$, und π als Limes aller dieser Eigenwerte wäre selbst ein Eigenwert $\pi = \omega_\pi$. Muß auch diese Frage als vorläufig unentschieden gelten, so wird sich doch im Folgenden nachweisen lassen, daß die beiden für die „Grenzzahl" aufgestellten Bedingungen auch *hinreichend* sind, daß nämlich jede beiden Bedingungen genügende Zahl π in der Tat als Charakteristik eines Normalbereiches auftreten kann.

§ 3. Die Entwickelung des Normalbereiches

Als „Normalbereich" bezeichneten wir jeden den ZF'-Axiomen genügenden Bereich von „Mengen" und „Urelementen". Ein solcher Normalbereich kann auch Teilbereiche besitzen, die selbst schon in Bezug auf die zwischen ihren Elementen geltende ε-Relation den Axiomen genügen, also Normalbereiche sind. Hierüber gilt nun zunächst der folgende

Hilfssatz. *Ein Teilbereich M eines Normalbereiches P ist selbst ein Normalbereich, wenn er 1) mit jeder seiner Mengen m zugleich deren Elemente enthält, und wenn er 2) jede Menge m des Normalbereiches P enthält, deren sämtliche Elemente x in M liegen. Umfaßt M außerdem die ganze „Basis" des Gesamtbereiches P, so ist er mit diesem identisch.*

---

[5] Vergl. *Baer* a. a. O. wie in [3]).

$\psi(\alpha) < \alpha$, just as asserted. Now, if $\pi < \psi(\pi) = \lim_{\alpha<\pi} \psi(\alpha)$, then there would already exist an $\alpha < \pi$, for which $\psi(\alpha) > \pi$, contrary to what has been proved. Thus we have as the *second condition*:

II) Every "boundary number", or "characteristic", of a normal domain is also an "eigenvalue", or "critical number", of the normal function $\psi(\xi)$ defined above.

These two conditions, which every "boundary number" must satisfy, are essentially independent of one another provided that only the character of a core number is postulated in I). It then follows immediately from the second condition that it cannot be a core number of the *first kind*: for, given two successive (transfinite) initial numbers ω_ν and $\omega_{\nu+1}$, we have $\omega_\nu < \omega_\nu + 1 < \omega_{\nu+1}$, and hence

$$\omega_{\nu+1} \leq \omega_\nu^* \leq \psi(\omega_\nu)^* = \psi(\omega_\nu + 1) < \psi(\omega_{\nu+1}) .$$

Thus, $\omega_{\nu+1}$ is certainly not an eigenvalue of the normal function. According to Cantor's conjecture, on the other hand, every "exorbitant number", every "core number of the second kind", would already satisfy the second condition.[7] For, in this case, we would have $\psi(\xi) = \omega_\xi$ for all transfinite ξ, and $\omega_\xi < \pi$ whenever $\xi < \pi$. The normal function ω_ξ would have eigenvalues $< \pi$, and π itself, being the limit of all these eigenvalues, would be an eigenvalue $\pi = \omega_\pi$. Although this question must be considered undecided for the time being, it is still possible to show in what follows that the two conditions specified for "boundary numbers" are also *sufficient*; that is, that every number π satisfying both conditions can in fact occur as the characteristic of a normal domain.

§ 3. The development of the normal domain

We called a "normal domain" any domain of "sets" and "urelements" that satisfies the ZF'-axioms. Such a normal domain can also have partial domains that themselves already satisfy the axioms with respect to the ε-relation obtaining among its elements, and that therefore are normal domains. Now, these partial domains are subject to the following

Lemma. *A partial domain M of a normal domain P is itself a normal domain if* 1) *it contains, along with each of its sets m, also the elements of m, and if* 2) *it contains every set m of the normal domain P all of whose elements x are in M. If M also includes the whole "basis" of the total domain P, then M is identical to P.*

---

[7] Compare *Baer 1929*.

In dem angenommen Falle sind nämlich die ZF'-Axiome, sofern sie für P gelten, auch für M erfüllt, namentlich wegen 2) auch U) und V). Das „Ersetzungsaxiom“ E) muß dabei natürlich so verstanden werden, daß die Elemente x', welche die Elemente x ersetzen sollen, wieder dem Teilbereich
36 M angehören müssen. Im | besonderen Falle, wo M die ganzen Basis umfaßt, enthält der Restbereich $R = P - M$ kein einziges Urelement, und jedes seiner Elemente wäre eine Menge r, deren Elemente, da sie nach der Annahme nicht alle in M liegen, wenigstens teilweise wieder in R auftreten müssen — im Widerspruch mit dem Axiom F).

Dagegen kann sehr wohl ein Normalbereich mit kleinerer Basis $Q' \subset Q$ im größeren enthalten sein. Ein solcher entsteht aus P durch Beschränkung auf alle solchen Mengen, deren rückschreitende Elementeketten $m \ni m_1 \ni m_2 \ni m_3 \dots$ gemäß F) ausschließlich in Urelementen aus Q' enden.

Erstes Entwickelungstheorem. *Jeder Normalbereich P von der Charakteristik π läßt sich zerlegen in eine nach dem Typus π wohlgeordnete* ⟦Folge⟧ *von nicht leeren und unter sich elementefremden „Schichten“ Q_α von der Beschaffenheit, daß jede Schicht Q_α alle in keiner früheren vorkommenden Elemente von P umfaßt, deren Elemente dem zugehörigen „Abschnitte“ P_α d. h. der Summe der vorangehenden Schichten angehören. Die erste Schicht Q_0 umfaßt dabei alle Urelemente.*

Die Teilbereiche *oder „Abschnitte“* P_α werden nämlich durch transfinite Induktion definirt vermöge der Festsetzungen:

1) $P_1 = Q_0 = Q$ umfasse die ganze Basis, die Gesamtheit der Urelemente.
2) $P_{\alpha+1} = P_\alpha + Q_\alpha$ soll alle in P_α „wurzelnden“ Mengen von P enthalten d. h. alle diejenigen, deren Elemente in P_α liegen.
3) Ist α eine Limeszahl, so bedeute P_α die Summe oder Vereinigung aller vorangehenden P_β mit kleineren Indizes $\beta < \alpha$.

Vermöge dieser Festsetzungen ist jedes P_α und schließlich auch $P_\pi = \sum_{\alpha<\pi} P_\alpha$ eindeutig bestimmt durch die vorangehenden und genügt den Forderungen des Theorems, sofern sich nachweisen läßt, daß P_π mit P identisch ist. Dabei enthält jede Schicht $Q_\alpha = P_{\alpha+1} - P_\alpha$ immer die Grundfolgen g_α vom gleichen Index, wie durch Induktion gezeigt werden kann. Denn $g_0 = u$ liegt in $Q_0 = P_1$, und gilt die Aussage für alle kleineren Indizes $\beta < \alpha$, so liegen alle Elemente von g_α, die ja selbst Grundfolgen g_β sind, in vorangehenden Schichten Q_β und damit auch in P_α, g_α selbst also jedenfalls in $P_{\alpha+1}$, aber nicht in P_α, da es sonst einer Schicht Q_β angehörte, die schon g_β, also ein Element von g_α enthält, im Widerspruch mit der Konstruktion. Also liegt auch g_α in Q_α, und diese Schicht ist nicht leer.
37 | Um nun den obigen „Hilfssatz“ auf den Unterbereich P_π von P anzuwenden, betrachten wir eine Menge r des Normalbereiches, deren Elemente sämtlich in P_π, etwa r_ν in der Schicht Q_{α_ν} liegen mögen. Diese Ordnungszahlen a_ν, die nicht alle verschieden zu sein brauchen, bilden dann eine wohlgeordnete Menge vom Typus $\varrho < \pi$, da ihre Mächtigkeit nicht größer als die

For, in the case assumed, the ZF'-axioms are satisfied also for M, provided that they hold for P; in particular, U) and V) are satisfied on account of 2). Of course, the "replacement axiom" E) is here to be understood as saying that the elements x' that are supposed to replace the elements x, must also belong to the partial domain M. As for the special case when M includes the whole basis, the remainder domain $R = P - M$ contains not a single urelement, and each of its elements would be a set r not all of whose elements are, by assumption, in M, and which, therefore, must have at least some elements occurring in R—in contradiction with axiom F).

On the other hand, it is quite possible that a normal domain with smaller base $Q' \subset Q$ is contained in the greater normal domain. Such a normal domain is obtained from P by a restriction to all those sets whose descending element chains $m \ni m_1 \ni m_2 \ni m_3 \dots$ end exclusively in urelements from Q', in accordance with F).

First development theorem. *Each normal domain P of characteristic π can be decomposed into a well-ordered* ⟦sequence⟧ *of type π of non-empty and disjoint "layers" Q_α, so that each layer Q_α includes all elements of P which occur in no earlier layer and whose elements belong to the corresponding "segment" P_α, that is, to the sum of the preceding layers. The first layer Q_α includes all the urelements.*

For the partial domains, *or "segments"*, P_α are defined by transfinite induction by virtue of the following stipulations:

1) $P_1 = Q_0 = Q$ shall include the whole basis, the totality of urelements.
2) $P_{\alpha+1} = P_\alpha + Q_\alpha$ shall contain all sets of P that are "rooted" in P_α, that is, all those sets whose elements lie in P_α.
3) If α is a limit number, then P_α shall be the sum or union of all preceding P_β with smaller indices $\beta < \alpha$.

It is by virtue of these stipulations that every P_α, and eventually also $P_\pi = \sum_{\alpha<\pi} P_\alpha$, is uniquely determined by the preceding ones and that it meets the demands of the theorem, provided that P_π can be shown to be identical to P. Here, each layer $Q_\alpha = P_{\alpha+1} - P_\alpha$ always contains the basic sequences g_α of the same index, as can be shown by induction. For $g_0 = u$ lies in $Q_0 = P_1$, and if the same statement holds for all smaller indices $\beta < \alpha$, then all elements of g_α, being basic sequences g_β themselves, are in preceding layers Q_β, and hence also in P_α. Thus, g_α lies in $P_{\alpha+1}$, but not in P_α, since otherwise it would belong to a layer Q_β that already contains g_β and, therefore, an element of g_α, contrary to the construction. Hence g_α, too, lies in Q_α, and this layer is not empty.

Now, in order to apply the "lemma" stated above to the subdomain P_π of P, we consider a set r of the normal domain whose elements are all in P_π, say r_ν in layer Q_{α_ν}. These ordinal numbers α_ν, not all of which have to be different, then form a well-ordered set of type $\varrho < \pi$, since its cardinality

von r sein kann. Da aber π als „Grenzzahl“ nach I) keiner kleineren konfinal ist, so besitzen alle diese a_ν eine obere Schranke $\alpha < \pi$, und alle r_ν, die Elemente von r, sind schon in P_α enthalten, r selbst also in $P_{\alpha+1}$ und damit auch in P_π. Der Unterbereich enthält also alle in ihm „wurzelnden“ Mengen von P, sowie alle Elemente seiner Elemente und ist, da er zugleich die ganze Basis umfaßt, mit dem zu entwickelnden Normalbereich identisch, womit der Beweis unseres Satzes vollendet ist.

Bezeichnen wir als „Einheitsbereich“ einen Normalbereich mit der „Basiszahl 1“, d. h. einen solchen, der einem *einzigen* Urelement entstammt, so gilt über seine Entwickelung der folgende Satz:

Zweites Entwickelungstheorem. *Bei der Entwickelung eines Einheitsbereiches hat jeder Abschnitt P_α die Mächtigkeit von $\psi(\alpha)$, enthält aber nur Mengen von kleinerer Kardinalzahl, während die entsprechende Schicht Q_α bereits Mengen dieser Mächtigkeit enthält. Jeder Abschnitt erster Art $P_{\beta+1}$ enthält als Mengen alle Unterbereiche des unmittelbar vorangehenden P_β, jeder Abschnitt zweiter Art alle vorangehenden Abschnitte und deren Unterbereiche. Der Einheitsbereich selbst hat die Mächtigkeit seiner Charakteristik π und enthält als Mengen alle seine Unterbereiche von kleinerer Mächtigkeit.*

Der Beweis wird wieder geführt durch transfinite Induktion unter der Annahme, daß die über P_α aufgestellte Behauptung zutreffe für alle kleineren Indizes $\beta < \alpha$, was für $\beta = 1$, $P_1 = Q$, $\psi(1) = 1$ sicher der Fall ist. Es sei nun $\alpha = \beta + 1$ von erster Art und nach der Annahme habe P_β die Mächtigkeit von $\psi(\beta) < \psi(\pi) = \pi$ und enthalte nur Mengen von kleinerer Kardinalzahl als $\psi(\beta)$, nämlich alle kleineren Abschnitte und deren Untermengen. Dann enthält $P_\alpha = P_\beta + Q_\beta$ gewiß *alle* Untermengen von P_β und auch nur solche, da jede Untermenge eines kleineren auch Teil des größeren ist. Also ist $P_\alpha = P_{\beta+1}$ von der Mächtigkeit $\mathfrak{p}_\alpha = 2^{\overline{\psi(\beta)}} = \overline{\psi}(\beta + 1) = \overline{\psi(\alpha)}$, enthält aber nur Mengen mit Kardinalzahlen $\leq \overline{\psi(\beta)} = \overline{\psi(\alpha)}$. Dagegen enthält die zugehö-
38 rige Schicht Q_α eine Menge von dieser | Kardinalzahl $\overline{\psi(\alpha)} < \overline{\pi}$, nämlich P_α selbst als Menge, die ja nach dem Ersetzungsaxiom in P vorhanden sein muß; aber auch keine größere, da Q_α nur aus Untermengen von P_α gebildet ist. Ist ferner α eine Limeszahl $< \pi$, also $P_\alpha = \sum_{\beta<\alpha} P_\beta$ die Summe aller kleineren Abschnitte P_β, die nach der Annahme die behaupteten Eigenschaften besitzen, so enthält auch P_α als Elemente nur Untermengen solcher Bereiche P_β und zwar *alle* diese Untermengen, da jede Untermenge von P_β schon im folgenden Abschnitte $P_{\beta+1}$ als Element enthalten ist. Alle diese Mengen haben dann Kardinalzahlen nicht größer als $\overline{\psi(\beta)} < \overline{\psi(\alpha)}$, und die Mächtigkeit des Abschnittes P_α selbst ist gegeben durch

$$\mathfrak{p}_\alpha = \lim_{\beta<\alpha} \mathfrak{p}_\beta = \lim_{\beta<\alpha} \overline{\psi(\beta)} = \overline{\psi(\alpha)} \leq \overline{\pi}.$$

cannot be greater than that of r. But since the boundary number π, in accordance with I), is cofinal with no smaller number, all the α_ν possess an upper bound $\alpha < \pi$, and all the r_ν, the elements of r, are already contained in P_α. Hence, r itself lies in $P_{\alpha+1}$, and, therefore, also in P_π. Hence, the subdomain contains all sets of P that are "rooted" in it, as well as all elements of its elements, and it is identical to the normal domain to be developed, since it also includes the whole basis, which completes the proof of our theorem.

Let us call by "unit domain" a normal domain with "basis number 1", that is, a normal domain that issues from a *single* urelement. Its development is subject to the following theorem:

Second development theorem. *In the development of a unit domain, each segment P_α has the cardinality of $\psi(\alpha)$ but contains only sets of smaller cardinal number, whereas the corresponding layer Q_α already contains sets of this cardinality. Each segment of the first kind $P_{\beta+1}$ contains as sets all subdomains of the immediately preceding P_β, and each segment of the second kind contains all preceding segments and their subdomains. As for the unit domain itself, it has the cardinality of its characteristic π and contains as sets all its subdomains of smaller cardinality.*

In order to prove the theorem we again use transfinite induction under the assumption that the claim made about P_α is correct for all smaller indices $\beta < \alpha$, which surely is the case for $\beta = 1$, $P_1 = Q$, $\psi(1) = 1$. Now, let $\alpha = \beta + 1$ be of the first kind. According to the assumption, P_β shall have the cardinality of $\psi(\beta) < \psi(\pi) = \pi$ and shall contain only sets of smaller cardinal number than $\psi(\beta)$; namely, it shall contain all smaller segments and their subsets. Then $P_\alpha = P_\beta + Q_\beta$ surely contains *all* subsets of P_β and only such, since each subset of a smaller segment is also part of a larger one. Hence, $P_\alpha = P_{\beta+1}$ is of cardinality $\mathfrak{p}_\alpha = 2^{\overline{\psi(\beta)}} = \overline{\psi(\beta+1)} = \overline{\psi(\alpha)}$,[8] but contains only sets with cardinal numbers $\leq \overline{\psi(\beta)} < \overline{\psi(\alpha)}$. On the other hand, the corresponding layer Q_α contains a set of this cardinal number $\overline{\psi(\alpha)} < \overline{\pi}$, namely P_α itself as a set which, after all, must be in P according to the replacement axiom, but none greater, since Q_α is formed only from subsets of P_α. If, furthermore, α is a limit number $< \pi$, and hence $P_\alpha = \sum_{\beta<\alpha} P_\beta$ the sum of all smaller segments P_β, which, by assumption, possess the stated properties, then P_α, too, contains as elements only subsets of those domains P_β, and indeed *all* such subsets, since each subset of P_β is already contained as an element in the following segment $P_{\beta+1}$. All these sets then have cardinal numbers not greater than $\overline{\psi(\beta)} < \overline{\psi(\alpha)}$, and the cardinality of the segment P_α itself is given by

$$\mathfrak{p}_\alpha = \lim_{\beta<\alpha} \mathfrak{p}_\beta = \lim_{\beta<\alpha} \overline{\psi(\beta)} = \overline{\psi(\alpha)} \leq \overline{\pi}\,.$$

---

[8] ⟦Zermelo erroneously writes "$\overline{\psi}(\beta+1)$" instead of "$\overline{\psi(\beta+1)}$".⟧

Für jedes $\alpha < \pi$ hat dann auch die Schicht Q_α die behauptete Eigenschaft, Mengen von der Kardinalzahl $\overline{\psi(\alpha)}$, aber keine größeren zu enthalten, nämlich P_α selbst und seine Untermengen. Für $\alpha = \pi$ aber ergibt sich ebenso als Mächtigkeit von P der Wert $\lim\limits_{\alpha<\pi} \overline{\psi(\alpha)} = \overline{\psi(\pi)} = \overline{\pi}$. Jedem Unterbereich von kleinerer Mächtigkeit als π entspricht dann in P eine äquivalente Grundfolge und daher nach E) auch eine Menge, die alle seine Elemente umfaßt. Für *Einheitsbereiche*, aber keineswegs für *beliebige* Normalbereiche gilt also das *v. Neumannsche* „Axiom", wonach *nur* solche Teilbereiche „zu groß" wären, um als „Mengen" auftreten zu können, welche von der gleichen Mächtigkeit sind wie der Gesamtbereich[6]. Durch die Beschränkung auf „Einheitsbereiche" würde aber die Mengenlehre ihre Anwendungsmöglichkeit zum größten Teile verlieren.

Aufgrund der gewonnenen Erkenntnis können wir jetzt die Entwickelung eines *beliebigen* Normalbereiches so abändern, daß in jede „Schicht" Q_α *nur* solche Mengen aus P aufgenommen werden, deren Kardinalzahl nicht größer ist als im Falle des Einheitsbereiches, nämlich $\leq \overline{\psi(\alpha)}$. Schließlich kommen sie doch alle an die Reihe, da für $\alpha < \pi$ auch immer $\psi(\alpha) < \psi(\pi) = \pi$ ist und wegen $\pi = \lim\limits_{\alpha<\pi} \psi(\alpha)$ jede Zahl $\varrho < \pi$ von einem Funktionswert $\psi(\alpha)$
39 über- | troffen wird. Die so entstehende „*kanonische*" Entwickelung hat nun den Vorzug, daß die Absonderung jeder einzelnen Schicht Q_α, unabhängig vom Gesamtbereiche und dessen Charakteristik, allein bestimmt wird durch ihren Index und durch die „Basis" des Normalbereiches, daß also bei gegebener Basis die Entwickelungen für die verschiedenen Grenzzahlen im Anfang immer übereinstimmen.

Drittes Entwickelungstheorem (Satz der „kanonischen" Entwickelung). *Jeder Normalbereich mit der Basis Q läßt sich zerlegen in eine mit Q beginnende wohlgeordnete Folge getrennter „Schichten" Q_α, wobei wieder jeder Schicht die Summe der vorangehenden als „Abschnitt" entspricht und jedes Q_α alle diejenigen Unterbereiche des zugehörigen Abschnittes P_α als Mengen enthält, die noch nicht im Abschnitte selbst liegen und keine größere Mächtigkeit haben als $\psi(\alpha)$. Diese letzte Einschränkung fällt weg im Falle des „Einheitsbereiches", wo „freie" und „kanonische" Entwickelung übereinstimmen. Bei der „kanonischen" Entwickelung ist jeder solche Abschnitt P_τ selbst ein Normalbereich, dessen Index τ den Bedingungen I, II einer „Grenzzahl" genügt.*

Der Beweis ist analog dem des ersten Entwickelungssatzes. Wie dort werden zunächst die Abschnitte P_α und Schichten Q_α induktiv definiert durch die Festsetzungen:

---

[6] *J. v. Neumann*, Die Axiomatisierung der Mengenlehre, Math. Zeitschr. Bd. 26 S. 669–752. 1928. Hier handelt es sich um das Axiom IV. 2, das a. a. O. insbesondere auf S. 677 f. erörtert wird.

For every $\alpha < \pi$, the layer Q_α has the stated property of containing sets of cardinal number $\overline{\psi(\alpha)}$ but none larger than that, namely P_α itself and its subsets. For $\alpha = \pi$, however, we likewise obtain as the cardinality of P the value $\lim_{\alpha<\pi} \overline{\psi(\alpha)} = \overline{\psi(\pi)} = \overline{\pi}$. To each subdomain of smaller cardinality than π there corresponds then in P an equivalent basic sequence, and hence, by E), also a set that includes all its elements. *Unit domains*, but not by any means *arbitrary* normal domains, are then subject to *v. Neumann's* "axiom" according to which the *only* partial domains "too large" to occur as "sets" are those of the cardinality of the total domain.[9] Set theory, however, would lose most of its applicability if it were restricted to "unit domains".

The knowledge we have gained enables us now to so modify the development of an *arbitrary* normal domain that *only* those sets enter from P into each "layer" Q_α whose cardinal number is not greater than would be the case with the unit domain, namely $\leq \overline{\psi(\alpha)}$. But eventually every set from P has its turn, since for $\alpha < \pi$ we always have $\psi(\alpha) < \psi(\pi) = \pi$ and since, because $\pi = \lim_{\alpha<\pi} \psi(\alpha)$, each number $\varrho < \pi$ is overtaken by a function value $\psi(\alpha)$. Thus arises a "*canonical*" development whose advantage is that the separation of each individual layer Q_α is determined only by its index and by the "basis" of the normal domain independently of the total domain and its characteristic; and hence that, for a given basis, the developments for the different boundary numbers are always in agreement in the beginning.

Third development theorem (Theorem of the "canonical" development). *Every normal domain with basis Q can be decomposed into a well-ordered sequence of separated "layers" Q_α that begins with Q, where again the sum of the preceding layers corresponds to each layer as a "segment", and where each Q_α contains as sets all those subdomains of the corresponding segment P_α that do not yet lie in the segment itself and that are not of greater cardinality than $\psi(\alpha)$. This last restriction drops out in the case of the "unit domain", where "free" and "canonical" development are in agreement. For the "canonical" development, each such segment P_τ is itself a normal domain whose index τ satisfies conditions I and II of a "boundary number".*

The proof is analogous to the proof of the first development theorem. Once again, we first define the segments P_α and layers Q_α inductively, using the stipulations:

9 *Von Neumann 1928d.* The axiom in question is axiom IV.2, discussed, in particular, in op. cit. pp. 677f.

1) $P_1 = Q_0 = Q$.
2) $P_{\alpha+1} = P_\alpha + Q_\alpha$ umfasse alle Mengen des Normalbereiches P, deren Elemente in P_α liegen und deren Kardinalzahlen $\leq \overline{\psi(\alpha)}$ sind.
3) Für jede Limeszahl α sei immer $P_\alpha = \sum\limits_{\beta<\alpha} P_\beta$ die Summe aller kleineren P_β.

Dann sind alle Abschnitte P_α und die entsprechenden Schichten $Q_\alpha = P_{\alpha+1} - P_\alpha$ eindeutig bestimmt für alle Indizes $\alpha \leq \pi$, und P_π ist ein wohlbestimmter Unterbereich von P, der wieder mit P identisch ist nach dem „Hilfssatze", weil er die ganze Basis, die Elemente seiner Elemente sowie alle in ihm „wurzelnden" Mengen von P enthält. Die letzte Bedingung ist auch hier erfüllt, da jede in P_α wurzelnde Menge r von der Kardinalzahl $\overline{\varrho} < \overline{\pi}$ spätestens in der Schicht Q_ϱ erscheint wegen $\varrho \leq \psi(\varrho)$.

Nun sei $\tau \leq \pi$ eine Zahl von Grenzzahl-Charakter und P_τ der ihr entsprechende Abschnitt der kanonischen Entwickelung von P. Dann enthält er nach
40 der Konstruktion lediglich Mengen mit Kar- | dinalzahlen $\overline{\varrho} \leq \overline{\psi(\alpha)} < \overline{\psi(\tau)} = \overline{\tau}$, da jede doch einer Schicht Q_α für $\alpha < \tau$ angehören muß. Umgekehrt muß auch jede in P_τ wurzelnde Menge r mit einer Kardinalzahl $< \overline{\tau}$, weil τ eine „Kernzahl" ist, bereits in einem kleineren Abschnitte P_α wurzeln und einem höheren P_γ angehören, wo γ nicht größer zu sein braucht als der Ordnungstypus ϱ von r, da ja $\varrho \leq \psi(\varrho)$ ist. Also enthält P_τ alle in ihm wurzelnden Mengen aus P, deren Kardinalzahlen kleiner sind als $\overline{\tau}$, insbesondere auch alle durch „Ersetzung" innerhalb P_τ gebildeten Mengen. Ist r eine beliebige Menge in P_τ und wohlgeordnet nach $\varrho < \tau$ so ist auch $\psi(\varrho) < \psi(\tau) = \tau$ sowie $2^{\overline{\varrho}} \leq 2^{\overline{\psi(\varrho)}} = \overline{\psi}(\varrho{+}1) < \overline{\tau}$ und mit r ist auch $\mathfrak{U}r$ Element von P_τ. Ist endlich r_ν eine nach dem Typus $\sigma < \tau$ wohlgeordnete Folge von Kardinalzahlen $< \overline{\tau}$, so haben sie eine obere Schranke $\overline{\mathfrak{t}}' < \overline{\tau}$, weil sonst τ konfinal σ wäre und keine Kernzahl, und es ist auch $\sum\limits_\nu \mathfrak{t}_\nu \leq \overline{\sigma}\mathfrak{t}' < \overline{\tau}$, d. h. auch das Axiom V) ist erfüllt im Abschnitte P_τ und dieser ist in der Tat ein Normalbereich. Damit ist zugleich erwiesen, daß die beiden für die Charakteristik eines Normalbereiches im § 2 aufgestellten Bedingungen I und II zugleich auch *hinreichend* sind, daß nämlich jede diesen Bedingungen genügende Ordnungszahl τ als „Grenzzahl" eines Normalbereiches auftreten kann. Vorausgesetzt wird dabei allerdings, daß diese Zahl τ überhaupt einem Bereiche angehört, für den die ZF'-Axiome erfüllt sind.

§ 4. Isomorphie und Automorphie der Normalbereiche

Zwei Normalbereiche P, P' heißen „isomorph", wenn sich die Elemente des einen ein-eindeutig abbilden lassen auf die Elemente x' des anderen, sodaß jede Grundrelation $a \in b$ in dem einen die entsprechende $a' \in b'$ in dem anderen nach sich zieht und umgekehrt. In isomorphen Bereichen entspricht also jedem Urelement u wieder ein Urelement u', jeder Menge m eine äquivalen-

1) $P_1 = Q_0 = Q$.
2) $P_{\alpha+1} = P_\alpha + Q_\alpha$ shall include all those sets of the normal domain P whose elements lie in P_α and whose cardinal numbers are $\leq \overline{\psi(\alpha)}$.
3) For every limit number α, let $P_\alpha = \sum_{\beta<\alpha} P_\beta$ the sum of all smaller P_β.

All segments P_α and the corresponding layers $Q_\alpha = P_{\alpha+1} - P_\alpha$ are then uniquely determined for all indices $\alpha \leq \pi$, and P_π is a well-determined subdomain of P, which, by the "lemma", is in turn identical to P, since it contains the whole basis, the elements of its elements, as well as all sets of P "rooted" in it. The last condition is also satisfied here since each set r of cardinal number $\overline{\varrho} < \overline{\pi}$ rooted in P_α appears in layer Q_ϱ at the latest, on account of $\varrho \leq \psi(\varrho)$.

Now let $\tau \leq \pi$ be a number of the character of a boundary number and let P_τ be the segment of the canonical development of P corresponding to it. According to the construction, it then contains only sets of cardinal numbers $\overline{\varrho} \leq \overline{\psi(\alpha)} < \overline{\psi(\tau)} = \overline{\tau}$, since, after all, each set must belong to some layer Q_α for $\alpha < \tau$. Conversely, every set r with a cardinal number $< \overline{\tau}$ rooted in P_τ must already be rooted in some smaller segment P_α since τ is a "core number", and it must belong to a higher P_γ, where γ does not need to be greater than the order type ϱ of r, since $\varrho \leq \psi(\varrho)$. Hence, P_τ contains all sets from P rooted in it and whose cardinal numbers are smaller than $\overline{\tau}$, particularly all those formed by means of "replacement" within P_τ. If r is an arbitrary set in P_τ well-ordered in type $\varrho < \tau$, then we have $\psi(\varrho) < \psi(\tau) = \tau$ as well as $2^{\overline{\varrho}} \leq 2^{\overline{\psi(\varrho)}} = \overline{\psi}(\varrho+1) < \overline{\tau}$, and, along with r, $\mathfrak{U}r$ is also an element of P_τ. If, finally, $\mathfrak{r}_\nu$ is a well-ordered sequence of cardinal numbers $< \overline{\tau}$ of type $\sigma < \tau$, then they have an upper bound $\mathfrak{r}' < \overline{\tau}$, since otherwise τ would be cofinal with σ and it would not be a core number. And we also have $\sum_\nu \mathfrak{r}_\nu \leq \overline{\sigma}\mathfrak{r}' < \overline{\tau}$; that is, axiom V), too, is satisfied in the segment P_τ, which is indeed a normal domain. Thus it is has also been demonstrated that the two conditions I and II stated for the characteristic of a normal domain in § 2 are also *sufficient* conditions, that is, that every ordinal number τ satisfying these conditions can occur as "boundary number" of a normal domain. What we assume here, however, is that this number τ belongs to some domain for which the ZF'-axioms are satisfied.

§ 4. Isomorphisms and automorphisms of the normal domains

Two normal domains P, P' are said to be "isomorphic" if the elements of the one can be mapped one-to-one onto the elements x' of the other such that each basic relation $a \,\epsilon\, b$ in the one gives rise to the corresponding basic relation $a' \,\epsilon\, b'$ in the other, and vice versa. Hence, in isomorphic domains there corresponds to each urelement u an urelement u', to each set m an equivalent

te Menge m', jeder Grundfolge g_α eine Grundfolge g'_α vom gleichen Index, der „Basis“ Q eine äquivalente Basis Q' und die „Grenzzahl“ oder „Charakteristik“ π sich selbst. Daß aber diese beiden letzten Bedingungen für die Isomorphie auch *hinreichend* sind, besagt das folgende Theorem.

Erster Isomorphiesatz. *Zwei Normalbereiche mit gleicher Charakteristik und äquivalenten Basen sind isomorph, und zwar ist die isomorphe Abbildung der bei den Bereiche auf einander* eindeutig *bestimmt durch die Abbildung ihrer Basen.*

41 | Zum Beweise bedienen wir uns der „Entwickelungssätze“, wobei wir beliebig die „freie“ oder die „kanonische Entwickelung“ zugrunde legen können. Wir denken uns also die Basen Q und Q' der beiden Normalbereiche eineindeutig auf einander abgebildet, so daß jedem Urelement u ein bestimmtes u' entspricht, und zeigen durch Induktion, daß auch jedem *Abschnitte* P_α der Entwickelung von P ein isomorpher Abschnitt P'_α in der Entwickelung des anderen zugeordnet werden kann, und zwar *eindeutig* für alle $\alpha \leq \pi$. Zwei Abschnitte P'_α, P'_β mit verschiedenen Indizes können schon deshalb nicht isomorph sein, weil der größere immer Grundfolgen enthält, denen im kleineren keine ähnlichen entsprechen. Nun nehmen wir an, es sei P_α isomorph abgebildet auf P'_α, was für $P_1 = Q$, $P'_1 = Q'$ vorausgesetzt ist. Dann werden gleichzeitig alle kleineren Abschnitte mit abgebildet auf solche von P', und bei allen diesen Abbildungen wird ein bestimmtes Element x immer auf *dasselbe* Element x' von P' abgebildet. Ist nun r irgend eine Menge aus Q_α, so liegen ihre Elemente r_ν alle in P_α, und die ihnen entsprechenden Elemente r'_ν in P'_α sind nach E) wieder die Elemente einer äquivalenten Menge r', da wegen $\pi = \pi'$ der Normalbereich P' Mengen dieser Mächtigkeit sicher enthält. Diese Menge r' muß auch im Falle der „kanonischen Entwickelung“ in $P_{\alpha+1}$ vorkommen, aber nicht schon in P'_α, da sonst wegen der Isomorphie auch die entsprechende Menge r schon in P_α läge und nicht in Q_α. Also entspricht jedem Element r von Q_α eindeutig ein solches r' von Q'_α und umgekehrt, und auch der Abschnitt $P_{\alpha+1} = P_\alpha + Q_\alpha$ ist eindeutig-isomorph abgebildet auf den Abschnitt $P'_{\alpha+1}$ des anderen Normalbereiches. Jetzt sei α eine Limeszahl, und es werde angenommen, daß für jedes kleinere $\beta < \alpha$ der Abschnitt P_β eindeutig-isomorph sei dem Abschnitte P'_β. Da hier jedes Element x von P_α sicher einem P_β angehört, so entspricht ihm bei allen diesen Abbildungen P_β, P'_β immer ein ganz bestimmtes Element x' von P'_α, und jedesmal, wo $a \in b$ ist, ist auch $a' \in b'$, da es stets einen Abschnitt P_β gibt, der beide Elemente enthält. Somit ist auch P_α eindeutig-isomorph P'_α für beliebige Limeszahlen $\alpha \leq \pi$, und wegen $P_\pi = P$, $P'_\pi = P'$ sind wie behauptet, die beiden Normalbereiche selbst eindeutig-isomorph.

Zweiter Isomorphiesatz. *Von zwei Normalbereichen mit äquivalenten Basen und verschiedenen Grenzzahlen π, π' ist stets der eine isomorph einem kanonischen Abschnitte des anderen.*

set m', to each basic sequence g_α a basic sequence g'_α with the same index, to the "basis" Q an equivalent basis Q', and to the "boundary number", or "characteristic", π there corresponds π itself. But that these last two conditions for the isomorphism are also *sufficient* is stated by the following theorem.

First isomorphism theorem. *Two normal domains with the same characteristic and with equivalent bases are isomorphic. In fact, the isomorphic mapping of the two domains onto one another is* uniquely *determined by the mapping of their bases.*

In order to prove this theorem we use the "development theorems", choosing as a starting point either the "free" or the "canonical development". We thus imagine the bases Q and Q' of the two normal domains mapped one-to-one onto one another such that there corresponds to each urelement u a particular u', and we show by induction that it is possible to associate with each *segment* P_α of the development of P an isomorphic segment P'_α in the development of the other, and that, in fact, it is possible to associate them *uniquely* for all $\alpha \leq \pi$. Two segments P'_α, P'_β with different indices cannot very well be isomorphic because the greater segment always contains basic sequences to which no similar basic sequences correspond in the smaller segment. Now let us assume that P_α is mapped isomorphically onto P'_α. This is presupposed for $P_1 = Q$, $P'_1 = Q'$. Then, at the same time, all smaller segments are also mapped onto such segments of P', and a given element x is always mapped onto the *same* element x' of P' under all these mappings. If now r is some set from Q_α, then all its elements r_ν are in P_α. According to E), the elements r'_ν in P'_α corresponding to the r_ν are, in turn, elements of an equivalent set r' since the normal domain P' certainly contains sets of this cardinality because $\pi = \pi'$. This set r' must also appear in $P'_{\alpha+1}$[10] in the case of the "canonical development", but not already in P'_α, since otherwise the corresponding set r would already lie in P_α, on account of the isomorphism, and not in Q_α. Thus to each element r in Q_α there corresponds uniquely such an r' in Q'_α, and vice versa, and the segment $P_{\alpha+1} = P_\alpha + Q_\alpha$, too, is uniquely-isomorphically mapped onto the segment $P'_{\alpha+1}$ of the other normal domain. Now let α be a limit number and let us assume that for each smaller $\beta < \alpha$ the segment P_β is uniquely-isomorphic to the segment P'_β. Since each element x in P_α certainly belongs to some P_β, there always corresponds to it under all these mappings P_β, P'_β a particular element x' in P'_α and whenever we have $a \in b$, we also have $a' \in b'$, since there always is a segment P_β containing both elements. Thus P_α, too, is uniquely-isomorphic to P'_α for arbitrary limit numbers $\alpha \leq \pi$, and, because $P_\pi = P$, $P'_\pi = P'$, the two normal domains themselves are uniquely-isomorphic, as was claimed.

Second isomorphism theorem. *Given two normal domains with equivalent bases and different boundary numbers π, π', one is always isomorphic to a canonical segment of the other.*

[10] ⟦Zermelo erroneously writes "$P_{\alpha+1}$" instead of "$P'_{\alpha+1}$".⟧

42 | Ist nämlich $Q \sim Q'$ und $\pi' > \pi$, so ist nach dem „dritten Entwickelungstheorem“ S. 39, der „kanonische Abschnitt“ P'_π, da π eine Grenzzahl ist, ein Normalbereich, der mit P die gleiche Charakteristik und eine äquivalente Basis hat, also nach dem vorigen Satze mit P isomorph.

Dritter Isomorphiesatz. *Von zwei Normalbereichen mit gleicher Charakteristik ist immer einer isomorph einem (echten oder unechten) Unterbereich des anderen.*

Es sei P ein Normalbereich und $Q' \subset Q$ ein Teil seiner Basis. Dann betrachten wir die Gesamtheit aller solchen Elemente von P, bei denen jede rückschreitende Kette von Elementen $m \ni m_1 \ni m_2 \ni m_3 \dots$ entsprechend dem Axiom F) mit einem Urelement in Q' endet, so erfüllt dieser Teilbereich P' von P alle Bedingungen unseres „Hilfssatzes“ im § 3. Er ist also selbst ein Normalbereich mit der gleichen Charakteristik π, weil er alle aus Q' entspringenden Grundfolgen von P enthält, und ist isomorph jedem Normalbereiche P'' von gleicher Charakteristik π, dessen Basis Q'' mit Q' äquivalent ist. Aus der vorausgesetzten Vergleichbarkeit beliebiger Mengen Q, Q'' folgt dann unmittelbar die Behauptung.

Die „Struktur“ eines Normalbereiches, d. h. das, was er mit allen isomorphen gemeinsam hat, oder sein „Modell-Typus“ wird nach dem hier Bewiesenen bestimmt durch *zwei* Zahlen, durch die Mächtigkeit seiner Basis $\mathfrak{q}$ und durch seine Charakteristik π, von denen die erste, die „Breite“ des Normalbereiches *beliebig* gewählt werden kann, während die andere, seine „Höhe“ die im § 2 angegebenen Eigenschaften einer „Grenzzahl“ besitzen muß. Diese „Modelltypen“ bilden also eine *zweifach wohlgeordnete Mannigfaltigkeit* von der Beschaffenheit, daß ein Modelltypus immer dann einem Bestandteile eines anderen isomorph ist, $\mu \leq \mu'$, wenn gleichzeitig $\mathfrak{q} \leq \mathfrak{q}'$ und $\pi \leq \pi'$ d. h. wenn *beide* bestimmenden Zahlen des einen kleiner oder gleich denen des anderen sind.

Da die isomorphe Abbildung zweier Normalbereiche auf einander, wo sie existiert, nach dem „ersten Isomorphiesatze“ *eindeutig* bestimmt ist durch die Abbildung ihrer Basen, so folgt, daß eine isomorphe Abbildung eines Normalbereiches auf sich *selbst*, also ein „Automorphismus“ nur möglich ist durch „Permutation“ seiner Basis, für „Einheitsbereiche“, die nur ein einziges
43 Urelement | enthalten, daher *unmöglich* ist. Ebenso entsteht eine isomorphe Abbildung eines Normalbereiches P auf einen *Teil* P' von sich aus jeder äquivalenten Abbildung der Basis Q auf einen ihrer Teile Q', wenn die Basis selbst *unendlich* ist. Es entspricht nämlich, wie wir beim Beweise des letzten Satzes sahen, jeder Teilbasis auch ein normaler Teilbereich P' von gleicher Charakteristik π, insbesondere auch jedem einzelnen Urelement u ein zugehöriger „Einheitsbereich“, und äquivalenten Teilbasen entsprechen isomorphe Teilbereiche, die man in Analogie mit der Körpertheorie als „konjugierte“ bezeichnen kann. Somit ergibt sich der folgende

For if $Q \sim Q'$ and $\pi' > \pi$, then, according to the "third development theorem" on p. 39, the "canonical segment" P'_π is a normal domain that shares the same characteristic and an equivalent basis with P since π is a boundary number. P'_π is therefore, according to the previous theorem, isomorphic to P.

Third isomorphism theorem. *Given two normal domains with the same characteristic, one always is isomorphic to a (proper or improper) subdomain of the other.*

Let P be a normal domain and let $Q' \subset Q$ be a part of its basis. We now consider the totality of all elements of P for which each descending chain of elements $m \ni m_1 \ni m_3 \ni \ldots$ ends in an urelement in Q' in accordance with axiom F). This partial domain P' of P then satisfies all the conditions of our "lemma" in § 3. Thus P' is itself a normal domain with the same characteristic π since it contains all basic sequences of P arising from Q', and it is isomorphic to every normal domain P'' with the same characteristic π whose basis Q'' is equivalent to Q'. The claim follows then immediately from the comparability of arbitrary sets Q, Q'', which we assumed.

According to what has been proved here, the "structure" of a normal domain, that is, what it has in common with all isomorphic normal domains, or its "model type", is determined by *two* numbers: the cardinality of its basis $\mathfrak{q}$ and its characteristic π. The first number, the "breadth" of the normal domain, can be chosen *arbitrarily*, whereas the second number, its "height", must possess the properties of a "boundary number" stated in § 2. These "model types" thus form a *doubly well-ordered manifold* such that one model type is isomorphic to a component of another, $\mu \leq \mu'$, whenever both $\mathfrak{q} \leq \mathfrak{q}'$ and $\pi \leq \pi'$, that is, whenever *both* determining numbers of the one are smaller than or equal to those of the other.

Since, according to the "first isomorphism theorem", the isomorphic mapping of two normal domains onto one another is, where it exists, *uniquely* determined by the mapping of its bases, it follows that an isomorphic mapping of a normal domain onto *itself*, that is, an "automorphism", is possible only by "permutation" of its basis, and, hence, *impossible* for "unit domains", which contain only one single urelement. Likewise, we obtain an isomorphic mapping of a normal domain P onto a *part* P' of itself from each equivalent mapping of the basis Q onto one of its parts Q', provided that the basis itself is *infinite*. For, as we have seen in the proof of the last theorem, there corresponds to each partial basis also a normal partial domain P' with the same characteristic π, and, in particular, to each individual urelement u also the associated "unit domain", and to equivalent partial bases correspond isomorphic partial domains, which may be called "conjugated" in analogy with field theory. Thus we obtain the following

Automorphiesatz. *Automorphismen d. h. isomorphe Abbildungen eines Normalbereiches auf sich selbst entsprechen eineindeutig den äquivalenten Abbildungen der Basis auf sich selbst und sind daher nur möglich bei einer Basiszahl $\mathfrak{q} > 1$; alle Einheitsbereiche sind „monomorph". Die Gruppe aller Automorphismen ist isomorph der zur Basis gehörenden Permutationsgruppe. Auch „Meromorphismen" d. h. isomorphe Abbildungen des Normalbereiches auf einen Teil von sich entsprechen den eineindeutigen Abbildungen der (unendlichen) Basis auf äquivalente Teile.*

§ 5. Existenzfragen, Widerspruchslosigkeit und Kategorizität

Unsere bisherigen Betrachtungen setzen die Existenz von „Normalbereichen" verschiedener Beschaffenheit voraus und gründen sich jedenfalls auf die Annahme von der Widerspruchlosigkeit der mengentheoretischen Axiome. Diese Widerspruchslosigkeit logisch-formal zu *beweisen*, soll auch hier nicht versucht werden. Dagegen soll unter der allgemeinen *Voraussetzung* dieser Widerspruchsfreiheit für die Mengenlehre überhaupt die (mathematische d. h. ideelle) *Existenz* der verschiedenen hier in Betracht kommenden Modelltypen geprüft werden. Wir setzen also die Existenz von Mengenbereichen, die den ZF-Axiomen genügen, für eine beliebige Basis voraus. Dann gibt es jedenfalls auch solche, die außerdem noch das „Fundierungsaxiom" F) erfüllen. Denn ist etwa M ein Mengenbereich von der vorausgesetzten Beschaffenheit, so bilden alle Elemente dieses Bereiches, die außerdem noch F) erfüllen, darunter natürlich auch alle Urelemente, einen wohldefinierten Unterbereich
44 | N von M, der für sich schon allen ZF'-Axiomen genügt und damit einen „Normalbereich" mit der gegebenen Basis Q darstellt.

Daß durch Verkleinerung der *Basis* wieder Normalbereiche als Teilbereiche des ersten entstehen, haben wir bereits im vorigen § 4 beim Beweise des „dritten Isomorphiesatzes" gezeigt. Dagegen ist noch nicht ohne weiteres klar, ob auch durch Verkleinerung oder Vergrößerung der *Charakteristik* neue Typen von Normalbereichen gewonnen werden können. Denn jede „Grenzzahl" muß ja den Bedingungen I) und II) des § 2 genügen und es steht noch in Frage, ob es überhaupt solche Zahlen von „Grenzzahl-Charakter" und wieviele solche es gibt. Nun ist aber ω, die Anfangszahl der zweiten Zahlenklasse, gewiß eine solche Zahl: ein „Eigenwert" der Funktion $\psi(\xi)$ und keiner kleineren konfinal. Und ω ist in der Tat auch die Charakteristik des niedersten Normalbereiches, der folgendermaßen entsteht: Wir lassen aus dem gegebenen Normalbereiche M alle diejenigen Mengen weg, für welche irgend eine gemäß F) gebildete „rückschreitende Elementenkette" der Form $m \ni m_1 \ni m_2 \ni m_3 \ldots$ eine „unendliche" Menge enthält. Dieser Bereich, der selbst nur noch *endliche* Mengen enthält, erfüllt alle Bedingungen eines Normalbereiches und ist zugleich der dem Index ω entsprechende Abschnitt P_ω der „kanonischen Entwickelung" des ursprünglichen Normalbereichs. Dieser „finitistische" Bereich, gegen den

Automorphism theorem. *Automorphisms, that is, isomorphic mappings of a normal domain onto itself, correspond one-to-one to the equivalent mappings of the basis onto itself, and are therefore possible only for basis number* $\mathfrak{q} > 1$*; all unit domains are "monomorphic". The group of all automorphisms is isomorphic to the group of permutations that belongs to the basis. Likewise, "meromorphisms", that is, isomorphic mappings of the normal domain onto a part of itself, correspond to the one-to-one mappings of the (infinite) basis onto equivalent parts.*

§ 5. Existence questions, consistency and categoricity

So far, our considerations have assumed the existence of different "normal domains" and have always been based on the assumption of the consistency of the set-theoretic axioms. We shall not attempt here to provide a logical formal *proof* of such consistency. Rather, making the general assumption of the consistency of set theory, we shall examine the (mathematical, that is, ideal) existence of the different model types relevant here. We therefore assume for an arbitrary basis the existence of domains of sets that satisfy the ZF-axioms. There then certainly also exist domains of sets that, in addition, satisfy the "foundation axiom" F). For if M is a domain of sets of the assumed constitution, then all those elements of this domain that also satisfy F), among them of course all urelements, form a well-defined subdomain N of M that already satisfies all ZF'-axioms and that therefore is a "normal domain" with the given basis Q.

Reducing the *basis* of a normal domain yields a normal domain which is a partial domain of the first, as was already shown in the proof of the "third isomorphism theorem" in § 4 above. By contrast, it is not clear yet, at least not without further ado, whether it is possible to obtain new types of normal domains by decreasing or increasing the *characteristic*. For after all, every "boundary number" must satisfy conditions I) and II) of § 2, and the question still remains to be answered whether such numbers of the "character of a boundary number" exist at all and if so, how many. Now ω, the initial number of the second number-class, certainly is such a number: it is an "eigenvalue" of the function $\psi(\xi)$ and cofinal with no smaller number. In fact, ω is also the characteristic of the lowest normal domain that arises as follows: we remove from a given normal domain M all those sets for which some "descending chain of elements", $m \ni m_1 \ni m_2 \ni m_3 \ni \ldots$, formed in accordance with F), contains an "infinite" set. This domain, which contains only *finite* sets, satisfies all conditions of a normal domain and is, at the same time, that segment P_ω of the "canonical development" of the original normal domain that corresponds to the index ω. This "finitistic" domain, against which even the "intuitionists"

trotz seiner eigenen Unendlichkeit selbst die „Intuitionisten" kaum etwas einzuwenden hätten, kann wenigstens dazu dienen, durch seine bloße Existenz die Widerspruchslosigkeit der ZF'-Axiome zu erweisen. Dagegen kann er, eben weil er keine unendlichen Mengen enthält, *nicht* als wahres „Modell" der Cantorschen Mengenlehre in Anspruch genommen werden. Aus ihm heraus führt erst mein früheres „Axiom des Unendlichen", das die Existenz wenigstens einer „unendlichen" Menge postuliert. Der *niederste* Normalbereich, der dieser Bedingung genügt und den ich als den „Cantorschen" bezeichnen möchte, hätte dann die Charakteristik π_1, nämlich den kleinsten Eigenwert der ψ-Funktion von Kernzahl-Charakter, also jedenfalls eine „reguläre Anfangszahl zweiter Art", wenn auch nicht notwendig die *kleinste* „exorbitante" Zahl überhaupt — wenigstens solange die Cantorsche Vermutung nicht bewiesen ist.

Aber gibt es überhaupt hinter ω solche Zahlen mit „Grenzzahl-Charak-
45 ter"? Gewiß, sofern es überhaupt eine „infinitistische" Men-| genlehre d. h. überhaupt Normalbereiche mit unendlichen Mengen gibt. Denn die *Gesamtheit* aller in einem solchen Bereiche vorkommenden „Grundfolgen" *hat* eben einen solchen Ordnungstypus π, wenn auch *innerhalb* des Bereiches keine Menge von diesem Typus π vorkommen kann. Und gibt es überhaupt „Grenzzahlen" $\pi > \omega$, so gibt es unter ihnen auch eine *kleinste* π_1. Freilich „beweisen" d. h. aus den allgemeinen ZF'-Axiomen ableiten läßt sich weder ihre Existenz noch ihre Nicht-Existenz, eben weil z. B. die Grenzzahl ω zwar im „Cantorschen" Bereich existiert, aber nicht im „finitistischen", weil m. a. W. die Frage in den verschiedenen „Modellen" der Mengenlehre *verschieden* beantwortet wird, also durch die Axiome allein noch nicht entschieden ist. Unser Axiomensystem ist eben *nicht-kategorisch*, was in diesem Falle kein Nachteil, sondern ein *Vorzug* ist. Denn gerade auf dieser Tatsache beruht die ungeheure Bedeutung und unbegrenzte Anwendbarkeit der Mengenlehre überhaupt. Natürlich kann man immer durch Hinzufügung weiterer „Axiome" die gewünschte Kategorizität künstlich *erzwingen*, aber immer nur auf Kosten der Allgemeinheit. Solche neuen Postulate, wie sie z. B. von *Fraenkel*[7], *Finsler*[8], *Neumann*[9], u. a. vorgeschlagen wurden, betreffen eben gar nicht die Mengenlehre *an sich*, sondern charakterisieren lediglich ein ganz spezielles vom jeweiligen Autor gewähltes *Modell*. In der Regel sind es „Einheitsbereiche", die bevorzugt werden — wodurch eigentlich, wie schon S. 38 bemerkt, die *Anwendbarkeit* der Mengenlehre preisgegeben würde. Außerdem pflegt man sich gewöhnlich auf den *niedersten* infinitistischen Bereich, den „Cantorschen" zu beschränken, worin ich ebenso wenig einen Vorteil erblicken kann. Vielmehr

---

[7] *Fraenkel*, Einleitung in die Mengenlehre, 3. Aufl. § 18. 5. S. 355. „Axiom der Beschränktheit".

[8] *Finsler*, Über die Grundlegung der Mengenlehre. Math. Zeitschr. 25, S. 683–713. Hierüber vergl. auch *R. Baer*, Über ein Vollständigkeitsaxiom in der Mengenlehre. Math. Zeitschr. Bd. 27, S. 536–539, 1928.

[9] *J. v. Neumann* wie oben [6]).

would hardly raise any objections, its infiniteness notwithstanding, may at least serve to demonstrate the consistency of the ZF'-axioms by dint of its mere existence. On the other hand, just because it does not contain infinite sets, it can *not* be claimed to be a true "model" of Cantorian set theory. It is only my previous "axiom of infinity", which postulates the existence of at least one "infinite" set, that leads us beyond this domain. The *lowest* normal domain satisfying this condition, which I shall call the "Cantorian" normal domain, would then have the characteristic π_1, namely the smallest eigenvalue of the ψ-function with the character of a core number, and thus certainly a "regular initial number of the second kind", even if not necessarily the *smallest* "exorbitant" number—at least as long as Cantor's conjecture has not been proved.

But are there any such numbers after ω that have the character of "boundary numbers"? The answer is certainly yes, provided that there is an "infinitistic" set theory at all, that is, provided that there are normal domains with infinite sets. For the *totality* of all "basic sequences" occurring in such domains simply *has* such an order type π, even though no set of this order type π can occur *within* the domain. And if there are any "boundary numbers" $\pi > \omega$ at all, then among them there is also a *smallest* one, π_1. It is of course not possible to "prove", that is, to deduce from the general ZF'-axioms, either its existence or its non-existence, simply because, for instance, the boundary number ω, even though it exists in the "Cantorian" domain, does not exist in the "finitistic" domain; because, in other words, the question receives *different* answers in different "models" of set theory, and is thus not decided merely by the axioms alone. Our axiom system is *non-categorical* after all, which, in this case, is not a disadvantage, but an *advantage*. For the enormous significance and unlimited applicability of set theory rests precisely on this fact. It is of course always possible to artificially *force* the desired categoricity by adding further "axioms", but only at the expense of generality. For new postulates such as those proposed by *Fraenkel*,[11] *Finsler*,[12] *Neumann*,[13] and others, simply do not concern set theory *as such* at all. They only characterize the special *model* chosen by the respective author. "Unit domains" are usually the preferred choice—which, in fact, would amount to sacrificing the *applicability* of set theory, as we have already noted on p. 38. Moreover, it is customary to restrict oneself to the *lowest* infinitistic domain, the "Cantorian" domain, which I find equally disadvantageous. Instead, set theory as a *science*

[11] *Fraenkel 1919*, 3rd ed., § 18.5. p. 355. "Axiom of restriction".

[12] *Finsler 1926*.

On this, cf. also *Baer 1928*.

[13] *Von Neumann 1928d*.

muß die Mengenlehre als *Wissenschaft* zunächst in vollster Allgemeinheit entwickelt werden, worauf die vergleichende Untersuchung der einzelnen *Modelle* als besonderes Problem vorgenommen werden kann.

46 Wodurch unterscheiden sich nun in der Mengenlehre tatsächlich | die verschiedenen Modelle mit gemeinsamer Basis, insbesondere die verschiedenen „Einheitsbereiche"? Wie wir sahen, durch ihre „Charakteristik", d. h. durch die Gesamtheit der in ihnen durch „Mengen" vertretenen Ordnungszahlen, oder durch die Gesamtheit der in ihnen enthaltenen „Grundfolgen" des nämlichen Urelementes. Da aber nur „Grenzzahlen" als „Charakteristik" dienen können, so ist jedes „Einheitsmodell" schon eindeutig bestimmt durch die Gesamtheit der in ihm vorhandenen (oder nicht vorhandenen) *Grundfolgen mit Grenzzahl-Typus*. Durch ihre Angabe, die in den verschiedenen Fällen mit Hilfe geeigneter Postulate erfolgen kann, ist dann der Modelltypus auch „kategorisch" festgelegt und zugleich mit seiner Charakteristik π (nach dem zweiten Entwickelungssatze des § 3) auch die Mächtigkeit $\overline{\pi}$ des entsprechenden Einheitsbereiches. Machen wir nun die allgemeine Hypothese, daß *jeder kategorisch bestimmte Bereich irgendwie auch als „Menge" aufgefaßt werden*, d. h. als Element eines (geeignet gewählten) Normalbereiches auftreten kann, so ergibt sich, daß jedem Normalbereiche ein höherer mit gleicher Basis, jedem Einheitsbereich ein höherer Einheitsbereich und damit auch jeder „Grenzzahl" π eine größere Grenzzahl π' entspricht. Ebenso entsteht aber auch aus jeder unendlichen Folge verschiedener Normalbereiche mit gemeinsamer Basis, die immer einer den andern als kanonische Abschnitte enthalten, durch Vereinigung und Verschmelzung ein kategorisch bestimmter Bereich von Mengen, der dann wieder zu einem Normalbereich von höherer Charakteristik ergänzt werden kann. Jeder kategorisch bestimmten Gesamtheit von „Grenzzahlen" folgt also wieder eine größere, und die Reihe „aller" Grenzzahlen ist ebenso unbegrenzt wie die Zahlenreihe selbst, sodaß auch jedem transfiniten Index eine bestimmte Grenzahl ein-eindeutig zugeordnet werden kann. „Beweisbar" aus den ZF'-Axiomen ist das natürlich wieder *nicht*, da das behauptete Verhalten aus jedem einzelnen Normalbereiche herausführt. Es muß vielmehr die *Existenz einer unbegrenzten Folge von Grenzzahlen* als neues *Axiom* für die „Meta-Mengenlehre" postuliert werden, wobei noch die Frage der „Widerspruchslosigkeit" einer näheren Prüfung bedarf. Wenn ich mich aber auch hier noch auf diese vorläufige Skizze beschränken und auf ihre spätere Ausführung verweisen muß, so dürfte doch Folgendes bereits ein-
47 leuchten, was als | das wesentliche Ergebnis der vorliegenden Untersuchung angesehen werden kann:

Die „ultrafiniten Antinomien der Mengenlehre", auf die sich wissenschaftliche Reaktionäre und Anti-Mathematiker in ihrem Kampfe gegen die Mengenlehre so eifrig und liebevoll berufen, diese scheinbaren „Widersprüche" beruhen lediglich auf einer Verwechselung der durch ihre Axiome nicht-kategorisch bestimmten *Mengenlehre selbst* mit den einzelnen sie darstellenden *Modellen*: was in einem Modelle als „ultrafinite Un- oder Übermenge"

must first be developed in greatest generality. The comparative investigation of individual *models* may then be approached as a special problem.

How, now, do the various models with a common basis, and in particular the various "unit domains", actually differ from one another in set theory? As we have seen, they differ with respect to their "characteristic", that is, the totality of ordinal numbers represented in them by "sets", or with respect to the totality of those "basic sequences" starting with the same urelement that are contained in them. But since only "boundary numbers" can serve as "characteristics", every "unit model" is uniquely determined already by the totality of the *basic sequences of boundary number type* that are (or are not) contained in it. Its specification then, which, in the different cases, can be effected by suitable postulates, "categorically" determines the model type and, together with its characteristic π (in accordance with the second development theorem of § 3) also the cardinality $\overline{\pi}$ of the corresponding unit domain. Let us now put forth the general hypothesis that *every categorically determined domain can also be conceived of as a "set" in one way or another*; that is, that it can occur as an element of a (suitably chosen) normal domain. It then follows that there corresponds to any normal domain a higher one with the same basis, to any unit domain a higher unit domain, and therefore also to any "boundary number" π a greater boundary number π'. Likewise, a categorically determined domain of sets arises through union and fusion from every infinite sequence of different normal domains with common basis, where one always contains the other as a canonical segment. This categorically determined domain of sets can then again be supplemented so as to become a normal domain of higher characteristic. Thus, to every categorically determined totality of "boundary numbers" there follows a greater one, and the sequence of "all" boundary numbers is as unlimited as the number series itself, allowing for the possibility of associating to every transfinite index a particular boundary number in one-to-one fashion. Once again, this is of course *not* "provable" on the basis of the ZF'-axioms, since the asserted behavior leads us beyond any individual normal domain. Rather, we must postulate the *existence of an unlimited sequence of boundary numbers* as a new axiom for the "meta-theory of sets", where the question of "consistency" still requires closer examination. While I have to confine myself here to this preliminary sketch and refer to its later elaboration, the following, which may be considered the essential result of the present investigation, should already be evident:

The "ultrafinite antinomies of set theory", to which scientific reactionaries and anti-mathematicians appeal in their fight against set theory with such eager passion, are only apparent "contradictions", due only to a confusion between *set theory itself*, which is non-categorically determined by its axioms, and the individual *models* representing it: what in one model appears as "ultrafinite non- or superset", is already a fully valid "set" with cardinal number

erscheint, ist im nächsthöheren bereits eine vollgültige „Menge" mit Kardinalzahl und Ordnungstypus und bildet selbst den Grundstein zum Aufbau des neuen Bereiches. Der unbegrenzten Reihe der Cantorschen Ordnungszahlen entspricht eine ebenso unbegrenzte Doppelreihe von wesentlich verschiedenen mengentheoretischen Modellen, in deren jedem die ganze klassische Theorie zum Ausdrucke kommt. Die beiden polar entgegengesetzten Tendenzen des denkenden Geistes, die Idee des schöpferischen *Fortschrittes* und die des zusammenfassenden *Abschlusses*, die auch den Kantischen „Antinomien" zugrunde liegen, finden ihre symbolische Darstellung und ihre symbolische Versöhnung in der auf den Begriff der Wohlordnung gegründeten transfiniten Zahlenreihe, die in ihrem schrankenlosen Fortschreiten keinen wahren Abschluß, wohl aber relative Haltpunkte besitzt, eben jene „Grenzzahlen", welche die höheren von den niederen Modelltypen scheiden. Und so führen auch die mengentheoretischen „Antinomien", richtig verstanden, statt zu einer Verengung und Verstümmelung vielmehr zu einer jetzt noch unübersehbaren Entfaltung und Bereicherung der mathematischen Wissenschaft.

Beim Abschluß dieser Arbeit ist es mir Bedürfnis, meinem Kollegen Herrn Dr. *Arnold Scholz*, der mich bei der Ausarbeitung dieser Untersuchung sowie bei den Korrekturen durch wertvolle Ratschläge auf das freundlichste unterstützte, meinen herzlichen Dank zu sagen.

Freiburg i. Br. den 13-ten April 1930.

Über die logische Form der mathematischen Theorien

1930b

Es wird der Begriff eines vollständigen Satzsystems eingeführt, d. i. eines Satzsystems, welches sämtliche logische Konsequenzen seiner Sätze enthält. Die Betrachtung derartiger Systeme kann in gewissen Fragen der axiomatischen Formulierung vorgezogen werden.

and ordinal type in the next higher model and, in turn, serves itself as the bed-stone in the construction of the new domain. To the unlimited series of Cantorian ordinal numbers there corresponds a likewise unlimited double series of essentially different set-theoretic models in each of which the entire classical theory finds its expression. The two diametrically opposed tendencies of the thinking mind, the ideas of creative *progress* and summary *completion*, which form also the basis of Kant's "antinomies", find their symbolic representation as well as their symbolic reconciliation in the transfinite number series, which rests upon the notion of well-ordering and which, though lacking in true completion on account of its boundless progressing, possesses relative way stations, namely those "boundary numbers", which separate the higher from the lower model types. Thus, instead of leading to constriction and mutilation, the set-theoretic "antinomies" lead, when understood correctly, to an as yet unforeseeable development and enrichment of the mathematical science.

In conclusion, I would like to express my sincere gratitude to my colleague Dr. *Arnold Scholz* for his kind support and valuable advice during the writing of this paper and the proof-reading of the manuscript.

Freiburg i. Br., on the 13th of April 1930.

On the logical form of mathematical theories

1930b

⟦The introductory note just before *s1929b* also addresses *1930b.*⟧

The concept of a complete propositional system is introduced, that is, of a propositional system containing all logical consequences of its propositions. The consideration of such systems may be preferred to the axiomatic formulation with respect to certain questions.

Introductory note to s1930d

Akihiro Kanamori

For the three years 1929 through 1931 Zermelo was supported by a fellowship provided by the Notgemeinschaft der Deutschen Wissenschaft, the Emergency Association of German Science. This association was founded in 1920 by the leading German academies and scientific organizations in order to avert a collapse of research in the economically disastrous time after the First World War. With a Swiss pension meager and inadequate, Zermelo had in 1929 applied for the fellowship with a research project "On the Nature and Foundations of Pure and Applied Mathematics and the Significance of the Infinite in Mathematics", and surely the fellowship was a significant factor in his return to the fray of foundational work. *s1930d* was a report, dated 3 December 1930, to the Notgemeinschaft for the continuation of the fellowship into 1931.[1]

Written for the purpose of apprising and explaining, the report is an informal account, well worth reading, of Zermelo's early career as well as of his *1930a* work and what was to be subsequent investigations (cf. *s1930e*). In what follows, we selectively make a few of the several possible points that can be made.

Zermelo makes informative remarks about the two axioms newly adopted in *1930a*. About the replacement axiom Zermelo writes: "... [the *1908b* axiom system] has received a valuable addition only once in the form of the 'replacement axiom' proposed by *Fraenkel*. On the other hand, it was only the concept of 'cofinality' introduced by *Hausdorff* that made possible a fruitful application of the new axiom." How Hausdorff had first ventured beyond Cantor into the higher uncountable is suitably noted; today however, one would mainly regard replacement as the provenance of formalized definitions by transfinite recursion and the sets they provide. As for the axiom of foundation, after articulating the problems "posed and solved" in his *1930a*, Zermelo newly highlights the axiom: "Using the new axiom it was possible to carry out a decomposition into layers, the 'development' of a 'normal domain' (that is, of a domain of sets satisfying the axiom system), and to answer the decisive main question in the 'isomorphism theorems'."

Having described how, in his *1930a* context, normal domains with equivalent bases are isomorphic to initial segments of each other, Zermelo writes: "From this already follows ... that *Cantor's* (generalized) conjecture [the generalized continuum hypothesis] ... does *not* depend on the choice of the model, but that it is decided (as true or false) once and for all by means of our axiom system." As Zermelo was evidently proceeding in second-order terms, this distinctly anticipates Georg Kreisel's observation (cf. *Kreisel 1971*) that

[1] See *Ebbinghaus 2007b* for the foregoing historical details.

the continuum hypothesis is not independent of second-order ZF although which way it is decided is not known.

Zermelo goes on to describe his *1930a* sequence of normal domains and their boundary numbers. The smallest boundary number is ω and the next, the least inaccessible cardinal, which for Zermelo corresponds to "'Cantor's domain', in which all of Cantor's set theory including the infinity axiom already finds it representation." But this, like any other normal domain, is capable of being extended to a higher domain, and so forth. With this, Zermelo draws the following distinction vis-à-vis Skolem:

> We thus arrive at a kind of "set-theoretic relativism", which, however, differs essentially from *Skolem's* "relativism", in which even the concepts of "partial sets" and that of "cardinalities" are being relativized: *Skolem* wishes to restrict the formation of subsets to special classes of defining functions, while I, in keeping with the true spirit of set theory, allow for the *free* division and postulate the existence of all partial sets formed in any arbitrary way. For *Skolem*, it is supposed to be possible to represent set theory *in its entirety* already in a *countable* model, and, e.g., the problem of the cardinality of the continuum already loses its real significance for him.

Concluding, Zermelo writes:

> The questions of the "existence" or "consistency" of the "higher" normal domains are of course not entirely settled in the present paper ⟦*1930a*⟧; but I believe to have found a procedure of making evident the consistency by means of the systematic construction of a "set-theoretic model" using "the unlimited number series".

The first would be described in *s1930e* and the second, in the notes *s1931e*. Zermelo continues with remarks that have a modern resonance in terms of reflection heuristics and large cardinal postulations:

> To this end, I need the "metamathematical" concept of "closed domains", which, e.g., corresponds to *Cantor's* "concept of sets" and which can be reduced to that of "categorical systems of postulates". Every "normal domain" is a "closed domain" and can therefore also be conceived of as a "set" in a "higher" ⟦normal domain⟧. ... No (closed) normal domain can represent set theory in its entirety since every "boundary number" corresponds to a segment of the number series, and hence no normal domain contains all boundary numbers. Set theory in its entirety is only representable in the "open" domain of all normal domains.

This distinction between open and closed domains would become the major demarcation for Zermelo's subsequent work in set theory (cf. *s1930e*).

Bericht an die Notgemeinschaft der Deutschen Wissenschaft über meine Forschungen betreffend die *Grundlagen der Mathematik*

s1930d

Schon vor 30 Jahren, als ich Privatdozent in Göttingen war, begann ich unter dem Einflusse *D. Hilberts*, dem ich überhaupt das meiste in meiner wissenschaftlichen Entwickelung zu verdanken habe, mich mit den Grundlagenfragen der Mathematik zu beschäftigen, insbesondere aber mit den grundlegenden Problemen der *Cantor*schen *Mengenlehre*, die mir in der damals so fruchtbaren Zusammenarbeit der Göttinger Mathematiker erst in ihrer vollen Bedeutung zum Bewußtsein kamen. Es war damals die Zeit, wo die "Antinomien", die scheinbaren "Widersprüche" in der Mengenlehre, die allgemeinste Aufmerksamkeit auf sich zogen und berufene wie unberufene Federn zu den kühnsten wie zu den ängstlichsten Lösungsversuchen veranlaßten. In der Überzeugung, daß in diesem Komplexe von Fragen auch die tiefsten Einblicke in das Wesen der Mathematik überhaupt zu gewinnen seien, wandte ich mich diesen Problemen zu in einer Reihe von Arbeiten, die u. a. die damals noch sehr umstrittene Möglichkeit der "Wohlordnung" betrafen und im Jahre 1907–8 in der Abhandlung "Über die Grundlagen der Mengenlehre" in den Mathematischen Annalen Bd. 65 zum vorläufigen Abschlusse kamen. Das von mir damals eingeführte Axiomen-System ist seitdem für die axiomatische Forschung auf diesem Gebiete im Wesentlichen maßgebend geblieben und hat inzwischen nur einmal durch das von *Fraenkel* vorgeschlagene "Ersetzungs-Axiom" eine wertvolle Erweiterung erfahren, während andererseits der von *Hausdorff* eingeführte Begriff der "Konfinalität" eine fruchtbare Anwendung des neuen Axioms erst ermöglichte. Mittlerweile war aber auch die Frage nach den "Grundlagen" aufs Neue wieder in Fluß gekommen durch das etwas geräuschvolle Auftreten der "Intuitionisten", die in temperamentvollen
2 Streitschriften eine | "Grundlagen-Krisis" der Mathematik verkündeten und so ziemlich der ganzen modernen Wissenschaft den Krieg erklärten — ohne selbst etwas Besseres an ihre Stelle setzen zu können. "Eine Mengenlehre als besondere mathematische Disziplin wird es nicht mehr geben," dekretierte einer ihrer eifrigsten Adepten — während gleichzeitig die neuen Lehrbücher der Mengenlehre nur so ins Kraut schossen. Diese Sachlage veranlaßte auch mich damals, den Grundlagen-Problemen wieder meine forschende Tätigkeit zuzuwenden, nachdem ich durch langwierige Krankheit und geistige Isolierung im Auslande der wissenschaftlichen Produktion schon fast entfremdet war. Ohne in dem proklamierten Streite zwischen "Intuitionismus" und "Formalismus" Parteigänger zu werden — ich halte diese Alternative überhaupt für eine logisch unzulässige Anwendung des "Tertium non datur" — glaubte ich doch zu einer Klärung der einschlägigen Fragen beitragen zu können: nicht als "Phi-

Report to the Emergency Association of German Science about my reasearch concerning the *foundations of mathematics*

s1930d

Already thirty years ago, as a *Privatdozent* in Göttingen, I began to concern myself with the foundational questions in mathematics under the influence of *D. Hilbert* to whom I owe the most for my scientific development. In particular, however, I concerned myself with the basic problems of *Cantor*'s *set theory* whose full significance I only realized during the so fruitful collaboration with the mathematicians in Göttingen. This was the time when the "antinomies", the apparent "contradictions" in set theory, attracted the widest attention and elicited attempts, both bold and timid, at their resolution from pens both skilled and not so skilled. Convinced that the deepest insights into the nature of mathematics altogether were to be found in this area of questions, I turned to these problems in a series of papers, which were concerned with, among other things, the possibility of "well-ordering", then still much-disputed, and which came to a preliminary conclusion in the 1907–8 paper "On the foundations of set theory" in the Mathematische Annalen vol. 65.[1] The axiom system I introduced at the time has essentially remained the standard for the axiomatic investigation in this field ever since. It received a valuable addition only once, in the form of the "replacement axiom" proposed by *Fraenkel*. On the other hand, it was only the concept of "cofinality" introduced by *Hausdorff* that made possible a fruitful application of the new axiom. In the meantime, however, the question about the "foundations" had gathered momentum once again as the "intuitionists" stridently proclaimed a "foundational crisis" in mathematics in fiery pamphlets and declared war on essentially all of modern science—without being able to substitute something better for it. "Set theory as a special mathematical discipline will no longer exist," decreed one of its most ardent disciples—when, at the same time, the new textbooks on set theory were springing up like mushrooms. At the time, this situation prompted me as well to take up my research on foundational problems after long illness and intellectual isolation abroad had almost estranged me from scientific production. While remaining neutral in the proclaimed dispute between "intuitionism" and "formalism"—a choice of alternatives, which, in my opinion, is but a logically illegitimate application of the "tertium non datur"—I believed that I could after all contribute to a clarification of the relevant questions: not as a "philosopher" who pronounces

---

1 ⟦*Zermelo 1908b*⟧.

losoph" durch Verkündung "apodiktischer" Prinzipien, welche durch Vermehrung der bestehenden Meinungen die Verwirrung nur noch zu steigern pflegen, sondern als Mathematiker durch Aufweisung objektiver mathematischer Zusammenhänge, die erst eine gesicherte Grundlage für alle philosophischen Theorien abgeben können. In der besonderen Frage der Mengenlehre nun, wo es sich vor allen Dingen um die Aufklärung der "Antinomien" handelt, stellte ich mir jetzt, dem aufgestellten Grundsatze entsprechend, die entscheidende Vorfrage: Wie muß ein "Bereich" von "Mengen" und "Urelementen" beschaffen sein, um den "allgemeinen" Axiomen der Mengenlehre zu genügen? Ist unser Axiomen-System "kategorisch" oder gibt es eine Vielheit wesentlich verschiedener "mengentheoretischer Modelle"? Ist der Begriff einer "Menge" im Gegensatz zu einer bloßen "Klasse" ein absoluter, durch logische Merkmale bestimmbarer oder nur ein relativer, abhängig von dem jeweils zugrunde geleg-
3 ten mengentheoretischen Modell? | Dieses Problem ist es, das ich mir in meiner 1930 erschienenen Fundamenta-Arbeit "Über Grenzzahlen und Mengenbereiche" gestellt und gelöst habe. Um es aber erfolgreich in Angriff nehmen zu können, ergab sich zunächst die Notwendigkeit, das "Zermelo-Fraenkel'sche Axiomensystem" durch ein weiteres, das "Fundierungs-Axiom" [[,]] zu ergänzen, das u. a. "sich selbst enthaltende" und "zirkelhafte Mengen" ausschließt und in allen praktisch wichtigen Fällen tatsächlich erfüllt ist. Mit Hilfe des neuen Axioms konnte eine schichtenförmige Zerlegung, die "Entwickelung" eines "Normalbereiches" (d. h. eines dem Axiomensystem genügenden Mengenbereiches) durchgeführt und in den "Isomorphie-Sätzen" die entscheidende Hauptfrage beantwortet werden. Zwei Normalbereiche sind dann und nur dann "isomorph", wenn 1) ihre "Basen" (d. h. die Gesamtheit ihrer Ur-Elemente) einander äquivalent und 2) ihre "Charakteristiken" (d. h. die oberen Grenzen der vorkommenden Alefs) einander gleich sind, wenn also auch jeder Menge des einen Bereiches mindestens eine äquivalente im anderen entspricht. Von zwei Bereichen mit äquivalenten Basen (aber verschiedenen Charakteristiken) ist immer der eine isomorph einem "kanonischen" Entwickelungs-Abschnitte des anderen. Hieraus folgt u. a. bereits, daß die (verallgemeinerte) *Cantor*sche Vermutung (wonach die Potenzmenge jeder Menge immer gerade die nächstfolgende Mächtigkeit haben soll) *nicht* von der Wahl des Modells abhängt, sondern durch unser Axiomensystem ein für allemal (als wahr oder als falsch) entschieden ist. Ein "Normalbereich" ist (bis auf isomorphe Abbildungen) bestimmt durch zwei Alefs, durch seine "Breite" (d. h. die Mächtigkeit seiner Basis) und durch seine "Höhe" (d. h. seine Charakteristik), und die Gesamtheit aller möglichen "Modell-Typen" wird also dargestellt durch eine (zweifach wohlgeordnete) Doppel-Reihe von Alefs, in welcher die "Breite" sämtliche Alefs durchläuft, die "Höhe" aber auf die Reihe der "Grenzzahlen" beschränkt ist. Um "Grenzzahl" oder "Charakteristik eines
4 | Normalbereiches" zu sein muß eine (transfinite) "Anfangszahl" zwei charakteristische Eigenschaften haben: sie muß "Eigenwert" oder "kritische Zahl" einer gewissen "Normalfunktion" sein, darf aber keiner kleineren Ordnungszahl "konfinal", muß vielmehr immer eine "reguläre Anfangszahl zweiter Art" im

"apodeictic" principles, which often add to the confusion by introducing yet another opinion, but as a mathematician who finds objective mathematical connections, which can, in turn, serve as a secure foundation for any philosophical theory. Considering the special question of set theory, where what matters most is the resolution of the "antinomies", I posed for myself the decisive preliminary question in accordance with the specified principle: How does a "domain" of "sets" and "urelements" have to be constituted so that it satisfies the "general" axioms of set theory? Is our axiom system "categorical" or are there a multitude of essentially different "set-theoretic" models? Is the concept of "set", as opposed to that of mere "class", an absolute one, capable of being determined by means of logical characteristics, or is it only a relative concept, dependent on the set-theoretic model upon which it happens to be based? This is the problem I posed for myself and solved in my paper "On boundary numbers and domains of sets" published in Fundamenta in 1930.[2] In order to tackle it successfully, however, I first had to supplement the "Zermelo-Fraenkel axiom system" by adding a further axiom, the "foundation axiom", which excludes, among other things, "self-containing" and "circular sets" and which, in all cases of practical significance, is in fact satisfied. By using the new axiom it was possible to carry out a decomposition into layers, the "development" of a "normal domain" (that is, of a domain of sets satisfying the axiom system), and to answer the decisive main question in the "isomorphism theorems". Two normal domains are "isomorphic" if and only if 1) their "bases" (that is, the totality of their urelements) are equivalent to one another and 2) their "characteristics" (that is, the upper limits of the occurring alephs) are equal, if, in other words, to each set of one domain there corresponds at least one equivalent one in the other domain. Of two domains with equivalent bases (but different characteristics) one is always isomorphic to a "canonical" development segment of the other. From this it already follows, among other things, that *Cantor*'s (generalized) conjecture (according to which the power set of any set is supposed to always be of the immediately succeeding cardinality) does *not* depend on the choice of the model, but that it is decided (as true or as false) once and for all by means of our axiom system. A "normal domain" is determined (up to isomorphic mappings) by two alephs, by its "breadth" (that is, the cardinality of its basis) and by its "height" (that is, its characteristic), and the totality of all possible "model types" is thus represented by a (doubly well-ordered) double sequence of alephs in which the "breadth" runs through all the alephs, but in which the "height" is restricted to the sequence of the "boundary numbers". For a (transfinite) "initial number" to be a "boundary number", or "characteristic of a normal domain" it must have the following two characteristic properties: it must be "eigenvalue" or "critical number" of a certain "normal function", but it must not be "cofinal" with any smaller ordinal number. Rather, it must

2 ⟦*Zermelo 1930a.*⟧

Sinne *Hausdorffs*, eine “exorbitante” Zahl sein. Die *Existenz* solcher “Grenzzahlen” kann nun freilich nicht aus dem Axiomensystem erwiesen sondern, weil sie eben nicht für *alle* Normalbereiche gilt, nur (für die höheren Bereiche metamathematisch) postuliert werden. Die (absolut) kleinste Grenzzahl ω begrenzt den “finitistischen Bereich”, der nur endliche Mengen enthält, die nächstfolgende π_1 den “Cantorschen Bereich”, in welchem bereits die ganze Cantorsche Mengenlehre einschließlich des Unendlichkeits-Axioms zur Darstellung gelangt. Aber auch dieser Bereich ist (wie jeder andere Normalbereich) noch erweiterungsfähig zu einem “höheren” Bereiche, welcher u. a. auch “Mengen” von der Mächtigkeit π_1 enthält. Ganz allgemein gesprochen lösen sich die “ultrafiniten Antinomien” bei dieser Betrachtungsweise dadurch, daß jeder Normalbereich N zwar Teilbereiche M besitzt, die “zu groß” sind, um in ihm “Mengen” zu sein, daß aber alle solchen Teilbereiche M wie auch N selbst bereits im nächstfolgenden Normalbereiche N' durch vollgültige “Mengen” vertreten sind. Jeder Normalbereich selbst ist “Menge” in allen höheren Bereichen, aber es gibt keinen höchsten Normalbereich, welcher *alle* Normalbereiche als Mengen enthielte. Charakterisiert wird der einzelne Normalbereich (abgesehen von seiner Basis) durch die in ihm vorkommenden “Mengen von Grenzzahl-Mächtigkeit”, der “Cantorsche” z. B. durch die Eigenschaft, außer den abzählbaren *keine* weiteren Mengen von Grenzzahl-Mächtigkeit zu enthalten. So gelangen wir also zu einer Art von “mengentheoretischem Relativismus”, der sich aber von dem Skolemschen “Relativismus”, in wel-
5 chem sogar die Begriffe von | “Teilmenge” und “Mächtigkeit” relativiert werden, grundsätzlich unterscheidet: *Skolem* will die Bildung der Untermengen auf besondere Klassen definierender Funktionen einschränken, während bei mir, dem wahren Geiste der Mengenlehre entsprechend, die *freie* Teilung zugelassen und die Existenz aller irgendwie gebildeten Teilmengen postuliert wird. Bei *Skolem* soll schon in einem *abzählbaren* Modell die *ganze* Mengenlehre dargestellt werden können, und für ihn verliert z. B. auch schon das Problem von der Mächtigkeit des Kontinuums seine eigentliche Bedeutung. Die Fragen nach der “Existenz” oder “Widerspruchslosigkeit” der “höheren” Normalbereiche sind freilich in der gedruckt vorliegenden Arbeit noch nicht völlig erledigt; aber ich glaube ein Verfahren gefunden zu haben, durch systematischen Aufbau eines “mengentheoretischen Modelles” und mit Hilfe der “unbegrenzten Zahlenreihe” diese Widerspruchslosigkeit einsichtig machen zu können. Ich brauche dazu den “metamathematischen” Begriff eines “geschlossenen Bereiches”, der etwa dem *Cantor*schen “Mengenbegriffe” entspricht und auf den eines “kategorischen Postulatsystems” zurückgeführt werden kann. Jeder “Normalbereich” ist ein “geschlossener Bereich” und kann daher in einem “höheren” auch als “Menge” aufgefaßt werden. Jeder *Abschnitt* der (transfiniten) Zahlenreihe ist ein geschlossener Bereich, aber auf jede “Menge”, jeden “geschlossenen” Bereich von Ordnungszahlen folgen immer noch weitere Ordnungszahlen: die “unbegrenzte” Zahlenreihe selbst ist ein “offener Bereich”. Kein (geschlossener) Normalbereich kann die ganze Mengenlehre darstellen, da jede “Grenzzahl” einem Abschnitt der Zahlenreihe entspricht und daher

always be a "regular initial number of the second kind" in *Hausdorff*'s sense, an "exorbitant" number. The *existence* of such "boundary numbers" cannot of course be shown from the axiom system but it can only be postulated (metamathematically for the higher domains) since it just does not hold for *all* normal domains. The (absolutely) smallest boundary number ω delimits the "finitistic domain", which only contains finite sets, the immediate successor π_1 ⟦delimits⟧ "Cantor's domain", in which all of Cantor's set theory including the infinity axiom already finds its representation. But (like any other normal domain) this domain, too, is still capable of being extended to a "higher" domain, which, among other things, also contains "sets" of cardinality π_1. Generally speaking, the "ultrafinite antinomies" are resolved on this view by the fact that, while every normal domain N possesses partial domains M which are "too big" to figure as "sets" in it, all such partial domains M as well as N itself are already represented in the immediately succeeding normal domain N' by fully valid "sets". Every normal domain is itself a "set" in all higher domains, but there is no highest normal domain containing *all* normal domains as sets. The individual normal domain is characterized (aside from its basis) by the "sets with boundary number cardinality" occurring in it; e.g., "Cantor's domain" by the property of containing *no* further sets of boundary number cardinality other than the countable ones. We thus arrive at a kind of "set-theoretic relativism", which, however, essentially differs from *Skolem*'s "relativism", in which even the concepts of "partial set" and that of "cardinality" are being relativized: *Skolem* wishes to restrict the formation of subsets to special classes of defining functions, while I, in keeping with the true spirit of set theory, allow for the *free* division and postulate the existence of all partial sets formed in any arbitrary way. For *Skolem*, it is supposed to be possible to represent set theory *in its entirety* already in a *countable* model, and, e.g., the problem of the cardinality of the continuum already loses its real significance for him. The questions of the "existence" or "consistency" of the "higher" normal domains are of course not entirely settled in the published paper; but I believe to have found a procedure of making evident this consistency by means of the systematic construction of a "set-theoretic model" and the "unlimited number series". To this end, I need the "metamathematical" concept of "closed domain", which, e.g., corresponds to *Cantor*'s "concept of sets" and can be reduced to the concept of "categorical system of postulates". Every "normal domain" is a "closed domain" and can therefore also be conceived of as a "set" in a "higher" ⟦normal domain⟧. Each *segment* of the (transfinite) number series is a closed domain, but each "set", each "closed" domain of ordinal numbers is always succeeded by still further ordinal numbers: the "unlimited" number series is itself an "open domain". No (closed) normal domain can represent set theory in its entirety, since every "boundary number" corresponds to a segment of the number series, and hence

kein Normalbereich alle Grenzzahlen enthält. Die ganze Mengenlehre ist allein darstellbar in der “offenen” Gesamtheit aller Normalbereiche.

6 | Anhang I

Zusammenstellung meiner bisherigen Fortschritte und Resultate zum Neuaufbau der Mengenlehre

1) Unterscheidung der allgemeinen Mengenlehre von den verschiedenen sie darstellenden Modellen, den “Normalbereichen”.
2) Ausscheidung “zirkelhafter” und ähnlicher Mengen durch Hinzufügung des “Fundierungs-Axioms”.
3) Einführung einer “Basis” von “Urelementen” anstelle der “Nullmenge”.
4) Vermeidung des *Skolem*schen “Relativismus” durch freie Bildung von Untermengen *ohne* “Definitheits-Beschränkung”.
5) Schichtenförmige “Entwickelung” eines “Normalbereiches” aus gegebener “Basis” mit Hilfe der Wohlordnung.
6) Begriff der “Grenzzahl” als “Charakteristik” eines Normalbereiches. “Finitistische” und “infinitistische” Normalbereiche.
7) Eigenschaften der “Grenzzahlen”: Sie sind “Eigenwerte” oder “kritische Zahlen” einer gewissen “Normalfunktion”, die gleichzeitig “Kernzahlen” d. h. “reguläre Anfangszahlen zweiter Art” sind.
8) “Isomorphie-Sätze” über Normalbereiche: jeder Modell-Typus ist bestimmt durch zwei Alefs, ⟦durch⟧ die Mächtigkeit seiner Basis und durch seine “Charakteristik” d. h. die obere Grenze der als Mengen vorkommenden Mächtigkeiten. Die Modelltypen bilden eine zweifach wohlgeordnete Doppelreihe. Die Gültigkeit der “*Cantor*schen Vermutung” ist *unabhängig* vom gewählten Modell.
9) Aufklärung der “ultrafiniten Antinomien” durch Unterscheidung der Modelltypen: Relativismus des Mengenbegriffs — in einem anderen als dem *Skolem*schen Sinne.

7 | Anhang II

Zusammenstellung weiterer in Vorbereitung begriffener Untersuchungen

1) Über die Konstruktion eines mengentheoretischen Modells und die Widerspruchslosigkeit der Mengenlehre.
2) Über “geschlossene” und “offene Bereiche” und den *Cantor*schen absoluten Mengenbegriff.
3) Über den mengentheoretischen Relativismus bei *Skolem* und mir und seine Bedeutung für das Kontinuum-Problem.
4) Über die Mathematik als die “Logik des Unendlichen” und die Unmöglichkeit einer “finitistischen Mathematik”.

no normal domain contains all boundary numbers. Set theory in its entirety is only representable in the "open" totality of all normal domains.

Appendix I

Synopsis of my previous advances and results concerning the new construction of set theory

1) Distinction between general set theory and the different models representing it, the "normal domains".
2) Exclusion of "circular" and similar sets by adding the "foundation axiom".
3) Introduction of a "basis" of "urelements" instead of the "null set".
4) Avoidance of *Skolem*'s "relativism" by means of the free formation of subsets *without* "definiteness restriction".
5) "Development" of a "normal domain" in layers from a given "basis" by means of well-ordering.
6) Concept of the "boundary number" as "characteristic" of a normal domain. "Finitistic" and "infinitistic" normal domains.
7) Properties of the "boundary numbers": They are "eigenvalues" or "critical numbers" of a certain "normal function" which are also "core numbers", that is, "regular initial numbers of the second kind".
8) "Isomorphism theorems" on normal domains: every model type is determined by two alephs: the cardinality of its basis and its "characteristic", that is, the upper limit of the cardinalities occurring as sets. The model types form a doubly well-ordered double sequence. The validity of "*Cantor*'s conjecture" is *independent* of the chosen model.
9) Resolution of the "ultrafinite antinomies" by distinguishing among the different model types: Relativism of the concept of set—in a sense other than *Skolem*'s.

Appendix II

Synopsis of further investigations currently in preparation

1) On the construction of a set-theoretic model and the consistency of set theory.
2) On "closed" and "open domains" and *Cantor*'s absolute set concept.
3) On the set-theoretic relativism in *Skolem*'s work and in my work, and its significance for the continuum problem.
4) On mathematics as the "logic of the infinite" and the impossibility of a "finitistic mathematics".

5) Über das Verhältnis der Mathematik zur *Anschauung*: erst mit der infinitistisch-logischen Verarbeitung eines anschaulich gegebenen Materiales beginnt die mathematische Wissenschaft und kann daher selbst *nicht* auf "Anschauung" gegründet werden. Auch in der Geometrie beruht der Vorzug der "Euklidischen Geometrie" *nicht* auf ihrer "anschaulichen Gegebenheit" sondern lediglich auf ihrer logisch-mathematischen Einfachheit.

5) On the relation of mathematics to *intuition*: mathematics begins only with the infinitistic-logical processing of material given in intuition. Thus it can *not* itself be based on "intuition". In geometry, too, the advantage of "Euclidean geometry" is *not* based on its "being given in intuition", but merely on its logico-mathematical simplicity.

Introductory note to s1930e

Akihiro Kanamori

In this note found in the *Nachlass*, Zermelo describes his ideas in 1930 on the construction of a model of set theory and, with it, the consistency of set theory. As such, the note provides details on the first two topics that Zermelo had listed in the appendix, "Synopsis of further investigations currently in preparation", of his report *s1930d* to the Notgemeinschaft der Deutschen Wissenschaft for the continuation of his fellowship into 1931. There are five carefully typed pages, the first four comprising a section "§ 1. The construction of the model" and dated 30 September 1930, and the last "On closed and open domains". At least the first four pages were probably intended to be at the beginning of a publication.[1]

What is most notable about *s1930e* is that it shows Zermelo to be undoubtedly the first to provide an analysis of set theory and a motivation for its axioms based on what we now call the *iterative conception.* This is a way of thinking about the subject matter for set theory as stratified in layers, a set at a layer consisting of elements from former layers and the layers indexed with transfinite numbers. As sources for this one has referred to *Zermelo 1930a* and its adoption of the axiom of foundation and to Gödel's work on L. In Gödel's writings one sees the iterative conception as an underlying motivation in terms of an extension into higher types (cf. his publication *1947,* 518ff and his lecture *1933*). Specific articulations of the iterative conception as itself motivating the axioms of set theory appeared in the 1970s, e.g. *Scott 1971*, *Shoenfield 1977*, *Wang 1974*, and *Boolos 1971.* In Zermelo's *s1930e* one now sees how the initial axiomatizer of set theory had already traveled down this path.

Zermelo first sets out his set-theoretic model, and it is his cumulative hierarchy from *1930a* based on a totality Q of urelements serving as a basis:

$$P_1 = Q\,;\ P_{\alpha+1} = P_\alpha \cup \mathcal{P}(P_\alpha)\,;\ \text{and } P_\alpha = \bigcup_{\beta<\alpha} P_\beta \text{ for limit ordinals } \alpha.$$

The difference here is that the presentation, with the goal of establishing the consistency of set theory, is more in the spirit of a schematic picture. In particular, Zermelo defines the "fundamental relation" $x \in y$ by first considering a subset M of a level P_α in an evidently informal sense, ascribing to it a corresponding member m of $P_{\alpha+1}$, and then specifying that $x \in m$ for all elements x of M. Zermelo himself did not take the linguistic turn, in that he did not develop an uninterpreted formalism, but here he comes closest in set theory to making a distinction between a semantic context and a syntactic counterpart.

[1] See *Ebbinghaus 2006* for historical details.

Zermelo goes on to argue for the axioms of set theory in terms of this schematic picture. As in his *1930a*, he argues for the separation axiom to be satisfied "in full generality and without the 'definiteness' restriction." He does not argue for the axiom of choice, for as in his *1930a* he takes it to be an underlying "general logical principle". Interestingly, he writes here "e.g., in the form of Hilbert's 'Aristides'." This refers to Hilbert's proof-theoretic τ operator, for which the scheme $A(\tau A) \rightarrow A(a)$ had been metaphorically described as: If Aristides the Just is corruptible, then everyone is corruptible. However antithetical Hilbert's finitary approach was for Zermelo, one finds him referring here explicitly to features of the former's proof theory. On the other hand, the connection between the τ operator and the axiom of choice is not a directly correlative one.

Zermelo's argument for the replacement axiom in terms of his schematic picture is weak—indeed, circular—in that he starts with a set m, replaces its members with sets from some P_β, and then argues that the result is a set, being in $P_{\beta+1}$. Zermelo writes: "Of course, the assumption that the replacing elements belong to a *segment* of the development, while being essential here, constitutes no real restriction." He however had emphasized in his *1930a* the importance of replacement in connection with cofinality, and so here he defeats its purpose by specifying in advance that the replacing members be from some fixed P_β. *Boolos 1971* pointed out how replacement is not well-motivated by the iterative conception.

Zermelo next gets to the question of when a P_π satisfies his axioms. He points out that the simpler axioms hold at any P_α and that the power set axiom holds at those with limit index α. He then focuses on the replacement axiom, and highlighting its import, argues that P_π satisfies all of the axioms exactly when π is a "boundary number [Grenzzahl]" in the sense of *1930a*, i.e. ω or a (strongly) inaccessible cardinal.

The last page of *s1930e* provides Zermelo's fullest articulation of his distinction between closed and open domains, broached at the end of *s1930d*. Zermelo begins:

> A "closed domain" is one which can be determined or ordered by means of a *categorical system of postulates.* It is precisely that which Cantor really meant by his well-known definition of "set".

Cantor 1895, 481 had "defined" a set as "any collection into a whole [Zusammenfassung zu einem Ganzen] of definite and separate objects of our intuition or our thought", and Zermelo is now specifying that sets are to be defined through a "categorical system of postulates". However, in what languages the definitions are to be given would never be adequately clarified.[2]

[2] See *Ebbinghaus 2003.*

Zermelo continues:

> An "open domain" is a *well-ordered sequence of domains successively comprising one another* constituted so that *every closed subdomain can always still be extended* in it. ... Furthermore, the *entire open domain can be well-ordered so that all elements of a preceding layer precede all elements of every subsequent one.*

Über das mengentheoretische Modell

s1930e

§1. Die Konstruktion des Modells

Es sei gegeben ein wohldefinierter "geschlossener" Bereich Q von Objekten "q", die wir als "Urelemente" bezeichnen, während der Bereich selbst die "Basis" der Konstruktion genannt werden soll. Wir bilden nun successive eine wohlgeordnete Reihe weiterer Bereiche, die wir "Schichten" nennen wollen, indem wir jeder Zahl α der transfiniten Zahlenreihe eine "Schicht" Q_α und einen zugehörigen "Abschnitt" P_α zuordnen nach der folgenden Vorschrift:

1) $P_1 = Q_0 = Q$
2) $P_{\alpha+1} = P_\alpha + Q_\alpha$
   $P_\alpha = \lim P_\beta$, wenn α eine Limeszahl ist und β alle kleineren Zahlen durchläuft.

Jedes Q_α soll dabei nur "neue" Elemente enthalten, d. h. solche, die in P_α nicht vorkommen. Zwischen den Elementen aller dieser Schichten führen wir nun eine Relation ein, die "Grundrelation" $a \in b$ in folgender Weise: Ist M irgend eine Untermenge von P_α, welche keinem kleineren Abschnitte P_β angehört, so entspricht ihr ein und nur ein Element m in Q_α, welches zu allen Elementen x von M in der Beziehung $x \in m$ steht und nur zu diesen, also sicher auch zu keinem Elemente derselben Schicht oder einer höheren Schicht. (Die Schichten Q_α selbst sind durch diese Bedingung nur nach ihrer Mächtigkeit bestimmt und im Übrigen willkürlich.)[1] Jede Schicht ist von höherer Mächtigkeit als

[1] ⟦The word "willkürlich" is underlined by pencil with a question mark on the margin.⟧

This alludes to the cumulative hierarchies of *1930a* and as there,

> Set theory *in its entirety can be completely* represented only in an "open model", and so it is in my "set-theoretic model" based on "absolute" or "canonical development".

While in *1930a* Zermelo had directly posited the set-theoretic universe as a sequence of domains given by an unending sequence of boundary numbers, here he is focusing on establishing the consistency of set theory through representing it with this "open model".

On the set-theoretic model*

s1930e

§1. The construction of the model

Assume a well-defined "closed" domain Q of objects "q", which we shall call "urelements", whereas the domain itself shall be called the "basis" of the construction. We now successively form a well-ordered sequence of further domains, which we shall call "layers", by correlating with each number α of the transfinite number series a "layer" Q_α and a "segment" P_α belonging to it according to the following instruction:

1) $P_1 = Q_0 = Q$
2) $P_{\alpha+1} = P_\alpha + Q_\alpha$
   $P_\alpha = \lim P_\beta$ where α is a limit number and β runs through all smaller numbers.

Each Q_α shall only contain "new" elements, that is, such ones that do not occur in P_α. We now introduce a relation among the elements of all these layers, the "fundamental relation" $\alpha \in \beta$, as follows: If M is some subset of P_α belonging to no smaller segment P_β, then to it there corresponds one and only one element m in Q_α such that the relation $x \in m$ holds for all elements x of M and only those, and hence certainly for no element of the same layer or a higher layer. (The layers Q_α themselves are determined by this condition only with respect to their cardinality. In any other regard, they are arbitrary.)

* ⟦The letters B, A, P, U, V, E, and F refer to the axioms of extensionality, separation, pairing, power set, union, replacement, and foundation, respectively. In his *1930a* Zermelo calls the system of these axioms, collectively referred to as BAPUVEF, the "supplemented *ZF*-system", or "*ZF'*-system".⟧

alle vorangehenden. Ist nämlich $\alpha = \beta + 1$ von erster Art, so enthält Q_α die Potenzmenge von Q_β, ist also mächtiger als Q_β (nach Cantor). Ist aber α Limeszahl und M eine "konfinale" Untermenge von P_α (die keinem kleineren Abschnitt angehört),[2] so kann man jedes ihrer Elemente durch ein beliebiges anderes aus derselben Schicht ersetzen und die entstehende Produktmenge, die in Q_α enthalten sein muß, ist nach König mächtiger als ihre Summe und jede von ihnen, sofern der Satz für alle vorausgehenden Schichten bereits zutrifft. Er ist also bewiesen durch Induktion.

2 | Das so entstehende Modell, das *unbegrenzt* fortgesetzt freilich einen "offenen" Bereich, keine "Menge" darstellen würde, während jeder seiner Abschnitte Mengencharakter besitzt, genügt nun, wie gezeigt werden soll, sämtlichen mengentheoretischen Axiomen in Bezug auf die eingeführte $\in$-Relation.

1) Das Axiom der "Bestimmtheit" B) ist erfüllt durch die Definition, da die Zuordnung von M und m ein-eindeutig ist.

2) Das Axiom der "Fundierung" F) ist erfüllt durch die Wohlordnung der Schichten, weil alle "Elemente" x einer "Menge" m immer vorangehenden Schichten Q_β angehören.

3) Das Axiom der "Paarung" P) gilt, weil irgend zwei Elemente stets einem Abschnittte P_α angehören, also einen Unterbereich dieses Abschnittes darstellen, dem spätestens in Q_α eine "Menge" entspricht.

4) Ist m eine Menge in Q_α und M der entsprechende Unterbereich von P_α, so entspricht auch jedem Unterbereich N von M eine "Menge" n spätestens in Q_α. Das "Aussonderungsaxiom" A) ist also in voller Allgemeinheit und ohne "Definitheits"-Beschränkung erfüllt.

5) Ist wieder m eine Menge aus Q_α, so liegen ihre Untermengen n sicher in $P_{\alpha+1}$, also ihre "Potenzmenge" Um in $Q_{\alpha+1}$. Ferner liegen die Elemente ihrer Elemente alle in P_α, also ihre "Vereinigungsmenge" Vm sicher in $P_{\alpha+1}$, nämlich entweder in Q_α oder in der unmittelbar vorangehenden Schicht. Mithin sind auch die Axiome U) und V) erfüllt.

6) Es sei wieder m eine Menge aus Q_α, deren Elemente in P_α ersetzt werden sollen durch andere x', die sämtlich einem und demselben Abschnitte P_β angehören. Dann erscheint die "ersetzende" Menge m' spätestens in Q_β, und das Axiom E) ist gleichfalls erfüllt. Die Voraussetzung, daß die ersetzenden Elemente einem *Abschnitte* der Entwickelung angehören müssen, ist dabei natürlich wesentlich, aber keine eigentliche Beschränkung.

3 | 7) Das "Axiom des Unendlichen" wird bereits im ersten transfiniten Abschnitte P_ω erfüllt.

8) Das "Auswahlaxiom" soll hier als allgemeines logisches Prinzip, etwa in der Form des Hilbertschen "Aristides", allen unseren Deduktionen zu grunde

---

[2] ⟦The text in brackets is underlined by pencil with a question mark on the margin.⟧

The cardinality of each layer is higher than that of any of its predecessors. For if $\alpha = \beta + 1$ is of the first kind, then Q_α contains the power set of Q_β, and hence is of a higher cardinality than Q_β (according to Cantor). If, however, α is a limit number and M a "cofinal" subset of P_α (belonging to no smaller segment), then we can replace each of its elements by an arbitrary different one from the same layer, and the resulting product set, which must be contained in Q_α, is, according to König, of a higher cardinality than both its sum and each of them, provided that the theorem already holds true for all preceding layers. It is therefore proved by induction.

The model so obtained, which, if it were continued *indefinitely*, would of course constitute an "open" domain, not a "set", while each of its segments possesses the character of sets, satisfies all set-theoretic axioms with respect to the introduced $\in$-relation, as shall be shown now.

1) The axiom of "definiteness" B) is satisfied by the definition since the correlation of M and m is one-to-one.

2) The axiom of "foundation" F) is satisfied by the well-ordering of the layers since all "elements" x of a "set" m always belong to preceding layers Q_β.

3) The axiom of "pairing" P) is valid since any two elements always belong to some segment P_α, and hence constitute a subdomain of this segment to which there corresponds a "set" no later than in Q_α.

4) If m is a set in Q_α and M the corresponding subdomain of P_α, then to each subdomain N of M there also corresponds a "set" n in no later layer than in Q_α. The "separation axiom" A) is therefore satisfied in full generality and without a "*definiteness*" restriction.

5) If m is again a set from Q_α, then its subsets n certainly lie in $P_{\alpha+1}$, and hence its "power set" Um in $Q_{\alpha+1}$. Furthermore, all elements of its elements lie in P_α, and hence its "union set" Vm certainly in $P_{\alpha+1}$, namely either in Q_α or in the immediately preceding layer. Thus, the axioms U) and V) are satisfied as well.

6) Again, let m be a set from Q_α whose elements are replaced in P_α by other x' all of which belong to one and the same segment P_β. The "replacing" set m' then appears in no later layer than in Q_β, and the axiom E) is satisfied as well. Of course, the assumption that the replacing elements belong to a *segment* of the development, while being essential here, constitutes no real restriction.

7) The "axiom of the infinite" is already satisfied in the first transfinite segment P_ω.

8) The "axiom of choice" shall underlie all our deductions as a general logical principle, e.g., in the form of Hilbert's "Aristides". Then, along with the

gelegt werden. Dann gilt also mit den übrigen Axiomen auch der Wohlordnungssatz und jede hier vorkommende "Menge" wie auch jeder "geschlossene" Bereich kann ohne Weiteres als *wohlgeordnet* angenommen werden.

Bezeichnen wir mit $f(\alpha)$ die Mächtigkeit des Abschnittes P_α oder genauer die Anfangszahl der entsprechenden Zahlenklasse, so erhalten wir eine "Normalfunktion" im Sinne von *Veblen* und *Hausdorff*. Nach unserer Konstruktion ist nämlich

$$f(\alpha+1) > f(\alpha),$$

weil der Abschnitt $P_{\alpha+1}$ alle Untermengen von P_α enthält. Außerdem ist stets $f(\alpha) = \lim f(x)$ für $x < \alpha$, wenn α eine Limeszahl ist, da ja P_α als Summe aller kleineren P_x erklärt ist. Dann ist die Funktion auch "eigentlich *monoton*"; ihr Anfangswert ist $f(1) = \kappa$ ⟦und⟧ ist die Mächtigkeit der Basis Q. Wie jede Normalfunktion besitzt auch diese "Eigenwerte" d. h. "kritische Zahlen" von der Eigenschaft $\lambda = f(\lambda)$, während sonst immer $\alpha < f(\alpha)$ ist.

Nun stellen wir die Frage: Wie muß ein *Abschnitt* P_τ unserer Entwickelung beschaffen sein, damit *in ihm* alle "konstituierenden" Axiome BAPUVEF der Mengenlehre erfüllt sind?

Hier ergibt sich zunächst: Die Axiome BFPAV sind in *jedem* Abschnitte von selbst erfüllt, das Axiom U) wenigstens in jedem Abschnitte mit *Limes*-Index. Dagegen erfordert das "*Ersetzungs-Axiom*" eine nähere Untersuchung. Ihm zufolge müßten alle Elemente x', die den Elementen x einer Menge m aus P_τ zugeordnet sind, sofern sie selbst diesem Abschnitte sämtlich angehören, wieder die Elemente einer Menge m' in P_τ bilden, also einem *kleineren* Abschnitte angehören, nämlich P_β, wenn m' in der Schicht Q_β liegen soll.

4 | Hieraus folgt zunächst, daß keine wohlgeordnete Menge m in P_τ einen Ordnungstypus $m \geq \tau$ haben kann. Denn sonst ließe sich ein Abschnitt m_1 von m ähnlich abbilden auf die wohlgeordnete Folge der Elemente q_α, wo jedes q_α ein bevorzugtes Element der Schicht Q_α sein soll für alle $\alpha < \tau$, und dieser Folge könnte innerhalb P_τ keine Menge m_1' entsprechen. Es sind also in P_τ immer nur Mengen kleinerer Ordnungszahl und kleinerer Mächtigkeit vertreten als τ. Insbesondere gilt dies auch von jeder Menge p_α in P_τ, welche alle Elemente des Abschnittes P_α enthält und, wie wir sahen, die Mächtigkeit $f(\alpha)$ besitzt. Mithin ist mit $\alpha < \tau$ auch stets $f(\alpha) < \tau$ und durch unsere Funktion wird der zu τ gehörige Abschnitt Z_τ der Zahlenreihe auf sich selbst abgebildet, und $\tau = f(\tau)$ ist "Eigenwert" der Funktion. Es ist aber kein innerhalb des Abschnittes selbst konstruierbarer Eigenwert, sondern eine "Kernzahl", eine "reguläre Anfangszahl", und zwar eine solche "zweiter Art", eine "exorbitante" Zahl. Wäre nämlich τ einer kleineren $\rho < \tau$ konfinal, so gäbe es im Abschnitte P_τ eine Menge m von der gleichen Mächtigkeit wie ρ und ihren Elementen ein-eindeutig entsprechend eine Folge von Elementen q_α aus Q_α, die keinem kleineren Abschnitte P_β angehören, also auch keine Menge m' innerhalb P_τ bestimmen könnte. Wäre aber τ eine Anfangszahl erster Art und $\sigma < \tau$ die unmittelbar vorangehende Anfangszahl, so enthielte P_τ eine Menge m von der Mächtigkeit σ, deren Potenzmenge Um wegen

other axioms, the theorem of well-ordering holds as well, and we can assume without any further ado that every "set" occurring here is *well-ordered*, and every "closed" domain as well.

If we use $f(\alpha)$ to refer to the cardinality of the segment P_α or, more precisely, the initial number of the corresponding number class, then we obtain a "normal function" in the sense of *Veblen* and *Hausdorff*. For according to our construction

$$f(\alpha + 1) > f(\alpha),$$

since the segment $P_{\alpha+1}$ contains all subsets of P_α. Furthermore, we always have $f(\alpha) = \lim f(x)$ for $x < \alpha$, assuming that α is a limit number, since P_α has been defined as the sum of all smaller P_x after all. Then the function is also "strictly *monotonous*"; its initial value is $f(1) = \kappa$ ⟦and⟧ the cardinality of the basis Q. This normal function, like any other, also has "eigenvalues", that is, "critical numbers" with the property $\lambda = f(\lambda)$, whereas in any other case we always have $\alpha < f(\alpha)$.

We now ask the question: How must a *segment* P_τ of our development be constituted so that all "constitutive" axioms BAPUVEF of set theory are satisfied *in it*?

At first, we have the following: The axioms BFPAV are automatically satisfied in *every* segment, and the axiom U) at least in every segment with *limit*-index. The "replacement axiom", on the other hand, requires closer investigation. According to it, all elements x' correlated with the elements x of a set m from P_τ would again have to form the elements of a set m' in P_τ, provided that they themselves all belong to this segment. In other words, they would have to belong to a *smaller* segment, namely P_β, if m' were to lie in the layer Q_β.

From this, at first, it follows that no well-ordered set m in P_τ can have an order type $m \geq \tau$. For, otherwise, there would be a similarity mapping of a segment m_1 of m onto the well-ordered sequence of elements q_α, where each q_α is supposed to be a designated element of the layer Q_α for all $\alpha < \tau$, and this sequence could have no set m_1' within P_τ corresponding to it. So the sets represented in P_τ are always only sets of smaller ordinal number and smaller cardinality than τ. In particular, this also holds true for every set p_α in P_τ, which contains all elements of the segment P_α and has, as we have seen, the cardinality $f(\alpha)$. Thus we always have $f(\alpha) < \tau$, if $\alpha < \tau$, and the segment Z_τ of the number series belonging to τ is mapped onto itself by virtue of our function, and $\tau = f(\tau)$ is an "eigenvalue" of the function. But it is not an eigenvalue that can be constructed within the segment itself. Rather, it is a "core number", or "regular initial number", and in particular one of the "second kind", an "exorbitant" number. For if τ were cofinal with a smaller $\rho < \tau$, then, in the segment P_τ, there would exist a set m of the same cardinality as ρ, and, corresponding to its elements one-to-one, a sequence of elements q_α from Q_α, which belong to no smaller segment P_β, and hence could determine no set m' within P_τ. But if τ were an initial number of the first kind and $\sigma < \tau$ the immediately preceding initial number, then P_τ would contain a set m of

ihrer Mächtigkeit $\geq \tau$ nicht mehr im Abschnitte P_τ enthalten wäre. Diese beiden Eigenschaften der "Grenzzahl" τ, gleichzeitig "Kernzahl" zu sein und "Eigenwert" der Funktion f, sind aber auch *hinreichend* für die Erfüllung des "Ersetzungsaxioms" in P_τ. Aus $\tau = f(\tau)$ folgt nämlich umgekehrt wieder, daß für jedes $\alpha < \tau$ auch $f(\alpha) < f(\tau) = \tau$ und daß daher jede in einer Schicht Q_α des Abschnittes P_τ enthaltene Menge m von kleinerer Mächtigkeit ist als τ. Ersetzt man nun die Elemente x_n einer solchen Menge m durch Elemente x'_n desselben Abschnittes P_τ, so liegen auch alle diese x'_n in einem kleineren Abschnitte P_β und bestimmen eine Menge m' in P_τ, weil sonst τ einer kleineren Zahl konfinal wäre gegen die Annahme.

5 | Über geschlossene und offene Bereiche

Ein "geschlossener" Bereich ist ein solcher, der durch ein *kategorisches Postulatsystem* bestimmt oder geordnet werden kann. Er ist genau das, was Cantor mit seiner bekannten Definition einer "Menge" eigentlich gemeint hat, und kann bei allen rein mathematischen Betrachtungen und Deduktionen überall und widerspruchsfrei als Menge behandelt werden. Jeder geschlossene Bereich kann wohlgeordnet werden und besitzt sowohl eine Mächtigkeit wie eine Ordinalzahl.

Ein "offener Bereich" ist eine *wohlgeordnete Folge successiv einander umschließender geschlossener Bereiche* von der Beschaffenheit, daß in ihm *jeder geschlossene Unterbereich noch erweitert werden kann.* Er kann also selbst gewiß kein "geschlossener" Bereich sein und durch kein *kategorisches* Postulatsystem geordnet werden. Dagegen ist jeder seiner *Abschnitte* geschlossen und wohlordnungsfähig, und auch der *ganze, offene Bereich* kann in der Weise *wohlgeordnet* werden, daß alle Elemente einer *vorangehenden Schicht* allen Elementen jeder folgenden vorangehen. Der offene Bereich hat *weder* eine Mächtigkeit *noch* eine Ordnungszahl.

Jeder *geschlossene* Bereich kann zu einem "Normalbereich" erweitert werden, er kann aber auch in einem passend gewählten Normalbereich als "Menge" erscheinen. Es gibt also *geschlossene* Bereiche, in denen die *klassische Mengenlehre* mit dem Axiomensystem BAPUVEF zur Darstellung kommt, darunter "finitistische" wie "infinitistische" Normalbereiche von beliebiger Charakteristik. Alle solchen Normalbereiche sind "geschlossene" Bereiche und gleichzeitig spezielle mengentheoretische "Modelle", die sich wesentlich von einander unterscheiden und die Mengenlehre mehr oder weniger *unvollständig* darstellen. Die *ganze* Mengenlehre kann *vollständig* nur in einem *"offenen Modell"* dargestellt werden, und sie wird es in meinem auf "absolute" oder auf *"kanonische Entwickelung"* gegründeten "mengentheoretischen Modell".

Die *unbegrenzte* Zahlenreihe ist ein *offener*, jeder ihrer *Abschnitte* ein *geschlossener* Bereich.

cardinality σ whose power set Um, since its cardinality $\geq \tau$, would no longer be contained in the segment P_τ. But these two properties of the "boundary number" τ, to be both "core number" and "eigenvalue" of the function f, are also *sufficient* for the satisfaction of the "replacement axiom" in P_τ. For from $\tau = f(\tau)$ it again follows conversely that for each $\alpha < \tau$ we also have $f(\alpha) < f(\tau) = \tau$, and hence that every set m contained in some layer Q_α of the segment P_τ is of cardinality smaller than τ. If we now replace the elements x_n of such a set m by elements x'_n of the same segment P_τ, then all these x'_n also lie in a smaller segment P_β and determine a set m' in P_τ, since otherwise τ would be cofinal with a smaller number, contrary to the assumption.

On closed and open domains

A "closed" domain is one such which can be determined or ordered by means of a *categorical system of postulates.* It is precisely that which Cantor really meant by his well-known definition of "set", and it can be treated as a set everywhere and without contradiction in all purely mathematical considerations and deductions. Every closed domain can be well-ordered and possesses both a cardinality and an ordinal number.

An "open domain" is a *well-ordered sequence of domains successively comprising one another* constituted so that *every closed subdomain can always still be extended* in it. Therefore, it certainly cannot be a "closed" domain, and it cannot be ordered by means of a *categorical* system of postulates. Each of its *segments*, on the other hand, is closed and capable of being well-ordered. Furthermore, the *entire open domain* can be *well-ordered* so that all elements of a *preceding layer* precede all elements of every subsequent one. The open domain has *neither* a cardinality *nor* an ordinal number.

Every *closed* domain can be extended to a "normal domain", but it can also occur as "set" in a suitably chosen normal domain. So there exist *closed* domains, among them "finitistic" and "infinitistic" normal domains of arbitrary characteristic, in which *classical set theory* with the axiom system BAPU-VEF finds its representation. All such normal domains are "closed" domains and, at the same time, special set-theoretic "models" that differ essentially from one another and represent set theory more or less *incompletely.* Set theory *in its entirety* can be *completely* represented only in an "*open model*", and so it is in my "set-theoretic model" based on "absolute" or "*canonical development*".

The *unlimited* number series is an *open* domain, and each of its *segments* is a *closed* domain.

Introductory note to 1930f

Albert Henrichs

In the bibliography of Zermelo's published work one item stands out and is certain to catch the attention of mathematicians and non-mathematicians alike. In 1930 Zermelo published a German translation of a portion of Book V of the *Odyssey* in a Viennese journal that catered to the interests of "the friends of antiquity" (*Zermelo s1930f*). It is remarkable that a mathematician in his late 50's should turn humanist, translate Homer into German verse, and present a specimen of his labor of love to the educated public. Such an enterprise required time and leisure and, even more, a reading knowledge of a complex poetic dialect of ancient Greek, a sense of rhythm and meter, and a command of the German language and its literature that went far beyond the capacities of most of his academic peers. In addition, Zermelo could not have succeeded without the support of friends and colleagues in the humanities who encouraged him in his endeavor and saw to it that a significant sample of his work as a translator would be published and thus brought to the attention of the world at large.

Many details surrounding Zermelo's translation remain unclear, including its exact date and even his original motivation for making this effort. Uncertainties like these leave plenty of room for conjecture; however we do have one valuable piece of information concerning the encouragement he received. It comes in the form of an editorial note that accompanies the published translation.[1] The note reads:

> A preeminent representative of the field of Classical Studies at the University of Freiburg i. Br. was kind enough to draw our attention to an unpublished translation of some books of the *Odyssey* by Ernst Zermelo, professor of mathematics at Freiburg University. Our readers can tell from the sample offered here that the author possesses a sure command of the (Greek) language, in addition to an uncommon sensitivity and power of expression. He did not intend a literal model translation of philological exactitude.[2]

The anonymous sponsor

It is regrettable that the editor of the *Wiener Blätter* chose to suppress the identity of the "preeminent" classicist who had brought Zermelo's translation

[1] The editor was Otto Barensfeld (1890–1985), who was known for attaching introductory appreciations to many of the "Samples of texts and translations" that appeared in the *Wiener Blätter*.

[2] The original German version of this note precedes the text of Zermelo's published translation as reproduced in this volume.

to his attention. He must have been a person of considerable distinction and would have been by definition a colleague of Zermelo's, since the latter had joined the Faculty of Natural Sciences and Mathematics at Freiburg as an adjunct professor in the fall of 1926.[3] At that time the Department of Classical Philology had two tenured professorships, one for Greek and the other for Latin. Between 1927 and 1934 in particular, several of the best classicists of the 20th century passed through Freiburg and briefly occupied these chairs. From 1927 to 1929 the senior Hellenist was Rudolf Pfeiffer (1889–1979), who left Freiburg after four semesters for a chair in Munich. Pfeiffer was intimately familiar with the Homeric poems and deeply interested in their ancient and modern reception.[4] In fact, he had written his Munich dissertation on the Augsburg Meistersinger Johannes Spreng (1524–1601) and his posthumous translation of the *Iliad* into rhyming German verse (1610).[5] It is not inconceivable that it was Pfeiffer who recommended Zermelo's translation, but the short duration of his tenure at Freiburg makes the identification unlikely. Pfeiffer's successor was Wolfgang Schadewaldt (1900–1974), who arrived in the fall of 1929 and moved from Freiburg to Leipzig in 1934, the year before Zermelo's dismissal from the university. Schadewaldt regarded translations of Greek and Latin texts "as the ultimate consummation of any philological endeavor."[6] A distinguished translator in his own right, his German prose version of the *Odyssey* appeared in 1958 to universal acclaim, followed posthumously in 1975 by his translation of the *Iliad.*[7] But despite his accelerated career, Schadewaldt's relatively young age and lack of seniority make it unlikely that he can be identified with the "preeminent classicist" who acted as a middleman between Zermelo and the Vienna journal.[8]

By far the likeliest candidate is Otto Immisch (1862–1936), who held the Latin chair from 1914 until his retirement in 1931. Virtually forgotten today, he was the most established and best known classicist at Freiburg during Zermelo's years on the faculty.[9] Equally well-versed and well-published in Greek

[3] Zermelo's name appears for the first time in the Freiburg course catalog for the winter semester 1926/27, where his German title is given as "ordentlicher Honorarprofessor" and its Latin equivalent as "Prof(essor) ord(inarius) hon(orarius)".

[4] See *Vogt 2001*; *Lloyd-Jones 1982.*

[5] *Pfeiffer 1914.*

[6] Hellmuth Flashar in *2005*, 496: "Als Summe der philologischen Bemühungen verstand S⟦chadewaldt⟧ die Übersetzung."

[7] See *Szlezák 2005.*

[8] Another obstacle to the identification is the brief interval between Schadewaldt's 1929 Freiburg appointment and the 1930 publication of Zermelo's translation.

[9] Immisch's successor in the Latin chair at Freiburg was Eduard Fraenkel (1888–1970), one of the greatest classicists of the 20th century, who came from Göttingen to Freiburg in the summer of 1931. He emigrated to England in 1934 after he had been stripped of his professorship in compliance with the new antisemitic laws by the then rector, the philosopher Martin Heidegger (1889–1976), who had succeeded Edmund Husserl in the fall of 1928. On Fraenkel see *Horsfall 1990.*

and Latin literature, Immisch was an all-around classicist of wide-ranging interests who fought publicly for humanistic values and for the survival of the traditional "humanistische Gymnasium" with its Latin and Greek curriculum.[10] He was co-founder and from 1919 until his death sole editor of "Das Erbe der Alten" ("The heritage of antiquity"), a series of monographs on topics relevant to "the nature and impact of antiquity". As a former rector, Immisch was a highly respected and visible member of the faculty and one of its most senior professors. Well-connected within the university, he apparently had ties to Lothar Heffter (1862–1962), a professor of mathematics who had been instrumental in securing the honorary professorship for Zermelo and whose Freiburg career overlapped almost exactly with that of Immisch.[11] Heffter mentions Immisch in his autobiography (*Heffter 1952*, 146), and it is a reasonable guess that Immisch met Zermelo through Heffter. With his commitment to making the world of antiquity accessible to a wider public, Immisch was clearly the kind of classicist who would have applauded Zermelo's translation project and recommended it to the editors of the *Wiener Blätter*.

The mathematician as humanist and hellenist

What attracted Zermelo to Homer, and why did he decide to translate the *Odyssey*? To the extent that we will ever know the answers to these questions, they can be found in his family background, his education, and ultimately in his desire to connect with the Greek roots of European culture, which he embraced wholeheartedly. After all, it was ancient Greece that had given the western world the foundations of literature and art, of democracy and political thought, as well as the bases of philosophy, mathematics, and logic.

Ernst Zermelo came from an academic family and had been exposed to things Greek at an early age (*Ebbinghaus 2007b*, Section 1.1). His father Theodor obtained a doctorate in 15th century French history and became a high school professor ("Gymnasialprofessor") whose specialties comprised history and geography, in addition to Greek and Latin as minor fields. The father and his academic interests were clearly a role model for Zermelo in his teenage years. When he was 13 he translated parts of the first book of Virgil's *Aeneid* into German verse, thus beginning the development of the skills he would use later as a translator of the *Odyssey*. As a young teacher his father had pointed the way for his son by compiling an anthology of his own translations of poems originally composed in French, Italian and

[10] For a succinct account of Immisch's life and career see *Becker 1974*.

[11] Both were born in June 1862; Heffter moved from Kiel to Freiburg in 1911, Immisch came there from Königsberg in 1914; both served as rectors of the university, Heffter in 1917/18 and Immisch in 1924/25; and both were members of the Heidelberg Academy.

English. He made sure that, starting in the fall of 1880, young Ernst attended the Luisenstädtisches Gymnasium in Berlin, where he learned Latin at the tender age of 9 and Greek at twelve. He was exposed to both languages on a regular basis until he graduated in March of 1889. He apparently did not enroll in any Greek or Latin courses while at the university, and one can only marvel at the fact that he managed to keep up his Greek on a very respectable level well into his late 50's and beyond.

In 1921 Zermelo left Zurich, where he had been living in early retirement, and moved to Freiburg i. Br. Unemployed for health reasons, he was unfulfilled and plagued by loneliness, even though he continued his mathematical research. He resumed teaching in 1926, when he was appointed to his honorary professorship. It was during that intervening period, between 1921 and 1926, that he embarked upon his translation project. In March 1926 he wrote to his two sisters that he could translate the entire *Odyssey* within a year or two if he had a publisher (*Ebbinghaus 2007b*, 144). He must have worked on the translation for some time. Probably by the winter of 1926/27 he had given typewritten copies of his translation of Books V and VI to a young protégé, the American philosopher Marvin Farber (1901–1980). Carbon copies of the original typescript survive in the Zermelo *Nachlass*. All in all, translations of three entire books (out of a total of 24) of the *Odyssey* are preserved in the *Nachlass* either in typewritten form (Books V and VI) or in shorthand (Book IX).[12] Translations of small portions of Books VII and VIII exist as well.[13] A specimen comprising lines 149–225 of Book V was published in 1930 (*Zermelo s1930f*). It omits six verses of the German translation, which correspond to lines 196–200 of the Homeric text. Apart from the omission and minor variations in the use of the letter "ß" and in punctuation the archived and published versions are completely identical.

The publication of 1930, as well as the full text of Book V, are reproduced in this volume as a permanent record of Zermelo's work as a translator. The omitted lines have been restored in the reproduction of the published sample. The typed translations of Books V and VI are highly polished versions that seem to represent the final form of his translation. In the winter semester of 1932/33, after an educational cruise to Greece, Italy and North Africa during the preceding spring, Zermelo gave a talk with the title "Selected samples of a modern translation of Homer (Odyssey V–VI)" for the Association of the Friends of the University (*Ebbinghaus 2007b*, 145). This suggests that translating the *Odyssey* was not a solipsistic exercise for him but that he intended to educate a wider public by introducing lay audiences to one of the two poems that mark the beginning of European literature.

---

[12] Preserved in the *Nachlass* as C 129/122 (Books V–VI) and C 129/220 (Book IX).

[13] The corresponding signatures in the *Nachlass* are C 129/221 (beginning of Book VII and end of Book VIII, typewritten) and C 129/222 (beginning of Book VII, additional typewritten copy).

Zermelo's *Odyssey* and Odysseus' return

Zermelo's selection of Books V–IX is hardly accidental. It must represent a deliberate choice, which has to do with the content of these books and their place in the poem's larger narrative. The five consecutive books that Zermelo chose do not constitute a self-contained narrative unit within the *Odyssey*, yet they are clearly set off from Books I–IV, the story of Odysseus' son Telemakhos, which are distinctly different in tone and structure. Books V–IX are part of a larger narrative entity that begins with Book V and ends with Book XII, and they comprise a succession of famous episodes that feature some of the most memorable moments in the entire epic. Books XIII–XXIV are set on Ithaca and deal with Odysseus' homecoming, his reunification with his family, and the revenge on the suitors, who had harassed his wife Penelope.

Book V opens with a council of the gods who decide that it is time to let Odysseus return to his native Ithaca. For seven long years, the goddess nymph Kalypso has held him against his will on her island of Ortygia, where he was washed ashore after a shipwreck during his journey from Troy and consigned to a life of sex and leisure. Hermes conveys Zeus' will to Kalypso, who agrees to release Odysseus. After five days of preparation, a lavish last meal and a final night together with Kalypso, Odysseus sets out to sea on a makeshift boat. But after seventeen days, Poseidon sends a gigantic wave, which destroys the raft. Shipwrecked once again, Odysseus is saved with Athena's help and finds himself stranded and naked on the shores of Scheria, the island of the seafaring and peace-loving Phaiakians. In Book VI Nausikaa finds him there, provides him with clothing and food and directs him to the royal palace. In Book VII he makes his way to the royal residence, where he is hospitably received by king Alkinoos and queen Arete. Book VIII is filled with athletic contests, dancing and the singing of songs. The blind bard Demodokos sings about the fall of Troy and makes Odysseus weep. In Book IX, Odysseus reveals his identity and embarks upon a lengthy account of his adventures during his journey from Troy, beginning with the Kikones, the Lotus-Eaters and the land of the one-eyed Kyklopes. The blinding of Polyphemos, the man-eating Kyklops, forms the narrative highlight of this book. Odysseus' tale goes on for three more books of fantastic episodes that include the divine sorceress Kirke who turns humans into pigs in Book X, a visit to the Underworld in Book XI, and finally in Book XII the Sirens, Skylla and Kharybdis, and the sacrilegious slaughter of the cattle of Helios. Book XII ends with Odysseus' retrospective account of his first shipwreck; this event, initially mentioned in Book V, now turns out to be the divine punishment for the slaughter. Odysseus emerges from the shipwreck as the sole survivor, only to land on Kalypso's island after nine days at sea. At this point, the story of Odysseus' wandering has come full circle and the narrative has reverted to its point of departure in Book V. In Book XIII Odysseus finally returns home to Ithaca.

The homecoming theme looms large in the *Odyssey;* it is in fact the principal poetic concern that drives the entire narrative. Its central importance explains why Zermelo chose Book V and the final encounter between Odysseus and Kalypso for publication. It is here that Odysseus emerges for the first time in the poem as an active character in the story, articulating his desire to leave Kalypso and to be united with his wife. In fact Odysseus uses the key term for the homecoming, *nóstos*, poignantly in the parting words that he addresses to the divine nymph who has kept him captive as her lover for so many years. In Mordaunt Barnard's English translation of 1876, which is reproduced below as a companion piece to Zermelo's German version, the lines in question read: "But even so I wish and long all day / For home, and my returning day to see" (lines 219–220). The day on which Odysseus returns to Ithaca is, in Greek, his *nóstimon êmar*, his "day (*êmar*) of homecoming (*nóstimon*)", the day of his *nóstos*.

In his own translation, Zermelo suppresses the entire phrase "my returning day" to meet metrical requirements and renders Odysseus' greatest wish as the desire "to come home" (*zur Heimat zu gelangen*), which is a literal but loaded translation of a related Greek phrase that precedes in the same verse (V 220 *oikade t' elthemenai*). The emotionally charged word *Heimat* captures the full force of the homecoming theme as effectively as the Greek *nóstos*. Zermelo's choice of *Odyssey* V 149–225 as his exhibition piece shows that he understood the thematic structure of the *Odyssey* extremely well. At the same time, his evocation of the *Heimatgefühl*, the "feeling for home", in rendering "the day of return" suggests that he deeply appreciated the Homeric characterization of Odysseus as a traveler in an alien world who, like the extraterrestrial ("E. T.") in the 1982 science fiction movie, simply but doggedly wants to "go home".

Zermelo as translator

This brings us to Zermelo's merits as a translator. He was certainly well prepared. He knew his Greek, and there is evidence that he had read up on Homer and his modern reception including translations, on Homeric poetics, on the textual criticism of the Homeric poems, and on the art of translation.[14] Did he put his talent and his tools to good use? The qualities one looks for in a translator of epic verse are an adequate degree of faithfulness to the Greek original; the ability to replicate its pace and rhythm creatively, especially if one translates into verse rather than prose; and finally, a congeniality of form and content, of diction and narration, that reminds the reader that what he is reading is world-class poetry of the epic, that is the narrative, variety.

[14] A list of relevant books is preserved in his *Nachlass* under C 129/220.

Zermelo deserves high marks on all three counts. His rendition is indeed accurate; that is, he fully understands the Greek in all its nuanced complexity and translates it accurately. That does not mean that his translations are always literal. They are not. In fact he often omits entire words and phrases found in the Greek text. The reason for his frequent liberties with the language of the original has to do with his choice of meter. Any translator who renders Homeric epic into a modern language has to choose between prose and verse before he even begins his task. The vast majority of modern renditions of the *Iliad* and *Odyssey* into German or English are translations into verse, especially into dactylic hexameters, the meter used in the Homeric poems. Prose translations are on the whole less common, but they include some outstanding specimens, especially Wolfgang Schadewaldt's German versions of both poems.

The *Iliad* and *Odyssey* are composed in dactylic hexameters, which consist of six metrical units in the form of dactyls (long-short-short) or spondees (long-long). Each verse comprises on average fifteen to sixteen syllables, which means that the lines are rather long and proceed at a stately pace that is compatible with the long narrative passages and frequent speeches that are typical of epic poetry. But English translators have often chosen other meters such as fourteen syllable verse, heroic couplets, and especially blank verse. Blank verse is an unrhymed iambic pentameter of ten or eleven syllables that was the preferred meter of the vast majority of English and German dramatists, including Shakespeare, Lessing and Goethe. Its use was equally widespread among non-dramatic poets writing in English from Milton via Keats and Shelley to Tennnyson, Browning and Robert Frost.

Given the choice between poetry and prose, Zermelo chose blank verse, presumably because this meter was well suited to achieve the "immediate liveliness" that he intended for his translation (*Ebbinghaus 2007b*, 144/145). Iambic rhythms occur naturally in German and in English, which explains why blank verse sounds more natural than the hexameter. But because of its comparative brevity and quick pace, the meter lacks the grandeur of the epic hexameter and is ultimately more suited to drama than to epic. Any translation of Homer into blank verse inevitably accelerates the pace of the narrative and intensifies the tone of the speeches. It thus affects the overall character of the poetry and changes its poetic essence. Because the structure of blank verse involves lines that are significantly shorter than those of the hexameter, its use poses problems for the translator, who either compromises by omitting Greek words from his translation or compensates by using more lines than the original text. Needless to say, Zermelo's translation pays a price for using shorter lines, and so does Barnard's, who translated the *Odyssey* into English blank verse some fifty years before Zermelo (*Barnard 1876*).

It is instructive to compare how the two translators deal with this problem. Miraculously, most of the time they manage to use the same number

of verses as in the original.[15] Barnard achieves this largely by taking advantage of the innate succinctness of the English language and by making clever use of condensed syntax. Zermelo is equally successful by regularly omitting words such as epithets that do not affect the basic meaning of the Greek verse. In epic poetry gods, mortals and even inanimate objects are often characterized by regularly recurring descriptive labels known as epithets. In the published passage alone, Odysseus is, in Barnard's translation, "brave" (149), "a man of great suffering" (171), "divine" (198), "wise" (203, 214) and has a "patient mind" (222); the sea is "barren" (158), "gloomy" (164) or "darksome" (221); wine is "red" (165), the sky is "broad" (169), words are "winged" (172), Kalypso is "divine" (180, 192), the gods are "blessed" (186), and Penelope is "chaste" (216). Of the 16 epithets found here, Barnard translates all, Zermelo a mere six. Widespread omission of epithets is common in Zermelo, and this practice necessarily reduces the epic flavor of his translation.

Zermelo ended his selection with a scene at nightfall—"the sun went down and darkness came" (Barnard). But this is not where the day ended for Kalypso and Odysseus in Homer. For the next two lines tell us how they spent the night: "They, going to the hollow cave's recess, / Each by the other stayed and joyed in love" (Barnard). In these brief lines, Kalypso and Odysseus make love for the last time, at least as far as we are told. As one would expect, Zermelo included this scene in the complete translation of Book V, which was never published in his lifetime and is made accessible in the Appendix below. But he stopped short of including the lines in question in his published selection even though they are inseparable from the verses that precede. One can only wonder why he chose to end his short version of the story of Odysseus and Kalypso with the falling night rather than with lovemaking.[16]

---

[15] However, Zermelo's translation of the 76 Homeric verses is 14% longer than Barnard's (92.5 versus 81 lines of blank verse). This difference has as much to do with the individual styles of the translators as with the languages into which they translate. Zermelo is decidedly more generous as a translator and often requires three lines where two suffice for Barnard. For instance, Zermelo needs 25 words to render the 16 Greek words of *Odyssey* V 160–161, whereas Barnard's version uses a mere 17 words. In Zermelo's translation of these lines, Calypso says: "Wenn du es denn begehrst, ich lass dich ziehn!" ("If you so desire, I let you go.") The phrase "if you so desire" has no equivalent in the Greek text, or in Barnard's translation.

[16] Heinz-Dieter Ebbinghaus suggests that the omission of the two lines that conclude the episode may have been mandated for reasons of space. If one assumes that the editor of the *Wiener Blätter* had allocated exactly two full pages to the sample of Zermelo's translation, any additional lines would have required an extra page. This explanation makes eminent sense.

Aus Homers Odyssee

1930f

V 149–225 übersetzt von *Univ.-Prof. Dr. Ernst Zermelo* in Freiburg i. Br.[*][1]

Die schöne Nymphe aber ging hinaus,
Nachdem sie den Befehl des Zeus vernommen,
Um den Odysseus draußen aufzusuchen.
Sie fand ihn wieder am Gestade sitzend,
Und nimmer wurden seine Augen trocken:
So floß sein Leben hin in Sehnsuchtstränen
Nach seiner fernen Heimat. Denn die Nymphe,
So schön sie war, gefiel ihm längst nicht mehr,
So daß er ungern nur und liebeleer
Das Lager teilte mit der Liebenden.
Bei Tage saß er immer auf dem Felsen
Und schaute weinend auf das weite Meer.
Jetzt trat sie freundlich zu ihm hin und sprach:
„Du Armer! Weine nicht und gräm dich nicht!
Du sollst nicht mehr dein Leben hier vertrauern!
Wenn du es denn begehrst, ich laß dich ziehn!
So hau dir Bäume flugs mit scharfer Axt
Und füge sie zum breiten Floß zusammen.
Bordbalken lege drauf so hoch und fest,
Dich sicher durch das wilde Meer zu tragen.
Dann will ich selber Wasser Dir und Brot
Und roten Wein als Reisezehrung geben,
Mit Kleidern dich versehn und günst'gen Wind
Dir in den Rücken senden, daß du sicher
Dein Heimatland erreichst, sofern die Götter
Die himmlischen es irgend dir gestatten;
Denn sie sind stärker, nur auf sie kommt's an!"
So sprach sie, doch der Held erschrak darüber

[*] Ein hervorragender Vertreter der Altertumswissenschaft an der Universität Freiburg i. Br. hat uns freundlichst die Bekanntschaft mit einigen Gesängen der bisher nicht veröffentlichten Odysseeübersetzung des Professors der Mathematik an der Freiburger Universität *Ernst Zermelo* vermittelt. Unsere Leser werden an der hier gebotenen Probe erkennen, daß der Verfasser nicht nur über sichere Sprachkenntnisse, sondern auch über eine nicht alltägliche Einfühlungsgabe und Ausdrucksfähigkeit verfügt; eine mit philologischer Akribie verfaßte wortgetreue Musterübersetzung hat er nicht beabsichtigt. ⟦English translation on p. 454⟧

[1] ⟦The line numbers given here correspond to the Greek text and are not found in Zermelo's published translation.⟧

From Homer's *Odyssey*

Translation *Barnard 1876,* 82-85

Od. V 149–225[1]

Soon as the nymph adorable had heard
Jove's message, she to brave Ulysses went.
She found him on the shore; his eyes from tears
Were never dry; his sweet life ebbed away,
In grief for home; the nymph no longer pleased,
Though by her in the cave at night he slept
Perforce, unwilling by a willing spouse.
By day he sat upon the rocks and beach,
Vexing his mind with tears, and groans and griefs,
And weeping looked upon the barren sea.
The goddess standing near him thus addressed:
"Weep not, ill-fated one! Nor let your life
Thus pine away. I'll freely let you go.
Come, cut long timbers with an axe and frame
A raft; upon it fix a lofty deck
That it may bear you o'er the gloomy sea.
Food, water and red wine I'll place therein,
To cheer your heart and hunger drive away.
I'll clothe you and will send a fav'ring breeze,
That you may unscathed to your country go,
If the gods want, who in the broad sky dwell,
Stronger than I to plan and execute."
She spoke. Ulysses, man of sufferings great,

[1] ⟦For the convenience of readers who lack German, Barnard's English version of *Odyssey* V 149–225 is reproduced here. Albert Henrichs chose this particular rendition because it closely reflects the spirit and the poetic diction of Zermelo's translation, including his use of blank verse. He also modified Barnard's version slightly to remove the most extreme archaisms such as "thou plannest" (line 174), "thou art" (182, 210), and "didst thou know" (206). The Reverend Mordaunt Roger Barnard (1828–1906) was an Anglican clergyman who translated not only Homer's *Odyssey* but also various works from Danish, Norwegian and Swedish.⟧

Und zweifelnd sprach er nun zu ihr die Worte:
„Da hast du sicher etwas andres vor,
Als jetzt mich heimzusenden, wie du sagst,
Wenn du mich heißest auf so schwachem Floß
Die ungeheure Meerflut zu befahren,
Die doch ein fest gebautes Ruderschiff
Bei günst'gem Winde kaum vermag zu zwingen!
Nein, nimmermehr besteig ich solch ein Floß,
Wofern du mir nicht feierlich gelobst
94 | Durch Eidschwur, Göttin, daß du bei dem Vorschlag
Nicht irgendeine Arglist führst im Schilde!“
So sprach er, doch die Göttin hört es lächelnd
Und streichelte den Helden mit der Hand,
Indem sie sprach: „Du bist mir doch ein Schelm,
Unübertrefflich stets an List und Vorsicht:
Wie klug und fein ist, was Du da geredet!
So hör' es denn die Erde und der Himmel,
Und in der Tiefe drunten hör's der Styx —
Was doch der stärkste Schwur ist bei den Göttern —,
Daß keine Arglist hier im Spiele ist!
Nur das allein hab' ich im Sinne hier,
Was ich auch für mich selbst beschließen würde,
Wär ich in gleicher Lage jetzt wie du!
Denn billig ist mein Sinn, und nicht von Stein
Ist doch mein Herz, nein voller Mitgefühl!“
So sprach die Göttin liebevoll und eilte
Voran zur Grotte und Odysseus folgte.
Sie traten ein, die Göttin und der Mann,
Und setzten sich, er auf denselben Sessel,
Von dem vorhin sich Hermes erst erhoben.[2]
Die Nymphe bot ihm liebreich zur Erquickung,
Was Menschenkinder essen, Speis und Trank,
Dann nahm sie selber Platz ihm gegenüber,
Von ihren Mägden mit der Götternahrung,
Ambrosia und Nektar, wohl versehn,
Und beide wandten sich dem Mahle zu.
Doch als sie sich an Speis und Trank gesättigt,
Begann Kalypso endlich das Gespräch:
„Laertes' edler Sohn, erfindungsreicher!
So willst du wirklich wieder fort von hier
Zur Heimat kehren? Nun, ich wünsch dir Glück!
Doch wüßtest du, wieviel dir noch an Leid

[2] ⟦Zermelo's published translation omitted the lines in italics, which have been reinserted from the full text of Book V as published in the appendix.⟧

Shuddered, and thus with winged words replied:
"Goddess! Surely something else than this
You plan and not at all my voyage home,
Who bid me on a raft the sea's vast depth,
Dread, difficult, to cross, which not e'en ships
Balanced and swift, rejoicing in Jove's breeze,
May pass. I would not 'gainst your will embark
Upon a raft, unless, o goddess, you
Would deign to swear to me a mighty oath
No other ill against me to devise."
Thus as he spoke divine Calypso smiled,
Caressed him with her hand, and thus replied:
"Sure you are crafty, skilled in no vain arts,
Who had the thought to offer such a speech.
Let earth attest, and heaven stretched above,
And water of the Styx which rolls below
(Oath greatest, strongest for the blessed gods),
That I no evil will against you plan.
But I will counsel and contrive for you
What, were there need, I'd purpose for myself.
My mind is upright, and my heart within
My breast not iron but compassionate."
The goddess spoke, and quickly led the way,
And he her footsteps followed as she went,
Entering the cave, the goddess and the man.
Upon the seat whence Hermes rose he sat,
And the nymph placed before him ev'ry food
To eat and drink, such food as mortals eat.
She then before divine Ulysses sat.
The maidens nectar and ambrosia brought
For her; and on the feast their hands they laid.
When they with meat and drink were satisfied,
Divine Calypso thus began to speak:
"Ulysses wise, Laertes' noble son!
Do you to home and your dear native land
Thus long to go? If so, may good betide!
But if you knew what woes it is your fate

Beschieden ist, bevor du sie erreichst:
Du zögst es vor, bei mir im Haus zu bleiben
Als ein Unsterblicher, und hörtest auf,
Nach deiner Gattin immer dich zu sehnen!
Denn doch nicht schlechter bin ich wohl als jene
An Antlitz und Gestalt. Wie könnte auch
Mit Göttern eine Sterbliche sich messen?"
Antwortend sprach Odysseus: „Edle Göttin!
Nicht zürne mir darum! Ich weiß ja selbst
Nur allzu wohl, wie sehr dir unterlegen
An Wuchs und Ansehn ist Penelope:
Ist sie doch sterblich nur, indessen dich
Unsterblichkeit und ew'ge Jugend ziert!
Allein was hilft's? Ich strebe Tag und Nacht
Und sehne mich zur Heimat zu gelangen!
Ja sollt' ein Gott mich auf dem Meer zerschmettern,
So mag er's tun: ich will auch das erdulden!
Denn standhaft ist das Herz mir in der Brust,
Nachdem ich ach! so vieles schon erfahren
An bitt'rem Leid im Krieg und auf dem Meere:
Jetzt mag es kommen, wie es eben will!"
So sprach er, und die Sonne sank ins Meer,
Und dunkel ward's. — — —

To suffer, all before arriving home,
Here would you stay with me and guard this house,
And be immortal, eager though to see
Your wife, whom you are always pining for.
And yet I boast that not inferior
Am I to her in form or countenance,
Nor is it fit in beauty or in shape
For mortals with immortals to contend."
The wise Ulysses answered her and said:
"Goddess adorable! Do not for this
Be angry with me. For I know full well
That chaste Penelope must yield to you
To look upon, in figure and in face,
She mortal, you immortal, ever young.
But even so I wish and long all day
For home, and my returning day to see.
If some god wreck me on the darksome wave
I will endure, and have a patient mind.
Already woes and toils I have endured
In waves and wars; let this to them be joined."
He spoke. The sun went down and darkness came.

Appendix:
Zermelo's translation of Book V of the *Odyssey*†

Kalypso. Das Floß

Als Eos sich erhob vom Rosenlager
Den Göttern und den Menschen Licht zu bringen,
Da saßen sie versammelt zur Beratung,
Die Götter alle und als mächtigster
Der Donnrer Zeus. Zu ihnen sprach Athene,
Der Leiden des Odysseus eingedenk,
Der immer noch im Haus der Nymphe weilte:
„Hört, Vater Zeus und all ihr andern Götter!
Bald wird doch keiner mehr von all den Fürsten,
Die jetzt das Szepter führen auf der Erde,
Gerecht und milde seines Amtes walten,
Nein hart und grausam werden sie hinfort
Nur Unrecht tun und wilde Frevel üben,
Wenn keiner des Odysseus mehr gedenkt
Von seinen Völkern, die er liebreich pflegte,
Wie je ein Vater seine Kinder nur!
Und er, in bitt'rem Harme sich verzehrend,
Er weilt noch immer auf Kalypsos Eiland,
Die ihn gefangen hält: denn ohne Schiffe
Und ohne Fahrgenossen, die ihn rudern,
Kann er die liebe Heimat nie erreichen.
Und nun beschloß man gar, ihm seinen Sohn
Zu töten, wenn er heimkehrt von der Reise,
Der jetzt nach Pylos zog und Lakedämon,
Um Kunde von dem Vater zu erlangen!"
Antwortend sprach zu ihr der Göttervater:
„Mein liebes Kind, wie soll ich das verstehn?
Ist's dir um den Odysseus so zu tun,
So hast Du selbst für ihn den Plan ersonnen,
Wie er heimkehrend Rache nehmen soll!
So führe denn zunächst, du kannst es ja,
Telemachos zurück in seine Heimat,
Daß er von dir behütet sicher sei
Und jene auf dem Schiff das Nachsehn haben!"
So sprach er, und zu seinem lieben Sohn,
Dem Hermes, dann sich wendend, sagte er:
„Nun, Hermes, da du doch mein Bote bist,

† ⟦Zermelo *Nachlass*, C 129/222. The lines published in *1930f* are given here in italics.⟧

So künde du sogleich der schönen Nymphe
Von mir den unabänderlichen Ratschluß,
Daß jetzt Odysseus nach der Heimat kehre,
Von Göttern nicht, von Menschen nicht geleitet,
Auf wohlgefügtem Floß und daß er dann
In zwanzig Tagen leidensreicher Fahrt
Nach Scheria der fruchtbaren gelange,
Wo die Phäaken nah den Göttern hausen.
Die werden dort wie einen Gott ihn ehren
Und ihn zu Schiff nach seiner Heimat senden,
Mit Erz und Gold und Kleidern reich beschenkt,
Wie er sie selbst vor Troja nicht gehabt,
Als er mit aller Beute heimwärts zog.
Denn ihm ist's vorbestimmt, zum Vaterlande
Und zu den Seinen endlich heimzukehren!"
So sprach er, und der Götterbote tat's:
Flugs band er sich die goldenen Sandalen,
Die ihn mit Windeseile über's Meer
Und weites Land hintrugen, an die Füße
Und nahm den Stab, womit er nach Gefallen
Der Menschen Augen bald in Schlaf versenken,
Bald aus dem Schlummer wieder wecken kann.
Mit diesem Stab in Händen flog er nun
Bis nach Pieria, dort ließ er sich
Auf's Wasser nieder, und dann schwebte er,
Hingleitend über's Meer wie eine Möwe,
Die in der salz'gen Flut nach Fischen tauchend
In leisem Aufschlag sich den Fittich netzt:
So glitt der Götterbote über's Meer.
Und als er dann die ferngeleg'ne Insel
In seinem Flug erreicht', stieg er ans Land,
Die schön gelockte Nymphe aufzusuchen.
Er fand sie auch daheim in ihrer Grotte:
Ein lustig Feuer brannte auf dem Herde,
Aus Zedernholz und Lebensbaum entfacht,
Und süßer Duft erfüllte rings den Raum.
Sie aber schritt, mit holder Stimme singend
Und bunte Fäden durch's Gewebe ziehend,
Am Webstuhl hin und her. Doch rings herum
Um ihre Grotte wuchs ein grüner Hain
Von Erlen, Pappeln, duftenden Zypressen,
Und Vögel nisteten dort im Gezweig:
Wie Eulen, Falken und geschwätz'ge Krähen,
Die auf dem Meere ihre Nahrung finden.

Doch um die Höhle selber rankte sich
Ein Weinstock üppig und von Trauben schwer.
Vier Quellen strömten nahe bei einander
Hierhin und dorthin silberklares Wasser.
Und schwellend Wiesenland umgab das Ganze,
Von Veilchen duftend und von würz'gem Eppich.
Hier staunte selbst ein Gott, der dies erblickte,
Und freute sich an all der Herrlichkeit.
So stand auch staunend jetzt der Götterbote.
Doch als er sich an allem satt gesehen,
Betrat er schnell die Grotte, und Kalypso
Erkannt' ihn auf den ersten Blick als Gott;
Denn leicht erkennen sich die Himmlischen,
Wenn einer noch so fern den andern wohnt.
Nur den Odysseus traf er nicht zu haus:
Denn wieder saß er, wie er immer pflegte,
Am Meeresstrand, im Kummer sich verzehrend,
Und schaute bitt'rer Tränen voll ins Weite.
Den Hermes aber fragte nun Kalypso,
Sobald er auf dem Sessel Platz genommen:
„Warum wohl, Hermes, kommst du jetzt zu mir,
Du hochgeehrter, lieber Götterbote?
Denn früher hast du mich doch nie besucht.
Was führt dich her? Ich will dir gerne dienen,
Wenn ich's vermag und wenn's erfüllbar ist!
Doch erst laß dich bewirten, lieber Gast!"
So sprechend rückte sie ihm einen Tisch
Mit Götterspeise vor und mischt' ihm Nektar,
Und schweigend aß und trank der Götterbote.
Nachdem er sich an Speis und Trank erfrischt,
Begann er nun die Rede zu der Nymphe:
„Wenn du als Göttin mich, den Gott, befragst,
Warum ich kam, so muß ich dir's verkünden,
Getreulich wie es ist, du willst es so!
Auf Zeus' Gebot, nicht eig'nen Antrieb komm ich.
Denn wer durchschritte gern die salz'ge Flut
Des öden Meers, wenn weit und breit kein Ort ist,
Wo man die Götter ehrt mit Opferfesten?
Doch keiner aller Götter könnte je
Dem Ägisträger sich entgegenstellen,
Um seinen festen Ratschluß zu vereiteln.
Nun sagt er, daß bei dir der Held verweile,
Der unglückseligste von allen denen,
Die einst um Troja stritten viele Jahre.

Den heißt dich Zeus jetzt ungesäumt entlassen:
Denn ihm ist's nicht bestimmt, im fremden Land
Zu sterben, sondern dies ist sein Geschick:
Zurückzukehren endlich in die Heimat,
Die Seinen und sein stattlich Haus zu sehn!"
So sprach er, und Entsetzen faßte sie,
Und schreckensbleich erwiderte die Göttin:
„Wie hart und grausam seid ihr doch, ihr Götter,
Daß ihr es einer Göttin nie erlaubt,
Sich einem Sterblichen als Weib zu schenken!
Als Eos den Orion sich gewann,
Da zürntet ihr der rosenfarbnen Göttin,
Bis auf Ortygia er dem Geschoß
Der jungfräulichen Artemis erlag.
Als Demeter einst dem Iasion
In zarter Herzensneigung sich gesellte,
Auf wohlgepflügtem Saatfeld ihn umarmend,
Ergrimmte Zeus und traf ihn mit dem Blitze!
Und so mißgönnt ihr auch jetzt mir, ihr Götter,
Mit einem Sterblichen den Bund zu schließen;
Und hab' ihn selber doch vom Tod errettet,
Als er auf schwachen Balken hilflos trieb,
Nachdem ihm mitten auf dem Meere Zeus
Mit seinem Feuerstrahl das Schiff zerschmettert:
Ich nahm ihn liebreich auf und pflegte ihn,
Versprach ihm auch, wenn er nur bei mir bliebe,
Unsterblichkeit und ew'ge Jugendfrische.
Doch da es keinem Gotte möglich ist,
Dem Ägisträger sich zu widersetzen
Und seinen festen Ratschluß zu vereiteln,
So geh er nur, wie jener es gebietet,
Auf's unbegrenzte Meer, zu seinem Unheil!
Zwar selber ihn begleiten kann ich nicht:
Hab ich doch Schiffe nicht noch Ruderknechte.
Doch will ich gern ihn fördern und beraten,
Wie er die Heimat unversehrt erreiche."
Antwortend sprach zu ihr der Götterbote:
„So send ihn denn von hinnen, wie Du sagst.
Doch hüte dich, den Donn'rer Zeus zu reizen
Und seine Rache dir auf's Haupt zu laden!"
So sprach er drohend und entfernte sich.
Die schöne Nymphe aber ging hinaus,
Nachdem sie den Befehl des Zeus vernommen,
Um den Odysseus draußen aufzusuchen.

Sie fand ihn wieder am Gestade sitzend,
Und nimmer wurden seine Augen trocken:
So floß sein Leben hin in Sehnsuchtstränen
Nach seiner fernen Heimat. Denn die Nymphe,
So schön sie war, gefiehl im längst nicht mehr,
Sodaß er ungern nur und liebeleer
Das Lager teilte mit der Liebenden.
Bei Tage saß er immer auf dem Felsen
Und schaute weinend auf das weite Meer.
Jetzt trat sie freundlich zu ihm hin und sprach:
„Du Armer! Weine nicht und gräm dich nicht!
Du sollst nicht mehr dein Leben hier vertrauern!
Wenn du es denn begehrst, ich laß dich ziehn!
So hau dir Bäume flugs mit scharfer Axt
Und füge sie zum breiten Floß zusammen.
Bordbalken lege drauf so hoch und fest,
Dich sicher durch das wilde Meer zu tragen.
Dann will ich selber Wasser Dir und Brot
Und roten Wein als Reisezehrung geben,
Mit Kleidern dich versehn und günst'gen Wind
Dir in den Rücken senden, daß du sicher
Dein Heimatland erreichst, sofern die Götter
Die himmlischen es irgend dir gestatten;
Denn sie sind stärker, nur auf sie kommt's an!"
So sprach sie, doch der Held erschrak darüber
Und zweifelnd sprach er nun zu ihr die Worte:
„Da hast du sicher etwas andres vor,
Als jetzt mich heimzusenden, wie du sagst,
Wenn du mich heißest auf so schwachem Floß
Die ungeheure Meerflut zu befahren,
Die doch ein fest gebautes Ruderschiff
Bei günst'gem Winde kaum vermag zu zwingen!
Nein, nimmermehr besteig ich solch ein Floß,
Wofern du mir nicht feierlich gelobst
Durch Eidschwur, Göttin, daß du bei dem Vorschlag
Nicht irgendeine Arglist führst im Schilde!"
So sprach er, doch die Göttin hört es lächelnd
Und streichelte den Helden mit der Hand,
Indem sie sprach: „Du bist mir doch ein Schelm,
Unübertrefflich stets an List und Vorsicht:
Wie klug und fein ist, was Du da geredet!
So hör' es denn die Erde und der Himmel,
Und in der Tiefe drunten hör's der Styx —
Was doch der stärkste Schwur ist bei den Göttern —,

Daß keine Arglist hier im Spiele ist!
Nur das allein hab' ich im Sinne hier,
Was ich auch für mich selbst beschließen würde,
Wär ich in gleicher Lage jetzt wie du!
Denn billig ist mein Sinn, und nicht von Stein
Ist doch mein Herz, nein voller Mitgefühl!"
So sprach die Göttin liebevoll und eilte
Voran zur Grotte und Odysseus folgte.
Sie traten ein, die Göttin und der Mann,
Und setzten sich, er auf denselben Sessel,
Von dem vorhin sich Hermes erst erhoben.
Die Nymphe bot ihm liebreich zur Erquickung,
Was Menschenkinder essen, Speis und Trank,
Dann nahm sie selber Platz ihm gegenüber,
Von ihren Mägden mit der Götternahrung,
Ambrosia und Nektar, wohl versehn,
Und beide wandten sich dem Mahle zu.
Doch als sie sich an Speis und Trank gesättigt,
Begann Kalypso endlich das Gespräch:
„Laertes' edler Sohn, erfindungsreicher!
So willst du wirklich wieder fort von hier
Zur Heimat kehren? Nun, ich wünsch dir Glück!
Doch wüßtest du, wieviel dir noch an Leid
Beschieden ist, bevor du sie erreichst:
Du zögst es vor, bei mir im Haus zu bleiben
Als ein Unsterblicher, und hörtest auf,
Nach deiner Gattin immer dich zu sehnen!
Denn doch nicht schlechter bin ich wohl als jene
An Antlitz und Gestalt. Wie könnte auch
Mit Göttern eine Sterbliche sich messen?"
Antwortend sprach Odysseus: „Edle Göttin!
Nicht zürne mir darum! Ich weiß ja selbst
Nur allzu wohl, wie sehr dir unterlegen
An Wuchs und Ansehn ist Penelope:
Ist sie doch sterblich nur, indessen dich
Unsterblichkeit und ew'ge Jugend ziert!
Allein was hilft's? Ich strebe Tag und Nacht
Und sehne mich zur Heimat zu gelangen!
Ja sollt' ein Gott mich auf dem Meer zerschmettern,
So mag er's tun: ich will auch das erdulden!
Denn standhaft ist das Herz mir in der Brust,
Nachdem ich ach! so vieles schon erfahren
An bitt'rem Leid im Krieg und auf dem Meere:
Jetzt mag es kommen, wie es eben will!"

So sprach er, und die Sonne sank ins Meer,
Und dunkel ward's. Da gingen sie zusammen
Ins Innerste der Grotte, ruhten dort
Und freuten sich der Liebe mit einander.
Als dann der Morgen kam und beide weckte,
Da zog Odysseus Rock und Mantel an,
Die schöne Nymphe aber hüllte sich
In schimmernde Gewänder zart und fein,
Schlang um die Hüften sich den gold'nen Gürtel
Und warf sich über's Haupt das Schleiertuch.
Nun sorgte sie für des Odysseus Heimkehr:
Sie gab ihm eine mächt'ge Doppelaxt,
Auf beiden Seiten scharf gewetzt und handlich
Mit einem Stiel aus schönem Oelbaumholz.
Dann gab sie ihm ein wohlgeschliff'nes Beil
Und führte ihn zu jenem Teil der Insel,
Wo hohe Bäume wachsen am Gestade,
Wie Erlen, Pappeln, himmelhohe Tannen,
Und viele trocken, dürr und gut zum Schiffbau.
Nachdem sie so die Bäume ihm gezeigt,
Ging sie allein zurück nach ihrer Wohnung.
Er aber fällte nun die hohen Bäume,
Das ging ihm hurtig, zwanzig an der Zahl,
Behaute sie darauf mit eh'rner Axt
Und schlichtete die Balken nach der Richtschnur.
Mit Bohrern, die ihm dann Kalypso brachte,
Durchbohrte er die Balken und verband sie
Mit Pflock und Klammer, wie es sich gehört.
So breit, wie ein geschickter Schiffbaumeister
Den Boden eines Lastschiffs machen würde,
So breit auch zimmert' sich ein Floß Odysseus.
Dann baute er den Bord mit dichten Rippen
Und schloß ihn ab mit aufgelegten Balken,
Errichtete den Mast mit Segelstangen
Und schuf zum Lenken sich ein Steuerruder.
Dann macht' er alles wasserdicht mit Flechtwerk
Und schüttete zuletzt den Ballast auf.
Nun brachte ihm Kalypso große Tücher,
Aus denen er geschickt sich Segel schnitt.
Drauf knüpfte er die Brassen und die Schoten
An Borde fest und wälzte dann das Floß
Mit Hebebäumen in die salz'ge Flut.
Am vierten Tage war das Werk vollendet.
Am fünften drauf entließ die Göttin ihn,

Mit Kleidern wohlversehn und frisch gebadet.
Zwei Schläuche gab sie ihm als Reisezehrung
Gefüllt mit Wasser und mit rotem Wein,
Und einen Ledersack mit Mundvorrat.
Dann ließ sie eine günst'ge Brise wehn,
Unschädlich und gelind. Da lachte ihm
Das Herz, und freudig spannte er die Segel
Und saß am Steuer, klug das Schiff zu lenken.
Ja, auch des Nachts schloß er die Augen nicht,
Nach den Plejaden schauend unverwandt
Und dem Böotus, der so spät erst schwindet,
Sowie dem Bären, der auch Wagen heißt
Und, dem Orion immer zugewendet,
Sich um sich selber dreht und der allein
Sich niemals badet im Okeanos.
Denn den zur Linken immer sich zu halten,
Dies hatte ihm Kalypso eingeschärft.
So fuhr er siebzehn Tage über's Meer.
Am achtzehnten schon tauchten wie ein Schild
Die nächsten Berge des Phäakenlandes
Als Schatten aus der nebelgleichen Flut.
Da kehrte vom Aethiopienland zurück
Der Erderschütt'rer und erblickte ihn,
Wie er da fuhr von den Bolymer Bergen,
Und stärker noch als je ergrimmte er
Und schüttelte sein lockig' Haupt und sprach:
„Was seh' ich? Ha! so haben denn die Götter,
Indeß ich fern bei den Aethiopen weilte,
Mit dem Odysseus anders es beschlossen:
Schon ist er nahe dem Phäakenlande,
Wo seiner Leiden Ende ihm bestimmt ist.
Doch vorher soll er noch, das schwör ich ihm,
Des Jammers vollgerüttelt' Maß erdulden."
So sprechend zog die Wolken er zusammen
Und nahm den Dreizack grimmig in die Hand,
Das ganze Meer in Aufruhr zu versetzen.
Von allen Seiten ließ er Stürme brausen
Und hüllte Land und Meer in Finsternis.
Da tobten denn zugleich der Ost und Süd,
Der Nord und West, gewalt'ge Wogen türmend!
Und das Entsetzen faßte bald Odysseus,
Und stille stand sein Herz, und stöhnend sprach er:
„Unsel'ger der ich bin! wie wird das enden?
So hätte doch die Göttin wahr gesprochen,

Die mir so bitt're Leiden auf dem Meere
Vorausgesagt, eh' ich zur Heimat käme!
Dies alles, fürcht' ich, wird sich hier erfüllen!
Wie hat doch Zeus den ganzen Himmel mir
Mit Wolken dicht bedeckt! Wie tobt das Meer!
Von allen Seiten brausen jetzt die Stürme,
Und unentrinnbar scheint mein Untergang!
Wie waren dreimal glücklich die Achäer,
Die auf dem weiten Feld vor Troja fielen
Im Kampf für die Atriden! Und ich selbst,
O hätt ich damals doch den Tod erlitten,
An jenem Tage mein Geschick erfüllt,
Als ich im wilden Sturm der Troerlanzen
Für den erschlagenen Peliden stritt!
Dort hätte man mich feierlich bestattet
Und meinen Ruhm verkündet; aber hier,
Da muß ich schmählich auf dem Meere enden!"
Wie er so sprach, da traf die Woge ihn
Mit fürchterlicher Wucht von oben her
Und riß das Floß herum, ihn selber aber
Ihn warf sie über Bord ins weite Meer.
Das Steuerruder glitt ihm aus den Händen,
Den Mastbaum brach der wilde Wirbelsturm,
Und weithin flogen Raa und Segelfetzen.
Er aber tief im Wasser brauchte lange,
Sich aus dem Wogenschwall emporzukämpfen,
Da ihn die Kleider lastend niederzogen,
Die ihm Kalypso mitgegeben hatte.
Da endlich taucht' er auf, und aus dem Munde
Quoll ihm das Wasser, strömt' ihm schwer vom Haupt.
In seiner Not vergaß er nicht das Floß:
Nachschwimmend faßt' er's noch und sprang hinauf,
Und so entrann er glücklich dem Verderben.
Doch hier- und dorthin trieb ihn nun die Flut.
So wie im Herbst der rauhe Nord die Diesteln
Weit über's Feld hintreibt in dichten Knäueln,
So trieben jetzt die Winde über's Meer
Hierhin und dorthin das gebroch'ne Floß.
Bald warf der Süd dem Nord es hin zum Spiele,
Bald überließ der Ost dem West die Beute.
Da sah ihn Ino, Kadmos' schöne Tochter,
Leukothea, die, einst ein Menschenkind,
In Meerestiefen jetzt als Göttin weilte.
Und voll Erbarmen sah sie den Odysseus,

Wie er dahintrieb so in seiner Not,
Und setzte sich auf's Floß zu ihm und sprach:
„Du Armer, ach! wie ist denn das gekommen,
Daß dir der Erderschütt'rer also zürnt,
Um soviel Schreckliches dir zu bereiten?
Doch soll er dich nicht ganz und gar verderben,
Wie er auch zürnt und tobt in seinem Grimm!
Nein, mach es so und merk', was ich dir sage:
Zieh erst die Kleider aus, die du da hast,
Und laß das Floß nur in den Winden treiben.
Dich selber aber wirf dann in die Flut
Und mit den Armen rudernd schwimm an's Land,
Wo dir die Rettung winkt bei den Phäaken.
Hier aber nimm den zauberkräft'gen Schleier
Und wind' ihn um den Leib: er rettet dich!
Doch wenn du dann das feste Land erreicht,
So lös ihn ab und wirf ihn hinter dich,
So weit du kannst, in's off'ne Meer hinaus!"
Mit diesen Worten gab sie ihm den Schleier
Und tauchte wie ein Wasserhuhn ins Meer,
Wo sie alsbald in dunkler Flut verschwand.
Odysseus aber blieb in bangem Zweifel,
Und überlegend sprach er so zu sich:
„Ach ist das nicht schon wieder solch ein Trug,
Wenn mich ein Gott verleiten will, das Floß,
Das mich hier birgt, freiwillig zu verlassen?
O nein, das tu ich nicht, denn schon erblick' ich
Von fern das Land, das doch mich retten soll,
Das beste wird wohl sein, ich mach es so:
Solang' das Floß mir noch zusammenhält,
Verharr ich drin, es komme, was da will.
Doch hat es erst die wilde Flut zerschmettert,
So bleibt mir nichts mehr übrig als zu schwimmen!"
Indem er dies noch zweifelnd überlegte,
Da ließ Poseidon eine mächt'ge Woge
Groß und gewaltig auf ihn niederstürzen.
Und wie ein Windstoß fährt in trock'ne Spreu
Und wirbelnd alles aus einander jagt,
So riß die Flut die Balken aus einander.
Odysseus aber schwang sich auf den einen
Und saß nun rittlings wie auf einem Pferde.
Kalypsos Kleider riß er sich herunter
Und wand sich um den Leib den Zauberschleier.
Dann stürzt' er sich kopfüber in die Flut,

Zum Schwimmen weit die Arme ausgebreitet.
Poseidon sah's und grimmig nickte er,
Und bei sich selber sprach der Erderschütt'rer:
„So treibe denn dahin in Not und Jammer,
Bis du die Menschen findest, die dich retten!
Doch sollst du vorher noch genug erdulden!"
So sprechend ließ er seine Rosse jagen,
Daß ihre Mähnen flogen, über's Meer,
Bis er nach Aega kam zu seinem Hause.
Jetzt griff Athene ein, die Tochter Zeus':
Erst hemmte sie den Lauf der andern Winde,
Bis sie gebändigt sich zur Ruhe legten.
Dann ließ sie einen steifen Nordwind wehen
Und brach ihm durch die Wellen glatte Bahn,
Damit Odysseus beim Phäakenvolke,
Dem Tod entronnen, endlich Rettung fände.
Zwei Tage und zwei Nächte trieb er schon
Im Wogenschwall umher, den Tod vor Augen.
Doch als der dritte Tag nun endlich anbrach,
Da legte sich der Wind, und glatte Stille
Umgab ihn rings, und nahe schien das Land,
Wie er mit scharfen Blicken vorwärts spähte.
Und wie die Kinder ihres Vaters Heilung
Mit heller Freude grüßen, der so lange
In schwerer Krankheit schmerzgefoltert lag,
Von einem bösen Dämon heimgesucht,
Doch von den Göttern glücklich noch gerettet:
So atmete Odysseus freudig auf,
Als er von Weitem Fels und Wald erblickte,
Und kräftig rudernd schwamm er nun voran,
Um endlich auf das feste Land zu steigen.
Doch als er schon so nah herangekommen,
Wie eines Menschen Stimme reicht, da hört' er
Ein fürchterlich' Getöse von den Klippen:
Denn mächtig schlug die Flut, aus Felsgestein
Aufspritzend, bis in Schaum sich alles löste.
Nicht Buchten gab's noch Reeden für die Schiffe,
Nur Klippen rings und starrend' Vorgebirg'!
Das fuhr auch dem Odysseus in die Glieder,
Und lähmendes Entsetzen packte ihn,
Sodaß er tief aufseufzend also sprach:
„Ach was ist das? Kaum hat mich Zeus das Land,
Das heiß ersehnte, endlich blicken lassen,
Und hab' mich glücklich durch die Flut gekämpft,

Da find' ich keinen Ausweg aus dem Meere:
Nur scharfe Klippen rings und glatte Felswand,
Und alles überschäumt von Brandungswellen!
Und nirgends flaches Land, um Fuß zu fassen
Und so sich dem Verderben zu entziehn!
Wag' ich mich hier heran, wird mich die Brandung,
So fürcht' ich, gegen scharfe Klippen schleudern,
Und jeder Widerstand ist da vergeblich.
Doch such' ich weiterschwimmend flache Stellen
Und Meeresbuchten auf, so fürcht' ich, wird
Auf's Neue mich der wilde Sturm erfassen
Und weit hinaus ins off'ne Meer mich treiben,
Erschöpft wie ich da bin und mühsam atmend!
Vielleicht sogar schickt mir ein Ungeheuer,
Wie doch so viele birgt der Meeresschlund,
Ein Gott, mich zu verschlingen, aus der Tiefe:
Weiß ich doch wohl, wie feindlich mir Poseidon!"
Indem er dies noch zweifelnd überdachte,
Da warf ihn eine Riesenwelle schon
Aufs zackige Gestein der Felsenküste
Und hätt' ihn dort zerschunden und zerschmettert,
Hätt' ihm Athene nicht ins Herz gegeben,
Den Fels mit beiden Händen zu umfassen.
So hing er keuchend an der spitzen Klippe,
Bis ihn die Unglückswelle überholt;
Und so entrann er diesmal dem Verderben.
Doch wiederkehrend traf sie ihn von vorn
Und schleudert' ihn zurück ins off'ne Meer.
Und wie ein Meerpolyp, aus seinem Lager
Herausgerissen, noch an seinen Armen
Saugnäpfe trägt gespickt mit vielen Steinchen,
So blieben auch am Fels von seinen Händen
Hautfetzen noch; und ihn verschlang die Flut.
So hätte denn Odysseus hier geendet
Auch gegen sein Geschick, hätt' ihm Athene
Besinnung nicht gewahrt zu seiner Rettung.
Jetzt aber, aus der Brandung aufgetaucht,
Schwamm er vorbei und spähte nach der Küste,
Ob er nicht flachen Strand und Buchten fände.
So kam er vor die Mündung eines Flusses:
Das schien ihm denn der rechte Ort zum Landen,
Von Felsen frei und sicher vor dem Winde,
Und an den Stromgott sandt' er das Gebet:
„Erhör mich Herr: denn dir vertrau ich mich,

Wer du auch seiest, heißersehnter Strom:
Ich flüchte mich vor des Poseidons Zorn!
Denn heilig ist auch den Unsterblichen,
Wer in der Not sich ihrem Schutz vertraut,
Wie ich jetzt dir in meinem Elend nahe:
So hab Erbarmen denn und rette mich!"
So sprach er, und schon hemmte seinen Lauf
Des Stromes Gott und glättete die Flut
Und nahm ihn rettend auf in seiner Mündung.
Dann sank er hin; es brachen ihm die Kniee,
Und auch die starken Arme lösten sich;
So matt war er und ganz erschöpft vom Schwimmen,
Die Glieder waren alle ihm geschwollen.
Und Wasser rann aus Mund und Nase ihm.
So lag er ohne Atem, ohne Stimme
Ohnmächtig da, von Schwäche ganz bewältigt.
Doch als er atmend wieder zu sich kam,
Da löste er sich ab den Zauberschleier
Und warf ihn nah der Mündung in den Fluß,
Wo ihn die Strömung schnell ins Meer entführte,
Und Ino fing ihn auf mit zarten Händen.
Odysseus aber warf sich in das Schilf
Und küßte weinend die geliebte Erde.
Doch seufzend sprach er zu sich selber dann:
„O weh! Jetzt geht's mir schlimm! Wie wird das enden?
Verbleib ich hier die ganze Nacht am Flusse,
So find ich noch den Tod in Frost und Nebel,
Entkräftigt wie ich bin und ohne Nahrung:
Denn in der Frühe weht ein eis'ger Wind!
Doch wenn ich hier den Hügel aufwärts klimme,
Um frostgeschützt im Busch zu übernachten,
Und dann erschöpft in tiefen Schlummer sinke,
So fall' ich wilden Tieren leicht zur Beute!"
Wie er so nachsann, schien es ihm das beste,
Zum Wald hinaufzusteigen, der da oben
Sich weithin sichtbar über'm Wasser zeigte.
Zwei Büsche fand er dort aus einer Wurzel,
Oelbaum und Wegdorn, engverwachsen vor.
Kein feuchter Wind durchdrang das grüne Laubwerk,
Noch je ein Sonnenstrahl und auch kein Regen:
So dicht war das Gestrüpp. Hier kroch er unter
Und machte sich mit eig'ner Hand ein Lager,
Bequem und warm aus aufgehäuften Blättern,
Die ringsumher den Boden ganz bedeckten,

Genug, zwei Männer oder drei zu schützen
Vor schärfstem Winterfrost. Da lachte ihm
Das Herz, wie er das Lager fertig sah,
Und tief aufseufzend warf er sich hinein
Und deckte sich mit Blättern ganz und gar.
So wie ein Landmann auf entleg'nem Feld,
Wo weit und breit kein Nachbar wohnt, das Feuer
Mit schwarzer Asche sorglich überdeckt,
Um über Nacht die Glut sich zu bewahren:
So hüllte sich Odysseus ganz in Blätter.
Athene aber sandt' ihm tiefen Schlaf,
Der neue Kraft dem Müden geben sollte,
Und schloß die Augen ihm mit leiser Hand.

Introductory note to s1931b, s1931c, Gödel 1931b, and s1931d

Heinz-Dieter Ebbinghaus

Through *Skolem 1930* Zermelo became aware of Thoralf Skolem's result (*1923*) that first-order set theory, if consistent, admits a countable model. Viewing axiomatic set theory as a foundation of Cantorian set theory with its unlimited progression of infinite cardinalities, Zermelo flatly rejected the basis of Skolem's argument, the first-order formulation of the axioms of separation and replacement.[1] In order to overcome the weakness of first-order logic, he started to realize a program that is charted in his theses concerning the infinite in mathematics, *s1921*: the development of infinitary languages and an infinitary logic as a means of giving mathematics a foundation which would preserve its "true" character. A second line he pursued consisted in developing set theory without Skolem's limitations on separation, namely by "keeping with the true spirit of set theory, [allowing] for the *free* division, and [postulating] the existence of all [subsets] formed in an arbitrary way" (*s1930d*); this resulted in his second-order axiom system of set theory and the cumulative hierarchy as developed in his *1930a*.

On 13 May 1931, about seven weeks after the appearance of *Gödel 1931a*, Reinhold Baer informed Zermelo about Kurt Gödel's first incompleteness theorem: Any consistent finitary axiom system of sufficient number-theoretic strength is incomplete in the sense that it admits propositions neither provable nor refutable in it. To put his considerations in concrete form, Gödel treated number theory in the system of *Principia mathematica*.

Zermelo viewed Gödel's result as further evidence for the inadequacy of any finitary approach to the foundation of mathematics. In mid-1931 his work on infinitary languages had progressed to such an extent that he decided to give a talk about it at the annual meeting of the Deutsche Mathematiker-Vereinigung (German Mathematical Union) in September 1931 in the small health resort of Bad Elster. When he learnt that Gödel had announced a talk on his incompleteness results there as well, Zermelo arranged that the two talks be scheduled together, his own following that of Gödel, and that there be a common discussion that should provide a forum for his struggle against the direction embodied by Gödel. Moreover, Zermelo tried to garner support. In a letter to an unknown addressee, perhaps Arnold Scholz,[2] he wrote:

> Will you really stay in your summer resort during the entire vacation or would you not prefer to come to Bad Elster for the meeting? It would be very important to me to have one or the other in the audi-

---

[1] Cf. *Ebbinghaus 2004* for Zermelo's immediate reaction and for further details.

[2] Carbon copy without first page in Universitätsarchiv Freiburg, Zermelo *Nachlass*, C 129/268.

ence who has read (and understood!) my Fundamenta paper ⟦*1930a*⟧. For I am sure that with my "infinitary logic" I will be faced with dissent from all sides: neither the "intuitionists" (of course, they are enemies of logic anyway) nor the formalists nor the Russellians will accept it. I only count on the young generation; their attitude towards such things is less prejudiced.

The letter *s1931c* to Reinhold Baer

Zermelo's letter *s1931c* to Baer was written about three weeks after the Bad Elster meeting. Above all, it gives a vivid impression of Zermelo's firm intention to work against the finitary approaches of Skolem and Gödel and also against intuitionism in a situation where "frivolous dilletantism [was] set to discredit the entire field". The letter reveals the measures he was going to take and shows that he drew strength from two points: a fallacious refutation of Gödel's argument (see the next section) and the convincing role of the axiom of foundation, "my 'principle of foundation' ": the axiom provides the basis for both the conception of the cumulative hierarchy of sets and for the hierarchy of his infinitary languages[3] and, hence, for the essentials of his anti-finitary program.

Zermelo reports on the two different notes he was publishing about his Bad Elster talk, an "*entirely* positive" one, *1932b*, which should be "intelligible to everybody" and was to appear in the scholarly magazine *Forschungen und Fortschritte*,[4] and a polemical one, *1932a*, for the *Jahresbericht der Deutschen Mathematiker-Vereinigung* that was intended to open the wanted discussion. Zermelo's statement that during the conference he avoided polemics against Gödel is supported by Olga Taussky-Todd; she speaks of a "peaceful meeting" (*1987*, 38) both had during a lunch break.

Zermelo complains that the deliberately planned common discussion after Gödel's and his talk in the afternoon of 15 September did not take place because of an "illicit" proposal made by Fraenkel. Ten years later, in his letter *s1941* to Paul Bernays, Zermelo would give a different reason, a "plot engineered by the Vienna Circle represented by Hahn and Gödel". Zermelo's 1931 view may have been influenced by the fact that by then the relationship between him and Fraenkel was suffering from differences about Fraenkel's contribution to the collected works of Cantor which Zermelo was editing (cf. *Ebbinghaus 2007b*, 158 ff).

One might wonder why Zermelo did not acknowledge Skolem's and Gödel's approaches at least for pragmatic reasons; for Skolem had demonstrated that first-order definiteness was sufficient to carry out all ordinary set-theoretic proofs, and Gödel had shown that finitary mathematics suffered from inherent weaknesses. Instead, Zermelo developed a strong feeling that

---

[3] Cf. (the introductory notes to) *1930a* and to *1932a etc.*

[4] Zermelo included a copy in the letter.

these approaches amounted to a severe attack against mathematics and that he, Zermelo, was in charge of fighting back in order to preserve mathematical science from damage.

The hint at "Schoenflies and his ilk" near the end of the letter may refer to widespread objections which Zermelo's first well-ordering proof and the axiom of choice had met.[5] It reveals that Zermelo viewed his struggle as a new version of that for the axiom of choice some 25 years earlier. However, whereas the dispute about the axiom of choice saw Zermelo as the victor, his struggle against a finitary foundation of mathematics failed entirely. There are several reasons. In the first place, he did not strive for a presentation of his counterarguments with a standard of precision as was exercised by Gödel and Skolem. Secondly, his epistemological engagement prevented him from considering the results of Gödel and Skolem in an unprejudiced way as mathematically impressive theorems, striving for their technical understanding, and only then pondering their epistemological meaning and utilizing the analyzing power they provide. As it was these results which shaped the discipline of mathematical logic in the 1930s, Zermelo placed himself outside the mainstream of mathematical foundations.

The Gödel correspondence

Before commenting on their letters, it may be useful to remember the basic differences between Zermelo and Gödel. For Zermelo a proposition or a proof of it is not a syntactic string of signs, but an ideal (infinitary) object. For Gödel's procedure the finitary representability of mathematical systems is indispensible. For only then can one code axioms, propositions, and proofs in an effective way by natural numbers, and the presupposed arithmetical power of the systems then allows provability to be treated in the systems themselves. By imitating the paradox of the liar on the syntactic level, thereby replacing truth by provability, one can finally infer the existence of undecidable propositions. For Zermelo the finitary point of view renders incompleteness a trivial fact: As the set of provable propositions in Gödel's sense is countable and as there are uncountably many true infinitary propositions,[6] there must be a true proposition which is not provable in Gödel's sense.

As described in the letter to Baer, Zermelo's plan for a common discussion after his and Gödel's talk at the Bad Elster meeting failed. But immediately afterwards Zermelo's attitude for fighting the finitary point of view was strengthened again when he thought he had found a gap in Gödel's argument. In the letter to Baer he writes:

> Once I have publicly declared Gödel's much-admired "proof" to be nonsense, the gentlemen will *have* to show their true colors; for, with

[5] With respect to Schoenflies, cf., for example, *Zermelo 1908a*, sect. d.
[6] The true propositions of his transfinite hierarchies of infinitary propositions.

it, Skolemism entirely collapses particularly in the form of Carnap's "PM-system".

The letter *s1931b* to Gödel, written on the weekend after the Bad Elster talks, describes a fallacious refutation which deviates from that indicated in the letter to Baer (and discussed below). Referring to *Gödel 1931a*, Zermelo makes the error of treating Gödel's $[R(n) : n]$, i.e. a name for a proposition, as a proposition and of conflating provability and truth. Having obtained the apparent contradiction, he attributes the failure to the "finitistic prejudice", namely the "erroneous assumption that every mathematically definable notion be expressible by a 'finite' combination of signs", and expresses his will to fight against the finitary point of view, at the same time trying to win Gödel as a witness and comrade.

Zermelo's error is a basic one. But such deficiencies were widespread among logicians of that time. Ivor Grattan-Guinness (*1979*, 296) reports that according to Barkley Rosser "it was only with Gödel's theorem that logicians realised how careful they needed to be with this matter." Zermelo never made a clear distinction between propositions and names of propositions and between propositions and their meaning. There is a single note in his *Nachlass* where he seems to draw a distinction between objects and variables for these objects, *s1931g*. However, the elaboration and extension *1935* of the thoughts on infinitary languages laid down in this note did not pursue this distinction.

Gödel's response, *Gödel 1931b*, is a ten-page letter, giving a detailed description of his method and explaining patiently where Zermelo went wrong. There is a point worth making here: On pages 7 and 8 Gödel explains that a weak form of his first incompleteness theorem can be established from the fact that the set of true propositions of the systems in question cannot be defined within the system. As the set of derivable propositions can be defined and as the latter set is contained in the first set, there are true sentences which are not provable. Gödel emphasizes that his *1931a* arguments yield more, namely a construction of undecidable propositions; moreover, the undecidable propositions are of a simple kind, namely arithmetical statements. At the end Gödel offers to provide Zermelo with information about doubts which had appeared to him when reading Zermelo's boundary number paper *1930a*.

Zermelo's answer, *s1931d*, written two weeks later, is short and definitive, without any reaction to Gödel's explanations and offer. Zermelo mainly discusses why he prefers his infinitary view: "[...] all you prove [...] amounts to what *I* always stress as well, namely that a 'finitistically restricted' proof schema *is not sufficient* for 'deciding' the propositions of an uncountable mathematical system." Pointing to Gödel's remark that a finitary proof system needs extension, he continues that in contrast to Gödel and at the very outset, he proceeded from a *more general* schema (his infinitary languages and logic), which did not *need* to be extended first, and—in a handwritten

addition to the typewritten body of the text—that in *his* system "really" ("auch wirklich") *any* proposition whatsoever was decidable.[7]

In the letter *s1931c* to Baer, Zermelo had reported on a further alleged contradiction in Gödel's work: The set of finitary sentences of a system is countable, but Gödel's diagonalization method, leading to propositions *in* the system, would show—in analogy to Cantor's diagonalization procedure—that there are uncountably many (finitary) propositions. The first part of Zermelo's *s1931d*[8] may be viewed as evidence that Zermelo had not yet obtained a clear picture of this situation; it may also be viewed as evidence that his idealistic view of languages was intruding into his considerations in an uncontrolled manner. In any case it shows that Zermelo was not yet willing or able to really follow Gödel's arguments. Gödel's impression of technical weakness, which had already emerged from Zermelo's first letter and may have led to the scholarly style of his *1931b*, became stronger and, together with Zermelo's sharp tone and uncompromising attitude in the Bad Elster report *1932a*, may have contributed to his decision to refrain from an answer.

The course of the correspondence may be considered unfortunate; for a discussion about Zermelo's *1930a* as Gödel had offered it at the end of his letter, could have provided valuable information about Gödel's early view of the cumulative hierarchy.

---

[7] For details about decidability in Zermelo's sense see the introductory note to *Zermelo 1932a*.

[8] It is, perhaps, influenced by Gödel's remark (p. 7 of Gödel's letter) that already according to Cantor's diagonalization method no single formal system could capture mathematics in its entirety.

Brief an Kurt Gödel vom 21. September 1931

s1931b

Freiburg i. Br. 21. IX 31
Karlstraße 60

Sehr geehrter Herr Gödel,

beifolgend sende ich Ihnen einen Abzug meiner Fundamenta-Arbeit und würde mich freuen, wenn ich Sie zu den wenigen zählen dürfte, die es wenigstens versucht haben, die dort entwickelten Gedanken und Methoden aufzunehmen

Despite Zermelo's antagonistic style a dispassionate analysis may lead to the conclusion that there was no irreconcilable gulf between him and Gödel. By ways which could have been mutually acknowledged, both had come to similar results: Gödel had proved in a strict mathematical sense that non-trivial finitary systems are inadequate to capture their richness, and he was open to considering stronger extensions. Zermelo had gained the insight into the incompleteness of finitary systems on grounds of his infinitary convictions and had conceived infinitary systems that according to his opinion were complete and adequate for mathematics. Kanamori (*2004*, 535) describes the situation with its inherent differences concisely as follows:

> For Gödel the incompleteness of formal systems is a crucial *mathematical* phenomenon to be reckoned with but also to transcend, whereas Zermelo, not having appreciated that logic has been submerged into mathematics, insisted on an infinitary logic that directly reflected transfinite reasoning.

The clash left both Gödel and Zermelo hurt. Zermelo's bitterness is still visible ten years later in his letter *s1941* to Paul Bernays. Gödel, unlike Zermelo not a man of controversies, felt completely misunderstood and attacked by scientifically unjustified arguments. Perhaps it was this impression that led him, in his later publications on the constructible hierarchy, to give less attention to Zermelo's boundary number paper *1930a* then could have been expected.[9]

[9] For details see the introductory note to *Zermelo 1930a* or *Ebbinghaus 2007b*, sect. 4.10.

Letter to Kurt Gödel of 21 September 1931

s1931b

Freiburg i. Br. 21. IX 31
Karlstraße 60

Dear Mr. Gödel,

Enclosed I send you a copy of my Fundamenta paper. It would please me if I could count you among those few who have at least tried to absorb the ideas and methods developed there and make them fruitful for their own research. It

und für die eigene Forschung fruchtbar zu machen. Während ich nun damit
beschäftigt war, von meinem in Elster gehaltenen Vortrag ein kurzes Referat
2 zu machen, und dabei auch | auf den Ihrigen Bezug nehmen mußte, kam es
mir nachträglich zum deutlichen Bewußtsein, daß Ihr Beweis für die Existenz
unentscheidbarer Sätze eine wesentliche *Lücke* aufweist. Um einen „unent-
scheidbaren" Satz aufzustellen, definieren Sie auf S. 175 ein „Klassenzeichen"
(eine Satzfunktion mit *einer* freien Variablen) $S = R(q)$ und zeigen dann,
daß weder $[R(q); q] = A$ noch seine Negation $\overline{A}$ „beweisbar" sei. Aber gehört
dann

$$S \equiv \overline{\mathrm{Bew}}\,[R(n); n]$$

wirklich Ihrem „System" an und sind Sie berechtigt, diese Funktion mit $R(q)$
3 zu iden-|tifizieren? Nur, weil es ein „Klassenzeichen" ist? Ich weiß, es folgt
nachher eine ausführliche Theorie der „Klassenzeichen". Aber zur Kritik ge-
nügt hier folgende Überlegung. Lassen Sie in Ihrer Formel (1) die Zeichen-
verbindung „Bew" fort und schreiben dafür

$$n \epsilon K^* \equiv \overline{[R(n); n]} \equiv S^*. \qquad (1^*)$$

Setzen Sie dann wieder $S^* = R(q^*)$, so folgt, daß der Satz

$$A^* = R(q^*; q^*)$$

weder „wahr" *noch* „falsch" sein kann, d. h. Ihre Annahme führt auf einen *Wi-
derspruch*, analog der Russel[[l]]'schen Antinomie. Der Fehler beruht — ebenso
wie in der Richard'schen und der Skolem'schen Paradoxie — auf der (irrigen)
4 | Voraussetzung, daß jeder mathematisch definierbare Begriff durch eine „end-
liche Zeichenverbindung" (nach einem *festen* System!) ausdrückbar sei, also
das, was ich das „finitistische Vorurteil" nenne. In Wirklichkeit steht es ganz
anders, und erst nach Überwindung dieses Vorurteils, die ich mir zur be-
sonderen Aufgabe gemacht habe, wird eine vernünftige „Meta-Mathematik"
möglich sein. Gerade Ihre Beweisführung würde, richtig gedeutet, sehr viel
dazu beitragen und damit der Wahrheit einen wesentlichen Dienst leisten
können. Aber so wie Ihr „Beweis" jetzt steht, kann ich ihn als bindend nicht
anerkennen. Das wollte ich Ihnen frühzeitig mitteilen, um Ihnen Zeit zur
Nachprüfung zu geben.

Mit bestem Gruß

E. Zermelo

was only while I was turning what I had presented in ⟦Bad⟧ Elster into a short paper, in the course of which I also had to refer to your ⟦presentation⟧, that I fully realized that your proof of the existence of undecidable propositions contains an essential *lacuna*. In order to set up an "undecidable" proposition, you define a "class sign" (a propositional function of *one* free variable) $S = R(q)$ on p. 175.[1] Then you show that neither $[R(q); q] = A$ nor its negation $\overline{A}$ are "provable". But does

$$S \equiv \overline{\text{Bew}}\,[R(n); n]$$

then really belong to your "system" and are you justified in identifying this function with $R(q)$? If so, just because it is a "class sign"? I know that you give an elaborate theory of "class signs" later on. But for a criticism the following consideration shall suffice here. Drop the combination of signs "Bew" from your formula (1) and instead write

$$n \epsilon K^* \equiv \overline{[R(n); n]} \equiv S^*. \tag{1*}$$

If you now again set $S^* = R(q^*)$, it then follows that the proposition

$$A^* = R(q^*; q^*)$$

can be *neither* "true" *nor* "false". That is, your assumption leads to a *contradiction*, analogously to Russel⟦l⟧'s antinomy. As with the Richard and Skolem paradoxes, the mistake is due to the (incorrect) assumption that it is possible to represent every mathematically definable concept by means of a "finite combination of signs" (in accordance with a *fixed* system!). In other words, ⟦it is due to⟧ what I call the "finitistic prejudice". Actually, things are quite different, and a reasonable "meta-mathematics" will only be possible once we have overcome this prejudice, a matter which I have made my special duty. Precisely your line of argument, when interpreted correctly, would contribute much to this end, thereby rendering a great service to the cause of truth. But I cannot recognize your "proof" as binding in its present state. I wanted to tell you this at an early stage so that you have time to check it over.

With best regards

E. Zermelo

---

[1] ⟦Zermelo refers to *Gödel 1931a*.⟧

Brief an Reinhold Baer vom 7. Oktober 1931

s1931c

Freiburg 7. X 31. Karlstr. 60

Lieber Herr Baer, beifolgend ein Durchschlag meines *neuen* Berichtes über den Elster-Vortrag, der auf Wunsch für die „Forschungen und Fortschritte“ hergestellt wurde, nach Möglichkeit „gemeinverständlich“, aber schon wegen der Kürze (4 Seiten) gewiß *nicht leicht* verständlich. Der ist natürlich ganz *ohne Polemik* oder historische Bezugnahme, enthält *nur* Positives. Der andere (für die Math. Vrgg⟦.⟧) *soll und muß* Widerspruch erregen bei denen, die es trifft. Denn ich bin meiner Sache ganz sicher und wünsche durchaus, daß die Frage einmal in Fluß kommt. In Bad Elster habe ich noch im Vortrage selbst und nachher eine direkte Polemik gegen Gödel vermieden: regsame Anfänger soll man nicht abschrecken. Ich habe extra Gödels Vortrag *vor* dem meinigen ansetzen lassen und eine *gemeinsame* Diskussion beantragt. Aber die Folge meiner Loyalität war einzig die, daß die ganze Diskussion auf den *un*berechtigten Vorschlag Fraenkels (bei *jeder* Gelegenheit fällt er mir in den Rücken!) *noch* weiter zurückgestellt wurde und dann im Sande verlief. Mein
2 eigener Vortrag fiel dabei ganz unter den Tisch. | Jetzt will ich es anders machen: die Herren *müssen* endlich Farbe bekennen, wenn ich öffentlich erkläre, daß der vielbewunderte Gödel'sche „Beweis“ Unsinn ist; denn mit ihm fällt der ganze Skolemismus insbesondere in der Form des Carnap'schen „PM-Systems“. Die Sache ist doch ganz einfach. Gödel geht von folgenden Voraussetzungen aus: 1) *Annahme A,* daß alle Satzfunktionen $R(n,x)$ des Systems (Gödel schreibt dafür $[R(n);x)]$ eine *abzählbare* Menge bilden, während auch die Variable x auf die *Zahlenreihe* beschränkt ist. Diese Annahme kommt also (verallgemeinert) darauf hinaus, daß die Menge der *Funktionen* dem Wertevorrat der *Variablen x ein-eindeutig zugeordnet* ist. 2) die *Annahme B,* daß man (analog dem Cantor'schen Diagonalverfahren) durch eine über *alle* Satzfunktionen erstreckbare Quantifikation *neue* Satzfunktionen *innerhalb des Systems* definieren könne. Diese *beiden Annahmen* stehen zueinander im *Widerspruch,* da nach Cantor die Menge der Funktionen von *höherer Mächtigkeit ist.* Gödel benutzt gleichzeitig *beide* Annahmen A, B und kann dann natürlich *alles* beweisen. Warum nur die Existenz „unentscheidbarer Sätze“, warum nicht die „Unsterblichkeit der Maikäfer“ oder gar die „Unfehlbarkeit der Fakultäten“? Auf meinen Brief hat er mir noch immer *nicht* geantwortet. Augenscheinlich hat er nichts mehr zu „meckern“.

3 | Könnten Sie nicht einmal in Ihrem Kolloquium in Halle die Frage zur Sprache bringen, indem Sie etwa *gleichzeitig* über Gödels und meinen Vortrag

Letter to Reinhold Baer of 7 October 1931

s1931c

⟦The introductory note just before *s1931b* also addresses *s1931c*.⟧

Freiburg 7. X 31. Karlstr. 60

Dear Mr. Baer, enclosed I send you a copy of my *new* report on the presentation given in ⟦Bad⟧ Elster, which I have been asked to prepare for "Forschungen und Fortschritte" in a form "intelligible to everybody", if possible. But, already on account of its brevity (4 pages), it is certainly *not easily* intelligible. Free of polemics or historical references, the report strikes an *entirely* positive tone. But the other one (for the Mathematical Union) *should and must* draw opposition from those at whom it is aimed. For I feel entirely sure of my ground and would certainly wish for the discussion of the question to at last get under way. During and after my presentation in Bad Elster, I still avoided any direct polemic against Gödel: eager beginners ought not to be deterred. I deliberately requested that Gödel's presentation be scheduled *earlier* than mine and that we have a *joint* discussion session. But the only consequence of my loyalty was that the entire discussion was put off *even* further on the *illicit* proposal by Fraenkel (he backstabs me *whenever* he can!). As a result, the discussion came to naught and my own presentation went completely by the board. This time, I will do things differently: once I have publicly declared Gödel's much-admired "proof" to be nonsense, the gentlemen will *have* to show their true colors; for, with it, Skolemism entirely collapses particularly in the form of Carnap's "PM-system". The matter is really very simple. Gödel starts from the following assumptions: 1) *Assumption A* that all propositional functions $R(n, x)$ of the system (or, with Gödel, $[R(n); x]$) form a *countable* set, where the variable x, too, is restricted to the *number series*. This assumption (when generalized) thus amounts to the assumption that the set of the *functions* is *correlated one-to-one* with the supply of values of the *variable* x. 2) the *assumption B* that (analogously to Cantor's diagonal procedure) *new* propositional functions can be defined *within the system* by means of a quantification extendable over *all* propositional functions. These *two assumptions* stand in *contradiction* with one another since, according to Cantor, the set of the functions is of *higher cardinality*. Gödel, using *both* assumptions A and B simultaneously, can of course prove *everything*. Why only settle for the existence of "undecidable propositions" when you can have the "immortality of the cockchafer" and even the "infallibility of the faculties"? He still has *not* responded to my letter. Obviously, he no longer has anything to "grouse" about.

Could you possibly raise the question in your colloquium in Halle some day by lecturing on Gödel's presentation and mine *simultaneously*. You may hold

referieren? Die beiden Durchschläge können Sie behalten, bis ich sie einmal selbst brauche und um Rücksendung bitte; ich habe ja noch ein weiteres Exemplar. Leider kann ich selbst nicht kommen — bei der 12stündigen Reise und den hohen Fahrkosten. In Berlin habe ich auch nichts mehr zu suchen und in Leipzig ebensowenig. Die Frage der Antinomie Richard und des Skolemismus *muß* endlich *ernsthaft* erörtert werden, nachdem leichtsinniger Dilettantismus wieder am Werke ist, das ganze Forschungsgebiet zu diskreditieren — wie Schoenflies und Genossen seinerzeit die Mengenlehre diskreditierten. In meinem „Fundierungsprinzip" glaube ich endlich die richtige Handhabe ge-
4 funden zu haben, um alles Erforderliche aufzuklären. Aber niemand | hat's verstanden, wie auch noch niemand auf meine Fundamenta-Arbeit reagiert hat — nicht einmal meine guten Freunde in Warschau. Hilbert sagte freilich einmal: eine Arbeit brauche 15 Jahre, um gelesen zu werden. Nun⟦,⟧ Hilbert selbst hat ja mit seiner „Geometrie" schnelle Erfolge gehabt; aber war das auch wirklich eine große Leistung? — Das Fundierungsprinzip beherrscht *alle* möglichen Satzsysteme, auch das Russell-Carnapsche, soweit es widerspruchsfrei ist. Die „finitistische" Annahme A) kommt auf eine *Beschränkung* der Quantifikationsstufe hinaus und ist dann natürlich mit einer „freien" Begriffsentwickelung im Sinne des Diagonalverfahrens B) *unvereinbar.* Also man entscheide sich für A) oder B): tertium non datur.

Mit besten Grüßen an Sie und die Ihrigen

E. Zermelo

Was macht Frl. Kröncke? Ich habe schon lange nichts mehr von ihr gehört.

Brief an Ernst Zermelo vom 12. Oktober 1931

Gödel 1931b

Wien 12./X. 1931

Sehr geehrter Herr Professor!

Besten Dank für Ihren Brief vom 21./IX. Ich konnte leider nicht sofort antworten, weil ich für einige Tage verreisen mußte, will dies aber jetzt umso ausführlicher tun.

Zunächst möchte ich feststellen, daß die ersten 3 Seiten meiner Arbeit natürlich kein bindender Beweis sein sollen. Auf S. 174 habe ich ja ausdrücklich gesagt, daß es sich vorerst bloß um die Skizzierung des Hauptgedankens eines

on to the two copies until I need them and ask you to return them to me; I still have another copy. I regret that the twelve-hour trip and high fares prevent me from coming myself. Also, I no longer have any business in Berlin, or in Leipzig for that matter. The issue of the Richard antinomy and that of Skolemism *must* finally be subjected to *earnest* discussion as frivolous dilettantism is set to discredit the entire field once again—just like Schoenflies and his ilk once discredited set theory. In my "principle of foundation" I believe to have finally found the right tool to clarify all that is necessary. But nobody has understood it. Nor has anybody reacted to my Fundamenta paper—not even my good friends in Warsaw. To be sure, Hilbert once said that it takes 15 years for a paper to be read. Now Hilbert himself achieved quick success with his "Geometry"; but was this really a great achievement?—The principle of foundation governs *all* possible propositional systems, including the one by Russell and Carnap, provided it is consistent. The "finitistic" assumption A) amounts to a *restriction* of the level of quantification and is then of course *incompatible* with a "free" development of concepts in the sense of the diagonal procedure B). Hence, one must choose either A) or B): tertium non datur.

My best regards to you and your family,

E. Zermelo

How is Ms Kröncke doing? I have not heard from her in a long time.

Letter to Ernst Zermelo of 12 October 1931

Gödel 1931b

⟦The introductory note just before *s1931b* also addresses *Gödel 1931b*.⟧

Vienna 12./X. 1931

Dear Professor,

Thank you for your letter of 21 September. Unfortunately, I was not able to respond immediately as I had to go away for a few days, but will do so all the more thoroughly now.

First, I would like to point out that the first 3 pages of my paper are of course not intended as a binding proof. As in fact explicitly stated on

Beweises handelt. Insbesondere wird natürlich die Lücke, auf die Sie hinweisen später ausgefüllt (vgl. S. 182 ff.) — ja man kann sagen, daß gerade dies der Hauptzweck der folgenden Überlegungen ist — u. zw. geschieht der Beweis in der Weise, daß die Definition der Klasse K auf einfache arithmetische Definitionsweisen (rekursive Def. etc.), welche in PM sicher formal ausdrückbar sind, zurückgeführt wird. — Ich glaube aber, daß man auch schon auf Grund
2 der ersten 3 Seiten meiner | Arbeit zu der Überzeugung von der Richtigkeit des Beweises kommen kann, wenn man die Sache genau durchdenkt.

Um dies auseinanderzusetzen knüpfe ich an Ihren Einwand an. Sie definieren eine Klasse K^* durch die Festsetzung: „n gehört zu K^*, wenn $[R(n); n]$ nicht *richtig* ist", während ich eine Klasse K definiere durch: „n gehört zu K, wenn $[R(n); n]$ nicht *beweisbar* ist". Die Annahme, daß K^* durch ein Klassenzeichen des gegebenen Systems ausdrückbar ist, führt dann auf einen Widerspruch (dies ist aber nicht meine, sondern Ihre Annahme). D. h. also man kann zeigen, daß die Klasse K^* in dem gegebenen System nicht vorkommt[1]. Von der Klasse K dagegen kann man dies *nicht* zeigen, sondern man kann im Gegenteil beweisen, daß sie mit einem Klassenzeichen des gegebenen Systems umfangsgleich ist, vorausgesetzt daß in dem gegebenen Syst. gewisse einfache
3 | arithmetische Begriffe (Addition u. Multiplikation) enthalten sind.

Um den Grund für dieses verschiedene Verhalten von K und K^* einzusehen, muß man zunächst die Definition von K^* in korrekter Form schreiben. Man kann nämlich *nicht* setzen:

$$n \,\epsilon\, K^* \equiv \overline{[R(n); n]}\,,$$

weil die Zeichenverbindung $\overline{[R(n); n]}$ keinen Sinn hat. Ein Negationsstrich hat ja nur Sinn über einer Zeichenverbindung, die eine Behauptung ausdrückt (über der Ziffer 5 etwa ist ein Neg.-Strich sinnlos). Die Zeichenverbindung „$[R(n); n]$" drückt aber *keine Behauptung* aus. „$[R(n); n]$" ist ja gleichbedeutend etwa mit folgenden deutschen Worten: „diejenige Formel der Princ. Math., welche aus dem n-ten Klassenzeichen bei Einsetzung der Zahl n für die Variable entsteht". „$[R(n); n]$" ist nicht etwa selbst diese Formel, „$[R(n); n]$" ist ja überhaupt keine Formel der Princ. Math., (denn das Zeichen [;] kommt
4 ja gar | nicht in den Princ. Math. vor), sondern „$[R(n); n]$" ist lediglich eine Abkürzung der unter Anführungszeichen stehenden deutschen Worte. Diese Worte drücken aber offenbar keine Behauptung aus, sondern sind die eindeutige Charakterisierung einer Formel (d. h. einer räumlichen Figur), ganz ebenso wie etwa die Worte „die erste Formel jenes Buches" keine Behauptung ausdrücken, wenn auch vielleicht die Formel, welche durch diese Worte charakterisiert wird, eine Behauptung ausdrückt. „$[R(n); n]$" ist also für jede bestimmte Zahl n *ein Name* (eine eindeutige Beschreibung) für eine bestimmte Formel (d. h. eine räumliche Figur) und ein Negationsstrich darüber hat daher ebensowenig Sinn, wie etwa über der Formel „$5 + n$", welche für jede

[1] daß es solche Klassen geben muß, folgt natürlich einfacher aus dem Diagonalverfahren oder Mächtigkeitsbetrachtungen.

p. 174, all I have done at the beginning is to sketch the main idea of a proof. In particular, the lacuna you have identified is of course filled later on (see pp. 182ff.)—in fact, one might say that precisely this is the main purpose of the subsequent considerations. In particular, the proof proceeds by reducing the definition of the class K to simple arithmetical methods of definition (definition by recursion, etc.), which certainly can be formally expressed in PM.—But I believe that it is even possible to convince oneself that the proof is correct just by considering the first 3 pages of my paper, provided the matter is thought through carefully.

In order to explain this, let me take up your objection. You define a class K^* by means of the stipulation: "n belongs to K^* if $[R(n); n]$ is not *correct*", whereas I define a class K by: "n belongs to K if $[R(n); n]$ is not *provable*". The assumption that K^* can be expressed by means of a class sign of the given system then leads to a contradiction (that, however, is not my assumption, but yours). In other words, it is therefore possible to show that the class K^* does not occur in the given system.[1] For the class K, on the other hand, it is *not* possible to show this. On the contrary, it is possible to prove that it is coextensive with a class sign of the given system, provided that the given system contains certain simple arithmetical notions (addition and multiplication).

In order to see why K and K^* behave differently in this respect, we must first put the definition of K^* in the correct form. For we can *not* set

$$n \,\epsilon\, K^* \equiv \overline{[R(n); n]}\,,$$

since the combination of signs $\overline{[R(n); n]}$ has no meaning. The negation bar is meaningful only when used above a combination of signs expressing an assertion (e.g., when used above the numeral 5, the negation bar is meaningless). But the combination of signs "$[R(n); n]$" does *not* express an *assertion*. For "$[R(n); n]$" is synonymous with, say, the following ⟦English⟧ words: "that formula of Principia Mathematica which results from the nth class sign by substituting the number n for the variable." But "$[R(n); n]$" itself is not this formula. After all, "$[R(n); n]$" is not a formula of Principia Mathematica (since the sign [;] does not occur in Principia Mathematica). Instead, "$[R(n); n]$" is but an abbreviation of the ⟦English⟧ words that are put in quotation marks. It is obvious, however, that these words do not express an assertion. Rather, they uniquely characterize a formula (that is, a spatial figure) just as, say, the words "the first formula of that book" do not express an assertion, even if the formula these words characterize happens to express an assertion. Thus, for any given number n, "$[R(n); n]$" is *a name* (a unique description) for a particular formula (that is, a spatial figure), and using a negation bar above it makes as much sense as using a negation bar above the formula "$5 + n$", which for every number n is a name for a particular natural number. The

[1] That classes of this kind must exist can of course be shown in a simpler fashion by means of the diagonal procedure or cardinality considerations.

Zahl n ein Name für eine bestimmte nat. Zahl ist. Die ganze Schwierigkeit rührt offenbar daher, daß es in der Metamathematik außer den Zeichen für Zahlen, Funktionen, etc. auch *Zeichen für Formeln* gibt und daß man ein
5 | Symbol, welches eine Formel bezeichnet, deutlich unterscheiden muß von dieser Formel selbst.

Die Definition für K^* muß man daher korrekt so schreiben:

$$n \,\epsilon\, K^* \equiv \overline{W}[R(n); n] \,, \qquad (1^*)$$

wobei $W(x)$ bedeuten soll: „x ist eine richtige Formel“ oder genauer: „x ist eine Formel, die eine wahre Behauptung ausdrückt.“[2] Jetzt zeigt sich ganz deutlich, daß in der Definition für K^* ein neuer Begriff, nämlich der Begriff „richtige Formel“ bzw. die Klasse der richtigen Formeln vorkommt. Dieser Begriff läßt sich aber *nicht* ohne weiteres auf eine kombinatorische Eigenschaft der Formeln zurückführen (sondern stützt sich auf die Bedeutung der Zeichen) u. läßt sich daher in der arithmetisierten Metamathematik nicht auf einfache arithmetische Begriffe zurückführen; oder anders ausgedrückt: Die
6 Klasse der richtigen Formeln | ist *nicht* durch ein Klassenzeichen des gegebenen Systems ausdrückbar[3] (daher auch nicht die daraus definierte Klasse K^*). Ganz anders steht es mit dem Begriff „beweisbare Formel“ (bzw⟦.⟧ der Klasse der beweisbaren Formeln, welche in der Definition von K vorkommt). Die Eigenschaft einer Formel, beweisbar zu sein, ist eine rein kombinatorische (formale), bei der es auf die Bedeutung der Zeichen *nicht* ankommt. Daß eine Formel A *in einem bestimmten System* beweisbar ist, heißt ja einfach, daß es eine endliche Reihe von Formeln gibt, welche mit irgendwelchen Axiomen des Systems beginnt u. mit A endet, und welche außerdem die Eigenschaft hat, daß jede Formel der Reihe aus irgendwelchen der vorhergehenden durch Anwendung einer der Schlußregeln hervorgeht; wobei als Schlußregeln
7 im wesentlichen nur die Einsetzungs- | und die Implikationsregel in Betracht kommen, welche lediglich auf einfache kombinatorische Eigenschaften der Formeln Bezug nehmen. Die Klasse der beweisbaren Formeln[4] läßt sich daher auf einfache arithmetische Begriffe zurückführen d. h. sie kommt unter den Klassenzeichen des gegebenen Systems vor u. ebenso die daraus abgeleitete Klasse K. Der ausführliche Beweis dafür findet sich auf S. 182 ff. meiner Arbeit.

Im Anschluß an das Gesagte kann man übrigens meinen Beweis auch so führen: Die Klasse W der richtigen Formeln *ist niemals* mit einem Klassenzeichen desselben Systems umfangsgleich (denn die Annahme, daß dies der Fall sei, führt auf einen Widerspruch). Die Klasse B der beweisbaren

---

[2] Jetzt ist die rechte Seite von (1^*) eine Behauptung geworden u. daher negierbar, ebenso wie etwa die Worte „die erste Formel jenes Buches ist richtig“ eine Behauptung ausdrücken.

[3] genauer gesprochen, handelt es sich natürlich um die Klasse derjenigen *Zahlen*, welche richtigen Formeln zugeordnet sind.

[4] genauer: die entsprechende Klasse natürlicher Zahlen.

whole difficulty appears to stem from the fact that in metamathematics we have, besides the signs for numbers, functions, etc., also *signs for formulas* as well as from the necessity to clearly distinguish a symbol denoting a formula from that formula itself.

The definition of K^* must therefore be written correctly as

$$n \epsilon K^* \equiv \overline{W}[R(n); n] , \tag{1*}$$

where $W(x)$ is supposed to mean: "x is a correct formula", or, more precisely: "x is a formula expressing a true assertion."[2] It has now become absolutely clear that the definition of K^* contains a new notion, namely the notion "correct formula", or, respectively, the class of correct formulas. But this notion can *not*, without any further ado, be reduced to a combinatorial property of the formulas (but depends on the meaning of the signs), and hence it cannot be reduced in the arithmetized metamathematics to simple arithmetical notions. Or, put differently: It is *not* possible to express the class of correct formulas by means of a class sign of the given system[3] (and hence, neither the class K^* defined from it). With the notion "provable formula" (or, respectively, the class of provable formulas occurring in the definition of K), things are entirely different. The property of a formula being provable is a purely combinatorial (formal) one, which does *not* depend on the meaning of the signs. After all, that a formula A is provable *in a certain system* simply means that there exists a finite sequence of formulas with some axioms of the system at its beginning and with A at its end, which has the property that each formula of the sequence results from some of the preceding ones by the application of one of the inference rules. Here, the substitution rule and the rule of implication, which only refer to simple combinatorial properties of the formulas, are really the only inference rules to be considered. The class of provable formulas[4] can therefore be reduced to simple arithmetical notions, that is, it occurs among the class signs of the given system, as does the class K derived from it. The detailed proof of this claim can be found on pp. 182ff. of my paper.

By the way, in connection with what has been said, it is also possible to carry out my proof as follows: The class W of correct formulas *is never* coextensive with a class sign of the same system (for the assumption that it is leads to a contradiction). The class B of provable formulas *is* coextensive with a class sign of the same system (as can be shown in detail); consequently,

---

[2] The right side of (1*) has now turned into an assertion, and hence is capable of being negated, just as, say, the words "the first formula of that book is correct" express an assertion.

[3] Put more precisely, what we are concerned with here is the class of those *numbers* that are correlated with correct formulas.

[4] More precisely: the corresponding class of natural numbers.

Formeln *ist* mit einem Klassenzeichen desselben Systems umfangsgleich (wie man ausführlich zeigen kann); folglich können B und W nicht miteinander
8 umfangsgleich sein. Weil aber $B \subseteq W$, so | gilt $B \subset W$ d. h. es gibt eine richtige Formel A, die nicht beweisbar ist. Weil A richtig ist, so ist auch non-A nicht beweisbar, d. h. A ist unentscheidbar. Dieser Beweis hat aber den Nachteil, daß er keine Konstruktion des unentscheidbaren Satzes liefert u. intuitionistisch nicht einwandfrei ist.

Ich möchte noch bemerken, daß ich den wesentlichen Punkt meines Resultates nicht darin sehe, daß man über jedes formale System irgend wie hinausgehen kann (das folgt schon nach dem Diagonalverfahren) sondern darin, daß es für jedes formale System der Mathem. Sätze gibt, die sich innerhalb dieses Systems *ausdrücken*, aber aus den Axiomen dieses Systems *nicht entscheiden* lassen, und daß diese Sätze sogar von rel. einfacher Art sind, nämlich der Theorie der pos. ganzen Zahlen angehören. Daß man die ganze Mathematik nicht in ein formales System einfangen kann, folgt schon nach dem
9 Cantorschen Diagonalverfahren, aber es blieb trotzdem denk- | bar, daß man wenigstens gewisse Teilsysteme der Mathematik vollständig (d. h. entscheidungsdefinit) formalisieren könnte. Mein Beweis zeigt, daß auch dies unmöglich ist, wenn das Teilsystem wenigstens die Begriffe der Addition u. Multiplikation ganzer Zahlen enthält. (Dabei ist unter Formalisierung zu verstehen: Zurückführung auf endlich viele Axiome und Schlußregeln.) Gewiß sind die rel. unentscheidbaren Sätze in höheren Systemen immer entscheidbar, worauf ich in meiner Arbeit auch ausdrücklich hingewiesen habe (vgl. S. 191 Fußnote 48a), aber auch in diesen höheren Systemen bleiben unentscheidbare Sätze derselben Art übrig u. s. w. in inf.

Es würde mich freuen, wenn meine Ausführungen Sie überzeugt hätten. Natürlich sollen auch diese kein „Beweis“ sein. Den Beweis finden Sie vielmehr
10 nur an den genannten Stellen meiner Arbeit. | Ich danke Ihnen bestens für die Zusendung Ihrer Fundamenta-Arbeit, doch ist dieselbe bisher leider nicht in meine Hände gelangt (vielleicht bei der Post verloren gegangen). Ich hatte Ihre Arbeit übrigens schon bald nach ihrem Erscheinen gelesen und es waren mir damals verschiedene Bedenken aufgetaucht, die ich Ihnen, falls sie Sie interessieren, nächstens gerne mitteilen will.

Mit den besten Grüßen

Ihr ergebener Kurt Gödel

Wien VIII. Josefstädterstrasse 43.

P.S. Gleichzeitig übersende ich Ihnen ein Separatum meiner Arbeit über den Funktionenkalkül, die ich in Bad Elster nicht bei mir hatte.

B and W cannot be coextensive with each other. But since $B \subseteq W$, it holds that $B \subset W$, that is, there is a correct formula A that is not provable. Since A is correct, non-A is not provable either, that is, A is undecidable. This proof, however, has the disadvantage that it does not provide a construction of the undecidable proposition and that it is not unobjectionable from an intuitionistic point of view.

I would like to add that, as I view it, the crucial point of my result is not that it is possible to somehow go beyond any formal system (that already follows according to the diagonal procedure), but that for every formal system of mathematics there are propositions which can be *expressed* within the system but *not decided* on the basis of the axioms of that system, and that these propositions are propositions of a relatively simple kind, namely propositions belonging to the theory of positive integers. That a single formal system cannot capture mathematics in its entirety already follows according to Cantor's diagonal procedure. But it nevertheless remained conceivable that at least certain partial systems of mathematics could be formalized completely (that is, in the sense of deductive completeness[5]). My proof shows that this, too, is impossible, provided that the partial system contains at least the notions of addition and multiplication of integers. (By formalization I here mean reduction to finitely many axioms and inference rules). To be sure, the relatively undecidable propositions are always decidable in higher systems, as I pointed out in my paper (see p. 191 footnote 48^a). But undecidable propositions of the same kind are left over in these higher systems as well, and so on in infinitum.

It would please me if my explanations convinced you. Of course, they are not supposed to be a "proof" either. Rather, you can only find the proof in the specified parts of my paper. I thank you very much for sending your Fundamenta paper, which, unfortunately, has not reached me yet (perhaps it got lost in the mail). By the way, I read your paper already soon after it was published and began to harbor various doubts about it, which I would be glad to impart to you in the near future, in case you are interested.

Best regards,

Yours sincerely, Kurt Gödel

Wien VIII. Josefstädterstrasse 43.

P.S. I am also sending you an offprint of my paper on the functional calculus, which I did not have with me in Bad Elster.

---

[5] ⟦"Deductively complete" ("entscheidungsdefinit"), also rendered "syntactically complete", means that for every proposition A of the system, either A or its negation is deducible.⟧

Brief an Kurt Gödel vom 29. Oktober 1931

s1931d

Freiburg i. Br. 29. Oktober 1931. Karlstraße 60.

Sehr geehrter Herr Gödel!

Ich danke Ihnen für Ihren freundlichen Brief, aus dem ich nun besser als aus Ihrer Abhandlung und Ihrem Vortrag entnehmen kann, wie Sie es eigentlich meinen. Also nur für die beweisbaren Sätze Ihres "PM-Systems", nicht für dessen Sätze *überhaupt* soll Ihre "finitistische Einschränkung" (wie ich es nenne) der Typenbildung zur Geltung kommen, während Sie für die Sätze des Systems freie Neubildungen nach Art des Cantorschen Diagonalverfahrens zulassen. Dann erhalten Sie natürlich ein nicht abzählbares System möglicher Sätze, unter denen nur eine abzählbare Teilmenge "beweisbar" wäre, und es muß sicherlich "unentscheidbare" Sätze geben. Daß diese "unentscheidbaren" Sätze in einem "höheren System" doch wieder "entscheidbar" werden, geben Sie ja zu. Aber dieses "höhere" System unterscheidet sich von dem ursprünglichen keineswegs durch Aufnahme neuer Sätze, wie man nach Ihren Formulierungen denken könnte, sondern lediglich durch neue Beweismittel, und alles, was Sie in der Abhandlung beweisen, kommt darauf hinaus, was auch *ich* immer betone, daß ein "finitistisch beschränktes" Beweis-Schema *nicht ausreicht*, um die Sätze eines *nicht-abzählbaren* mathematischen Systems zu "entscheiden". Oder können Sie etwa "beweisen", daß *Ihr* Schema das "einzig mögliche" ist? Das geht doch wohl nicht; denn was ein "Beweis" eigentlich ist, ist nicht selbst wieder "beweisbar" sondern muß in irgend einer Form angenommen, vorausgesetzt werden. Und darum handelt es sich hier eben: was versteht man unter einem Beweis? Ganz allgemein versteht man darunter ein System von Sätzen derart, daß unter Annahme der Prämissen die Gültigkeit der Behauptung *einsichtig* gemacht werden kann. Und es ist nur noch die Frage, was alles "einsichtig" ist? Jedenfalls nicht bloß, und das zeigen Sie gerade selbst, die Sätze irgend eines finitistischen Schemas, das ja auch in Ihrem Falle immer wieder erweitert werden kann. Aber damit wären wir eigentlich einig: nur daß ich eben von vorn herein ein allgemeineres Schema, das nicht erst erweitert zu werden *braucht*, zugrunde lege. Und in *diesem* System sind dann auch wirklich *alle* Sätze "entscheidbar"! Soviel für heute; ich bin jetzt leider sehr in Anspruch genommen.

Mit bestem Gruß ergebenst E. Zermelo

Letter to Kurt Gödel of 29 October 1931

s1931d

⟦The introductory note just before *s1931b* also addresses *s1931d.*⟧

Freiburg i. Br. 29 October 1931. Karlstrasse 60.

Dear Mr. Gödel!

I'd like to thank you for your friendly letter. I am able to gain a better understanding from it about what you really want to say than before from your paper and your lecture. Your "finitistic restriction" (as I call it) on the formation of types is thus only meant to apply to the *provable* propositions of your "PM-system", and not to its *propositions in general.* At the same time, you allow for new *propositions* of the system to be *freely* formed in accordance with Cantor's diagonal procedure. Of course, you then obtain an *uncountable* system of possible propositions only a *countable* partial set of which would be "provable", and there would certainly have to exist "undecidable" propositions. That these "undecidable" propositions still become "decidable" in a "higher system" you freely admit. But this "higher" system does not differ at all from the original one in the admission of new *propositions*, as your formulations might lead one to believe, but only in new *means of proof*, and all you prove in your paper amounts to what I, too, always stress, namely that a "finitistically restricted" proof schema *is not sufficient* for "deciding" the propositions of an uncountable mathematical system. Or can you perhaps "prove" that *your* schema is the "only possible one"? But that would hardly work; for what a "proof" really is is itself not "provable" but must, in some form or another, be *assumed*, be *presupposed.* And this is the whole point, What do we mean by "proof"? Generally speaking, by "proof" we mean a system of propositions so constituted that, under the assumption of the premisses, the validity of the claim can be made *obvious.* The only remaining question is, What all is considered "obvious"? Certainly *not merely*, as you yourself have shown after all, the propositions of some finitistic schema, which, in your case, too, can always be *extended.* But then we would actually agree, except that I, at the very outset, proceed from a *more general* schema, which does not *need* to be extended first. And in *this* system now really *any* proposition whatsoever is decidable! So much for today; I am unfortunately much occupied at present.

With best regards, your most sincerely, E. Zermelo

Introductory note to s1931e and s1933b

Akihiro Kanamori

In this series of short notes found in the *Nachlass*, Zermelo articulates his late views on and approach to the ordinal numbers. These views encompass the boundary numbers of *1930a* and are related to his project of making evident the consistency of set theory "by means of the systematic construction of a 'set-theoretic model' using 'the unlimited number series' ", as stated at the end of his *s1930d*. Like his *s1930e* "On the set-theoretic model", the notes here are carefully typed. On the other hand, they are disparate sketches with redundancies and different emphases.[1] Although Zermelo had sketched the rudiments of the von Neumann ordinals perhaps as early as 1913[2] and had applied this specific representation of the Cantorian ordinal numbers in his *1930a*, he does not appeal to this reduction but pursues the ordinal numbers *ab initio* as autonomous concept.

Note 1 establishes the existence of exorbitant numbers based on the existence of normal domains satisfying ZF. The exorbitant numbers are the uncountable, regular limit cardinal numbers, now also called the weakly inaccessible cardinals. The term "exorbitant" derives from *Hausdorff 1914*, 131, where these cardinals were first considered and so regarded. As for normal domains, these are the models of set theory described in *1930a*. The argument is based on uncountable boundary numbers, now called the (strongly) inaccessible cardinals, being exorbitant numbers. The least boundary number of a normal domain satisfying ZF is uncountable, as the axiom of infinity is included in ZF; the domain is "categorically determined" by minimality and hence a set (cf. *1930a*); and consequently, its cardinality is an exorbitant number, being the boundary number. Zermelo here is focusing on exorbitant numbers, which he gets as boundary numbers, and he is well aware of the difference. He notably points that "t[T]here is something to be said" for the hypothesis that the cardinality of the continuum is an exorbitant number—a hypothesis antithetical to Hausdorff's original description as "exorbitant".

---

[1] See *Ebbinghaus 2006* for historical details.
[2] See *Ebbinghaus 2007b*, 133.

Note 2 discusses a property of the ultrafinite number series. The term is distinctive and refers to Zermelo's extension of the ordinal numbers as in *1930a*. As he begins:

> The "ultrafinite number series" is characterized by the fact that every "set", that is, every *categorically defined domain* of numbers, is always succeeded by yet other ones. The "transfinite" number series already ends with the "boundary number" of the "Cantor domain".

However, in succeeding notes "ultrafinite" is not used again; there is "unlimited", a reversion to "transfinite", and also, "open". The property discussed in the note involves cofinality, but is not separately distinctive nor further pursued.

Note 3 is an incisive, axiomatic presentation of the ordinal numbers as sets, much as one would see nowadays with the von Neumann ordinals. Proceeding in terms of "Z-segments" and "Z-elements" Zermelo however does not actually specify extensionally what these sets are. He could have worked with the representative basic sequences of *1930a* built from an urelement, but maintains a certain abstraction. The totality of all Z-elements is not a set but an open domain (cf. *s1930e*).

Note 4 specifies, for "an (open) domain Z" as in the previous note, the "fundamental relation" $a < b$. This specification is in the spirit of that of the membership relation $x \in y$ in *s1930e*, Zermelo's closest approach in set theory to making a distinction between a semantic context and a syntactic counterpart.

Note 5 takes a synthetic approach to the boundary numbers. The domain of all ordinal numbers is confronted by the Burali-Forti paradox, and so "we must proceed *differently* in order to get 'set domains' that we can use." To this end the replacement axiom serves, together with closure under power sets, to get boundary numbers. Zermelo thus approaches boundary numbers axiomatically without appeal to the normal domains of *1930a*.

Note 6 motivates having arbitrarily large boundary numbers, again in terms of replacement and categorical determination. Note 7 is similar, but proceeds more locally in terms of an axiomatization of boundary numbers.

The fragment *s1933b* seems to be a final, succinct statement about the ordinal numbers, one that puts together several features of the *s1931e* notes.

Sieben Noten über Ordinalzahlen und große Kardinalzahlen *

s1931e

Note 1: Beweis für die Existenz der *exorbitanten* Zahlen

Angenommen, es gäbe keine "regulären Anfangszahlen zweiter Art", aber es gäbe eine (widerspruchsfreie) Mengenlehre, welche die ZF'-Axiome erfüllt, es gäbe also "Normalbereiche" N auf einer gegebenen "Basis" Q. Dann kann kein Teilbereich N' von N, welcher Q enthält, ebenfalls ein Normalbereich sein, weil er dann ein "kanonischer Abschnitt" in der Entwickelung von N wäre und eine Grenzzahl $\pi > \omega$ zur "Charakteristik" hätte. Der Bereich N enthielte also Mengen N' von "Grenzzahl-Mächtigkeit" — gegen unsere Annahme. Also bestimmt unser Axiomensystem zusammen mit der Basis Q *kategorisch* den Bereich N, dieser wäre ein *geschlossener Bereich*, d. h. eine *Menge*, und der Ordnungstypus der "Entwickelung" von N wäre selbst eine "Grenzzahl" und mithin auch eine "exorbitante Zahl".

Noch einfacher geht der Beweis folgendermaßen: Es sei N ein beliebiger Normalbereich und u ein Element seiner Basis. Dann enthält N als Unterbereich einen "Einheitsbereich" E mit u als Basis, der dann wieder, sofern er keine Abschnitte von "exorbitantem" Index besitzt, eindeutig und kategorisch bestimmt, also eine *Menge* ist, deren Mächtigkeit als Charakteristik eines Normalbereiches notwendig *exorbitant* ist.

Die *kleinste exorbitante Zahl* braucht *nicht* notwendig eine Grenzzahl zu sein; sie könnte sogar schon *unter* der Mächtigkeit des *Kontinuums* liegen oder gerade *selbst* von dieser Mächtigkeit sein, eine Hypothese, die manches für sich zu haben scheint. Hierüber ließe sich vielleicht noch einiges feststellen — nachdem ein mal die Existenz exorbitanter Zahlen überhaupt gesichert ist.

Note 2: Eine Eigenschaft der *ultrafiniten* Zahlenreihe

Die "ultrafinite Zahlenreihe" ist dadurch ausgezeichnet, daß auf jede "Menge" d. h. auf jeden *kategorisch definierten Bereich* von Zahlen immer noch weitere folgen. Die *"transfinite"* Zahlenreihe endet bereits mit der "Grenzzahl" des "Cantorschen Bereiches". Sie ist dadurch charakterisiert, daß 1) auf jede Zahl noch eine solche von der Mächtigkeit ihrer Potenzmenge folgt, und daß 2) jeder Abschnitt, der einem kleineren konfinal ist, noch fortsetzbar ist.

Ein *"Zahlenabschnitt"* d. h. ein Abschnitt der Zahlenreihe heiße *"abgeschlossen"*, wenn er ein letztes Element enthält, also von "erster Art" ist, sonst heißt er *"offen"*.

* ⟦The notes are written on separate pages and unnumbered.⟧

Seven notes on ordinal numbers and large cardinals

s1931e

Note 1: Proof of the existence of *exorbitant* numbers

Suppose that "regular initial numbers of the second kind" did not exist, but that there existed a (consistent) set theory satisfying the ZF'-axioms, and hence that there existed "normal domains" N on a given "basis" Q. Then no partial domain N' of N containing Q can also be a normal domain, since otherwise it would then be a "canonical segment" in the development of N and would have a boundary number $\pi > \omega$ as its "characteristic". Hence, the domain N would contain sets N' of "boundary-number-cardinality"—contrary to our assumption. Together with the basis Q, our axiom system therefore *categorically* determines the domain N, which would be a *closed* domain, that is, a *set*, and the order type of the "development" of N would itself be a "boundary number", and hence also an "exorbitant number".

An even simpler proof runs as follows: Let N be an arbitrary normal domain and u an element of its basis. Then N contains as subdomain a "unit domain" E with u as basis, which then, provided it has no segments of "exorbitant" index, is again uniquely and categorically determined, that is, it is a *set* whose cardinality must be *exorbitant* because it is the characteristic of a normal domain.

The *smallest exorbitant number* need *not* necessarily be a boundary number; it could even already lie *below* the cardinality of the *continuum*, or it could *itself* be of this cardinality. There is something to be said for this hypothesis, or so it seems. A thing or two could be stated about this matter—after the existence of an exorbitant number has been ascertained in the first place.

Note 2: A property of the *ultrafinite* number series

The "ultrafinite number series" is characterized by the fact that every "set", that is, every *categorically defined domain* of numbers, is always succeeded by yet other ones. The "*transfinite*" number series already ends with the "boundary number" of the "Cantor domain". It is characterized by the fact that 1) every number is succeeded by one with the cardinality of its power set, and that 2) every segment cofinal with a smaller one is still capable of being continued.

A "*number segment*", that is a segment of the number series, is said to be "*closed*" if it contains a last element, and hence is of the "first kind". Otherwise, it is said to be "*open*".

Eine Teilmenge M eines Zahlenabschnittes A heiße *"konvergent"*, wenn sie einem kleineren Abschnitte, einem *echten* Abschnitte B von A angehört; im anderen Falle heißt sie *"divergent"*. Eine Teilmenge M von A heiße *"abgeschlossen"*, wenn sie mit jeder in A konvergenten Teilfolge M' von M auch deren Limeszahl enthält. Sie braucht aber, wenn A selbst "offen" ist, selbst kein letztes Element zu besitzen, da sie dann selbst in A divergieren kann. Eine *divergente abgeschlossene* Teilmenge M *ohne letztes Element* kann keinem *echten* Abschnitte A' von A ähnlich sein, weil sie dann auch das Limes-Element enthielte, das diesem Abschnitte entspräche, und damit wieder doch ein letztes Element hätte. Eine solche Teilfolge ist also dem *ganzen* Abschnitte A ähnlich, und diese Abbildung ist nichts anderes als eine *"Normalfunktion"* $g(x)$, welche der ursprünglichen Funktion $f(x) = x'$ zugeordnet ist, die jedes Element von M auf das nächstfolgende abbildet. Jede "Normalfunktion" besitzt aber *"Eigenwerte"* oder "Fixzahlen", die dann selbst wieder eine divergente Teilfolge M' bilden und eine neue Normalfunktion bestimmen, u. s. f. ins Unbegrenzte.

Ein Zahlenabschnitt heiße *"reduzibel"*, wenn er eine divergente Teilfolge von kleinerem Ordnungstyp enthält, oder m. a. W. wenn ein kleinerer Ab-
2 schnitt einer divergenten Teilfolge ähnlich ist. Im anderen | Falle heißt der Abschnitt *"irreduzibel"* oder ein *"Kern-Abschnitt"*. Ein Abschnitt ist "reduzibel" oder "irreduzibel", je nachdem seine Ordnungszahl einer kleineren *"konfinal"* ist oder nicht; zu den irreduziblen Abschnitten gehören *"Kernzahlen"* oder *"reguläre Anfangszahlen"* als Ordnungstypen. Alle *"Anfangszahlen erster Art"* sind bekanntlich *"regulär"*; dagegen war es bisher noch zweifelhaft, ob es außer der ersten transfiniten Zahl ω noch weitere "Anfangszahlen zweiter Art", oder *"exorbitante Zahlen"* gibt.

Note 3: Konstruktion der Zahlenreihe *Z*

Gegeben sei ein System von "Z-Abschnitten" von folgender Beschaffenheit:

1) Jeder "Z-Abschnitt" ist eine *wohlgeordnete Menge*.
2) Von je zwei "Z-Abschnitten" ist immer der eine ein *Abschnitt* des anderen.
3) Jeder "Menge" von "Z-Abschnitten" entspricht ein "Z-Abschnitt", der sie alle als Abschnitte enthält.
4) Jeder Abschnitt eines "Z-Abschnittes" ist selbst ein "Z-Abschnitt".

Hieraus folgt nun, wie leicht zu sehen:

5) Jedem Z-Abschnitt A entspricht ein und nur ein weiterer A', der genau *ein* Element mehr enthält, das auf A "unmittelbar folgende" Element a.
6) Jeder Menge T von Z-Abschnitten entspricht als weiterer Z-Abschnitt seine "Vereinigung" ST, der alle Elemente von T zu Abschnitten hat, der auf T "unmittelbar folgende" Z-Abschnitt.

A partial set M of a number segment A is said to be "convergent" if it belongs to a smaller segment, a *proper* segment B of A; otherwise, it said to be "divergent". A partial set M of A is said to be "closed" if along with every partial sequence M' of M convergent in A it also contains its limit number. However, if A itself is "open", it does not need to contain a last element itself since it then can itself diverge in A. A *divergent closed* partial set M *lacking a last element* cannot be similar to any *proper* segment A' of A, since otherwise it would also contain the limit element corresponding to this segment, and hence it would possess a last element after all. Such a partial sequence is therefore similar to the *entire* segment A, and this mapping is but a "*normal function*" $g(x)$ assigned to the original function $f(x) = x'$, which maps each element of M onto its immediate successor. But every "normal function" possesses "*eigenvalues*" or "fixed numbers", which themselves again form a divergent partial sequence M' and determine a new normal function, and so on indefinitely.

A number segment is said to be "*reducible*" if it contains a divergent partial sequence of smaller order type, or, in other words, if a smaller segment is similar to a divergent partial sequence. Otherwise, the segment is said to be "*irreducible*" or a "*core segment*". A segment is "reducible" or "irreducible", depending on whether or not its ordinal number is "*cofinal*" with a smaller one; to the irreducible segments there belong as order types "*core numbers*", or "*regular initial numbers*". All "*initial numbers of the first kind*" are "*regular*", as is well-known; up to now it has, however, been uncertain whether there exist, besides the first transfinite number ω, further "initial numbers of the second kind", or "*exorbitant numbers*".

Note 3: Construction of the number series Z

Suppose a system of "Z-segments" is constituted as follows:

1) Every "Z-segment" is a *well-ordered set*.
2) Of any two "Z-segments" one is always a *segment* of the other.
3) To every "set" of "Z-segments" there corresponds a "Z-segment" containing them all as segments.
4) Every segment of a "Z-segment" is itself a "Z-segment".

From this, as can be seen easily, it nows follows that:

5) To every Z-segment A there corresponds one and only one further A' containing exactly *one* more element, the element a "immediately succeeding" A.
6) To every set T of Z-segments there corresponds as a further Z-segment its "union" ST which has all elements of T as segments, the Z-segment "immediately succeeding" T.

7) Jede *Menge* M von Elementen aus Z-Abschnitten ist *wohlgeordnet*. Denn ihr entspricht die Menge T der Z-Abschnitte, denen sie angehören, und M ist dann Untermenge von ST.
8) Die Gesamtheit *aller* Z-Elemente (d. h. *aller* Ordnungszahlen) ist gewiß *keine Menge*, da *jeder* Z-Menge M nach 7) ein sie alle enthaltender Z-Abschnitt A entspricht und nach 5) ein auf sie alle "folgendes" Element a.
9) Dieser *"offene Bereich"* aller Z-Elemente ist gleichwohl *"wohlgeordnet"* in dem Sinne, daß jeder Teilbereich Z_1 von Z ein *erstes* Element besitzt, nämlich das erste Element a_1 unter allen denen, die einem gegebenen a vorangehen und die als Abschnitt eines Z-Abschnittes selbst eine Menge bilden.
10) In diesem "offenen Bereiche" sind "Quantifikationen", d. h. "universale" und "partikulare" Aussagen sinnvoll, sofern sie auf Quantifikationen in *"geschlossenen"* Teilbereichen, d. h. in "Mengen" zurückgeführt werden können.

Note 4: *Axiomatik der Ordinalzahlen* in der *unbegrenzten* Zahlenreihe

In einem (offenen) Bereiche Z von Elementen $a, b, c, \ldots$ gelten für die "Grundrelation" $a < b$ die Postulate:

1) Jedes Element a außer e bestimmt als "Abschnitt" ein-eindeutig eine "Menge" d. h. einen "geschlossenen" Unterbereich A, welcher alle Elemente $x < a$ umfaßt und a selbst nicht enthält.
2) Jede Menge T von Abschnitten enthält einen "kleinsten" Abschnitt d. h. einen solchen, der in allen übrigen als (echter) Teil enthalten ist. Insbesondere gilt dies auch von je zwei Elementen a, b und es ist immer $a < b$, wenn A der "kleinere" Abschnitt ist. Das "Anfangselement" e, das keinen Abschnitt bestimmt, geht allen übrigen voran⟦:⟧ $e < a$.
3) Jede Menge T von Abschnitten erzeugt durch "Vereinigung" wieder einen Abschnitt ST, der alle diese Abschnitte als Teile enthält, mithin auch ein zugehöriges Element s, dem alle Elemente $a < s$ "vorangehen".

Aus 1) und 2) folgt bereits, daß die Relation $a < b$ eine *Wohlordnung* darstellt: denn unter beliebig vielen Abschnitten ist immer einer der "kleinste", und diese Eigenschaft überträgt sich dann auf die zugehörigen "bestimmenden" Elemente. Daß alle diese Abschnitte A selbst eine "Menge" bilden, braucht dabei nicht vorausgesetzt zu werden; denn diese Bedingung ist wegen 1) erfüllt für alle $b < a$, die einem gegebenen a vorangehen, und das kleinste Element unter diesen ist wegen der "Transitivität" unserer Relation zugleich das kleinste unter allen des Teilbereiches T.

7) Every *set* M of elements from Z-segments is *well-ordered.* For to it there corresponds the set T of the Z-segments to which they belong, and M then is a subset of ST.
8) The totality of *all* Z-elements (that is, *all* ordinal numbers) is certainly *no set*, since to *each* Z-set M there corresponds, according to 7), a Z-segment A containing them all and an element a "succeeding" them all, according to 5).
9) This "*open domain*" of all Z-elements is nevertheless "*well-ordered*" in the sense that every partial domain Z_1 of Z possesses a *first* element, namely the first element a_1 among all those preceding a given a which themselves, as a segment of a Z-segment, form a set.
10) In this "open domain", "quantifications", that is, "universal " and "particular" assertions, are meaningful, provided they can be reduced to quantifications in "*closed*" partial domains, that is, in "sets".

Note 4: *Axiomatics of the ordinal numbers* in the *unlimited* number series

In an (open) domain Z of elements $a, b, c, \ldots$, the "fundamental relation" $a < b$ is subject to the postulates:

1) Every element a, besides e, determines one-to-one as a "segment" a "set", that is, a "closed" subdomain A that comprises all elements $x < a$ and that does not contain a itself.
2) Every set T of segments contains a "smallest" segment, that is, one contained as (proper) part in all others. In particular, this holds true also for any two elements a, b, and we always have $a < b$, if A is the "smaller" segment. The "initial element" e, which does not determine any segments, precedes all others⟦:⟧ $e < a$.
3) Every set T of segments in turn generates a segment ST by means of "union", which contains as parts all these segments, and hence also a corresponding element s "preceded" by all elements $a < s$.

From 1) and 2) it already follows that the relation $a < b$ constitutes a *well-ordering*: for among an arbitrary number of segments there is always a "smallest" one, and this property is then passed on to the corresponding "determining" elements. We need not assume here that all these segments A themselves form a "set"; for this condition is satisfied, on account of 1), for all $b < a$ which precede a given a, and the smallest element among these is also the smallest among all those of the partial domain T because of the "transitivity" of our relation.

Aus 3) folgt weiter, daß jeder Abschnitt A sowie jede "Summe" S von Abschnitten, die einer *"Menge"* entspricht, noch *"fortsetzbar"* ist, der *ganze* Bereich Z, d. h. die *unbegrenzte Zahlenreihe* gewiß *keine Menge* sein kann. Sie ist vielmehr der Prototyp eines *"offenen Bereiches"*.

Note 5: Über die transfinite Zahlenreihe und die Antinomie der Ordnungszahlen

Ein *"Bereich"* von Ordnungszahlen muß folgende Eigenschaften haben:

I) Er muß *wohlgeordnet* sein, d. h. jeder Unterbereich M von B muß ein *erstes* Element haben.
II) Er muß *fortsetzbar* sein, also etwa
   a) Jedem Element a folgt ein weiteres $a + 1$, *nächstfolgendes*,
   b) auf jede *"Menge"* von Ordnungszahlen folgt eine weitere: $\lim a_n = b$.

Wollte man nun, um den Bereich *aller* Ordnungszahlen zu erhalten, die Bedingung IIb) so auffassen, daß auf jeden *Unterbereich* noch weitere Zahlen folgen sollen, so gälte dies auch von B selbst, und es käme ein *Widerspruch*. Wir müssen also *anders* verfahren, um zu brauchbaren "Mengenbereichen" zu kommen. Hierzu dient am besten das *"Ersetzungsprinzip"*. Als *"Menge"* soll jeder solche Unterbereich M von B gelten, der einem *echten Abschnitte* von B *äquivalent* ist. Nur auf *solche* Unterbereiche brauchen weitere Elemente zu folgen, also auch nicht mehr auf B selbst, sofern der Ordnungstypus von B eine *"Anfangszahl"* ist. Es darf aber auch keine Anfangszahl sein, die einer kleineren *konfinal* ist, sondern es muß eine *"reguläre Anfangszahl"*, eine *"Kernzahl"* sein. Dieser Bedingung genügte jede "Anfangszahl *erster Art"*, die eine unmittelbar vorangehende hat. Solche werden aber *ausgeschlossen* durch die weitere Bedingung

III) Jedem (echten) Abschnitte von B entspricht ein solcher von größerer *Mächtigkeit*, er enthält keine Ordnungszahl von *größter Mächtigkeit*:
   a) mit jeder Zahlenklasse enthält er auch die *nächstfolgende*,
   b) mit jedem Abschnitt enthält der Bereich auch einen von der Mächtigkeit seiner *Potenzmenge*.

Aus III a) folgt unmittelbar, daß der O-Typus von B eine *Kernzahl zweiter Art*, eine *"exorbitante Zahl"* sein muß, aus III b) aber weiter, daß er außerdem eine *Fixzahl der Normalfunktion* $g(x)$, also eine *"Grenzzahl"* sein muß.

From 3), furthermore, it follows that each segment A as well as each "sum" S of segments corresponding to a "*set*" is still "*capable of being continued*", [that] the *entire* domain Z, that is, the *unlimited number series*, can certainly be *no set*. Rather, it is the prototype of an "*open domain*".

Note 5: On the transfinite number series and the antinomy of the ordinal numbers

A "*domain*" of ordinal numbers must have the following properties:

I) It must be *well-ordered*, that is, each subdomain M of B must have a *first* element.
II) It must be *capable of being continued*, that is, e.g.,
   a) Each element a is *immediately succeeded* by a further $a + 1$,
   b) each "*set*" of ordinal numbers is succeeded by a further one: $\lim a_n = b$.

If now, in order to obtain the domain of *all* ordinal numbers, we were to understand the condition IIb) so that each *subdomain* is supposed to be succeeded by still further numbers, then this would hold true also for B itself, and we would have a *contradiction*. Thus, we must proceed *differently* in order to get "set domains" that we can use. For this purpose the "*principle of replacement*" serves us best. Every subdomain M of B *equivalent* to a *proper segment* of B shall be considered a "*set*". It is only *such* subdomains that need to be succeeded by further elements, and hence no further elements need to succeed B itself any longer, provided that the order type of B is an "*initial number*". On the other hand, it must not be an initial number that is *cofinal* with a smaller one either. Instead, it must be a "*regular initial number*", a "*core number*". This condition is satisfied by every "initial number *of the first kind*" that is immediately preceded by such a number. Those, however, are *excluded* by the further condition

III) To each (proper) segment of B there corresponds one such with *greater cardinality*, it does not contain an ordinal number with the *greatest cardinality*:
   a) along with every number class it also contains its *immediate successor*,
   b) along with every segment the domain also contains one with the cardinality of its *power set*.

From III a) it immediately follows that the o[rder] type of B must be a *core number of the second kind*, an "*exorbitant number*", but from III b), furthermore, that it also must be a *fixed number of the normal function* $g(x)$, and hence a "*boundary number*".

Note 6: Die Grundeigenschaft der *offenen* Zahlenreihe

Ersetzt man jedes Element eines echten *Abschnittes* der Zahlenreihe durch ein anderes, so gehören alle diese Zahlen x' wieder einem *Abschnitte* der Zahlenreihe an, d. h. auf ihre Gesamtheit folgen immer noch weitere Zahlen. Nehmen wir insbesondere eine solche Ersetzung an, daß der kleineren Zahl $x < y$ auch immer die kleinere $x' < y'$ entspricht, so haben wir eine *ähnliche Abbildung* auf einen *Teil* eines *größeren Abschnittes.* Also:

Jeder *Teil* der Zahlenreihe, der einem *Abschnitte* (oder einem Teil eines solchen) *ähnlich* ist, gehört immer selbst (als Teil) einem (echten) *Abschnitte* an. Oder: jede Limeszahl ist *konfinal* einer *größeren*, während umgekehrt bekanntlich *nicht* jede Ordnungszahl einer *kleineren konfinal* zu sein braucht, z. B. die *"regulären Anfangszahlen erster Art"*, nämlich die Anfangszahlen aller Zahlenklassen vom Index $n+1$, wie überhaupt alle *Kernzahlen.* Ist nun aber A ein Zahlenabschnitt von *Kernzahl-Typus*, also keinem kleineren konfinal, so gilt für *ihn* bereits die *gleiche Grundeigenschaft* wie für die *ganze* Zahlenreihe, aber auch *nur* für solche Abschnitte. Gäbe es keine Kernzahlen, so wäre ein unserer Bedingung genügender Zahlenabschnitt *kategorisch* bestimmt, es wäre eine *Menge* vom *Kernzahlypus.* Es muß also jedenfalls Kernzahlen geben, und zwar in *jedem Reste* der Zahlenreihe und *hinter* jeder gegebenen Kernzahl. Da aber jede Kernzahl zugleich eine *Anfangszahl* ist, so gibt es auch immer *höhere Zahlenklassen* d. h. höhere *Mächtigkeiten (Alefs)* zu jeder gegebenen. Gäbe es hinwieder nur *Kernzahlen erster Art*, so wäre ein Zahlenabschnitt von der genannten Eigenschaft, der *mit* jeder Anfangszahl die *nächstfolgende* enthielte, wieder *kategorisch bestimmt*, also eine *Menge*, und sein Ordnungstypus wäre eine *Kernzahl zweiter Art*; also gibt es auch *exorbitante Zahlen.* Soll außerdem der Abschnitt mit jeder Mächtigkeit auch die der *Potenzmenge* enthalten, so folgt ebenso die *Existenz der Grenzzahlen.*

Note 7: Die *"Hauptabschnitte"* der Zahlenreihe

Ein "Hauptabschnitt" H hat folgende Eigenschaften:

1) Jedem Abschnitte A von H entspricht ein anderer A' von der Mächtigkeit seiner Potenzmenge UA.
2) Jede Teilmenge M von H, die einem *echten Abschnitte* H_1 von H *äquivalent* ist, gehört selbst einem *echten Abschnitte* von H an.

Ein solcher Abschnitt ist z. B. der kleinste transfinite Abschnitt vom Typus ω, und im übrigen alle durch *"Grenzzahlen"* gebildeten Abschnitte. Gäbe es außer ω keine weiteren Grenzzahlen, so wäre außer Z_ω die *ganze* Zahlenreihe der *einzige* "Hauptabschnitt" und wäre durch unser Postulatsystem *kategorisch*

Note 6: The basic property of the *open* number series

If each element of a proper *segment* of the number series is replaced by another one, then all these numbers x' again belong to a *segment* of the number series, that is, its totality is always succeeded by still further numbers. In particular, if we assume a replacement such that to the smaller number $x < y$ there always also corresponds a smaller one $x' < y'$, then we have a *similar mapping* onto a part of a *greater segment*. Hence:

Every *part* of the number series *similar* to a *segment* (or a part of one such) always belongs itself (as a part) to a (proper) *segment*. Or: every limit number is *cofinal* with a *greater one*, whereas conversely, as is well-known, *not* every ordinal number needs to be *cofinal* with a *smaller one*, e.g., the "*regular initial numbers of the first kind*", namely the initial numbers of all number classes of index $n+1$, and all *core numbers* in general. But now if A is a number segment of the *type of core numbers*, and hence cofinal with no smaller one, then the *same basic property* holding of the *entire* number series already holds of *it*, but *only* of such segments. If there were no core numbers, then a number segment satisfying our condition would be *categorically* determined, it would be a *set* of *the type of core numbers*. Hence, there have to be core numbers in any case, namely in *every remainder* of the number series and *after* every given core number. But since every core number is also an *initial number*, then there always are also *higher number classes*, that is, higher *cardinalities (alephs)* for each given one. If, on the other hand, there only existed *core numbers of the first kind*, then a number segment with the specified property containing *along with* each initial number its *immediate successor* would again be *categorically determined*, and hence a set, and its order type would be a *core number of the second kind*; so there are also *exorbitant numbers*. If, furthermore, the segment is to contain along with each cardinality also that of the *power set*, then the *existence of the boundary numbers* follows as well.

Note 7: The "*principal segments*" of the number series

A "principal segment" H has the following properties:

1) To every segment A of H there corresponds another A' with the cardinality of its power set UA.
2) Every partial set M of H *equivalent* to a *proper segment* H_1 of H belongs itself to a *proper segment* of H.

The smallest transfinite segment, of type ω, for instance, is a segment of this kind and, incidentally, all segments formed by means of "*boundary numbers*". If there were no further boundary numbers, besides ω, then, besides Z_ω, the number series *in its entirety* would be the *only* "principal segment", and

bestimmt, also ein "*geschlossener Bereich*", eine "*Menge*" und daher nach der Grundeigenschaft der *"offenen"* Zahlenreihe doch noch *fortsetzbar*, und ihr Ordnungstypus π wäre in der Tat eine *Grenzzahl*. Also gibt es in der Zahlenreihe *"tatsächlich"* Hauptabschnitte *oberhalb* ω sowie auch oberhalb *jeder weiteren* Grenzzahl.

Durch die Normalfunktion $\psi(x)$, welche der Bedingung $\psi(x+1) = 2^{\psi(x)}$ genügt,[1] wird wegen der Eigenschaft 1) jeder Hauptabschnitt auf *sich selbst* abgebildet, und daher ist jede Grenzzahl eine *Fixzahl*, ein *Eigenwert* unserer Normalfunktion. Zugleich darf wegen 2) kein Hauptabschnitt einem kleineren *"konfinal"* sein, jede Grenzzahl muß eine *"Kernzahl"* und zwar wegen 1) eine solche "zweiter Art", eine *"exorbitante Zahl"* sein. Aus unserer Betrachtung folgt also unter anderem auch die *Existenz der exorbitanten Zahlen*, und zwar *oberhalb* jeder Grenze.

[1] ⟦The ψ-formula is only in a carbon copy and written there in pencil.⟧

it would be *categorically determined* by means of our system of postulates. Hence, it would be a "*closed domain*", a "*set*", and it would therefore, according to the basic property of the "*open*" number series, yet still be *capable of being continued*, and its order type π would indeed be a *boundary number*. So there "*really*" are principal segments *above* ω in the number series, as well as above *any further* boundary number.

Every principal segment is mapped *onto itself* by means of the normal function $\psi(x)$ satisfying the condition $\psi\,(x+1) = 2^{\psi(x)}$ because of property 1). Hence, every boundary number is a *fixed number*, an *eigenvalue* of our normal function. At the same time, because of 2), no principal segment can be "*cofinal*" with a smaller one, every boundary number must be a "*core number*" and, in particular, because of 1), one such of the "second kind", an "*exorbitant number*". So from our consideration also follows, among other things, the *existence of the exorbitant numbers*, and in particular ⟦their existence⟧ *above* every limit.

Introductory note to s1931f and s1932d

Heinz-Dieter Ebbinghaus

In his Warsaw talks *s1929b* Zermelo had proposed a new concept of set with the intention of separating sets from classes in a way more coherent than that of von Neumann. Without providing details, Zermelo takes as sets those classes which are domains of structures allowing for a categorical definition, i. e., a definition which up to isomorphism has exactly one model. There is no explanation about the additional relations allowed for definitions; moreover, there is no comment on the language in which the definitions are to be given. Examples such as the Peano axioms for the natural numbers and Hilbert's axioms for the real numbers suggest that second-order definitions should be allowed. As stated in *s1930e*, Zermelo is fully convinced that a set in the new sense "is precisely that which Cantor really meant by his well-known definition of 'set', and it can be treated as a set everywhere and without contradiction in all purely mathematical considerations and deductions" (ibid., 5).

As argued by Taylor in *1993*, the 1908 axiomatization already shows that "an intuitive concept of mathematical definability is at the root of Zermelo's concept of set" (p. 549) and that "absolutely all mathematical objects are capable of (predicative) 'determination' and, hence, are more or less definable" (p. 551). Seen in the light of his new approach, definability has become *the* dominant property of a set.

The new concept comes with a shift concerning the role of axioms. Hitherto, Zermelo had emphasized that the universe of sets should be described or given in an axiomatic way. Now, the axiomatic determination is shifted from the universe to the individual sets: he *defines* the universe as the totality of sets obeying the condition of categorical definability. In order to ensure that the set-theoretic arguments remain valid, he therefore has to show that the universe of categorically definable sets is an inner model of the set-theoretic axioms inside the domain of background classes.

Early in 1931 he turned to this task. In the carefully typed note *s1931f* and in the fragment *s1932d*, probably part of a version for a textbook on set theory, he gives a more detailed elaboration. In the following we comment on these, finishing with remarks on their further perspective, in particular on that concerning the axiom of separation.[1]

The note *s1931f* and the universe of closed domains

Taking up the terminology from *s1930e*, Zermelo speaks of categorically defined sets as "closed domains" in contrast, for example, to the "open domain" of ordinal numbers. Closedness in this sense, however, must not be identified

[1] For further details, see *Ebbinghaus 2003*.

with closedness in some topological sense. Zermelo uses the German term "geschlossen" which is different from the German term "abgeschlossen" for "closed" in the topological sense. His terminology evokes the impression of "finished" and may refer to the cumulative hierarchy: a "closed" domain is one delimited by the hierarchy, being contained in some level, in contrast to the "open" domain of ordinals.[2]

Zermelo sketches proofs showing that the universe of categorically definable sets satisfies the axioms of power set (in 1)), union (in 2)), and separation (in 3)). The axiom of replacement is treated only in the fragment *s1932d*. The axiom of infinity may have seemed trivial to him, as the introduction states that the set of natural numbers can be categorically defined by the Peano axioms, or it may have been skipped—like in the boundary number paper *1930a*—as not belonging to general set theory. Moreover, like in that paper and as explicitly mentioned in *s1930e*, the axiom of choice is taken for granted as a "general logical principle", leaving us with the question whether or why choice sets are sets in the new sense.

The proofs given suffer not only from their sketchy character, but also from more essential deficiencies: There is no hint as to questions of absoluteness and no comment on the principles for the background domain of classes used in the arguments. Moreover, there is no comment on the nature of the isomorphisms between models of categorical definitions. Apparently they have to be considered as belonging to the background universe of classes. As already mentioned above, there are no remarks on the language to be used for the categorical definitions.

When treating the power set axiom in 1), Zermelo emphasizes that he is working with full power sets and, with respect to Skolem's first-order version of definiteness, argues that "allness" and "quantification" are fundamental logical categories which cannot be restricted, as any restriction would have to refer to quantifications as well, thus leading to a *regressus in infinitum*. The passage is evidence for his belief that for Skolem the power set of a set consists only of first-order definable subsets. Apparently Reinhold Baer had not been able to convince him that he was wrong when interpreting Skolem in this way. For as early as on 27 May 1930 Baer had written to him that "a subset of a set belonging to some domain of sets may exist even if there is no definite function according to the axiom of separation; it is only not forced to exist in this case."[3]

The criticism of Skolem is taken up again in part 3). The allusions to David Hilbert and Felix Bernstein are to their papers *Hilbert 1926* and *Bernstein 1905a* where both claim to have proved the continuum hypothesis by showing in essence that the functions over the set of natural numbers can be enumerated by the ordinal numbers of the second number class. Already in 1907, in a letter to Leonard Nelson, Zermelo had criticized Bernstein for

[2] Cf. also (the introductory note to) *s1930e*.

[3] Cf. *Ebbinghaus 2004*, 78.

his "peculiar behaviour concerning the continuum hypothesis (where he [...] against his better judgement claimed to have a proof of Cantor's well-known theorem[4])."[5] The hint at Fraenkel refers to *Fraenkel 1923a*; on p. 208 Fraenkel states that there is no conjecture even about the direction where a possible solution of the continuum problem could be found.

The fragment *s1932d* and the axiom of separation

Between 1928 and 1942 Zermelo worked on a monograph on set theory. There is strong evidence that the manuscript was never completed. The *Nachlass* contains only some drafts of smaller parts, among them the present fragment *s1932d*. It is typed less carefully than others. Evidently it was written down as a collection of thoughts and far from being in final shape. However, it does indicate that the monograph was to be based on the new notion of set.

According to the intended purpose and in contrast to *s1931f*, the presentation is not concerned with criticism of the finitary approach of Skolem. Zermelo starts forthwith by setting out his view of (infinite) sets. Infinite sets, in his conviction, are incapable of being presented empirically; they exist as Platonic ideas, open only to an axiomatic description as exemplified by the notion of categorical definability. He then shows that the universe of categorically defined sets satisfies the axioms of separation, power set, union, and replacement. By reasons commented on above, the axioms of infinity and choice are not taken into consideration. Despite being somewhat more detailed, the presentation suffers from the same deficiencies as that in *s1931f*.

The axiom of separation or, more exactly, the theorem corresponding to it, has the following succinct formulation:

Every well-defined part T of a set S is itself a set.

Here, the well-definedness of T is to mean that the categorical definition of T is compatible with the categorical definition of S in the sense that isomorphisms between models of the categorical definition of S induce isomorphisms between the models of the categorical definition of T. In *s1931f*, well-defined parts of a set S were defined in a similar way, but less explicitly, namely

[4] I. e. of the continuum hypothesis.
[5] Cf. *Ebbinghaus 2007b*, 75.

as given by a definition (Satzfunktion) which is *definit* with respect to the categorical definition δ of S in the sense that it is uniquely determined by means of the fundamental relations making up the relations in δ.

In the lecture notes of Zermelo's last course on set theory given in Freiburg in the winter semester 1933/34, the theorem of separation has the following form:

Every (categorically defined) part of a set is itself a set.

This is then re-formulated in the even simpler form:

Every *part* of a set is itself a set.

This is the last version of separation which can be found in the *Nachlass*. So in the end, the notion of definiteness has been absorbed into the notion of set.

At first glance there is a similarity to the formulation of the axiom of separation in the boundary number paper *1930a*:

To each part of a set there corresponds a set
that contains all elements of that part.

The difference lies in the sets forming the universe. Whereas in *1930a* we have a free "unfolding" of sets and full power sets, the lecture notes presuppose the universe to consist only of the categorically definable sets.

There now seem to be two concepts of the set-theoretic universe: a dynamic one where the universe evolves along the inexhaustible series of the ordinal numbers and a static one where the universe is given by an appropriate notion of language. A reconciliation between these concepts would have been possible by working definitions into a cumulative hierarchy. Gödel's constructible hierarchy accomplishes this task for first-order definability. However, it is beyond question that Zermelo had neither the inclination nor the technical competence to go in such a direction.

The fragment ends with elementary considerations about equivalent sets, i. e., sets that can be mapped onto each other by one-to-one functions. In contrast to the precise treatment of equivalence in *1908b* and probably because of the textbook character, the mappings ensuring equivalence need not be sets, but may be taken from the background universe of classes. There are no documents that give information about what this would finally lead to.

Sätze über *geschlossene Bereiche*

s1931f

Ein "geschlossener" Bereich ist ein solcher, der durch ein *kategorisches Postulatsystem* oder sonstige Festsetzungen "kategorisch" d. h. bis auf *"isomorphe"* Abbildungen bestimmt ist. So ist z. B. die "Zahlenreihe", wie sie etwa durch die *"Peanoschen Axiome"* bestimmt ist, der einfachste Typus eines *unendlichen geschlossenen Bereiches.* Für solche geschlossenen Bereiche gelten nun folgende Sätze:

1) Ist M ein geschlossener Bereich, so ist auch UM, die Gesamtheit *aller Teilbereiche* von M, ein geschlossener Bereich. Denn seien M und M' irgend zwei demselben Postulatsystem genügende, also unter sich *äquivalente* Bereiche. Dann entspricht auch jeder Teilmenge M_1 von M, also jedem Element von UM, in M' eine *äquivalente* Teilmenge M_1' und damit auch ein ganz bestimmtes Element von UM', die beiden Bereiche UM und UM' sind also isomorph aufeinander abgebildet. Hierbei ist natürlich wesentlich, daß sowohl UM wie UM' auch wirklich *alle* Teilmengen von M bzw. M' enthalten soll, nicht nur die durch besonders konstruierte Satzfunktionen definierten. Der Begriff der "Allheit" oder der "Quantifikation" muß überhaupt als nicht weiter analysierbare logische Grundkategorie jeder mathematischen Betrachtung zugrunde gelegt werden. Wollte man im besonderen Falle die Allheit einschränken durch besondere Bedingungen, so müßte auch dies wieder geschehen in Gestalt von Quantifikationen, und wir kämen so zu einem regressus in infinitum.

2) Ist T ein geschlossener Bereich und jedem seiner Elemente t eineindeutig zugeordnet ein geschlossener Bereich M_t derart, daß alle diese Bereiche M_t unter sich elementefremd sind, so bildet die "Vereinigung" aller dieser Bereiche M_t wieder einen geschlossenen Bereich ST.

Denn seien S und S' irgend zwei Modelle dieser "Vereinigung" und m irgend ein Element von M_t, so entspricht ihm auch ein ganz bestimmtes Element t' von T' und im zugehörigen M_t' auch ein ganz bestimmtes Element m', weil ja die Bereiche M_t und M_t' einander äquivalent sind. Also sind auch S und S' äquivalent und die "Vereinigung" ST ist *kategorisch bestimmt.*

2 | 3) Aus einem geschlossenen Bereiche M wird ein geschlossener Teilbereich M_f *ausgesondert* durch jede *Satzfunktion* $f(x)$, welche in dem M *definierenden* Postulatsystem P *"definit"* d. h. durch seine *"Grundrelationen"* eindeutig bestimmt ist. Denn unter dieser Voraussetzung werden aus je zwei isomorphen äquivalenten Modellen M und M' auch wieder *isomorphe* Teilbereiche M_f und M_f' ausgesondert, und die Definition des Teilbereiches ist daher ebenfalls *kategorisch.* Diese "Definitheits-Bedingung" betrifft aber lediglich die *Definition* einzelner Teilbereiche, nicht aber ihre (objektive) *Existenz* in

Theorems on *closed domains*

s1931f

A "closed" domain is one that is "categorically" determined, that is, up to "*isomorphic*" mappings, by means of a *categorical system of postulates* or other stipulations. The "number series", for instance, as determined by means of "*Peano's axioms*", is the simplest type of an *infinite closed domain.* Such closed domains are subject to the following theorems:

1) If M is a closed domain, then UM, the totality of *all partial domains* of M, is also a closed domain. For let M and M' be any two domains which satisfy the same system of postulates, and hence which are mutually *equivalent.* Then there also corresponds to each partial set M_1 of M, that is, to each element of UM, in M' an *equivalent* partial set M_1', and thus also a specific element of UM'. Hence, the two domains UM and UM' are isomorphically mapped onto one another. In this case it is of course essential that both UM and UM' are really supposed to contain *all* partial sets of M and M' respectively, not only those defined by means of specially constructed propositional functions. In general, the concept of "allness", or "quantification", must lie at the foundation of any mathematical consideration as a basic logical category incapable of further analysis. If we were to restrict the allness in a particular case by means of special conditions, then we would have to do so using quantifications, which would lead us to a regressus in infinitum.

2) If T is a closed domain and if with each of its elements t a closed domain M_t is correlated one-to-one so that all these domains M_t are mutually exclusive, then the "union" of all these domains M_t is again a closed domain ST.

For let S and S' be any two models of this "union" and m some element of M_t. Then to it there also corresponds a specific element t' of T' and also a specific element m' in the corresponding $M_{t'}'$ since the domains M_t and $M_{t'}'$ are mutually equivalent.[1] Hence, S and S', too, are equivalent, and the "union" ST is *categorically determined.*

3) From a closed domain M we may separate out a closed partial domain M_f by means of any *propositional function* $f(x)$ that is "*definite*" in the system of postulates *defining* M, that is, which is uniquely determined by means of its "*fundamental relations*". For, on this assumption, ⟦the two⟧ partial domains M_f and M_f' separated from two isomorphic equivalent models M and M' are again *isomorphic*, and the definition of the partial domain is therefore also *categorical.* This "*condition of definiteness*", however, only concerns the *definition* of individual partial domains, but not its (objective)

[1] ⟦Zermelo erroneously writes "M_t'" instead of "$M_{t'}'$".⟧

der Gesamtheit *aller* Teilbereiche, wie sie unter 1) in Betracht gezogen wurde, sie kann also auch *nicht*, wie dies z. B. bei *Skolem* geschieht, zu einer willkürlichen *Einschränkung des Mengenbereiches* und zu einer sinnwidrigen *Relativierung* des Mengen- und Mächtigkeitsbegriffes verwendet werden. Natürlich ist es immer möglich, durch solche willkürlichen Einschränkungen der zugelassenen Teilmengen z. B. den Begriff des *Kontinuums* derart umzudeuten oder zu verwässern, daß es nach Belieben von *erster* oder von *zweiter Mächtigkeit* wird. Für das eigentliche *"Kontinuums-Problem"*, das sich eben unweigerlich auf die Gesamtheit *aller* Teilmengen der Zahlenreihe bezieht, also in gewissem Sinne dem *"Vollständigkeitsaxiome"* genügt, wird durch solche Umdeutungen m. Er. *nicht das geringste* gewonnen. Solche Betrachtungen könnten, wenn überhaupt etwas, nur das eine beweisen, daß das Kontinuum *mindestens* von erster bzw. von zweiter Mächtigkeit ist — was doch ohnehin schon bewiesen ist. Wenn selbst *Hilbert* nach dem Vorgange von *Bernstein* eine Zeit lang ähnliche Wege gegangen ist (vestigia terrent!), so ist das vielleicht als Alterserscheinung erklärlich. Bei *Skolem* ist es aber schon ein kaum noch entschuldbarer Mangel an Selbstkritik: an der ganzen Mengenlehre interessiert ihn überhaupt *nur* noch sein *Pseudo-Relativismus!* Zur Freude aller Reaktionäre und Antimathematiker! *Fraenkel* ist auch hier völlig hilflos. —

existence in the totality of *all* partial domains as considered in 1). Hence, it can *not* be used for an arbitrary *restriction of the domain of sets* and for an absurd *relativization* of the concepts of set and cardinality, as is the case, e.g., with *Skolem*. It is of course always possible to re-interpret or to dilute, e.g., the concept of the *continuum* by means of such arbitrary restrictions of the permitted partial sets so that it is left to one's discretion whether it has *first* or *second cardinality*. In my opinion, *nothing at all* is really gained by such re-interpretations for the "*continuum problem*", which is inevitably concerned with the totality of *all* partial sets of the number series, and which therefore, in a certain sense, satisfies the "*completeness axiom*". If anything, such considerations can only prove that the continuum is *at least* of first or of second cardinality—but this has already been proved. That even *Hilbert*, following *Bernstein*, trod similar paths for some time (vestigia terrent!) we may perhaps explain as a sign of old age. But in *Skolem*'s case this lack of self-criticism is beyond pardon: in all of set theory he finds *nothing* of interest *but* his *pseudo-relativism*! To the delight of all reactionaries and anti-mathematicians! And once again, *Fraenkel* is completely helpless here.—

Introductory note to s1931g

R. Gregory Taylor*

Zermelo published three papers outlining his theory of systems of infinitely long propositions. From 1931 there is *1932a*, in four pages, based on Zermelo's presentation at the September 1931 conference in Bad Elster. From the same year comes *1932b*, in two pages, which summarizes *1932a* for a general audience. Finally, there is the fuller exposition of *1935*. In addition, Zermelo's *Nachlass* contains several related documents from the same period. Most interesting among them is handwritten *s1931g*, whose title is almost identical to that of *1935*. One notable feature of *s1931g* concerns Zermelo's notion of definiteness. That notion, familiar from *1908b* and *1929a*, plays no role in *1932a*, *1932b*, or *1935*. In contrast, definiteness arises in *s1931g* within a context that is almost that of systems of infinitely long propositions. This suggests that definiteness never ceased to figure in Zermelo's theorizing about mathematical systems.

Dated July 1931, *s1931g* was drafted only two months before *1932a*. Since there is much overlap between *s1931g* and *1932a*, the former must be regarded as a preliminary step in Zermelo's development of his theory of systems of infinitely long propositions. Its notational ambiguity, among other things, suggests that *s1931g* presents first formulations of ideas published in *1932a*, *1932b*, and *1935*. There are some differences, however.

First, whereas what Zermelo emphasizes in *1932a*, *1932b*, and *1935* are well-founded hierarchies of propositions, by contrast, systems of propositions figure not at all in *s1931g*. Nor is Zermelo's topic the infinitary character of propositions. Rather, Zermelo's focus in *s1931g* is elementary model theory together with his notion of symmetric proposition. (Note that if domain $\mathfrak{G}$ is infinite, then the disjunctions presented at the end of *s1931g* are infinite, however.)

Second, whereas the systems of *1932a*, *1932b*, and *1935* are variable-free, already the first sentence of *s1931g* introduces variables $\xi, \eta, \ldots$ ranging over domain elements $\alpha, \beta, \ldots$. Consequently, only *s1931g* suggests a theory of relations or propositional functions complementing Zermelo's theory of proposition systems.

What might seem to constitute a third difference is perhaps mere appearance. Zermelo's examples in *1932a*, *1932b*, and *1935* involve a minimal signature comprising a single dyadic relation. In contrast, the first sentence of *s1931g* may indicate a signature consisting of several fundamental relations of distinct degrees. If so, however, then that greater generality is not maintained in what follows. Also, one cannot rule out the possibility that $p_{\xi,\eta,\ldots}$, $q_{\xi,\eta,\ldots}$,

---

* The author of this introductory note wishes to thank H.-D. Ebbinghaus and A. Kanamori for helpful comments and J. Stanton for editorial assistance.

and $r_{\xi,\eta,...}$ indicate only that variables $\xi, \eta, \ldots$ have been assigned different values in the case of one and the same, not necessarily dyadic, fundamental relation R.

With respect to semantics, *s1931g* is likewise indeterminate. Whereas "truth distributions" in *1932a*, *1932b*, and *1935* are representable as two-dimensional Boolean matrices whose order is given by the cardinality of an assumed domain of individuals, what Zermelo terms a "complete matrix" in *s1931g* may or may not signify something more general, e.g., a p-tuple $\langle M_1, \ldots, M_p\rangle$ of Boolean matrices of fixed dimensions $n_1, \ldots, n_p$, respectively, but all of one and the same order. Alternatively, what Zermelo here calls a "complete matrix" would, in fullest generality, be a p-tuple $\langle f_1, \ldots, f_p\rangle$ of Boolean-valued functions with f_ℓ assigning value 1 to arguments $\xi_1, \ldots, \xi_{n_\ell}$ just in case relation $R_\ell \xi_1 \ldots \xi_{n_\ell}$ holds. This second alternative is closer to what is suggested in the second sentence of *s1931g*. Note, however, that that sentence points in two directions. On the one hand, describing $u_\zeta = u_{\xi,\eta,...}$ as a single function suggests the minimal context of *1932a*, *1932b*, and *1935*. On the other hand, describing u_ζ as assigning a Boolean value "to every system $\zeta = (\xi, \eta, \ldots)$ appearing in the fundamental relations" suggests that a multiplicity of degrees is at issue. (If a unit signature is in fact involved, then the qualifier "appearing in the fundamental relations" adds nothing whatsoever.)

In the second paragraph of *s1931g*, definite propositions are characterized, in fullest generality, as those whose truth value is expressible as a function $f(u)$ of an arbitrary complete matrix $u = u_\zeta^{(1)}, u_\zeta^{(2)}, \ldots$, where each of $u_\zeta^{(1)}, u_\zeta^{(2)}, \ldots$ is a Boolean entry within u, one supposes. The truth value of s might then be the Boolean product or sum of $u_\zeta^{(1)}, u_\zeta^{(2)}, \ldots$ or the result of applying some more complicated truth function. In a sense, definiteness and completeness are complementary notions in Zermelo's lexicon: definite propositions are precisely those whose truth values are determined without ambiguity by any complete matrix, and complete matrices are just those structures that determine the truth value of any definite proposition.

A certain conflation of propositions with their semantic values is characteristic of *s1931g*, as when Zermelo writes $s = f(u)$. Similarly,

$$s \equiv U_s = \sum_\lambda s_\lambda u^{(\lambda)} = \sum_\lambda s_\lambda \prod_\zeta u_\zeta^{(\lambda)} \tag{\dagger}$$

is problematic on multiple grounds, although Zermelo's likely intentions can be made out. (In what follows, we introduce notation found elsewhere in Zermelo's writings as well as some of our own.)

Let ${}^{n\mathfrak{G}}2$ denote the set of all n-ary Boolean-valued functions with arguments in domain $\mathfrak{G}$. We let U denote ${}^{n_1\mathfrak{G}}2 \times \ldots \times {}^{n_p\mathfrak{G}}2$. Where s is a proposition over $\mathfrak{G}$ and signature $\Sigma = \langle R_1, \ldots, R_p\rangle$ with R_ℓ an n_ℓ-adic relation and $u =: \langle f_1, \ldots, f_p\rangle \in U$, we define the *semantic value* val(s, u) *of s at u* as follows. If s is elementary, say, $R_\ell \mathfrak{b}_1 \ldots \mathfrak{b}_{n_\ell}$ with $1 \le \ell \le p$ and $\mathfrak{b}_1, \ldots, \mathfrak{b}_{n_\ell} \in \mathfrak{G}$,

then $\mathrm{val}(s,u)$ is $f_\ell(\mathfrak{b}_1,\ldots,\mathfrak{b}_{n_\ell})$. If s is of the form $\overline{t}$ with t a proposition over $\mathfrak{G}$ and Σ, then $\mathrm{val}(s,u) = 1$ if and only if $\mathrm{val}(t,u) = 0$. If s is a disjunction $\mathfrak{D}(K)$ with K a set of propositions over $\mathfrak{G}$ and Σ, then $\mathrm{val}(s,u) = 1$ if and only if $\mathrm{val}(t,u) = 1$ for some $t \in K$. We set $U_s =: \{u \in U \mid \mathrm{val}(s,u) = 1\}$. Propositions s and t are *logically equivalent* provided that $U_s = U_t$, and we write $s \equiv t$.

Let domain $\mathfrak{G}$ and signature $\Sigma = \langle R_1,\ldots,R_p\rangle$ be given. As a description, modulo fundamental relation R_ℓ, of $f \in {}^{n_\ell\mathfrak{G}}2$, we offer the length-$|\mathfrak{G}|^{n_\ell}$ conjunction $\mathrm{eldg}_{\mathfrak{G},R_\ell}(f)$ of literals

$$\prod \left\{R_\ell\mathfrak{b}_1\ldots\mathfrak{b}_{n_\ell} \;\middle|\; \mathfrak{b}_1,\ldots,\mathfrak{b}_{n_\ell} \in \mathfrak{G} \text{ and } f(\mathfrak{b}_1,\ldots,\mathfrak{b}_{n_\ell}) = 1\right\} \cup \left\{\overline{R_\ell\mathfrak{b}_1\ldots\mathfrak{b}_{n_\ell}} \;\middle|\; \mathfrak{b}_1,\ldots,\mathfrak{b}_{n_\ell} \in \mathfrak{G} \text{ and } f(\mathfrak{b}_1,\ldots,\mathfrak{b}_{n_\ell}) = 0\right\}.$$

Finally, setting $\mathrm{eldg}_{\mathfrak{G},\Sigma}(\langle f_1,\ldots,f_p\rangle) =: \prod_{1\le\ell\le p}\mathrm{eldg}_{\mathfrak{G},R_\ell}(f_\ell)$, we can perhaps recast (†) as

$$s \equiv \sum_{\langle f_1,\ldots,f_p\rangle \in U_s} \mathrm{eldg}_{\mathfrak{G},\Sigma}(\langle f_1,\ldots,f_p\rangle),$$

which says that s may be brought into a certain normal form, as Zermelo observes. (We have made no explicit use of Zermelo's s_λ, the characteristic function of U_s.)

In the definition of symmetric proposition given in the final paragraph of *s1931g*, Zermelo fails to distinguish two notions of symmetry. One is syntactic: proposition s is symmetric if permuting individual constituents invariably leaves s itself unchanged. The other notion of symmetry is semantic: s is symmetric if U_s is permutation-invariant. Zermelo seems to assume that the two notions are equivalent, which they are not. (For details, see the introductory note to *1932a*.)

The definition of symmetric proposition S^* in terms of given proposition s provides a method of generating symmetric propositions over given $\mathfrak{G}$ and Σ. Namely, form the disjunction S^* of all images of s under permutations of $\mathfrak{G}$. Alternatively, choose an arbitrary Boolean matrix of order $|\mathfrak{G}|$ and permute it in all possible ways. Next, form the disjunction of the elementary diagrams of (the functions represented by) the resulting matrices. (Foundational consid-

erations mandate that $\mathfrak{G}$ not be too big, although Zermelo does not mention this in *s1931g*.)

Within algebraic number theory, a polynomial $p(x_1, \ldots, x_n)$ is said to be *symmetric* if it is unchanged by arbitrary permutations of variables $x_1, \ldots, x_n$. Examples with $n = 3$ are $x^2 + y^2 + z^2$ and $xy + xz + yz$. It can then be shown that any symmetric polynomial in variables $x_1, \ldots, x_n$ over a field F can be expressed as a polynomial over F in the elementary symmetric functions $\sigma_1, \ldots, \sigma_n$, where σ_k with $1 \leq k \leq n$ is the sum of $\binom{n}{k}$ distinct terms, each the product of k distinct variables from among $x_1, \ldots, x_n$. For instance, with $n = 3$ we have $x^2 + y^2 + z^2 = \sigma_1^2 - 2\sigma_2$.

In *s1931g* Zermelo begins to develop an analogy between his symmetric propositions and the symmetric polynomials of algebraic number theory. Within the theory of symmetric propositions, the analogue of elementary symmetric functions is what Zermelo, in *1932a* and *1935*, calls "categorical propositions", where proposition s is categorical if U_s is nonempty and permutation-invariant and no nonempty proper subset of U_s is permutation-invariant. The final sentence of *s1931g* alludes to Zermelo's result, stated in *1932a* but proved only in *1935*, that any symmetric proposition is expressible as a disjunction of categorical propositions. The reference to linearity would indicate logical sums of categorical propositions *qua* elementary symmetric functions raised to power $n = 1$.

Finally, we observe that, unlike Zermelo's symmetric propositions, symmetric polynomials may be defined either syntactically or semantically with identical results. To begin, polynomial $p(x_1, \ldots, x_n)$ over $\Re$ is symmetric if it is taken into itself, as an inscription, by any permutation of variables $x_1, \ldots, x_n$. Equivalently, $p(x_1, \ldots, x_n)$ is symmetric if $J =: \{\langle a_1, \ldots, a_n\rangle \in \mathbb{C}^n \mid p(a_1, \ldots, a_n) = 0\}$ is permutation-invariant in the sense that, where $\pi \in S_n$ is arbitrary, $\langle a_{\pi^{-1}(1)}, \ldots, a_{\pi^{-1}(n)}\rangle \in J$ whenever $\langle a_1, \ldots, a_n\rangle \in J$. Thus, the analogy between symmetric polynomials and symmetric propositions has clear limits. That having been said, symmetric polynomials had proved useful, from Leibniz' day forth, in connection with integer partitions (see *Dickson 1919–23*, vol. 2, ch. III). Much later, they figured in the classical theory of algebraic number fields (see *Landau 1918*, §§ 1–5). Zermelo expected—or at least hoped—that symmetric propositions would be similarly efficacious within a new, infinitary logic.

Allgemeine Theorie der mathematischen Systeme

s1931g

July 1931.

Jeder mathematischen Theorie liegt zugrunde ein „Grundbereich“ $\mathfrak{G}$ von Elementen $\alpha, \beta, \ldots, \xi, \eta \ldots$ und von „Grundrelationen“ R

$$p_{\xi,\eta\ldots} \quad q_{\xi,\eta\ldots} \quad r_{\xi,\eta\ldots}$$

zwischen diesen Elementen, welche „gelten“ oder „nicht gelten“ können z. B.

$$r_{\xi\eta} \quad \text{oder} \quad \overline{r_{\xi\eta}}$$

sobald die „Variablen“ $\xi, \eta, \ldots$ „spezifiziert“ sind

$$\xi = \alpha, \ \eta = \beta.$$

Eine „vollständige Matrix“ ist eine Funktion

$$u_\zeta = u_{\xi,\eta,\ldots}$$

welche *jedem* Wertesystem $\zeta = (\xi, \eta, \ldots)$ das bei den Grundrelationen vorkommt, einen der beiden Werte 1 oder 0 zuordnet, je nachdem die Relation $u_{\xi,\eta\ldots}$ *gilt* oder *nicht gilt*, und es ist

$$u = \prod_\zeta u_\zeta$$

wahr oder falsch, es „gilt“ u oder $\overline{u}$, je nachdem u_ζ für „alle“ ζ wahr *oder* u_ζ für „mindestens ein“ ζ falsch ist.

2 | Ein „Satz“ s heißt „definit“ in Bezug auf „Grundbereich“ und „Grundrelationen“, wenn seine „Gültigkeit“ (d. h. sein „Gelten“ oder „Nichtgelten“) durch jede „Matrix“ u vollständig *bestimmt* ist, wenn also sein „Wahrheitswert“ als „Funktion“ von u dargestellt werden kann.

$$s = f(u) = f(u_\zeta^{(1)}, u_\zeta^{(2)}, \ldots)$$

Jedem „definiten Satz“ s ist demgemäß zugeordnet 1) die Gesamtheit U_s der Matrizen u, für die er „wahr“ $s_\lambda = 1$ und 2) die Gesamtheit $V_s = U - U_s$, für die er „falsch“ d.h. $s_\lambda = 0$ ist.

$$s \equiv U_s = \sum_\lambda s_\lambda u^{(\lambda)} = \sum_\lambda s_\lambda \prod_\zeta u_\zeta^{(\lambda)}$$

General theory of mathematical systems

s1931g

July 1931.

As basis of every mathematical theory we have a "fundamental domain" $\mathfrak{G}$ of elements $\alpha, \beta, \ldots, \xi, \eta \ldots$ and "fundamental relations" R

$$p_{\xi,\eta\ldots} \quad q_{\xi,\eta\ldots} \quad r_{\xi,\eta\ldots}$$

between these elements, which can "hold" or "not hold", e.g.

$$r_{\xi\eta} \quad \text{or} \quad \overline{r_{\xi\eta}}\,,$$

once the "variables" $\xi, \eta, \ldots$ are specified $\xi = \alpha,\ \eta = \beta$. A "complete matrix" is a function

$$u_\zeta = u_{\xi,\eta,\ldots}\,,$$

which assigns one of the two values 1 or 0 to *every* system of values $\zeta = (\xi, \eta, \ldots)$ appearing in the fundamental relations, depending on whether the relation $u_{\xi,\eta\ldots}$ *holds* or *does not hold.* And

$$u = \prod_\zeta u_\zeta$$

is true or false, u or $\overline{u}$ "holds", depending on whether u_ζ is true for "all" ζ *or* u_ζ is false for "at least one" ζ.

A "proposition" s is termed "definite" with respect to "fundamental domain" and "fundamental relations" if its validity (i.e. its "holding" or "not holding") is completely *determined* by every "matrix" u, in other words, if its value can be represented as a "function" of u.

$$s = f(u) = f(u_\zeta^{(1)}, u_\zeta^{(2)}, \ldots)$$

Accordingly, to every "definite proposition" s there corresponds (1) the totality U_s of all matrices u for which it is "true", i.e. $s_\lambda = 1$ and (2) the totality $V_s = U - U_s$ for which it is "false", i.e. $s_\lambda = 0$.

$$s \equiv U_s = \sum_\lambda s_\lambda u^{(\lambda)} = \sum_\lambda s_\lambda \prod_\zeta u_\zeta^{(\lambda)},$$

wo das Σ-Zeichen die „Disjunktion“ („oder“) ausdrückt und der Index λ *alle* Matrizen $u^{(\lambda)}$ durchläuft, wobei nur die Elemente von U_s von 0 verschieden sind. Jeder „definite“ Satz ist also „äquivalent“ einem Satze der Form

$$\sum s_\lambda u^{(\lambda)} ,$$

er läßt sich *eindeutig* auf diese „Normalform“ bringen, sodaß je zwei „äquivalenten“ Sätzen *dieselbe* Normalform entspricht.

Ein Satz heißt „invariant“ oder „symmetrisch“, wenn er bei allen „Permutationen“ π der Elemente $\xi, \eta \ldots$ des Grundbereiches „in sich übergeht“, d. h. „wahr“ oder „falsch“ bleibt nach Ausführung jeder Permutation π. Aus einem beliebigen Satze s entsteht ein symmetrischer, wenn man *alle* Sätze $s^{(\pi)}$, welche durch Permutationen aus ihm hervorgehen, (wobei π die „Gruppe“ aller Permutationen von $\mathfrak{G}$ durchläuft) durch Disjunktion mit einander verbindet

$$S^* = \sum_\pi s^{(\pi)} ,$$

wobei S mit s äquivalent ist, wenn schon s symmetrisch ist. Insbesondere entsteht aus jeder *Matrix* $u = \sum_\zeta u_\zeta$ ein symmetrischer Satz

$$u^* = \sum_\pi u^{(\pi)} ,$$

wobei $u^{(\pi)}$ immer die „Klasse“ der durch Permutationen aus u hervorgehenden („homologen“?) Matrizen enthält. So entspricht jeder „Klasse“ homologer Matrizen als „elementare symmetrische Funktion“ als „Klasseninvariante“ ein symmetrischer Satz s^* und *jeder symmetrischer* Satz s ist darstellbar als Disjunktion (lineare Verbindung) dieser „elementaren symmetrischen Funktionen“.

where the symbol Σ expresses "disjunction" ("or") and the index λ ranges through *all* matrices $u^{(\lambda)}$, whereby only the elements of U_s are not canceled. Every "definite" proposition is thus "equivalent" to a proposition of the form

$$\sum s_\lambda u^{(\lambda)}.$$

Any proposition may be brought into this normal form *in a unique manner* so that any two "equivalent" propositions correspond to the *same* normal form.

A proposition is termed "invariant" or "symmetric" if it goes over into itself under any "permutation" π of the elements $\xi, \eta, \ldots$ of the fundamental domain, i.e. remains "true" or "false" after application of any permutation π. From an arbitrary proposition s one obtains a symmetric proposition if one disjoins *all* propositions $s^{(\pi)}$ obtained from s by permutation (whereby π ranges over the "group" of all permutations of $\mathfrak{G}$)

$$S^* = \sum_\pi s^{(\pi)},$$

whereby S^*[1] is equivalent to s if s is already symmetric. In particular, a symmetric proposition

$$u^* = \sum_\pi u^{(\pi)}$$

is obtained from every *matrix* $u = \prod_\zeta u^{(\zeta)}$, where $u^{(\pi)}$ contains the class of all matrices obtainable from u by permutation ("homologous"?). Thus to every "class" of homologous matrices there corresponds a symmetric proposition s^* as "elementary symmetric function" as "class invariant", and *every symmetric* proposition s is expressible as the disjunction (linear combination) of these "elementary symmetric functions".

[1] ⟦Zermelo erroneously writes "S" instead of "S^*".⟧

Introductory note to 1932a, 1932b, and 1935

R. Gregory Taylor*

Zermelo's final publications concerning mathematical logic are *1932a*, *1932b*, and *1935*. The three articles cover much the same ground, indicating that the ideas involved date from 1931 or before. Zermelo's presentation in *1932a* of his "infinitary proof theory" is compressed, but that paper offers the most with regard to philosophical motivation and bears the clearest relation to the five theses of *s1921*.

1. Regarding *1932a*

In his opening sentences Zermelo attacks the "finitistic prejudice" that he has come to associate with Skolem. Relativistic consequences of the downward Löwenheim–Skolem theorem constitute a *reductio* of "Skolemism" *ad absurdum* in Zermelo's eyes, including, it seems, the very concept of a formal language that was fast becoming the standard within the discipline. The so-called Skolem paradox is taken, obscurely, to reinstate the paradox of Richard.

The last half of the first paragraph elaborates Thesis IV of *s1921* with its call for a "Platon[ist]ic logic" based on "infinitary intuition". Like Theses II and IV of *s1921*, the long fourth sentence concerning "conceptually ideal relations" has its source in an endnote of *Cantor 1883b*—endnote 1 in the case of Theses II and IV, endnote 2 in the present case. The reference to sign combinations is an allusion to what Hilbert called proof theory, whereas Zermelo's pledge to accommodate mathematics' entire legacy implies disapproval, consonant with Thesis IV, of the intuitionist's refusal to do so. Zermelo's discussion is informed by Kant's distinction between concepts and intuitions. However, the yet more emphatic Kantian tone of the inspirational passages from Cantor—multiplicity is given whereas unity must be forged—is muted, both here and in *s1921*, in favor of considerations derived from Zermelo's debate with his own contemporaries. Finally, in the sentence concerning conceptually ideal relations it is clear that Zermelo has in mind, principally, entire hierarchies of sets or propositions rather than infinite sets or propositions, taken individually, as in *s1921*. (The former are the focus of Cantor's endnote as well.)

The scientific warrant for Zermelo's philosophical remarks is his description of a well-founded hierarchy of propositions. (Throughout our discussion we preserve Zermelo's algebraically suggestive notation, some of which is first introduced only in *1935*.) Given *basis* Q of elementary propositions over

* The author of this introductory note wishes to thank H.-D. Ebbinghaus and A. Kanamori for comments on earlier drafts and J. Stanton for editorial advice.

fixed domain $G = \{x, y, z, \ldots\}$, we close under unary negation, infinitary conjunction, and infinitary disjunction. Relative to fixed G, disjoint union $S := \sum_{\alpha \in \mathrm{On}} Q_\alpha$ is a *system of propositions* well-founded with respect to the immediate proper subcomponent relation, where each Q_α is defined by

$$Q_\alpha = \begin{cases} Q & \text{if } \alpha = 0 \\ \{\overline{s} \mid s \in Q_\beta\} + \big(\{\mathfrak{K}(T) \mid T \subseteq \sum_{\gamma \leq \beta} Q_\gamma\} + \\ \qquad \{\mathfrak{D}(T) \mid T \subseteq \sum_{\gamma \leq \beta} Q_\gamma\}\big) - \sum_{\gamma \leq \beta} Q_\gamma & \text{if } \alpha = \beta + 1 \\ \big(\{\mathfrak{K}(T) \mid T \subseteq \sum_{\gamma < \alpha} Q_\gamma\} + \\ \qquad \{\mathfrak{D}(T) \mid T \subseteq \sum_{\gamma < \alpha} Q_\gamma\}\big) - \sum_{\gamma < \alpha} Q_\gamma & \text{otherwise}. \end{cases} \qquad (\dagger)$$

Here $\mathfrak{D}(\cdot)$ and $\mathfrak{K}(\cdot)$ denote logical operations of disjunction and conjunction, respectively, applied to sets of propositions. (Following Zermelo, we write $s+t$ for $\mathfrak{D}(\{s,t\})$ and st for $\mathfrak{K}(\{s,t\})$.) Each proposition of system S thereby possesses a unique rank or "level of quantification". If G happens to be infinite, then already Q_1 contains propositions of infinite length.

Zermelo's propositions, like those of Russell, have both individual and relational constituents. One important difference between Russell and Zermelo concerns individuals. Russell's individuals are nonmathematical, and mathematical objects are higher-type objects in his system. Zermelo countenances individuals of every sort. However, his focus in *1932a*, *1932b*, and *1935* is mathematical propositions whose individual constituents are numbers or sets, and relational constituents are irreducibly mathematical (cf. *1935*, §3). A second difference concerns semantics. Zermelo's propositions have truth values only relative to interpretations of constituent relations (truth distributions). They are not timelessly true or false, as are Russell's propositions. Indeed, Husserl's concept of a manifold, comprising a domain of objects together with relations that may or may not hold of those objects, surely influenced the development of Zermelo's theory of systems of infinitely long propositions more directly than did anything to be found in *Principia* (see *Husserl 1922*, §§ 69–70).

In his fifth paragraph, Zermelo assumes a Boolean-matrix-based semantics for propositions of a system S starting from 1)–3) of *1935*, §4. He writes of a set U_s of truth distributions (equivalently, Boolean matrices) making proposition s true. In the minimal setting that Zermelo assumes (that of a single dyadic relation), U_s will be a set of two-dimensional Boolean matrices of order $|G|$.

In his sixth paragraph, Zermelo introduces a sequence of metalogical notions in quick succession, where each notion is defined in terms of truth distributions. Thus we are treated to characterizations of "absolutely true (false) with respect to Q", "satisfiable with respect to Q", "reconcilable with respect to Q", and so forth. It is unclear that the relativizations to basis Q add anything in the case of those metalogical notions. Happily, a certain

domain-relativity is obvious in the case of the two metalogical notions Zermelo introduces next.

In the same long paragraph, Zermelo presents definitions of "symmetric proposition" and "categorical proposition" (cf. below). This proximity is significant. It indicates that Zermelo regards symmetry and categoricity as metalogical properties on a par with validity, satisfiability, and so forth. In particular, to the extent that the symmetry concept constitutes Zermelo's analysis of "general proposition", Zermelo's proof theory is self-characterizing, which is in keeping with his description, elsewhere, of logic, the science of generality, as a "reflexive science" (see *Farber 1926/27a* and the final paragraphs of the introductory note to *s1921*).

Zermelo's definition of "symmetric proposition" is structural in character, but, within the same sentence, he goes on to introduce semantic considerations. (Incidentally, an urelement is any member of basis G; no hierarchy of urelements and sets is at issue here.)

> If a proposition s is "symmetric", i.e. *invariant* under the *permutations* of the urelements $x, y, z, \ldots$ of the domain G, then the same holds of the corresponding set U_s.

As defined in the antecedent here, a symmetric proposition s is one such that, for any domain permutation π, image $\pi(s)$ is s itself. On this account, symmetry would pertain to the way in which elementary propositions are collected together in a particular case (cf. *s1921*, Thesis V): elementary subcomponents of symmetric s are organized, within s, in such a way that, holding relational constituents fixed, s remains unchanged when individual constituents are interchanged *ad libitum.*

If we take symmetry to be defined structurally, then the consequent of the quoted sentence might appear to give an equivalent, semantic, characterization of symmetry. It does not. Symmetry implies semantic invariance; but the converse fails since a contradiction such as $xRy\ \overline{xRy}$ is semantically invariant but not symmetric. Nor do contingent propositions fare better: $(xRx + \overline{xRx})\ xRx\ yRy$ with $G = \{x, y\}$ is semantically invariant but not symmetric. Since semantic invariance, but not symmetry in the structural sense, is preserved under logical equivalence, the former seems the better notion. Zermelo's remarks regarding axiom systems at the very end of the sixth paragraph hint at his reason, fully revealed only in *1935*, for introducing symmetry. With regard to the issue of domain-relativity, note that $xRy + yRx$ is symmetric (in either sense) just in case $G = \{x, y\}$.

As defined in the same final paragraph, a "categorical proposition" s is one such that (1) U_s is nonempty and permutation-invariant and (2) no nonempty proper subset of U_s is permutation-invariant. Thus any categorical proposition is both satisfiable and symmetric. Further, as demonstrated in *1935* but merely asserted here, any symmetric proposition is equivalent to a disjunction of categorical propositions. (Zermelo's terminology harkens to Kant's distinction between categorical and disjunctive judgments.) In this connec-

tion, Zermelo mentions an analogy with the manner in which the elementary symmetric functions constitute a basis for the set of symmetric functions of n real variables, a result associated with Newton (see the introductory note to *s1931g*). Zermelo's emphasis upon the subcategory of categorical propositions, defined semantically but not structurally, provides another, aesthetic, reason to prefer the semantic notion of symmetry.

In his third paragraph, Zermelo offers a semantic characterization of proof relative to some assumed system S of propositions with basis Q. Suppose that proposition s in S has (nested) conditional form. Starting with antecedent(s) and consequent, we close under the proper subformula relation, which means that, by well-foundedness, certain members of Q will be introduced. Now the question is, Does any truth distribution with respect to the resulting set or "system" make antecedent(s) true but consequent false? If not, then this second system constitutes a proof of s. Otherwise, s is refutable. On one reading, presentation of a proof in Zermelo's sense amounts to an infinite thought-experiment (presentation of an infinite truth table or semantic tableau), as does determining whether s is provable. The proof itself is a certain infinite mathematical object—perhaps an ordered set or a tree.

Zermelo asserts that any indirect proof, in his sense, of a given proposition is readily transformed into a direct proof of that same proposition (cf. *1932b* and *1935*). Zermelo would intend something such as the following. The set $A = \{s \vee t, \neg s, \neg t\}$ amounts to an indirect proof of t from premises $\{s \vee t, \neg s\}$ in that no row within a certain four-row truth table makes all three members of A true. Further, just one row makes the first two members of A true, namely, $s = \text{false}, t = \text{true}, \neg s = \text{true}, \neg t = \text{false}, s \vee t = \text{true}$. If each false member of A is replaced by its negation, we obtain $A' = \{s \vee t, \neg s, t\}$, which, by the same row, counts as a direct proof, in Zermelo's sense, of t from premises $s \vee t$ and $\neg s$. Zermelo's interest in this feature of his notion of proof may be explained by the fact that, within any standard deductive system for propositional logic such as that of Hilbert and Ackermann, it is no straightforward matter to transform an indirect proof, however that notion is characterized, into a direct proof of the same proposition.

As indicated in his fourth paragraph, systems of infinitely long propositions together with semantic proof are Zermelo's answer to the challenge of Hilbert's *non ignorabimus*. In the first instance, this means that no proposition involving appropriate relations is independent of the adopted axioms. One might wonder whether, in addition, semantic proof constitutes a decision procedure for logical consequence on Zermelo's view. Whereas it is hard to see how an infinite thought-experiment or infinite truth table could count as a decision procedure, it is also true that, in the period before Turing clarified the concept, there was no consensus regarding this matter.

Semantic proof constitutes Zermelo's response to Gödel as well (same paragraph). In *s1931d* Zermelo writes in effect that, in one of his own systems of infinitely long propositions, each and every (conditional) proposition is decidable in the sense that, "under the validity of [its antecedent], the validity

of [its consequent] can be made *obvious*". The discussion there suggests that, indeed, a decision procedure for logical consequence is what Zermelo offers despite its involving potentially infinitely many steps.

Cardinality considerations figuring in the fifth paragraph of *1932a* are clearly related to the discussion, one paragraph up, of Gödel's results. According to Zermelo, Gödel's incompleteness theorem is, properly understood, unsurprising and, by implication, uninteresting (cf. *s1931d*). Let fundamental domain G be the set of natural numbers and let Q be the collection of all fundamental relations involving dyadic "is the successor of", monadic "is zero", and dyadic "is equal to". Let S be the system of infinitely long propositions based on Q. Consider the truth distribution that models identity and makes $\ulcorner$1 is the successor of 0$\urcorner$, $\ulcorner$2 is the successor of 1$\urcorner$, ..., and $\ulcorner$0 is zero$\urcorner$ true and all other elementary propositions false. Let us say that a proposition of S is true* if it is true according to this distribution. Zermelo's dismissal of Gödel's proof would then be based on the indisputable fact that cardinality alone dictates that there exists in S a (true*) proposition that is not the expansion of any (provable) sentence of first-order Peano arithmetic (if that theory is consistent). (Universally [existentially] quantified formulæ expand, recursively, to denumerably infinite conjunctions [disjunctions].)

Gödel presented his recent incompleteness results to the annual meeting of the German Mathematical Union held at Bad Elster in September 1931. Zermelo was present and spoke the same day, and *1932a* is a report of his talk, to which he had now added the critique of young Gödel's ideas that constitutes his fourth paragraph.

2. Regarding *1932b*

The German scholarly magazine *Forschungen und Fortschritte* was published in Berlin between 1925 and 1967 under the auspices of the scientific academies of Berlin, Göttingen, Heidelberg, Leipzig, Munich, and Vienna. Zermelo's two-page article appears in the eighth volume, sandwiched between an article concerning Northern European painting of the early sixteenth century and another on the elements of nuclear physics. In harmony with the generalist character of the host journal, Zermelo opens with a sequence of five general questions concerning the nature of mathematics:

(1) What is a mathematical proposition?
(2) What is a mathematical proof?
(3) What is a mathematical theory?
(4) What is a mathematical discipline?
(5) Finally, what are the general logical laws common to all mathematical systems?

Zermelo's goal is to show that an investigation of systems of infinitely long propositions provides interesting answers to all five questions.

Regarding the first question, systems of infinitely long propositions are described in a manner consistent with more detailed presentations in *1932a* and *1935*. Thus mathematical propositions are elements of certain sorts of structures.

Zermelo answers his second question in his fifth paragraph: a (mathematical) proof of proposition s (from premises $\emptyset$) within system S is any truth distribution to basis Q of S such that s comes out true as dictated by "well-known syllogistic rules"—a reference to the classical semantics of propositional logic. (Clauses for disjunction and conjunction are given in the third paragraph, and the meaning of negation can be inferred from the later reference to the law of excluded middle (see also 1)–3) in *1935*, §4).) These rules together with the law of contradiction (what implies a contradiction is false) constitute Zermelo's answer to his fifth and last question.

As for the third question, a theory based on "axiom system" s in S would be the smallest semantically closed subset of S containing s (cf. Zermelo's penultimate sentence). Said theory is mathematical if basis Q of S involves mathematical relations such as quadratic "$a^2 + b^2 + c^2 = r^2$" or dyadic "a lies on line g" exclusively. In *1932a* and *1935* Zermelo emphasizes more restrictive notions by adding the requirement that s be symmetric in his technical sense. (*1932b* omits any mention of symmetry or categoricity.)

It might appear that Zermelo has failed to answer his fourth question. In his fifth paragraph, Zermelo reiterates his claim that all propositions of a system S are decidable, where the decision procedure involves reasoning about something such as an infinite truth table or semantic tableau. With this remark concerning decidability, Zermelo would likely see himself as having characterized mathematical disciplines. To see why, we must turn to Husserl and begin with the latter's notion of a "definite manifold" (*Husserl 1928*, §72):

> [A "*definite*" *manifold* or "*mathematical manifold in the pregnant sense*"] is characterized by the fact that a *finite number of concepts and propositions*, arising in any given case from the nature of the domain in question, *determines completely and unambiguously the totality of all possible formations and does so in accordance with purely analytical necessity*, with the result that, *in principle, nothing further remains open within the domain.*

Further, according to Husserl, a definite manifold is axiomatically characterizable or "definable" in such manner that "the concepts 'true' and 'derivable from the axioms' are equivalent, as are the concepts 'false' and 'refutable based on the axioms' ", which would indicate that Husserl assumes a categorical axiomatization. Finally, any deductive discipline based on such axioms is what Husserl calls a "mathematical discipline".

The operant notion of completeness appears in Husserl's earliest writings on mathematics and greatly influenced Hilbert and all who followed. It is an entirely syntactic notion or an entirely semantic notion, depending upon what one takes "purely analytical necessity" and "derivable from the axioms" to

mean. Gödel, who, like Hilbert, took derivability to refer to some given formal system of mathematical axioms together with underlying logical axioms and rules of inference, proved that entire ranges of theories are not "syntactically complete", or not "decidable", assuming their consistency. (Here a theory T is syntactically complete if, for any sentence ϕ of the language of T, either ϕ or $\neg\phi$ is derivable from the axioms of T.) All those theories fail to be mathematical disciplines in Husserl's sense, assuming that "derivable" does mean "formally derivable". In contrast, according to Zermelo, derivability amounts to semantic consequence. Hence, any mathematical theory T, taken to be a set of propositions of some given system S, is doubtless decidable in the sense that, for any proposition s in S, either $\bigcap_{t\in T} U_t \subseteq U_s$ holds or fails to hold. Thus, if derivability is taken, following Zermelo, to mean semantic consequence, then any axiomatic theory whatever qualifies as a mathematical discipline. This seems to be Zermelo's answer to his fourth question. (As for Husserl, he surely intended a higher standard.)

Zermelo's opening remarks in *s1931c* suggest that both *1932a* and *1932b* were put into final form during September 1931. Whereas *1932a* takes aim at Gödel, Skolem, and unnamed others along the way, *1932b*, in contrast, is free of polemics.

3. Regarding *1935*

The axioms for set theory presented in *1930a* include the axiom of foundation asserting that any nonempty set a contains an element b such that $a \cap b = \emptyset$. Zermelo generalizes this feature of the membership relation in *1935*, which is widely recognized as the source of the theory of well-founded relations (see *Levy 1979*, 63 and 67). In fact, Zermelo himself applies the adjective "well-founded" to domains only. In particular, hierarchies of (urelements and) sets and hierarchies of (fundamental relations and) propositions are two important genres of well-founded domains that he cites.

In 1.1 Zermelo introduces "founding relation" as a generalization of Cantor's notion of well-ordering: dyadic relation $\mathfrak{f}$ *well-founds* set S if every nonempty subset T of S has an $\mathfrak{f}$-minimal element. It is clear from 1.5 that an irreflexive relation is intended. Zermelo allows that $\mathfrak{f}$ be a linear ordering of S, however, in which case $\mathfrak{f}$ is transitive and well-orders S, since any $\mathfrak{f}$-minimal element of $T \subseteq S$ is now an $\mathfrak{f}$-minimum element (cf. 1.6).

Zermelo had exploited hierarchies of urelements and sets to great effect, within set theory, in *1930a*. His objective in *1935* is to match that achievement within general logic. To begin, he formulates and proves "development" Theorem 1.7, a general result to the effect that a well-founded domain S can be stratified: its elements can be distributed over a well-ordered sequence of mutually disjoint "layers" of a (possibly transfinite) hierarchy whereby (1) lowest layer Q_0 (or *basis*) is the collection of all $\mathfrak{f}$-minimal elements of S and (2) layer Q_α with $\alpha > 0$ contains all and only those elements of S that

are "rooted" in layers below Q_α but not contained therein. (An element b of S is rooted in $\bigcup_{\beta<\alpha} Q_\beta$, writes Zermelo, if $a\mathfrak{f}b$ implies that a is an element of $\bigcup_{\beta<\alpha} Q_\beta$.)

In subsequent sections of *1935* Zermelo describes two applications of well-foundedness within logic. Ultimately, he characterizes his investigation as providing the elements of an infinitary proof theory, and proof, the focus of §2, is the first application Zermelo considers:

(1) If $\{a, b, c, \ldots\}$ is a set of propositions whose collective truth implies that of proposition p, then Zermelo says that p follows from $a, b, c, \ldots$ and writes $a\,b\,c \ldots \rightarrow p$, where $\rightarrow$ is a dyadic relation holding between the possibly infinite conjunction of $a, b, c, \ldots$ and proposition p.
(2) Further, if $a\,b\,c \ldots \rightarrow p$ holds, then $a\mathfrak{f}p$, $b\mathfrak{f}p$, $c\mathfrak{f}p$, ... all hold, where $\mathfrak{f}$ denotes the dyadic *partial justification* relation, and each of $a, b, c, \ldots$ is said to "justify" p.

Zermelo's usage indicates that (1) constitutes only a necessary condition for $a\,b\,c \ldots \rightarrow p$ holding; otherwise, $\rightarrow$ would well-found no domain whatever since $p \rightarrow p$ always. Zermelo's intentions with respect to (1) are clarified, we think, if particular applications of some fixed collection of inference rules based on 1)–3) in §4 provide context (more below). Having introduced $\rightarrow$ for (1) and $\mathfrak{f}$ for (2), Zermelo proceeds to use the more perspicuous $\rightarrow$ for (2) as well (cf. the last sentence of the proof in §2). Within the statement of his theorem, Zermelo in effect defines a proof to be any system S of propositions well-founded by $\rightarrow$. The official definition of *(direct) proof* that follows is marred unnecessarily by its overemphasis of truth.

Regarding the issue of inference rules, consider the following. Any system S well-founded by $\mathfrak{f}$ counts as a proof of any $\mathfrak{f}$-maximal proposition p^* in S, let us suppose. However, as described by Zermelo in terms of truth only, dyadic $\mathfrak{f}$ is transitive. This means, in turn, that the basis Q_0 of S and p^*, taken together, constitute an alternative one-step proof of p^*, which cannot be what Zermelo intends. (In §4 Zermelo questions whether his generalized notion of proof is viable since infinitely many intermediate propositions may be involved; if, in fact, $\mathfrak{f}$ is transitive so that any multi-step proof is replaceable by a one-step proof, then Zermelo's fear is hard to fathom.) Instead, as indicated already, we prefer to read Zermelo as assuming a fixed collection of inference rules or syllogisms: $a\mathfrak{f}p$ holds if a is among the premises and p is the conclusion of a specific application of some rule. On this syntactic reading, $\mathfrak{f}$ is nontransitive assuming rules based on 1)–3) of §4.

The syntactic reading does not comport well with Zermelo's brief discussion of mathematical induction at the end of §2, however. There his focus is the well-foundedness of ω-sequence $p_1, p_2, \ldots$ of propositions asserting that some property holds of natural numbers $1, 2, \ldots$, respectively. Meanwhile, $p_{n-1} \rightarrow p_n$ is assumed to be "proved". Even if we take $p_{n-1} \rightarrow p_n$ to indicate the partial ground relation, what rule or syllogism licenses the inference from

"alternating group A_{45} is generated by 3-cycles" (plus lemmata) to "A_{46} is generated by 3-cycles"? Only semantic proof makes sense here.

Zermelo's second application of well-foundedness is to hierarchies of propositions. In §3 a system S of propositions over domain G and some finite collection of explicitly given relations is described as a hierarchy of mutually disjoint layers $Q_1, Q_2, \ldots, Q_\omega, \ldots$ (cf. (†) in Sect. 1. Cumulative level P_α is then defined as $\sum_{\beta<\alpha} Q_\beta$. Such an $S = \sum_{\alpha\geq 1} P_\alpha$ is well-founded by dyadic relation $\mathfrak{f}$, where $a\mathfrak{f}b$ holds just in case proposition a is an immediate proper subcomponent of proposition b. As was once common in technical German, Zermelo uses the term *abgeleitet* both for the (inverse of the) proper subcomponent relation holding between two propositions (cf. 1)–3) at the end of the first paragraph of §3) as well as for derivability in the inferential sense (cf. *s1921*, Thesis III). More often, Zermelo writes *deduzierbar* or *syllogistisch ableitbar* for the latter.

Zermelo's viewpoint is not the model-theoretic one of Gödel and Tarski according to which a mathematical theory comprises a set of formulæ of an initially uninterpreted language. Rather, Zermelo's starting point is invariably some domain of "urelements", fixed in advance and relative to which the relations of the theory are interpreted; in this sense, mathematical theories are domain-related according to him. Consequently, whereas a given mathematical theory will in general have multiple interpretations, those interpretations can be expected to fall within a predetermined range. In effect, the individuals associated with a given theory live behind a veil of ignorance regarding their properties and the relations they bear to one another. Although Zermelo does not say so explicitly, it is as if any theory is associated with a certain domain cardinality—$\aleph_0$ in the case of arithmetic and strongly inaccessible θ in the case of any set theory. (The "boundary numbers" that figure in *1930a* are strongly inaccessible ordinals including ω.)

Zermelo envisions rational reconstructions of a variety of mathematical theories within systems of propositions erected above bases of elementary propositions determined by appropriate domains and relations. Examples of such theories are mentioned at the end of §3, where a distinction is made between theories of a particular structure and general theories instantiated by families of structures. In the case of set theory, this distinction is that between (1) the theory of a particular strongly inaccessible initial segment (normal domain) of some given cumulative hierarchy of urelements and sets as described in *1930a* and (2) the theory of all strongly inaccessible initial segments of that hierarchy. With respect to (1), basis Q would comprise all propositions of the form $a \in b$, where a and b are members of the initial segment in question. As for (2), Zermelo says only that the situation is different.

Whereas Zermelo introduces his notion of symmetric proposition at the very end of *1932a*, its significance becomes clear only in *1935*. Kant held that mathematical axioms must be general in character. But which propo-

sitions of Zermelo's systems count as general? His answer in §5 seems to be that the general propositions are precisely the symmetric ones, although he formulates this answer, elliptically, by saying only that "the axioms are symmetric". Zermelo's comment, in his next sentence, that the fundamental relations need not be symmetric is a vast understatement: in the general case, an elementary proposition over domain G is symmetric only if G is a singleton.

Further remarks regarding elementary propositions invoke the structural notion of symmetry (cf. Sect. 1). In contrast to symmetric propositions, elementary propositions are not fixed but, rather, are taken to one another by the several permutations of domain G. The action of these permutations induces a certain "principal group" [*Hauptgruppe*] $\mathfrak{H}$ constituting a subgroup of the group of all permutations of basis Q. Propositions fixed by members of $\mathfrak{H}$ are termed "symmetric". As an example, let $G = \{a, b, c\}$ and let dyadic relation R be given. Then basis Q comprises nine elementary propositions. In this case, principal group $\mathfrak{H}$ is an order-6 subgroup of the order-9! group of all permutations of Q. As noted in *s1931g*, $\mathfrak{H}$ is isomorphic to the symmetric group $S(G)$. In particular, it does not depend on R.

Zermelo does nothing with $\mathfrak{H}$ either in *1935* or elsewhere. However, its introduction is of some interest, nonetheless, since it indicates the foundational thrust of the theory of symmetric propositions. Zermelo does not mention Felix Klein, but his use of the term *Hauptgruppe* doubtless points to the latter's group-theoretic characterizations of geometries—here we rely on authoritative *Wussing 1984*—and is evidence that Zermelo is drawn to a group-theoretic characterization of logic(s). As we read him, Zermelo is saying that any system $\sum_{\alpha\geq 1} P_\alpha$ of propositions is characterized by the principal group $\mathfrak{H}$ of permutations of its elementary propositions. And just as, on Klein's account, the principal group of transformations associated with any geometry determines a collection of features of the Euclidean plane preserved by those transformations, $\mathfrak{H}$ similarly fixes a collection of symmetric propositions and hence determines a notion of general proposition characteristic of $\sum_{\alpha\geq 1} P_\alpha$.

The twenty-fifth volume of *Fundamenta mathematicæ*, in which *1935* appears, commemorates the founding of that Polish journal, which was the first devoted to set theory and one with which Zermelo had an association extending over several years (see the first paragraph of *Fraenkel 1935*). Its title indicates that *1935* is but a first installment. The three-page manuscript *s1931g*, cited one paragraph back, bears essentially the same title and somewhat extends the discussion published in *1935*. Plausibly, the envisioned sequel to *1935* would have drawn on ideas sketched there. (Regarding chronology, see the opening paragraph of Sect. 1.) Zermelo's use of Boolean matrices (essentially Boolean-valued functions) to model relational structures suggests what is done in *Mautner 1946*.

4. Concluding remarks

Strangely, the results of Zermelo's infinitary investigations, as set forth in *1932a*, *1932b*, and *1935*, had no influence on the subsequent development of infinitary logic as carried out by Tarski's students, among others, starting in the late fifties. One must ask why. Hilbert's influence is blamed at least indirectly in *Barwise 1981*, and the overly general character of Zermelo's proposal regarding proof comes in for criticism in *van Rootselaar 1976*. We find neither suggestion convincing. More probably, one should look to the cultural and scientific discontinuities that befell central Europe, otherwise occupied, at the end of Zermelo's career. There is also the fact that Zermelo's late papers have remained, until now, untranslated. Both *1932a* and *1935* are included in the bibliography *Church 1936* but without an abstract or an introductory sentence that would inform potential readers of their content. A footnote on page 52 of *Lakatos 1976* indicates that Zermelo's late papers

Über Stufen der Quantifikation und die Logik des Unendlichen

1932a

Von der Voraussetzung ausgehend, daß alle mathematischen Begriffe und Sätze durch ein *festes endliches Zeichensystem* darstellbar sein müßten, gerät man schon beim arithmetischen Kontinuum unausweichlich in die bekannte „*Richardsche Antinomie*", wie sie neuerdings, nachdem sie schon lange erledigt und begraben schien, im *Skolemismus*, der Lehre, daß *jede* mathematische Theorie, auch die Mengenlehre, in einem *abzählbaren Modell* realisierbar sei, ihre fröhliche Auferstehung gefunden hat. Aus widerspruchsvollen Prämissen kann man bekanntlich alles beweisen, was man will; aber auch die seltsamsten Konsequenzen, die *Skolem* und andere aus ihrer Grundannahme gezogen haben, z. B. die „Relativität" des Teilmengen- wie des Äquivalenzbegriffes, scheinen noch nicht genügt zu haben, um gegen eine Lehre bedenklich zu stimmen, die für manche bereits die Kraft eines über alle Kritik erhabenen *Dogmas* angenommen zu haben scheint. Eine gesunde „Metamathematik", eine wahre „Logik des Unendlichen", wird aber erst möglich sein durch eine *grundsätzliche Abkehrung* von der oben charakterisierten Voraussetzung, die ich als das „*finitistische Vorurteil*" bezeichnen möchte. Überhaupt sind *nicht* wie manche annehmen, „Zeichenverbindungen" der wahre

were well known in British philosophy circles during the fifties. In contrast, their omission from the bibliography of *Tarski 1958* almost surely indicates that Tarski did not know of them in 1958. (H. Jerome Keisler, a student of Tarski from 1959 to 1961, recalled no mention of Zermelo's infinitary investigations during his Berkeley years.) Bernays did know Zermelo's papers, however, and recommended the work on infinitary logic to his own student, Erwin Engeler. Engeler, in turn a colleague of Tarski in Berkeley from 1961 to 1963, remembered bringing Zermelo's papers to the attention of Tarski, and colleagues Leon Henkin and Robert Vaught, at that time.

Finally, we turn from logic to the philosophy of logic. It is argued in *Taylor 2009* that symmetric propositions, as described in *1932a* and *1935*, constitute Zermelo's analysis of the notion of a general proposition, and in *Taylor 2008* it is shown that this analysis is equivalent to the analysis of logical terms presented in *Tarski 1986*. If this is correct, then Zermelo has standing as a philosopher of logic.

On levels of quantification and the logic of the infinite

1932a

Proceeding from the assumption that it should be possible to represent all mathematical concepts and theorems by means of a *fixed finite system of signs*, we inevitably run into the well-known "*Richard antinomy*" already in the case of the arithmetical continuum. This antinomy, which seemed long buried, has recently celebrated a merry resurrection in the form of *Skolemism*, i.e. the doctrine that *every* mathematical theory, including set theory, can be realized in a *countable model*. As is well known, everything can be proved from contradictory premises; but even the oddest conclusions *Skolem* and others have drawn from their basic assumption, such as the "relativity" of the concept of partial sets and that of equivalence, have apparently not sufficed to raise doubts about a theory which, for some, has already attained the status of a *dogma* beyond all criticism. But a healthy "metamathematics", a true "logic of the infinite", will only become possible once we have *definitively renounced* the assumption characterized above, which I would like to call the "*finitistic prejudice*". Mathematics, generally speaking, is *not* really concerned with "combinations of signs", as some assume, but with *conceptually ideal*

Gegenstand der Mathematik, sondern *begrifflich-ideale Relationen* zwischen den Elementen einer begrifflich gesetzten *unendlichen Mannigfaltigkeit*, und unsere Zeichensysteme sind dabei immer nur *unvollkommene* und von Fall zu Fall wechselnde *Hilfsmittel* unseres *endlichen* Verstandes, des Unendlichen, das wir *nicht unmittelbar* und *intuitiv* „überblicken" oder erfassen können, wenigstens in schrittweiser Annäherung Herr zu werden. Im folgenden soll nun versucht werden, die Grundlagen einer „mathematischen Logik" zu entwickeln, die *frei* vom „finitistischen Vorurteil" und von inneren Widersprüchen Raum genug bieten soll für die *gesamte bisherige Mathematik* und ihre fruchtbare Weiterentwicklung unter *Verzicht* auf alle *willkürlichen Verbote* und *Einschränkungen.*

86 | Jeder mathematischen Theorie liegt zugrunde ein im allgemeinen unendlicher „*Urbereich*" G von Elementen $x, y, z, \ldots$, zwischen denen „*Grundrelationen*" $q(x, y, z, \ldots)$ bestehen können, welche zusammen, d. h. für alle denkbaren Kombinationen der $x, y, z, \ldots$ wieder einen unendlichen Bereich Q möglicher Relationen konstituieren. Aus den Grundrelationen werden weitere „*abgeleitete*" Relationen oder „*Sätze*" gebildet durch die *logischen Elementaroperationen* der „*Negation*", „*Konjunktion*" und „*Disjunktion*", wobei die beiden letztgenannten „verbindenden" Operationen, die wir mit der Bezeichnung „*Quantifikation*" zusammenfassen wollen, beliebig über endliche und unendliche Bereiche bereits definierter Sätze erstreckt werden können. Auf diese Weise entstehen „*Satzsysteme*" S, die wieder unendlich sein können, aber, zur Vermeidung eines „*circulus in definiendo*" oder eines „*regressus in infinitum*", im Sinne meines „Fundierungsaxioms" (Fundamenta Mathematicae T. 16) „*wohlfundiert*" sein müssen in bezug auf die definierende Operation der Quantifikation. Ein Satzsystem S heißt „*wohlfundiert*" in bezug auf eine „erzeugende Operation" f, wenn jedes (echte oder unechte) Teilsystem T von S mindestens einen Satz t enthält, der von *keinem* weiteren Satze t aus T „abhängt", d. h. zu ihm in der Beziehung f steht. Dann enthält auch das Gesamtsystem S notwendig Sätze q, die von *keinem* Satze s aus S „abhängen" und als „Ursätze" seine „*Basis*" Q konstituieren. Wir sagen dann auch, es sei S „wohlfundiert *auf die Basis* Q". Es soll also hier das System der „abgeleiteten Sätze" *wohlfundiert* sein *auf den Bereich* Q *der Grundrelationen.* Nun gilt aber für jeden wohlfundierten Bereich S das allgemeine „*Entwicklungstheorem*" (vgl. a. a. O. § 3), demzufolge jeder solche Bereich *eindeutig* in eine *wohlgeordnete Folge von „Schichten"* Q_α zerlegt werden kann derart, daß die Elemente s einer Schicht Q_α immer nur von solchen *vorangehender* Schichten abhängen und dabei immer der *niedersten* Schicht von dieser Beschaffenheit angehören. Dabei bildet dann die „Basis" Q die unterste Schicht überhaupt $Q = Q_0$, und jedem „abgeleiteten" Satze s entspricht hier eine ganz bestimmte (endliche oder transfinite) Ordnungszahl α als Schicht-Index oder als „*Stufe der Quantifikation*", wobei den „Grundrelationen" q immer die Stufe 0 zukommt.

Werden nun die Grundrelationen q in beliebiger Weise *verteilt* in „wahre" und „falsche": $Q = W + V$, wobei jeder Untermenge u von Q eine solche „*Wahrheitsverteilung*" entspricht, die im Falle einer *binären* Grundrelation

relations among the elements of a conceptually posited *infinite manifold.* Our systems of signs are but *imperfect expedients* of our *finite* mind, which we, adapting them to the circumstances at hand, employ in order to at least gradually get a hold on the infinite, which we can *neither* "survey" *nor* grasp *immediately* and *intuitively.* In what follows I will now try to develop the foundations of a "mathematical logic" that is *free* from both the "finitistic prejudice" and inner contradictions and that is supposed to allow for *all of mathematics as it currently exists* and permit its fruitful further development *without arbitrary prohibitions* and *restrictions.*

Every mathematical theory is based on a generally infinite "*urdomain*" G of elements $x, y, z, \ldots$ between which "*fundamental relations*" $q\,(x, y, z, \ldots)$ may obtain, which, when taken together, i.e. for all conceivable combinations of the $x, y, z, \ldots$, again form an infinite domain Q of possible relations. Proceeding from the fundamental relations, further, "*derived*" relations or "*propositions*" are formed by means of the *logical elementary operations* of "*negation*", "*conjunction*" and "*disjunction*". Here, it is possible to arbitrarily extend the latter two "combining" operations, for which I shall summarily use the term "*quantification*", over finite and infinite domains of propositions already defined. In this way, we obtain "*propositional systems*" S, which may be infinite again but which must be "*well-founded*" in the sense of my "axiom of foundation" (*Zermelo 1930a*) with respect to the defining operation of quantification in order to avoid a "*circulus in definiendo*" and a "*regressus in infinitum*". A propositional system S is called "*well-founded*" with respect to a "generating operation" f if each (proper or improper) partial system T of S contains at least one proposition t which "depends" on, i.e., bears the relation f to, *no* further proposition t from T. Then the total system S, too, necessarily contains propositions q which do *not* "depend" on *any* proposition s from S and which, as "urpropositions", form its "*basis*" Q. In this case, we also say that S is "well-founded *on the basis* Q". Hence, let the system of the "derived propositions" be *well-founded on the basis Q of fundamental relations.* But now, every well-founded domain S is subject to the general "*development theorem*" (see op. cit. § 3) according to which it is possible to *uniquely* decompose every such domain into a *well-ordered sequence of "layers" Q_α* so that the elements s of a layer Q_α always depend only on those of *preceding* layers and, at the same time, always belong to the *lowest* layer so constituted. In this case, the "basis" Q is then the lowest possible layer $Q = Q_0$, and to each "derived" proposition s there corresponds a specific (finite or transfinite) ordinal number α as layer-index or as "*level of quantification*", where the "fundamental relations" q are always of level 0.

Now let us suppose that each fundamental relation q is *classified* as either "true" or "false" in some arbitrary fashion: $Q = W + V$, where to each subset u of Q there corresponds one such "*truth distribution*", which, in the case of

die Form einer zweidimensionalen „*Matrix*“ annimmt, so *überträgt sich* diese Einteilung nach den *allgemeinen syllogistischen Regeln* auch auf jede folgende „Schicht“ und damit, sofern unser System „*wohlfundiert*“ ist, auf das *ganze System* S der „*abgeleiteten Relationen*“. Denn das Restsystem R der noch unverteilten Sätze enthielte immer einen Satz r_1, der, da er nicht zur Basis gehört, von bereits verteilten Sätzen aus $S - R$ durch Quantifikation „abhinge“ und damit auch syllogistisch „*bestimmbar*“ wäre. Es gilt also auch für das ganze wohlfundierte Satzsystem S der „*Satz des Widerspruchs*“ und der des „*ausgeschlossenen Dritten*“, insofern seine Gültigkeit für den Bereich Q der „Grundrelationen“ vorausgesetzt wird. Wird dabei ein Satz s von S als „wahr“ bestimmt, so ist er damit auch „*bewiesen*“ unter Voraussetzung der zugrunde gelegten „Matrix“ der Grundrelationen, und zwar auf der gleichen
87 „Quantifikationsstufe“, auf | der er definiert ist. Denn ein mathematischer „*Beweis*“ ist überhaupt nichts anderes als ein *durch Quantifikation wohlfundiertes System von Sätzen, das ohne Verletzung der syllogistischen Regeln nicht in zwei Klassen, „wahre“ und „falsche“ zerlegt werden kann, wobei die Voraussetzungen zur ersten und der zu beweisende Satz zur zweiten gehörte.* Ersetzt man die in unserem „Beweis“ als „falsch“ bestimmten Sätze überall durch ihre Negate, wodurch sie in „wahre“ verwandelt werden, so entsteht ein neuer, nur aus „wahren“ Sätzen gebildeter „Beweis“, d. h. mit anderen Worten: jeder „*indirekte Beweis*“ (in unserem Sinne) läßt sich durch einen „*direkten*“ ersetzen, und zwar wieder auf der gleichen Quantifikationsstufe, in welcher der Satz selbst definiert ist.

Für unseren Standpunkt ist also jeder „wahre“ Satz zugleich auch „*beweisbar*“, sowie jeder durch ein wohlfundiertes Satzsystem S „definierbare“ Satz zugleich auch „*entscheidbar*“, *und zwar ohne* daß ein Übergang zu einer höheren Quantifikationsstufe erforderlich wäre. Es gibt *keine* (objektiv) „*unentscheidbaren*“ Sätze. Demgegenüber versuchte Herr *Gödel* (Wiener Monatshefte Bd. 38, S. 173) das *Gegenteil* zu beweisen, indem er für ein „*PM*-System“ von beschränkter (nämlich *endlicher*) Quantifikationsstufe einen Satz A herzuleiten suchte, der *nachweislich* (wenigstens in diesem System) *unentscheidbar* sein soll. Der Gödelsche Beweis kommt aber nur dadurch zustande, daß bei ihm die „finitistische“ Einschränkung lediglich auf die „beweisbaren“ Sätze des Systems, nicht auf *alle* dem System angehörigen Sätze angewendet wird. So bilden nur die ersteren, nicht die letzteren eine *abzählbare* Menge, und es muß natürlich in *diesem* Sinne „unentscheidbare“ Sätze geben. Aber gerade der von G. als Beispiel eines „unentscheidbaren“ konstruierte Satz erweist sich, wie er selbst feststellt, nachher doch wieder als „beweisbar“, wenn auch nicht im Sinne der ursprünglichen Definition. Diese ganze Argumentation kann also m. E. nur ⟦dazu⟧ dienen, die Unzulänglichkeit *jeder* „finitistischen“ Beweistheorie zu erhärten, ohne doch ein Mittel zur Behebung dieses Übelstandes an die Hand zu geben. Die eigentliche Frage, ob es absolut-unentscheidbare Sätze, absolut-unlösbare Probleme in der Mathematik gibt, wird durch solche relativistischen Betrachtungen in keiner Weise berührt.

a *binary* fundamental relation, takes the form of a two-dimensional "*matrix*". Then this classification also *carries over* to every succeeding "layer" according to the *general syllogistic rules*, and hence to the *entire system* S of the "*derived relations*", provided that our system is "*well-founded*". For otherwise the remainder system R of the propositions not yet classified would always contain a proposition r_1 which, since it does not belong to the basis, would "depend" on propositions from $S - R$ already classified by dint of quantification. Hence, it, too, would be syllogistically "*determinable*". The "*law of contradiction*" and that of the "*excluded middle*" are therefore valid for the entire well-founded propositional system S as well, provided that its ⟦sic!⟧ validity has been assumed for the domain Q of the "fundamental relations". Once a proposition s of S is determined as "true", it is also "*proved*", under the assumption of the underlying "matrix" of the fundamental relations, namely on that "level of quantification" on which it is defined. For a mathematical "*proof*" is nothing but *a system of propositions well-founded by means of quantification which cannot be decomposed into two classes, "true" and "false", without violating the syllogistic rules, where the assumptions belong to the first ⟦class⟧ and the proposition to be proved to the second one.* If we replace all the propositions which, in our "proof", are determined as "false" by their negations, thereby transforming them into "true" ⟦propositions⟧, then we obtain a new "proof" consisting of "true" propositions only. In other words, every (in our sense) "*indirect proof*" can be replaced by a "*direct*" one, namely, once again, on that level of quantification on which the proposition itself is defined.

From our point of view, every "true" proposition is therefore also "*provable*", and every proposition "definable" by means of a well-founded propositional system S is also "*decidable*", *without* ascent to a higher level of quantification being necessary. There are *no* (objectively) "*undecidable*" propositions. On the other hand, Mr. *Gödel* (*1931a*) has tried to prove the *opposite.* In order to do so, he tried to derive a proposition A, given a "PM-system" of limited (namely *finite*) level of quantification, which is supposed to be *demonstrably* (at least in this system) *undecidable.* But the only reason Gödel's proof works is because he applies the "finitistic" restriction only to the "provable" propositions of the system and not to *all* propositions belonging to the system. Thus only the former, but not the latter, form a *countable* set, and of course, when understood in *this* sense, there must exist "undecidable" propositions. But, as G. himself states, the very proposition constructed by him as an example of an "undecidable" proposition later turns out to be "decidable" after all, even if not in the sense of the original definition. This whole argument, in my opinion, only serves as evidence for the inadequacy of *any* "finitistic" proof theory without, however, providing the means to remove this ill. Such relativistic considerations in no way touch on the real question as to whether there are absolutely undecidable propositions or absolutely unsolvable problems in mathematics.

Da der „Wahrheitswert" (ob wahr oder falsch) eines „wohldefinierten" d. h. einem wohlfundierten Satzsysteme entnommenen Satzes s ausschließlich bestimmt ist durch die Gültigkeitsmatrix seiner *Basis*, so ist jedem solchen Satze s zugeordnet die Menge U_s aller derjenigen Matrizen (oder Wahrheitsverteilungen der Basis), für welche s „wahr" ist. Somit ist der Satz selbst *„logisch äquivalent", d. h. gleichzeitig wahr und falsch* mit der über alle in U_s enthaltenen Verteilungen u erstreckten *Disjunktion.* Ist k die Mächtigkeit von Q, für eine einzige binäre Relation zwischen den Elementen eines Urbereiches von der Mächtigkeit m also $k = m^2$, so ist 2^k die Mächtigkeit für die Menge aller Verteilungen oder Matrizen und

$$2^{2^k} = 2^{2^{m^2}}$$

die Anzahl aller „wesentlich verschiedenen", d. h. unter sich nicht äquivalenten Sätze, die aus den Grundrelationen entspringen.

88 | Ist $U_s = U$ die *alle* Matrizen umfassende Menge, so gilt der Satz s für alle möglichen Wahrheitsverteilungen, er ist „absolut wahr" für die zugrunde gelegte Basis. Ist dagegen $U = 0$, so gilt er für *keine* Verteilung, er ist dann „absolut falsch" oder „widerspruchsvoll" für unsere Basis; in jedem anderen Falle ist er „möglich" oder „widerspruchsfrei". Ist $U_s = U_t$, so gilt t für jede Verteilung, für welche s gilt, und t *„folgt aus s"*, ist aus ihm „deduzierbar" oder „beweisbar". Für $U_s = U_t$ sind beide Sätze *„äquivalent"*, d. h. *gleichzeitig* wahr oder falsch. Ist der Durchschnitt $U_sU_t = 0$, so sind sie miteinander „unvereinbar" oder „im Widerspruch", in jedem anderen Falle „vereinbar". Ist ein Satz *„symmetrisch"*, d. h. *invariant* gegenüber den *Permutationen* der Urelemente $x, y, z, \ldots$, des Bereiches G, so gilt das gleiche auch von der zugehörigen Menge U_s, und wenn *kein echter Teil* dieser Menge die gleiche Eigenschaft hat, sondern alle zugehörigen Matrizen durch *Permutation* auseinander hervorgehen, so ist jeder weitere *symmetrische* Satz t des Systems *entweder* eine *Folge* von s *oder* aber mit ihm *unvereinbar.* Wir haben dann ein *„kategorisches Axiomensystem"*. *Jeder symmetrische Satz s* ist äquivalent einer über lauter *kategorische* Sätze k_α erstreckten *Disjunktion* — in Analogie mit dem bekannten algebraischen Theorem von den *„elementaren symmetrischen Funktionen"*. In gewissem Sinne lassen sich also *beliebige symmetrische Axiomensysteme* auf *„kategorische"* zurückführen. Diese Beispiele mögen genügen, um zu zeigen, was für verschiedenartige Fragen für unseren Standpunkt der metamathematischen Untersuchung zugänglich werden.

Eine ausführliche Darstellung des Gegenstandes ist für die „Math. Annalen" in Aussicht genommen.

Since the "truth value" (whether it is true or false) of a proposition s which is "well-defined", i.e., which belongs to a well-founded propositional system, is solely determined by the validity matrix of its *basis*, there corresponds to each such proposition s the set U_s of all those matrices (or truth distributions of the basis) for which s is "true". Thus the proposition itself is "*logically equivalent*", *i.e. true and false at the same time*, to the *disjunction* extending over all distributions u contained in U_s. Let k be the cardinality of Q for a single binary relation among the elements of an urdomain of cardinality m, i.e., $k = m^2$. Then 2^k is the cardinality of the set of all distributions or matrices, and we have

$$2^{2^k} = 2^{2^{m^2}}$$

as the number of all "essentially different", i.e. mutually nonequivalent, propositions arising from the fundamental relations.

Let $U_s = U$ be the set of *all* matrices. Then the proposition s holds for all possible truth distributions. It is "absolutely true" for the underlying basis. If, on the other hand, $U = 0$, then it holds for *no* distribution. It is, in this case, "absolutely false", or "contradictory", for our basis; in any other case, it is "possible", or "consistent". If $U_s \subseteq$ [1] U_t, then t holds for every distribution for which s holds, and t "*follows from s*". It is "deducible" from it, or "provable". In the case of $U_s = U_t$, both propositions are "*equivalent*", i.e., *simultaneously* true or false. If the intersection $U_sU_t = 0$, then they are "incompatible", or "in contradiction", with one another. In any other case, they are "compatible". If a proposition is "*symmetric*", i.e., *invariant* under the *permutations* of the urelements $x, y, z, \ldots$ of the domain G, then the same holds of the corresponding set U_s. And if *no* ⟦nonempty⟧ *proper* part of this set possesses the same property but if all corresponding matrices arise from one another by *permutation*, then any further *symmetric* proposition t of the system is *either* a *consequence* of s *or incompatible* with it. We then have a "*categorical axiom system*". *Every symmetric proposition s* is equivalent to a *disjunction* extending over propositions k_α all of which are *categorical*—in analogy with the well-known algebraic theorem of the "*elementary symmetric functions*". Hence, in a certain sense, it is possible to reduce *arbitrary symmetric axiom systems* to "*categorical*" ones. These examples may suffice to show the great variety of questions that, from our point of view, become amenable to metamathematical investigation.

I intend to elaborate on the topic in the "Math. Annalen".

[1] ⟦Zermelo erroneously writes "=" instead of "⊆".⟧

Über mathematische Systeme und die Logik des Unendlichen

1932b

Was ist ein „mathematischer Satz“, ein „mathematischer Beweis“, eine „mathematische Theorie“, eine „mathematische Disziplin“? Um eine allgemeine Theorie der mathematischen Satz-Systeme, wie sie allen mathematischen Disziplinen zugrunde liegen, handelt es sich in den nachfolgenden Betrachtungen, die hier in kurzem Auszuge wiedergegeben werden sollen. Ein mathematischer „Satz“ hat nur Sinn und Bedeutung innerhalb eines mathematischen Systems, einer Theorie oder einer (umfassenden) Disziplin, wie z. B. der „Euklidischen Geometrie“ oder der „Arithmetik der reellen Zahlen“. Welches aber sind die charakteristischen Merkmale, welches die allgemeinen logischen Grundgesetze, die allen „mathematischen Systemen“ gemeinsam sind?

Den Ausgangspunkt jeder mathematischen Theorie bildet ein „Grundbereich“ von „Urelementen“ $a, b, c, \ldots$, zwischen denen Grundrelationen, wie z. B. $a < b$, $a + b = c$, $a^2 + b^2 + c^2 = r^2$ oder: „der Punkt a liegt auf der Geraden g“, definiert sind, so daß sie zwischen je zwei, drei oder mehr Urelementen bestehen oder nicht bestehen können. Dieser Grundbereich G wird im allgemeinen „unendlich“ sein, d. h. unendlich viele Urelemente $a, b, c, \ldots$ umfassen, und dementsprechend wird auch die Gesamtheit Q der (möglichen) Grundrelationen im allgemeinen unendlich sein. Handelt es sich z. B. um eine „binäre Relation“, d. h. um eine solche zwischen je zwei Elementen, und um einen Grundbereich von der „Kardinalzahl“ m, so wird die Gesamtheit der zwischen ihnen möglichen Grundrelationen die (endliche oder transfinite) Kardinalzahl oder „Mächtigkeit“ m^2 besitzen.

Aus den „Grundrelationen“ werden nun weitere „Sätze“ „abgeleitet“ durch die logischen Elementaroperationen der „Negation“, „Konjunktion“ und „Disjunktion“, die im Deutschen durch die Wörter „nicht“, „und“ und „oder“ ausgedrückt werden: jedem Satz a entspricht ein Satz $-a$ („a nicht“) sowie jedem Paar a, b das weitere Paar „a und b“ und „a oder b“. Die beiden letzten „verbindenden Operationen“ oder „Quantifikationen“ sind aber nicht auf „Paare“ beschränkt, sondern können auf drei, mehr, ja auf beliebige endliche oder unendliche Bereiche von Sätzen ausgedehnt werden, wobei die so erweiterte „Konjunktion“ und „Disjunktion“ die Bedeutung annehmen, daß im ersten Falle „alle“, im zweiten „mindestens ein“ Satz des betrachteten Bereiches B, über den „quantifiziert“ wird, gelten soll. Durch fortgesetzte Anwendung dieser Operationen der Negation und Quantifikation auf die bereits gewonnenen Sätze entstehen immer weitere „abgeleitete Sätze“ s, die dann zusammen mit den ursprünglichen „Grundrelationen“ q ein „Satz-System“ S konstituieren. Bei einer solchen „successiven Definition“ muß aber natürlich jeder „regressus in infinitum“ sowie jeder „circulus in definiendo“ vermieden werden, und dafür ergibt sich als „notwendige und hinreichende Bedingung“ die folgen-

On mathematical systems and the logic of the infinite

1932b

⟦The introductory note just before *1932a* also addresses *1932b*.⟧

What is a "mathematical proposition", a "mathematical proof", a "mathematical theory", a "mathematical discipline"? A general theory of propositional systems as it underlies all mathematical disciplines is the subject of the following considerations outlined briefly here. A mathematical "proposition" makes sense and has a meaning only within a mathematical system, a theory or a (comprehensive) discipline as, e.g., "Euclidean geometry" or the "arithmetic of real numbers". But what are the characteristic features, what are the general basic laws of logic common to all "mathematical systems"?

What forms the starting point of any mathematical theory is a "basic domain" of "urelements" $a, b, c, \ldots$ among which fundamental relations, such as $a < b$, $a + b = c$, $a^2 + b^2 + c^2 = r^2$, and "point a lies on line g", are defined so that they obtain or do not obtain between any two, three or more urelements respectively. In general, this basic domain G is "infinite", that is, it comprises infinitely many urelements $a, b, c, \ldots$, and, accordingly, the totality Q of the (possible) fundamental relations is, in general, infinite as well. Consider, e.g., a "binary relation", that is, one obtaining between two elements, and a basic domain with "cardinal number" m. In this case, the totality of the fundamental relations possibly obtaining between them possesses the (finite or transfinite) cardinal number or "cardinality" m^2.

From the "fundamental relations" we now "derive" further "propositions" by means of the elementary logical operations of "negation", "conjunction" and "disjunction", which are expressed in English by use of the words "not", "and" and "or": to every proposition a there corresponds a proposition $-a$ ("not a") and to every pair a, b the further pair "a and b" and "a or b". The last two "combining operations", or "quantifications", however, are not restricted to "pairs", but can also be extended to three or more, and even to arbitrary finite and infinite domains of propositions, where the "conjunction" and "disjunction" so extended assume such a meaning that, in the first case, "all" propositions of the considered domain B over which we "quantify" are supposed to be valid and, in the second case, "at least one" proposition is valid. Repeated application of these operations of negation and quantification to the propositions already obtained always generates further "derived propositions" s, which, together with the original "fundamental relations" q then form a "propositional system" S. However, when using such a "successive definition" it is of course necessary to avoid any "regressus in infinitum" and any "circulus in definiendo". To this end, we have the following as "necessary and sufficient condition": the system S defined by means of our operations

de: das durch unsere Operationen definierte System S muß „wohlfundiert" sein auf die Gesamtheit Q der Grundrelationen als „Basis", d. h. es muß jeder Teilbereich T von S mindestens einen Satz t_1 enthalten, der nicht aus weiteren Sätzen t von T durch Negation oder Quantifikation abgeleitet ist, also entweder selbst zu Q gehört oder aus Sätzen $a, b, c, \ldots$ des Restsystems $R = S - T$ abgeleitet ist. Diese Bedingung der „Fundierung" ist die denkbar allgemeinste und setzt insbesondere *nicht* voraus, daß die Gesamtheit der „erzeugenden Quantifkationen" eine *endliche* sein müsse. Dagegen folgt in der Tat aus unserer Bedingung, daß man, von einem Satze s des Systems ausgehend und in der Erzeugungsreihe rückwärtsschreitend, immer nach einer endlichen Anzahl von Schritten notwendig bei einer Grundrelation q endet.

Weiter ergibt sich, daß ein in dieser Weise „wohlfundiertes System" sich (eindeutig) in eine „wohlgeordnete" Folge getrennter „Schichten" zerlegen läßt derart, daß die Sätze jeder Schicht Q_r ausschließlich von Sätzen der vorangehenden Schichten „abhängen", d. h. durch Quantifikation aus ihnen hervorgehen und zugleich der niedersten Schicht angehören, für welche dies der Fall ist. Unter einer „wohlgeordneten Folge" versteht man dabei eine solche, in welcher jede Teilfolge wie auch die ganze immer ein *erstes*, wenn auch *nicht* notwendig ein *letztes* Element enthält. So ist z. B. die Reihe der natürlichen Zahlen $1, 2, 3, \ldots$ eine wohlgeordnete Folge, ebenso auch die Folge

$$1, 2, 3, \ldots 1', 2', 3', \ldots 1'', 2'', 3'' ,$$

nicht aber eine Folge der Form

$$\ldots 3, 2, 1, 1', 2', 3', \ldots 1'' ,$$

welche *kein erstes* Element besitzt. Die Schicht, welcher ein gegebener Satz s angehört, oder den sie charakterisierenden Index r nennen wir auch seine „Quantifikationsstufe". Die „Basis" Q hat dabei den Index 0 und umfaßt alle „Grundrelationen" q.

Diese Grundrelationen q, die aus allen Kombinationen der Urelemente entstehen, werden nicht alle zugleich gelten, sondern sie zerfallen in zwei Klassen, in „wahre" und „falsche", und jeder beliebigen Einteilung der Basis S entspricht eine „Matrix" p als mögliche Wahrheitsverteilung. Jede solche „Wahrheitsverteilung" überträgt sich aber nach den bekannten syllogistischen Regeln auf alle Sätze s des ganzen wohlfundierten Systems S. Denn der „Wahrheitswert" jedes aus den Sätzen $a, b, c \ldots$ durch Negation und Quantifikation „abgeleiteten" Satzes t ist immer eindeutig bestimmt durch die Wahrheitswerte der Sätze $a, b, c \ldots$, von denen er „unmittelbar abhängt", und jedes etwa noch unverteilte Restsystem T enthält mindestens einen Satz t_1, der von Sätzen $a, b, c \ldots$ außerhalb T unmittelbar abhängt, also von bereits verteilten, und somit durch diese mitbestimmt ist. Jeder Wahrheitsverteilung der Basis

must be "well-founded" on the totality Q of the fundamental relations as "basis", that is, each partial domain T of S must contain at least one proposition t_1 which is not derived from further propositions t of T by means of negation or quantification, and hence which either itself belongs to Q or is derived from propositions $a, b, c, \ldots$ of the remainder system $R = S - T$. This condition of "foundation" is the most general one conceivable. In particular, it does *not* assume that the totality of the "generating quantifications" must be a *finite* one. On the other hand, it in fact follows from our condition that, by necessity, we always reach a fundamental relation q in a finite number of steps when starting out from a proposition s of the system and going backwards in the sequence of generation.

Furthermore, it follows that we can (uniquely) decompose a thus "well-founded system" into a "well-ordered" sequence of separated "layers" so that the propositions of each layer Q_r exclusively "depend" on propositions of the preceding layers, that is, that they arise from them by means of ⟦negation or⟧ quantification and, at the same time, belong to the lowest layer for which this is the case. By a "well-ordered sequence" we mean one in which every subsequence as well as the entire ⟦sequence⟧ always contains a *first* if *not* necessarily a *last* element. Thus, e.g., the sequence of the natural numbers $1, 2, 3, \ldots$ is a well-ordered sequence, as is also the sequence

$$1, 2, 3, \ldots 1', 2', 3', \ldots 1'', 2'', 3'' ,$$

but not a sequence of the form

$$\ldots 3, 2, 1, 1', 2', 3', \ldots 1'' ,$$

which contains *no first* element. The layer to which a given proposition s belongs, or the index r characterizing it, is said to be its "quantification level". The "basis" Q has here the index 0 and comprises all "fundamental relations" q.

These fundamental relations q arising from all combinations of the urelements do not all obtain simultaneously but fall into two classes, "true" and "false", and to each arbitrary classification of the basis Q[1] there corresponds a "matrix" p as a possible truth distribution. Each such "truth distribution" is, however, being passed on to all propositions s of the entire well-founded system S in accordance with the well-known syllogistic rules. For the "truth value" of every proposition t "derived" from the propositions $a, b, c \ldots$ by means of negation and quantification is always uniquely determined by the truth values of the propositions $a, b, c \ldots$, on which it "immediately depends", and any remainder system T which may not have been distributed yet contains at least one proposition t_1 that ⟦either belongs to basis Q or⟧ immediately depends on propositions $a, b, c \ldots$ outside of T, and hence on propositions already distributed, and thus also determined by those propositions.

---

1 ⟦Zermelo erroneously writes "S" instead of "Q".⟧

Q entspricht also eine solche des ganzen Systems S, und diese syllogistische Wahrheitsbestimmung eines in dieser Weise auf die Basis „wohlfundierten" Satzes s als „wahr" oder „falsch" kann auch als der „Beweis" des Satzes s oder seines Gegenteils $-s$, als seine „Entscheidung" bezeichnet werden. In diesem Sinne ist jeder Satz eines wohlfundierten Systems durch die Wahrheitsverteilung der Basis entschieden und somit „entscheidbar" innerhalb des definierenden Systems. Die Sätze „vom Widerspruch" und vom „ausgeschlossenen dritten" gelten für jedes wohlfundierte System, insofern ihre Gültigkeit für die „Basis des Systems", die Gesamtheit der Grundrelationen vorausgesetzt wird. Jeder „wahre" Satze ist auch „beweisbar", und zwar auf der gleichen Quantifikationsstufe, auf der er „definiert" ist. Des weiteren zeigt sich, daß
7 ein solches wohlfundiertes | System noch wohlfundiert bleibt, wenn man jeden „falschen" Satz b durch seine Negation $-b$ ersetzt, so daß nur „wahre" Sätze übrig bleiben. Jeder „wahre" Satz ist also syllogistisch ableitbar ausschließlich mit Hilfe „wahrer" Sätze oder: jeder „indirekte Beweis" ist durch einen „direkten" ersetzbar.

Da die „Wahrheit" oder „Falschheit" eines wohlfundierten Satzes eindeutig bestimmt ist durch die Wahrheitsverteilung seiner Basis, so ist jedem solchen Satze s zugeordnet die Gesamtheit U_s aller möglichen Verteilungen p, für welche er wahr ist, er ist also „logisch äquivalent" (d. i. gleichzeitig wahr und falsch) mit der über alle Sätze von U_s erstreckten Disjunktion. Umfaßt U_s alle möglichen Verteilungen, $U_s = U$, so ist s „absolut wahr"; ist $U_s = 0$, so ist s „absolut falsch" oder „widerspruchsvoll" für diese Basis. Ist für zwei Sätze s und t, U_s enthalten in U_t, so ist t immer wahr, wenn s wahr ist, oder „t folgt aus s". Betrachtet man s als „Axiomensystem", so konstituieren die sämtlichen „aus s folgenden" Sätze t die „auf das Axiomensystem s gegründete mathematische Theorie". — Diese Beispiele und Andeutungen mögen hier genügen, um zu zeigen, in welcher Weise die verschiedenen auf Form und Wesen der Mathematik gerichteten „metamathematischen" Fragen und Probleme von der hier entwickelten Grundlage aus einer exakten wissenschaftlichen Untersuchung zugänglich werden.

Thus, to every truth distribution of the basis Q there corresponds one of the entire system S, and this syllogistic truth determination as "true" or "false" of a proposition s thus "well-founded" on the basis can also be said to be the "proof" of the proposition s or of its opposite $-s$, its "decision". In this sense, each proposition of a well-founded system is decided by means of the truth distribution of the basis, and is hence "decidable" within the defining system. The laws "of contradiction" and of "the excluded middle" are valid for every well-founded system insofar as their validity is assumed for the "basis of the system", the totality of the fundamental relations. Every "true" proposition is also "provable", namely on the same quantification level on which it is "defined". Furthermore, it becomes evident that such a well-founded system remains well-founded if every "false" proposition b is replaced by its negation $-b$ so that we are only left with "true" propositions. Every "true" proposition is thus syllogistically derivable by means of "true" propositions alone, or: every "indirect proof" can be replaced by a "direct" one.

Since the "truth" or "falsity" of a well-founded proposition is uniquely determined by means of the truth distribution of its basis, correlated to every such proposition s there is the totality U_s of all possible distributions p for which it is true, and hence it is "logically equivalent" (that is, simultaneously true or false) to the disjunction extended over all propositions of U_s. If U_s comprises all possible distributions, $U_s = U$, then s is "absolutely true"; if $U_s = 0$, then s is "absolutely false" or "contradictory" for this basis. If, for two propositions s and t, U_s is contained in U_t, then t is true whenever s is true, or "t follows from s". If we consider s as an "axiom system", then all of the propositions t "following from s" together form "the mathematical theory based on the axiom system s".—These examples and suggestions shall suffice here to show the way in which the various "metamathematical" questions and problems concerning the form and nature of mathematics become amenable to precise scientific investigation once approached from the foundation developed here.

Introductory note to 1932c

Heinz-Dieter Ebbinghaus

Zermelo's preface to the edition of Cantor's collected works provides information in two directions, information by what it says and information by what it does not say. To explain the second, we start with some part of the prehistory of the edition which concerns the collaboration with Abraham Fraenkel. We then comment on the preface.

The Fraenkel affair

At the end of 1926 Zermelo approached Springer Verlag, asking them whether they were willing to publish the collected works of Georg Cantor with he himself taking on the editorship. Springer agreed, informing him that Fraenkel had asked them whether they would publish a biography of Cantor which Fraenkel was preparing for the Deutsche Mathematiker-Vereinigung (*Fraenkel 1930*). In March 1927 Zermelo offered Fraenkel a collaboration either by taking part in the editorship or by contributing his Cantor biography in a suitable form. Fraenkel answered immediately, expressing his pleasure and emphasizing that he felt honoured by the possibility of a collaboration. Eventually, the second alternative was agreed upon.

In October Fraenkel sent a first version of the biography. Zermelo's answer, a typewritten letter of five pages, was to become the origin of a growing estrangement. With a mixture of objective comments, ironic side remarks, and harsh criticism which nevertheless shows an inherent sensibility, Zermelo urged Fraenkel to adopt his view of how to write a biography. As late as December 1929 Fraenkel wrote that with regard to the factual and in many ways stylistic changes Zermelo had proposed, he was not able to agree to the majority of them, because he did not want to disown either his scientific views or his style. The relationship between the two pioneers of set theory was never to recover, even when after the appearance of the Cantor edition the letters lost their frostiness. Instead of Fraenkel, Reinhold Baer and Arnold Scholz became the main partners in the Cantor enterprise.

Annotations

The first sentences of the preface bear witness to Zermelo's high estimation for Cantor's set-theoretic achievements. As regards style and contents, Zermelo showers praise on Cantor's work. This admiration goes together with his general intention to defend without compromise what he considered as Cantor's heritage, being totally unwilling to relinquish Cantorian set theory to the finitary approaches of Thoralf Skolem and Kurt Gödel and to the principles of intuitionism. At the end of the preface, he will come back to this point.

When describing the contents and commenting on the choice of papers, Zermelo emphasizes that Cantor's papers on trigonometric series (*1871a, 1871b, 1872*) lead step by step with inherent necessity to the conception of transfinite ordinal numbers; they thus could be considered the birthplace of Cantorian set theory.[1] Cantor himself emphasizes this point in a footnote of *1880*, 358. Strangely enough, the Cantor edition does not contain the footnotes of this paper.

The description of the contents that the preface provides is rather detailed. It is therefore astonishing at first glance that Fraenkel's biography, forming the last section and comprising more than thirty pages, is not mentioned. The reasons are be found in the historical circumstances described above. There is an early, but undated, draft of the preface where Zermelo did praise the cooperation with Fraenkel as follows: "The concluding part consists of a biography of Cantor from the experienced pen of the mathematical author who is well-known to wide circles because of his 'Einleitung in die Mengenlehre' [2]" ("Den Schluß bildet eine Biographie Cantors aus der bewährten Feder des durch seine 'Einleitung in die Mengenlehre' weiteren Kreisen bekannten Autors").

As Zermelo explains, the appendix of the Cantor edition contains *fragments* of hitherto unpublished letters of the Cantor-Dedekind correspondence. In particular, all parts concerned with personal matters have been left out. This feature corresponds to one of Zermelo's criticisms of Fraenkel's Cantor biography: Personal information such as that from letters provided, for example, by family members should only be used to widen the perspective of the writer and published, if at all, only very reluctantly.

The transcriptions of the letters had been provided by the philosopher Jean Cavaillès (1903–1944), probably during a stay in Germany in 1930. Cavaillès had shown a strong interest in the foundations of set theory in general and in Zermelo's views in particular, which resulted in two major publications (*1938a, 1938b*). He intended to edit all his transcriptions himself, considering a publication in the *Jahresbericht der Deutschen Mathematiker-Vereinigung*.[3] Five years later he published them together with Emmy Noether in a separate edition (*Noether and Cavaillès 1937*). In May 1930 Emmy Noether, then working on the edition of Dedekind's collected papers, proposed to Zermelo incorporating the correspondence into the Cantor collection because "it is Cantor who makes the new scientific statements." Zermelo rejected her proposal, selecting only those parts of the correspondence which formed "an essential and indispensable supplement" (*Cantor 1932*, 451). The passages chosen concern Cantor's "final conceptions about the system of *all* ordinals and the system of *all* cardinals together with consistent and inconsistent totalities" (ibid., 451), thus throwing light on the

[1] Cf. also *Dauben 1990*, 43f, *Hallett 1984*, 3f, and *Purkert-Ilgauds 1987*, 39.

[2] *Fraenkel 1919.*

[3] Letter to Zermelo of 12 July 1932; Universitätsarchiv Freiburg, sign. C 129/21.

history of the paradoxes. The selection comprises parts of the letters which Cantor wrote to Dedekind on 28 July, 3 August, 28 August, 30 August, and 31 August 1899 and Dedekind wrote to Cantor on 29 August 1899. Cantor's second letter is given as a part of the first one. In addition to the points mentioned above, the selection provides in print Dedekind's now classical proof of the equivalence theorem, i.e., the corresponding part of Dedekind's letter, and Cantor's answer of 30 August 1899.[4] Dedekind's proof is the most elegant one among the variety of proofs of the equivalence theorem, written down by him already in July 1887, but published for the first time only here and, at the same time, in Dedekind's collected papers (*1932*, 447–449). Zermelo, who had found a similar proof in June 1905 and published it in his axiomatization paper *1908b*, wonders (*Cantor 1932*, 451, fn. 2) "why neither Dedekind nor Cantor decided then to publish this at any rate not unimportant proof".[5]

Despite his admiration for Cantor's work, Zermelo does not abstain from criticism. In particular, he criticizes Cantor's intuitive use of successive choices (*Cantor 1932*, 352, fn. 6, and 451, fn. 1). The extensive second footnote refers to Cantor's letter to Dedekind of 3 August 1899. In this letter Cantor sketches an intuitive proof of the well-ordering principle, making use of successive choices along all ordinals, thereby speaking of a "projection" of ordinals into sets. The key of the proof consists in showing that the cardinal of an infinite set is an aleph. (Then infinite sets are equivalent to well-ordered sets and so are well-orderable as well.) If the set S were a counterexample, i.e. a set without an aleph, then a projection of the inconsistent multiplicity of all ordinals into S (by choosing different elements $s_0, s_1, \ldots, s_\omega, \ldots$ of S) would show that S itself encompassed an inconsistent multiplicity and, hence, could not be a set—a contradiction. In the footnote Zermelo criticizes this procedure as being based on a "vague 'intuition'":

> Here temporal intuition is applied to a process which transcends all of intuition, and there is faked a being that should be able to make *successive* arbitrary choices [... The proof works] only by applying the "axiom of choice" which postulates the possibility of *simultaneous* choices and which is used by Cantor instinctively without being conscious of it, but which he nowhere formulates explicitly.

[4] The edition of the letters is criticized by Ivor Grattan-Guinness; cf., for example, *Grattan-Guinness 2000*, 548; cf. also *Cantor 1991*, 406. For other corrections concerning Zermelo's edition, cf. *Grattan-Guinness 1974*, 134–136.—For the overall reception of the edition, cf. *Ebbinghaus 2007b*, 162–163.

[5] For details concerning the proof history of the equivalence theorem, cf. *Kanamori 2004*, §4, *Ferreirós 1999*, sect. VII, or (the introductory notes to) *Zermelo 1901* and *Zermelo 1908b*.

During the editorial work Zermelo became involved in a serious foundational controversy. On the one hand, it was rooted in intuitionism and its endeavour to propagate a kind of mathematics which had been deprived of actual infinity and classical logic.[6] On the other hand, it resulted from Gödel's and Skolem's finitary approaches to mathematics with their inherent limitations: Gödel's first incompleteness theorem, based on finitary axiomatizations of mathematical theories and reasoning, and Skolem's result about the existence of countable models of set theory, a consequence of axiomatizing set theory in first-order logic. In what became a personal campaign, Zermelo fought for what he called "true mathematics", Cantor's set theory being an essential part of it.[7] The last paragraph of the preface addresses this matter of concern. It takes up a remark of Cantor's about Antoine Gombaud, named Chevalier de Méré, in Cantor's historical notes on probability theory (*Cantor 1873*, 36). In 1654 Gombaud turned to Blaise Pascal for intuitive justifications of and with doubts about some facts of probability dealing with dice games. Pascal discussed the matter with Pierre de Fermat in a series of letters, in this way enhancing the foundation of probability theory. Having reported on these events, Cantor (ibid.) continues:

> I believe that the Chevalier de Meré can be taken as a cautionary example by all opponents to exact research—such opponents exist at any time and any place; for it may easily occur that just at the spot where they try to give science its deadly wound, a new branch, more beautiful, if possible, and more promising than all former ones, will burst into bloom before their eyes—like probability theory before the eyes of the Chevalier de Meré.

Zermelo's preface ends quoting this passage, blaming the opponents of Cantorian set theory as starting their attacks in timid weakness. In an annotation to Cantor's passage he writes (*Cantor 1932*, 367):

> Even more justified, Cantor's concluding remark about the Chevalier de Meré could be related to the fate of set theory and to its opponents: a strange prophecy at a time where the future pioneer had merely started his genuine lifework. [...] The present remark characterizes the fate of any reactionary direction in a striking way and will be relevant again and again.

---

[6] Zermelo's aversion to intuitionism becomes visible again and again, starting in the early 1920s in conversations with the philosopher Marvin Farber (cf. *Ebbinghaus 2007b*, sect. 4.3.2) and ending with *s1937*. Cf., in particular, (the inroductory note to) *s1929b*.

[7] Cf. (the introductory notes to) *s1931b–d, 1932a,* and *s1937*.

Vorwort zu *Cantor 1932*

1932c

In der Geschichte der Wissenschaften ist es gewiß ein seltener Fall, wenn eine ganze wissenschaftliche Disziplin von grundlegender Bedeutung der schöpferischen Tat eines einzelnen zu verdanken ist. Dieser Fall ist verwirklicht in der Schöpfung *Georg Cantor*s, der Mengenlehre, einer neuen mathematischen Disziplin, die während eines Zeitraumes von etwa 25 Jahren in einer Reihe von Abhandlungen ein und desselben Forschers in ihren Grundzügen entwickelt, seitdem zum bleibenden Besitze der Wissenschaft geworden ist, so daß alle späteren Forschungen auf diesem Gebiete nur noch als ergänzende Ausführungen seiner grundlegenden Gedanken aufzufassen sind. Aber auch abgesehen von dieser ihrer historischen Bedeutung sind die *Cantor*schen Originalabhandlungen noch für den heutigen Leser von unmittelbarem Interesse, in ihrer klassischen Einfachheit und Präzision ebenso zur ersten Einführung geeignet und darin noch von keinem neueren Lehrbuch übertroffen, wie auch für den Fortgeschrittenen durch die Fülle der zugrunde liegenden Gedanken eine genußreich anregende Lektüre. Der immer noch wachsende Einfluß der Mengenlehre auf alle Zweige der modernen Mathematik und vor allem ihre überragende Bedeutung für die heutige Grundlagenforschung haben bei Mathematikern wie bei Philosophen den Wunsch entstehen lassen, die in verschiedenen Zeitschriften zerstreuten und teilweise schwer zugänglichen Abhandlungen in ihrem natürlichen Zusammenhange lesen und studieren zu können. Diesem Bedürfnisse zu entsprechen ist die hier vorliegende Gesamtausgabe bestimmt, welche aber außer den rein mengentheoretischen auch alle übrigen wissenschaftlichen Abhandlungen *Cantor*s mathematischen und philosophischen Inhalts umfaßt, einschließlich der (lateinisch geschriebenen) Dissertation und Habilitationsschrift sowie insbesondere auch der zuerst in der „Zeitschrift für Philosophie und philosophische Kritik" erschienenen Aufsätze, in denen *Cantor* im Briefwechsel mit verschiedenen Mathematikern und Philosophen seinen Unendlichkeitsbegriff entwickelt und gegen philosophische und theologische Einwände verteidigt. Während nun die Aufnahme dieser mit der Mengenlehre im engsten Zusammenhange stehenden philosophischen Abhandlungen keiner besonderen Begründung bedarf, ist die Aufnahme der nicht sehr umfangreichen zahlentheoretischen Arbeiten hauptsächlich in biographischem Interesse erfolgt, um dem Leser auch die ersten Anfänge dieses Forscherlebens vor Augen zu führen. Endlich sind die funktionentheoretischen
IV Untersuchungen *Cantor*s, welche hauptsächlich die | Theorie der trigonometrischen Reihen betreffen und noch heute von gegenständlichem Interesse sind, schon deshalb unentbehrlich, weil sich an diesen Problemen zuerst die grundlegenden neuen Ideen, die dann zur Mengenlehre führten, Schritt für Schritt entwickelt haben.

Preface to *Cantor 1932*

1932c

It is certainly rarely the case in the history of the sciences that an entire scientific discipline of fundamental significance is due to the creative act of a single individual. This is the case with *Georg Cantor*'s creation, set theory, a new mathematical discipline, which was developed in its outlines over the course of about 25 years in a series of publications by one and the same researcher and of which science has permanently seized possession since then so that all later research in this field can only be considered as supplementary development of his fundamental ideas. But even aside from their historical significance, *Cantor*'s original papers are still of immediate interest to today's reader. Their classic simplicity and precision commend them as a first introduction, and no more recent textbook has yet surpassed them in this respect. Moreover, the advanced reader will be pleasantly stimulated by the abundance of the underlying ideas. The still growing influence of set theory on all branches of modern mathematics, and in particular its great significance for today's foundational research, has instilled the desire in both mathematicians and philosophers to read and study the papers in their natural order, which are now scattered over various journals and not all of which are easily accessible. The present complete edition is meant to meet this need. Besides the purely set-theoretic papers, it also contains all other scientific treatises by *Cantor* of mathematical or philosophical content, including the dissertation (in Latin) and the habilitation, and in particular also the papers published first in the "Zeitschrift für Philosophie und philosophische Kritik", in which *Cantor*, in correspondence with various mathematicians and philosophers, develops his concept of infinity and defends it against philosophical and theological objections. Now, there is no need to justify the inclusion of these philosophical treatises, which are intimately related to set theory. By contrast, the few number-theoretic papers have been included mainly for biographical interest in order to bring to the reader's attention the early stages in the life of this researcher. Finally, *Cantor*'s function-theoretic investigations, which are mainly concerned with the theory of trigonometric series and are still of substantial interest today, are indispensable, especially since it was in connection with these problems that those basic new ideas that then led to set theory were first developed step by step.

Die Abhandlungen sind in der vorliegenden Ausgabe nach ihren Stoffgebieten in vier Hauptabschnitte eingeteilt, je nachdem sie die Zahlentheorie und Algebra, die Funktionentheorie, die Mengenlehre oder die Geschichte der Mathematik und die Philosophie betreffen, und in den einzelnen Abteilungen nach der Zeit ihres Erscheinens geordnet. Der Abdruck erfolgte originalgetreu, unter sorgfältiger Verbesserung aller nachweisbaren Versehen und Druckfehler des Originals und unter Einführung der heutigen Rechtschreibung; kürzere Zusätze des Herausgebers im Texte sind durch [eckige] Klammern kenntlich gemacht. Die „Anmerkungen" hinter den einzelnen Abhandlungen sind teils erläuternder, teils kritischer Natur und enthalten u. a. auch Hinweise auf die Bedeutung der betreffenden Arbeiten und auf die sich anschließende spätere Literatur. Doch sind diese Anmerkungen durchaus für den *mathematisch*, insbesondere mengentheoretisch interessierten Leser bestimmt; von einer spezifisch *philosophischen* Würdigung der einschlägigen Aufsätze wurde hier abgesehen. — Ein terminologischer Index der mengentheoretischen Grundbegriffe soll dazu dienen, das Studium der vielfach aufeinander bezugnehmenden Arbeiten nach Möglichkeit zu erleichtern.

Zur Ergänzung der Abhandlungen wurden als „Anhang" aus dem bisher unveröffentlichten Briefwechsel zwischen *Cantor* und *Dedekind* unter Ausscheidung alles rein Persönlichen einige längere und kürzere Stücke aufgenommen, die mir für den Leser des *Cantor*schen Lebenswerkes von besonderem Interesse zu sein schienen. Diese Ausführungen beziehen sich größtenteils auf die „inkonsistenten" Gesamtheiten *aller* Ordnungszahlen und *aller* Alefs und damit auf die später so sattsam diskutierten „Antinomien der Mengenlehre", die also, wie aus den Briefstellen deutlich hervorgeht, *Cantor* längst bekannt waren und von ihm bereits zutreffend aufgefaßt und gewertet wurden. In diesen Briefen findet sich aber auch *Dedekind*s bisher unbekannter Beweis des „Äquivalenzsatzes", der, mit Hilfe der „Ketten"-Theorie geführt, den Beweis des Herausgebers von 1908 antizipiert und bei dieser Gelegenheit zum ersten Male im Druck erscheint. Die Abschriften nach den Originalbriefen hat Herr *Cavaillès* (Paris) während seines Studienaufenthaltes in Göttingen herstellen lassen und mir für diese Ausgabe freundlichst zur Verfügung gestellt. Ich möchte nicht verfehlen, ihm hierfür auch an dieser Stelle meinen verbindlichsten Dank auszusprechen.

Ganz besonderen Dank schulde ich vor allem Herrn Dr. *Reinhold Baer*, Privatdozent in Halle, der alle Korrekturen des Umbruches mitgelesen und mich durch vielfache Ratschläge und Literaturnachweise bei meiner Arbeit
V | wesentlich unterstützt hat, sowie auch meinen hiesigen Freunden und Kollegen, den Herren Dr. *Arnold Scholz* und Dr. *Robert Breusch*, die mir häufig bei Korrektur und Kommentar halfen, endlich auch Herrn Prof. Dr. *Oskar Becker* (Bonn) für seine freundlichen Auskünfte bezüglich der philosophischen Literatur. Aber auch der Verlagsbuchhandlung, die meinen Wünschen in bezug auf Anlage und Ausstattung des Werkes immer bereitwillig entgegenkam, sowie der Druckerei, die alle nicht immer leichten Korrekturen mit der größten Sorgfalt ausführte, bin ich zu größtem Danke verpflichtet.

The present edition groups the papers in four main parts according to subject matter: number theory and algebra, complex function theory, set theory, and history of mathematics and philosophy. In each part, the papers appear in the order of the date of their publication. The text is true to the original. All obvious slip-ups and misprints in the original have been meticulously corrected. Modern orthography has been introduced; [square] brackets indicate brief supplementary notes by the editor. The "notes" following each paper are in part elucidatory, in part critical. They also contain, among other things, indications of the significance of the respective papers and references to relevant later literature. But these notes are certainly intended for the reader with *mathematical*, and in particular set-theoretic, interests; a specifically *philosophical* assessment of the relevant papers is not given there.—A terminological index of the set-theoretic concepts is intended to facilitate as much as possible the study of the papers, which often contain cross-references.

Several longer and shorter pieces from the hitherto unpublished correspondence between *Cantor* and *Dedekind*, which, I assume, the reader of *Cantor*'s life-work will find of particular interest, were included as "appendix" in order to supplement the papers, with the exception of anything on matters of purely personal concern. For the most part, these writings are concerned with the "inconsistent" totalities of *all* ordinal numbers and of *all* alephs, and hence with the "antinomies of set theory", which later became the subject of so much debate. As can be clearly seen from the passages in the letters, *Cantor* had long known about them and had already properly grasped and evaluated them. But, in these letters, we also find the hitherto unknown proof by *Dedekind* of the "equivalence theorem", which is carried out by means of the "chain"-theory and which anticipates the proof by the editor from 1908. It appears in print here for the first time. The copies of the original letters were made by Mr. *Cavaillès* (Paris) during the time of his studies in Göttingen. I do not want to miss the opportunity to express my gratitude to him for having kindly made them available to me for this edition.

I am especially indebted, first and foremost, to Dr. *Reinhold Baer*, *Privatdozent* in Halle, who read all corrections to the galley proofs and who lent crucial support to my work with much advice and references to the literature. I am also indebted to my friends and colleagues here, Dr. *Arnold Scholz* and Dr. *Robert Breusch*, who often provided assistance with the corrections and the commentary, and, finally, to Prof. Dr. *Oskar Becker* (Bonn), who kindly provided information on relevant philosophical literature. I am also greatly indebted to the publisher who always readily accepted my wishes regarding design and presentation of the volume, and to the printer who carried out the sometimes difficult corrections with greatest diligence.

Möge das Werk in der Form, wie sie hier vorliegt, recht viele Leser finden und in weiten Kreisen der Kenntnis und dem Verständnis des *Cantor*schen Lebenswerkes dienen im Sinne seines Urhebers und im Geiste echter Wissenschaft, unabhängig von Zeit- und Modeströmungen und unbeirrt durch die Angriffe derer, die in ängstlicher Schwäche eine Wissenschaft, die sie nicht mehr meistern können, zur Umkehr nötigen möchten. Diesen aber, sagt *Cantor*, „kann es leicht begegnen, daß genau an jener Stelle, wo sie der Wissenschaft die tödliche Wunde zu geben suchen, ein neuer Zweig derselben, schöner, wenn möglich, und zukunftsreicher als alle früheren, rasch vor ihren Augen aufblüht — wie die Wahrscheinlichkeitsrechnung vor den Augen des Chevalier de Meré“.

Freiburg i. Br., 22. März 1932 *E. Zermelo.*

Mengenlehre 1932

s1932d

Die Mengenlehre hat es zu tun mit den mathematisch definierten unendlichen Gesamtheiten oder Bereichen, die als “Mengen” bezeichnet werden, und unter denen die “endlichen” nur als besonderer Grenzfall auftreten. Da eine unendliche Gesamtheit niemals empirisch gegeben oder vorgelegt werden kann, so kann die Definition solcher Bereiche immer nur *axiomatisch* erfolgen durch Angabe eines Systems von Bedingungen, welchen dieser ideell gesetzte und nur im Sinne einer platonischen Idee existierende unendliche Bereich genügen soll. Beispiele solcher axiomatisch definierten unendlichen Gesamtheiten oder Mengen sind etwa das System der natürlichen Zahlen im Sinne der Peanoschen Postulate oder das System der reellen Zahlen im Sinne der Hilbertschen Axiome.

Aber nicht jedes willkürlich aufgestellte System von Postulaten oder Axiomen definiert auch eine “Menge” im Sinne der Mengenlehre, sondern es müssen dazu besondere Bedingungen erfüllt sein, damit dies möglich ist; und zwar muß

1) das Postulatensystem widerspruchsfrei sein, d. h. die an den Bereich gestellten Bedingungen müssen miteinander vereinbar sein, und

2) das Postulatensystem muß außerdem “kategorisch” sein, nämlich so beschaffen, daß je zwei Bereiche, welche den Postulaten genügen, isomorf d. h. unter Erhaltung der Grundrelationen ein-eindeutig aufeinander abgebildet werden können.

So ist z. B. das System *aller* Gesamtheiten, das sich selbst wie auch jedes seiner Teilsysteme enthalten müßte, ein sich selbst widersprechender Be-

May this volume find a wide audience and may it serve to both broaden and deepen the understanding of *Cantor*'s life-work according to the intention of its creator, in the spirit of true science, independently of changing times and fashions, and resolutely in the face of assaults by those who wish to reverse the direction of a science they are no longer able to master, weak and anxious as they are. But they, *Cantor* says, "may easily find that at the very point where they had sought to inflict upon science a deadly wound, a fresh branch thereof will quickly begin to blossom before their eyes more beautifully, if that is possible, and with a future more promising than that of all the past ones—as the probability calculus did before the eyes of the Chevalier de Meré."

Freiburg i. Br., March 22, 1932 *E. Zermelo.*

Set theory 1932

s1932d

⟦The introductory note just before *s1931f* also addresses *s1932d.*⟧

Set theory is concerned with those mathematically defined infinite totalities or domains which are called "sets" and among which the "finite" ones only occur as a special borderline case. Since an infinite totality can never be given or presented empirically, the definition of such a domain can always only proceed *axiomatically* through the specification of a system of conditions that this ideally posited infinite domain, which only exists as an idea in Plato's sense, is supposed to satisfy. Examples of such axiomatically defined infinite totalities or sets are the system of the natural numbers in the sense of Peano's postulates and the system of the reals in the sense of Hilbert's axioms.

But not every system of postulates or axioms specified arbitrarily also defines a "set" in the sense of set theory. Rather, for this to be possible, special conditions must be satisfied; namely,

1) the system of postulates must be consistent, that is, the conditions to which a domain is subjected must be jointly compatible, and

2) the system of postulates must also be "categorical", namely so constituted that any two domains satisfying the postulates are isomorphic, that is, it must be possible to map one onto the other one-to-one while preserving the fundamental relations.

Thus, e.g., the system of *all* totalities, which would have to contain itself as well as each of its partial systems, is a concept that contradicts itself,

griff, also gewiß keine "Menge". Ebensowenig wird durch die arithmetischen Grundgesetze der Addition und Multiplikation, also durch die "Körper-Axiome" allein schon eine "Menge" definiert, weil dieses Postulatensystem durch wesentlich verschiedene d. h. unter sich unisomorfe "Körper" wie den der rationalen, den der reellen und den der komplexen Zahlen erfüllt wird, also sicher nicht kategorisch ist.

Dagegen bildet das System der Peanoschen Axiome ein besonders einfaches und wichtiges Beispiel eines kategorischen Postulatsystems und definiert damit auch die Gesamtheit der "natürlichen Zahlen" als eine unendliche *Menge* besonderer Art, nämlich als eine "abzählbar unendliche" Menge. Auch die Hilbertschen Axiome der reellen Zahlen (einschließlich der Anordnungs- und Stetigkeits-Gesetze) definieren kategorisch einen unendlichen mathematischen Bereich, also eine "unendliche Menge", die aber, wie wir sehen werden, von der soeben genannten "abzählbaren" wesentlich verschieden ist.

2 | Wir gelangen somit zu folgender Festsetzung:

Eine "Menge" ist ein durch ein kategorisches Postulatensystem charakterisierbarer endlicher oder unendlicher Bereich von wohlunterscheidbaren Dingen oder Objekten.

Hieraus ergeben sich nun die folgenden Folgerungen:

1) Jeder wohldefinierte Teil einer Menge ist selbst eine Menge.

Damit ein Teilbereich T eines kategorisch charakterisierten Bereiches B als "wohldefiniert" gelten kann, muß er "invariant" definiert sein, d. h. die Entscheidung darüber, ob ein Element x von B zu T oder nicht zu T gehört, muß so beschaffen sein, daß wenn der Bereich B isomorf abgebildet wird auf einen anderen B', dem Teil T auch ein ganz bestimmter Teil T' von B' entspricht, welcher der gleichen Bedingung genügt. In diesem Falle ist aber auch der Bereich T durch das T charakterisierende Postulatensystem einerseits und die den Teil als solchen bestimmende neu hinzukommende Bedingung andererseits kategorisch charakterisiert.

2) Die Gesamtheit aller Teilmengen einer gegebenen Menge ist selbst eine Menge.

Wird nämlich der kategorisch bestimmte Bereich B isomorf abgebildet auf einen anderen B', so wird gleichzeitig auch jeder Teilbereich T mitabgebildet auf einen ganz bestimmten Teilbereich T' von B', und damit ist die Gesamtheit dieser Teilbereiche T ebenfalls kategorisch charakterisiert.

3) Ist T eine Menge (zweiter Stufe), deren Elemente M sämtlich Mengen sind, so bilden auch die Elemente ihrer Elemente selbst eine Menge S, die als die "Vereinigung" dieser Mengen M bezeichnet wird.

Wir denken uns die Menge T isomorf abgebildet auf eine andere T' und gleichzeitig jedes Element M von T isomorf auf das entsprechende Element M' von T' — es wird hierbei von dem allgemeinen "Auswahlprinzip" Gebrauch gemacht. Dann wird auch jedes Element x von S abgebildet auf ein
3 ganz bestimmtes Element x' des entsprechenden | Bereiches S', der die Ele-

and hence certainly no "set". Likewise, the basic laws of arithmetic governing addition and multiplication, that is, the "field axioms", do not define a "set" by themselves since this system of postulates is satisfied by essentially different, that is, mutually nonisomorphic "fields" such as that of the rational numbers, that of the reals and that of the complex numbers, and hence is certainly not categorical.

The system of Peano's axioms, on the other hand, forms a particularly simple and important example of a categorical system of postulates and thus also defines the totality of the "natural numbers" as an infinite *set* of a special kind, namely as a "countably infinite" set. Hilbert's axioms of the reals (including the laws of order and continuity), too, categorically define an infinite mathematical domain, and hence an "infinite set", which, however, as we shall see, essentially differs from the "countable" one just mentioned.

We thus arrive at the following stipulation:

A "set" is a finite or infinite domain of well-distinguishable things or objects that can be characterized by means of a categorical system of postulates.

Thence, now, the following consequences arise:

1) Every well-defined part of a set is itself a set.

In order for a partial domain T of a categorically characterized domain B to be considered "well-defined", it must be defined "invariantly", that is, the decision as to whether or not an element x of B belongs to T must be so constituted that if the domain B is mapped isomorphically onto another one B', then there also corresponds to the part T a specific part T' of B' satisfying the same condition. In this case, however, the domain T is also categorically characterized by means of the system of postulates characterizing T, on the one hand, and by means of the new condition, which determines the part as such, on the other.

2) The totality of all partial sets of a given set is itself a set.

For if the categorically determined domain B is mapped isomorphically onto another one B', then, at the same time, each partial domain T is also mapped isomorphically onto a specific partial domain T' of B'. Hence, the totality of these partial domains T, too, is categorically characterized.

3) If T is a set (of second order) all of whose elements M are sets, then the elements of its elements themselves form a set S, which is called the "union" of these sets M.

Let us assume that the set T is mapped isomorphically onto another one T' and, at the same time, each element M of T onto the corresponding element M' of T'—in this case, we make use of the general "principle of choice". Then, each element x of S is also mapped onto a specific element x'

mente der Elemente von T' umfaßt; durch diese Festsetzungen ist also auch der Bereich S kategorisch charakterisiert.

4) Ersetzt man die Elemente x einer Menge M nach einer "invariant" definierten Vorschrift durch andere Elemente y, so bilden auch diese y die Elemente einer Menge N.

Wird nämlich M isomorf abgebildet auf M', so entspricht auch jedem Element y von N ein bestimmtes Element y', und identischen y entsprechen auch identische y'. Somit bilden diese y' die Elemente eines mit N isomorfen Bereiches N', und auch dieser Bereich N ist kategorisch charakterisiert, mithin eine "Menge".

5) Zwei Mengen M und N sollen dann und nur dann als "gleich", d. h. als identisch gelten, wenn jedes Element der einen zugleich auch Element der anderen ist. Ist zwar jedes Element m von M zugleich Element von N, aber nicht umgekehrt, so ist M eine (echte) "Teilmenge" von N. Soll aber die Möglichkeit der Identität zugelassen werden, so heißt M eine "Untermenge" von N.

Zwei Mengen, M und N, welche demselben (kategorischen) Postulatensystem genügen, brauchen deswegen noch keineswegs mit einander identisch zu sein. Das Postulatensystem als solches genügt noch nicht, um eine Menge als solche vollständig zu charakterisieren, es bestimmt vielmehr immer nur eine ganze Klasse von Mengen, zu der unendlich viele Einzel-Mengen gehören können. Und zwar haben alle Mengen einer solchen "Klasse" eine bestimmte *Eigenschaft* mit einander gemein, die wir jetzt charakterisieren wollen.

Definition. Zwei Mengen M und N heißen mit einander "äquivalent", $M \sim N$, wenn es möglich ist, jedem Element m der einen Menge M ein bestimmtes Element n der anderen N so zuzuordnen, daß verschiedenen Elementen m, m' auch immer verschiedene n, n' entsprechen und die zugeordneten n auch die Elemente der Menge N erschöpfen, die Zuordnung also ein-eindeutig ist.

4 | Aus dieser Definition ergeben sich unmittelbar die Folgerungen:

1) Jede Menge ist sich selbst äquivalent.
2) Ist M äquivalent N, so ist auch umgekehrt N äquivalent M.
3) Sind zwei Mengen M und R einer dritten N äquivalent: $M \sim N$ und $N \sim R$, so sind sie auch untereinander äquivalent: $M \sim R$.

Die Relation der Äquivalenz ist also 1) reflexiv, 2) symmetrisch, und 3) transitiv. Aus diesen drei Eigenschaften der Relation ergibt sich aber, daß alle Mengen in bezug auf diese Relation in getrennte Klassen von unter sich äquivalenten Mengen zerfallen, daß also zwei Mengen dann und nur dann zur selben "Klasse" gehören, wenn sie einander äquivalent sind. Jede Menge M gehört dann einer und nur einer solchen "Äquivalenz-Klasse" an, und alle Angehörigen einer solchen Klasse haben ein Gemeinsames, eine "Klassen-Invariante", die wir in unserem Falle mit Cantor als "Mächtigkeit" oder "Kardinalzahl" bezeichnen, die aber eben nichts weiter ist als die Zugehörigkeit

of the corresponding domain S' comprising the elements of the elements of T'; hence, the domain S, too, is categorically characterized by these stipulations.

4) If the elements x of a set M are replaced by other elements y in accordance with an "invariantly" defined instruction, then these y, too, form the elements of a set N.

For if M is mapped isomorphically onto M', then there also corresponds to each element y of N a particular element y', and to identical y there also correspond identical y'. These y' thus form the elements of a domain N' isomorphic with N, and this domain N, too, is categorically characterized, and hence a "set".

5) Two sets M and N are to be considered "equal", that is, identical, if and only if each element of one is also an element of the other. If each element m of M is also an element of N, but not vice versa, then M is a (proper) "partial set" of N. But if we want to allow for the possibility of identity, then we call M a "subset" of N.

Two sets, M and N, need certainly not be identical to one another just because they satisfy the same (categorical) system of postulates. The system of postulates as such is not sufficient in order to completely characterize a set as such. Rather, it always determines only an entire class of sets to which infinitely many individual sets may belong. In particular, all sets of such a "class" have a particular *property* in common, which we will characterize now.

Definition. Two sets M and N are said to be mutually "equivalent", $M \sim N$, if it is possible to assign to each element m of the one set M a particular element n of the other N so that to different elements m, m' there always also correspond different n, n' and the assigned n also exhaust the elements of the set N, and hence the assignment is one-to-one.

From this definition the following consequences immediately arise:

1) Every set is equivalent to itself.
2) If M is equivalent to N, then, conversely, N is also equivalent to M.
3) If two sets M and R are equivalent to a third one N: $M \sim N$ and $N \sim R$, then they are also equivalent to one another: $M \sim R$.

The relation of equivalence is therefore 1) reflexive, 2) symmetric and 3) transitive. But from these three properties of the relation it follows that, with respect to this relation, all sets fall into separate classes of mutually equivalent sets so that consequently two sets belong to the same "class" if and only if they are equivalent to one another. Every set M then belongs to one and only one such "equivalence class", and all members of such a class have something in common, a "class invariant", which, in our case, is called "cardinality" or "cardinal number" following Cantor, which, however, is

zu einer bestimmten Äquivalenz-Klasse. Jeder Menge M kommt also eine und nur eine Kardinalzahl als "Mächtigkeit" zu, und zwei Mengen M und N besitzen dann und nur dann dieselbe Kardinalzahl, wenn sie äquivalent sind. Diese Kardinalzahlen wollen wir zur Unterscheidung von den Mengen selbst und ihren Elementen in der Regel mit kleinen deutschen Buchstaben bezeichnen, z. B. die von M mit $\mathfrak{m}$, die von N mit $\mathfrak{n}$, u. s. w.

So bilden alle aus einem einzigen Element gebildeten Mengen, alle "Einheitsmengen" zusammen eine einzige Klasse, der die Zahl "Eins" als Kardinalzahl zukommt. Ebenso bilden die aus je zwei, drei, u. s. w. gebildeten "endlichen" Mengen besondere Klassen mit den entsprechenden "endlichen Kardinalzahlen" Zwei, Drei, u. s. w. Ein "Beweis" für diese Tatsache läßt sich freilich erst liefern auf grund einer genauen Definition der "endlichen Menge", die erst auf einer höheren Stufe unserer Theorie, etwa mit Hilfe der "Wohlordnung" gegeben werden kann.

5 | Da je zwei isomorfe Bereiche natürlich auch äquivalent sind, so wird durch ein kategorisches Postulatensystem auch die Mächtigkeit oder Kardinalzahl des charakterisierten Bereiches eindeutig bestimmt, d. h. die Mächtigkeit ist eine invariante Eigenschaft jeder Menge.

Zu jeder Menge M gehört als *"Potenzmenge"* eine weitere Menge $U(M)$, welche alle "Untermengen" M_1 von M (einschließlich der Menge selbst und der "Nullmenge") und nur diese als Elemente enthält. Zu zwei äquivalenten Mengen M und N gehören auch äquivalente Potenzmengen $U(M)$ und $U(N)$, da bei jeder ein-eindeutigen Abbildung von M auf N auch jeder Untermenge M_1 von M auch eine äquivalente Untermenge N_1 von N entspricht und zwei verschiedenen Untermengen M_1 und M_2 auch immer verschiedene N_1 und N_2. Also ist die Mächtigkeit der Potenzmenge $U(M)$ eindeutig bestimmt durch die Mächtigkeit der Menge M selbst.

Sind die Mengen M und N unter sich elementenfremd, ebenso auch die Mengen M' und N' und ist $M \sim M'$ sowie $N \sim N'$, so ist auch die Menge $M+N$, welche alle Elemente von M sowie alle Elemente von N und nur diese enthält und als ihre Summe oder Vereinigung bezeichnet wird, äquivalent der entsprechend gebildeten Summe $M' + N'$. Denn da jedes Element der einen Summe im betrachteten Falle entweder zu M oder zu N gehört, so entspricht ihm bei jeder solchen Doppel-Abbildung von M auf M' und von N auf N' ein ganz bestimmtes Element s' von M' oder von N', und verschiedenen s auch immer verschiedene s', wobei auch alle Elemente von $M' + N'$ zur Verwendung kommen. Die Kardinalzahl der (exklusiven) Summe ist also durch die Kardinalzahlen $\mathfrak{m}$ und $\mathfrak{n}$ der beiden Summanden eindeutig bestimmt und kann daher mit $\mathfrak{m} + \mathfrak{n}$ bezeichnet werden. Hiermit ist also die "Addition der Kardinalzahlen" erklärt als eine sinngemäße Erweiterung der gewöhnlichen "Addition" endlicher Anzahlen.

Auch für die so erweiterte "Addition" gilt sowohl das "assoziative" wie das "kommutative" Grundgesetz:

$$\mathfrak{m} + (\mathfrak{n} + \mathfrak{r}) = (\mathfrak{m} + \mathfrak{n}) + \mathfrak{r}, \;\; \mathfrak{m} + \mathfrak{n} = \mathfrak{n} + \mathfrak{m}\,.$$

but the membership to a particular equivalence class. To every set M there belongs therefore one and only one cardinal number as ⟦its⟧ "cardinality", and two sets M and N possess the same cardinal number if and only if they are equivalent. In order to distinguish them from the sets themselves and their elements we will usually refer to these cardinal numbers by lower-case Gothic letters, e.g., to that of M by $\mathfrak{m}$, to that of N by $\mathfrak{n}$, etc.

Thus, all sets formed from a single element, all "unit sets", together form a single class to which the number "One" belongs as ⟦its⟧ cardinal number. Likewise, the "finite" sets formed from two, three, etc., elements form special classes with the corresponding "finite cardinal numbers" Two, Three, etc. Of course, only a precise definition of "finite sets" furnishes a "proof" of this fact, a definition, which can be given only on a higher level of our theory, e.g., by means of the "well-ordering".

Since any two isomorphic domains are of course also equivalent, the cardinality, or cardinal number, of the characterized domain is uniquely determined by means of a categorical system of postulates as well, that is, the cardinality is an invariant property of every set.

To every set M there belongs as "power set" a further set $U(M)$ containing as elements all "subsets" M_1 of M (including the set itself and the "null set") and only these. To two equivalent sets M and N there also belong equivalent power sets $U(M)$ and $U(N)$ since in every one-to-one mapping of M onto N there also corresponds to each subset M_1 of M an equivalent subset N_1 of N, and to two different subsets M_1 and M_2 always also two different N_1 and N_2. The cardinality of the power set $U(M)$ is therefore uniquely determined by the cardinality of the set M itself.

If two sets M and N are disjoint, and so are the sets M' and N', and if both $M \sim M'$ and $N \sim N'$, then the set $M + N$ containing all elements of M and all elements of N and only those, which is called their sum or union, is also equivalent to the sum $M' + N'$ formed accordingly. Since, in the case under consideration, each element of one sum belongs either to M or to N, there also corresponds to it in every such double-mapping of M onto M' and of N onto N' a specific element s' of M' or of N', and to different s also always different s', where all elements of $M' + N'$ are being used. The cardinal number of the (exclusive) sum is therefore uniquely determined by the cardinal numbers $\mathfrak{m}$ and $\mathfrak{n}$ of the two summands, and hence we can refer to it by $\mathfrak{m}+\mathfrak{n}$. Thus, by expanding "addition" of finite numbers in keeping with its original meaning, we have explained the "addition of cardinal numbers".

The thus expanded "addition" is also subject to the basic laws of "associativity" and "commutativity":

$$\mathfrak{m} + (\mathfrak{n} + \mathfrak{r}) = (\mathfrak{m} + \mathfrak{n}) + \mathfrak{r}, \ \mathfrak{m} + \mathfrak{n} = \mathfrak{n} + \mathfrak{m}\,.$$

Die unbegrenzte Zahlenreihe und die exorbitanten Zahlen

s1933b

§1. Die Axiome der Zahlenreihe

Die transfinite Zahlenreihe Z ist ein Bereich von Elementen, welche vermöge der Relation $a < b$ geordnet sind gemäß den folgenden Postulaten:

1) Die Relation bestimmt eine *lineare* Anordnung und zwar eine *Wohlordnung:* unter beliebig vielen Elementen $[\![t]\!]$, die einem Teilbereich $T \subseteq Z$ angehören, gibt es immer ein kleinstes $t_0 \leq t$.

2) Jedem *echten* Abschnitt der Zahlenreihe, d. h. jedem Unterbereich von Elementen $< a$, entspricht ein Abschnitt von *größerer Mächtigkeit.*

3) Jeder Abschnitt, der einem kleineren *konfinal* ist, ist "fortsetzbar", ist ein *echter* Abschnitt.

4) Jeder Abschnitt, der durch ein *kategorisches* Postulatsystem charakterisiert werden kann, ist *fortsetzbar*, ist eine "Menge".

The unlimited number series and the exorbitant numbers

s1933b

⟦The introductory note just before *s1931e* also addresses *s1933b*.⟧

§1. The axioms of the number series

The transfinite number series Z is a domain of elements ordered by means of the relation $a < b$ according to the following postulates:

1) The relation determines a *linear* ordering, in particular a *well-ordering*: among arbitrarily many elements ⟦t⟧ belonging to a partial domain $T \subseteq Z$ there is always a least $t_0 \leq t$.

2) To each *proper* segment of the number series, that is, to each subdomain of elements $< a$, there corresponds a segment of *greater cardinality*.

3) Every segment *cofinal* with a smaller one is "capable of being continued", is a *proper* segment.

4) Every segment that can be characterized by means of a *categorical* system of postulates is *capable of being continued*, is a "set".

Introductory note to 1934

Dieter Wolke

Zermelo's article *1934*, his only one of purely number-theoretic character, contains two remarks on elementary prime number theory. The first, shorter remark is of particular interest.

The so-called fundamental theorem of arithmetic, i.e. unique factorization in $\mathbb{N}$, was first explicitly stated and proved by Carl Friedrich Gauß (*Gauß 1801*). The proof of the uniqueness rests upon Euclid's Theorem 30 in book VII of the *Elements*: If a prime p divides a product, then it divides at least one of the factors. This is derived by means of properties of the greatest common divisor and uses Euclid's algorithm. A proof that dispenses with Euclid's algorithm can be found in Edmund Landau's *Elementare Zahlentheorie* (*Landau 1927*).

Zermelo provides a simple uniqueness proof, which proceeds by induction and does not use the fact $p \mid ab \Rightarrow p \mid a \vee p \mid b$. Let $m > 1$ be the smallest number with two different product representations, $m = qm_0 = pm'$, q and p primes, $q < p$, q and p the smallest primes in the two products, resp. (Equality $p = q$ is excluded by the induction hypothesis.) Then consideration of $m - q\frac{m}{p}$ leads to a contradiction.

Zermelo states that in 1912 he shared a version of his proof with number theorists like Adolf Hurwitz and Edmund Landau, who assured him that his proof had not been published at this point.

In a letter written on 1 August 1926, Kurt Hensel asked Zermelo for permission to use his idea in his lectures (*Ebbinghaus 2007b*, 178, fn. 144). After a conversation with Kurt Hensel, Helmut Hasse used Zermelo's idea for the proof of the unique factorization property in integral domains (*Hasse 1928*). In his article, Hasse refers to Leopold Kronecker's *Vorlesungen über Zahlentheorie* (*Kronecker 1901*), which state a uniqueness proof without Euclid's algorithm. This proof can be considered a precursor of Zermelo's proof. Kronecker's idea is stated concisely by Gerhard Klappauf (*1935*): Let m be the smallest number with two factorizations

$$m = p_1 \cdot \ldots \cdot p_s = q_1 \cdot \ldots \cdot q_t,$$

and let q_1 be the smallest of the q_i's and p_i's. Write

$$p_i = q_1 Q_i + r_i, \quad 0 < r_i < q_1,$$

then

$$m = p_1 \cdot \ldots \cdot p_s = r_1 \cdot \ldots \cdot r_s + q_1 Q \quad (Q \in \mathbb{N})$$

immediately gives a contradiction. Klappauf names Zermelo and Frederick Alexander Lindemann (see below) as the authors who first created new, brief uniqueness proofs. Klappauf's method can also be found in Ernst Trost's book *Primzahlen* (*Trost 1953*).

In his later textbooks (*1949, 1950*) Hasse always used Zermelo's "beautiful idea" (*1949,* 3) to prove the "Hauptsatz".

On 2 November 1934, Zermelo submitted his article to the Wissenschaftliche Gesellschaft zu Göttingen, where he was listed as corresponding member since 18 December 1931 (*Ebbinghaus 2007b,* 177).

In *1933*, Frederick Alexander Lindemann (= Lord Cherwell) published a uniqueness proof which is strikingly similar to Zermelo's. Instead of Zermelo's $m - q\frac{m}{p}$, Lindemann chose $m - pq$. For all that is known, both authors have worked independently.

The first textbooks in which Zermelo's idea occurred, were those by Arnold Scholz (*1939*) and by Richard Courant and Herbert Robbins (*1941*). In the first, Zermelo's name is mentioned, in the second not.

The Lindemann-Zermelo method has been described in various textbooks, e.g. in *Bundschuh 1980, Davenport 1952, Hardy and Wright 1938, Rademacher 1964, Remmert and Ullrich 1987.* A detailed discussion of the different approaches can be found in John W. Dawson's article *2006.*

The second part of Zermelo's paper refers to chapter 22 of the booklet *Von Zahlen und Figuren* by Hans Rademacher and Otto Toeplitz (*Rademacher and Toeplitz 1933*). The authors show that the number 30 is the largest integer n with the property that all positive integers smaller than n and coprime to n are primes. The proof mainly rests upon the following elementary relation. If p_n denotes the n'th prime, then, for $n \geq 4$, we have

$$p_{n+1}^2 < p_1 \cdot \ldots \cdot p_n.$$

Zermelo provides a shorter, but not fundamentally different proof. He indicates that in a similar manner the more general inequality

$$p_{n+1}^k < p_1 \cdot \ldots \cdot p_n \quad \text{for} \quad k \leq 2, \quad n \geq N_k$$

can be derived.

Elementare Betrachtungen zur Theorie der Primzahlen

1934

Vorgelegt in der Sitzung am 2. November 1934

Die folgenden Betrachtungen sind angeregt durch die für „Liebhaber der Mathematik“ bestimmte Schrift von *H. Rademacher* und *O. Toeplitz* „Von Zahlen und Figuren“ (Berlin 1930) und sollen an zwei Beispielen zeigen, wie es auch in der elementaren Zahlentheorie durchaus möglich ist, die Beweise unbeschadet ihres elementaren Charakters wesentlich einfacher zu führen, als dies herkömmlich und auch in der genannten Schrift geschehen ist.

1. Läßt sich eine Zahl nur auf eine Weise in Primfaktoren zerlegen? (*Rademacher–Toeplitz* Nr. 11.)

Der Satz von der eindeutigen Zerlegung aller natürlichen Zahlen in Primteiler wird im Buche wie überhaupt in den meisten mir bekannten Darstellungen der Zahlentheorie auf die Begriffe des „größten gemeinsamen Teilers“ bzw. des „kleinsten gemeinsamen Vielfachen“ zweier Zahlen gegründet. Sind diese beiden Begriffe auch unentbehrlich für den systematischen *Aufbau* der elementaren Zahlentheorie, so läßt sich doch der *Beweis* des vorliegenden Satzes, solange er für sich betrachtet wird, wie ich hier zeigen möchte, auch *ohne* diese Hilfsmittel in ganz einfacher Weise durchführen. Der Beweis wird durch „Induktion“ geführt, d. h. es wird die eindeutige Zerlegung bewiesen für eine Zahl m unter der *Annahme*, daß der Satz für alle *kleineren* Zahlen $m_1 < m$ bereits zutreffe, und gründet sich im Wesentlichen auf die Haupteigenschaft der Zahlenreihe, „wohlgeordnet“ zu sein. Unter den von 1 verschiedenen Teilern von m gibt es einen *kleinsten* $q \leqq m$, der sicher eine Primzahl sein muß; denn jeder echte Teiler von q wäre auch Teiler von m und noch kleiner. Dann ist aber

$$m_0 = \frac{m}{q} < m$$

44 | nach unserer Annahme *eindeutig* in Primfaktoren zerlegbar, also ist auch $m = qm_0$ jedenfalls ein Primzahlprodukt, und diese Darstellung ist auch die *einzige*, in welcher q als Faktor vorkommt. Bei jeder *anderen* Zerlegung müßte der *kleinste* auftretende Primfaktor $p > q$ sein. Dann wäre aber

$$m_1 = m - q\frac{m}{p} = (p-q)\frac{m}{p} < m$$

nach der Annahme wieder *eindeutig* zerlegbar, und q als Teiler von m und somit auch von m_1 müßte unter den Primfaktoren der rechten Seite vorkom-

Elementary considerations concerning the theory of prime numbers

1934

Presented at the meeting of 2 November 1934

The following considerations were prompted by the book "Von Zahlen und Figuren" written for the "mathematical enthusiast", *Rademacher and Toeplitz 1930*. Using two examples, it shall be shown how it is also possible in elementary number theory to carry out the proofs without loss of their elementary character in a fashion much simpler than that in which this has usually been done, and also in the book just mentioned.

1. Is there only one way of decomposing a number into prime factors? (*Rademacher and Toeplitz 1930,* No. 11.)

In the book, as in most other accounts of number theory known to me, the theorem on the unique decomposition of all natural numbers into prime divisors is based on the concepts of the "greatest common divisor" and "least common multiple" of two numbers. Even though these two notions are indispensable to the systematic *construction* of elementary number theory, it is also possible, as I will show here, to carry out the *proof* of the present theorem *without* these expedients in a very simple fashion provided that it is considered in isolation. The proof proceeds by "induction", that is, the unique decomposition is proved for a number m on the *assumption* that the theorem already holds for all *smaller* numbers $m_1 < m$, and is essentially based on the basic property of the number series of being "well-ordered". Among the various divisors of m different from 1 there is a *least one* $q \leq m$, which certainly has to be a prime number; for every proper divisor of q would also be a divisor of m and an even smaller one. But then

$$m_0 = \frac{m}{q} < m$$

is, according to our assumption, *uniquely* decomposable into prime factors, and hence $m = qm_0$ is certainly a product of prime numbers, and this representation is also the *only one* in which q occurs as a factor. In any *other* decomposition, the *least* prime factor occurring in it would have to be $p > q$. But then

$$m_1 = m - q\frac{m}{p} = (p-q)\frac{m}{p} < m$$

would, according to the assumption, again be *uniquely* decomposable. And q, being divisor of m, and hence also of m_1, would have to occur among

men, während doch $p - q$ durch q nicht teilbar ist und $\frac{m}{p}$ nur Primfaktoren $p' \geqq p > q$ enthalten soll. Also enthält *jede* Primzahlzerlegung von m den Faktor q und mit $\frac{m}{q} < m$ ist auch $m = q\frac{m}{q}$ nur auf *eine* Weise zerlegbar.

Der vorstehende Beweis ist eine leichte Abänderung eines früheren Beweises, den ich bereits um 1912 verschiedenen Arithmetikern wie *A. Hurwitz* und *E. Landau* brieflich mitgeteilt hatte, ohne daß sie sich seiner Präexistenz in der Literatur erinnert hätten.

Wie ich einer freundlichen Mitteilung des Herrn *Herglotz* entnehme, ist auf meinen damaligen Beweis inzwischen bereits Bezug genommen worden. Vgl. *H. Hasse*, Über eindeutige Zerlegung in Primelemente oder in Primhauptideale in Integritätsbereichen, Crelle 159 (1928), S. 3, Anm. 1 und S. 6d)., Derselbe, Aufgabensammlung zur höheren Algebra, 2, I, § 1, Aufg. 25, S. 72/73.

2. Eine Eigenschaft der Zahl 30 und ihre Verallgemeinerung.
(a. a. O. Nr. 22.)

Es handelt sich hier um einen elementaren Beweis der Tatsache, daß von $n = 4$ ab das Produkt der n ersten Primzahlen immer größer ist als das Quadrat der nächstfolgenden Primzahl:

$$P_n = p_1 p_2 \dots p_n > p_{n+1}^2\,, \tag{1}$$

und seine Verallgemeinerung kommt darauf hinaus, daß auch jeder höheren Potenz p_{n+1}^k eine Zahl N_k zugeordnet werden kann, sodaß für $n \geqq N_k$ stets

$$P_n > p_{n+1}^k \qquad (1)_k$$

ausfällt.

45 | Der Beweis von (1) wird a. a. O. von *Rademacher* und *Toeplitz* geführt im Anschluß an eine Mitteilung von *H. Bonse* aus dem Jahre 1907 (Archiv der Math. u. Phys., dritte Reihe Bd. 12, S. 292–295), während bezüglich seiner Verallgemeinerung $(1)_k$ nur auf die Arbeit von *R. Remak* (in der gleichen Zeitschrift Bd. 15, S. 186–193, 1908) hingewiesen wird. Beide Beweise lassen sich aber wesentlich einfacher und durchsichtiger gestalten als bei *Bonse* selbst und in der Darstellung von *Rademacher–Toeplitz*.

Für ein beliebig gegebenes m werden die p_m Ausdrücke betrachtet

$$M_x = xP_{m-1} - 1\,, \tag{2}$$

wo x die Zahlen $1, 2, \dots p_m$ durchläuft, sodaß augenscheinlich immer

$$M_x \leqq M_{p_m} = P_m - 1 < P_m \tag{3}$$

ausfällt. Alle diese Ausdrücke sind relativ prim zu P_{m-1} und untereinander inkongruent nach jeder Primzahl $p_{m+t} \geqq p_m$. Wird also jedem M_x die *kleinste* in ihm aufgehende Primzahl q_x zugeordnet, so ist jedes $q_x \geqq p_m$ und alle

the prime factors on the right-hand side, whereas $p - q$ is not divisible by q, and $\frac{m}{p}$ is supposed to contain only prime factors $p' \geqq p > q$. Thus *every* decomposition of m into prime numbers contains the factor q, and, since $\frac{m}{q} < m$, $m = q\frac{m}{q}$, too, can be decomposed in *one* way only.

The proof stated here is a slightly modified version of an earlier proof which I communicated to various arithmeticians such as *A. Hurwitz* and *E. Landau* in writing as early as around 1912 without evoking from them any recollection as to whether it already existed in the literature.

Mr. *Herglotz* was kind enough to communicate to me references to my earlier proof made in the meantime. See *Hasse 1928*, p. 3, remark 1 and p. 6 d)., *Hasse 1934*, 2, I, § 1, exercise 25, p. 72/73.

2. A property of the number 30 and its generalization.
(*Rademacher and Toeplitz 1930*, No. 22.)

At issue here is an elementary proof of the fact that from $n = 4$ onwards the product of the first n prime numbers is always greater than the square of the immediately succeeding prime number:

$$P_n = p_1 p_2 \ldots p_n > p_{n+1}^2 \,, \tag{1}$$

and its generalization amounts to this: with every higher power p_{n+1}^k, too, it is possible to associate a number N_k such that for $n \geq N_k$ we always have

$$P_n > p_{n+1}^k \,. \tag*{$(1)_k$}$$

Rademacher and *Toeplitz* (op. cit.) provide a proof of (1) following the 1907 report by *H. Bonse* (*1907*, pp. 292–295). Concerning its generalization $(1)_k$, however, they only refer to the paper *Remak 1908*. But it is possible to realize both proofs in a much simpler and much more transparent fashion than has been done by *Bonse* himself and by *Rademacher* and *Toeplitz* in their account.

Let us consider, for an arbitrary m, the p_m expressions

$$M_x = xP_{m-1} - 1 \,, \tag{2}$$

where x runs through the numbers $1, 2, \ldots p_m$ so that we evidently always have

$$M_x \leqq M_{p_m} = P_m - 1 < P_m \tag{3}$$

All these expressions are relatively prime with respect to P_{m-1} and are mutually incongruent modulo every prime number $p_{m+t} \geqq p_m$. If we therefore associate with each M_x the *least* prime number dividing it, then each $q_x \geqq p_m$

diese q_x sind von einander *verschieden*. Sie können daher nicht alle der Reihe $p_m, p_{m+1}, \dots p_{m+p_m-2}$ angehören, sondern mindestens eine von ihnen muß von der Form p_{m+s} sein, wo

$$m + s \geqq m + p_m - 1 \tag{4}$$

ist, sodaß wir wegen (3) haben

$$P_m > p_{m+s} \geqq p_{m+p_m-1} \, . \tag{5}$$

Nun ist von $m = 3$ an stets $p_m - m \geqq 2$, also $m + p_m - 1 \geqq 2m + 1$ und von $m = 4$ an stets $p_m - m \geqq 3$, also $m + p_m - 1 \geqq 2m + 2$. Daher wird für jedes $2m \geqq 6$ stets

$$P_{2m} > P_m^2 > p_{2m+1}^2 \tag{6}$$

und für jedes $2m + 1 \geqq 9$ stets

$$P_{2m+1} > P_m^2 > p_{2m+2}^2 \, , \tag{7}$$

d. h. (1) gilt für alle geraden Zahlen $\geqq 6$ und für alle ungeraden $\geqq 9$. Da aber (1) numerisch auch für $n = 4, 5, 7$ verifiziert werden kann, so gilt die Formel (1) allgemein für *jedes* $n \geqq 4$.

In ähnlicher Weise läßt sich auch die *Remak*'sche Beweisführung für die Verallgemeinerung $(1)_k$ vereinfachen.

Für ein beliebiges $k \geqq 2$ sei $n = mk + r$, wo $0 \leqq r \leqq k - 1$, also

$$n + 1 \leqq (m + 1)k \tag{8}$$

ist.

46 | Nach einem bekannten und mit elementaren Mitteln leicht zu beweisenden Satze der Primzahl-Theorie (vgl. *E. Landau*, Handbuch der Lehre von den Primzahlen Bd. I § 15) strebt der Quotient $\frac{m}{p_m}$ mit wachsendem m der Grenze 0 zu. Man kann daher für jedes k eine Zahl $m_k \geqq k$ so bestimmen, daß für alle $m \geqq m_k$ wegen (8) immer

$$p_m > mk \geqq (m + 1)k - m \geqq n + 1 - m$$

und somit

$$m + p_m - 1 \geqq n + 1 \tag{9}$$

wird.

Dann wird aber wegen (5)

$$P_m > p_{n+1}$$

und weiter

$$P_n \geqq P_{mk} > P_m^k > p_{n+1}^k \tag{10}$$

für alle $n \geqq km_k$, d. h. die zu beweisende Ungleichheit $(1)_k$.

Freiburg i. Br., 1. September 1934.

and all these q_x are *distinct* from one another. Thus they cannot all belong to the sequence $p_m, p_{m+1}, \ldots p_{m+p_m-2}$, but at least one of them must be of the form p_{m+s}, where

$$m + s \geqq m + p_m - 1\,, \tag{4}$$

so that, on account of (3), we have

$$P_m > p_{m+s} \geqq p_{m+p_m-1}\,. \tag{5}$$

Now from $m = 3$ onwards we always have $p_m - m \geqq 2$, and hence $m + p_m - 1 \geqq 2m + 1$, and from $m = 4$ onwards always $p_m - m \geqq 3$, and hence $m + p_m - 1 \geqq 2m + 2$. Thus for every $2m \geqq 6$ we always have

$$P_{2m} > P_m^2 > p_{2m+1}^2 \tag{6}$$

and for every $2m + 1 \geqq 9$ always

$$P_{2m+1} > P_m^2 > p_{2m+2}^2\,, \tag{7}$$

i.e., (1) holds for all even numbers $\geqq 6$ and for all odd ones $\geqq 9$. But since it is possible to numerically verify (1) also for $n = 4, 5, 7$, the formula (1) generally holds for *every* $n \geqq 4$.

It it is also possible to simplify *Remak*'s proof of the generalization $(1)_k$ in a similar way.

For an arbitrary $k \geqq 2$ assume that $n = mk + r$, where $0 \leqq r \leqq k - 1$, and hence that

$$n + 1 \leqq (m + 1)\,k\,. \tag{8}$$

According to a well-known theorem of the theory of prime numbers (*Landau 1909,* § 15), which can easily be proved by elementary means, the quotient $\frac{m}{p_m}$ approaches 0 as a limit as m increases. It is therefore possible to determine for every k a number $m_k \geqq k$ so that, for all $m \geqq m_k$, on account of (8), we always have

$$p_m > mk \geqq (m + 1)\,k - m \geqq n + 1 - m\,,$$

and hence

$$m + p_m - 1 \geqq n + 1\,. \tag{9}$$

But then, on account of (5), we have

$$P_m > p_{n+1}$$

and, furthermore,

$$P_n \geqq P_{mk} > P_m^k > p_{n+1}^k \tag{10}$$

for all $n \geqq km_k$, i.e., the inequality $(1)_k$, which was to be proved.

Freiburg i. Br., September 1, 1934.

Grundlagen einer allgemeinen Theorie der mathematischen Satzsysteme
(Erste Mitteilung)

1935

Die folgenden Betrachtungen bringen eine nähere Ausführung der im September 1932 auf der Mathematiker-Versammlung in Bad Elster (vgl. D. Math. Vg. Bd. 41, S. 85–88) von mir vorläufig und andeutungsweise vorgetragenen Gedanken und gründen sich im Wesentlichen auf den Begriff der „fundierenden Relationen", den ich schon im 1930 in meiner Arbeit über „*Grenzzahlen und Mengenbereiche*" (diese Zeitschrift Bd. 16, S. 29–47) im besonderen Falle der ε-Beziehung (der Einordnung eines Elementes in eine Menge) verwendet habe. Ich beginne daher mit einer allgemeinen Charakterisierung dieses Begriffes, der in gewissem Sinne eine Erweiterung der Cantor'schen „Wohlordnung" darstellt, um dann zu seiner Anwendung auf die Syllogistik der Satzsysteme überzugehen.

§ 1. Fundierende Relationen und wohlgeschichtete Mengen

(1) **Definition.** Durch eine binäre Relation $x\mathfrak{f}y$ wird ein Bereich S „wohlfundiert", wenn jeder Unterbereich $T \subset S$ (der nicht verschwindet) mindestens ein „Anfangselement" t_0 enthält, das zu *keinem* Elemente t von T in der Beziehung $t\mathfrak{f}t_0$ steht.

(2) In einem wohlfundierten Bereiche S gibt es *keine unbegrenzt rückschreitende* Kette der Form

$$a_1\mathfrak{f}a\,, \quad a_2\mathfrak{f}a_1\,, \quad a_3\mathfrak{f}a_2\,, \ldots\,,$$

da für den aus $a, a_1, a_2, \ldots$ gebildeten Teilbereich A unsere Forderung (1)
137 nicht erfüllt wäre. Diese Bedingung ist aber auch *hinrei- | chend* für die Fundierung von S. Wäre nämlich $T \subset S$ ein Teilbereich *ohne* Anfangselement und t ein beliebiges Element von T, so gäbe es in T ein $t_1\mathfrak{f}t$, ein $t_2\mathfrak{f}t_1$, ein $t_3\mathfrak{f}t_2$ in infinitum entgegen unserer Annahme, da sie alle auch in S liegen.

(3) Mit einem Bereiche S ist auch jeder seiner *Teilbereiche* T wohlfundiert. Denn enthielte T einen Teilbereich T_1 *ohne* Anfangselement, so wäre das auch ein Teilbereich von S, und jedem Elemente t_1 von T_1 entspräche mindestens ein $t\mathfrak{f}t_1$ in T_1 und S wäre nicht wohlfundiert.

Foundations of a general theory of the mathematical propositional systems (First notice)

1935

[The introductory note just before *1932a* also addresses *1935*.]

The following considerations elaborate ideas whose preliminary outlines I surveyed in September 1932[1] at the meeting of mathematicians in Bad Elster (see *Zermelo 1932a*). They essentially rest on the concept of the "founding relations" of which I already made use in 1930 in my paper on "*boundary numbers and domains of sets*" (*Zermelo 1930a*) in the special case of the ε-relation (the insertion of an element into a set). I therefore begin with a general characterization of this concept, which, in a certain sense, extends Cantor's "well-ordering", and then turn to its application to the syllogistic of propositional systems.

§ 1. Founding relations and well-layered sets

(1) **Definition.** A domain S is "well-founded" by means of a binary relation $x \mathfrak{f} y$ if every (nonempty) subdomain $T \subset S$ contains at least one "initial element" t_0 such that *no* element t of T bears the relation $t \mathfrak{f} t_0$ to t_0.

(2) In a well-founded domain S there exists *no infinite descending* chain of the form

$$a_1 \mathfrak{f} a\,, \quad a_2 \mathfrak{f} a_1\,, \quad a_3 \mathfrak{f} a_2\,, \ldots\,,$$

since the partial domain A formed from $a, a_1, a_2, \ldots$ would not meet our requirement (1). But this condition also is *sufficient* for the founding of S. For if $T \subset S$ were a partial domain *without* initial element and t an arbitrary element of T, then there would be in T a $t_1 \mathfrak{f} t$, a $t_2 \mathfrak{f} t_1$, a $t_3 \mathfrak{f} t_2$ in infinitum contrary to our assumption, since all of them lie also in S.

(3) Along with a domain S each of its *partial domains* T is well-founded as well. For if T contained a partial domain T_1 *without* initial element, then this would also be a partial domain of S, and to each element t_1 of T_1 there would correspond at least one $t \mathfrak{f} t_1$ in T_1, and S would not be well-founded.

[1] [The conference took place in September 1931.]

(4) Jeder wohlfundierte Bereich S enthält mindestens ein Ur-Anfangselement s_0, das zu keinem weiteren Elemente s in der Beziehung $s\mathfrak{f}s_0$ steht. Die Gesamtheit aller solchen Elemente s_0 bezeichnen wir als die „Basis“ des Bereiches S.

(5) In einem wohlfundierten Bereiche S ist *niemals* $a\mathfrak{f}a$, *niemals* gleichzeitig $a\mathfrak{f}b$ und $b\mathfrak{f}a$, oder gleichzeitig $a\mathfrak{f}b$, $b\mathfrak{f}c$ und $c\mathfrak{f}a$, d. h. es sind alle *zyklischen* Relationen dieser Art

$$a_1\mathfrak{f}a\,, \quad a_2\mathfrak{f}a_1\,, \ldots\,, a\mathfrak{f}a_n$$

ausgeschlossen, da sonst für die aus a, aus a, b, bzw. aus a, b, c u. s. w. gebildeten Teilbereiche kein Anfangselement existierte.

(6) Gilt in einem wohlfundierten Bereiche S zugleich die „Trichotomie“, wonach für zwei beliebige Elemente a, b immer mindestens eine der Relationen $a\mathfrak{f}b$, $a = b$, $b\mathfrak{f}a$ bestehen muß, so ist der Bereich durch unsere Relation zugleich „linear geordnet“ und zwar „wohlgeordnet“. Denn aus $a\mathfrak{f}b$, $b\mathfrak{f}c$ folgt dann immer $a\mathfrak{f}c$, d. h. es gilt das Gesetz der „Transitivität“, weil wegen (5) weder $a = c$ noch $c\mathfrak{f}a$ gelten kann. Ist $T \subset S$ ein beliebiger Teilbereich und t_0 ein Anfangselement von T, so ist für jedes andere $t \in T$ stets $t_0\mathfrak{f}t$, d. h. t_0 ist das *einzige* Anfangselement, das „erste“ Element von T in der linearen Anordnung.

(7) **Satz.** Jeder wohlfundierte Bereich S läßt sich „entwickeln“ in eine wohlgeordnete Folge von „Schichten“ $Q_0, Q_1, Q_2, \ldots$ derart, daß die Elemente jeder Schicht Q_α in den vorangehenden Q_β „wurzeln“, daß nämlich, wenn a ein Element von Q_α und $x\mathfrak{f}a$, immer x einer *niederen* Schicht Q_β angehört und daß ferner jede Schicht Q_α *alle* in den vorangehenden wurzelnden, aber noch nicht ihnen angehörenden Elemente a enthält.

138 | **Beweis.** Die Gesamtheit Q der Ur-Anfangselemente q, die „Basis“ des Bereiches, bildet die erste Schicht $Q_0 = P_1$, die Gesamtheit der in Q wurzelnden Elemente die nächste Schicht Q_1, die Gesamtheit der in $P_2 = Q_0 + Q_1$ wurzelnden, aber nicht in P_2 liegenden Elemente, die Schicht Q_2 u. s. w. Allgemein, wenn α irgend eine endliche oder transfinite Ordnungszahl ist und jeder vorangehenden $\beta < \alpha$ eine unseren Bedingungen genügende Schicht Q_β entspricht, so sei $P_\alpha = \Sigma Q_\beta$, d. h. der zu Q_α gehörende „Abschnitt“, die Vereinigung aller *vorangehenden* Schichten und Q_α enthalte alle in P_α wurzelnden Elemente von S, die *nicht* in P_α liegen. Dann ist in der Tat

$$S = Q_0 + Q_1 + Q_2 + \cdots + Q_\alpha + \ldots$$

Wäre nämlich R die Gesamtheit der in *keiner* Schicht Q_α vorkommenden Elemente und r_0 ein Anfangselement von R, so könnte es kein Basiselement sein, weil es sonst zu Q_0 gehörte, und jedes Element x, das zu r_0 in der Beziehung $x\mathfrak{f}r_0$ stände, gehörte einer Schicht und damit auch einem Abschnitte P_α an, und die Vereinigung aller dieser P_α wäre selbst ein Abschnitt P_ϱ, in

(4) Every well-founded domain S contains at least one ur-initial element s_0 such that no further element s bears the relation $s \mathfrak{f} s_0$ to it. We call the totality of all such elements s_0 the "basis" of the domain S.

(5) In a well-founded domain S, we *never* have $a \mathfrak{f} a$, we *never* have both $a \mathfrak{f} b$ and $b \mathfrak{f} a$, nor do we have $a \mathfrak{f} b$, $b \mathfrak{f} c$ and $c \mathfrak{f} a$ at the same time. That is, all *cyclical* relations of the kind

$$a_1 \mathfrak{f} a\,, \quad a_2 \mathfrak{f} a_1\,, \ldots\,, a \mathfrak{f} a_n$$

are excluded, since otherwise the partial domains formed from a, from a, b and from a, b, c etc. would lack an initial element.

(6) If "trichotomy", too, holds in a well-founded domain S, according to which for any two elements a, b at least one of the relations $a \mathfrak{f} b$, $a = b$, $b \mathfrak{f} a$ must always obtain, then the domain is also "linearly ordered" by means of our relation, namely "well-ordered". For from $a \mathfrak{f} b$, $b \mathfrak{f} c$ it then always follows that $a \mathfrak{f} c$, that is, the law of "transitivity" holds, since by (5) neither $a = c$ nor $c \mathfrak{f} a$ can hold. If $T \subset S$ is an arbitrary partial domain and t_0 an initial element of T, then, for any other $t \in T$, we always have $t_0 \mathfrak{f} t$, that is, t_0 is the *only* initial element, the "first" element of T in the linear ordering.

(7) **Theorem.** Every well-founded domain S can be "developed" into a well-ordered sequence of "layers" Q_0, Q_1, $Q_2, \ldots$ so that the elements of each layer Q_α are "rooted" in the preceding Q_β, that is, if a is an element of Q_α and $x \mathfrak{f} a$, then x always belongs to a *lower* layer Q_β, and, furthermore, each layer Q_α contains *all* elements a rooted but not yet contained in the preceding layers.

Proof. The totality Q of ur-initial elements q, the "basis" of the domain, is the first layer $Q_0 = P_1$, the totality of the elements rooted in Q is the next layer Q_1, the totality of the elements that are rooted in $P_2 = Q_0 + Q_1$ but not contained in P_2 is the layer Q_2 etc. Generally, if α is any finite or transfinite ordinal number and if to each preceding $\beta < \alpha$ there corresponds a layer Q_β satisfying our conditions, then let $P_\alpha = \sum Q_\beta$, that is, the "segment" belonging to Q_α, the union of all *preceding* layers, and let Q_α contain all elements from S rooted in P_α which do *not* lie in P_α. Then we in fact have

$$S = Q_0 + Q_1 + Q_2 + \ldots + Q_\alpha + \ldots$$

For if R were the totality of elements occurring in *no* layer Q_α and r_0 an initial element of R, then it could not be a basis element, since otherwise it would belong to Q_0, and every element x bearing the relation $x \mathfrak{f} r_0$ to r_0 would belong to some layer, and thus to some segment P_α as well, and the union of all these P_α would itself be a segment P_ϱ in which r_0 would be

welchem r_0 wurzelte, also müßte r_0 selbst der zugehörigen Schicht Q_ϱ angehören — gegen die Annahme.

Eine durch $\mathfrak{f}$ wohlfundierte Menge S heißt daher auch eine „wohlgeschichtete Menge"

$$S = Q_0 + Q_1 + Q_2 + \ldots = \sum_{\alpha \geq 0} Q_\alpha .$$

(8) Besteht eine Relation $b\mathfrak{f}a$, wo b der Schicht Q_β und a der Schicht Q_α angehört, so ist immer $\beta < \alpha$, da jedes a nur in P_α, d. h. in vorangehenden Schichten Q_β wurzelt.

(9) **Satz.** Ist ein Bereich S wohlfundiert auf eine Basis Q und diese wieder durch Hinzufügung neuer zwischen ihren Elementen bestehender Relationen $q_1\mathfrak{f}q_2$ wohlfundiert auf eine neue Basis $Q^* \subset Q$, so ist nach der Hinzufügung auch S wohlfundiert auf Q^*.

Ist $T \subset S$ ein nicht verschwindender Teilbereich von S und t_0 ein Anfangselement von T in der ursprünglichen Fundierung, so braucht dies noch kein Anfangselement im neuen Sinne zu sein, sofern t_0 ein Element von Q ist. Die Gesamtheit aller solchen Elemente $t_0 = q$ von T bilden dann eine (nicht verschwindende) Teilmenge Q' von Q, und jedes Anfangselement q^* dieser Teilmenge bei der neuen Fundierung $\mathfrak{f}'$ ist zugleich Anfangselement von T,
139 da | kein weiteres Element t von T, ob es nun zu Q' oder zu $T - Q'$ gehört, zu t_0 in der Relation $t_0\mathfrak{f}'q^*$ stehen kann. Also ist auch S wohlfundiert auf Q^*.

Anwendungen

1) Auf die ε-Relation des Enthaltenseins $x \epsilon m$, wenn x Element der Menge m sein soll. Die Elemente eines „Normalbereiches" sind dann wohlfundiert auf die Gesamtheit der „Urelemente" als Basis.

2) Auf die „echte" Subsumption $m \subset n$ (wobei die Identität ausgeschlossen ist) zweier Mengen. Die (nicht verschwindenden) Untermengen einer *endlichen* Menge sind dann wohlfundiert auf die Gesamtheit der „Einheitsmengen" $\{a\}$, welche nur je *ein* Element enthalten. Diese Eigenschaft kann als *Definition* der endlichen Menge verwendet werden, wie dies durch *A. Tarski*, Fund. Math. Bd. 6, S. 54–55, ausgeführt wurde.

§ 2. Anwendung auf die Beweistheorie

Die fundierende Relation, um die es sich hier handelt, ist die zwischen „Grund" und „Folge", welche zwischen „Sätzen" $a, b, c, \ldots$ bestehen oder nicht bestehen kann. Wir sagen „p folgt aus $a, b, c, \ldots$" und schreiben $a, b, c, \ldots \to p$, wenn mit der Wahrheit von $a, b, c, \ldots$ auch die von p gesetzt sein soll, und nennen den Komplex $a, b, c, \ldots$ den „Grund", den Satz p die „Folge" und den die-

rooted. Hence, r_0 itself would have to belong to the corresponding layer Q_ϱ —contrary to the assumption.

A set S well-founded by means of $\mathfrak{f}$ is therefore also called a "well-layered set"

$$S = Q_0 + Q_1 + Q_2 + \ldots = \sum_{\alpha \geq 0} Q_\alpha .$$

(8) If a relation $b\,\mathfrak{f}\,a$ obtains with b belonging to the layer Q_β and a to the layer Q_α, then we always have $\beta < \alpha$, since each a is only rooted in P_α, that is, in preceding layers Q_β.

(9) **Theorem**. If a domain S is well-founded on a basis Q and if the latter is again well-founded on a new basis $Q^* \subset Q$ when new relations $q_1\,\mathfrak{f}\,q_2$ obtaining between its elements are added, then S, too, is well-founded on Q^* subsequent to the addition.

If $T \subset S$ is a nonempty partial domain of S and t_0 an initial element of T in the original founding, then the latter need not yet be an initial element in the new sense, provided that t_0 is an element of Q. The totality of all such elements $t_0 = q$ of T then form a (nonempty) partial set Q' of Q, and each initial element q^* of this partial set in the new founding $\mathfrak{f}'$ is also an initial element of T, since no further element t of T, may it belong to Q' or to $T - Q'$, can bear the relation $t\mathfrak{f}'q^*$ to q^*.[2] Hence, S, too, is well-founded on Q^*.

Applications

1) To the ε-relation of containment $x \,\epsilon\, m$, if x is supposed to be an element of the set m. In this case, the elements of a "normal domain" are well-founded on the totality of the "urelements" as the basis.

2) To the "proper" subsumption $m \subset n$ of two sets (where identity is excluded). In this case, the (nonempty) subsets of a *finite* set are well-founded on the totality of the "unit sets" $\{a\}$ each of which contains only *one* element. This property can be used as a *definition* of finite sets, as elaborated in *Tarski 1924a*, pp. 54–55.

§ 2. Applications to proof theory

Our concern here is with the founding relation between "ground" and "consequence", which may or may not obtain among "propositions" $a, b, c, \ldots$. We say "p follows from $a, b, c, \ldots$" and write $a, b, c, \ldots \rightarrow p$ if along with the truth of $a, b, c, \ldots$ also that of p is supposed to be posited. And we call the complex $a, b, c, \ldots$ the "ground", the proposition p the "consequence", and the sense

[2] ⟦Zermelo erroneously writes "t_0" instead of "t".⟧

ser Formel innewohnenden Sinn die „Begründung“ des Satzes p. Diese Relation ist „binär“, wenn entweder nur ein einziger Satz zur Begründung ausreicht, oder aber alle begründenden Sätze a, b, c zu einem „Komplex“ $A = a\,b\,c \ldots$ zusammengefaßt werden. In jedem Falle heiße der Komplex A, die „Hypothese“, das „Vorderglied“, der Satz p die „These“, das „Hinterglied“ der Relation. Ist a ein einzelner zur Begründung von p erforderlicher oder verwendeter Satz, so sagen wir „a begründet p“ und schreiben $a\mathfrak{f}p$, auch wenn a zur Begründung von p nicht ausreichen sollte. In diesem Sinne der „teilweisen Begründung“ gilt nun das

Theorem. Ist ein System S von Sätzen p „wohlfundiert“ durch die „Begründungs-“ oder „Folgerelation“ $a \to p$, so sind alle Sätze des Systems „wahr“, insofern die Sätze seiner „Basis“ es sind. Die Sätze sind aus denen der Basis, die man hier die „Voraussetzung“ nennt, „abgeleitet“, „bewiesen“, und das System selbst ist der „Beweis“.

140 | **Beweis des Theorems.** Angenommen, ein Teil U der fraglichen Sätze (die dann gewiß nicht zur Basis Q gehören) wäre falsch, so enthielte U mindestens einen Satz u_0, der als „Anfangselement“ zu keinem weiteren Satze u von U in der „Folgerelation“ $u \to u_0$ steht. Da aber u_0 kein Basis-Element ist, so gibt es einen Satz v_1 in $V = S - U$, für den $v_1 \to u_0$ wäre, ja *alle* zur Begründung von u_0 dienenden Sätze $v_1, v_2, v_3, \ldots$ wären in V enthalten, wären „wahr“ und damit wäre auch u_0 selbst wahr — entgegen unserer Annahme.

Definition. Ein (direkter) *Beweis* ist ein durch Folgerung (Begründung) wohlfundiertes System von Sätzen, zu denen der zu beweisende gehört, und dessen Basis aus lauter wahren Sätzen besteht, welche die Voraussetzung des Satzes bilden.

Das typische Beispiel eines solchen „Beweises“ bildet das Schlußverfahren der „vollständigen Induktion“. Das System S besteht hier aus Sätzen p_n, welche den Zahlen der natürlichen Zahlenreihe eineindeutig entsprechen, wobei für jedes n die Folgerung $p_{n-1} \to p_n$ erwiesen sei und die Gültigkeit des Satzes p_1 (des einzigen der Basis) angenommen wird. Dieses System ist „wohlfundiert“, denn jedes Teilsystem T enthält ein p_n mit *kleinstem* Index, sodaß das entsprechende p_{n-1} nicht mehr zu T gehört. Wesentlich für die Gültigkeit dieses Schlußverfahrens ist also die Eigenschaft der Zahlenreihe, daß unter beliebig vielen Zahlen eine die *kleinste* ist und daß *jeder* Zahl n (außer der ersten) eine unmittelbar vorangehende $n - 1$ entspricht.

§ 3. Grundrelationen und abgeleitete Sätze

Jede mathematische Theorie bezieht sich auf einen (im Allgemeinen unendlichen) Bereich von Elementen oder Gegenständen (z. B. Zahlen, Punkten,

inherent to this formula the "justification" of the proposition p. This relation is "binary" if either only a single proposition suffices for the justification or all justifying propositions a, b, c are combined into a "complex" $A = a\,b\,c\ldots$. In each case, the complex A is said to be the "hypothesis", or "antecedent", and the proposition p the "thesis", or "consequent", of the relation. If a is some proposition required or used for the justification of p, then we say that "a justifies p" and write $a \mathfrak{f} p$, even if a were not to suffice for the justification of p. In this sense of a "partial justification", we now have the

Theorem. If a system S of propositions p is "well-founded" by means of the "justification" or "consequence relation" $a \to p$, then all propositions of the system are "true" provided that the propositions of its "basis" are true. The propositions are "derived", or "proved", from those of the basis, which is called here the "presupposition", and the system itself is the "proof".

Proof of the theorem. Let us assume that some part U of the propositions in question (which then, to be sure, do not belong to the basis Q) is false. U would then contain at least one proposition u_0 as an "initial element" such that no further proposition u of U bears the "consequence relation" $u \to u_0$ to u_0. But since u_0 is not a basis element, there is a proposition v_1 in $V = S - U$ for which we would have $v_1 \to u_0$. In fact, *all* propositions $v_1, v_2, v_3, \ldots$ which serve to justify u_0 would be contained in V, they would be "true", and thus u_0 itself would be true as well—contrary to our assumption.

Definition. A (direct) *proof* is a system of propositions well-founded by means of inference (justification) to which the proposition to be proved belongs and whose basis consists only of true propositions forming the presupposition of the proposition.

The method of inference known as "mathematical induction" provides the typical example of such a "proof". In this case, the system S consists of propositions p_n corresponding one-to-one to the numbers of the natural number series, where we suppose that for each n the consequence $p_{n-1} \to p_n$ has been demonstrated, and where the validity of the proposition p_1 (the only one of the basis) is assumed. This system is "well-founded" since each partial system T contains some p_n with the *least* index such that the corresponding p_{n-1} no longer belongs to T. Hence, what is indispensable for the validity of this inference method is the property of the number series that among arbitrarily many numbers one is the *least* and that to *each* number n (besides the first) there corresponds an $n - 1$ immediately preceding it.

§ 3. Fundamental relations and derived propositions

Every mathematical theory is concerned with a (generally infinite) domain of elements or objects (e.g., numbers, points, figures, etc.), among which certain

Figuren u. s. w.), zwischen denen gewisse „Grundrelationen“ z. B.

$$a < b\,, \qquad a + b = c\,, \qquad a \text{ liegt auf der Geraden } b\,c$$

bestehen oder nicht bestehen können. Aus diesen Grundrelationen werden nun weitere Relationen abgeleitet durch die logischen Operationen der Konjunktion und Disjunktion („und“ und „oder“) in Verbindung mit der Negation ($\overline{s}$ = Nicht-s), angewendet auf endliche oder unendliche Gesamtheiten von Sätzen in endlicher oder unendlicher Wiederholung. So bedeutet die Konjunk-
141 tions Aussage | $\mathfrak{K}(S_1)$, daß *alle* Sätze des Teilbereiches S_1 von S *gleichzeitig* gelten sollen, die Disjunktions-Aussage $\mathfrak{D}(S_1)$, daß von allen Sätzen von S_1 *mindestens einer* gelten soll. Damit die so abgeleiteten Sätze „wohldefiniert“ seien, jeder „circulus in definiendo“ vermieden werde, machen wir die Annahme, daß das ganze so entstehende Satzsystem S durch die definierenden Relationen „wohlfundiert“ sei auf die Gesamtheit Q der Grundrelationen als „Basis“. Die „fundierende“ Relation $a\mathfrak{f}b$ kann hier von dreierlei Art sein: 1) a ist einer der Sätze, deren *Konjunktion* b ist, 2) a ist ein Glied der durch b dargestellten *Disjunktion* oder 3) b ist die *Negation* von $a, b = \overline{a}$.

Im letzteren Falle ist natürlich auch a die Negation von b, aber es ist nicht als solche *definiert*, und darauf kommt es hier an. Andernfalls wäre ja gleichzeitig $a\mathfrak{f}b$ und $b\mathfrak{f}a$, was dem Wesen der Fundierung widerstreitet, da dann der Teilbereich $\{a, b\}$ „auf sich selbst beruhte“. Die definierenden Relationen dürfen in dieser Hinsicht eben nicht ohne weiteres durch logisch äquivalente andere ersetzt werden.

Nun sei Q irgend ein „geschlossener Bereich“, eine „Menge“ von Grundrelationen q. Dann bilden wir die Menge aller durch die definierenden Operationen der Konjunktion, Disjunktion und Negation aus ihnen hervorgehenden „abgeleiteten Sätze erster Stufe“ Q_1. Setzen wir hier $P_2 = Q + Q_1$, so können wir mit P_2 ebenso verfahren und erhalten weitere „Schichten“ $Q_2, Q_3, \dots Q_\omega, \dots, Q_\alpha$, wobei immer jede Schicht Q_α alle die aus der Gesamtheit der vorangehenden $P_\alpha = \sum_{\beta<\alpha} Q_\beta$ unmittelbar abgeleiteten Sätze umfaßt, die noch nicht in P_α enthalten sind. Indem nun α die ganze transfinite Zahlenreihe durchläuft, entsteht successive ein „wohlfundiertes“ System aller aus Q mittelbar oder unmittelbar abgeleiteten Sätze, die zunächst alle als *verschieden* betrachtet werden, auch wo sie wie s und $\overline{\overline{s}}$ logisch äquivalent sein sollten.

Das so gebildete Satzsystem S ist in der Tat „wohlfundiert“ auf diese Grundrelationen, mit Q als „Basis“. Ist nämlich T irgend ein Unterbereich von S, so gibt es unter den in ihm vertretenen Schichten Q_α eine solche von niederstem Index α_0, und jeder dieser Schicht angehörige Satz t_0 hängt dann, sofern er nicht selbst der Basis Q angehört, unmittelbar nur von Sätzen aus niederen Schichten Q_β ($\beta < \alpha_0$) ab, die in T nicht vorkommen. Nach oben hin
142 kann dieses | „wohlgeschichtete“ Satzsystem beliebig, z. B. mit einer wohldefinierten „Grenzzahl“ π (vgl. meine Note in Fund. Math. Bd. 16), abgeschlossen werden und besitzt dann alle Eigenschaften einer „Menge“. In dieser allgemei-

"fundamental relations" such as

$$a < b\,, \qquad a + b = c\,, \qquad a \text{ lies on the line } b\,c$$

may or may not obtain. From these fundamental relations further relations are derived now by means of the logical operations of conjunction and disjunction ("and" and "or") in combination with negation ($\overline{s}$ = not-s) as they are applied to finite or infinite totalities of propositions in finite or infinite repetition. Thus, the assertion of the conjunction $\mathfrak{K}(S_1)$ means that *all* propositions of the partial domain S_1 of S are supposed to hold *simultaneously*, and the assertion of the disjunction $\mathfrak{D}(S_1)$ means that among all propositions of S_1 *at least one* is supposed to hold. In order for the propositions so derived to be "well-defined", in order to avoid any "circulus in definiendo", we assume that the entire system of propositions S so obtained is "well-founded" by means of the defining relations on the totality Q of the fundamental relations as the "basis". The "founding" relation $a\,\mathfrak{f}\,b$ can be one of three kinds here: 1) a is one of the propositions whose *conjunction* is b, 2) a is a term of the *disjunction* represented by means of b, or 3) b is the *negation* of a, $b = \overline{a}$.

In the last case, to be sure, a is also the negation of b. But it is not *defined* as such, which is what matters here. For otherwise we would have both $a\,\mathfrak{f}\,b$ and $b\,\mathfrak{f}\,a$, which runs counter to the nature of founding since the partial domain $\{a, b\}$ would then "rest on itself". In this regard, the defining relations simply must not be replaced with logically equivalent different ones.

Now let Q be some "closed domain", a "set" of fundamental relations q. We then form the set of all "derived propositions of first rank" Q_1 obtained from them by means of the defining operations of conjunction, disjunction and negation. If we set here $P_2 = Q + Q_1$, then we can treat P_2 in the same way, obtaining further "layers" $Q_2, Q_3, \ldots Q_\omega, \ldots, Q_\alpha$, where each layer Q_α comprises all propositions that are immediately derived from the totality of the preceding $P_\alpha = \sum_{\beta<\alpha} Q_\beta$ and that are not yet contained in P_α. By letting α run through the entire transfinite number series, a "well-founded" system arises successively of all propositions derived mediately or immediately from Q, all of which are, at first, considered *different*, even if they should be logically equivalent such as s and $\overline{\overline{s}}$.

The system of propositions S so formed is in fact "well-founded" on these fundamental relations with Q as the "basis". For if T is some subdomain of S, then among those layers Q_α represented in it there is one of lowest index α_0, and each proposition t_0 belonging to this layer immediately depends then only on propositions from lower layers Q_β ($\beta < \alpha_0$) not occuring in T, provided that $[\![t_0]\!]$ itself does not belong to the basis Q. This "well-layered" system of propositions can be arbitrarily closed off upwardly, e.g., by means of a well-defined "boundary number" π (cf. *Zermelo 1930a*). It then possesses all the properties of a "set". Any mathematical discipline that rests on a par-

nen Form läßt sich jede auf einen bestimmten Urbereich von Elementen und bestimmte Grundrelationen gegründete mathematische Disziplin darstellen, so die Arithmetik der rationalen, algebraischen Zahlen und Funktionen, die Analysis der reellen und komplexen Zahlen und Funktionen, ebenso die Geometrie jedes Raumes von gegebener Dimensionenzahl, auch die Mengenlehre eines gegebenen „Normalbereiches" (a. a. O., S. 46). Etwas anders verhält es sich dagegen mit der Theorie der „offenen" Bereiche, wie der „*allgemeinen* Körpertheorie", der *allgemeinen* Geometrie oder der *allgemeinen*, über *alle* Normalbereiche ausgedehnten Mengenlehre.

§ 4. Die Wahrheitsverteilung in Satzsystemen

Die eigentliche Logik und damit die mathematische Wissenschaft beginnt erst mit der Verteilung der Sätze eines Systemes in „wahre" und „falsche", in solche, die als „gültig" oder „ungültig" angesehen werden. Wir gehen aus von einer willkürlichen „Wahrheitsverteilung" der „Basis" $Q = V_0 + U_0$ und übertragen sie auf das ganze auf Q fundierte System $S = V + U$ nach folgenden syllogistischen Regeln:

1) Jede Konjunktion und jede Disjunktion wahrer Sätze ist wahr, jede Konjunktion und Disjunktion falscher Sätze ist falsch.
2) Jede „gemischte" Konjunktion (von teils wahren, teils falschen Sätzen) ist falsch, jede „gemischte" Disjunktion ist wahr.
3) Jede Negation eines wahren Satzes ist falsch, jede Negation eines falschen Satzes ist wahr.

Gäbe es Sätze t des Systems S, die bei einer Wahrheitsverteilung nicht eindeutig mitverteilt würden, so bildeten sie einen Teilbereich T mit einem „Anfangselement" t_0, einem Satze, der nach der gemachten Annahme kein Basiselement wäre und unmittelbar nur von bereits verteilten Sätzen des Bereiches $S - T$ abhinge, also nach den Regeln 1)–3) in bezug auf seine Zugehörigkeit doch wieder eindeutig bestimmt wäre.

Von zwei Sätzen s, r des Systems sagen wir, sie seien „äquivalent", wenn sie bei *jeder* Wahrheitsverteilung zur selben „Hälfte" gehören, immer *gleichzeitig*
143 wahr oder falsch sind; wir sagen, sie seien „kon- | travalent" oder „widerstreiten" einander, wenn sie bei *jeder* Verteilung immer zu *verschiedenen* Hälften gehören.

Alle unter sich äquivalenten Sätze $a, a', a'', \ldots$ bilden eine „Klasse" A, der eine andere Klasse $\overline{A}$ der widerstreitenden Sätze $b, b', b'', \ldots$ entspricht.

Sind zwei Sätze a, b einem dritten c äquivalent, so sind sie unter einander äquivalent. Jeder Satz ist sich selbst äquivalent und kein Satz widerstreitet sich selbst. (Ein „widerspruchsvoller" Satz ist vielmehr ein solcher, der bei jeder Verteilung zu den „falschen" gehört, ein „absolut falscher Satz"; vgl. S. 144).

ticular urdomain of elements and on particular fundamental relations can be represented in this general way, such as the arithmetic of rational, algebraic numbers and functions, the analysis of real and complex numbers and functions, the geometry of any space of a given number of dimensions, and also the set theory of a given "normal domain" (op. cit., p. 46). Matters are somewhat different, on the other hand, with regard to the theory of "open" domains such as the "*general* theory of fields", *general* geometry or *general* set theory extended to *all* normal domains.

§ 4. The distribution of truth in propositional systems

Logic proper, and hence mathematical science, begins only with the classification of the propositions of a system into "true" and "false" propositions, with their distribution into propositions considered "valid" and "invalid" respectively. Our starting point is an arbitrary "truth distribution" of the "basis" $Q = V_0 + U_0$, which we then carry over into the entire system $S = V + U$ founded on Q according to the following syllogistic rules:

1) Every conjunction and every disjunction of true propositions is true, every conjunction and disjunction of false propositions is false.
2) Every "mixed" conjunction (of propositions some of which are true and some false) is false, every "mixed" disjunction is true.
3) The negation of any true proposition is false, the negation of any false proposition is true.

Propositions t of the system S that are not uniquely distributed along with the others in a truth distribution would, if they existed, form a partial domain T with some "initial element" t_0, a proposition, which, by our assumption, would not be a basis element and which would immediately depend only on already distributed propositions of the domain $S - T$. Hence, the rules 1)–3) would uniquely determine where this proposition belongs after all.

We call two propositions s, r of the system "equivalent" if they both belong to the same "half" in *every* truth distribution, if they are always *both* true or *both* false; we say that they are "contravalent", or that they "oppose" one another, if they belong to *different* halves in *every* distribution.

All mutually equivalent propositions $a, a', a'', \ldots$ form a "class" A to which there corresponds another class $\overline{A}$ of the opposing propositions $b, b', b'', \ldots$.

If two propositions a, b are equivalent to a third c, then they are mutually equivalent. Every proposition is equivalent to itself, and no proposition opposes itself. (A "contradictory" proposition, by contrast, is one that belongs to the "false" propositions in every distribution, an "absolutely false proposition"; cf. p. 144.)

Jeder Satz a widerstreitet seiner Negation $\overline{a}$ und ist äquivalent seiner doppelten Negation $\overline{\overline{a}}$. Die Konjunktion $a\overline{a}$ (a und $\overline{a}$) ist immer falsch, d. h. in *jeder* Wahrheitsverteilung, „absolut falsch"; die Disjunktion $a + \overline{a}$ (a oder $\overline{a}$) ist immer, bei jeder Wahrheitsverteilung, wahr, „absolut wahr". Der „Satz vom Widerspruch" und der vom „ausgeschlossenen Dritten" sowie der von der „doppelten Verneinung" ergeben sich also hier als einfache Folgen der „syllogistischen Regeln".

Alle „absolut wahren" Sätze sind unter einander äquivalent, ebenso alle „absolut falschen" Sätze unter einander.

Weiter ergibt sich aus denselben Regeln das

Theorem. Die Negation einer aus den Sätzen $a, b, c, \ldots$ gebildeten Konjunktion k ist immer äquivalent der aus den Negaten $\overline{a}, \overline{b}, \overline{c}, \ldots$ gebildeten Disjunktion l'; ebenso das Negat einer aus $a, b, c, \ldots$ gebildeten Disjunktion l äquivalent der Konjunktion k' der Negate $\overline{a}, \overline{b}, \overline{c}, \ldots$

Denn *entweder* 1) sind bei einer Verteilung $a, b, c, \ldots$ *alle wahr*, dann sind es auch k und l; zugleich sind dann *alle* $\overline{a}, \overline{b}, \overline{c}, \ldots$ falsch und damit auch k' und l' falsch und $\overline{k'}$ und $\overline{l'}$ sind wahr. *Oder* 2) sind alle $a, b, c, \ldots$ falsch und damit auch k und l falsch, zugleich alle $\overline{a}, \overline{b}, \overline{c}, \ldots$ wahr und k' und l' wahr, also $\overline{k'}$ und $\overline{l'}$ falsch. *Oder* 3) sind einige $a, b, c, \ldots$ wahr, andere falsch; dann ist k falsch und l wahr, zugleich aber auch k' falsch und l' wahr. D. h. in allen Fällen ist immer k wahr, wenn l' falsch ist, und l wahr, wenn k' falsch ist, und umgekehrt.

Als „Wahrheitsbereich" V bezeichnen wir jeden Teilbereich des Satzsy-
144 stemes S, der bei einer „Verteilung" die „wahren" | Sätze enthält; entsprechend als „Falschheitsbereich" einen U, der bei einer Verteilung die falschen Sätze umfaßt. Der *Durchschnitt* $V^*(s)$ *aller* V*-Bereiche*, die einen gegebenen Satz s enthalten, umfaßt alle Sätze t, welche immer wahr sind, wenn s wahr ist, also alle aus s *folgenden* Sätze und heiße der „Folgebereich von s".

Der Durchschnitt $U^*(s)$ aller Falschheitsbereiche U, welche s enthalten, umfaßt alle Sätze, die immer falsch sind, wenn s falsch ist, d. h. alle Sätze, *aus denen* s *folgt*, und heiße der „Ursprungsbereich von s". Der Durchschnitt aller Wahrheits- *und* Falschheitsbereiche, welche s enthalten, umfaßt alle Sätze des Systems, die mit s zugleich wahr und falsch sind, also alle mit s *äquivalenten* Sätze; er ist zugleich der Durchschnitt des Folge- und des Ursprungsbereiches von s und heiße der „Äquivalenzbereich von s": $A^*(s) = U^*(s)V^*(s)$.

Ist z. B. s ein „Axiomensystem" einer auf die Grundrelationen gegründeten mathematischen Theorie, z. B. der Arithmetik oder der euklidischen Geometrie, so umfaßt $A^*(s)$ alle äquivalenten Axiomensysteme, $V^*(s)$ alle Sätze der aus s fließenden Theorie, insbesondere alle *allgemeineren* Axiomensysteme, und $U^*(s)$ alle *spezielleren* Axiomensysteme.

Liegt ein Satz t im Folgebereich $V^*(s)$, „folgt" also aus s, so ist er auch „syllogistisch ableitbar" aus s, und zwar bereits innerhalb des gemeinsamen

Every proposition a opposes its negation $\overline{a}$ and is equivalent to its double negation $\overline{\overline{a}}$. The conjunction $a\,\overline{a}$ (a and $\overline{a}$) is always false, that is, ⟦false⟧ in *every* truth distribution, "absolutely false"; the disjunction $a + \overline{a}$ (a or $\overline{a}$) is always, ⟦that is,⟧ in every truth distribution, true, "absolutely true". In other words, the "law of contradiction" and that of the "excluded middle" as well as the one of "double negation" arise here simply as consequences of the "syllogistic rules".

All "absolutely true" propositions are mutually equivalent, as are all "absolutely false" propositions.

Furthermore, the same rules give rise to the

Theorem. The negation of a conjunction k formed from the propositions $a, b, c, \ldots$ is always equivalent to the disjunction l' formed from the negations $\overline{a}, \overline{b}, \overline{c}, \ldots$; likewise, the negation of the disjunction l formed from $a, b, c, \ldots$ is equivalent to the conjunction k' of the negations $\overline{a}, \overline{b}, \overline{c}, \ldots$.

For *either* 1) $a, b, c, \ldots$ are *all true* in some distribution. In this case, k and l are true as well; at the same time, $\overline{a}, \overline{b}, \overline{c}, \ldots$ are *all* false, and hence k' and l', too, are false, and $\overline{k'}$ and $\overline{l'}$ are true. *Or* 2) $a, b, c, \ldots$ are all false, and hence k and l are false as well. At the same time, $\overline{a}, \overline{b}, \overline{c}, \ldots$ are all true as are k' and l', and hence $\overline{k'}$ and $\overline{l'}$ are false. *Or* 3) some of the $a, b, c, \ldots$ are true, and others are false; in this case, k is false and l is true but, at the same time, k' is false and l' true as well. That is, in all cases, k is always true when l' is false, and l is always true when k' is false, and vice versa.

Any partial domain of the system S of propositions containing the "true" propositions in some "distribution" is called a "truth domain" V; correspondingly, a "falsity domain" will be a U comprising the false propositions in some distribution. The *intersection* $V^*(s)$ of *all V-domains* containing a given proposition s comprises all those propositions t that are true if s is true, and hence all propositions *following* from s. It will be called the "consequence domain of s".

The intersection $U^*(s)$ of all falsity domains U containing s comprises all those propositions that are false whenever s is false, that is, all those propositions *from which s follows.* It shall be called the "origin domain of s". The intersection of all truth *and* falsity domains containing s comprises all propositions of the system that are true and false along with s, that is, all propositions *equivalent* to s; it is also the intersection of the consequence and origin domains of s and shall be called the "equivalence domain of s": $A^*(s) = U^*(s)\,V^*(s)$.

If, e.g., s is a "system of axioms" of a mathematical theory that rests on the fundamental relations of, e.g., arithmetic or of Euclidean geometry, then $A^*(s)$ comprises all equivalent axiom systems, and $V^*(s)$ comprises all propositions of the theory that follows from s, and in particular all *more general* axiom systems, and $U^*(s)$ comprises all *more particular* axiom systems.

If a proposition t lies in the consequence domain $V^*(s)$, that is, if it "follows" from s, then it also is "syllogistically derivable" from s. In particular,

„Wurzelbereiches“ oder „Definitionsbereiches“ $W(s,t)$ von s und t. Es ist dies der Durchschnitt aller s und t enthaltenden „Wurzelbereiche“ W, nämlich aller solchen, die mit jedem in ihnen enthaltenen abgeleiteten Satze auch seine sämtlichen „Wurzeln“ enthalten, d. h. mit jeder Negation $\overline{a}$ den verneinten Satz a, und mit jeder Konjunktion oder Disjunktion ihre sämtlichen Glieder. Dann entspricht jeder beliebigen Wahrheitsverteilung des Gesamtsystems auch eine solche von $W(s,t)$, und jeder Durchschnitt V' mit einem Wahrheitsbereiche V enthält entweder s nicht oder er enthält zugleich auch t. Der Bereich W enthält alle zur Ableitung von t erforderlichen Zwischensätze, und ihre „Wertung“, ihre Wahrheitsverteilung, erfolgt in ihm nach den syllogistischen Regeln.

Demnach wäre also jeder aus s folgende Satz auch „beweisbar“, aber zunächst nur im absoluten, „infinitistischen“ Sinne. Ein solcher „Beweis“ enthält zumeist *unendlich viele* Zwischensätze, und es ist noch nicht gesagt, in wie weit und durch welche Hilfsmittel er auch unserem *endlichen* Verstande einleuchtend gemacht werden kann. Im Grunde ist *jeder* mathematische
145 Beweis, z. B. das Schluß- | verfahren der „vollständigen Induktion“ durchaus „infinitistisch“ und doch vermögen wir ihn einzusehen. Feste Grenzen der Verständlichkeit gibt es hier augenscheinlich nicht.

§ 5. Symmetrie und Kategorizität

Die Urelemente, die den mathematischen Theorien zugrunde gelegt werden und zwischen denen die „Grundrelationen“ gelten (oder nicht gelten) sollen, sind in der Regel *gleichberechtigt*. Die „Axiome“ bleiben bestehen, wenn die Urelemente unter einander vertauscht werden, sie sind „symmetrisch“, d. h. gegen diese Permutationen „invariant“, und das gleiche gilt auch von einem Teil der abgeleiteten Sätze. Dagegen brauchen die „Grundrelationen“ durchaus nicht selbst „invariant“ oder „symmetrisch“ zu sein. Vielmehr gehen sie durch diese Permutationen $\mathfrak{P}$ *in einander* über, sie erfahren eine gewisse Gruppe $\mathfrak{H}$ von Permutationen unter einander, die eine Untergruppe *aller* Permutationen der Basis darstellt und als „Hauptgruppe“ bezeichnet werden soll. „Symmetrisch“ sind dann alle solche Sätze des Systems, die bei Permutationen der Hauptgruppe in sich übergehen. Z. B. ist jede Konjunktion und jede Disjunktion symmetrisch, wenn sie über alle solchen Sätze (z. B. Grundrelationen) erstreckt wird, die durch Permutationen von $\mathfrak{H}$ in einander übergehen. Nun kann es vorkommen, daß ein symmetrischer Satz (z. B. Axiomensystem) die Eigenschaft hat, daß *alle* Wahrheitsbereiche, in denen er vorkommt, durch Permutationen der Hauptgruppe aus einander hervorgehen. Dann heißt der Satz „kategorisch“.

Theorem. Jeder symmetrische Satz ist äquivalent einer Disjunktion von kategorischen Sätzen.

it is so derivable already within the common "root domain", or "definition domain", $W(s,t)$ of s and t. This is the intersection of all "root domains" W containing s and t, namely of all those which along with every derived proposition contained in them also contain all of its "roots", that is, along with every negation $\overline{a}$ the negated proposition a, and along with every conjunction or disjunction all of their terms. There corresponds then to any arbitrary truth distribution of the entire system also one such of $W(s,t)$, and every intersection V' with a truth domain V either fails to contain s or contains it along with t. The domain W contains all intermediate propositions required for the derivation of t, and their "valuation", their truth distribution, proceeds in it according to the syllogistic rules.

Accordingly, every proposition that follows from s would therefore be "provable" as well, but, at first, only in the absolute, "infinitistic" sense. Such a "proof" for the most part contains *infinitely many* intermediate propositions, and it has yet to be determined to what extent and by what means it can be rendered evident to our *finite* mind. *Every* mathematical proof, such as the inference method of mathematical induction, is actually quite "infinitistic", and yet we are capable of grasping it. Firm limits on comprehensibility do not seem to exist here.

§ 5. Symmetry and categoricity

The urelements upon which mathematical theories rest and among which the "fundamental relations" are supposed to hold (or not to hold) are, in general, *on a par*. The "axioms" remain in effect when the urelements are interchanged. They are "symmetric", that is, they are "invariant" under these permutations, and the same also holds for a portion of the derived propositions. On the other hand, the "fundamental relations" do not have to be "invariant" or "symmetric" themselves at all. Rather, they are turned *into one another* by means of these permutations $\mathfrak{P}$, they are subject to a certain group $\mathfrak{H}$ of permutations, which is a subgroup of *all* permutations of the basis and which shall be called the "principal group". "Symmetric" are then all such propositions of the system which turn into themselves under the permutations of the principal group. E.g., every conjunction and every disjunction is symmetric when extended to all such propositions (e.g., fundamental relations) which are turned into one another by means of permutations of $\mathfrak{H}$. Now, it is possible that some symmetric proposition (e.g., axiom system) has the property that *all* truth domains in which it occurs arise from one another by means of permutations of the principal group. In this case, the proposition is called "categorical".

Theorem. Every symmetric proposition is equivalent to a disjunction of categorical propositions.

Beweis. Jeder Wahrheitsbereich V enthält einen für ihn charakteristischen „erzeugenden“ Satz v, der keinem anderen (vollständigen) Wahrheitsbereiche angehört. Es ist dies die Konjunktion aller ihm angehörenden Grundrelationen q und der Negate $\overline{q}$ aller ihm *nicht* angehörenden, und durch ihn ist nach § 4 die ganze Wahrheitsverteilung eindeutig bestimmt. Gehört nun ein Satz s gleichzeitig den Wahrheitsbereichen $V, V', V'' \ldots$ und nur diesen an, so ist er dann und nur dann wahr, wenn mindenstens einer der „Erzeugenden“ $v, v', v'' \ldots$ wahr ist. Er ist also äquivalent der *Disjunktion* aller dieser Erzeu-
146 genden. Ist der Satz selbst *symmetrisch,* | so gehört er mit jedem V gleichzeitig immer den durch Permutation aus V hervorgehenden Wahrheitsbereichen an, er enthält also mit v zugleich die Disjunktion v^* aller aus v durch Permutation hervorgehenden Sätze, und diese Disjunktion v^* ist selbst „kategorisch“. Sollte es außer den so aus V hervorgehenden Wahrheitsbereichen noch weitere $V_\alpha, V_\beta, \ldots$ geben, denen s angehört, so entsprechen auch diesen weitere „elementar-symmetrische“ oder „kategorische“ Sätze $v^*_\alpha, v^*_\beta, v^*_\gamma, \ldots$ und s ist dann und nur dann wahr, wenn einer dieser Sätze wahr ist, d. h. s ist „äquivalent“ der Disjunktion aller dieser kategorischen Sätze, w. z. b. w.

Die vorstehenden Ausführungen bilden erst den *Anfang* einer noch nicht abgeschlossenen Untersuchung, welche die Begründung einer „infinitistischen“ echt mathematischen Syllogistik und Beweistheorie zum Ziele hat. Einer ehrenvollen Einladung der Redaktion folgend, habe ich hier meine vorläufigen Ergebnisse für diesen Festband zusammengestellt in der Hoffnung, in einer weiteren Mitteilung die erforderlichen Ergänzungen nachholen zu können.

Proof. Every truth domain V contains some characteristic "generating" proposition v that belongs to no other (complete) truth domain. This is the conjunction of all fundamental relations q belonging to it and of the negations $\overline{q}$ of all those *not* belonging to it. The entire truth distribution is, according to §4, uniquely determined by means of it. If now a proposition s belongs to all truth domains $V, V', V'' \ldots$ and only to those, then it is true if and only if at least one of the "generating" $v, v', v'' \ldots$ is true. It hence is equivalent to the *disjunction* of all these generating propositions. If the proposition itself is *symmetric*, then along with each V it always also belongs to the truth domains arising from V by means of permutation. Hence, it contains, along with v, also the disjunction v^* of all propositions arising from v by means of permutation, and this disjunction v^* is itself "categorical". In case there are further $V_\alpha, V_\beta, \ldots$ to which s belongs, besides the truth domains thus arising from V, then to these, too, there correspond further "elementary symmetric" or "categorical" propositions $v^*_\alpha, v^*_\beta, v^*_\gamma, \ldots$, and s is true if and only if one of these propositions is true, that is, s is "equivalent" to the disjunction of all these categorical propositions, q.e.d.

The above elaborations are only the *beginning* of a not yet completed investigation that seeks the foundation of an "infinitistic", genuinely mathematical syllogistic and proof theory. Upon the editors' honorary invitation, I have put together my preliminary results for the present celebratory volume, hoping that I shall be able to hand in the required supplements in a later notice.

Introductory note to s1937

Dirk van Dalen

At the Scandinavian mathematics congress of 1922, Skolem drew an astonishing conclusion from a theorem on models of first-order theories that he had established in 1920. This theorem, now known as the Löwenheim-Skolem theorem, states that if a countable first-order theory has an infinite model at all, then it has a countable model. Löwenheim's earlier paper *1915* had handled the "one-sentence" case, and Skolem's generalization allowed arbitrary countable sets of sentences. Modulo Gödel's completeness theorem, the theorem is also expressed as "a consistent and countable first-order theory has a countable model". We will pass over the finer details of Skolem's proof and the formulations involved, and concentrate on the application made by Skolem in the context of set theory.

Since the existing axiomatization of set theory, laid down by Zermelo and augmented by Fraenkel, was regarded by Skolem as a first-order theory with the axioms of separation and of replacement taken to be schemes (the latter being suggested by Skolem, independently of Fraenkel, in his 1922 talk), Skolem was fully justified in asserting that if ZF (Zermelo-Fraenkel set theory) has a model, it has a countable model. (Historical note: Skolem communicated this consequence to Felix Bernstein in Göttingen in 1915/16; see *Skolem 1923*.) A striking consequence is that a first-order set theory, which proves Cantor's theorem that there are uncountable sets, if consistent, has a countable model. This phenomenon has since become known as *Skolem's paradox*. The paradox puzzled, but did not overly worry Skolem's contemporaries. Fraenkel, in his authoritative expositions, remained sceptical about the significance and the correctness of Skolem's results, at least until the end of the twenties. The strongest opposition came from the father of the axiomatization of set theory, Zermelo; he was well-aware of the foundational issue: how to handle quantification in set theory—what is meant by "all sets".

In axiomatized theories, one interprets "for all" in a model-theoretic setting. Hilbert's *Foundations of geometry* had set the standard for mathematics, and the mathematical community knew very well that in statements like "all squares of numbers are non-negative", it is crucial to specify the domain of "numbers". Skolem had accepted the natural view that for set theory, with its first-order formalization, quantification came to specifying inside a model (in this case a set with a binary relation) the interpretation of the objects and the relation. For set theory this spelled trouble, as there are two positions to be considered: the "outside"-position, where "element of" and "set" have their intuitive meaning, and the "inside"-position, where the meanings need not have any resemblance to their cousins outside. The observation made by Skolem was that—from the inside—the countable model behaved, according

to ordinary logic, as a perfect instance of the universe of sets (following ZF), but that—seen from the outside—it was in flagrant contradiction with all we had learned. For instance, one might take a set-object p_M in the model M and look at all set-objects q_M that obey $q_M \in_M p_M$. These set-objects form a set $p_{outside}$ from the outside view. Now it is not difficult to see that, e.g. p_M may be uncountable in M, whereas $p_{outside}$ is countable. This can indeed, as Skolem observed, take place in a completely natural setting: the element-relation in the model may be the same as the outside one. In model-theoretic terms, the smaller model may be an elementary submodel of the big model. Hence cardinality is *relative to the model.*

At this point Zermelo refused to follow Skolem. He insisted that "for all" should be the natural second-order notion.[1] In *s1930d* he expressed his position as "Skolem wishes to restrict the formation of subdomains to special classes of defining functions, while I, in keeping with the true spirit of set theory, allow for the *free* division and postulate the existence of all partial domains formed in any arbitrary way." Returning to the issue in *s1931f*, he gave further motivation for turning down the Skolem interpretation of "all" and "exists": "The concept of 'allness' or 'quantification' must lie at the foundation of any mathematical consideration as a basic logical category incapable of further analysis. If we were to restrict the allness in a particular case by means of special conditions, then we would have to do so using quantifications, which would lead us to a regressus in infinitum."

Indeed, he had in his *Fundamenta mathematicae* paper *1935*, as well as in the papers in the *Jahresbericht*, *1932a*, and in the *Forschungen und Fortschritte*, *1932b*, pointed to new ways to create an infinitary logic that would allow him to obtain an alternative formulation of set theory. The fact that no such comprehensive new foundation of set theory was presented by Zermelo suggests that the task of setting up an adequate infinitary language and logic was at the time beyond him. The intricacies were indeed formidable. Zermelo's second-order approach to set theory has been revisited in modern times. For models of this theory see *Uzquiano 1999.*

Not satisfied with criticizing the "finitistic" set theory with its relativism, Zermelo wanted to refute relativism in set theory. The present note in his *Nachlass*, dated October 4, 1937, claimed to reduce relativism to absurdum. The note was not published, and Zermelo did not return to it. In the paper *van Dalen and Ebbinghaus 2000* the note is analyzed in the context of the period. The argument in the note, as one would expect, foundered at the interpretation of "all" (ibid., 156):

> To see where the refutation breaks down, the modern reader might visualize the argument by thinking of a countable elementary submodel V' of a standard model V of ZF, he would take M to be ω, and pick for K and K' the respective continua in V and V'. Now,

[1] For the second-order approach see also *Zermelo 1930a*.

> translated to this situation, Zermelo's proof makes use of *arbitrary* intersections and unions of subsets of K', [...] where for the argument "arbitrary" would have to mean "in the sense of V'". The strong closure property, however, is not available. Should we allow it, then elementary equivalence could not apply.

So, Zermelo's refutation amounts to a proof of the set-theoretic statement that, given a denumerable set M, any subset of the powerset K of M that is closed under arbitrary unions and intersections (and under complements with respect to M) and whose union is M, is either finite or of the same cardinality as K.

It should be noted that Zermelo had another *bête noire*: intuitionism. Zermelo's references to this topic are not systematic enough to allow us to reach definite conclusions on his views. It seems that on the whole Zermelo did not wish to see intuitionism as a serious foundational program. It appears that on this topic he was not particularly well-informed, as one can see from his remark below on banning the infinite from mathematics. In *s1929b* he seems to see intuitionism as a school that practices mathematics in "disjunctive

Der Relativismus in der Mengenlehre und der sogenannte Skolem'sche Satz

s1937

Zermelo, 4. X. 37

Das „Kontinuum" wird in der Mengenlehre gewöhnlich definiert durch die Menge P *aller* Untermengen M_1 einer „abzählbaren" Menge M. Aber ist diese Definition auch *eindeutig*, ist der Begriff *aller* Untermengen nicht zu unbestimmt? Könnte es nicht *verschiedene* Grade dieser Allheit und damit verschiedene „Modelle" des Kontinuums geben, könnte nicht vielleicht dieses Kontinuum unbeschadet seiner formalen Eigenschaften sogar (entsprechend dem „Skolemschen Satze") durch ein *abzählbares* Modell dargestellt werden? Dann müßte es möglich sein, ein solches „mageres" Kontinuum K' in ein „fetteres" K so einzubauen, daß sich alle formalen Eigenschaften auch auf dieses übertragen lassen.

2 | Wir betrachten eine *abzählbare* Menge M und aus der Gesamtheit K *aller* ihrer Untermengen N (im Sinne des „fetteren" Kontinuums) eine Teilgesamtheit K' von folgender Eigenschaft:

systems", i.e. that restricts its attention to those parts where the principle of the excluded middle is valid, and thus misses even the most elementary parts of mathematics. In the "Bericht an die Notgemeinschaft", *s1930d*, Zermelo mentions the formalism-intuitionism dispute, and declares himself to be completely neutral on the issue. His infinitary approach to logic, however, makes Hilbert's finitist views a better target for technical foundational criticism, which moreover ties in with Zermelo's opposition to Skolemism.

It is worthwhile to note here that Brouwer had, independently of Zermelo, embraced a form of infinitary logic in his paper in the Riemann volume, *Brouwer 1927*. Apparently this had escaped Zermelo's attention.

There is an interesting passage in the letter from Reinhold Baer to Zermelo of July 12, 1930 (cf. *Ebbinghaus 2004*, 81), where Baer points out that at least Brouwer's "urintuition" offers a direct access to the uncountable, thus going beyond Skolem, and shoring up Zermelo's views on the transcending the countable (without supporting Cantor's higher cardinalities).

On the whole, Zermelo seems to have been inadequately informed about Brouwer's doctrines; he accuses Brouwer, for example, of banning the infinite from mathematics. This may explain his harsh view of intuitionism.

Relativism in set theory and the so-called Skolem theorem

s1937

Zermelo, 4. X. 37

The "continuum" is usually defined in set theory by means of the set P of *all* subsets M_1 of a "countable" set M. But is this definition also *univocal*? Isn't the concept of *all* subsets too indeterminate? Isn't it possible that there are *different* degrees of this allness, and hence different "models" of the continuum? Isn't it perhaps even possible to represent this continuum by means of a countable model (in accordance with "Skolem's theorem") without loss of its formal properties? If so, then it should be possible to build such a "meager" continuum K' into a "fatter" continuum K so that all formal properties can be transferred also onto the latter.

Consider a *countable* set M and a partial totality K' from the totality K of *all* its subsets N (in the sense of the "fatter" continuum) with the following property:

1) Ist R eine Untermenge von K', so sind sowohl die zu R gehörige Vereinigungsmenge S_R wie ihr Durchschnitt D_R wieder Elemente von K'. M. a. W. Summe und Durchschnitt beliebig vieler (auch ∞ vieler) Mengen aus K' gehören wieder zu K', die Elemente von K' bilden einen „Mengenring".
2) Jedes Element m von M ist in mindestens einer Menge aus K' enthalten.

Dann sei N_m der *Durchschnitt* aller Mengen N aus K', welche m enthalten, also selbst ein Element von K', und zwei solche Durchschnitte N_a und N_b sind entweder identisch oder elementenfremd. Dadurch wird die Gesamtmenge M zerlegt in eine Summe von elementenfremden Bestandteilen N_m, welche einzeln entweder endlich oder abzählbar sein müssen. Ferner wird *jede* Menge N aus K' gleichfalls in solche Teile N_m zerspalten und jede Vereinigung, jede Summe solcher N_m muss nach 1) wieder ein Element von K' sein.

3 | Bezeichnet man also mit T die Menge aller (von einander verschiedenen) M_a, so entspricht jedes Element von K' ein-eindeutig einer Untermenge von T und K' ist äquivalent der Menge $\mathfrak{U}T$ aller Untermengen von T. Nun ist die Menge der Teile, in die eine *abzählbare* Menge M zerspalten werden kann, entweder *endlich* oder selbst *abzählbar*, d. h. der Menge M äquivalent, und das Pseudo-Kontinuum K' daher entweder selbst endlich oder äquivalent $K' \sim \mathfrak{U}T \sim \mathfrak{U}M = K$, d. h. äquivalent dem ursprünglichen (nicht abzählbaren) Kontinuum K. Es ist daher unmöglich, das Kontinuum in einem *abzählbaren* Modell darzustellen, es müßte dann endlich sein. Der „Skolem'sche Satz" führt also zu der interessanten Folgerung, daß sich *unendliche* Mengen in endlichen Modellen realisieren lassen — eine Folgerung, die nicht paradoxer wäre als manche andere aus diesem schönen Satze bereits gezogene Konsequenzen. Hiermit wäre denn auch das Ideal des „Intuitionismus", die Abschaffung des Unendlichen aus der Mathematik, der Verwirklichung nahe gebracht zugleich mit dem des „Formalismus", der bekanntlich den Beweis der Widerspruchsfreiheit erstrebt. Denn aus absurden Prämissen kann man bekanntlich *alles* beweisen. Also auch die Widerspruchslosigkeit eines beliebigen Satzsystems.

1) If R is a subset of K', then both the union set S_R belonging to R and its intersection D_R are again elements of K'. In other words, sum and intersection of arbitrarily many (also ∞ many) sets from K' again belong to K'. The elements of K' form a "ring of sets".
2) Every element m of M is contained in at least one set from K'.

Then, let N_m be the *intersection* of all sets N from K' containing m. In other words, let it be itself an element of K'. Two such intersections N_a and N_b are either identical or disjoint. The total set M is thus being decomposed into a sum of disjoint components N_m each of which must either be finite or countable. Furthermore, *each* set N from K' is also being split into such parts N_m, and every union, every sum of such N_m must, according to 1), again be an element of K'.

If we therefore refer to the set of all (mutually distinct) N_α[1] as T, then every element of K' corresponds one-to-one to a subset of T, and K' is equivalent to the set $\mathfrak{U}T$ of all subsets of T. Now, the set of the parts into which a *countable* set M can be split is either *finite* or itself *countable*, that is, equivalent to the set M, and the pseudo-continuum K' is therefore either itself finite or equivalent $K' \sim \mathfrak{U}T \sim \mathfrak{U}M = K$, that is, equivalent to the original (not countable) continuum K. It is therefore impossible to represent the continuum in a *countable* model, for it would have to be finite. "Skolem's theorem" thus leads to the interesting consequence that it is possible to realize *infinite* sets in finite models—a consequence which would be no more paradoxical than many a consequence already drawn from this beautiful theorem. Thus, we would have also come closer to realizing the ideal of "intuitionism", the abolition of the infinite from mathematics, as well as that of "formalism", which, as is well-known, seeks the proof of consistency. For from absurd premisses *anything* can be proved, as is well-known. Hence, also the consistency of an arbitrary propositional system.

---

[1] ⟦Zermelo erroneously writes "M_a" instead of "N_a".⟧

Introductory note to s1941

Heinz-Dieter Ebbinghaus

Zermelo's letter to Paul Bernays, an answer to Bernays' 70th birthday congratulations, has been included as the last entry of the present collection as it vividly conveys the feeling of loneliness and the fear of falling into oblivion which dominated Zermelo's last years. There are several reasons that contributed to this.

Scientifically, his fight against the predominance of finitary approaches in the foundations of mathematics had failed. Even more, he was considered as a researcher of the previous generation unable to play an active part in shaping mathematical logic in the 1930s. According to John W. Dawson in his biography of Kurt Gödel (*Dawson 1997*, 75), Zermelo's fight was a "reactionary" one.

Concerning his career, his unstable state of health caused by a nervous constitution and tuberculosis of the lungs had delayed the first offer of a university position—a full professorship at the University of Zurich—until he was nearly 40 years old and had led to his compulsory retirement only six years later. A second teaching position, an honorary one at the University of Freiburg which he had obtained ten years after his retirement, had been lost for political reasons and under deplorable circumstances.

Not one for systematic development and exploration, Zermelo would have needed a group to discuss and develop the consequences of his ideas. In Göttingen he had enjoyed such a group. However, his later circumstances had not readily offered an opportunity to build up a school. Furthermore, he was not a man with a penchant for cooperation, but one with strong reservations against "cliques" and with an urge for scientific freedom which despised all opportunism, a man of solitude who liked controversial debates and only in individual cases opened himself up to others.

Zermelo was now living in a remote part of a suburban residential area of Freiburg and hardly had any contact with the Institute of Mathematics. There was only the infrequent exchange of letters on special occasions with Wilhelm Süss, one of the two full professors in the Institute and later founder of the Mathematical Research Institute Oberwolfach. The other full professor, Gustav Doetsch, had been the driving force behind the denunciation that had led to the loss of Zermelo's honorary professorship. Arnold Scholz, his main and perhaps only scientific partner and friend during the late 1930s, was far away in Kiel.

Annotations

The invitation to Göttingen concerned a colloquium on the occasion of Zermelo's 70th birthday. It was initiated by and organized with the help of Arnold Scholz and took place on 19 July 1941. Among the speakers were Konrad Knopp and Bartel Leendert van der Waerden. Zermelo gave three talks: "Rubber ball and lampshade" ("Gummiball und Lampenschirm") on the folding and bending of flexible surfaces, a topic he had already discussed in a talk given to the Polish Mathematical Society in June 1929; "Building roads in the mountains" ("Straßenbau im Gebirge") on shortest lines of limited steepness, following *1902d*; and "How to break a piece of sugar?" ("Wie zerbricht man ein Stück Zucker?") on lines of fracture of a rectangle, based on *1933a*. Later he wrote to Süss that "such opportunities have become so rare that they are seized with great pleasure."

Zermelo's statement that he had never claimed priority for the axiom of choice displays his conviction that mathematical principles are common property, that only the proofs based on them can be property of an individual mathematician (*1908a*, 118, first footnote).

The foundational congress in Zurich to which Zermelo refers is presumably the congress *Les fondements et la méthode des sciences mathématiques* in December 1938; see *Gonseth 1941*.

The Bad Elster meeting is the 1931 annual meeting of the Deutsche Mathematiker-Vereinigung; it took place in September 1931 at Bad Elster, a small spa in Saxony. It was at the time when Zermelo had started to work out his theory of infinitary languages in order to overcome the weakness of finitary systems which Thoralf Skolem and Gödel had revealed. Having become aware that Gödel had announced a talk about his results at the Bad Elster meeting, he made up his mind to use this encounter as an opportunity to fight for his convictions. As described in the letter *s1931c* to Reinhold Baer, he deliberately had Gödel's lecture put on *before* his and asked that they be discussed *together*. The account of the failure of this strategy given there is quite different from that in the present letter to Bernays: "But [...] the entire discussion was put off *even* further on the *illicit* proposal by Fraenkel (he backstabs me *whenever* he can!). As a result, the discussion came to naught and my own presentation went completely by the board."

Brief an Paul Bernays vom 1. Oktober 1941

s1941

Freiburg 1. Okt. 1941
Günterstal Schauinslandstr. 99

Lieber Herr Bernays!

Haben Sie vielen Dank für Ihren liebenswürdigen Brief und die freundlichen Glückwünsche zu meinem Altersjubiläum. Ich freue mich sehr darüber, daß immer noch einige meiner früheren Kollegen und Mitarbeiter sich meiner erinnern, während ich schon so manche meiner Freunde durch den Tod verloren habe. Man wird eben immer einsamer, ist aber umso dankbarer für jedes freundliche Gedenken. Wenn ich auch immer noch wissenschaftlich interessiert und beschäftigt bin, so vermisse ich doch allzu sehr jeden wissenschaftlichen Gedankenaustausch, der mir früher, namentlich in meiner Göttinger Zeit so reichlich zuteil geworden war. Umso mehr freute ich mich über eine Einladung nach Göttingen für die letzte Woche des Sommersemesters, wo ich über einige kleine Fragen der angewandten Mathematik (Straßenbau im Gebirge als Variationsproblem, Gummiball und Lampenschirm als Knickungsaufgabe, und Wie zerbricht man ein Stück Zucker?) vortragen durfte und dabei alte Freunde wie Herglotz und Scholz wiedersehen konnte. Über die Aus-
2 wirkung meiner eigentlichen Lebensarbeit, | soweit sie die „Grundlagen“ und die Mengenlehre betrifft, mache ich mir freilich keine Illusionen mehr. Wo mein Name noch genannt wird, geschieht es immer *nur* in Verbindung mit dem „Auswahlprinzip“ auf das ich *niemals* Prioritätsansprüche gestellt habe. So war das auch bei dem Grundlagen-Kongress in Zürich, zu dem ich nicht eingeladen wurde, dessen Bericht Sie mir aber freundlicherweise zugesandt haben. In *keinem* der Vorträge oder Diskussionen, soweit ich mich bisher überzeugte, wurde eine meiner seit 1904 erschienenen Arbeiten (namentlich die beiden Annalen-Noten von 1908 und die beiden in den „Fundamenta“ erschienenen im letzten Jahrzehnt) angeführt oder berücksichtigt, während die fragwürdigen Verdienste eines Gödel oder Skolem reichlich breitgetreten wurden. Dabei erinnere ich mich, daß schon bei der Mathematiker-Tagung in Bad Elster mein Vortrag über Satz-Systeme durch eine Intrige der von Hahn und Gödel vertretenen Wiener Schule von der Diskussion ausgeschlossen wurde, und habe seitdem die Lust verloren, über Grundlagen vorzutragen. So geht es augenscheinlich jedem, der keine „Schule“ oder Klique hinter sich hat. Aber vielleicht kommt noch eine Zeit, wo auch meine Arbeiten wieder entdeckt und gelesen werden. Ihnen jedenfalls sage ich nochmals herzlichen Dank für Ihre freundliche Bemühung. Ergebenst Ihr

E. Zermelo

Letter to Paul Bernays of 1 October 1941

s1941

Freiburg Oct 1, 1941
Günterstal Schauinslandstr. 99

Dear Mr. Bernays,

I thank you for your delightful letter and the kind congratulations on the anniversary of my birthday. Having already lost many a friend to death, I am very happy that some of my former colleagues and collaborators still remember me. For as one finds oneself growing more and more lonely, one is all the more grateful for anyone cherishing one's memory. While I am still interested in and concerned with scientific matters, I dearly miss that scientific exchange of ideas in which I partook so amply in the past, especially during my days in Göttingen. I was all the more delighted to receive an invitation to visit Göttingen for the last week of the summer semester, where I could lecture on several minor questions in applied mathematics (mountain road construction as a variation problem, rubber balls and lamp shades as kink problems, and, How to break a piece of sugar?) and see old friends again, such as Herglotz and Scholz. Of course, I no longer harbor any illusions about the impact of my real life-work, as far as "foundations" and set theory are concerned. Every mention of my name is invariably connected *only* with the "principle of choice", to which I have *never* laid any special claim. This is also what happened at the congress on the foundations in Zurich, to which I had not been invited, but whose proceedings you were kind enough to send to me. As far as I have been able to ascertain, *none* of the presentations and discussions refer to or consider any of my papers published since 1904 (namely the two 1908 papers in *Annalen* and the two papers in "Fundamenta" published in the past decade).[1] The questionable contributions of a Skolem and a Gödel, on the other hand, have been expatiated upon at great length. And I recall that my presentation on propositional systems had already been excluded from a discussion session during the meeting of mathematicians in Bad Elster due to a plot engineered by the Vienna Circle represented by Hahn and Gödel. Ever since, I have lost all interest in speaking publicly on foundational matters. This apparently is the lot of anyone who has no "school" or clique behind him. But perhaps a time will come when even my papers will be rediscovered and read again. At any rate, allow me to express, once again, my gratitude to you for your kind efforts. Yours sincerely,

E. Zermelo

---

[1] ⟦*1908a, 1908b* and *1930a, 1935,* respectively.⟧

Bibliography

Ackermann, Wilhelm

See Hilbert, David and Wilhelm Ackermann.

Alexandrow, Waldemar

1915 *Elementare Grundlagen für die Theorie des Maßes*, Ph. D. Thesis, University of Zurich.

Baer, Reinhold

1928 Über ein Vollständigkeitsaxiom der Mengenlehre, *Mathematische Zeitschrift* **27**, 536–539.

1929 Zur Axiomatik der Kardinalzahlarithmetik, *Mathematische Zeitschrift* **29**, 381–396.

1952 Endlichkeitskriterien für Kommutatorgruppen, *Mathematische Annalen* **124**, 161–177.

Banaschewski, Bernhard and Guillaume Brümmer

1986 Thoughts on the Cantor-Bernstein theorem, *Quaestiones mathematicae* **9**, 1–27.

Barnard, Mordaunt

1876 *The Odyssey of Homer, rendered into English blank verse*, Williams and Norgate: London.

Barwise, Jon

1981 Infinitary logics, in Evandro Agazzi (ed.), *Modern logic: A survey*, D. Reidel: Dordrecht, 93–112.

Becker, Carl

1974 Immisch, Otto, in *Neue deutsche Biographie*, vol. 10, Duncker & Humblot: Berlin, 164-165.

Bell, John

1995 Type reducing correspondences and well-orderings: Frege's and Zermelo's constructions re-examined, *The Journal of Symbolic Logic* **60**, 209–221.

2009 The axiom of choice, in Edward N. Zalta (ed.), *The Stanford encyclopedia of philosophy (spring 2009 edition)*, http://plato.stanford.edu/archives/spr2009/entries/axiom-choice/.

Bernays, Paul

1913 Zur elementaren Theorie der Landauschen Funktion $\phi(\alpha)$, *Vierteljahrsschrift der Naturforschenden Gesellschaft in Zürich* **58**, 203–238; Habilitation Thesis, University of Zurich.

1941 A system of axiomatic set theory, *The Journal of Symbolic Logic* **6**, 1–17.
1942 A system of axiomatic set theory. III, *The Journal of Symbolic Logic* **7**, 65–89.

Bernstein, Felix
1901 *Untersuchungen aus der Mengenlehre*, Halle a. S., Ph. D. Thesis, University of Göttingen; reprinted as *Bernstein 1905c*.
1904 Bemerkung zur Mengenlehre, *Nachrichten von der Königlichen Gesellschaft der Wissenschaften zu Göttingen, Mathematisch-physikalische Klasse* **1904**, 557–560.
1905a Die Theorie der reellen Zahlen, *Jahresbericht der Deutschen Mathematiker-Vereinigung* **14**, 447–449.
1905b Über die Reihe der transfiniten Ordnungszahlen, *Mathematische Annalen* **60**, 187–193.
1905c Untersuchungen aus der Mengenlehre, *Mathematische Annalen* **61**, 117–155.

Bilharz, Herbert
1951 Über die Gaußsche Methode zur angenäherten Berechnung bestimmter Integrale. Ernst Zermelo zum 80. Geburtstag gewidmet, *Mathematische Nachrichten* **6**, 171–192.

Birkhoff, Garrett and Saunders Mac Lane
1965 *A survey of modern algebra*, 3rd edition, Macmillan: New York.

Blumberg, Henry
1913 Über eine von Herrn Maillet vorgeschlagene Definition der ganzen transzendenten Zahlen, *Archiv der Mathematik und Physik*, 3rd series, **20**, 53–57.

Bois-Reymond, Paul du *see* du Bois-Reymond, Paul

Boltzmann, Ludwig
1896 Entgegnung auf die wärmetheoretischen Betrachtungen des Hrn. E. Zermelo, *Annalen der Physik und Chemie*, new series, **57**, 773–784.
1897 Zu Hrn. Zermelo's Abhandlung „Ueber die mechanische Erklärung irreversibler Vorgänge", *Annalen der Physik und Chemie*, new series, **60**, 392–398.

Bolza, Oskar
1909 *Vorlesungen über Variationsrechnung*, B. G. Teubner: Leipzig.

Bolzano, Bernard
1827/44 *Philosophische Tagebücher 1827–1844*, complete edition, series II-B, vol. 18/2, Frommann-Holzboog: Stuttgart.

1837 *Wissenschaftslehre*, 4 vols., J. E. v. Seidelsche Buchhandlung: Sulzbach.
1851 *Paradoxien des Unendlichen*, C. H. Reclam: Leipzig.
1975 *Einleitung zur Größenlehre und erste Begriffe der allgemeinen Größenlehre*, complete edition, vol. II A 7, edited by Jan Berg, Frommann-Holzboog: Stuttgart.

Bonse, H.
1907 Über eine bekannte Eigenschaft der Zahl 30 und ihre Verallgemeinerung, *Archiv der Mathematik und Physik* **12**, 292–295.

Boole, George
1847 *The mathematical analysis of logic*, Macmillan: Cambridge.

Boolos, George S.
1971 The iterative concept of set, *Journal of Philosophy* **68**, 215–231; reprinted in Richard Jeffrey (ed.), *Logic, logic, and logic*, Harvard University Press: Cambridge, Massachusetts.

Borel, Émile
1898 *Leçons sur la théorie des fonctions,* Gauthier-Villars: Paris; 4th edition 1950.
1905a Quelques remarques sur les principes de la théorie des ensembles, *Mathematische Annalen* **60**, 194–195; reprinted in *Borel 1972*, vol. III, 1251–1252.
1905b Cinq lettres sur la théorie des ensembles, *Bulletin de la Societé Mathématique de France* **33**, 261–273 (Hadamard to Borel, Baire to Hadamard, Lebesgue to Borel, Hadamard to Borel, Borel to Hadamard); reprinted in *Borel 1972*, vol. III, 1253–1265.
1905c *Leçons sur les fonctions de variables réelles*, Gauthier-Villars: Paris.
1921 La théorie du jeu et les équations intégrales à noyau symétrique, *Comptes Rendus de l'Académie des Sciences, Série 1, Mathématique* **173**, 1304–1308; reprinted in *Borel 1972*, vol. II, 901–904.
1924 Sur les jeux où interviennent l'hasard et l'habileté des joueurs, in *Éléments de la théorie des probabilités*, 3rd edition, Librairie Scientifique, J. Hermann: Paris, 204–224.
1927 Sur les systèmes de formes linéaires à determinant symétrique gauche et la théorie générale de jeu, *Comptes Rendus de l'Académie des Sciences, Série 1, Mathématique* **184**, 52–54; reprinted in *Borel 1972*, vol. II, 1123–1124.
1972 *Œuvres de Émile Borel*, vol. I–IV, Centre national de la recherche scientifique: Paris.

Brouwer, Luitzen Egbertus Jan

1927 Über Definitionsbereiche von Funktionen, *Mathematische Annalen* **97**, 60–75.

Brümmer, Guillaume

See Banaschewski, Bernhard and Guillaume Brümmer.

Bundschuh, Peter

1988 *Einführung in die Zahlentheorie*, Springer-Verlag: Berlin, Heidelberg, et al.

Burali-Forti, Cesare

1897 Una questione sui numeri transfiniti, *Rendiconti del Circolo Matematico di Palermo* **11**, 154–164.

Cantor, Georg

1871a Notiz zu dem Aufsatze: Beweis, daß eine für jeden reellen Wert von x durch eine trigonometrische Reihe gegebene Funktion $f(x)$ sich nur auf eine einzige Weise in dieser Form darstellen läßt, *Journal für die reine und angewandte Mathematik* **73**, 294–296; reprinted in *Cantor 1932*, 84–86.

1871b Ueber trigonometrische Reihen, *Mathematische Annalen* **4**, 139–143; reprinted in *Cantor 1932*, 87–91.

1872 Ueber die Ausdehnung eines Satzes aus der Theorie der trigonometrischen Reihen, *Mathematische Annalen* **5**, 123–132; reprinted in *Cantor 1932*, 92–102.

1873 Historische Notizen über die Wahrscheinlichkeitsrechnung, *Sitzungsberichte der Naturforschenden Gesellschaft zu Halle* **1873**, 34–42; reprinted in *Cantor 1932*, 357–367.

1878 Ein Beitrag zur Mannigfaltigkeitslehre, *Journal für die reine und angewandte Mathematik* **84**, 242–258; reprinted in *Cantor 1932*, 119–133.

1879 Ueber unendliche, lineare Punktmannichfaltigkeiten. 1., *Mathematische Annalen* **15**, 1–7; reprinted in *Cantor 1932*, 139–145.

1880 Ueber unendliche, lineare Punktmannichfaltigkeiten. 2., *Mathematische Annalen* **17**, 355–358; reprinted in *Cantor 1932*, 145–148.

1882 Ueber unendliche, lineare Punktmannichfaltigkeiten. 3., *Mathematische Annalen* **20**, 113–121; reprinted in *Cantor 1932*, 149–157.

1883a Ueber unendliche, lineare Punktmannichfaltigkeiten. 4., *Mathematische Annalen* **21**, 51–58; reprinted in *Cantor 1932*, 157–164.

1883b Ueber unendliche, lineare Punktmannichfaltigkeiten. 5., *Mathematische Annalen* **21**, 545–591; reprinted in *Cantor 1932*, 165–208.

1884 Ueber unendliche, lineare Punktmannichfaltigkeiten. 6., *Mathematische Annalen* **23**, 453–488; reprinted in *Cantor 1932*, 210–244.

1887 Mitteilungen zur Lehre vom Transfiniten I, *Zeitschrift für Philosophie und philosophische Kritik* **91**, 81–125; reprinted in *Cantor 1932*, 378–419.

1888 Mitteilungen zur Lehre vom Transfiniten II, *Zeitschrift für Philosophie und philosophische Kritik* **92**, 240–265; reprinted in *Cantor 1932*, 420–439.

1890 *Zur Lehre vom Transfiniten. Gesammelte Abhandlungen aus der Zeitschrift für Philosophie und philosophische Kritik von Georg Cantor, ord. Professor an der Universität Halle – Erste Abteilung*, C. E. M. Pfeffer (Norbert Stricker): Halle.

1890/91 Ueber eine elementare Frage der Mannigfaltigkeitslehre, *Jahresbericht der Deutschen Mathematiker-Vereinigung* **1**, 75–78; reprinted in *Cantor 1932*, 278–280.

1895 Beiträge zur Begründung der transfiniten Mengenlehre (Erster Artikel), *Mathematische Annalen* **46**, 481–512; reprinted in *Cantor 1932*, 282–311.

1897 Beiträge zur Begründung der transfiniten Mengenlehre (Zweiter Artikel), *Mathematische Annalen* **49**, 207–246; reprinted in *Cantor 1932*, 312–351.

1932 *Gesammelte Abhandlungen mathematischen und philosophischen Inhalts. Mit erläuternden Anmerkungen sowie mit Ergänzungen aus dem Briefwechsel Cantor–Dedekind herausgegeben von Ernst Zermelo*, Springer-Verlag: Berlin; reprinted with biographical supplement 1980 and 1990.

1991 *Briefe,* edited by Herbert Meschkowski and Winfried Nilson, Springer-Verlag: Berlin, Heidelberg, New York et al.

Carathéodory, Constantin

1904 *Über die diskontinuierlichen Lösungen in der Variationsrechnung*, Ph. D. Thesis, University of Göttingen; reprinted in *Carathéodory 1954/57*, vol. I, 3–79.

1914 Über das lineare Maß von Punktmengen – Eine Verallgemeinerung des Längenbegriffs, *Nachrichten der Königlichen Gesellschaft der Wissenschaften zu Göttingen, Mathematisch-physikalische Klasse aus dem Jahre 1914,* 404–426; reprinted in *Carathéodory 1954/57,* vol. IV, 249–275.

1918 *Vorlesungen über reelle Funktionen*, B. G. Teubner: Leipzig and Berlin.

1927 *Vorlesungen über reelle Funktionen*, 2nd edition, B. G. Teubner: Leipzig and Berlin.

1935 *Variationsrechnung und partielle Differentialgleichungen erster Ordnung*, B. G. Teubner: Leipzig.

1954/57 *Gesammelte Mathematische Schriften*, vol. I–V, Beck: Munich.

Cauchy, Augustin-Louis

1821 *Cours d'analyse de l'école royale polytechnique. Première partie: Analyse algébrique*, DeBure Frères: Paris; reprinted in *Œuvres complètes d'Augustin Cauchy*, 2nd series, vol. 3, Gauthier-Villars: Paris, 1897.

Cavaillès, Jean

1938a *Remarques sur la formation de la théorie abstraite des ensembles*, Hermann: Paris.

1938b *Méthode axiomatique et formalisme. Essai sur le problème du fondement des mathématiques*, Hermann: Paris.

See also Noether, Emmy and Jean Cavaillès.

Church, Alonzo (ed.)

1936 A bibliography of symbolic logic (1666–1935), *The Journal of Symbolic Logic* **1**, 121–218; first version of *Church 1984*.

1984 *A bibliography of symbolic logic: Revised and expanded edition*, Association for Symbolic Logic.

Courant, Richard and Herbert Robbins

1941 *What is mathematics?*, Oxford University Press: London, New York, Toronto.

Dauben, Joseph Warren

1990 *Georg Cantor: His mathematics and philosophy of the infinite*, Princeton University Press: Princeton, New Jersey.

Davenport, Harold

1952 *The higher arithmetic*, Hutchinson: London.

David, Herbert A.

1988 *The method of paired comparisons*, 2nd edition, Oxford University Press: Oxford; 1st edition Charles Griffin & Company Ltd.: London.

Dawson, John W., Jr.

1997 *Logical dilemmas: The life and work of Kurt Gödel*, A. K. Peters: Wellesley, Massachusetts.

2006 Why do mathematicians re-prove theorems?, *Philosophia mathematica* **14**, 269–286.

Dedekind, Richard

1872 *Stetigkeit und irrationale Zahlen*, Vieweg: Braunschweig.

1888 *Was sind und was sollen die Zahlen?*, Vieweg: Braunschweig.

1932 *Gesammelte mathematische Werke*, vol. III, edited by Robert Fricke, Emmy Noether and Øystein Ore, Vieweg: Braunschweig.

Deiser, Oliver

2004 *Einführung in die Mengenlehre*, 2nd edition, Springer-Verlag: Berlin, Heidelberg, et al.

2005 Der Multiplikationssatz der Mengenlehre, *Jahresbericht der Deutschen Mathematiker-Vereinigung* **107**, 89–109.

2006 *Orte, Listen, Aggregate*, Habilitation Thesis, Freie Universität Berlin.

200? On the development of the notion of a cardinal number; in preparation.

de la Vallée Poussin, Charles-Jean *see* la Vallée Poussin, Charles-Jean, Baron de

Dick, Auguste

1970 *Emmy Noether, 1882–1935*, Birkhäuser Verlag: Basel.

Dickson, Leonard E.

1919–1923 *History of the theory of numbers*, vols. 1–3, Carnegie Institution: Washington, D.C.

Dirichlet, Peter Gustav Lejeune

1871 *Vorlesungen über Zahlentheorie*, edited by Richard Dedekind, 2nd edition, Vieweg: Braunschweig.

du Bois-Reymond, Paul

1873 Eine neue Theorie der Convergenz und Divergenz von Reihen mit positiven Gliedern, *Journal für die reine und angewandte Mathematik* **76**, 61–91.

1879 Fortsetzung der Erläuterungen zu den Anfangsgründen der Variationsrechnung, *Mathematische Annalen* **15**, 564–576.

Ebbinghaus, Heinz-Dieter

2003 Zermelo: Definiteness and the universe of definable sets, *History and Philosophy of Logic* **24**, 197–219.

2004 Zermelo in the mirror of the Baer correspondence 1930–1931, *Historia mathematica* **31**, 76–86.

2006 Zermelo: Boundary numbers and domains of sets continued, *History and Philosophy of Logic* **27**, 285–306.

2007a Zermelo and the Heidelberg congress 1904, *Historia mathematica* **34**, 428–432.

2007b (in cooperation with Volker Peckhaus) *Ernst Zermelo. An approach to his life and work*, Springer-Verlag: Berlin and Heidelberg.

See also van Dalen, Dirk and Heinz-Dieter Ebbinghaus.

Farber, Marvin

1926/27a Undated handwritten notes of conversations with Zermelo concerning the philosophy of logic and mathematics, Marvin Farber Papers on Philosophy and Phenomenology, University Archives, State University of New York at Buffalo, 22/5F/768, folders 33.10–11.

1926/27b Multiple outlines of an envisioned book entitled variously "Logic" or "The foundations of logic: Studies concerning the structure and function of logic", Marvin Farber Papers on Philosophy and Phenomenology, University Archives, State University of New York at Buffalo, 22/5F/768, folders 33.10–11.

1927 *The foundations of logic: Studies concerning the structure and function of logic,* typescript, initial draft of the first three chapters of a book on the philosophy of logic, Marvin Farber Papers on Philosophy and Phenomenology, University Archives, State University of New York at Buffalo, 22/5F/768, folder 28.12.

Fechner, Gustav Theodor

1875 *Kleine Schriften*, Breitkopf und Härtel: Leipzig.

Felgner, Ulrich

2002a Der Begriff der Kardinalzahl, in *Hausdorff 2002*, 634–644.

2002b Ein Brief Gödels zum Fundierungsaxiom, in Bernd Buldt et al. (eds.), *Kurt Gödel. Wahrheit & Beweisbarkeit*, vol. 2: *Kompendium zum Werk*, öbv & hpt: Vienna, 205–213.

2002c Der Begriff der Funktion, in *Hausdorff 2002*, 621–633.

2009 Dichtung und Wahrheit – zur Geschichte der Antinomie vom Lügner, in Günther Löfflath (ed.), *Mathematik-Logik-Philosophie, Ideen und ihre historischen Wechselwirkungen*, Harry Deutsch: Frankfurt.

Ferreirós, José

1993 On the relations between Georg Cantor and Richard Dedekind, *Historia mathematica* **20**, 343–363.

1999 *Labyrinth of thought. A history of set theory and its role in modern mathematics*, Birkhäuser: Basel, Boston, and Berlin 1999; 2nd edition 2007.

Finsler, Paul

1926 Über die Grundlegung der Mengenlehre. I. Teil: Die Mengen und ihre Axiome, *Mathematische Zeitschrift* **25**, 683–713.

Flashar, Hellmut

2005 Schadewaldt, Wolfgang Otto Bernhard, in *Neue deutsche Biographie*, vol. 22, Duncker & Humblot: Berlin, 495-496.

Fraenkel, Abraham A.

1919 *Einleitung in die Mengenlehre*, Springer-Verlag: Berlin; 2nd edition *see Fraenkel 1923a*; 3rd edition *see Fraenkel 1928*.

1921 Über die Zermelosche Begründung der Mengenlehre, *Jahresbericht der Deutschen Mathematiker-Vereinigung* **30**, 2nd section, 97–98.

1922a Der Begriff »definit« und die Unabhängigkeit des Auswahlaxioms, *Die Preussische Akademie der Wissenschaften. Sitzungsberichte. Physikalisch-Mathematische Klasse* **1922**, 253–257.

1922b Zu den Grundlagen der Cantor-Zermeloschen Mengenlehre, *Mathematische Annalen* **86**, 230–237.

1923a *Einleitung in die Mengenlehre*, 2nd enlarged edition of *Fraenkel 1919*, Springer-Verlag: Berlin.

1923b Die Axiome der Mengenlehre, *Scripta atque Bibliothecae Hierosolymitanarum* **1**, 1–8.

1925 Untersuchungen über die Grundlagen der Mengenlehre, *Mathematische Zeitschrift* **22**, 250–273.

1927 *Zehn Vorlesungen über die Grundlegung der Mengenlehre*, B. G. Teubner: Leipzig and Berlin.

1928 *Einleitung in die Mengenlehre*, 3rd expanded edition of *Fraenkel 1919*.

1930 Georg Cantor, *Jahrebericht der Deutschen Mathematiker-Vereinigung* **39**, 189–266.

1935 Zum Diagonalverfahren Cantors, *Fundamenta mathematicae* **25**, 45–50.

1961 *Abstract set theory*, 2nd revised edition, North-Holland Publishing Company: Amsterdam.

1967 *Lebenskreise. Aus den Erinnerungen eines jüdischen Mathematikers*, Deutsche Verlags-Anstalt: Stuttgart.

Fraenkel, Adolf *see* Fraenkel, Abraham A.

Franchella, Miriam

1997 On the origins of Dénes König's infinity lemma, *Archive for History of Exact Sciences* **51**, 3–27.

Fréchet, Maurice

1915 Sur l'intégrale d'une fonctionnelle étendue à un ensemble abstrait, *Bulletin de la Société Mathématique de France* **43**, 249–265.

1923 Des familles et fonctions additives d'ensembles abstraits, *Fundamenta mathematicae* **4**, 329–365.

1924 Des familles et fonctions additives d'ensembles abstraits (suite), *Fundamenta mathematicae* **5**, 206–251.

Frege, Gottlob

1879 *Begriffsschrift, eine der arithmetischen nachgebildete Formelsprache des reinen Denkens*, Louis Nebert: Halle an der Saale.

1892 Review of *Cantor 1890*, *Zeitschrift für Philosophie und philosophische Kritik* **100**, 269–272; reprinted in *Frege 1967*, 163–166.
1967 *Kleine Schriften*, edited by Ignacio Angelelli, Olms: Hildesheim.
1976 *Nachgelassene Schriften und wissenschaftlicher Briefwechsel*, vol. II: *Wissenschaftlicher Briefwechsel*, edited by Gottfried Gabriel et al., Felix Meiner Verlag: Hamburg.
1986 *Grundlagen der Arithmetik*, centenary edition by Christian Thiel, Meiner: Hamburg.

Friedman, Harvey
1971 Higher set theory and mathematical practice, *Annals of Mathematical Logic* **2**, 326–357.

Fritsch, Rudolf and Gerda Fritsch
2001 Ansätze zu einer wissenschaftlichen Biographie von Arthur Schoenflies (1853–1928), in Menso Folkerts et al. (eds.), *Florilegium Astronomicum, Festschrift für Felix Schmeidler*, *Algorismus Studien*, issue 37, Institut für Geschichte der Naturwissenschaften: Munich, 141–186.

Gauß, Carl Friedrich
1801 *Disquisitiones arithmeticae*, Gerh. Fleischer, Jun.: Leipzig.

Genocchi, Angelo
1899 *Differentialrechnung und Grundzüge der Integralrechnung*, edited by Guiseppe Peano, German translation by Georg Bohlmann and Adolf Schepp, B. G. Teubner: Leipzig.

Gericke, Helmuth
1970 *Geschichte des Zahlbegriffs*, Bibliographisches Institut: Mannheim.

Gibbs, Josiah Willard
1902 *Elementary principles in statistical mechanics, developed with especial reference to the rational foundation of thermodynamics*, Scribner: New York.
1905 *Elementare Grundlagen der statistischen Mechanik entwickelt besonders im Hinblick auf eine rationelle Begründung der Thermodynamik*, German edition of *Gibbs 1902* by Ernst Zermelo, J. A. Barth: Leipzig.

Givant, Steven and Alfred Tarski
1977 Peano arithmetic and the Zermelo-like theory of sets with finite ranks, Abstract 77T-E51, *Notices of the American Mathematical Society* **24**, A-437.
See also Tarski, Alfred and Steven Givant.

Glazebrook, Richard Tedley

1894 *Light, an elementary textbook theoretical and practical*, Cambridge University Press: Cambridge.

1897 *Das Licht. Grundriß der Optik, für Studierende und Schüler*, German edition of *Glazebrook 1894* by Ernst Zermelo, S. Calvary & Co.: Berlin.

Gödel, Kurt

1931a Über formal unentscheidbare Sätze der *Principia mathematica* und verwandter Systeme I, *Monatshefte für Mathematik und Physik* **38**, 173–198; reprinted with English translation in *Gödel 1986*, 144–195.

1931b Letter to E. Zermelo of 12 October 1931, Universitätsarchiv Freiburg, Zermelo *Nachlass*, C129/36; reprinted with English translation in *Gödel 2003*, 422-429.

1933 The present situation in the foundation of mathematics, invited lecture delivered to the meeting of the American Mathematical Society, December 29–30, 1933 in Cambridge, Massachusetts; reprinted in *Gödel 1995*, 45–53.

1939 The consistency of the generalized continuum hypothesis, *Bulletin of the American Mathematical Society* **45**, 93; reprinted in *Gödel 1990*, 27.

1940 *The consistency of the axiom of choice and of the generalized continuum hypothesis with the axioms of set theory*, Annals of Mathematics Studies, vol. 3, Princeton University Press: Princeton, New Jersey; reprinted in *Gödel 1990*, 33-101.

1944 Russell's mathematical logic, in Paul Arthur Schilpp (ed.), *The philosophy of Bertrand Russell*, The Open Court Publishing Co.: La Salle, Illinois, 125–153; reprinted in *Gödel 1990*, 119–141.

1947 What is Cantor's continuum problem?, *American Mathematical Monthly* **54**, 515–525, errata **55**, 151; reprinted in *Gödel 1990*, 176–187.

1986 *Collected works*, vol. I, edited by Solomon Feferman et al., Oxford University Press: New York and Clarendon Press: Oxford.

1990 *Collected works*, vol. II, edited by Solomon Feferman et al., Oxford University Press: New York and Oxford.

1995 *Collected works*, vol. III, edited by Solomon Feferman et al., Oxford University Press: New York and Oxford.

2003 *Collected works*, vol. V, edited by John W. Dawson, Jr., et al., Oxford University Press: New York and Oxford.

Gonseth, Ferdinand (ed.)

1941 *Les entretiens de Zurich sur les fondements et la méthode des sciences mathématiques, 6–9 décembre 1938. Exposés et discussions*, Leemann: Zürich.

Grattan-Guinness, Ivor

1974 The rediscovery of the Cantor-Dedekind-correspondence, *Jahresbericht der Deutschen Mathematiker-Vereinigung* **76**, 104–139.

1979 In memoriam Kurt Gödel: His 1931 correspondence with Zermelo on his incompletability theorem, *Historia mathematica* **6**, 294–304.

2000 *The search for mathematical roots, 1870–1940: Logic, set theories and the foundations of mathematics from Cantor through Russell to Gödel*, Princeton University Press: Princeton and Oxford.

Grelling, Kurt

1910 *Die Axiome der Arithmetik, mit besonderer Berücksichtigung der Beziehungen zur Mengenlehre*, Ph. D. Thesis, University of Göttingen.

Hahn, Hans and Ernst Zermelo

1904 Weiterentwicklung der Variationsrechnung in den letzten Jahren, *Enzyklopädie der mathematischen Wissenschaften mit Einschluss ihrer Anwendungen*, vol. 2: *Analysis*, part 1, 1st half, Teubner: Leipzig, 1899–1916, 626–641.

Hajek, Petr

See Vopěnka, Petr and Petr Hajek.

Hallett, Michael

1979a Towards a theory of mathematical research programmes (I), *British Journal for the Philosophy of Science* **30**, 1–25.

1979b Towards a theory of mathematical research programmes (II), *British Journal for the Philosophy of Science* **30**, 135–159.

1984 *Cantorian set theory and the limitation of size*, Clarendon Press: Oxford.

2008 The 'purity of method' in Hilbert's *Grundlagen der Geometrie*, in Paolo Mancosu (ed.), *The philosophy of mathematical practice*, Clarendon Press: Oxford, 198–255.

2010 Frege and Hilbert, in Michael Potter and Thomas Ricketts (eds.), *The Cambridge companion to Frege*, Cambridge University Press: Cambridge, forthcoming.

Hallett, Michael and Ulrich Majer (eds.)

2004 *David Hilbert's lectures on the foundations of geometry, 1891–1902*, *Hilbert's lectures on the foundations of mathematics and physics*, vol. 1, Springer-Verlag: Berlin, Heidelberg, New York.

Halpern, J. Daniel and Paul E. Howard

1970 Cardinals m such that $2m = m$, *Proceedings of the American Mathematical Society* **26**, 487–490.

Hamel, Georg

1905 Eine Basis aller Zahlen und die unstetigen Lösungen der Funktionalgleichung $f(x+y) = f(x) + f(y)$, *Mathematische Annalen* **60**, 459–462.

Hardy, G. Harold

1904 A theorem concerning the infinite cardinal numbers, *The Quarterly Journal of Pure and Applied Mathematics* **35**, 87–94.

Hardy, G. Harold and E. Maitland Wright

1938 *An introduction to the theory of numbers*, Clarendon Press: Oxford.

Hartogs, Friedrich

1915 Über das Problem der Wohlordnung, *Mathematische Annalen* **76**, 438–443.

Hasse, Helmut

1928 Über eindeutige Zerlegung in Primelemente oder in Primhauptideale in Integritätsbereichen, *Journal für die reine und angewandte Mathematik* **159**, 3–12.

1934 *Aufgabensammlung zur höheren Algebra*, Walter de Gruyter: Berlin.

1949 *Zahlentheorie*, Akademie-Verlag: Berlin.

1950 *Vorlesungen über Zahlentheorie*, Springer-Verlag: Berlin, Göttingen, Heidelberg.

Hauschild, Kurt

1966 Bemerkungen, das Fundierungsaxiom betreffend, *Zeitschrift für mathematische Logik und Grundlagen der Mathematik* **12**, 51–56.

Hausdorff, Felix

1908 Grundzüge einer Theorie der geordneten Mengen, *Mathematische Annalen* **65**, 435–505.

1909 Die Graduierung nach dem Endverlauf, *Abhandlungen der Königlich-Sächsischen Gesellschaft der Wissenschaften, mathematisch-physikalische Klasse* **31**, 297–334.

1914 *Grundzüge der Mengenlehre*, Verlag von Veit & Comp.: Leipzig; reprinted in *Hausdorff 2002*, 93–476; 2nd edition *see Hausdorff 1927*; 3rd edition *see Hausdorff 1935*.

1927 *Mengenlehre*, thoroughly revised 2nd edition of *Hausdorff 1914*, Walter de Gruyter: Berlin; reprinted in *Hausdorff 2008*, 42-307.

1935 *Mengenlehre*, 3rd edition of *Hausdorff 1914*, Walter de Gruyter: Berlin.

2002 *Gesammelte Werke*, vol. II: *Grundzüge der Mengenlehre,* edited by Egbert Brieskorn et al., Springer-Verlag: Berlin, Heidelberg, et al.
2008 *Gesammelte Werke*, vol. III: *Deskriptive Mengenlehre und Topologie,* edited by Egbert Brieskorn et al., Springer-Verlag: Berlin, Heidelberg, et al.

Heath, Thomas
1931 *A manual of Greek mathematics*, Oxford University Press: Oxford.

Heffter, Lothar
1952 *Beglückte Rückschau auf neun Jahrzehnte. Ein Professorenleben,* Schulz: Freiburg i. Br.

Hertz, Paul
1922 Über Axiomensysteme für beliebige Satzsysteme, I. Teil: Sätze ersten Grades (Über die Axiomensysteme von der kleinsten Satzzahl und den Begriff des idealen Elementes), *Mathematische Annalen* **87**, 246–269.
1923 Über Axiomensysteme für beliebige Satzsysteme, Teil II: Sätze höheren Grades, *Mathematische Annalen* **89**, 76–102.
1929 Über Axiomensysteme für beliebige Satzsysteme, *Mathematische Annalen* **101**, 457–514.

Hessenberg, Gerhard
1906 *Grundbegriffe der Mengenlehre*, *Abhandlungen der neuen Friesschen Schule*, new series, **1**, 479–706; reprint Vandenhoeck & Ruprecht: Göttingen 1906.

Hilbert, David
1897 Die Theorie der algebraischen Zahlkörper, *Jahresbericht der Deutschen Mathematiker-Vereinigung* **4**, 175–546.
1899 *Grundlagen der Geometrie*, in *Festschrift zur Feier der Enthüllung des Gauss-Weber-Denkmals in Göttingen*, B. G. Teubner: Leipzig.
1900a Über den Zahlbegriff, *Jahresbericht der Deutschen Mathematiker-Vereinigung* **8**, 180–184.
1900b Mathematische Probleme. Vortrag, gehalten auf dem internationalen Mathematiker-Kongreß zu Paris 1900, *Nachrichten von der königlichen Gesellschaft der Wissenschaften zu Göttingen, mathematisch-physikalische Klasse aus dem Jahre 1900*, 253–297.
1902 *Grundlagen der Geometrie*, lecture notes by August Adler, summer semester 1902, Mathematical Institute Göttingen; published as Ch. 6 in *Hallett and Majer 2004*.

1920 *Probleme der Mathematischen Logik*, lecture notes by Moses Schönfinkel and Paul Bernays, summer semester 1920, Mathematical Institute Göttingen.

1926 Über das Unendliche, *Mathematische Annalen* **95**, 161–190.

Hilbert, David and Wilhelm Ackermann

1928 *Grundzüge der theoretischen Logik*, Springer-Verlag: Berlin.

Hobson, Ernest William

1907 *The theory of functions of a real variable and the theory of Fourier's series*, 1st edition in one volume, Cambridge University Press: Cambridge; 2nd edition vol. I 1921, vol. II 1926; 3rd edition vol. I 1927.

Horsfall, Nicholas

1990 Eduard Fraenkel, in Ward W. Briggs and William M. Calder III (eds.), *Classical scholarship: A biographical encyclopedia*, Garland Publishing Inc.: New York and London, 61–67.

Howard, Paul E.

See Halpern, J. Daniel and Paul E. Howard, and the next entry.

Howard, Paul E. and Jean Rubin

1998 *Consequences of the axiom of choice*, American Mathematical Society: Providence, Rhode Island.

Husserl, Edmund

1891 *Philosophie der Arithmetik. Psychologische und logische Untersuchungen*, vol. 1, C. E. M. Pfeffer (Robert Stricker): Halle a. S.; critical edition *Husserl 1975.*

1922 *Logische Untersuchungen*, vol. I: *Prolegomena zur reinen Logik*, Max Niemeyer: Halle; reprinted in *Husserl 1975.*

1928 *Ideen zu einer reinen Phänomenologie und phänomenologischen Philosophie*, first book: *Allgemeine Einführung in die reine Phänomenologie*, Max Niemeyer: Halle; reprinted in *Husserl 1950.*

1950 *Husserliana*, vol. 3, edited by Walter Biemel, Nijhoff: The Hague.

1956 *Husserliana*, vol. 7, edited by Rudolf Boehm, Nijhoff: The Hague.

1974 *Husserliana*, vol. 17, edited by Paul Janssen, Nijhoff: The Hague.

1975 *Husserliana*, vol. 18, edited by Elmar Holenstein, Nijhoff: The Hague.

1979 *Aufsätze und Rezensionen 1890–1910*, *Husserliana*, vol. 22, edited by Bernhard Rang, Nijhoff: The Hague.

Ilgauds, Hans Joachim

See Purkert, Walter and Hans Joachim Ilgauds.

Jech, Thomas

1973 *The axiom of choice*, North-Holland Publishing Company: Amsterdam.

2003 *Set theory: The third millennium edition*, Springer: Berlin, Heidelberg, et al.

Jensen, Ronald B.

1966 Independence of the axiom of dependent choices from the countable axiom of choice (abstract), *The Journal of Symbolic Logic* **31**, 294.

Jones, Antonia J.

2000 *Game theory: Mathematical models of conflict*, Horwood Publishing Limited: Chichester.

Jourdain, Philip Edward Bertrand

1904a On the transfinite cardinal numbers of well-ordered aggregates, *Philosophical Magazine*, series 6, **7**, 61–75.

1904b On the transfinite cardinal numbers of number classes in general, *Philosophical Magazine*, series 6, **7**, 294–303.

1905a On transfinite cardinal numbers of the exponential form, *Philosophical Magazine*, series 6, **9**, 42–56.

1905b On a proof that every aggregate can be well-ordered, *Mathematische Annalen* **60**, 465–470.

Kaemmel, Thomas

2006 *Arthur Schoenflies – Mathematiker und Kristallforscher. Eine Biographie mit Aufstieg und Zerstreuung einer jüdischen Familie*, Projekte-Verlag: Halle.

Kalmár, Lásló

1928/29 Zur Theorie der abstrakten Spiele, *Acta litterarum ac scientiarum Regiae Universitatis Hungaricae Francisco-Josephinae, sectio scientiarum mathematicarum* **4**, 65–85.

Kanamori, Akihiro

1997 The mathematical import of Zermelo's well-ordering theorem, *The Bulletin of Symbolic Logic* **3**, 273–298.

2003 *The higher infinite*, 2nd edition, Springer-Verlag: Berlin, Heidelberg, et al.

2004 Zermelo and set theory, *The Bulletin of Symbolic Logic* **10**, 487–553.

2006 Levy and set theory, *Annals of Pure and Applied Logic* **140**, 233–252.

Kechris, Alexander S.
1995 *Classical descriptive set theory*, Springer-Verlag: New York, Berlin, Heidelberg, et al.

Kirby, Laurence
2009 Finitary set theory, *Notre Dame Journal of Formal Logic* **50**, forthcoming.

Klappauf, Gerhard
1935 Beweis des Fundamentalsatzes der Zahlentheorie, *Jahresbericht der Deutschen Mathematiker-Vereinigung*, 2nd section, **45**, 130.

Knaster, Bronisław
1928 Un théorème sur les fonctions d'ensembles, *Annales de la Societé Polonaise de Mathématique* **6**, 133–134.

Kneser, Adolf
1900 *Lehrbuch der Variationsrechnung*, Vieweg: Braunschweig.
1904 Variationsrechnung, in *Encyklopädie der mathematischen Wissenschaften mit Einschluss ihrer Anwendungen*, vol. 2: *Analysis*, part 1.1, 571–625, B. G. Teubner: Leipzig, 1899–1916.

Kneser, Hellmuth
1940 Der Fundamentalsatz der Algebra und der Intuitionismus, *Mathematische Zeitschrift* **46**, 287–302.

König, Dénes
1926 Sur les correspondances multivoques des ensembles, *Fundamenta mathematicae* **8**, 114–134.
1927a Über eine Schlussweise aus dem Endlichen ins Unendliche, *Acta litterarum ac scientiarum Regiae Universitatis Hungaricae Francisco-Josephinae, sectio scientiarum mathematicarum* **3**, 121–130.
1927b Zusatz zu §5, in *D. König 1927a,* 129–130.

König, Julius
1905a Zum Kontinuum-Problem, in Adolf Krazer (ed.), *Verhandlungen des Dritten Internationalen Mathematiker-Kongresses in Heidelberg vom 8. bis 13. August 1904*, Teubner: Leipzig, 144–147.
1905b Zum Kontinuum-Problem, *Mathematische Annalen* **60**, 177–180.
1905c Über die Grundlagen der Mengenlehre und das Kontinuumproblem, *Mathematische Annalen* **61**, 156–160.
1906 Sur la théorie des ensembles, *Comptes Rendus Hebdomadaires des Séances de l'Académie des Sciences* **143**, 110–112.
1907 Über die Grundlagen der Mengenlehre und das Kontinuumproblem (Zweite Mitteilung), *Mathematische Annalen* **63**, 217–221.

Korselt, Alwin
1911 Über einen Beweis des Äquivalenzsatzes, *Mathematische Annalen* **70**, 294–296.

Kreisel, Georg

1971 Observations on popular discussions of foundations, in Dana S. Scott (ed.), *Axiomatic set theory*, Proceedings of Symposia in Pure Mathematics, vol. 13, part 1, American Mathematical Society: Providence, Rhode Island.

Kreiser, Lothar

1995 Die Hörer Freges und sein Briefpartner Alwin Korselt, electronically published in *Wittgenstein Studies* **1**, file 24-1-95.txt, Vienna.

Kronecker, Leopold

1882 *Grundzüge einer arithmetischen Theorie der algebraischen Grössen*, Reimer: Berlin; reprinted in *Kronecker 1897*, 237–387.

1897 *Werke*, vol. 2, edited by Kurt Hensel, B. G. Teubner: Leipzig.

1901 *Vorlesungen über Zahlentheorie*, ed. by Kurt Hensel, B. G. Teubner: Leipzig.

Kuhn, Harold W.

1953 Extensive games: The problem of information, in *Kuhn and Tucker 1953*, 193–216.

Kuhn, Harold W. and Albert W. Tucker (eds.)

1953 *Contributions to the theory of games*, vol. 2, Annals of Mathematical Studies 28, Princeton University Press: Princeton, New Jersey.

Kuratowski, Kazimierz

1921 Sur la notion de l'ordre dans la théorie des ensembles, *Fundamenta mathematicae* **2**, 161–171.

1922 Une méthode d'élimination des nombres transfini des raisonnements mathématiques, *Fundamenta mathematicae* **3**, 76–108.

Lakatos, Imre

1976 *Proofs and refutations: The logic of mathematical discovery*, edited by John Worrall and Elie Zahar, Cambridge University Press: Cambridge.

Landau, Edmund

1909 *Handbuch der Lehre von der Verteilung der Primzahlen*, vol. I, B. G. Teubner: Leipzig and Berlin.

1917a R. Dedekind. Gedächtnisrede, gehalten in der öffentlichen Sitzung der Königlichen Gesellschaft der Wissenschaften zu Göttingen am 12. Mai 1917, *Nachrichten der Königlichen Gesellschaft der Wissenschaften zu Göttingen. Geschäftliche Mitteilungen* **1917**, 50–70; reprinted in *Landau 1986*, 449–465.

1917b Abschnitt 3, in *Landau 1917a*, 55–58.

1918 *Einführung in die elementare und analytische Theorie der algebraischen Zahlen und der Ideale*, B. G. Teubner: Leipzig and Berlin.
1927 *Elementare Zahlentheorie*, Hirzel: Leipzig.
1986 *Collected works*, vol. 6, edited by Paul T. Bateman et al., Thales Verlag: Essen.

la Vallée Poussin, Charles-Jean, Baron de

1909 *Cours d'analyse infinitésimale*, Uystpuyst-Dieudonné: Louvain, Paris; vol. I, 2nd edition 1909, 3rd edition 1914; vol. II, 2nd edition 1912.
1916 *Intégrales de Lebesgue, fonctions d'ensemble, classes de Baire*, Gauthier-Villars: Paris.

Lebesgue, Henri

1907 Sur les transformations ponctuelles transformant les plans en plans qu'on peut définir par des procédés analytiques, *Atti della Reale Accademia di Scienze di Torino* **42**, 2–10; reprinted in *Lebesgue 1972*, 218–226.
1972 *Œuvres scientifiques*, vol. III, L'Enseignement Mathématiques: Geneva.

Leibniz, Gottfried Wilhelm

1856 *Mathematische Schriften*, vol. 3, edited by Carl Immanuel Gerhardt, H. W. Schmidt: Halle.

Levi, Beppo

1902 Intorno alla teoria degli aggregati, *Reale Istituto lombardo di scienze e lettere, Rendiconti*, 2nd series, **35**, 863–868.

Levy, Azriel

1963 Independence results in set theory by Cohen's method, *Notices of the American Mathematical Society* **10**, 592–593.
1969 The definability of cardinal numbers, in Jack J. Bulloff et al. (eds.), *Foundations of mathematics. Symposium papers commemorating the sixtieth birthday of Kurt Gödel*, Springer-Verlag: Berlin, Heidelberg, et al., 15–38.
1979 *Basic set theory*, Springer-Verlag: Berlin, Heidelberg, et al.

Lindemann, Frederick Alexander

1933 The unique factorization of a positive integer, *Quarterly Journal of Mathematics* **4**, 319–320.

Lloyd-Jones, Hugh

1982 Rudolf Pfeiffer, in Hugh Lloyd-Jones, *Blood for the ghosts: Classical influences in the nineteenth and twentieth centuries*, Duckworth: London, 261–270.

Löwenheim, Leopold

1915 Über Möglichkeiten im Relativkalkül, *Mathematische Annalen* **76**, 447–470.

Luce, R. Duncan and Howard Raiffa

1957 *Games and decisions: A study of the behavioral models project*, Bureau of Applied Social Research: Columbia University.

Mac Lane, Saunders

See Birkhoff, Garrett and Saunders Mac Lane.

Maillet, Edmond

1906 *Introduction à la théorie des nombres transcendants et des propriétés arithmétiques des fonctions*, Gauthier-Villars: Paris.

Majer, Ulrich

See Hallett, Michael and Ulrich Majer.

Martin, D. Anthony

1975 Borel determinacy, *Annals of Mathematics* **102**, 363–371.

Mathias, Adrian R. D.

2001 Slim models of Zermelo set theory, *The Journal of Symbolic Logic* **66**, 487–496.

Mautner, Felix I.

1946 An extension of Klein's Erlanger Programm: Logic as invariant theory, *American Journal of Mathematics* **68**, 345–384.

Mayer, Adolph

1899 Über die Aufstellung der Differentialgleichungen für reibungslose Punktsysteme, die Bedingungsgleichungen unterworfen sind, *Berichte über die Verhandlungen der Königlich-Sächsischen Gesellschaft der Wissenschaften zu Leipzig* **51**, 224–244.

Mendelson, Elliott

2004 *Introducing game theory and its applications*, Chapman & Hall/CRC: Boca Raton, Florida.

Meschkowski, Herbert

1967 *Probleme des Unendlichen, Werk und Leben Georg Cantors*, Vieweg: Braunschweig.

Moore, Gregory H.

1976 Ernst Zermelo, A. E. Harward, and the axiomatization of set theory, *Historia mathematica* **3**, 206–209.

1978 The origins of Zermelo's axiomatization of set theory, *Journal of Philosophical Logic* **7**, 307–329.
1982 *Zermelo's axiom of choice: Its origins, development and influence*, Springer-Verlag: New York, Heidelberg, and Berlin.
2002 Hilbert on the infinite: The role of set theory in the evolution of Hilbert's thought, *Historia mathematica* **29**, 40–64.

Morgenstern, Oskar
See von Neumann, John and Oskar Morgenstern.

Neumann, Johann von and Neumann, John von *see* von Neumann, John

Noether, Emmy
1916 Die allgemeinsten Bereiche aus ganzen transzendenten Zahlen, *Mathematische Annalen* **77**, 103-128; corrections ibid. **81**, 30.

Noether, Emmy and Jean Cavaillès (eds.)
1937 *Briefwechsel Cantor–Dedekind*, Hermann: Paris.

Parsons, Charles
1987 Developing arithmetic in set theory without infinity: Some historical remarks, *History and Philosophy of Logic* **8**, 201–213.
2008 *Mathematical thought and its objects*, Cambridge University Press: Cambridge.

Peano, Guiseppe
1897 *Formulaire de mathématiques*, vol. 2.1: *Logique mathématique*, Bocca: Turin.
1906a Super theorema de Cantor-Bernstein, *Rendiconti del Circolo Matematico di Palermo* **21**, 360–366; with the addition *Peano 1906b* also in *Revista de mathematica* **8**, 136–143; reprinted in *Peano 1957*, 337–344.
1906b Additione, *Revista de mathematica* **8**, 143–157; reprinted in *Peano 1957*, 344–358.
1957 *Opere scelte*, vol. 1, Cremonese: Rome.
1990 *Arbeiten zur Analysis und mathematischen Logik*, reprint of the appendices I–IV composed by Guiseppe Peano in *Genocchi 1899*, edited by Günter Asser, Teubner-Archiv zur Mathematik, vol. 13, B. G. Teubner: Leipzig and Springer-Verlag: Wien and New York.

Peckhaus, Volker
1990a *Hilbertprogramm und Kritische Philosophie. Das Göttinger Modell interdisziplinärer Zusammenarbeit zwischen Mathematik und Philosophie*, Vandenhoeck & Ruprecht: Göttingen.
1990b 'Ich habe mich wohl gehütet, alle Patronen auf einmal zu verschießen'. Ernst Zermelo in Göttingen, *History and Philosophy of Logic* **11**, 19–58.

1993 Kurt Grelling und der logische Empirismus, in Rudolf Haller and Friedrich Stadler (eds.), *Wien-Berlin-Prag: Der Aufstieg der wissenschaftlichen Philosophie*, Hölder-Pichler-Tempsky: Wien, 362–385.

1994 Von Nelson zu Reichenbach: Kurt Grelling in Göttingen und Berlin, in Lutz Dannenberg, Andreas Kamlah, and Lothar Schäfer (eds.), *Hans Reichenbach und die Berliner Gruppe*, Vieweg: Braunschweig and Wiesbaden, 53–73.

See also *Ebbinghaus 2007b*.

Pfeiffer, Rudolf

1914 *Der Augsburger Meistersinger und Homerübersetzer Johannes Spreng*, Ph. D. Thesis, University of Munich, P. J. Pfeiffer: Augsburg.

Plessner, Abraham I.

See Schlesinger, Ludwig and Abraham I. Plessner.

Plotinus

1964 Von den Zahlen, in Richard Harder (ed.), *Plotins Schriften*, vol. III, Felix Meiner: Hamburg, 162–217.

Poincaré, Henri

1893 Le mécanisme et l'expérience, *Revue de métaphysique et de morale* **1**, 534–537.

1905 Les mathématiques et la logique, *Revue de métaphysique et de morale* **13**, 815–835.

1906a Les mathématiques et la logique, *Revue de métaphysique et de morale* **14**, 17–34.

1906b Les mathématiques et la logique, *Revue de métaphysique et de morale* **14**, 294–317.

1906c A propos de la logistique, *Revue de métaphysique et de morale* **14**, 866–868.

1909a Réflexions sur les deux notes précédents, *Acta mathematica* **32**, 195–200.

1909b Le logique de l'infini, *Revue de métaphysique et de morale* **17**, 462–482.

1910a *Sechs Vorträge über ausgewählte Gegenstände aus der reinen Mathematik und mathematischen Physik*, B. G. Teubner: Leipzig and Berlin.

1910b Über transfinite Zahlen, in *Poincaré 1910a*, 45–48.

1912 Le logique de l'infini, *Scientia* **12**, 1–11.

Previale, Flavio

1994 Induction and foundation in the theory of hereditarily finite sets, *Archive for Mathematical Logic* **33**, 213–241.

Purkert, Walter

2002 Grundzüge der Mengenlehre – Historische Einführung, in *Hausdorff 2002*, 1–89.

Purkert, Walter and Hans Joachim Ilgauds

1987 *Georg Cantor 1845–1918*, Birkhäuser: Basel, Boston, and Stuttgart.

Pycior, Helena M.

1981 George Peacock and the British origins of symbolic algebra, *Historia mathematica* **8**, 23–45.

Quine, Willard Van Orman

1963 *Set theory and its logic*, Belknap Press: Cambridge, Massachusetts.

Rademacher, Hans

1964 *Lectures on elementary number theory*, Blaisdell Publishing Company: New York et al.

Rademacher, Hans and Otto Toeplitz

1930 *Von Zahlen und Figuren*, Springer-Verlag: Berlin.

Radon, Johann

1913 Theorie und Anwendungen der absolut additiven Mengenfunktionen, *Sitzungsberichte der Kaiserlichen Akademie der Wissenschaften in Wien, Mathematisch-naturwissenschaftliche Klasse* **122**, 1295–1438; reprinted in Johann Radon: *Gesammelte Abhandlungen*, vol. I, Österreichische Akademie der Wissenschaften: Wien 1987, and Birkhäuser: Basel and Boston 1987, 45–188.

Raiffa, Howard

See Luce, R. Duncan and Howard Raiffa.

Ramsey, Frank Plumpton

1926 The foundations of mathematics, *Proceedings of the London Mathematical Society*, 2nd series, **25**, 338–384.

Rang, Bernhard and Wolfgang Thomas

1981 Zermelo's discovery of the Russell paradox, *Historia mathematica* **8**, 15–22.

Rautenberg, Wolfgang

1987 Über den Cantor-Bernsteinschen Äquivalenzsatz, *Mathematisch-Physikalische Semesterberichte* **34**, 71–88.

Remak, Robert Erich

1909 Elementare Verallgemeinerung einer bekannten Eigenschaft der Zahl 30, *Archiv der Mathematik und Physik* **15**, 186–193.

Remmert, Reinhold

1991 The fundamental theorem of algebra, in Heinz-Dieter Ebbinghaus et al. (eds.), *Numbers*, Springer-Verlag: New York, Berlin, et al., 97–122.

Remmert, Reinhold and Peter Ullrich

1987 *Elementare Zahlentheorie*, Birkhäuser: Basel and Boston.

Riesenfeld, Ernst and Ernst Zermelo

1909 Die Einstellung der Grenzkonzentration an der Trennungsfläche zweier Lösungsmittel, *Physikalische Zeitschrift* **10**, 958–961.

Ritzberger, Klaus

2002 *Foundations of non-cooperative game theory*, Oxford University Press: Oxford.

Robbins, Herbert

See Courant, Richard and Herbert Robbins.

Rosenthal, Artur

See Zoretti, Ludovic and Artur Rosenthal.

Rubin, Jean

See Howard, Paul E. and Jean Rubin.

Russell, Bertrand

1903 *The principles of mathematics*, vol. I, Cambridge University Press: Cambridge.

1906a On some difficulties in the theory of transfinite numbers and order types, *Proceedings of the London Mathematical Society*, 2nd series, **4**, 29–53.

1906b Les mathématiques et la logique, *Revue de métaphysique et de morale* **14**, 627–650.

1911 Sur les axiomes de l'infini et du transfini, *Société mathématique de France, comptes rendus des séances* **2**, 22–35; reprinted in *Russell 1992*, 399–408.

1918/19 The philosophy of logical atomism, *The Monist* **28**, 495–527, **29**, 32–63, 190–222, 345–380; reprinted in *Russell 1956*, 177–281.

1956 *Logic and knowledge: Essays 1901–1950*, George Allen & Unwin: London.

1992 *The collected papers of Bertrand Russell*, vol. 6: *Logical and philosophical papers, 1909–13*, edited by John Slater, Routledge: London and New York.

See also Whitehead, Alfred North and Bertrand Russell.

Schappacher, Norbert

2005 Felix Bernstein, *International Statistical Review* **73**, 3–7.

Schlesinger, Ludwig and Abraham I. Plessner

1926 *Lebesguesche Integrale und Fouriersche Reihen*, Walter de Gruyter: Berlin and Leipzig.

Schoenflies, Arthur

1898 Mengenlehre, *Enzyklopädie der mathematischen Wissenschaften mit Einschluss ihrer Anwendungen*, vol. 1: *Arithmetik und Algebra*, part I A 5, B. G. Teubner: Leipzig, 1898–1904, 184–207.

1900 Die Entwickelung der Lehre von den Punktmannigfaltigkeiten. Bericht, erstattet der Deutschen Mathematiker-Vereinigung, *Jahresbericht der Deutschen Mathematiker-Vereinigung* **8**(2), 3–250.

1905 Über wohlgeordnete Mengen, *Mathematische Annalen* **60**, 181–186.

1911 Über die Stellung der Definition in der Axiomatik, *Jahresbericht der Deutschen Mathematiker-Vereinigung* **20**, 222–255.

1913 *Entwickelung der Mengenlehre und ihrer Anwendungen. Erste Hälfte: Allgemeine Theorie der unendlichen Mengen und Theorie der Punktmengen. Umarbeitung des im VIII. Bande der Jahresberichte der Deutschen Mathematiker-Vereinigung erstatteten Berichts*, B. G. Teubner: Leipzig.

Scholz, Arnold

1939 *Einführung in die Zahlentheorie*, Walter de Gruyter: Berlin.

Schröder, Ernst

1890 *Vorlesungen über die Algebra der Logik (Exakte Logik)*, vol. 1, Teubner: Leipzig.

1896 Über G. Cantorsche Sätze, *Jahresbericht der Deutschen Mathematiker-Vereinigung* **5**, 81–82.

1898 Über zwei Definitionen der Endlichkeit und G. Cantorsche Sätze, *Nova Acta Academiae Caesareae Leopoldino-Carolinae Germanicae Naturae Curiosorum* **71**, 303–362.

Schröder-Heister, Peter

2002 Resolution and the origin of structural reasoning. Early proof-theoretic ideas of Hertz and Gentzen, *The Bulletin of Symbolic Logic* **8**, 246–265.

Schwalbe, Ulrich and Paul Walker

2001 Zermelo and the early history of game theory, *Games and Economic Behavior* **34**, 123–137.

Scott, Dana S.

1971 Axiomatizing set theory, in Dana S. Scott (ed.), *Axiomatic set theory*, Proceedings of Symposia in Pure Mathematics vol. 13, part 1, American Mathematical Society: Providence, Rhode Island, 207–214.

Sebestik, Jan

2000 Bolzanos Paradoxien des Unendlichen, in Edgar Morscher (ed.), *Bolzanos geistiges Erbe für das 21. Jahrhundert*, series *Beiträge zur Bolzano-Forschung*, vol. 11, Academia Verlag: St. Augustin bei Bonn, 231-249.

Segal, Sanford L.

2003 *Mathematicians under the Nazis*, Princeton University Press: Princeton and Oxford.

Serret, Joseph Alfred

1899 *Lehrbuch der Differential- und Integralrechnung*, German edition by Axel Harnack, 2nd revised edition, vol. 2: *Integralrechnung*, edited by Georg Bohlmann with the assistance of Heinrich Liebmann and Ernst Zermelo, B. G. Teubner: Leipzig.

1904 *Lehrbuch der Differential- und Integralrechnung*, German edition by Axel Harnack, 2nd revised edition, vol. 3: *Differentialgleichungen und Variationsrechnung*, edited by Georg Bohlmann and Ernst Zermelo, B. G. Teubner: Leipzig.

Shapiro, Stewart and Gabriel Uzquiano

2008 Frege meets Zermelo: A perspective on ineffability and reflection, *The Review of Symbolic Logic* **1**, 241–266.

Shegalkin, Ivan Ivanovich

1907 *Transfinite numbers* (Russian), University Press: Moscow.

Shepherdson, John C.

1952 Inner models for set theory—Part II, *The Journal of Symbolic Logic* **17**, 225–237.

Shoenfield, Joseph

1977 Axioms of set theory, in Barwise, Jon (ed.), *Handbook of Mathematical Logic*, North-Holland: Amsterdam, 321–344.

Sierpiński, Wacław and Alfred Tarski

1930 Sur une propriété caractéristique des nombres inaccessibles, *Fundamenta mathematicae* **15**, 292–300.

Skolem, Thoralf

1923 Einige Bemerkungen zur axiomatischen Begründung der Mengenlehre, *Wissenschaftliche Vorträge gehalten auf dem Fünften Kongress der Skandinavischen Mathematiker in Helsingfors vom 4. bis 7. Juli 1922*, Akademische Buchhandlung: Helsingfors 1923, 217–232; reprinted in *Skolem 1970*, 137–152.

1930 Einige Bemerkungen zu der Abhandlung von E. Zermelo: "Über die Definitheit in der Axiomatik", *Fundamenta mathematicae* **15**, 337–341; reprinted in *Skolem 1970*, 275–279.

1970 *Selected works in logic*, edited by Jens Erik Fenstad, Universitetsforlaget: Oslo, Bergen, Tromsø.

Stäckel, Paul

1907 Zu H. Webers Elementarer Mengenlehre, *Jahresbericht der Deutschen Mathematiker-Vereinigung* **16**, 425-428.

Steinitz, Ernst

1910 Algebraische Theorie der Körper, *Journal für die reine und angewandte Mathematik* **137**, 167–305.

Stewart, Frank M.

See Gale, David and Frank M. Stewart.

Straffin, Philip D.

1993 *Game theory and strategy*, Mathematical Association of America: Providence, Rhode Island.

Szlezák, Thomas Alexander (ed.)

2005 *Wolfgang Schadewaldt und die Gräzistik des 20. Jahrhunderts*, Olms: Hildesheim.

Tait, William W.

1998 Zermelo's conception of set theory and reflection principles, in Matthias Schirn (ed.), *The philosophy of mathematics today*, Oxford University Press: Oxford, 469–483.

Tarski, Alfred

1924a Sur les ensembles finis, *Fundamenta mathematicae* **6**, 45–95.

1924b Sur quelques théorèmes qui équivalent à l'axiom du choix, *Fundamenta mathematicae* **5**, 147–154.

1925 Quelques théorèmes sur les alephs, *Fundamenta mathematicae* **7**, 1–14.

1938 Über unerreichbare Kardinalzahlen, *Fundamenta mathematicae* **30**, 68–89.
1948 Axiomatic and algebraic aspects of two theorems on sums of cardinals, *Fundamenta mathematicae* **35**, 79–104.
1955 A lattice-theoretical fixpoint theorem and its applications, *Pacific Journal of Mathematics* **5**, 285–309.
1958 Remarks on predicate logic with infinitely long expressions, *Colloquium mathematicum* **6**, 171–176.
1986 What are logical notions?, *History and Philosophy of Logic* **7**, 143–154; text, edited with introduction by John Corcoran, of a lecture first presented in London in 1966 and repeated in Buffalo in 1973.

See also Givant, Steven and Alfred Tarski.

Tarski, Alfred and Steven Givant

1987 A formulation of set theory without variables, *American Mathematical Society Colloquium Publications* **41**, American Mathematical Society: Providence, Rhode Island.

Taussky-Todd, Olga

1987 Remembrances of Kurt Gödel, in Rudolf Gödel et al., *Gödel remembered: Salzburg 10–12 July 1983*, Bibliopolis: Naples, 29–41.

Taylor, R. Gregory

1993 Zermelo, reductionism, and the philosophy of mathematics, *Notre Dame Journal of Formal Logic* **34**, 539–563.
2002 Zermelo's Cantorian theory of systems of infinitely long propositions, *The Bulletin of Symbolic Logic* **8**, 478–515.
2008 Symmetric propositions and logical quantifiers, *Journal of Philosophical Logic* **37**, 575–591.
2009 Zermelo's analysis of 'general proposition', *History and Philosophy of Logic* **30**, 141–155.

Thomas, Wolfgang

See Rang, Bernhard and Wolfgang Thomas.

Toeplitz, Otto

See Rademacher, Hans and Otto Toeplitz.

Trost, Ernst

1953 *Primzahlen*, Birkhäuser: Basel and Stuttgart.

Tucker, Albert W.

See Kuhn, Harold W. and Albert W. Tucker.

Uffink, Jos

2004 Boltzmann's work in statistical physics, in Edward N. Zalta (ed.), *The Stanford encyclopedia of philosophy*, http://plato.stanford.edu/entries/statphys-Boltzmann/.

Ullrich, Peter

See Remmert, Reinhold and Peter Ullrich.

Uzquiano, Gabriel

1999 Models of second-order Zermelo set theory, *The Bulletin of Symbolic Logic* **5**, 289–302.

See also Shapiro, Stewart and Gabriel Uzquiano.

van Dalen, Dirk and Heinz-Dieter Ebbinghaus

2000 Zermelo and the Skolem paradox, *The Bulletin of Symbolic Logic* **6**, 145–161.

van Heijenoort, Jean (ed.)

1967 *From Frege to Gödel: A source book in mathematical logic, 1879–1931*, Harvard University Press: Cambridge, Massachusetts.

van Rootselaar, Bob

1976 Ernst Friedrich Ferdinand Zermelo, in Charles Coulston Gillispie et al. (eds.), *Dictionary of scientific biography*, vol. 14, Scribner: New York, 613–616.

Vitali, Guiseppe

1904 Sui gruppi di punti, *Rendiconti del Circolo Matematico di Palermo* **18**, 116–126; reprinted in Vitali, Guiseppe: *Opere sull'analisi reale e complessa. Carteggio*, Edizioni Cremonese: Rome, 1984, 139–149.

Vogt, Ernst

2001 Pfeiffer, Rudolf Carl Franz Otto, in *Neue deutsche Biographie*, vol. 20, Duncker & Humblot: Berlin, 323–324.

von Neumann, John

1923 Zur Einführung der transfiniten Zahlen, *Acta litterarum ac scientiarum Regiae Universitatis Hungaricae Francisco-Josephinae, sectio scientiarum mathematicarum* **1**, 199–208; reprinted in *von Neumann 1961*, 24–33.

1925 Eine Axiomatisierung der Mengenlehre, *Journal für die reine und angewandte Mathematik* **154**, 219–240; reprinted in *von Neumann 1961*, 35–56.

1928a Über die Definition durch transfinite Induktion und verwandte Fragen der allgemeinen Mengenlehre, *Mathematische Annalen* **99**, 373–391; reprinted in *von Neumann 1961*, 320–338.

1928b Ein System algebraisch unabhängiger Zahlen, *Mathematische Annalen* **99**, 134–141; reprinted in *von Neumann 1961*, 312–319.

1928c Zur Theorie der Gesellschaftsspiele, *Mathematische Annalen* **100**, 295–320.

1928d Die Axiomatisierung der Mengenlehre, *Mathematische Zeitschrift* **27**, 669–752; reprinted in *von Neumann 1961*, 339–422.

1929 Über die Widerspruchsfreitheitsfrage in der axiomatischen Mengenlehre, *Journal für die reine und angewandte Mathematik* **160**, 227–241; reprinted in *von Neumann 1961*, 494–508.

1961 *Collected Works*, vol. 1, edited by Abraham Haskel Taub, Pergamon Press: Oxford et al.

von Neumann, John and Oskar Morgenstern

1943 *Theory of games and economic behavior*, Princeton University Press: Princeton, New Jersey.

Vopěnka, Petr and Petr Hajek

1963 Über die Gültigkeit des Fundierungsaxioms in speziellen Systemen der Mengentheorie, *Zeitschrift für mathematische Logik und Grundlagen der Mathematik* **9**, 235–241.

Vorob'ev, Nicolai N.

1994 *Foundations of game theory: Noncooperative games*, Birkhäuser: Basel.

Walker, Paul

See Schwalbe, Ulrich and Paul Walker.

Wang, Hao

1974 The concept of set, in Hao Wang, *From mathematics to philosophy*, Humanities Press: New York, 181–223; reprinted in Paul Benacerraf and Hilary Putnam (eds.), *Philosophy of mathematics. Selected readings*, 2nd edition, Cambridge University Press: Cambridge, 1983, 530–570.

Weber, Heinrich

1895 *Lehrbuch der Algebra*, vol. 1, Vieweg: Brauschweig.

1906 Elementare Mengenlehre, *Jahresbericht der Deutschen Mathematiker-Vereinigung* **15**, 173–184.

1912 *Lehrbuch der Algebra*, combined edition, Vieweg: Braunschweig.

Weyl, Hermann

1910 Über die Definition der mathematischen Grundbegriffe, *Mathematisch-naturwissenschaftliche Blätter* **7**, 93–95, 109–113; reprinted in *Weyl 1968*, 298–304.

1918 *Das Kontinuum*, Verlag von Veit: Leipzig; reprinted in *Weyl 1966*.

1966 *Das Kontinuum und andere Monographien*, Chelsea: New York.
1968 *Gesammelte Abhandlungen*, vol. 1, edited by Komaravolu Chandrasekharan, Springer-Verlag: Berlin, Heidelberg, et al.

Whitehead, Alfred North and Bertrand Russell
1912 *Principia mathematica*, vol. II, Cambridge University Press: Cambridge.
1925 *Principia mathematica*, 2nd edition, Cambridge University Press: Cambridge.

Wright, E. Maitland
See Hardy, G. Harold and E. Maitland Wright.

Wussing, Hans
1984 *The genesis of the abstract group concept: A contribution to the history of the origin of the abstract group concept*, MIT Press: Cambridge, Massachusetts; English translation of Hans Wußing, *Die Genesis des abstrakten Gruppenbegriffs*, Deutscher Verlag der Wissenschaften: Berlin, 1969.

Zagare, Frank C.
1984 *Game theory: Concepts and applications*, Quantitative Applications in the Social Sciences, Sage Publications: Thousand Oaks, California.

Zermelo, Ernst
1894 *Untersuchungen zur Variations-Rechnung*, Ph. D. Thesis, University of Berlin.
1896a Ueber einen Satz der Dynamik und die mechanische Wärmetheorie, *Annalen der Physik und Chemie,* new series, **57**, 485–494.
1896b Ueber mechanische Erklärungen irreversibler Vorgänge. Eine Antwort auf Hrn. Boltzmann's „Entgegnung", *Annalen der Physik und Chemie,* new series, **59**, 793–801.
1899a Ueber die Bewegung eines Punktsystems bei Bedingungsungleichungen, *Nachrichten von der Königlichen Gesellschaft der Wissenschaften zu Göttingen. Mathematisch-physikalische Klasse. Aus dem Jahre 1899,* 306–310.
s1899b Wie bewegt sich ein unausdehnbarer materieller Faden unter dem Einfluß von Kräften mit dem Potentiale $W(x, y, z)$?, manuscript, 3pp., Niedersächsische Staats- und Universitätsbibliothek Göttingen, Cod. Ms. D. Hilbert 722.
1900 Über die Anwendung der Wahrscheinlichkeitsrechnung auf dynamische Systeme, *Physikalische Zeitschrift* **1**, 317–320.
1901 Ueber die Addition transfiniter Cardinalzahlen, *Nachrichten von der Königlichen Gesellschaft der Wissenschaften zu Göttingen. Mathematisch-physikalische Klasse aus dem Jahre 1901*, 34–38.

1902a Hydrodynamische Untersuchungen über die Wirbelbewegungen in einer Kugelfläche, *Zeitschrift für Mathematik und Physik* **47**, 201–237.

s1902b Hydrodynamische Untersuchungen über die Wirbelbewegungen in einer Kugelfläche (Zweite Mitteilung), manuscript with pages 61–117, Universitätsarchiv Freiburg, Zermelo *Nachlass*, C 129/225.

s1902c §5. Die absolute Bewegung, manuscript with pages 119–131, Universitätsarchiv Freiburg, Zermelo *Nachlass*, C 129/253.

1902d Zur Theorie der kürzesten Linien, *Jahresbericht der Deutschen Mathematiker-Vereinigung* **11**, 184–187.

1903 Über die Herleitung der Differentialgleichung bei Variationsproblemen, *Mathematische Annalen* **58**, 558–564.

1904 Beweis, daß jede Menge wohlgeordnet werden kann, *Mathematische Annalen* **59**, 514–516.

1906 Review of *Gibbs 1902* and *Gibbs 1905*, *Jahresbericht der Deutschen Mathematiker-Vereinigung* **15**, 232–242.

1908a Neuer Beweis für die Möglichkeit einer Wohlordnung, *Mathematische Annalen* **65**, 107–128.

1908b Untersuchungen über die Grundlagen der Mengenlehre. I, *Mathematische Annalen* **65**, 261–281.

1909a Sur les ensembles finis et le principe de l'induction complète, *Acta mathematica* **32**, 185–193.

1909b Ueber die Grundlagen der Arithmetik, in G. Castelnuovo (ed.), *Atti del IV Congresso Internazionale dei Matematici (Roma, 6–11 Aprile 1908)*, vol. 2: *Comunicazioni delle sezioni I e II*, Tipografia della R. Accademia dei Lincei: Rome, 8–11.

1913 Über eine Anwendung der Mengenlehre auf die Theorie des Schachspiels, in Ernest William Hobson and Augustus Edward Hough Love (eds.): *Proceedings of the 5th international congress of mathematicians, Cambridge 1912*, vol. 2, Cambridge University Press: Cambridge, 501–504.

1914 Über ganze transzendente Zahlen, *Mathematische Annalen* **75**, 434–442.

s1921 Thesen über das Unendliche in der Mathematik, typescript, 1 p., Universitätsarchiv Freiburg, Zermelo *Nachlass*, part of C 129/278.

1927 Über das Maß und die Diskrepanz von Punktmengen, *Journal für die reine und angewandte Mathematik* **158**, 154–167.

1928 Die Berechnung der Turnier-Ergebnisse als ein Maximumproblem der Wahrscheinlichkeitsrechnung, *Mathematische Zeitschrift* **29**, 436–460.

1929a Über den Begriff der Definitheit in der Axiomatik, *Fundamenta mathematicae* **14**, 339–344.

s1929b Vortrags-Themata für Warschau 1929, typescript, 9 pp., Universitätsarchiv Freiburg, Zermelo *Nachlass*, C 129/288.

1930a Über Grenzzahlen und Mengenbereiche. Neue Untersuchungen über die Grundlagen der Mengenlehre, *Fundamenta mathematicae* **16**, 29–47.

1930b Über die logische Form der mathematischen Theorien, *Annales de la Société Polonaise de Mathématique* **9**, 187.

1930c Über die Navigation in der Luft als Problem der Variationsrechnung, *Jahresbericht der Deutschen Mathematiker-Vereinigung* **39**, 2nd section, 44–48.

s1930d Bericht an die Notgemeinschaft der Deutschen Wissenschaft über meine Forschungen betreffend die *Grundlagen der Mathematik,* typescript, 5 pp., with appendices, 2pp., dated 3 December 1930, Universitätsarchiv Freiburg, Zermelo *Nachlass*, part of C 129/140.

s1930e Über das mengentheoretische Modell, typescript, 5 pp., Universitätsarchiv Freiburg, Zermelo *Nachlass*, part of C 129/292. Copy of first 4 pages, dated September 1930, in C 129/212.

1930f Aus Homers Odyssee, *Wiener Blätter für die Freunde der Antike* **6**, 93–94.

1931a Über das Navigationsproblem bei ruhender oder veränderlicher Windverteilung. *Zeitschrift für angewandte Mathematik und Mechanik* **11**, 114–124.

s1931b Letter to Kurt Gödel of 21 September 1931, Gödel *Nachlass*, Institute for Advanced Study, 013270. Draft in Universitätsarchiv Freiburg, Zermelo *Nachlass*, part of C 129/36; reprinted in *Gödel 2003*, 420–423.

s1931c Letter to Reinhold Baer of 7 October 1931, in the possession of Ulrich Felgner and given to him by Reinhold Baer.

s1931d Letter to Kurt Gödel of 29 October 1931, Gödel *Nachlass*, Institute for Advanced Study, 013271. Carbon copy in Universitätsarchiv Freiburg, Zermelo *Nachlass*, part of C 129/36; reprinted in *Gödel 2003*, 430–431.

s1931e Seven notes on ordinal numbers and large cardinals, typescript, 8 pp., undated, ca. 1931, Universitätsarchiv Freiburg, Zermelo *Nachlass*, C 129/236.

s1931f Sätze über *geschlossene Bereiche*, typescript, 2 pp., undated, ca. 1931, Universitätsarchiv Freiburg, Zermelo *Nachlass*, C 129/236.

s1931g Allgemeine Theorie der mathematischen Systeme, manuscript, dated July 1931, Universitätsarchiv Freiburg, Zermelo *Nachlass*, C 129/294.

1932a Über Stufen der Quantifikation und die Logik des Unendlichen, *Jahresbericht der Deutschen Mathematiker-Vereinigung* **41**, 2nd section, 85–88.

1932b Über mathematische Systeme und die Logik des Unendlichen, *Forschungen und Fortschritte* **8**, 6–7.

1932c Vorwort. Preface to *Cantor 1932*, in *Cantor 1932,* iii–v.

s1932d Mengenlehre 1932, typescript, 5 pp., Universitätsarchiv Freiburg, Zermelo *Nachlass*, part of C 129/290.

1933a Über die Bruchlinien zentrierter Ovale. Wie zerbricht ein Stück Zucker?, *Zeitschrift für angewandte Mathematik und Mechanik* **13**, 168–170.

s1933b Die unbegrenzte Zahlenreihe und die exorbitanten Zahlen, typescript, 1 p., November 1933, Universitätsarchiv Freiburg, Zermelo *Nachlass*, part of C 129/239.

1934 Elementare Betrachtungen zur Theorie der Primzahlen, *Nachrichten von der Gesellschaft der Wissenschaften zu Göttingen, Fachgruppe I; Mathematik*, new series, **1**, 43–46.

1935 Grundlagen einer allgemeinen Theorie der mathematischen Satzsysteme (Erste Mitteilung), *Fundamenta mathematicae* **25**, 136–146.

s1937 Der Relativismus in der Mengenlehre und der sogenannte Skolemsche Satz, manuscript, dated 4 October 1937, 3 pp., Universitätsarchiv Freiburg, Zermelo *Nachlass*, part of C 129/217.

s1941 Letter to Paul Bernays of 1 October 1941, ETH-Bibliothek Zürich, *Nachlass* Bernays, Hs. 975.5259.

1967a Proof that every set can be well-ordered, English translation of *Zermelo 1904*, in *van Heijenoort 1967*, 139–141.

1967b A new proof of the possibility of a well-ordering, English translation of *Zermelo 1908a*, in *van Heijenoort 1967*, 183–198.

1967c Investigations in the foundations of set theory I, English translation of *Zermelo 1908b*, in *van Heijenoort 1967*, 200–215.

See also Hahn, Hans and Ernst Zermelo, and Riesenfeld, Ernst and Ernst Zermelo.

Zermelo, Theodor

1875 *August Ludwig Schlözer, ein Publicist im alten Reich*, W. Weber: Berlin.

Zita, Kurt

1931 *Beiträge zu einem Variationsproblem von Zermelo*, Ph. D. Thesis, University of Breslau.

Zoretti, Ludovic and Artur Rosenthal

1924 Neuere Untersuchungen über Funktionen reeller Veränderlichen. Die Punktmengen, in: *Enzyklopädie der mathematischen Wissenschaften mit Einschluss ihrer Anwendungen*, vol. II: *Analysis*, part C 9a, B. G. Teubner: Leipzig, 1899–1924, 851–1030.

Index

MIX
Papier aus verantwortungsvollen Quellen
Paper from responsible sources
FSC® C105338

If you have any concerns about our products,
you can contact us on
ProductSafety@springernature.com

In case Publisher is established outside the EU,
the EU authorized representative is:
Springer Nature Customer Service Center GmbH
Europaplatz 3, 69115 Heidelberg, Germany

Printed by Libri Plureos GmbH
in Hamburg, Germany